Advance in Nanoparticles as Vaccine Adjuvants

Advance in Nanoparticles as Vaccine Adjuvants

Collection Editors

Eduardo Gomez-Casado
Sohrab Ahmadivand

MDPI

Basel • Beijing • Wuhan • Barcelona • Belgrade • Novi Sad • Cluj • Manchester

Collection Editors

Eduardo Gomez-Casado
Department of Biotechnology
INIA-CSIC
Madrid
Spain

Sohrab Ahmadivand
Faculty of
Veterinary Medicine
Ludwig-Maximilians-Universität
München
München
Germany

Editorial Office
MDPI AG
Grosspeteranlage 5
4052 Basel, Switzerland

This is a reprint of the Topical Collection, published open access by the journal *Vaccines* (ISSN 2076-393X), freely accessible at: https://www.mdpi.com/journal/vaccines/topical_collections/Nanoparticles_Vaccine.

For citation purposes, cite each article independently as indicated on the article page online and as indicated below:

Lastname, A.A.; Lastname, B.B. Article Title. *Journal Name* **Year**, *Volume Number*, Page Range.

ISBN 978-3-7258-7122-3 (Hbk)
ISBN 978-3-7258-7123-0 (PDF)
https://doi.org/10.3390/books978-3-7258-7123-0

Contents

vaccines

MDPI

Editorial

Advances in Nanoparticles as Vaccine Adjuvants

Sohrab Ahmadivand [1] and Eduardo Gomez-Casado [2,*]

1 Faculty of Veterinary Medicine, Ludwig-Maximilians University Munich, 80539 Munich, Germany; sohrab.ahmadivand@lmu.de
2 Department of Biotechnology, Instituto Nacional de Investigación y Tecnología Agraria y Alimentaria, Consejo Superior de Investigaciones Científicas (INIA-CSIC), 28040 Madrid, Spain
* Correspondence: casado@inia.csic.es

1. Introduction

The development of safe, effective, and durable vaccines remains a central goal in preventing infectious diseases in humans and animals. Modern non-replicating platforms offer advantages in safety, antigen specificity, and manufacturing scalability; however, they often exhibit limited intrinsic immunogenicity. As a result, the identification and optimization of effective adjuvants have become critical determinants of vaccine performance [1].

Adjuvants enhance immune responses primarily by activating innate pathways, including pattern recognition receptors (PRRs), and can also promote germinal center formation, T follicular helper (Tfh) cell differentiation, and the magnitude, quality, and durability of adaptive immunity, including Th1/Th2 polarization and antibody titers [2].

Classical adjuvants, including aluminum salts and oil-based emulsions (e.g., MF59®, Montanide), are widely used and generally effective but may have limited capacity to elicit robust cellular or mucosal immunity. Aluminum salts primarily induce antibody (Th2) responses [3], whereas oil-emulsion adjuvants can cause side effects, including local reactogenicity and, occasionally, systemic reactions [4]. Furthermore, they provide limited flexibility to fine-tune immune responses and may be suboptimal for vaccines targeting intracellular pathogens, rapidly mutating viruses, or complex antigens, highlighting the need for next-generation adjuvant systems with greater design precision.

Recent advances in nanotechnology have significantly enhanced vaccine adjuvant design by enabling precise control over antigen delivery and immune activation. Nanoparticles can be engineered with defined size, shape, surface chemistry, and degradation kinetics, which influence lymphatic trafficking, antigen uptake, and immune recognition [5]. These properties allow nanoparticles to act as active immunostimulatory platforms rather than passive carriers, integrating targeted antigen presentation, innate immune engagement, and controlled antigen release within a single construct. Such attributes drive enhanced humoral and cellular responses compared with traditional adjuvants and are a foundational rationale for next-generation vaccine design [6].

For instance, inorganic nanoparticles, including nano-sized aluminum salts, silica, and clay, improve upon conventional adjuvants by enhancing cellular uptake and activating innate immune pathways. Lipid-based nanoparticles and oil-in-water emulsions (e.g., MF59®, AS03®) further augment immune responses while maintaining favorable safety profiles, and their conversion to well-defined nanoscale formulations may enhance antigen delivery and immunogenicity [7]. Organic systems, such as chitosan nanoparticles, offer increased stability and immunogenicity. Lipid nanoparticles (LNPs) possess intrinsic adjuvant activity, activating innate immunity and promoting robust antibody responses. Self-assembling protein nanocages (SAPNs), such as ferritin, mimic viral size and structure, enabling multivalent antigen display, improved stability, and enhanced immunogenicity. Protein-based adjuvants like flagellin, which engage TLR5 [8], can be incorporated into SAPNs to further boost adaptive immunity and allow co-display and co-delivery with antigens [9].

This Special Issue, "Advances in Nanoparticles as Vaccine Adjuvants", compiles original research and comprehensive reviews that collectively delineate the scope, mechanistic diversity, and translational relevance of nanoparticle-based adjuvant systems. The studies presented underscore that rationally designed nanoparticles can integrate structural organization, targeted delivery, and immune modulation to enhance vaccine efficacy, safety, and translational potential. Together, they establish nanoparticle-based adjuvants as a cornerstone of next-generation vaccine design for infectious diseases.

2. An Overview of Advances in Nanoparticles as Vaccine Adjuvants

A substantial body of work in this Special Issue highlights self-assembling protein nanocages (SAPNs) with pathogen-mimicking architectures as advanced vaccine platforms that enable multivalent antigen display and co-delivery of adjuvants. Bacterial self-assembling protein nanocages exploit nanoscale symmetry and repetitive antigen display to promote strong B-cell receptor crosslinking and effective adaptive immune responses, thereby enhancing the stability and immunogenicity of subunit vaccines, as reviewed by Lamontagne et al. [10].

SAPNs such as ferritin represent versatile, multifunctional platforms with broad applicability and have been shown to enhance antigen persistence, facilitate efficient uptake by antigen-presenting cells (APCs), and amplify immune activation in subunit, DNA, and mRNA vaccines. As noted by Ahmadivand [9], beyond structural antigen display and RNAi therapeutics, SAPNs can also co-deliver antigens and adjuvants. They achieve this either through multivalent surface display or internal encapsulation of protein or small-molecule adjuvants, promoting efficient APC activation, sustained immune stimulation, and reduced off-target toxicity. Flexible design strategies allow single or dual adjuvant configurations and can tailor Th1/Th2 immune bias. This SAPN-adjuvant approach offers controlled and modular immune modulation, improving vaccine safety and efficacy, particularly for viral infections.

Protein nanoparticles also support multivalent vaccine designs in which structurally distinct antigens can be presented simultaneously on a single scaffold. For example, Xia et al. [11]. developed a chimeric viral nanoparticle, P24-αTSR, that self-assembles into a 24-meric octahedral structure displaying both the norovirus protruding (P) domain and the Plasmodium αTSR domain. This dual presentation retains the functional conformations of both antigens and elicits strong pathogen-specific antibody responses in mice, demonstrating the potential of SAPNs to co-deliver multiple antigens while promoting potent adaptive immunity.

Structural stability is an important factor in enhancing adjuvant performance. In this context, Wong et al. [12]. developed thermally stable protein nanoparticles that maintain immunogenicity under elevated temperatures and induce durable humoral immune responses after a single immunization, suggesting that similarly stable SAPN platforms could achieve comparable benefits.

Polymeric nanoparticles are a well-established and versatile class of vaccine adjuvants, with immunological outcomes determined by material composition, degradation kinetics, and mechanical properties. As demonstrated by Cui et al. [13], biodegradable PLGA–PEG 25% nanoparticles exhibit antigen-dependent adjuvanticity: nanovaccines conjugated with the S. aureus exotoxin rHlaH35L elicited more robust humoral, cellular, and innate immune responses, as well as superior protective efficacy, compared with those carrying

 https://doi.org/10.3390/vaccines14020159

the cell-wall protein rSpam, or both antigens combined. This underscores the critical importance of rational antigen–nanoparticle pairing in optimizing vaccine potency. The influence of physical nanoparticle properties was further shown by Song et al. [14], who reported that nanoparticle stiffness and degradation kinetics regulate antigen accessibility and cellular uptake in a route-dependent manner in an S. aureus mouse model, resulting in enhanced humoral and cellular immunity compared with conventional alum adjuvants. In addition, Su et al. [15] demonstrated that encapsulation of ginseng stem-leaf saponins within PLGA nanoparticles improves adjuvant stability, promotes dendritic cell-mediated antigen presentation, and amplifies both mucosal and systemic immune responses while maintaining a favorable safety profile. Overall, these findings illustrate that antigen selection, nanoparticle physicochemical properties, and bioactive payloads directly govern the magnitude, quality, and durability of vaccine-induced immune responses, providing a rational framework for next-generation polymeric adjuvants.

Nanoparticles also play a pivotal role in the adjuvantation of nucleic acid and mucosal vaccines, settings in which effective immune stimulation is particularly challenging. Sousa de Pinho et al. [16] showed that yeast-derived β-glucan capsules function as biologically derived nanoparticles that enhance DNA vaccine uptake by macrophages and APCs, improving immune priming while maintaining minimal cytotoxicity. In mucosal vaccination, polymeric caffeic acid acts primarily as a nanoparticle-based antigen delivery system rather than a classical immune stimulant, promoting targeted transport to mucosal dendritic cells and eliciting both local and systemic immune responses [17]. Chitosan-based nanoparticles further illustrate this dual adjuvant–carrier functionality: their mucoadhesive properties, intrinsic immunostimulatory activity, and sustained antigen-release capacity make them highly effective for mucosal immunization [18]. Consistent with these findings, Yang et al. [19] demonstrated that surface-modified liposomes incorporating lipidated chitosan derivatives, such as oleoyl-quaternized chitosan (OTMC), function as self-adjuvanting intranasal vaccine platforms, eliciting robust systemic and mucosal immune responses and promoting functional antibody and cytokine production.

Lipid nanoparticles (LNPs) carrying mRNA have also been shown to actively stimulate innate immune pathways in addition to delivering their cargo, shaping downstream adaptive responses. However, formulation features such as PEGylation, while improving particle stability, can trigger complement activation and hypersensitivity reactions, highlighting the need for careful optimization to balance safety and immunostimulatory efficacy [20].

Several review articles place nanoparticle adjuvants within broader translational and disease-specific contexts. Comprehensive overviews of nanovaccine platforms demonstrate how nanoparticles enhance antigen protection, immune targeting, and response durability across bacterial and viral infections [21]. Beyond immunological effects, the clinical translation of nanoparticle adjuvants also requires consideration of manufacturability, scalability, and regulatory compliance, as systematically analyzed by Filipić et al. [7]. Disease-specific applications further illustrate the value of these platforms. In tuberculosis vaccination, Wang et al. [22] emphasized that nanoparticle adjuvants improve antigen presentation kinetics and promote sustained cellular immunity, which is essential for protection against intracellular pathogens. Similarly, in aquaculture, nanoparticle-based formulations offer welfare-conscious alternatives to traditional oil- or aluminum-based adjuvants, enhancing vaccine efficacy while minimizing adverse effects in teleost fish [23].

Innovative strategies expanding the functional capabilities of nanoparticle adjuvants have been demonstrated through controlled antigen release systems. Brubaker et al. [24] showed that antigen-containing powders coated via atomic layer deposition enabled tunable delayed-release kinetics, providing precise temporal control over immune stimulation

https://doi.org/10.3390/vaccines14020159

and supporting the development of single-administration vaccination strategies. Complementing this approach, Shalash et al. [25] highlighted that nanoparticle-compatible formulations can enhance peptide-based vaccines, which often suffer from low intrinsic immunogenicity. Optimized peptide antigens, particularly when conjugated to T-helper epitopes such as PADRE, elicited robust and functional antibody responses, including strong opsonic activity, even under complex co-infection conditions, reinforcing the role of nanoparticles in overcoming fundamental limitations of peptide vaccines.

3. Conclusions

This Special Issue demonstrates that nanoparticles function as active, designable vaccine adjuvants rather than passive carriers. Platforms including self-assembling protein nanocages, polymeric systems, lipid-based carriers, and nucleic acid formulations enhance antigen stability, enable multivalent and targeted presentation, and overcome intrinsic limitations of subunit, peptide, mucosal, and nucleic acid vaccines. Rational design, considering antigen selection, structural stability, physicochemical properties, and co-delivered adjuvants, directly governs the magnitude, quality, and durability of immune responses. Collectively, these studies underscore the potential of nanoparticle-based platforms to advance next-generation vaccines for both human and veterinary applications, with future efforts focusing on precision engineering, controlled or single-dose delivery, mucosal targeting, and integration with emerging technologies to develop safe, scalable, and highly protective vaccines.

Author Contributions: S.A. drafted the manuscript and E.G.-C. participated in manuscript writing-editing. All authors have read and agreed to the published version of the manuscript.

Acknowledgments: We thank all authors for their contributions and everyone who assisted in preparing this special issue. We also gratefully acknowledge the research grant from the Spanish Ministry of Science and Universities (Project PID2022-140624OR-I00) and the German Research Foundation (DFG, Project Number 557327539).

Conflicts of Interest: The author declares no conflicts of interest.

References

1. Pollard, A.J.; Bijker, E.M. A guide to vaccinology: From basic principles to new developments. *Nat. Rev. Immunol.* **2021**, *21*, 83–100. [CrossRef] [PubMed]
2. Ahmadivand, S.; Fux, R.; Palić, D. Role of T Follicular Helper Cells in Viral Infections and Vaccine Design. *Cells* **2025**, *14*, 508. [CrossRef]
3. Kool, M.; Fierens, K.; Lambrecht, B.N. Alum adjuvant: Some of the tricks of the oldest adjuvant. *J. Med. Microbiol.* **2012**, *61*, 927–934. [CrossRef]
4. Petrovsky, N. Comparative safety of vaccine adjuvants: A summary of current evidence and future needs. *Drug Saf.* **2015**, *38*, 1059–1074. [CrossRef]
5. Moni, S.S.; Abdelwahab, S.I.; Jabeen, A.; Elmobark, M.E.; Aqaili, D.; Gohal, G.; Oraibi, B.; Farasani, A.M.; Jerah, A.A.; Alnajai, M.M.A.; et al. Advancements in Vaccine Adjuvants: The Journey from Alum to Nano Formulations. *Vaccines* **2023**, *11*, 1704. [CrossRef]
6. Fan, J.; Jin, S.; Gilmartin, L.; Toth, I.; Hussein, W.M.; Stephenson, R.J. Advances in Infectious Disease Vaccine Adjuvants. *Vaccines* **2022**, *10*, 1120. [CrossRef]
7. Filipić, B.; Pantelić, I.; Nikolić, I.; Majhen, D.; Stojić-Vukanić, Z.; Savić, S.; Krajišnik, D. Nanoparticle-Based Adjuvants and Delivery Systems for Modern Vaccines. *Vaccines* **2023**, *11*, 1172. [CrossRef]
8. Terron-Exposito, R.; Dudognon, B.; Galindo, I.; Quetglas, J.I.; Coll, J.M.; Escribano, J.M.; Gomez-Casado, E. Antibodies against *Marinobacter algicola* and *Salmonella typhimurium* flagellins do not cross-neutralize TLR5 activation. *PLoS ONE* **2012**, *7*, e48466. [CrossRef]
9. Ahmadivand, S. Innovation in mRNA Vaccines and RNAi via Protein Nanocages. *Vaccines* **2025**, *13*, 653. [CrossRef] [PubMed]
10. Lamontagne, F.; Khatri, V.; St-Louis, P.; Bourgault, S.; Archambault, D. Vaccination Strategies Based on Bacterial Self-Assembling Proteins as Antigen Delivery Nanoscaffolds. *Vaccines* **2022**, *10*, 1920. [CrossRef] [PubMed]

 https://doi.org/10.3390/vaccines14020159

11. Xia, M.; Huang, P.; Vago, F.S.; Jiang, W.; Jiang, X.; Tan, M. A Nanoparticle Comprising the Receptor-Binding Domains of Norovirus and *Plasmodium* as a Combination Vaccine Candidate. *Vaccines* **2025**, *13*, 34. [CrossRef] [PubMed]
12. Wong, T.-T.; Liou, G.-G.; Kan, M.-C. A Thermal-Stable Protein Nanoparticle That Stimulates Long-Lasting Humoral Immune Response. *Vaccines* **2023**, *11*, 426. [CrossRef] [PubMed]
13. Cui, M.; Xi, J.; Shi, Z.; Zhu, Y.; Ma, Z.; Li, M.; Yang, Q.; Song, C.; Fan, L. Antigen-Dependent Adjuvanticity of Poly(lactic-co-glycolic acid)-Polyethylene Glycol 25% Nanoparticles for Enhanced Vaccine Efficacy. *Vaccines* **2025**, *13*, 317. [CrossRef]
14. Song, C.; Hu, J.; Liu, Y.; Tian, Y.; Zhu, Y.; Xi, J.; Cui, M.; Wang, X.; Zhang, B.-Z.; Fan, L.; et al. Vaccination-Route-Dependent Adjuvanticity of Antigen-Carrying Nanoparticles for Enhanced Vaccine Efficacy. *Vaccines* **2024**, *12*, 125. [CrossRef] [PubMed]
15. Su, F.; Xu, L.; Xue, Y.; Xu, W.; Li, J.; Yu, B.; Ye, S.; Yuan, X. Immune Enhancement of Nanoparticle-Encapsulated Ginseng Stem–Leaf Saponins on Porcine Epidemic Diarrhea Virus Vaccine in Mice. *Vaccines* **2022**, *10*, 1810. [CrossRef]
16. Sousa de Pinho, S.; Invenção, M.C.V.; Silva, A.J.D.; de Macêdo, L.S.; Espinoza, B.C.F.; Leal, L.R.S.; da Gama, M.A.T.M.; de Moura, I.A.; dos Santos Silva, M.E.; de Souza, D.V.S.; et al. *Pichia pastoris*-Derived β-Glucan Capsules as a Delivery System for DNA Vaccines. *Vaccines* **2024**, *12*, 1428. [CrossRef]
17. Tada, R.; Nagai, Y.; Ogasawara, M.; Saito, M.; Ohshima, A.; Yamanaka, D.; Kunisawa, J.; Adachi, Y.; Negishi, Y. Polymeric Caffeic Acid Acts as an Antigen Delivery Carrier for Mucosal Vaccine Formulation by Forming a Complex with an Antigenic Protein. *Vaccines* **2024**, *12*, 449. [CrossRef]
18. Gaglio, S.C.; Perduca, M.; Zipeto, D.; Bardi, G. Efficiency of Chitosan Nanocarriers in Vaccinology for Mucosal Immunization. *Vaccines* **2023**, *11*, 1333. [CrossRef]
19. Yang, J.; Boer, J.C.; Khongkow, M.; Phunpee, S.; Khalil, Z.G.; Bashiri, S.; Deceneux, C.; Goodchild, G.; Hussein, W.M.; Capon, R.J.; et al. The Development of Surface-Modified Liposomes as an Intranasal Delivery System for Group A Streptococcus Vaccines. *Vaccines* **2023**, *11*, 305. [CrossRef]
20. Jo, H.; Jeoung, J.; Kim, W.; Jeoung, D. Regulating Immune Responses Induced by PEGylated Messenger RNA–Lipid Nanoparticle Vaccine. *Vaccines* **2025**, *13*, 14. [CrossRef]
21. Saleh, M.; El-Moghazy, A.; Elgohary, A.H.; Saber, W.I.A.; Helmy, Y.A. Revolutionizing Nanovaccines: A New Era of Immunization. *Vaccines* **2025**, *13*, 126. [CrossRef]
22. Wang, J.; Zhao, Z.; Wang, Q.; Shi, J.; Wong, D.W.-C.; Cheung, J.C.-W. Advancements in Nanoparticle-Based Adjuvants for Enhanced Tuberculosis Vaccination: A Review. *Vaccines* **2024**, *12*, 1335. [CrossRef]
23. Tammas, I.; Bitchava, K.; Gelasakis, A.I. Advances in Vaccine Adjuvants for Teleost Fish: Implications for Aquatic Welfare and the Potential of Nanoparticle-Based Formulations. *Vaccines* **2024**, *12*, 1347. [CrossRef]
24. Brubaker, S.W.; Walters, I.R.; Hite, E.M.; Antunez, L.R.; Palm, E.L.; Funke, H.H.; Steadman, B.L. Demonstration of Tunable Control over a Delayed-Release Vaccine Using Atomic Layer Deposition. *Vaccines* **2024**, *12*, 761. [CrossRef]
25. Shalash, A.O.; Sun, H.; Cui, Y.; Wang, J.; Arnts, B.; Bauer, J.; Hussein, W.M.; Khalil, Z.G.; Skwarczynski, M.; Toth, I. Evaluation of Peptide-Based Vaccines against Group A Streptococcus in *Staphylococcus aureus*-Infected Mice. *Vaccines* **2025**, *13*, 632. [CrossRef]

https://doi.org/10.3390/vaccines14020159

vaccines

MDPI

Article

Evaluation of Peptide-Based Vaccines Against *Group A Streptococcus* in *Staphylococcus aureus*-Infected Mice

Ahmed O. Shalash [1,*], Haolan Sun [1], Yiru Cui [1], Jingwen Wang [1], Barb Arnts [2], Jannah Bauer [2], Waleed M. Hussein [3], Zeinab G. Khalil [2,3,4], Mariusz Skwarczynski [1] and Istvan Toth [1,4]

1 School of Chemistry and Molecular Biosciences, The University of Queensland, Brisbane, QLD 4072, Australia; m.skwarczynski@uq.edu.au (M.S.)
2 Australian Institute for Bioengineering and Nanotechnology, The University of Queensland, Biological Resources, Brisbane, QLD 4072, Australia
3 Institute for Molecular Bioscience, The University of Queensland, Brisbane, QLD 4072, Australia
4 School of Pharmacy and Pharmaceutical Sciences, The University of Queensland, Brisbane, QLD 4072, Australia
* Correspondence: a.shalash@uq.edu.au

Abstract: Background: *Group A Streptococcus* (GAS) is a major human pathogen associated with serious diseases. Evaluating immune responses against GAS vaccines—immunogenicity, quality, and efficacy—is complicated by interference from co-infections, like *Staphylococcus aureus* (*S. aureus*). We aimed to evaluate peptide-based GAS vaccines in mice for antisera efficacy against standard and mutant GAS strains and to assess immunological methods under co-infection conditions. Methods: Female C57BL/6 mice were infected with *S. aureus* and immunized with various M-protein-derived peptide antigens: J8, J8i, J8i-J8i, and the native p145 sequence. Two novel, conserved M-protein-derived antigens (NTD and CTD2) were also evaluated. Enzyme-linked immunosorbent assays (ELISAs) were used to assess immunogenicity and GAS-specific antibody responses. Peptide antigens were either conjugated to or physically mixed with the PADRE T-helper epitope and tested for enhanced antisera immunogenicity and opsonic efficacy. Result: ELISA against the immunizing peptides as coating antigens reflected the immunogenicity, while p145-based ELISA correlated with GAS-specific antibody titres without *S. aureus* interference for J8-based vaccines. Immunogenicity ranked J8 > J8i ≈ J8i-J8i > p145. NTD and CTD2 antisera demonstrated opsonic activity, indicating protective potential. PADRE–J8 conjugates significantly enhanced antibody magnitude and quality, producing strong opsonic bactericidal responses against both standard and p145-mutant GAS strains. PADRE–J8i was effective only against standard strains. This is the first report to suggest at least two B-cell epitopes within the J8i peptide. Conclusion: These findings support the diagnostic utility of p145, NTD, and CTD2 under co-infection settings, and the vaccine potential of J8, NTD, and CTD2, particularly when conjugated to a T helper for enhanced antigen presentation.

Keywords: *Group A Streptococcus*; vaccines; immunogenicity; immune response quality; novel antigens; Mutant bacterial strains; *S. aureus*-infected mice

1. Introduction

Group A streptococcus (GAS) infection causes a wide range of human diseases, from mild to severe life-threatening illnesses, like invasive GAS, affecting over 600,000 patients each year, with a 25% mortality rate [1–3]. As beta-lactam antibiotics continue to lose their efficacy, GAS infection remain among the top ten infectious diseases causing mortality worldwide, and there is an urgent need to develop an effective vaccine and immunological evaluation methods that are broadly applicable to suit various research settings [2,4,5].

A successful vaccine must induce high-quality, opsonic antibodies capable of recognizing GAS during infection, while immunological assays should reliably measure these responses, even in the context of co-existing or prior infections [3,6–8].

Among the streptococcal surface antigens, the M-protein is the most prominent vaccine target for GAS vaccine development. The M-protein is a major virulence factor of GAS and a well-characterized cell wall-anchored surface antigen (Figure 1). Structurally, it is an α-helical coiled-coil dimer protein, which is anchored to the bacterial cell wall via a conserved *C*-terminal LPXTG motif [9]. The coiled-coil dimeric structure of the M-protein facilitates immune evasion [10]. M-protein further supports host immune modulation by binding to host plasmin(ogen) and fibrinogen, inhibiting complement activation, degrading a key opsonin (C3b), and thus aiding in opsonization and phagocytosis escape, and it plays a role in adhesion to host cell glycans [11–14].

M-protein's *N*-terminal region (50 residues) is the emm-defining segment, and it is hypervariable [15]. The *N*-terminal region is immunogenic and surface exposed; however, it is a risky antigen source due to limited strain coverage. In contrast, the *C*-terminal membrane-proximal region is highly conserved, and thus a good source of vaccine antigens with high coverage against many GAS strains. However, full-length M-protein vaccines risk eliciting autoimmune responses due to sequence similarities with human myosin [16–19]. In addition, with over 250 serotypes exhibiting M-protein sequence variations, the usage of a single homolog of M-protein, 34 to 85% conserved (51.3 $\pm$ 17.4 average, Figure 1), or its variable fragments as a vaccine antigen will allow mutant escape and provide limited coverage against diverse GAS strains [3].

To overcome these challenges, two strategies were adopted to develop M-protein-based vaccine antigens against GAS. These strategies aimed to broaden vaccine coverage, to prevent mutant escape, and to avoid eliciting cross-reactive autoimmune responses. These strategies are either using (1) a minimal and conserved *C*-terminal-derived immunogenic peptide antigen (like J8 peptide), or (2) a modified chimeric protein combining multiple strain-specific *N*-terminal fragments (like the StreptAnova® vaccine antigen) [20,21].

The J8 peptide antigen (**QAEDKVKQ**-SREAKKQVEKAL-**KQLEDKVQ**, bold residues are derived from yeast GCN-4 protein [22]) was designed to include a highly conserved sequence; however, its M-protein-derived sequence content is only 58 to 100% conserved (80.3 $\pm$ 19.8 average) and reported to be found in only 37% of GAS genomes [3]. J8 was also designed to avoid a shared epitope with human myosin in the native p145 epitope sequence (**LRRDLDA**-SREAKKQVEKALE, bold residues are shared with human myosin, 60 to 100% conserved, 80.0 $\pm$ 19.5 average). J8 is also flanked with GCN-4 sequences at both termini of the minimalist epitope J8i (SREAKKQVEKAL, 58 to 100% conserved, 80.3 $\pm$ 19.8 average) to impart an equivalent helicity to the p145 original native sequence's conformation. The J8 peptide is known to induce broadly protective opsonic antibodies. M-protein is completely helical in structure, and thus imparting helicity to M-protein-derived antigens such as J8 in peptide vaccines significantly improves their immunogenicity and efficacy [6,23]. To further assist the presentation of J8 to antigen-presenting cells, various approaches have been adopted, such as (a) conjugation to a carrier protein, like diphtheria toxoid [24], or (b) conjugation of a hydrophobic moiety, with adjuvanting properties, to enable self-assembly into an immunogenic nanovaccine, such self-adjuvanting moieties including polyhydrophobic amino acids, like polyleucine, or lipid core peptides [6,7].

```
GCA_900983385.1_emm221.0          MVRKDTNRQYSLRKLKKGTASVAVALTVVGAGLASQTEVKADQPVDHHRY
SERO_012                          MAKNTTNRHYSLRKLKTGTASVAVALTVVGAGLVAGQTVRADHSDLVAEK
GCA_900983395_emm221.0            MAKNTTNRHYSLRKLKTGTASVAVALTVVVAGHTVKANDDITSLTPILSG
SERO_06                           MAKNNTNRHYSLRKLKKGTASVAVALSVIGAGLVVNTNEVSARVFPRGTV
SERO_05                           MARENTNKHYWLRKLKKGTASVAVALSVLGAGLVVNTNEVSAAVT-RGTI
SERO_24                           MTKNNTNRHYSLRKLKTGTASVAVALTVLGAGLVVNTNEVSAVAT-RSQT
GCA_900983435_emm18.12 NS1944     MARKDTNKQYSLRKLKTGTASVAVAVAVLGAGFANQTTVKANSKTPAPAP
SERO_02F                          MARKDTNKQYSLRKLKTGTASVAVAVAVLGAGFANQTTVKANSKNPVP--
GCA_900983335_emm221.0            MTRQQTKKNYSLRKLKTGTASVAVALTVLGAGFANQTEVKAEEAKKTVPY
GCA_900995335_emm18.12 NS5582     MTRQQTKKNYSLRKLKTGTASVAVALTVLGAGFANQTEVKAEGVQFDWEK
SERO_02                           MARQQTKKNYSLRKLKTGTASVAVALTVLGAGFANQTEVRADEAKKMEVK
SERO_M49                          MARQQTKKNYSLRKLKTGTASVAVALTVLGAGFANQTEVRAEGVNATTS-
                                  *.:: *:::* *****.********::*: ** .

GCA_900983385.1_emm221.0          L RRDLDASREAKKQVEKALEEANSKLAALEKLNKELEESKKLTEKEKAELQ
SERO_012                          L RRDLDASREAKKQVEKALEEANSKLAALEKLNKDLEESKKLTEKEKAELQ
GCA_900983395_emm221.0            L RRDLDASREAKKQVEKALEEANSKLAALEKLNKELEESKKLTEKEKAELQ
SERO_06                           L RRDLDASREAKKQVEKALEEANSKLAALEKLNKELEESKKLTEKEKAELQ
SERO_05                           L RRDLDASREAKKQVEKALEEANSKLAALEKLNKELEESKKLTEKEKAELQ
SERO_24                           L RRDLDASREAKKQVEKALEEANSKLAALEKLNKELEESKKLTEKEKAELQ
GCA_900983435_emm18.12            L SRDLEASREAKKKVEADLAEANSKLQALEKLNKELEEGKKLSEKEKAELQ
SERO_02F                          L SRDLEASREAKKKVEADLAEANSKLQALEKLNKELEEGKKLSEKEKAELQ
GCA_900983335_emm221.0            L SRDLEASREANKKVTSELTQAKAQLSALE-------ESKKLSEKEKAELQ
GCA_900995335_emm18.12            L SRDLEASREANKKVTSELTQAKAQLSAFE-------ESKKLSEKEKAELQ
SERO_02                           L SRDLEASREANKKVTSELTQAKAQLSALE-------ESKKLSEKEKAELQ
SERO_M49                          L SRDLEASREANKKVTSELTQAKAQLSALE-------ESKKLSEKEKAELQ
                                  *  ***:*****:*:*    * :*:::* *:*       *.***:********

GCA_900983385.1_emm221.0          AKLEAEAKALKEKLAKQAEELAKLRAEKASDSQTPDAKPGNKAVPGKGQA
SERO_012                    41%   AKLEAEAKALKEQLAKQAEELAKLRAGKASDSQTPDAKPGNKAVPGKGQA
GCA_900983395_emm221.0      34%   AKLEAEAKALKEQLAKQAEELAKLRAGKASDSQTPDTKPGNKAVPGKGQA
SERO_06                     51%   AKLEAEAKALKEQLAKQAEELAKLRAGKASDSQTPDAKPGNKVVPGKGQA
SERO_05                     85%   AKLEAEAKALKEQLAKQAEELAKLRAGKASDSQTPDTKPGNKAVPGKGQA
SERO_24                     38%   AKLEAEAKALKEKLAKQAEELAKLRAGKASDSQTPDAKPGNKAVPGKGQA
GCA_900983435_emm18.12      41%   ARLEAEAKALKEQLAKQAEELAKLKGNQTPN-------------------
SERO_02F                    44%   AKLEAEAKALKEQLAKQAEELAKLKGNQTPN-------------------
GCA_900983335_emm221.0      34%   AKLDAQGKALKEQLAKQAEELAKLRAEKAAGSKTPATKPANKERSGR--A
GCA_900995335_emm18.12      47%   AKLDAQGKALKEQLAKQAEELAKLRAEKAAGSKTPATKPANKERSGR--A
SERO_02                     73%   AKLDAQGKALKEQLAKQTEELAKLRAEKAAGSKTPATKPANKERSGR--A
SERO_M49                    71%   AKLDAQGKALKEQLAKQTEELAKLRAEKAAGSKTPATKPANKERSGR--A
                                  *:*:*:.*****:****:******:.  ::..
```

Figure 1. Partial sequences from full-length M-protein alignment using Clustal-w, from different GAS serotypes, including serotypes 2, 2F, 4, 5, 6, 12, 24, M49, M221.0, and M18.12 [GCA_900983435 (NS1944) and GCA_900995335 (NS5582) mutant strains], obtained from Uniprot and the protein database. Derived peptides are NTD (LRKLKKGTASVAVAL; highlighted in green colour, top); p145 (LRRDLDASREAKKQVEKALE, enclosing J8i (SREAKKQVEKAL), highlighted in pink, middle); and CTD2 (KALKEQLAKQAEELAKLR, highlighted in blue, bottom), asterisks represent highly conserved residues among GAS strains.

In contrast, the other approach employs the hypervariable *N*-terminal segment of M-protein, but due to its inter-strain variability, the StreptAnova® vaccine combines 30-valent antigens, each of 50 amino acids, from different GAS strains [20,21], into a single chimeric protein sequence that protects against these strains, imparting broader coverage. However, protection is limited mostly to the 30 GAS strains, from which antigens were derived [8,21]. Therefore, in either strategy, the antigens are intentionally abbreviated or have minimalist immunogenic components to prevent the autoantigen sequences located at the *C*-terminus of M-protein from inciting autoimmune responses, thus preventing the generation of cross-reactive antibodies [8]. Thus, when developing GAS vaccines using such derived and non-native antigen sequences, e.g., GCN-4 sequences or carrier proteins, the quality of the elicited immune responses and the absence of cross-reactivity are constantly being investigated [17–19], and further improvements in the quality of immune responses could result in dramatic efficacy improvements (Figure 2) [25]. For example, a significant proportion of generated antibodies against J8 has been hypothesized to be directed against the irrelevant GCN-4 flanking sequences (Figure 2A).

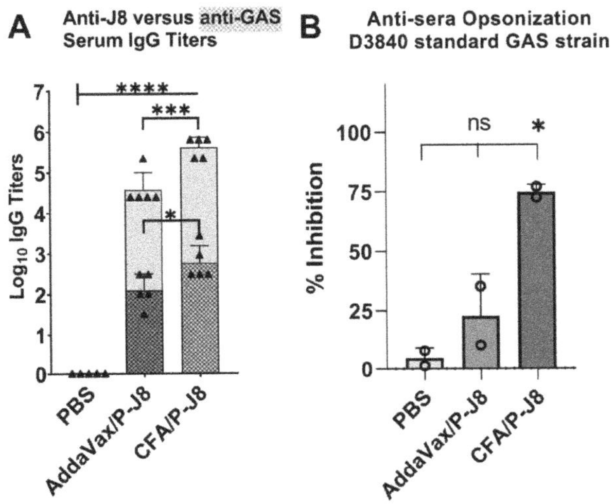

Figure 2. Immune responses and efficacy of J8, adapted from [25]. (**A**) Serum antigen-specific IgG generated against PADRE-J8 (P-J8) peptide antigen in C57BL/6 mice (n = 5), where the shaded portion represents anti-GAS titres (GAS bacteria coated on ELISA plates) compared to the anti-J8 peptide titres, while the gaps represent 300- and 600-fold for AddaVax and CFA vaccines, respectively, between generated anti-peptide titres and anti-GAS high-quality antibodies. (**B**) The opsonic capacity of those 10-fold diluted antisera generated against J8, with a small difference between the two vaccines, resulted in a dramatic change in opsonization bactericidal efficacy. One-way ANOVA with Dunnett's multiple comparisons test was used for statistical analysis. Significance was set to $p < 0.05$ (*), $p < 0.001$ (***), and $p < 0.0001$ (****).

Ongoing research is being conducted to discover novel, highly conserved, and immunogenic peptide antigens, derived from various surface proteins, to be employed in developing next-generation GAS vaccines, like FNBR and S2 [26]. Such antigens could be useful in triggering broader immune responses and reducing mutant escape.

Gram-positive and immobile *Staphylococcus aureus* (*S. aureus*) is a type of cocci bacteria, which infects all mammalian species and is commonly harboured in the nasopharynx of 30% of healthy humans. Human infections cause septicaemias and localised inflammatory diseases, but the majority of infected animals and humans are asymptomatic. Thus, infected immunocompetent mice do not exhibit symptoms of illness. Infection is mucosal and predominantly localizes to the epidermis, gastrointestinal system, and nasopharynx, which are the primary sites of *S. aureus* colonization [27,28]. Further, the infection is easily transmissible via several different routes, including contaminated aerosol or direct contact with fomites, from infected animals, where animals housed/caged together are expected to share the infection through breathing contaminated aerosols or via contact during grooming. Such a common infection will serve to establish potentially interfering or cross-reactive immune responses against GAS that originally result from a pre-existing *S. aureus* infection.

In this study, we aim to (1) discover novel, conserved, and minimalist surface-derived peptide antigens, and (2) establish immunological assays that can evaluate the magnitude, type, and quality of generated antibodies against native streptococcal surface antigens in the presence of a co-infection. We seek develop robust immunological assays in the presence of potential interference from prior or co-existing infections, more similar to clinical settings. Thus, we initially evaluated M-proteins from over 12 diverse GAS strains (Figure 1) and identified two highly conserved sequences, NTD (LRKLKKGTASVAVAL, 87 to 100% conserved, 93.3 ± 4.2 on average) and CTD2 (KALKEQLAKQAEELAKLR, 89 to 100% conserved, 97.0 ± 3.5 on average), compared to J8i or p145. These peptides were selected based on their surface exposure, immunogenic potential, and ability to induce bactericidal antibodies across GAS strains (strain coverage), including those with p145 sequence mutations. To this end, we synthesized J8-derived peptide antigens, including J8, J8i, p145, and J8i-J8i dimer, as well as the new antigens, characterized their secondary structure, and adjuvanted with either complete Freund's adjuvant (CFA) or 15-mer polyleucine to immunize C57BL/6 mice, which were infected, since birth, with *S. aureus*. We also compared the outcomes to immunized, non-infected mice, with CFA-immunized J8 or saline as controls (Figure 2). We conducted ELISA against each peptide (J8, J8i, J8i-J8i, p145, NTD, and CTD2) and used carvacrol-inactivated whole-GAS bacteria as ELISA coating antigens. Finally, we conducted a bactericidal/opsonization assay for the antisera. Ultimately, the immune responses against each antigen correlated with the recognition of whole GAS bacteria or the overall bactericidal efficacy. In other words, our goal was to investigate which of the ELISA coating antigens (J8, J8i, p145, NTD, CTD2, or whole inactivated GAS) best represented each of the following: (1) the magnitude, (2) quality/avidity, or (3) bactericidal efficacy of generated antigen-specific antibodies against GAS. Moreover, we also evaluated the usefulness of the two newly discovered peptide antigens.

2. Materials and Methods

2.1. Materials

Analytical-grade reagents were used throughout the experiments. Dichloromethane (DCM), *N,N*-dimethylformamide (DMF), chromatography-grade methanol (MeOH) and acetonitrile (MeCN), and peptide-grade piperidine were obtained from Merck (Kilsyth, VIC, Australia). *N,N,N′,N′*-tetramethyl-O-(7-azabenzotriazol-1-yl)uroniumhexafluorophosphate (HATU) was obtained from Mimotopes (Melbourne, VIC, Australia). Trifluoroacetic acid (TFA) and diisopropylethylamine (DIPEA) were acquired from Novachem (Heidelberg West, VIC, Australia). 2,2,2-Trifluoroethanol (TFE) was obtained from ThermoFisher Scientific (Banyo, QLD, Australia). Fluoronylmethoxycarbonyl (FMOC)-protected amino acids and Rink amide resin (100–200 mesh, substitution value of 0.56 mmol/g) were acquired from Novabiochem (Läufelfingen, Switzerland). Horseradish peroxidase (HRP)-goat anti-mouse IgG, phosphate buffer (PBS), O-phenylenediamine dihydrochloride (OPD) substrate, and triisopropylsilane (TIPS) were obtained from Sigma Aldrich (Castle Hill, NSW, Australia). Microplates (96-well, flat base, polystyrene, high binding) were obtained from Sarstedt (Mawson Lakes, SA, Australia).

2.2. Equipment

Reverse-phase analytical liquid chromatography (RP)-HPLC was employed for measuring peptide purity using a Shimadzu HPLC system (SPD-M10A VP, LC-20AB, DGU-20A5, SIL-20AC HT, Shimadzu, Kyoto, Japan) equipped with a Grace Vydac 214TP C4 5 μm column or Vydac 218TP C18 5 μm column. The mobile-phase flow rate was set to 1 mL/min with gradient elution using a mixture of solvent A (100% Milli-Q water and 0.1% TFA) and solvent B (90% MeCN in Milli-Q water and 0.1% TFA). Preparative HPLC was employed for peptide purification on a Shimadzu HPLC system (SPD-20A, LC-20AP, CBM-20A, FRC-10A, Shimadzu, Kyoto, Japan) equipped with a C4 10 μm (Grace Vydac,

214TP1022) or C18 10 μm (Grace Vydac, 218TP1022) column. The mobile-phase flow rate was set to 20 mL/min using a gradient of solvents A and B. Synthesized peptide identity was confirmed via electrospray ionization mass spectrometry (ESI-MS) (PerkinElmer-Sciex API3000, Concord, ON, Canada). ESI-MS solvent was a 1:1 mixture of solvent C (100% Milli-Q water and 0.1% AcOH) and solvent D (90% MeCN in Milli-Q water and 0.1% AcOH) and was set to a flow rate of 0.5 mL/min. Particle size distribution of self-assembled peptide vaccines was measured using dynamic light scattering (DLS) (zeta sizer 3000TM, Malvern Instruments, Malvern, UK). The secondary structure of the peptide antigens was analysed using a Jasco-720 spectropolarimeter (Jasco Corp., Tokyo, Japan). Optical density (OD) values of immunological assays were measured by the SpectraMAX 250 plate reader (Molecular Devices, San Jose, CA, USA).

2.3. Peptide Synthesis

Peptide antigens (Figure 3 and Table 1) were synthesized using standard Fmoc solid-phase peptide synthesis (SPPS) as previously described [6,29]. Briefly, a 0.1 mmole scale of Rink amide resin was weighed for each peptide and allowed to swell in DMF overnight. After swelling, DMF was removed and a solution of 20% piperidine in DMF was added twice (5 min × 2) to remove Fmoc protection groups on the functionalized resin reaction sites. After deprotection of resin, the first C-terminal FMOC-protected amino acid in the sequence was weighed (4.2 equivalents) and activated using HATU (4.0 equivalents) and DIPEA (6.2 equivalents) in DMF, then added to the resin for coupling twice (20 min × 2). The resin was washed five times with DMF and deprotected again as mentioned above, followed by coupling the following fmoc-protected amino acid in the peptide sequence. Cycles of deprotection and coupling were continued till the entire peptide antigen sequence was complete. The peptidyl-resin was deprotected from N-terminal Fmoc once more, washed with DMF (×3), DCM (×3), and MeOH (×3), and left to dry overnight under vacuum at room temperature (RT). Each peptide antigen was cleaved from the resin using a 10 mL mixture of TFA, Milli-Q water, and TIPS (95:2.5:2.5) while stirring (250v RPM) for 4 h at RT. Afterwards, the resin was removed by filtration, and the TFA-containing peptide solution was dried under a vacuum. The dried peptide was washed with ice-cold diethyl ether: *N*-hexane three times, to remove the scavengers. The precipitated peptide antigen was filtered under vacuum, redissolved in a suitable volume of acetonitrile: water (1:1) solution, and purified using preparative HPLC.

Table 1. Sequence of all synthesized peptides.

Peptide	Peptide Sequence	Analytical Data	Purity	Yield
J8	QAEDKVKQSREAKKQVEKALKQLEDKVQ Molecular weight: 3282.6 g/mol.	ESI-MS: $[M + 3H]^{+3}$ m/z 1095.8 (cal. 1095.3), $[M + 4H]^{+4}$ m/z 821.0 (cal. 821.5); HPLC, t_R = 20.0 min (C18 column, 0–100% solvent B, 53 min)	98%	53%
J8i	SREAKKQVEKAL Molecular weight: 1385.6 g/mol.	ESI-MS: $[M + H]^{+}$ m/z 1387.9 (cal. 1387.6), $[M + 2H]^{+2}$ m/z 693.5 (cal. 694.5); HPLC, t_R = 15.4 min (C18 column, 0–100% solvent B, 53 min)	97%	55%
J8i-J8i	SREAKKQVEKALSREAKKQVEKAL Molecular weight: 2754.2 g/mol.	ESI-MS: $[M + 3H]^{+3}$ m/z 920.0 (cal. 919.1), $[M + 4H]^{+4}$ m/z 690.6 (cal. 689.6), $[M + 5H]^{+5}$ m/z 552.0 (cal. 551.8); HPLC, t_R = 19.4 min (C18 column, 0–100% solvent B, 53 min)	93%	57%

Table 1. *Cont.*

Peptide	Peptide Sequence	Analytical Data	Purity	Yield
P145	LRRDLDASREAKKQVEKALE Molecular weight: 2354.7 g/mol.	ESI-MS: $[M + 2H]^{+2}$ m/z 1177.9 (cal. 1178.4), $[M + 3H]^{+3}$ m/z 785.1 (cal. 785.9), $[M + 4H]^{+4}$ m/z 588.7 (cal. 589.7); HPLC, t_R = 19.3 min (C4 column, 0–100% solvent B, 53 min)	95%	52%
NTD	LRKLKKGTASVAVAL Molecular weight: 1553.9 g/mol.	ESI-MS: $[M + 2H]^{+2}$ m/z 777.7 (cal. 778.0), $[M + 3H]^{+3}$ m/z 518.5 (cal. 519.0); HPLC, t_R = 18.9 min (C18 column, 0–100% solvent B, 53 min)	96%	55%
CTD2	KALKEQLAKQAEELAKLR Molecular weight: 2066.5 g/mol.	ESI-MS: $[M + 2H]^{+2}$ m/z 1034.8 (cal. 1034.3), $[M + 3H]^{+3}$ m/z 685.6 (cal. 689.8); HPLC, t_R = 19.9 min (C18 column, 0–100% solvent B, 53 min)	97%	53%
P-J8	AKFVAAWTLKAAA- QAEDKVKQSREAKKQVEKALKQLEDKVQ Molecular weight: 4653.4 g/mol.	ESI-MS: $[M + 4H]^{+4}$ m/z 1164.8 (cal. 1164.4), $[M + 5H]^{+5}$ m/z 932.2 (cal. 931.7), $[M + 6H]^{+6}$ m/z:776.9 (cal. 776.6); HPLC, t_R = 21.2 min (C18 column, 0–100% solvent B, 53 min)	96%	43%
P-J8i	AKFVAAWTLKAAA-SREAKKQVEKAL Molecular weight: 2715.2 g/mol.	ESI-MS: $[M + 2H]^{+2}$ m/z 1357.9 (cal. 1358.6), $[M + 3H]^{+3}$ m/z:907.0 (cal. 906.1), $[M + 4H]^{+4}$ m/z:681.4 (cal. 679.8); HPLC, t_R = 28.0 min (C18 column, 0–100% solvent B, 53 min)	98%	45%
P-J8i-J8i	AKFVAAWTLKAAA- SREAKKQVEKALSREAKKQVEKAL Molecular weight: 4083.8 g/mol.	ESI-MS: $[M + 3H]^{+3}$ m/z 1363.1 (cal. 1362.3), $[M + 4H]^{+4}$ m/z: 1023.1 (cal. 1022.0), $[M + 5H]^{+5}$ m/z: 818.5 (cal. 817.8); HPLC, t_R = 21.9 min (C18 column, 0–100% solvent B, 53 min)	97%	46%
L_{15}-P-J8	LLLLLLLLLLLLLLL- AKFVAAWTLKAAA- QAEDKVKQSREAKKQVEKALKQLEDKVQ Molecular weight: 6350.8 g/mol.	ESI-MS: $[M + 5H]^{+5}$ m/z 1269.8 (cal. 1271.2), $[M + 6H]^{+6}$ m/z: 1059.7 (cal. 1059.5), $[M + 7H]^{+7}$ m/z: 908.5 (cal. 908.3); HPLC, t_R = 37.2 min (C4 column, 0–100% solvent B, 53 min)	96%	23%
L_{15}-P-J8i	LLLLLLLLLLLLLLL- AKFVAAWTLKAAA-SREAKKQVEKAL Molecular weight: 4454.7 g/mol.	ESI-MS: $[M + 4H]^{+4}$ m/z: 1115.2 (cal. 1114.7), $[M + 5H]^{+5}$ m/z: 892.5 (cal. 891.9); HPLC, t_R = 38.4 min (C4 column, 0–100% solvent B, 53 min)	95%	18%
L_{15}-P-J8i-J8i	LLLLLLLLLLLLLLL- AKFVAAWTLKAAA- SREAKKQVEKALSREAKKQVEKAL Molecular weight: 5783.1 g/mol.	ESI-MS: $[M + 5H]^{+5}$ m/z 1157.2 (cal. 1157.6), $[M + 6H]^{+6}$ m/z: 964.8 (cal. 964.9), $[M + 7H]^{+7}$ m/z: 826.9 (cal. 827.2); HPLC, t_R = 35.5 min (C4 column, 0–100% solvent B, 53 min)	95%	21%
L_{15}-P-p145	LLLLLLLLLLLLLLL- AKFVAAWTLKAAA- LRRDLDASREAKKQVEKALE Molecular weight: 5381.7 g/mol.	ESI-MS: $[M + 4H]^{+4}$ m/z 1346.8 (cal. 1346.4), $[M + 5H]^{+5}$ m/z: 1076.5 (cal. 1077.3), $[M + 6H]^{+6}$ m/z: 897.5 (cal. 898.0); HPLC, t_R = 34.2 min (C4 column, 0–100% solvent B, 53 min)	96%	19%

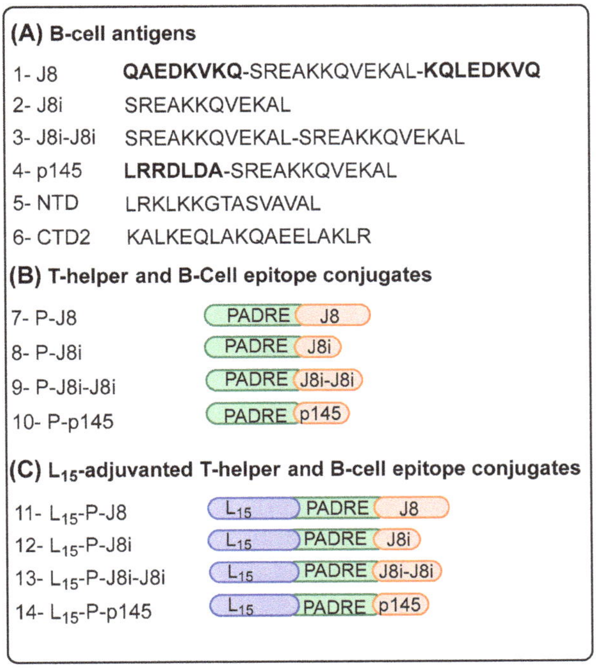

Figure 3. Synthesized peptides: (**A**) B-cell antigens, (**B**) T-helper epitope conjugates with B-cell antigens, and (**C**) 15-mer polyleucine (L_{15}, blue oval), T-helper epitopes (PADRE: (P) green oval), and B-cell epitopes (orange oval).

2.4. Preparation of Vaccines

The vaccines included CFA-adjuvanted peptide antigens or 15-mer polyleucine adjuvanted compounds and were all adjusted to a final antigen concentration of 0.5 mg/mL, to deliver 25 µg per 50 µL dose/mouse. CFA vaccines were prepared by vortexing the adjuvant (bacterium suspension in mineral oil) for 1 min and taking the required volume of adjuvant and mixing 1:1 v/v with the peptide antigen solution in PBS. Afterwards, CFA emulsification was conducted using an ultrasound probe homogenizer by sonicating at 40% power and 40% pulser for 5 min in an ice bath. Meanwhile, polyleucine compounds were simply dissolved in PBS, to achieve an antigen concentration of 0.5 mg/mL in PBS, and vortexed for 1 min.

2.5. Characterization of Vaccines: Size, Shape, and Structural Conformation

The particle size and polydispersity (PDI) of self-assembled polyleucine vaccines were evaluated using DLS at a scattering angle of 173° and 25 °C using disposable cuvettes. The vaccine candidates were dissolved at a concentration of 0.1 mg/mL using PBS, and the size was presented as an average of 5 measurements.

Transmission electron microscopy (TEM) was used to confirm the self-assembled polyleucine vaccines' sizes and investigate their shapes. The samples (0.1 mg/mL concentration) were dropped onto a carbon-coated grid and were stained using uranyl acetate and dried before imaging via a Joel electron microscope (JEM-1010, JEOL Ltd., Tokyo, Japan) operated at 100 kV.

The secondary structure of the peptide antigens was analysed using circular dichroism. Each peptide was dissolved at a concentration of 0.25 mg/mL in 10 mM PBS, a volume of 200 µL was added to a 1 mm quartz microcuvette, and we performed measurements, then a further 25% TFE was added, mixed well, and remeasured. Quantitative secondary structural determination was performed on the resulting molar ellipticity data, where the resulting composite spectra were fitted to yield the contribution of each pure spectral component, i.e., β-sheet, random coil, or α-helix spectrum. The individual pure spectra were previously reported using polylysine peptide [30], and the quantitative secondary structure was determined as described previously [6].

2.6. *In Vivo Assessment in Mice*

The mice were obtained from Australian Bio Resources (ABR, New South Wales, Australia), and animal experimentation procedures were approved by the University of Queensland Animal Ethics Committee (AEC), ethics number 2021/AE000311. Naïve (uninfected) or *S. aureus*-infected C57BL/6 female mice (6–8 weeks old) were randomly housed in groups of four. After acclimatization for a week, pre-immunization, 10 μL of blood was collected from each mouse two days before immunization (-2d) from the tail tip and diluted 10-fold in PBS. CFA-immunized mice were subcutaneously immunized, and blood was collected 42 days post-immunization (42d) after euthanasia. Polyleucine vaccine-immunized mice were immunized on days 0, 14, and 28, then sera were collected 42 days post-immunization after euthanasia. Mice were generally euthanized using CO_2 on the 42nd day post-immunization, and blood was collected via heart puncture. The collected blood (undiluted or diluted) was centrifuged at 3600 rpm for 10 min, and sera were transferred and stored in centrifuge tubes at −80 °C for further immunological analyses.

Mice were infected since birth by *S. aureus*, and infection was confirmed by capillary electrophoresis. Faecal extracts were tested for *Staphylococcus aureus* through 3 different PCR primer pairs that specifically target the *S. aureus* DNA sequence of the Sa422, GyrA, and Spa genes, while excluding the DNA sequence of other bacterial pathogens. This ensured that all *S. aureus* species would be detected if present; it also allowed for any false positives to be identified and ruled out. The testing was performed on faecal samples pooled from each cage. Bacterial DNA extraction was performed using a Qiagen QIAamp Power Fecal Pro extraction kit. Extracts were tested with generic bacterial 16S primers to verify successful extraction and that the negative results were valid. The testing was performed using fluorescently labelled PCR primers, and then the PCR products were detected via capillary electrophoresis on an Applied Biosystems 3730 DNA analyser. The results were then viewed as fluorescence chromatograms in specialised software. Initial positive PCR results, as well as positive PCR results from a ZymoBIOMICS Microbial Community Standard control (including *S. aureus*), were also sequenced by Sanger sequencing to validate the test results.

2.7. *Evaluation of Serological Immune Responses: Immunogenicity (Against Peptides), Avidity (Against GAS), and the Cross-Reactivity Between the Two Bacteria*

The enzyme-linked immunosorbent assay (ELISA) was performed to determine the antigen-specific IgG antibody titre. Coating was performed using each peptide antigen or whole inactivated GAS bacteria, as described previously [31], to determine the magnitude, type, quality, or potential cross-reactive immune responses. To this end, serum antibody titres against the same peptide antigen used in vaccination represented immunogenicity and/or type of triggered immune response. Meanwhile, serum antibody titres against native M-protein derived sequences (p145, J8i, NTD, and CTD2) and against whole GAS bacteria represented immune responses of high-quality and cross-reactive ones between GAS and *S. aureus*. Microplates (96-well) were coated with peptide antigen at a concentration of 0.3 μg/well by dissolving peptide in a pH 9.6 carbonate coating buffer (CCB) and incubating at 37 °C for 90 min. Alternatively, for a whole inactivated GAS bacterium coating, bacterium suspension at a concentration of 2.5×10^8 CFU/mL in CCB was added to each well and left to bind overnight at 4 °C. The plate contents were discarded and

250 μL/well of 5% *w*/*v* skim milk solution was added and incubated at 37 °C for 90 min, then plates were washed. Serum samples were added to the first row at a 1:100 final dilution and 3-fold serially diluted down the columns in 0.5% *w*/*v* skim milk solution. and plates were incubated at 37 °C for 90 min followed by washing. HRP-goat anti-mouse IgG secondary antibodies were diluted 1:3000 in 0.5% *w*/*v* skim milk solution and added to the plates. Plates were incubated to bind the secondary antibodies, then they were washed. Finally, OPD substrate was added (100 μL/well) and left to react for 20 min, then the stop solution (1N sulphuric acid) was added (100 μL/well), and plates were measured for absorbance using a plate reader.

2.8. Opsonization Assay

An opsonization assay was employed to establish the bactericidal inhibition of vaccine antisera, as reported previously [6,7] using '10-fold diluted sera' to detect more significant opsonic antisera and thus the most promising antigens for future vaccine development. Two clinical GAS isolates were used: D3840 (nasopharynx swabs) and ACM-2727 (Royal Brisbane Hospital). These strains have highly conserved p145/J8i in their M-protein sequence. In addition, another two strains, NS1944 and NS5582, which have mutant J8i sequences, are only 75% and 58.3% conserved compared to the original J8i sequence, respectively, where mutations are located at the C-terminus of J8i (Figure 1). Briefly, high-quality antibodies recognize and bind to GAS and induce antibody-dependent phagocytosis of bacteria by macrophages. For each GAS strain, bacteria were cultured by transferring a single colony to Todd–Hewitt broth with 5% yeast extracts in a 50 mL Falcon tube and incubating at 37 °C for 24 h to obtain a concentration of 10^7 CFU/mL. The culture is then diluted 100-fold in PBS. In a 96-well microplate, a 10 μL volume of the diluted culture was transferred, followed by adding 10 μL of heat-inactivated (50 °C, 10 min) murine sera and 80 μL of horse blood, the contents were mixed, and the microplates were incubated. The opsonization activity was calculated as the percentage reduction in CFU compared to the PBS negative control group sera. The assay was performed in duplicate. The bactericidal/inhibitory effect against the GAS strains was presented as a percentage compared to the non-inhibited control (Equation (1)). In addition, cross-clade inhibition can be calculated using variance of inhibition % among the four tested strains, and opsonization efficacy represented as log reduction in GAS bacteria can be normalized by the titres to produce an opsonic capacity index to rank the tested peptide vaccines in their potential opsonic capacity, regardless of their current immunogenicity in the current experiment (Equation (2)). The seeded count of GAS bacteria per test replicate is 10^3 CFU, and the percentage inhibition can be converted to log reduction first before normalization by serum anti-peptide antibody titres (Equation (2)).

$$\textbf{Inhibition \%} = \left[1 - \frac{\textbf{CFU}_{\textbf{Sera}}}{\textbf{CFU}_{\textbf{Medium}}}\right] \times 100 \quad (1)$$

$$\text{Opsonic Capacity Index} = \frac{Log_{10}1000 - Log_{10}[10 \times (100 - inh\%)]}{Log_{10}anti - peptide\ IgG\ titers} \quad (2)$$

2.9. Data Analysis

The main statistical test employed was one-way ANOVA, followed by Dunnett's multiple comparisons test. Significance was set to $p < 0.05$ (*), $p < 0.01$ (**), $p < 0.001$ (***), and $p < 0.0001$ (****). In addition, OriginPro 2020 (OriginLab Corporation, Northampton, MA, USA) and GraphPad Prism v8 (GraphPad Software Inc., San Diego, CA, USA) were used for the mathematical and statistical presentation of the data.

3. Results

3.1. Peptide Synthesis and Characterization

Peptides were synthesized using Fmoc-based SPPS (Figure 4, Table 1) and their purity and identity were confirmed using RP-HPLC and ESI-MS, respectively. The synthesized peptides included four self-adjuvanted polyleucine-based vaccines (L_{15}-P-J8, L_{15}-P-J8i, L_{15}-P-p145, and L_{15}-P-J8i-J8i). In addition, the synthesized peptides also included antigens either conjugated to PADRE T-helper epitope (P-J8i, P-J8, P-J8i-J8i, P-p145) or those physically mixed with it (P+J8i, P+J8, P+J8i-J8i, P+p145, P+NTD, P+CTD2).

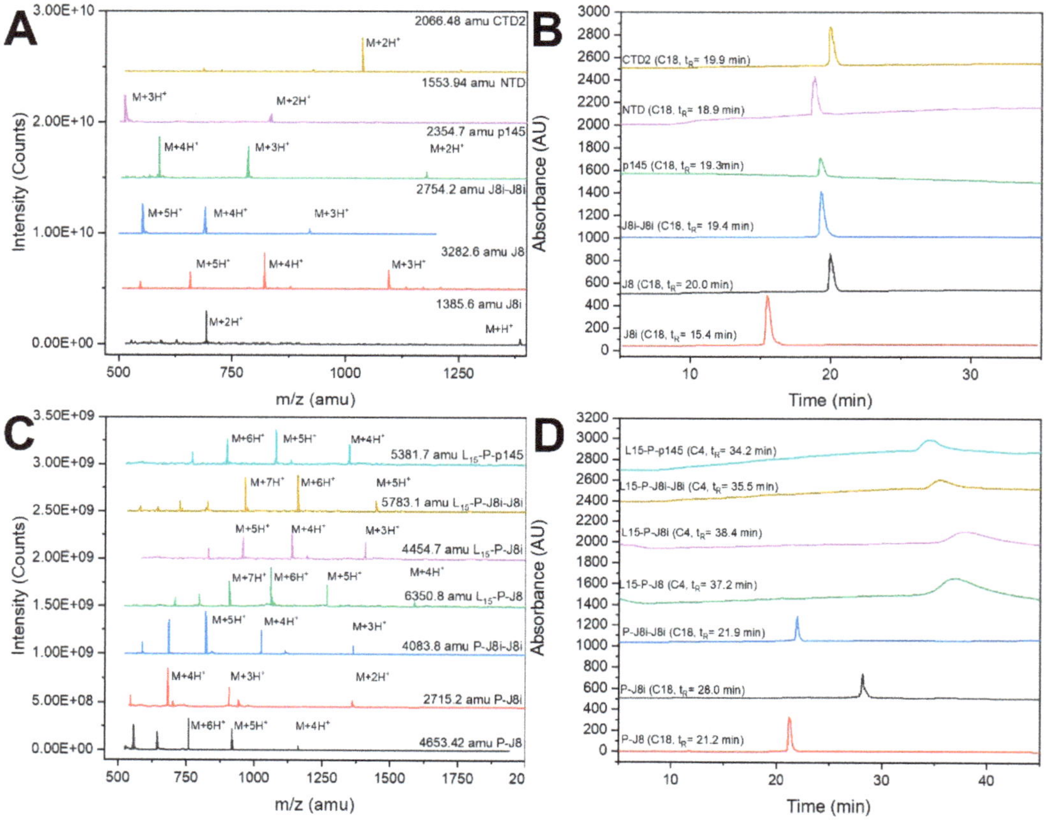

Figure 4. Synthesized peptides: (**A**,**B**) ESI-MS spectra and HPLC chromatograms of B-cell antigens, (**C**,**D**) ESI-MS spectra and HPLC chromatograms of T-helper epitope conjugates with B-cell antigens, with or without 15-mer polyleucine.

3.2. Size and Shape of Self-Assembled Nanovaccines

Peptides were dissolved in PBS, and the size distribution of self-assembled polyleucine peptide nanovaccines was evaluated via DLS. L_{15}-P-J8i self-assembled into microparticles with a peak diameter of 2.2 ± 0.3 μm, including aggregates at a larger size, as the polydispersity index was very high, 1.0, and this size was significantly larger than the other polyleucine peptide vaccines. This was attributed to the small size of the hydrophilic J8i peptide compared to the hydrophobic sequence of PADRE and L_{15} at the construct's *N*-terminus (Figure 5). L_{15}-P-J8, L_{15}-P-p145, and L_{15}-P-J8i-J8i self-assembled mainly into nanovaccines with peak sizes of 235 ± 13 nm, 87 ± 4 nm, and 81 ± 4 nm, with polydispersity indices of 0.62 ± 0.08, 0.87 ± 0.14, and 0.73 ± 0.05. L_{15}-P-p145, and L_{15}-P-J8i-J8i also distinctly included aggregates at microparticles of 1.8 ± 0.8 μm and 1.3 ± 0.2 μm in diameter. L_{15}-P-J8, L_{15}-P-p145, L_{15}-P-J8i, and L_{15}-P-J8i-J8i self-assembled into micro-/nanoparticles with z-average values of 484 ± 38, 427 ± 158, 2818 ± 650, and 135 ± 3 nm, respectively. Moreover, the TEM micrographs showed sizes that matched the DLS measurements (Figure 6).

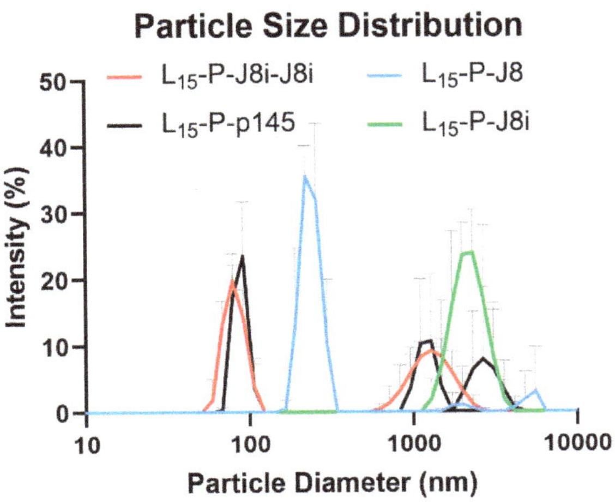

Figure 5. Size distribution of self-assembled 15-mer polyleucine vaccines in PBS via DLS.

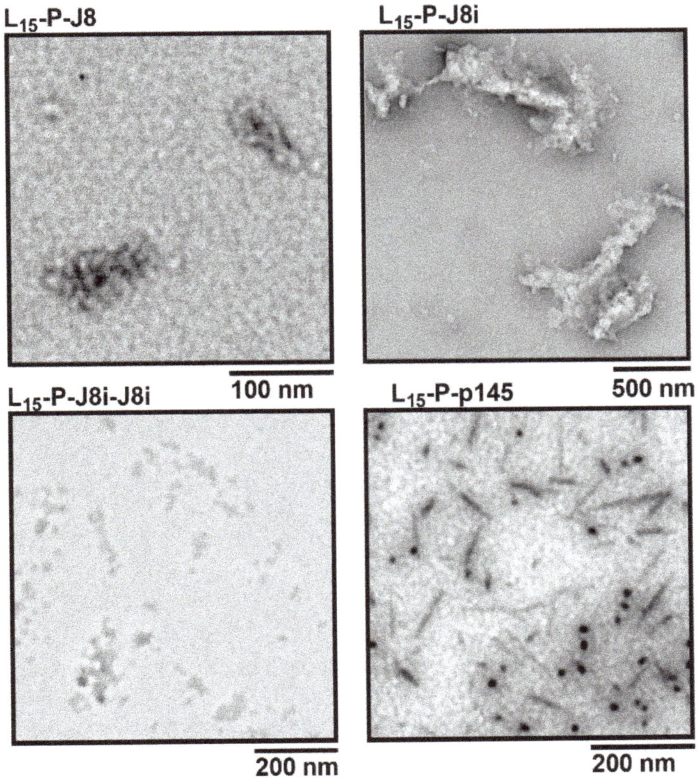

Figure 6. TEM electromicrographs of self-assembled 15-mer polyleucine vaccines in PBS.

3.3. Secondary Structural Evaluation of Peptide Antigens and Vaccines

The secondary structures of the synthesized peptides were evaluated using CD spectropolarography (Figure 7 and Table S1). J8-derivative B-cell antigens (J8, J8i, J8i-J8i, and p145) were mainly random coil (73 to 87%) with some helical structures (25 to 13%). The random coil or conformation-less content followed the length of the B-cell antigens, where the minimalist peptide J8i had only a 13% α-helix content, compared to p145, J8i-J8i, and J8, which had 17.25, 25%, and 27%, respectively. The conjugation of PADRE universal T helper also increased the helical content of the whole construct, where P-J8, P-J8i, and P-J8i-J8i had 31%, 25%, and 23% helical contents, respectively.

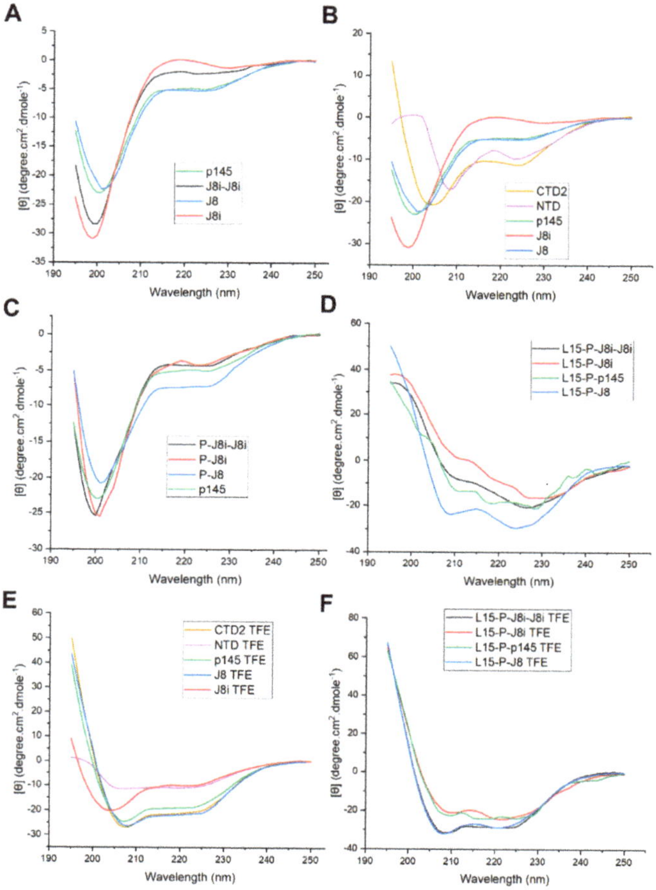

Figure 7. Circular dichroism measurements of the peptide vaccines and antigens, including (**A**) new and reported B-cell antigens, (**B**) J8-derivatives of B-cell antigens, (**C**) T-helper epitope conjugates to B-cell antigens, (**D**) polyleucine-adjuvanted vaccines, and (**E**,**F**) peptide antigens and vaccines in PBS including trifluoroethanol.

Recently, it has been reported that peptides adopt a native conformation in the vicinity of lipid membranes in vivo. Researchers attempted to mimic such an effect on peptide secondary structures by including small unilamellar liposomes or by including a cosolvent such as TFE. The secondary structures were evaluated in the presence of TFE for comparison [32,33]. The inclusion of TFE increased helical contents of the epitopes, where J8i, p145, J8i-J8i, and J8 had 41%, 67%, 71%, and 74%, respectively. Moreover, the conjugation to PADRE rendered P-J8, P-J8i, and P-J8i-J8i peptides of equivalent and high helical contents, ranging from 74 to 77%, in the presence of TFE. CTD2 and NTD had relatively high α-helical contents of 45% and 55% in PBS, which increased to 55% and 65% in TFE/PBS, respectively. Polyleucine vaccine constructs had relatively high helical contents >75% in PBS and increased to completely α-helical (>92%) in TFE/PBS.

These peptide antigens were adjuvanted with complete Freund's adjuvant (CFA) and investigated for their immunogenicity, quality of generated antibodies, and antisera's opsonization efficacy, following single immunization of C57BL/6 mice or three immunizations for polyleucine immunized C57BL/6 mice, followed by immunological analyses of the murine sera. This aimed to investigate the impacts of altering the B-cell epitope (J8-derivatives), the effect of T-helper conjugation on the B-cell epitope, and employing polyleucine as an adjuvant on the immunogenicity, quality, and efficacy of generated immune responses in the presence of another infection, i.e., *S. aureus*.

3.4. Immunological Evaluation of Mouse Sera

Initially, due to the existing postnatal infection of the C57BL/6 mice with *S. aureus* (Figure S1), serum IgG titres against GAS bacteria were detected; however, no serum IgG titres were detected against the peptide antigens (Figures 8 and S2). The mice were immunized with each peptide vaccine (n = 4), and sera were collected and analysed for immune responses after 42 days post-immunization, against various coating antigens J8, J8i, p145, NTD, and CTD2, and whole inactivated GAS (Figure 8A). Anti-peptide ELISAs were performed among the different mouse groups, where each mouse group's sera were evaluated against the peptide antigen with which it was immunized. This represented an evaluation and means of comparison among the antigens' immunogenicities.

The CFA/P-J8 immunized mouse group was the most immunogenic in generating anti-peptide serum IgG titres (Figure 8B), compared to the other groups, followed by CFA/P+J8, CFA/P-J8i, CFA/P+J8i-J8i, and CFA/P+J8i; these vaccines generated lower serum anti-peptide IgG titres ($p < 0.01$) than CFA/P-J8 but were similar to each other ($p > 0.05$). This demonstrated the superior immunogenicity of J8 compared to the other antigens, and improved immunogenicity due to the conjugation of the helper T-cell epitope, as CFA/P+J8-immunized group sera had lower anti-J8 IgG titres than those in the sera of the CFA/P-J8-immunized group ($p < 0.01$). The newly tested peptides CFA/P+NTD and CFA/P+CTD2 generated fewer serum anti-peptide IgG titres ($\log_{10}$ IgG titres of 3 to 3.5) compared to the CFA/P+J8- and CFA/P+J8i-immunized groups (Log_{10} anti-peptide IgG titres of 4.5 to 5), suggesting that NTD and CTD2 are less immunogenic than J8 or J8i. The rank order of absolute immunogenicity was as follows: P-J8 > P+J8 ≈ P-J8i ≈ P+J8i ≥ P+J8i-J8i > P+p145 ≈ CTD2 ≈ NTD, with either adjuvant polyleucine or CFA.

Anti-J8i ELISA was conducted on immunized mouse sera. As J8i is the native minimalist peptide sourced from p145 within the GAS bacterium M-protein sequence, it was expected that anti-J8i titres would be highly relevant to the immunogenicity of J8i-based vaccines (Figure 8C). Although it is a completely native peptide derived from M-protein, its short minimalist sequence may prevent it from adopting its native conformation. Anti-J8i serum IgG titres were highest for CFA/P-J8i-immunized mouse group sera, similar to those generated by CFA/P+J8i-immunized mice ($p > 0.05$). The CFA/P+J8i-J8i- and L_{15}-P-J8i-J8i-immunized mouse groups had slightly lower anti-J8i IgG titres in mouse sera ($p > 0.05$ and $p < 0.05$, respectively) compared to CFA/P-J8i. Meanwhile, p145 vaccines (CFA/P+p145 and L_{15}-P-p145) generated significantly lower anti-J8i IgG titres in the sera ($p < 0.01$ and $p <$ 0.05) compared to CFA/P-J8i and L_{15}-P-J8i, respectively. In sum, the anti-J8i serum IgG titres generally increased with the usage of more minimalistic antigens (in the rank order of P-J8i ≈ P+J8i > P+J8i-J8i > P+p145 > J8), and most of the generated anti-J8i IgG did not recognize J8 but recognized p145 to some extent. The Shapiro–Wilk test was conducted, at

$\alpha = 0.05$, to assess the normality of variances for the anti-J8i IgG immunological responses of the mouse groups, and they were normally distributed except for anti-J8i IgG for the CFA/P+J8i-J8i and L_{15}-P-p145 vaccines.

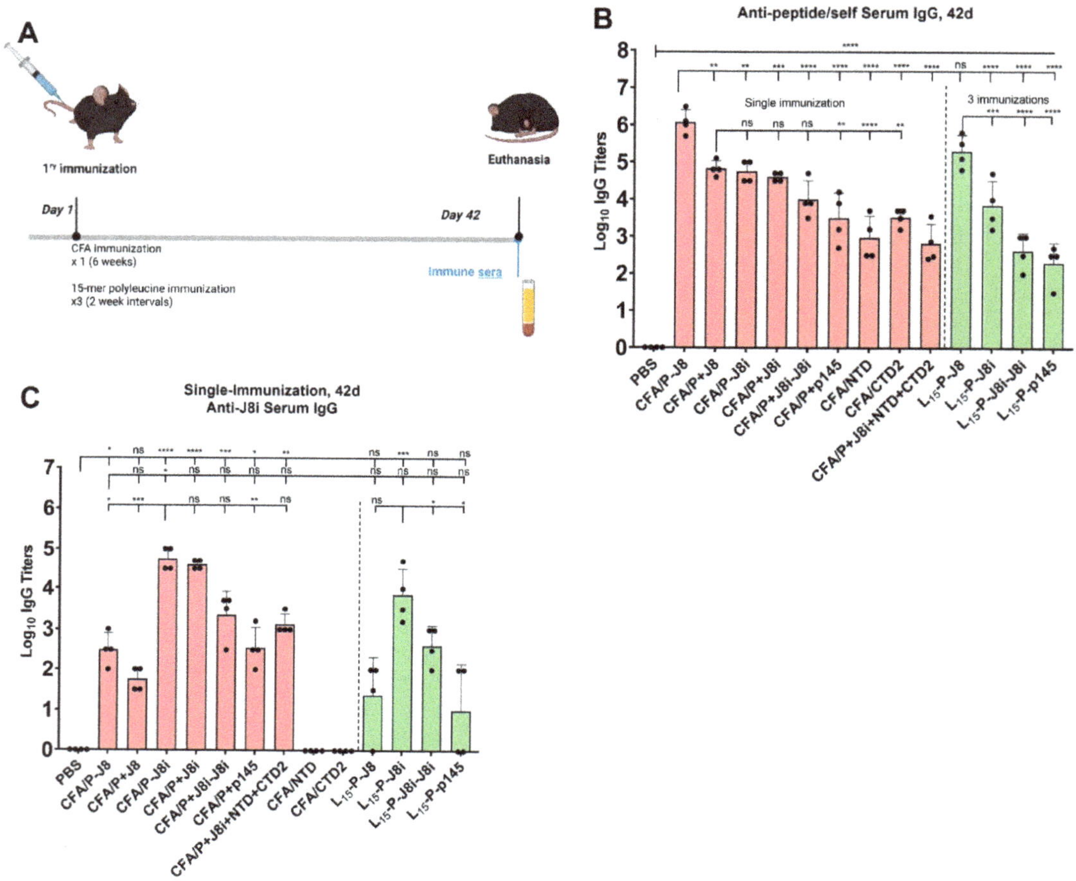

Figure 8. Immunization of C57BL/6 mice and immunological analyses of sera: (**A**) immunization schedule, (**B**) serum anti-peptide IgG titres, each against the intended peptide vaccine antigen, and (**C**) serum anti-J8i using ELISA. One-way ANOVA with Dunnett's multiple comparisons test was used for statistics. Significance was set to $p < 0.05$ (*), $p < 0.01$ (**), $p < 0.001$ (***), and $p < 0.0001$ (****).

Anti-J8 ELISAs were performed among the different mouse groups compared to PBS and to CFA/P-J8 as negative and positive control groups, respectively (Figure 9A). The CFA/P+J8-immunized mouse group was slightly less immunogenic compared to CFA/P-J8, $p < 0.05$. Three immunizations with L_{15}-P-J8 triggered equivalent immune responses to CFA/P-J8, $p > 0.05$, and all other mouse groups were significantly less immunogenic compared to CFA/P-J8. Furthermore, CFA/P-J8i was significantly less immunogenic than CFA/P-J8 ($p < 0.001$ and $p < 0.0001$, respectively). CFA/P+J8i was significantly less immunogenic than CFA/P+J8 ($p < 0.05$). In addition, CFA/P+J8i-J8i-immunized mouse groups had significantly lower anti-J8 serum IgG titres compared to CFA/P-J8 ($p < 0.0001$) but were similar to those generated by the CFA/P+J8-immunized mouse group ($p > 0.05$), whereas CFA/P+p145 generated significantly lower anti-J8 serum IgG titres compared to CFA/P-J8 or CFA/P-J8i ($p < 0.0001$ and $p < 0.01$, respectively). Finally, L_{15}-P-J8i and

L_{15}-P-J8i-J8i generated similar serum anti-J8 IgG titres to one another ($p > 0.05$), while L_{15}-P-p145 was significantly less immunogenic ($p < 0.001$). In sum, the generation of anti-J8 IgG titres in the serum was in the following rank order with either adjuvant system: P-J8 > P+J8 ≈ P-J8i > P+J8i-J8i > P+J8i >> P+p145. Further, the PBS group of infected mice, as well as CFA/NTD and CFA/CTD2 immunized mice, did not generate detectable anti-J8, which shows J8-specific responses and suggests that the peptide is not shared with *S. aureus* bacterial surface antigens. The Shapiro–Wilk test was conducted, at $\alpha = 0.05$, to assess the normality of variances for the anti-J8 IgG immunological responses of the mouse groups, and they were normally distributed except for anti-J8 for the L_{15}-P-J8i-J8i vaccine.

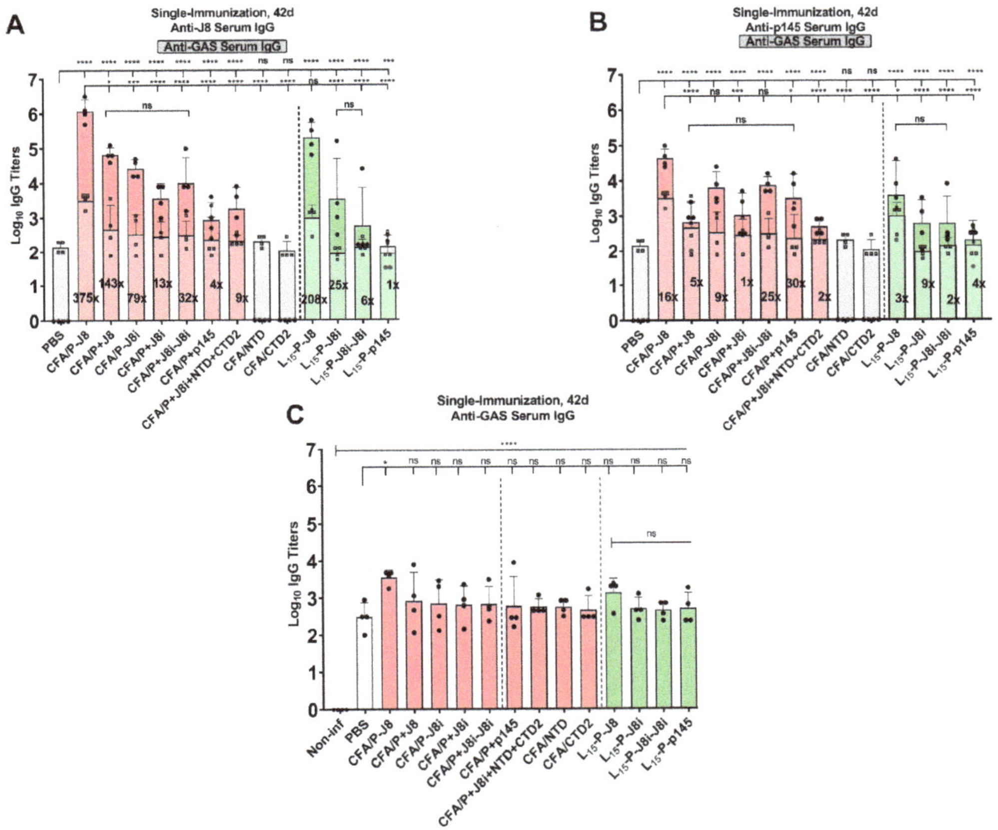

Figure 9. Immunological analyses of C57BL/6 mice sera. (**A**) Serum anti-J8 and anti-GAS IgG titres, where shaded bars represent the anti-GAS titres, determined using the ELISA. The numerical factor at the base of each column represents the value of the gap between anti-J8 and anti-GAS IgG titres. (**B**) Serum anti-p145 and anti-GAS IgG titres, where shaded bars represent the anti-GAS titres, determined using ELISA. The numerical factor at the base of each column represents the value of the gap between anti-p145 and anti-GAS IgG titres, and (**C**) serum anti-GAS using the ELISA. One-way ANOVA with Dunnett's multiple comparisons test was used for statistics. Significance was set to $p < 0.05$ (*), $p < 0.001$ (***), and $p < 0.0001$ (****).

Anti-p145 ELISA was conducted on immunized mouse sera. As p145 is the native epitope source (of J8 and J8i) in the GAS M-protein, it was expected that anti-p145 titres would be highly relevant to antibody quality or the avidity of the generated immune responses (Figure 9B). Anti-p145 serum IgG titres were the highest for CFA/P-J8 immunized mouse group sera, potentially on account of the immunogenicity of the P-J8 peptide. The

CFA/P+p145- and L_{15}-P-p145-immunized mouse groups had slightly lower anti-p145 IgG titres in mouse sera ($p < 0.05$) compared to CFA/P-J8. Meanwhile, CFA/P+J8, CFA/P-J8i, CFA/P+J8i, and CFA/P+p145 generated similar serum titres of anti-p145 IgG ($p > 0.05$). In addition, the L_{15}-P-J8-, L_{15}-P-J8i-, and L_{15}-P-J8i-J8i-immunized mouse groups also had similar serum anti-p145 IgG titres ($p > 0.05$). Surprisingly, when comparing CFA/P-J8- and CFA/P+J8-immunized group sera, a dramatic reduction in anti-p145 was observed when PADRE was not conjugated ($p < 0.0001$). The Shapiro–Wilk test was conducted, at $\alpha = 0.05$, to assess the normality of variances for the anti-p145 IgG immunological responses of the mouse groups, and they were normally distributed except for anti-p145 IgG for the L_{15}-P-J8i-J8i vaccine.

An Anti-GAS ELISA was conducted on mouse immune sera (Figure 9C). CFA/P-J8 generated the highest serum anti-GAS IgG titres. Nonetheless, L_{15}-P-J8 generated anti-GAS serum IgG titres similar to those generated by CFA/P-J8. However, all the other groups, including the CFA/P+J8-immunized mouse group, generated lower anti-GAS titre levels compared to CFA/P-J8 ($p < 0.05$). Polyleucine-adjuvanted mouse groups generated similar anti-GAS IgG titres to each other, and L_{15}-P-J8 generated similar anti-GAS IgG titres to the CFA/P-J8 immunized mouse group, after three immunizations. However, L_{15}-P-J8i, L_{15}-P-J8i-J8i, and L_{15}-P-p145 generated lower but non-significantly different anti-GAS serum IgG compared to CFA/P-J8 or L_{15}-P-J8 ($p > 0.05$). In addition, the NTD- and CTD2-immunized mouse groups generated significant but modest anti-GAS titres, which suggests that the epitopes are not as immunogenic as J8 (or its derivatives) but represent a promising approach for developing a multivalent vaccine against GAS. It was also observed that the PBS group of infected mice generated some anti-GAS IgG titres ($\log_{10}$ serum IgG of 2.0 to 3.0) compared to naïve sera from the non-infected PBS group (Figure S2). This suggests that there are some shared epitopes between surface antigens of GAS and *S. aureus* bacteria. The Shapiro–Wilk test was conducted, at $\alpha = 0.05$, to assess the normality of variances for the anti-GAS IgG immunological responses of the mouse groups, and they were normally distributed except for anti-GAS IgG for the CFA/P+J8i and CFA/P+J8i-J8i vaccines.

3.5. *Opsonization Efficacy Evaluation of Mouse Sera*

Sera of the immunized mouse groups were evaluated for their opsonic bactericidal inhibition activity against four GAS clinically isolated strains, two of which were standard strains with a conserved p145/J8i sequence (D3840 and ACM2727), while the other two clinical strains were mutant virulent strains with mutated p145/J8i sequences (NS1944 and NS5582) (Figure 1). Pre-immunization sera did not significantly impact the opsonization efficacy of any of the tested strains, $p > 0.05$ (Figure S3).

The standard strains were effectively opsonized by many antisera, >80%, against many vaccine antigens, including CFA/P-J8, CFA/P+J8, CFA/P-J8i, and L_{15}-P-J8, compared to PBS control group sera ($p < 0.0001$) (Figure 10A). The T-helper-conjugated vaccine CFA/P-J8i generated more bactericidal antisera against the standard GAS strains compared to antisera against CFA/P+J8i, $p < 0.001$ (Figure 10B). In addition, the antisera against L_{15}-P-J8i vaccines did not exhibit opsonic efficacy against the standard GAS strains. In contrast, antisera against CFA/P+CTD2, CFA/P+NTD, CFA/P+J8i, and CFA/P+J8i-J8i significantly but modestly generated opsonic antisera against GAS standard strains ($\leq$50%, $p < 0.05$, Figure 10A), with CFA/P+J8i-J8i and CFA/P+J8i vaccines generating more significant opsonic antibodies ($p < 0.01$) compared to PBS control group sera. In contrast, the performance of the antisera generated by the vaccines was very different against mutant GAS strains. The two mutant strains NS1944 and NS5582 resisted opsonization by the antisera generated by the vaccines with many of the tested antigens: CFA/P+J8, CFA/P+J8i, CFA/P-J8i, CFA/P+J8i-J8i, CFA/P+p145, CFA/P+NTD, and CFA/P+CTD2

($p > 0.05$, Figures 10 and 11). Only antisera against P-J8 vaccines, adjuvanted with either CFA or 15-mer polyleucine, elicited effective opsonic bactericidal activity against all tested GAS strains, including the mutant virulent GAS strains ($p < 0.001$, Figures 10 and 11).

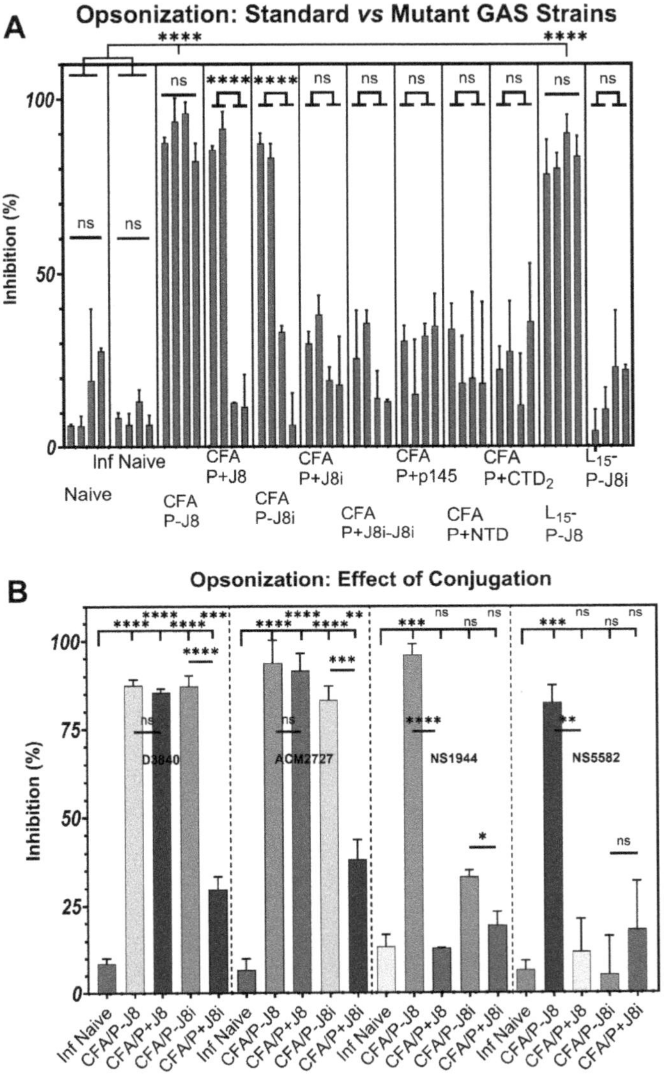

Figure 10. GAS opsonization inhibition via (**A**) antisera of immunized mouse groups against four clinical strains—two standard strains (D3840 and ACM2727) and two mutant virulent strains (NS1944 and NS5582)—where each bar for a given vaccine represents bactericidal inhibition against a specific GAS strain from left to right in the order of D3840, ACM2727, NS1944, and NS5582, respectively. (**B**) The influence of B-cell antigen conjugation to the T helper on GAS opsonization inhibition of immunized mouse groups against four clinical strains—two standard strains (D3840 and ACM2727) and two mutant virulent strains (NS1944 and NS5582). Statistics were compared for bactericidal inhibition between the two standard and the two mutant strains and compared to the PBS negative control group. One-way ANOVA with Dunnett's multiple comparisons test was used for statistics. Significance was set to $p < 0.05$ (*), $p < 0.01$ (**), $p < 0.001$ (***), and $p < 0.0001$ (****).

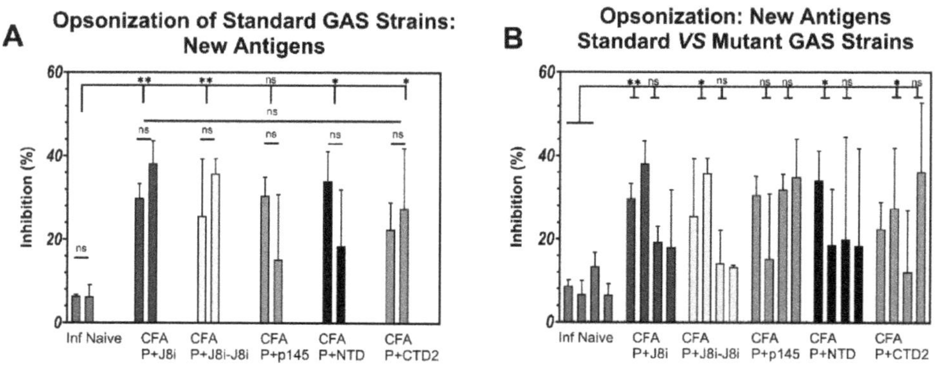

Figure 11. GAS opsonization inhibition of immunized mouse groups, comparing vaccine groups with new antigens and J8-derivative antigens against (**A**) the standard two GAS strains or (**B**) four standard and mutant GAS strains, respectively. Statistics compare bactericidal inhibition across all tested clinical GAS strains between standard and mutant strains and between immunized groups and the PBS negative control group. One-way ANOVA with Dunnett's multiple comparisons test was used for statistical analysis. Significance was set to $p < 0.05$ (*), and $p < 0.01$ (**).

The new B-cell antigens NTD and CTD2 (CFA/P+NTD and CFA/P+CTD2) generated antisera that were moderately opsonic against the standard GAS strains ($p < 0.05$ Figure 11A); however, they had no significant opsonic activity against the mutant strains, despite being conserved in their M-protein sequences ($p > 0.05$, Figures 1 and 11B). Notably, the J8i and J8i-J8i antigens' (CFA/P+J8i and CFA/P+J8i-J8i) antisera were more opsonically effective ($p < 0.01$) against standard GAS strains compared to NTD and CTD2 ($p < 0.05$) when either was compared to the PBS negative control group, but they were all more or less at similar modest inhibition levels ranging between 20% and 40% (Figures 10A and 11A).

4. Discussion

4.1. Size and Secondary Structural Characterization of Synthesized Peptide Constructs

Peptide antigens were successfully synthesized. The 15-mer polyleucine self-assembled peptide vaccines were characterized for their size distribution using DLS and TEM. The most stable nanovaccine size, with a smaller aggregation extent, was observed for L_{15}-P-J8, and this was attributed to the hydrophilicity of J8 B-cell antigen, which represented a significant portion of the sequence, compared to J8i-J8i or p145. Therefore, the hydrophilicity of B-cell antigens at C-termini played a major role in the self-assembled size of the peptide constructs, where a longer hydrophilic sequence resulted in smaller, more monodisperse, and less aggregating nanovaccines; the rank order of peptides' hydrophilicity was J8 > J8i-J8i > p145 > J8i.

The peptide constructs were dissolved in PBS and characterized for their secondary structural conformation using CD spectroscopy, both in the presence and absence of TFE co-solvent. It was observed that the helical content increased with the length of the B-cell antigen employed. The helical content demonstrated that J8 was precisely designed to match the p145 native peptide structure, after omitting the autoimmune sequence at its *N*-terminus. In addition, this also shows that J8i-J8i dimer and J8i minimalist peptides lack the necessary residues to convey a matching native helical structure to that observed for the p145 peptide. In addition, the conjugation of PADRE T helper to B-cell antigens increased the helical conformation content of the constructs. The impact of PADRE on the short J8i sequence was higher in the induction of helicity compared to the J8i-J8i dimer, probably due to the greater proportion of the short epitope being influenced by neighbouring residues to PADRE. The addition of TFE to peptide solutions increased the helical conformation contents for many peptides; however, the helical content increase in J8i peptide was modest compared to the increase observed for the other B-cell antigens J8, J8i-J8i, and p145. This demonstrated that J8i still lacked the necessary residues to adopt a native helical conformation; however, p145, J8i-J8i dimer, and J8 epitopes were of equivalent helical contents ranging from 67 to 74%. Further, the conjugation of PADRE increased the helical content of J8i, in the presence of TFE; however, it did not significantly increase J8 or J8i-J8i dimer, compared to constructs without PADRE, in the presence of TFE. This demonstrated that conjugation of PADRE not only helps with generating robust, long-lasting immune responses as a universal helper T-cell epitope for C57BL/6 mice [6,7] but also assists in adopting a native structural conformation for short neighbouring epitopes like J8i. However, it remains to be tested whether the peptide structure influenced their immunogenicity and the quality of triggered responses in the mice.

4.2. The Immunological Evaluation of Mouse Sera: Immunogenicity and Immune Response Quality

The peptide vaccines were prepared and used to immunize female C57BL/6 mice. Murine sera were collected and analysed for immunological responses using ELISA against various antigens, including the immunizing peptides themselves (Figure 8B), J8, J8i, and p145, as well as the inactivated GAS bacteria. ELISA against the immunizing peptide antigens provided the highest lgo_{10} titre responses, thus supporting that they are most suited to evaluating vaccine immunogenicity. In terms of absolute immunogenicity (anti-peptide IgG titres), conjugation to PADRE T helper enhanced the immunogenicity over a physical mixture between PADRE and B-cell antigen groups. In addition, the immunogenicity of J8-derivative vaccine J8 was the greatest, followed by J8i, J8i-J8i, and then p145.

ELISA against J8i peptide demonstrated higher responses for J8i-based vaccines over those employing J8 or p145 antigens (Figure 8B). However, there are gaps of 0.9 and 1.6 log_{10} units between anti-J8i and anti-p145 IgG titres of CFA/P-J8i and CFA/P+J8i, respectively. This suggests that the immunogenicity of J8i is higher than p145, close in ELISA unit values to anti-J8 IgG titres for the CFA/P-J8-immunized group; however, the high-quality antibodies generated by J8i-vaccines that recognized p145 were only 12.5% or 2.5%, for PADRE conjugated to, or physically mixed with, J8i vaccine antigens, respectively. This also demonstrated that the conjugation of PADRE to J8i improved the quality of generated antibodies rather than immunogenicity, potentially by supporting a more native helical structure for J8i. In sum, the anti-J8i serum IgG titres generally increased with the usage of more minimalistic antigens, and most of the generated anti-J8i IgG did not recognize J8 but recognized p145 more slightly. Further, the conjugation of PADRE to short J8i improved the quality of generated anti-J8i serum IgG antibodies in terms of recognition of p145 antigen, compared to a CFA/P+J8i physical mixture.

The ELISA against the J8 peptide demonstrated higher responses for CFA/P-J8 and CFA/P-J8i vaccines over CFA/P+J8 and CFA/P+J8i, respectively (Figure 9A). This demonstrated the importance of conjugating the PADRE T-helper epitope to short peptide antigens for enhancement of their native structure and for delivery to the same lymph node, which was often reported to enhance immunogenicity, despite the presence of freely mixed PADRE T-helper epitope. Although CFA contains inactivated mycobacteria, which form a source of numerous helper T-cell epitopes, and thus the enhanced immunogenicity could be attributed to the increased adoption of a helical secondary structure of the B-cell epitope induced by the conjugated T helper. In addition, serum anti-J8 IgG titres generated by the vaccines were highest for CFA/P+J8, followed by P+J8i-J8i, P+J8i, and then P+p145. Surprisingly, the gap value between anti-GAS titres and anti-J8 titres was smaller for mixed non-conjugated T-helper-based vaccines, like CFA/P+J8 and CFA/P+J8i, compared to their conjugated counterparts, CFA P-J8 and CFA/P-J8i, by 3- to 5-fold, and the gap was wider with shorter B-cell antigen sequences (Figure 9A). This suggests that recognition of GAS bacteria might be slightly better with the physical mixture, thus producing higher-quality antibodies; however, due to the interference of *S. aureus* infection in anti-GAS titres, it is difficult to provide definitive conclusions without the opsonization assay outcomes.

The ELISA against the p145 peptide demonstrated the highest responses for the CFA/P-J8 vaccine (Figure 9B). The anti-p145 serum IgG titres generally increased with the increase in peptide immunogenicity; however, the conjugation of PADRE to short J8i or using the dimer J8i-J8i, i.e., CFA/P-J8i- and CFA/P+J8i+J8i-immunized mouse groups, improved the magnitude of triggered immune responses compared to P+J8i or P+p145. Further, the conjugation of PADRE to J8 dramatically improved its anti-p145 titres, despite the fact that J8 is relatively long and equipped with GCN-4 at both termini. In addition, the gaps between anti-p145 titres and anti-GAS titres (Figure 9C) were noticeably smaller than any other ELISA peptide-antigen coat; the gaps varied between 2- and 30-fold for all p145-based vaccine derivatives and only varied 1- to 16-fold for J8- and J8i-based vaccines, which suggests that p145 can be employed for ELISA antigen coating as a measure of antibody quality or immune response avidity.

Anti-GAS IgG titres often correlate best with opsonization and represent the overall quality of generated antibodies and their ability to recognize GAS (Figure 9C). The anti-GAS IgG serum titres were similar among various antigens; however, the conjugation of PADRE was essential to enhance the anti-GAS titres of the J8 and J8i peptides. In addition, there was an observed interference by the existing *S. aureus* infection. Thus, solutions to this issue include either blank subtraction (pre-immunization from post-immunization sera) or the usage of alternative distinctive (non-shared) peptides or proteins from GAS as ELISA coating antigens, to restore the specificity of the assay. Alternatively, NTD and CTD generated significant but very low serum anti-GAS IgG titres; therefore, they should be subjected to further improvements in future research.

4.3. *In Vitro Opsonization Efficacy of Mouse Sera Against Standard and Mutant GAS Strains*

The murine antisera against vaccines were evaluated for their opsonic bactericidal efficacy in vitro against two standard and two mutant GAS strains. The conjugation of PADRE to J8 was not necessary to generate effective opsonic antibodies against the standard GAS strains, as CFA/P+J8 was effective in generating opsonic antibodies against standard GAS strains (Figure 10B). In contrast, the conjugation of PADRE to J8i was necessary to generate opsonic antibodies against standard GAS strains, as CFA/P+J8i did not generate effective opsonic antibodies compared to CFA/P-J8i, thus PADRE conjugation at the *N*-terminus and CFA lipids potentially contributed to restoring the native conformation of the short J8i epitope. Surprisingly, L_{15}-P-J8i was also ineffective in generating opsonic antibodies against the standard GAS strains, which suggests that the β-like protein conformation structure induced by the polyleucine at one end, and random coil residues at the *C*-terminus of the short J8i peptide at the other end, may have obscured the structure of the short J8i sequence.

The opsonic efficacy of immunized mice antisera was very different from that of the mutant GAS strains. Only L_{15}-P-J8 and CFA/P-J8 generated highly opsonic antisera against mutant GAS strains, compared to the antisera against other vaccines, including CFA/P+J8. In the context of generating opsonic antibodies using J8-derivative antigens against the mutant virulent GAS strains, it is apparent that not only is J8 antigen, including GCN-4 sequences, necessary for the opsonic activity but also its conjugation at the B-cell antigen's *N*-terminus to the PADRE T-helper epitope, which aids in helical conformation induction and antigen presentation. Thus, these modifications to the vaccine antigen supported both the vaccine antigen's immunogenicity and efficacy, especially against mutant GAS strains;

the same effect could hold for T-helper carrier protein sources like diphtheria toxins. Since the mutations in NS1944 and NS5582 in p145/J8i sequences are located at its *C*-terminus (Figure 1), this suggests that the conjugation of PADRE and GCN-4 sequences to J8 was necessary for effective antigen presentation of the conserved *N*-terminal residues of J8i (Figures 10 and 11), which could be the main source that generated opsonic antibodies against these mutant GAS strains. Meanwhile, conjugation of PADRE and use of lipid-based adjuvants were adequate to generate effective opsonic antibodies against 'standard GAS strains' where *C*-terminus residues of p145/J8i remain conserved in their M-protein sequences (Figure 1). In contrast, antisera against CFA/P+NTD and CFA/P+CTD2 were only moderately opsonic against standard GAS strains and were not effective against the mutant GAS strains (Figure 10), despite the fact that their sequences are conserved in both GAS strain types, potentially due to their modest immunogenicity. Thus, further development is required for these new antigens to boost their immunogenicity and the quality of the generated antibodies to be as effective as J8; among such modifications are conjugation to PADRE, flanking with helical conformation-inducing sequences, or using a carrier T-helper source protein. These modifications could help promote the use of these new antigens in future vaccines against GAS infection. However, the new antigens NTD and CTD2 proved their usefulness as GAS-specific antigens in the presence of *S. aureus* infection, and thus could also be of diagnostic value.

The opsonic bactericidal inhibition of a given antisera is dependent on vaccine immunogenicity, i.e., the magnitude of generated antibody titres. Thus, normalization of the opsonic efficacy, or log reduction in seeded GAS bacteria, normalized by antigen-specific antibody titres was necessary to compare among the employed vaccine antigens (Figure 12), herein termed the opsonization capacity index. When comparing the opsonization capacity indices of the antisera against standard GAS strains (Figure 12A,C), it was observed that the antisera of CFA/P-J8-, CFA/P-J8i-, and CFA/P+J8-immunized groups were similar to each other, $p > 0.05$, and very significantly higher than all the other immunized groups, including L_{15}-P-J8, $p < 0.0001$. This means that when the average generated anti-peptide antibody titres were taken into account, the antibodies of P-J8, P+J8, and P-J8i became equivalent in their opsonic activity against 'the standard GAS strains', which suggests that the role of PADRE conjugation to J8 supported the antigen's immunogenicity and antigen presentation to immune cells rather than restoring the native structural conformation. However, the latter observation was noticeably more relevant for the J8i minimalist epitope (CFA/P-J8i), as it significantly reduced its antisera's capacity index when it was not conjugated to PADRE (CFA/P+J8i). Overall, J8 antisera had the highest opsonization capacity index among physically mixed B-cell antigens (Figure 12C). However, when comparing opsonization capacity indices of antisera against the mutant strains, only antisera with J8 conjugated to PADRE T helper (CFA/P-J8 or L_{15}-P-J8 vaccines) had very high and significant opsonic capacity indices compared to all other vaccines, $p < 0.0001$, and this opsonic capacity was comparable to that observed for the standard strains (Figure 12B). This is a very valuable finding because few-residue mutations, but not complete deletion, in J8i sequences did not prevent the generation of effective opsonic antibodies against these mutant GAS strains. Surprisingly, when opsonization was normalized by immunogenicity for 'the physically mixed vaccines', i.e., non-conjugated T-helper epitopes, and new antigen CFA/P+p145, CFA/P+CTD2, and CFA/P+NTD vaccines, their antisera had a significantly higher opsonization capacity, against the mutant strains, than J8, J8i, and J8i-J8i, $p < 0.01$, in the rank order of p145 >> CTD2 $\approx$ NTD >> J8i > J8i-J8i > J8 (Figure 12D). This suggests that these new B-cell antigens could be very promising vaccine antigens but require further development to boost their immunogenicity, thereby potentially achieving a similar potent opsonic efficacy to P-J8.

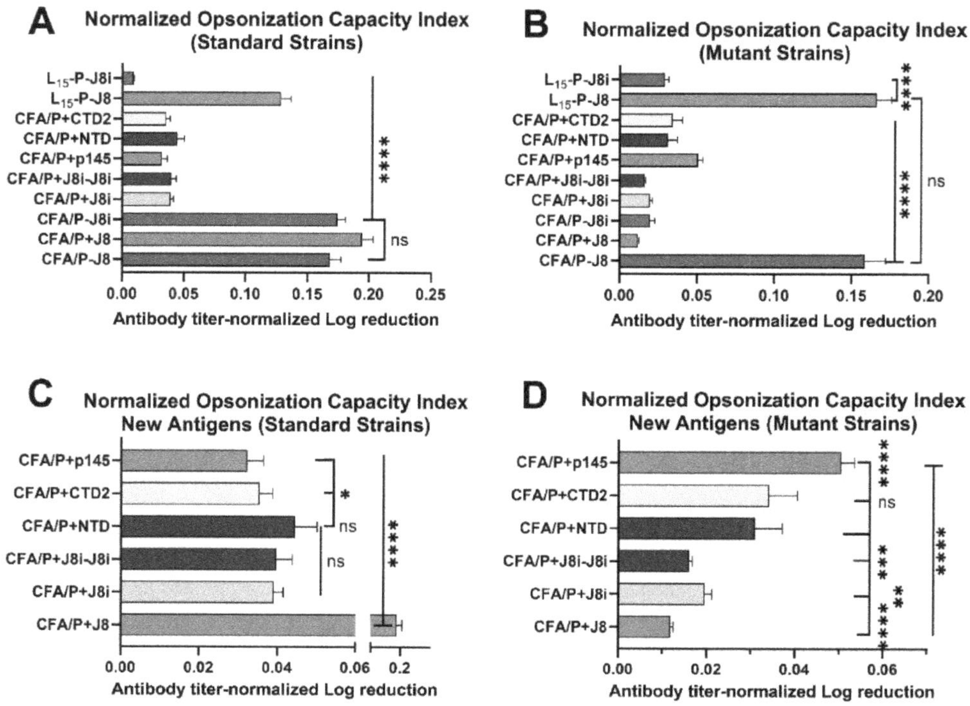

Figure 12. Opsonization capacity indices of antisera from immunized mouse groups (**A**,**C**), comparing vaccinated mouse groups and new antigens' opsonization capacity indices against the two standard GAS strains, respectively. (**B**,**D**) Vaccinated mouse groups and new antigens' opsonization capacity indices against the two mutant GAS strains, respectively. Statistics comparing the opsonization capacity indices among the antisera of the immunized groups against standard (left panels (**A**,**C**)) or mutant GAS strains (right panels (**B**,**D**)). One-way ANOVA with Dunnett's multiple comparisons test was used for statistics. Significance was set to $p < 0.05$ (*), $p < 0.01$ (**), $p < 0.001$ (***), and $p < 0.0001$ (****).

4.4. Relationships Among Immunological Responses and Opsonic Bactericidal Activity

The relationships among the immune responses were further investigated; anti-J8 serum IgG titres were plotted against anti-J8 IgG titres and against anti-p145 serum IgG titres (Figure 13A,B). A clear gap/x-offset (log_{10} of 1.9, ca. 80-fold, Figure 13A) was observed only for J8-based vaccine immunized groups between anti-J8i and anti-J8 IgG titres. The gap represented an 80-fold increase in anti-J8 IgG, for J8-based vaccines, at the expense of anti-J8i antibodies generated by the same vaccines, which could be due to antibodies generated against the non-native GCN-4 sequences. However, the gap/x-offset between J8 and J8i vaccines was almost completely closed when p145 was used as a coat for ELISA, i.e., anti-p145 serum IgG (Figure 13B), which demonstrates that 'the extra anti-J8 IgG titres' generated by J8-based vaccines, which caused the gap in Figure 13A, were recognized by p145, thus these are not irrelevant antibodies against the non-native GCN-4 sequence residues. This outcome suggests that p145 and J8 are more related to each other than J8i is related J8, and that the addition of GCN-4 sequences transformed the structure of J8i peptide, bring it closer in conformation to p145's native structure within M-protein, and potentially restored the structure of unknown B-cell epitopes located within the J8i sequence. Thus, J8i potentially contains more than one B-cell epitope, whose structure was lost, especially when the minimalist J8i was used 'unconjugated' as a vaccine antigen.

It remains to be seen whether further improvements over GCN-4 sequences can bring J8i even closer to p145, thereby closing the gap completely and enhancing the quality of generated antibodies, while boosting the antigen's immunogenicity.

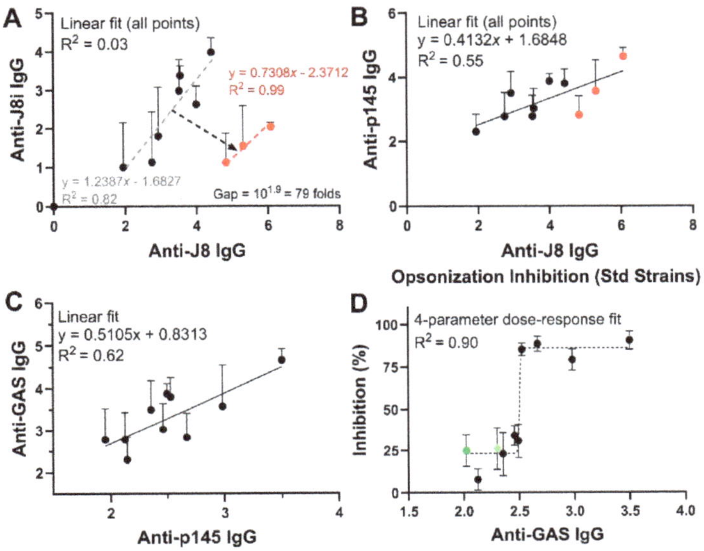

Figure 13. Correlations among the generated immune responses and opsonisation inhibition of standard GAS strains: between (**A**) serum anti-J8 and serum anti-J8i IgG titres (red circles represent J8-based vaccines datapoints), (**B**) serum anti-p145 and serum anti-J8 IgG titres (red circles represent J8-based vaccines datapoints), (**C**) serum anti-GAS and serum anti-p145 IgG titres, and (**D**) opsonization inhibition and serum anti-GAS IgG titres (green circles represent NTD and CTD2 datapoints).

The relationships between efficacy and antigen-specific antibody titres were further investigated. The relationships between anti-GAS IgG titres and anti-p145 IgG titres and between anti-GAS IgG titres and opsonization efficacy against 'the standard strains' were explored (Figure 13C,D). A linear fit between anti-p145 and anti-GAS IgG titres gave an R^2 of 0.62, supporting the usage of p145 to assess the quality of generated antibodies, even in the presence of a *S. aureus*-related immune responses. In addition, the anti-GAS titres correlated with opsonization efficacy against standard GAS strains, using 10-fold diluted sera, in a four-parameter dose–response sigmoidal pattern, gave an R^2 of 0.9, and 50% inhibitory anti-GAS $\log_{10}$ IgG titres (IC_{50}) of 2.5 (or 1.5 for undiluted sera) against standard GAS strains, which is equivalent to anti-J8 $\log_{10}$ IgG titres of about 4.85 (or 3.85 for undiluted sera).

To further investigate whether multiple B-cell epitopes are present within J8i/p145 peptide (LRRDLDA-**SREAKKQVEKAL**-E, bold sequence represents J8i within p145), p145 was synthesized but split into two smaller peptides; the first was 95% conserved in all employed, mutant and standard, GAS strains [p145-p1: RDLDA**SREAK**], and the second was significantly changed, and thus only 30% conserved [p145-p2: **KKQVEKAL**]. The immunized mouse group sera against CFA/P-J8, CFA/P+J8, CFA/P-J8i, CFA/P+J8i, L_{15}-P-J8, and L_{15}-P-J8i were tested for their IgG antibody titres against p145-P1 and p145-P2 peptides and the inactivated mutant NS1944 GAS strain (Figure 14). The total anti-p145 titres were approximately equal to the sum of anti-p145-P1 and anti-p145-P2 IgG titres (Figure 14A). Thus, IgG titres against the split peptides p145-P1 and p145-P2 support the hypothesis that at least two B-cell epitopes are located within the p145 sequence. Further, the antibodies generated against p145-P1 conserved peptide were well-correlated with opsonic bactericidal inhibition of 'the mutant NS1944 GAS' strain; the fit gave $R^2 = 0.93$, with anti-p145-P1 IC_{50} corresponding to log_{10} of 3.45, at a 10-fold serum dilution (Figure 14B,C). This suggested that (a) antibodies generated against the p145-P1 conserved peptide fragment are mainly responsible for the opsonic activity against mutant GAS strains, and (b) that mutant GAS strains require the generation of higher log_{10} antibody titres of the high-quality anti-GAS_{NS1944} antibodies for effective opsonic activity (Figure 14B,C) compared to anti-GAS IgG titres required for opsonic activity against standard GAS strains (Figure 12D). To further support this finding, the ratio of anti-p145-P1 to anti-p145-P2 IgG titres was well-correlated with opsonization capacity against the mutant NS1944 GAS strain, normalized by anti-p145 IgG titres (Figure 14D), and ratio values above unity, i.e., vaccines that generated more anti-p145-P1 than anti-p145-P2 IgG titres (CFA/P-J8 and L_{15}-P-J8), were highly opsonic against the mutant strains. The ratio of anti-p145-P1 to anti-p145-P2 IgG titres was higher in the vaccines that employed (a) J8 as the B-cell antigen, i.e., GCN-4 sequence flanking of J8i, and (b) conjugation of PADRE to the *N*-terminus of J8. The absence of those two factors focused the generated serum high-quality antibody responses toward the less conserved *C*-terminus portion of the p145 peptide, i.e., p145-P2, and rendered these responses less effective against the mutant GAS strains, such as those generated by CFA/P+J8, CFA/P-J8i, or CFA/P+J8i vaccines (Figure 14D). Thus, we recommend evaluating immune responses against p145-P1 and p145-P2 peptides in future J8-based vaccine development endeavours to assess immune responses against the conserved versus mutant/altered sequences of the p145 peptide. These responses may provide powerful immune correlates of protection against mutant hypervirulent GAS strains.

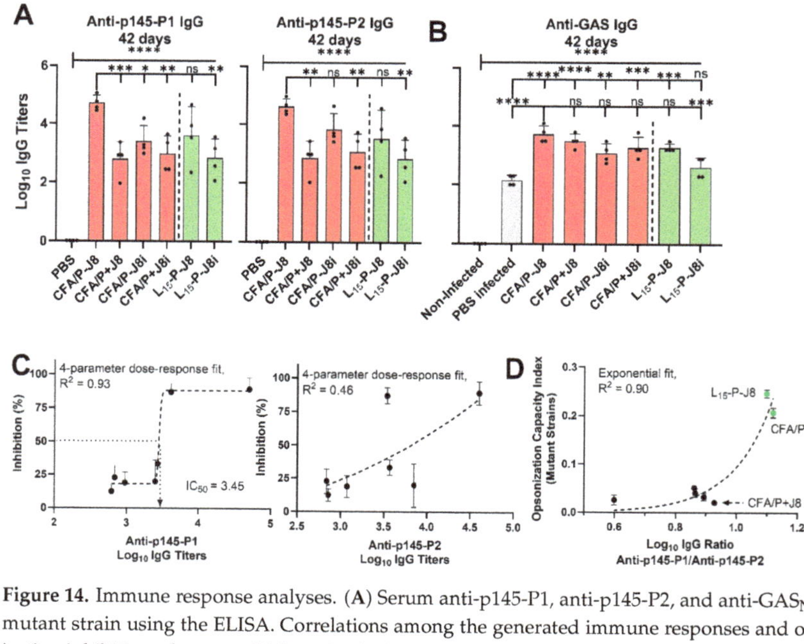

Figure 14. Immune response analyses. (**A**) Serum anti-p145-P1, anti-p145-P2, and anti-GAS_{NS1944} mutant strain using the ELISA. Correlations among the generated immune responses and opsonisation inhibition of mutant GAS strains (NS1944 and NS5582): correlation between opsonization bactericidal activity and (**B**) serum anti-p145-P1 or serum anti-p145-P2 IgG titres, and (**C**) the ratio between serum log_{10} anti-p145-P1 and anti-p145-P2 IgG, where green data points represent the sera of mouse groups immunized with L_{15}-P-J8 and CFA/P-J8. Data points were fitted using either a 4-parameter dose–response Boltzmann (panel (**C**)) or an exponential fit (panel (**D**)). One-way ANOVA with Dunnett's multiple comparisons test was used for statistical analysis. Significance was set to $p < 0.05$ (*), $p < 0.01$ (**), $p < 0.001$ (***), and $p < 0.0001$ (****).

Although functional antibody responses were demonstrated using in vitro opsonization assays [20,21]—an established correlate of protection—future studies will include live GAS infection challenge studies to assess the overall vaccine-induced protection, bacterial clearance, and survival outcomes. The immunological assays validated here will help optimize and interpret such in vivo challenge studies, particularly in the context of co-infection.

5. Conclusions

In sum, P-J8 proved to be a valuable vaccine antigen against both standard and mutant GAS strains with altered p145 sequences in their M-proteins. Conjugation of PADRE T helper to J8 proved valuable to opsonic bactericidal efficacy against mutant GAS strains, while P-J8i proved useful as a vaccine antigen only against standard GAS strains, and conjugation to PADRE was necessary for generating opsonic antibodies. While helicity plays a major role in the activity of M-protein-derived antigens, it is not the sole player in generating high-quality opsonic antibodies against GAS, as our results suggest the presence of at least two B-cell epitopes within the J8i sequence, which are dramatically affected by the neighbouring conjugated peptides, like PADRE and/or flanking GCN-4 sequences, thus significantly impacting their generated antisera's opsonic activity.

CTD2, NTD, and p145 peptides are of diagnostic value for GAS infection, more appropriate than whole pathogens or whole proteins, which have shared antigens among different bacterial pathogens, and thus may indicate infection by other pathogenic bacteria than GAS. These novel peptide antigens CTD2 and NTD generated significantly opsonic sera, although not as powerful as J8, thus they require further development to impart helicity, e.g., via flanking with GCN-4 and conjugation to a carrier protein, a T-helper cell epitope, or both. Such novel and highly conserved antigens will prove useful if their antisera opsonisation is further enhanced in future studies.

Immunological assays vary in their usability under clinical and 'advanced' pre-clinical conditions, and immune sera cannot be evaluated only with anti-GAS ELISA to detect GAS-specific antibody titres if prior infections may have occurred. Thus, testing pre-immunisation sera is important, and conducting ELISAs against GAS-specific peptide or protein antigens could be useful to establish GAS-specific immunogenicity and immune response avidity, respectively. In contrast, the opsonisation assay seems to be less affected by co-infection; however, the interfering antibody titres against shared epitopes were modest, and thus the effect on opsonic efficacy may vary if proper anti-*S. aureus* titres are to develop.

Supplementary Materials: The following supporting information can be downloaded at https://www.mdpi.com/article/10.3390/vaccines13060632/s1, Table S1: Quantitative secondary structure estimates of peptide antigens; Figure S1: Capillary electrophoresis analyses examples of pooled faecal samples from infected mouse groups; Figure S2: Pre-immunization serum IgG titers; Figure S3: GAS opsonisation inhibition of non-infected naïve mouse sera and preimmunization infected mouse sera.

Author Contributions: Conceptualization, A.O.S., M.S., and I.T.; methodology, A.O.S., H.S., Y.C., J.W., B.A., J.B., W.M.H., and Z.G.K. (design and execution of opsonization assays, and GAS inactivation); software, A.O.S.; validation, A.O.S., W.M.H., Z.G.K. (optimization and replication of opsonization assays for NS strains), M.S., and I.T.; formal analysis, A.O.S., Z.G.K., M.S., and I.T.; investigation, A.O.S., H.S., Y.C., J.W., B.A., J.B., W.M.H., and Z.G.K. (experimental support for key efficacy data); resources, Z.G.K., M.S., and I.T.; data curation, A.O.S., M.S., and I.T.; writing—original draft preparation, A.O.S.; writing—review and editing, A.O.S., W.M.H., Z.G.K., M.S., and I.T.; visualization, A.O.S.; supervision, A.O.S., M.S., and I.T.; project administration, I.T.; funding acquisition, Z.G.K., M.S., and I.T. All authors have read and agreed to the published version of the manuscript.

Funding: This recsearch was financially supported through the Australian Research Council Discovery Project, titled "Poly(amino acids) as immune stimulators" (DP210101280). In addition, Z.G.K. was supported by the ARC Future Fellowship (FT230100468).

Institutional Review Board Statement: Animal study procedures were approved by the University of Queensland Animal Ethics Committee (AEC), ethics number 2021/AE000311 (17 December 2021).

Informed Consent Statement: Not applicable.

Data Availability Statement: Data Available upon request.

Acknowledgments: We acknowledge the facilities and scientific and technical assistance provided by the Australian Microscopy & Microanalysis Research Facility at the Centre for Microscopy and Microanalysis, the University of Queensland. We would also like to acknowledge the University of Queensland Genetic Research Services, which confirmed *S. aureus* infection in mice via PCR. Furthermore, we would like to acknowledge funding support through the Australian Research Council Discovery Project, titled "Poly(amino acids) as immune stimulators" (DP210101280). Z.G.K. acknowledges support by the ARC Future Fellowship (FT230100468), and we would also like to acknowledge the roles of David Paterson and Patrick Harris (UQ Centre for Clinical Research) for the D3840 GAS strain, Robert Capon (IMB, UQ) for the ACM-2727 GAS strain, and Mark Walker for the NS1944 and NS5582 isolates of GAS.

Conflicts of Interest: The authors declare no conflicts of interest.

References

1. Sanyahumbi, S. Streptococcus pyogenes: Basic Biology to Clinical Manifestations. In *Streptococcus pyogenes: Basic Biology to Clinical Manifestations*; Ferretti, J.J., Stevens, D.L., Fischetti, V.A., Eds.; University of Oklahoma Health Sciences Center: Oklahoma City, OK, USA, 2016.
2. Vekemans, J.; Gouvea-Reis, F.; Kim, J.H.; Excler, J.L.; Smeesters, P.R.; O'Brien, K.L.; Van Beneden, C.A.; Steer, A.C.; Carapetis, J.R.; Kaslow, D.C. The Path to Group A *Streptococcus* Vaccines: World Health Organization Research and Development Technology Roadmap and Preferred Product Characteristics. *Clin. Infect. Dis. Off. Publ. Infect. Dis. Soc. Am.* **2019**, *69*, 877–883. [CrossRef] [PubMed]
3. Davies, M.R.; McIntyre, L.; Mutreja, A.; Lacey, J.A.; Lees, J.A.; Towers, R.J.; Duchêne, S.; Smeesters, P.R.; Frost, H.R.; Price, D.J.; et al. Atlas of group A streptococcal vaccine candidates compiled using large-scale comparative genomics. *Nat. Genet.* **2019**, *51*, 1035–1043. [CrossRef] [PubMed]
4. Musser, J.M.; Beres, S.B.; Zhu, L.; Olsen, R.J.; Vuopio, J.; Hyyryläinen, H.L.; Gröndahl-Yli-Hannuksela, K.; Kristinsson, K.G.; Darenberg, J.; Henriques-Normark, B.; et al. Reduced In Vitro Susceptibility of *Streptococcus pyogenes* to β-Lactam Antibiotics Associated with Mutations in the *pbp2x* Gene Is Geographically Widespread. *J. Clin. Microbiol.* **2020**, *58*, 10-1128. [CrossRef] [PubMed]
5. Tedijanto, C.; Olesen, S.W.; Grad, Y.H.; Lipsitch, M. Estimating the proportion of bystander selection for antibiotic resistance among potentially pathogenic bacterial flora. *Proc. Natl. Acad. Sci. USA* **2018**, *115*, E11988–E11995. [CrossRef]
6. Azuar, A.; Li, Z.; Shibu, M.A.; Zhao, L.; Luo, Y.; Shalash, A.O.; Khalil, Z.G.; Capon, R.J.; Hussein, W.M.; Toth, I.; et al. Poly(hydrophobic amino acid)-Based Self-Adjuvanting Nanoparticles for Group A *Streptococcus* Vaccine Delivery. *J. Med. Chem.* **2021**, *64*, 2648–2658. [CrossRef]
7. Skwarczynski, M.; Zhao, G.; Boer, J.C.; Ozberk, V.; Azuar, A.; Cruz, J.G.; Giddam, A.K.; Khalil, Z.G.; Pandey, M.; Shibu, M.A.; et al. Poly(amino acids) as a potent self-adjuvanting delivery system for peptide-based nanovaccines. *Sci. Adv.* **2020**, *6*, eaax2285. [CrossRef]
8. Walkinshaw, D.R.; Wright, M.E.E.; Mullin, A.E.; Excler, J.-L.; Kim, J.H.; Steer, A.C. The *Streptococcus pyogenes* vaccine landscape. *npj Vaccines* **2023**, *8*, 16. [CrossRef]
9. Russo, B.T.; Ayinuola, Y.A.; Singh, D.; Carothers, K.; Fischetti, V.A.; Flores-Mireles, A.L.; Lee, S.W.; Ploplis, V.A.; Liang, Z.; Castellino, F.J. The M Protein of *Streptococcus pyogenes* Strain AP53 Retains Cell Surface Functional Plasminogen Binding after Inactivation of the Sortase A Gene. *J. Bacteriol.* **2020**, *202*, 10-1128. [CrossRef]
10. Brouwer, S.; Rivera-Hernandez, T.; Curren, B.F.; Harbison-Price, N.; De Oliveira, D.M.P.; Jespersen, M.G.; Davies, M.R.; Walker, M.J. Pathogenesis, epidemiology and control of Group A *Streptococcus* infection. *Nat. Rev. Microbiol.* **2023**, *21*, 431–447. [CrossRef]
11. Kehoe, M.A.; Miller, L.; Poirier, T.P.; Whitnack, E.; Beachey, E.H.; Robinson, J.H.; Pinkney, M. Molecular Biology of Group A Streptococcal M Proteins. In *Pathogenesis of Wound and Biomaterial-Associated Infections*; Wadström, T., Eliasson, I., Holder, I., Ljungh, Å., Eds.; Springer: London, UK, 1990; pp. 47–54.

12. Caparon, M.G.; Stephens, D.S.; Olsén, A.; Scott, J.R. Role of M protein in adherence of group A streptococci. *Infect. Immun.* **1991**, *59*, 1811–1817. [CrossRef]
13. Ryan, P.A.; Pancholi, V.; Fischetti, V.A. Group A streptococci bind to mucin and human pharyngeal cells through sialic acid-containing receptors. *Infect. Immun.* **2001**, *69*, 7402–7412. [CrossRef] [PubMed]
14. Walker, M.J.; Barnett, T.C.; McArthur, J.D.; Cole, J.N.; Gillen, C.M.; Henningham, A.; Sriprakash, K.S.; Sanderson-Smith, M.L.; Nizet, V. Disease manifestations and pathogenic mechanisms of Group A *Streptococcus*. *Clin. Microbiol. Rev.* **2014**, *27*, 264–301. [CrossRef] [PubMed]
15. Li, Z.; Sakota, V.; Jackson, D.; Franklin, A.R.; Beall, B. Array of M protein gene subtypes in 1064 recent invasive group A streptococcus isolates recovered from the active bacterial core surveillance. *J. Infect. Dis.* **2003**, *188*, 1587–1592. [CrossRef] [PubMed]
16. Massell, B.F.; Honikman, L.H.; Amezcua, J. Rheumatic fever following streptococcal vaccination. Report of three cases. *JAMA* **1969**, *207*, 1115–1119. [CrossRef]
17. Sekuloski, S.; Batzloff, M.R.; Griffin, P.; Parsonage, W.; Elliott, S.; Hartas, J.; O'Rourke, P.; Marquart, L.; Pandey, M.; Rubin, F.A.; et al. Evaluation of safety and immunogenicity of a group A streptococcus vaccine candidate (MJ8VAX) in a randomized clinical trial. *PLoS ONE* **2018**, *13*, e0198658. [CrossRef]
18. Hayman, W.A.; Brandt, E.R.; Relf, W.A.; Cooper, J.; Saul, A.; Good, M.F. Mapping the minimal murine T cell and B cell epitopes within a peptide vaccine candidate from the conserved region of the M protein of group A streptococcus. *Int. Immunol.* **1997**, *9*, 1723–1733. [CrossRef]
19. Pruksakorn, S.; Currie, B.; Brandt, E.; Phornphutkul, C.; Hunsakunachai, S.; Manmontri, A.; Robinson, J.H.; Kehoe, M.A.; Galbraith, A.; Good, M.F. Identification of T cell autoepitopes that cross-react with the C-terminal segment of the M protein of group A streptococci. *Int. Immunol.* **1994**, *6*, 1235–1244. [CrossRef]
20. Pastural, É.; McNeil, S.A.; MacKinnon-Cameron, D.; Ye, L.; Langley, J.M.; Stewart, R.; Martin, L.H.; Hurley, G.J.; Salehi, S.; Penfound, T.A.; et al. Safety and immunogenicity of a 30-valent M protein-based group a streptococcal vaccine in healthy adult volunteers: A randomized, controlled phase I study. *Vaccine* **2020**, *38*, 1384–1392. [CrossRef]
21. Dale, J.B.; Penfound, T.A.; Chiang, E.Y.; Walton, W.J. New 30-valent M protein-based vaccine evokes cross-opsonic antibodies against non-vaccine serotypes of group A streptococci. *Vaccine* **2011**, *29*, 8175–8178. [CrossRef]
22. Brandt, E.R.; Hayman, W.A.; Currie, B.; Carapetis, J.; Wood, Y.; Jackson, D.C.; Cooper, J.; Melrose, W.D.; Saul, A.J.; Good, M.F. Opsonic human antibodies from an endemic population specific for a conserved epitope on the M protein of group A streptococci. *Immunology* **1996**, *89*, 331–337. [CrossRef]
23. Pandey, M.; Wykes, M.N.; Hartas, J.; Good, M.F.; Batzloff, M.R. Long-Term Antibody Memory Induced by Synthetic Peptide Vaccination Is Protective against *Streptococcus pyogenes* Infection and Is Independent of Memory T Cell Help. *J. Immunol.* **2013**, *190*, 2692–2701. [CrossRef] [PubMed]
24. Batzloff, M.R.; Hayman, W.A.; Davies, M.R.; Zeng, M.; Pruksakorn, S.; Brandt, E.R.; Good, M.F. Protection against group A streptococcus by immunization with J8-diphtheria toxoid: Contribution of J8- and diphtheria toxoid-specific antibodies to protection. *J. Infect. Dis.* **2003**, *187*, 1598–1608. [CrossRef] [PubMed]
25. Shalash, A.O.; Hussein, W.M.; Nahar, U.J.; Wang, J.; Lu, L.; Azuar, A.; Kiong, J.J.E.; Koirala, P.; Khalil, Z.G.; Toth, I. Rational Development of a Novel Emulgel Adjuvant for Single-Shot Effective Vaccination: A Multivariate Analysis Approach. *Adv. Mater.* **2025**, 2506496. [CrossRef] [PubMed]
26. Pandey, M.; Powell, J.; Calcutt, A.; Zaman, M.; Phillips, Z.N.; Ho, M.F.; Batzloff, M.R.; Good, M.F. Physicochemical characterisation, immunogenicity and protective efficacy of a lead streptococcal vaccine: Progress towards Phase I trial. *Sci. Rep.* **2017**, *7*, 13786. [CrossRef]
27. Schulz, D.; Grumann, D.; Trübe, P.; Pritchett-Corning, K.; Johnson, S.; Reppschläger, K.; Gumz, J.; Sundaramoorthy, N.; Michalik, S.; Berg, S.; et al. Laboratory Mice Are Frequently Colonized with *Staphylococcus aureus* and Mount a Systemic Immune Response-Note of Caution for In vivo Infection Experiments. *Front. Cell. Infect. Microbiol.* **2017**, *7*, 152. [CrossRef]
28. Tong, S.Y.; Davis, J.S.; Eichenberger, E.; Holland, T.L.; Fowler, V.G., Jr. *Staphylococcus aureus* infections: Epidemiology, pathophysiology, clinical manifestations, and management. *Clin. Microbiol. Rev.* **2015**, *28*, 603–661. [CrossRef]
29. Shalash, A.O.; Becker, L.; Yang, J.; Giacomin, P.; Pearson, M.; Hussein, W.M.; Loukas, A.; Toth, I.; Skwarczynski, M. Development of a peptide vaccine against hookworm infection: Immunogenicity, efficacy, and immune correlates of protection. *J. Allergy Clin. Immunol.* **2022**, *150*, 157–169.e110. [CrossRef]
30. Greenfield, N.J. Using circular dichroism spectra to estimate protein secondary structure. *Nat. Protoc.* **2007**, *1*, 2876–2890. [CrossRef]
31. Wijesundara, N.M.; Lee, S.F.; Cheng, Z.; Davidson, R.; Rupasinghe, H.P.V. Carvacrol exhibits rapid bactericidal activity against *Streptococcus pyogenes* through cell membrane damage. *Sci. Rep.* **2021**, *11*, 1487. [CrossRef]

32. Toniolo, C.; Bonora, G.M.; Lüscher, I.F.; Schneider, C.H. Chain-length dependence for secondary structure formation of homo-oligopeptides from epsilon-tert.-butyloxycarbonyl-L-lysine with a lipophilic C-terminal group. *Int. J. Pept. Protein Res.* **1984**, *23*, 47–54. [CrossRef]
33. Gazzara, J.A.; Phillips, M.C.; Lund-Katz, S.; Palgunachari, M.N.; Segrest, J.P.; Anantharamaiah, G.M.; Snow, J.W. Interaction of class A amphipathic helical peptides with phospholipid unilamellar vesicles. *J. Lipid Res.* **1997**, *38*, 2134–2146. [CrossRef]

vaccines

Article

Antigen-Dependent Adjuvanticity of Poly(lactic-co-glycolic acid)-polyethylene Glycol 25% Nanoparticles for Enhanced Vaccine Efficacy

Minxuan Cui [1,†], Jiayue Xi [1,†], Zhuoyue Shi [1], Yupu Zhu [2], Zhengjun Ma [1], Muqiong Li [1], Qian Yang [3], Chaojun Song [4,*] and Li Fan [1,*]

1 Shaanxi Key Laboratory of Chiral Drug and Vaccine Adjuvants, Department of Pharmaceutical Chemistry and Analysis, School of Pharmacy, Air Force Medical University, 169th Changle West Road, Xi'an 710032, China; cuiminxuan2521@fmmu.edu.cn (M.C.); 0518xjy@fmmu.edu.cn (J.X.); shizhuoyue2021@163.com (Z.S.); mazhengjun@fmmu.edu.cn (Z.M.); limuqiong1981@fmmu.edu.cn (M.L.)

2 Department of Occupational & Environmental Health, The Ministry of Education Key Lab of Hazard Assessment and Control in Special Operational Environment, School of Public Health, Air Force Medical University, 169th Changle West Road, Xi'an 710032, China; zhuyupupapa@fmmu.edu.cn

3 Department of Chinese Materia Medical and Natural Medicines, School of Pharmacy, Air Force Medical University, 169th Changle West Road, Xi'an 710032, China; yangqian@fmmu.edu.cn

4 School of Life Science, Northwestern Polytechnical University, 127th Youyi West Road, Xi'an 710072, China

* Correspondence: cj6005@nwpu.edu.cn (C.S.); xxfanny@fmmu.edu.cn (L.F.)

† These authors contributed equally to this work.

Abstract: Background: A key component in modern vaccine development is the adjuvant, which enhances and/or modulates the antigen-specific immune response. In recent years, nanoparticle (NP)-based adjuvants have attracted much research attention owing to their ability to enhance vaccine potency. Nonetheless, how the selection of different antigens influences the overall vaccine efficacy when combined with the same nanoparticle adjuvant is less discussed, which is important for practical applications. **Methods**: Non-toxic mutants of exotoxin Hla (rHlaH35L) and cell-wall-anchored protein SpA(rSpam) were covalently conjugated to Poly(lactic-co-glycolic acid)-polyethylene glycol (PLGA-PEG) 25% NPs (25% NPs) as antigens to prepare nanovaccines. Antibody titers, cytokine secretion levels, and the antibody bacteriolytic capacity were tested to investigate immune activation. To evaluate the protective efficacy of the nanovaccine, immunized mice were challenged with *S. aureus* ATCC 25923 at three different lethal doses: $1 \times LD_{100}$, $2 \times LD_{100}$, and $4 \times LD_{100}$. **Results**: We showed that 25% NP-rHlaH35L nanovaccines were associated with more efficient humoral, cellular, and innate immune responses and protection potency compared with 25% NP-rSpam. Moreover, the overall vaccine potency of 25% NP-rHlaH35L was even better than the combination vaccination of both 25% NP-rHlaH35L and 25% NP-rSpam. In comparison to the clinically used aluminum (alum) adjuvant, the 25% NP adjuvants were found to stimulate humoral and cellular immune responses efficiently, irrespective of the antigen type. For antigens, either exotoxins or cell-wall-anchored proteins, the 25% NP-based vaccines show excellent protection for mice from *S. aureus* infection with survival rates of 100% after lethal challenge, which is significantly superior to the clinically used alum adjuvant. Moreover, due to the superior immune response elicited by 25% NP-rHlaH35L, the animals inoculated with this formulation survived even after two times the lethal dose of *S. aureus* administration. **Conclusions**: We demonstrated that the type of antigen plays a key role in determining the overall vaccine efficacy in the immune system when different kinds of antigens are conjugated with a specific nanoparticle adjuvant, paving a new way for vaccine design based on 25% NP adjuvants with enhanced potency and reduced side effects.

Keywords: antigen; nanovaccine; *Staphylococcus aureus*; nanoparticle adjuvant; adjuvanticity; PLGA; subunit vaccine

1. Introduction

Staphylococcus aureus (*S. aureus*) is a Gram-positive commensal bacterium that can cause severe opportunistic infections, such as bacteremia, sepsis, and endocarditis [1]. In recent years, *S. aureus* has become a public health concern due to the emergence and spread of methicillin-resistant *S. aureus* (MRSA) strains that are resistant to most antibiotic treatments [2]. Vaccination has been envisioned to be an effective solution [3], but the development of *S. aureus* vaccines via the conventional approach of generating antibodies against surface antigens has appeared to be insufficient [4], due to immunoevasive adaptations and the lack of cellular response activation [5]. Extensive research supports the notion that T-cell-mediated immunity is a pivotal factor in *S. aureus* vaccine efficacy. For instance, while B-cell-deficient mice show no increased vulnerability to systemic *S. aureus* infection compared to wild-type mice [6], mice deficient in T cells [7], IFN-γ [8,9], TNF [10], or dual IL-17A/F [11] demonstrate heightened susceptibility to such infections. This evidence highlights the necessity for novel vaccine designs and approaches to achieve protection against *S. aureus*. At its core, a successful vaccine must integrate pathogen-specific antigens with immunostimulatory adjuvants to trigger a protective adaptive immune response. An effective immune response against *S. aureus* necessitates a multifaceted approach, involving antibodies that directly neutralize bacterial viability and toxicity, enhance opsonophagocytic activity, and promote T-cell-mediated immunity to recruit phagocytic cells to the site of infection [12].

The selection of *S. aureus*-specific antigens and adjuvants for immunostimulation are two critical components in developing a protective *S. aureus* vaccine. Antigens determine the targeting of the immune protection of vaccines and the protective spectrum of vaccines. A principle for selecting an antigen should be its conservation among different strains. Exotoxins and cell-wall-anchored (CWA) proteins are usually selected as antigens, which are widely expressed in most *S. aureus* strains for the following reasons: (1) Functionally, the antigen modified by exotoxins can block the process of inflammation and target cell lysis [13] and the antigen using CWA proteins can block bacterial adhesion colonization, nutrient acquisition, and immune evasion [14]. (2) In terms of secretion and expression characteristics, exotoxins and CWA proteins can simulate the process from bacterial adhesion to virulence, which has the advantage of biomimicry.

On the other hand, the adjuvant determines the production time, titer, and maintenance time of neutralizing antibodies, which influence the protective efficacy of the bacterial vaccine [15,16]. In fact, in animal models, antigen-specific antibody titers, antigen-specific T cell responses, and bacterial clearance during infection all increased with the addition of adjuvants [17]. In recent years, nanoparticles have been widely used in the research of new vaccine adjuvants due to their ease of modification and high efficiency [18]. Adjusting the size and shape of nanoparticles can alter the type and degree of vaccine immune activation [19]. Spherical polystyrene nanoparticles with a size of 193 nm produced a TH1-biased immune response, while larger rod-shaped polystyrene particles (1530 nm) induce a stronger TH2-biased immune response [20]. The rigidity of the nanoparticles also appears to change their pharmaceutical response characteristics, and it has been reported that the doping ratio of polyethylene glycol(PEG) in poly(lactic-co-glycolic acid)(PLGA) nanoparticles modulates the rigidity of PLGA nanoparticles. The rigidity of PLGA nanoparticle adjuvants affects the phagocytic ability of dendritic cells (DCs) and the rate of antigen release, ultimately leading to differences in vaccine protection. In short, a nanovaccine prepared with 25% PEG-doped PLGA nanoparticles combined with the *S. aureus* antigen rEsxB provided 100% protection in a mouse *S. aureus* infection model during subcutaneous immunization [21]. Wang et al. prepared negatively charged core-shell nanomicelles with different rigid cores using PLGA and PCL with varying proportions of LA and GA as soft cores and PSt as hard cores. They investigated the effect of core rigidity on the uptake efficiency of tumor cells. The results showed that the rigidity of the nanomicelle core significantly affected the uptake efficiency of the tumor cells. PLGA (LA/GA, 48/48, mole ratio) nanoparticles with a soft core structure exhibited the highest uptake efficiency by tumor cells [22]. Gao et al. prepared core–shell PLGA–liposome nanoparticles with different rigidities and studied their transport ability in tissues covered with biohydrogel. The results showed that the moderately rigid nanoparticles are deformed into ellipsoids, which promotes the rapid diffusion of nanoparticles in the hydrogel mesh. The encapsulation of doxorubicin by moderately rigid nanoparticles significantly increased the bioavailability of the drug in vivo [23].

Moreover, adjuvants have a clear selective preference for specific antigens [24]. For example, polyvinylimine and chitosan can enhance the binding level of antigens to mucosal surfaces for pathogens that invade through mucosal pathways [25]. The complex life cycle of *Plasmodium* and the variety of antigens put forward higher requirements for the development of malaria vaccines [26], and accordingly, multi-functional adjuvants are needed. The RTS,S vaccine is a malaria vaccine candidate that targets the spore phase of *Plasmodium* infection [27]. AS01 was selected as the vaccine adjuvant, consisting of 3-O-deacyl-4′-monophospholipid A (MPL) and saponin QS-21. MPL is a toll-like receptor-4(TLR4) agonist that induces mouse T cells to produce IFN-γ and convert antibodies to IgG 2a/c, while QS-21 induces the production of neutralizing antibodies and cytotoxic T cell responses [28]. Animal studies have shown that the RTS,S vaccine induces a rapid, instantaneous innate immune response in the injection site, draining lymph nodes, and produces antibody titers 20 times higher than natural exposure [29]. These studies suggested that adjuvants may also have a selective preference for different types of *S. aureus* antigens.

Herein, as illustrated in Figure 1, in order to mimic *S. aureus*, one exotoxin (Hla) and one CWA protein (SpA) were selected as vaccine antigens. PLGA nanoparticles with a 25% PEG doping ratio were chosen as the adjuvant. rHlaH35L and rSpam, the non-toxic mutants of Hla and the CWA protein SpA of *S. aureus*, were first prepared as recombinant antigens. Covalent bonding was also chosen as the conjugation method between the antigen and the adjuvant. The immunoactivation efficacy, protection, and biosafety issues have been well-studied. Moreover, the antigen selection preference of the 25% PLGA-PEG nanoparticles (25% NPs) adjuvant was also tested. We aimed to obtain an optimized nanoparticle vaccine formulation with the most protective effect to pave the way for future clinical trials.

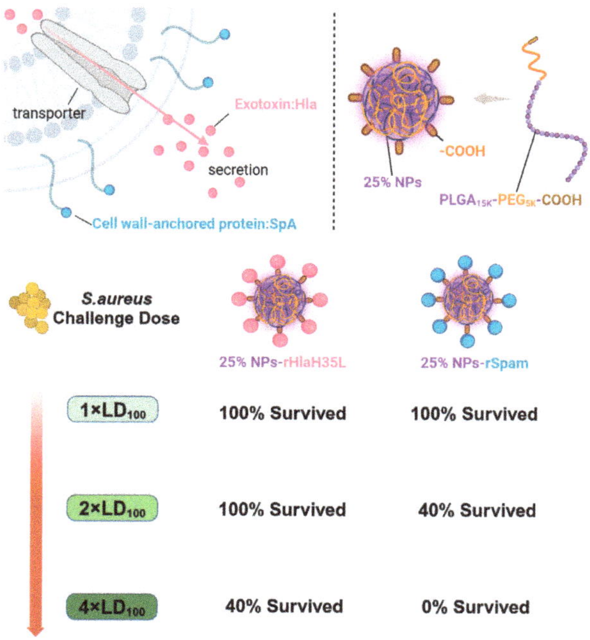

Figure 1. A graphic illustration of the immune efficacy of antigen-dependent adjuvants. Nanovaccines prepared with 25% NPs with either the exotoxin Hla or the CWA protein SpA exhibited different protection at different challenge doses. When the challenge dose was 1 × LD_{100}, both nanovaccines provided 100% protection. However, when the challenge dose increased to 2 × LD_{100}, the survival rate of 25% NP-rHlaH35L remained 100%, while the survival rate of the 25% NP-rSpam group was 40%. As the challenge dose continued to increase, the survival rates of both groups decreased, and the protective efficacy of 25% NP-rHlaH35L was superior to that of the 25% NP-rSpam group. Created in BioRender (https://BioRender.com).

2. Materials and Methods

2.1. Expression and Purification of rHlaH35L and rSpam Antigen

α-hemolysin (Hla), a critical exotoxin produced by *S. aureus*, plays a significant role in the pathogen's infection process. Hla binds to the ADAM-10 receptor on target cell membranes, leading to cell lysis. To eliminate its hemolytic toxicity, the histidine at position 35 in the *hla* sequence was substituted with leucine, resulting in the HlaH35L plasmid [30–32]. Protein A (SpA), a cell-wall-anchored protein expressed by *S. aureus*, is essential for immune evasion. The immunoglobulin-binding domains (E, D, A, B, and C) of SpA are frequently employed as vaccine antigens. In wild-type SpA, two consecutive glutamines (QQ) at positions 7–8 and two consecutive aspartic acids (DD) at positions 34–35 within its five structural domains are crucial for immune escape. To generate a SpA variant, glutamines at positions 7–8 in each domain were replaced with lysine, and aspartic acids at positions 35–36 were substituted with alanine [33,34]. The expression plasmid was transformed into BL21 (DE3) *E. coli* (Tsingke Biotechnology Co., Ltd, Beijing, China), and recombinant proteins rHlaH35L and rSpam, both carrying a 6 × His-tag, were produced using standard IPTG induction [35]. These proteins were purified from the lysate of IPTG-induced BL21 (DE3) *E. coli* using Ni-Sepharose column chromatography. The Ni-Sepharose column (QIAGEN, Shanghai, China) was washed with phosphate buffer (PBS) and eluted with 100 mmol/L, 300 mmol/L, and 500 mmol/L imidazole gradients(Merck, Shanghai, China). The purified proteins were dialyzed into PBS buffer, and endotoxins were removed using a High-Capacity Endotoxin Removal column (Xiamen Bioendo Technology Co., Ltd.,

Xiamen, China). Protein concentrations were quantified using the BCA assay(Thermo Fisher Scientific, Waltham, MA, USA), and purity was assessed by SDS-PAGE.

2.2. Preparation and General Characterizations of 25% NPs Conjugated with Antigen

The preparation of PLGA-PEG 25% nanoparticles (25% NPs) followed a precipitation/solvent diffusion method, adapted from our previously established procedures with slight modifications [21]. In detail, 100 mg of $PLGA_{15K}$-PEG_{5K}-COOH polymers (50:50 lactic acid to glycolic acid ratio, Xi'an Ruixi Biological Technology Co., Ltd., Xi'an, China) were dissolved in an organic solvent mixture containing 1.5 mL of dichloromethane and 1 mL of acetone. This organic solution was slowly added to 10 mL of a 5% aqueous PVA solution (Shanghai Macklin Biochemical Technology Co., Ltd., Shanghai, China) under constant stirring at 600 rpm. The mixture was emulsified using an ultrasonic probe to form an oil-in-water emulsion. The emulsion was then transferred into 50 mL of DI water and stirred overnight. The 25% NP suspension was filtered through a 0.45 μm (Merck, Shanghai, China) filter to remove aggregates and centrifuged at 18,000× *g* at 4 °C for 1 h. The resulting precipitate was resuspended in DI water and centrifuged again to yield the final 25% NPs. Then, 25% NPs were resuspended in 10 mL of MES buffer, followed by the addition of EDC solution and NHS solution(Merck Millipore, Billerica, MA, USA). The mixture was stirred at 600 rpm for 3 h. The activated 25% NPs were collected by centrifugation at 18,000× *g* for 1 h and then resuspended in a solution containing recombinant proteins. The pH of the solution was adjusted to 8.0, and the mixture was stirred vertically overnight at 4 °C. The hydrodynamic diameter and zeta potential of 25% NPs, 25% NP-rHlaH35L, and 25% NP-rSpam were determined using dynamic light scattering (Delsa™ Nano, Beckman-Coulter, High Wycombe, UK). Fourier Transform Infrared Spectroscopy (Thermo Nicolet IS50, Thermo Fisher Scientific, Waltham, MA, USA) was utilized to verify successful crosslinking. The Bradford protein assay (Beyotime, Shanghai, China) was employed to measure the total recombinant protein concentration loaded onto 25% NPs, and the loading efficiency was calculated using the following formula:

$$\text{Loading effiency(mass ratio)} = \frac{m_{\text{recombinant protein on the 25\%NPs(mg)}}}{m_{\text{25\%NPs(mg)}}} \times 100\% \quad (1)$$

2.3. Evaluation of Biocompatibility of 25% NP Antigen In Vitro and In Vivo

Cell counting kit-8 (CCK-8) was performed using the L929 fibroblast cell line to assess the cytotoxicity of nanovaccines. In brief, cells were plated at a concentration of 5×10^3 cells/well in 96-well plates and allowed to adhere overnight. Gradient concentrations of nanovaccines, antigens, and nanoparticle solutions (ranging from 1000 μg/mL to 7.8125 μg/mL) were added to the wells, with each concentration tested in six replicate wells. After 24 h of incubation, CCK-8 solution was added to each well, followed by incubation at room temperature in the dark for 1 h. Absorbance was measured at 450 nm. Cell viability was determined using the following formula:

$$\text{Cell Viability} = \frac{A_{\text{test}} - A_{\text{background}}}{A_{\text{control}} - A_{\text{background}}} \times 100\% \quad (2)$$

Female BALB/C mice, aged 8 to 10 weeks, were randomly assigned to 6 groups and immunized subcutaneously with nanovaccines or antigens on day 0, followed by boost immunizations on day 14 and day 28 to evaluate the in vivo biocompatibility of the nanovaccines. The body weight and physical condition (including activity level, appetite, coat condition, and skin integrity) of the mice were recorded on days 0, 7, 14, 21, 28, and 35. On day 35, the mice were euthanized, and tissues from the heart, liver, spleen, lung, and kidney were collected. Hematoxylin and eosin (HE) staining, following the standard method described in our earlier publication [36], was performed to examine tissue toxicity caused by the nanovaccine.

2.4. Animal Immunization

BALB/C mice used in the experiments were purchased from the Animal Experiment Center of Air Force Medical University. All animal experiments complied with the National Research Council's Guide for the Care and Use of Laboratory Animals, and the animal experiments were approved by the Animal Care and Ethics Committee of Fourth Military Medical University (Approval NO. KY20213144-1). Female BALB/C mice of 8–10 weeks of age were used for immunization, and all animals were healthy before immunization. Animals were divided into 7 groups and inoculated with PBS buffer, rHlaH35L, rSpam, rHlaH35L + rSpam, clinically used alum mixed with rHlaH35L (alum-rHlaH35L), rSpam (alum-rSpam), and alum-rHlaH35L + alum-rSpam were set as positive control, and 25% NP antigens were set as test groups. Mice were immunized subcutaneously on days 14 and 28 after the first immunization for a boost with the same formulation (antigen concentration of 25 μg/mice). Based on the average antigen loading efficiency and antigen dosage, the concentration of nanoparticles in 25% NP-rHlaH35L was 2272.7 μg/mice; the concentration of nanoparticles in 25% NP-rSpam was 1388.9 μg/mice; the concentration of nanoparticles in the antigen combine group was 3661.6 μg/mice.

2.5. ELISA Assay

Blood was collected from the tail tip vein of mice on the 17th and 35th days from the initial immunization and incubated at 37 °C for 1 h to allow the serum to precipitate. The serum was centrifuged and stored at −80 °C. The antigen was diluted to 10 μg/mL with coating buffer and added to a 96-well plate and stored at 4 °C overnight. Mouse serum was diluted to different gradients with antibody diluent solution. The gradient-diluted mouse serum was added to a 96-well plate and incubated at 37 °C for 1 h. Then, the plate was washed 5 times with PBST. HRP-labeled secondary antibody was added and incubated at 37 °C for 45 min, and then the plate was washed with PBST 5 times, and the plate was incubated at room temperature for 20 min with the addition of ABTS chromogenic solution. The absorbance at 405 nm was detected by a microplate reader (Biotek SYNERGY LX Instruments, Santa Clara, CA, USA). The AUC value of the antibody titer was calculated as we reported before.

2.6. ELISPOT Assay

Mice spleens were isolated and ground to prepare mice splenocyte homogenates after the mice were euthanized in an aseptic environment. The splenocyte homogenate was centrifuged, mixed with erythrocyte lysate, left at room temperature for 5 min, centrifuged again, and counted, and the cell density was adjusted to 1×10^7 cells/mL. The 96-well plate was washed 5 times using sterile PBS and then incubated with 1640 medium containing 10% FCS at room temperature for 30 min. The antigen was diluted with medium to 4 μg/mL, and 100 μL of antigen solution was added to each well as an immunostimulant. Finally, 100 μL of cell suspension was added, and the plate was incubated for 24 h at 37 °C. The plate was washed 5 times with sterile PBS. The detection antibody was diluted to working concentration with 0.5% FCS in PBS. A total of 100 μL of detection antibody was added to each well and incubated for 2 h at room temperature. The plate was washed 5 times with sterile PBS, and 100 μL of ACP solution diluted with 0.5% FCS-PBS was added to each well and incubated for 1 h at room temperature. The plate was washed five times with sterile PBS, and 100 μL of TMB solution was added to each well until unique spots appeared. The plates were rinsed with plenty of water to terminate the color development and counted in an ELISPOT reader (Cellular Technology Limited, Cleveland, OH, USA).

2.7. Lethal Challenge

A total of 100 μL of *S. aureus* (strain ATCC 25923) at different concentrations (for the $1 \times LD_{100}$ lethal challenge group, the concentration of *S. aureus* was 2.56×10^9 CFU/mL; for the $2 \times LD_{100}$ lethal challenge group, the concentration of *S. aureus* was 5.12×10^9 CFU/mL; and for the $4 \times LD_{100}$ lethal challenge group, the concentration of *S. aureus* was 1.024×10^{10} CFU/mL) were administered intravenously to the mice on the 35th day after prime immunization. Then, the physical condition and mortality of mice in each group were observed and recorded daily. At the end point of the experiment, all remaining animals were euthanized.

2.8. Serum Bactericidal Test

The Neisser–Wechsberg tube lysis reaction was used to determine the neutralizing activity of antibodies in the sera of immunized mice. *S. aureus* ATCC 25923 was cultured to OD 600 nm = 0.5. Healthy guinea pig sera were isolated and prepared. Bacterial liquid, guinea pig serum, and mouse serum immunized with nanovaccine were mixed with sterile PBS at a dilution ratio of 1:1000,000 and incubated at 37 °C for 1 h. The solution was spread on Luria–Bertani solid plates, and single clones on the plates were counted on the next day. Mouse sera from PBS and antigen groups were used as a negative control for simultaneous experiments. The lysis rate was calculated using the following formula:

$$\text{Percentage of lytic bacteria}(\%) = 1 - \frac{\text{The CFU of treated group}}{\text{The CFU of control group}} \times 100\% \quad (3)$$

2.9. Statistical Analysis

All results were expressed as mean ± standard deviation (SD) and statistically analyzed using GraphPad Prism 9.0 software. Data distribution was assessed using the Shapiro–Wilk test, which confirmed normality. Analysis of variance (one-way ANOVA) was used to determine the statistical significance of the differences. Log-rank test was used to determine the statistical significance of survival rates. Statistical difference was defined as * $p < 0.05$, ** $p < 0.01$, *** $p < 0.001$, and **** $p < 0.0001$, and ns (not significant) for $p \geq 0.05$ with $p < 0.05$ considered statistically significant.

3. Results

3.1. General Characterizations of 25% NPs Conjugated with Antigen

The molecular weight and purity of rHlaH35L and rSpam were confirmed by SDS-PAGE (Figure S1). The successful covalent binding of the two antigens to 25% NPs was verified by the presence of amide bonds in the Fourier Transform Infrared Spectroscopy (FTIR) spectra of conjugated NP samples (Figure 2a). The loading efficiencies of rHlaH35L and rSpam on 25% NPs were approximately 1~2% (mass ratio), respectively (Figure 2b).

The determination results of the hydrodynamic diameter and zeta potential for various batches of nanoparticles and nanovaccines can be found in the Supporting Information (Tables S1 and S2). The representative average size of 25% NPs is 172.9 ± 6.963 nm (Figure 3a). As shown in Figure 3b,c, a small increase of ~20 nm in the average size of the NPs was observed after the antigen conjugation (25% NP-rHlaH35L:189.9 ± 9.531 nm; 25% NP-rSpam:201.5 ± 8.834 nm). Since the synthesized nanoparticles are being carboxyl-terminated, the zeta potential of 25% NPs is −17.917 ± 1.107 mV (Figure 3d). Additionally, after antigen conjugation, the zeta potential of both 25% NP-rHlaH35L (−8.897 ± 0.956 mV) and 25% NP-rSpam (−22.074 ± 1.898 mV) remains negative (Figure 3e,f).

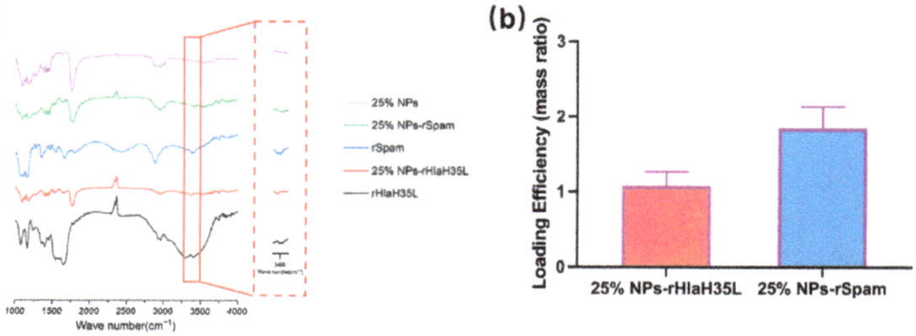

Figure 2. Nanovaccine via condensation reaction as enhanced *S. aureus* vaccine adjuvant. (**a**) FTIR spectra taken from 25% NPs, 25% NP-rHlaH35L, rHlaH35L, 25% NP-rSpam, and rSpam samples. (**b**) Antigen loading efficiency of 25% NP antigen measured by Bradford.

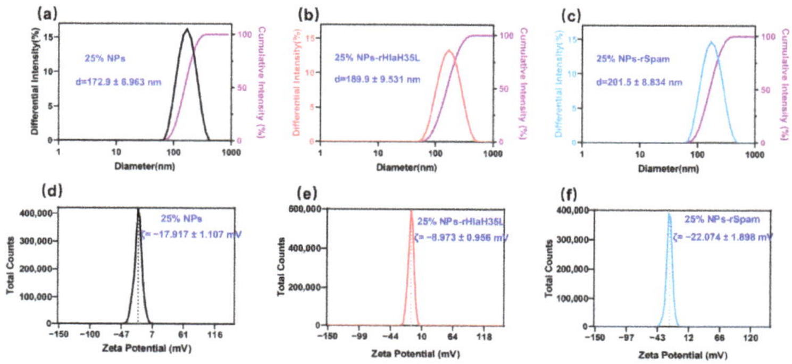

Figure 3. Hydrodynamic diameter of (**a**) 25% NPs, (**b**) 25% NP-rHlaH35L, and (**c**) 25% NP-rSpam. Zeta potential of (**d**) 25% NPs, (**e**) 25% NP-rHlaH35L, and (**f**) 25% NP-rSpam.

3.2. Biocompatibility of Nanovaccines

The vaccine demonstrated good safety both in vitro and in vivo. In the CCK-8 experiments, the survival rate of the L929 cell line remained above 95% at nanoparticle concentrations ranging from 7.8 ng/mL to 1 mg/mL (Figure S2a). The body weights of the immunized mice did not show any significant decrease during the monitoring period (Figure S2b). The immunized mice were in good physiological condition, exhibiting normal appetites and high levels of vitality. Their skin and fur were in excellent condition. No significant damage was observed in the major organs of the immunized mice (Figure S2c).

3.3. The Efficacy Evaluation of Nanovaccines

After S.C. vaccination of the antigen, alum/antigen and 25% NP/antigen conjugates, the antigen-specific IgG levels were monitored for all treatment groups. In our research, mice were first inoculated at day 0, then boosted at day 14 and day 28. Blood samples were collected on day 17 and day 35 in order to evaluate the specific IgG levels in all of the treatment groups (Figure 4a). Day 17 was selected as the sampling point because IgM, the dominant antibody subtype before the second vaccination, exhibits a transient peak during the early immune response and typically declines as the immune system shifts to IgG production. According to the immune response rules, protective IgG antibody levels should rise after the second vaccination. As expected, and as shown in Figure 4b, the PBS buffer group and antigen groups (rHlaH35L, rSpam, and rHlaH35L + rSpam) did not show significant antibody growth. On the contrary, the positive alum adjuvant control groups, the alum-rHlaH35L and alum-rSpam treatment groups, showed significant IgG increases compared with the corresponding antigen ones. The results above confirmed that the evaluation system runs normally and that the results are logical. Then, we analyzed the results of the experimental groups, including the combination of the alum adjuvants with antigen and the antigen combined with 25% NP adjuvants. The alum-rHlaH35L + alum-rSpam treatment group did not show any significant difference compared to the groups receiving alum adjuvant with antigen alone. However, in the 25% NP-adjuvanted immune groups, either combined with the single antigen (rHlaH35L or rSpam) or the antigen combination (rHlaH35L and rSpam), antigen-specific IgG titers significantly increased. Moreover, the 25% NP adjuvant conjugated with the antigen combination presented double IgG AUC compared with the conjugated with a single antigen. The results above indicated that even in the early stage in the immunization period, only 3 days after the second vaccination, 25% NPs serving as a vaccine adjuvant could elicit a high antibody titer level.

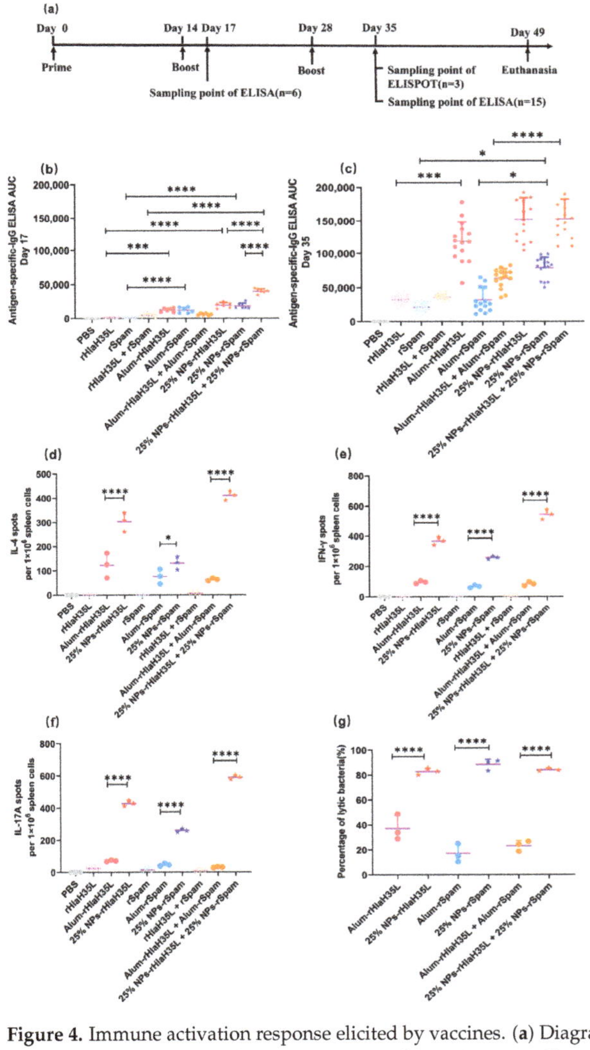

Figure 4. Immune activation response elicited by vaccines. (**a**) Diagram of immunization procedure and sampling time points for immunoassay. (**b**) Antigen-specific serum IgG titer at day 17 ($n = 6$). (**c**) Antigen-specific serum IgG titer at day 35 ($n = 15$). (**d**) IL-4 secretion level revealed by ELISPOT at day 35 ($n = 3$). (**e**) IFN-γ secretion level revealed by ELISPOT at day 35 ($n = 3$). (**f**) IL-17A secretion level revealed by ELISPOT at day 35 ($n = 3$). (**g**) Percentage of lytic bacteria of serum antibody at day 35. All data are expressed as mean ± S.D. Statistical significance is indicated by $p < 0.05$ (* $p < 0.05$; *** $p < 0.001$; **** $p < 0.0001$).

As time went on, on day 35, the 7th day after the third vaccination, the PBS group and negative free antigen groups (including the single antigen and antigen combinations) showed no significant IgG increase (Figure 4c), which is comparable with existing research [37]. In the positive-alum control groups, alum could significantly increase the rHlaH35L and the combined antigen (rHlaH35L and rSpam)-specific IgG titer; however, alum-rSpam did not present a higher antibody titer level compared with the antigen groups. The absolute IgG AUC value of alum-rSpam on day 35 is comparable with the one on day 17, indicating that the alum adjuvant is not able to elicit a strong rSpam-induced humoral immune response. In the 25% NP adjuvant vaccine groups, 25% NPs conjugated with single rHlaH35L could elicit a similar IgG titer level with alum-rHlaH35L. Notably, it is pleasing that the 25% NP adjuvant could strongly elicit an rSpam-specific antibody titer level, which the alum adjuvant could not. The IgG level of 25% NP-rSpam is significantly higher than that of the alum-rSpam and pure rSpam vaccination groups. Consequently, the titer level of 25% NPs conjugated with the rHlaH35L and rSpam antigen combination is much higher than that of the alum ones.

As the antibody titer is representative of the humoral immune response, the secretion levels of IL-4 in the spleen of immunized mice exhibit a similar trend (Figure 4d). Both in the single-antigen vaccine group and in the antigen-combination vaccine group, the number of IL-4 spots in the 25% NP adjuvant vaccine groups is much higher than that in the alum ones, indicating that for *S. aureus* recombinant protein rHlaH35L and rSpam, the adjuvanticity for eliciting a humoral immune response of 25% NPs is higher than that of alum.

On the other hand, the cellular immune response activated by specific vaccine groups was also evaluated by an ELISPOT assay. IFN-γ spots were employed for measuring cellular immune response activation. As shown in Figure 4e, vaccines with adjuvant groups including alum and 25% NPs could elicit significant cellular immune responses. Furthermore, all 25% NP vaccine groups exhibited a significantly higher number of IFN-γ spots compared to the alum-adjuvanted groups, irrespective of the antigen formulation (single antigen or antigen combination). The results indicate that in the cellular immune response activation aspect, 25% NPs showed stronger adjuvanticity.

Apart from the enhancement of humoral and cellular immune response, Th17-associated immune responses are another consideration in the establishment of an effective vaccine design against *S. aureus* infections. Given that IL-17A produced by Th17 cells is an essential cytokine in the primary defense against *S. aureus* infections, an ELISPOT assay of IL-17A was also carried out to determining whether the 25% NP vaccine could eradicate *S. aureus* in the host. As shown in Figure 4f, as the vaccine adjuvant, 25% NPs could elicit five to six times the IL-17A-specific spots compared with the alum adjuvant when combined with a single antigen. Moreover, 25% NPs conjugated with the antigen combination (rHlaH35L and rSpam) could induce an IL-17A level in the serum 19 times higher than that of the alum group, indicating that 25% NPs as the vaccine adjuvant could effectively induce Th17-associated immune responses.

Although the evidence of humoral, cellular, and Th17-associated immune response activation was found in the 25% NP vaccine groups, the neutralizing capability of the antibody is the crucial factor that determines the vaccine protection ability. Bacteriolysis evaluation was employed to investigate the bacteriolysis effects of the antibody produced by vaccination of the 25% NP single or combined antigen (with the corresponding alum vaccine groups as the control groups). The results (Figure 4g) reveal the significantly higher bacteria-neutralizing capability of antibodies produced by mice vaccinated with 25% NP vaccines over the alum ones. The results are consistent with the mice survival rates in the lethal challenge experiment.

3.4. Lethal Challenge

Before the lethal challenge, all of the mice were healthy. For the groups of mice treated with pure antigen/the antigen combination or PBS, all of the mice died following intravenous administration of a lethal dose of *S. aureus*. (Figure S4a–c). A total of 20% of the mice survived in the alum-rSpam-treated group and 50% of the mice survived in the alum-rHlaH35L- and alum-rHlaH35L + alum-rSpam-treated groups. In comparison, the survival rates of the 25% NP vaccine groups with pure antigen and the antigen combination were much higher (about 100%) than those of the alum vaccine groups in general (Figure 5b–d). No significant difference was identified at this lethal challenge dose using the single-antigen or antigen-combination 25% NP vaccines. However, in the efficacy evaluation above, we did observe significant differences in terms of humoral, cellular, or Th17-associated immune response activations between the three 25% NP vaccine groups. The results suggest that the lethal dose is too low to differentiate the efficacy of these three nanovaccines. Thus, we increased the challenge dose to a 2 × LD_{100} and 4 × LD_{100} lethal dose.

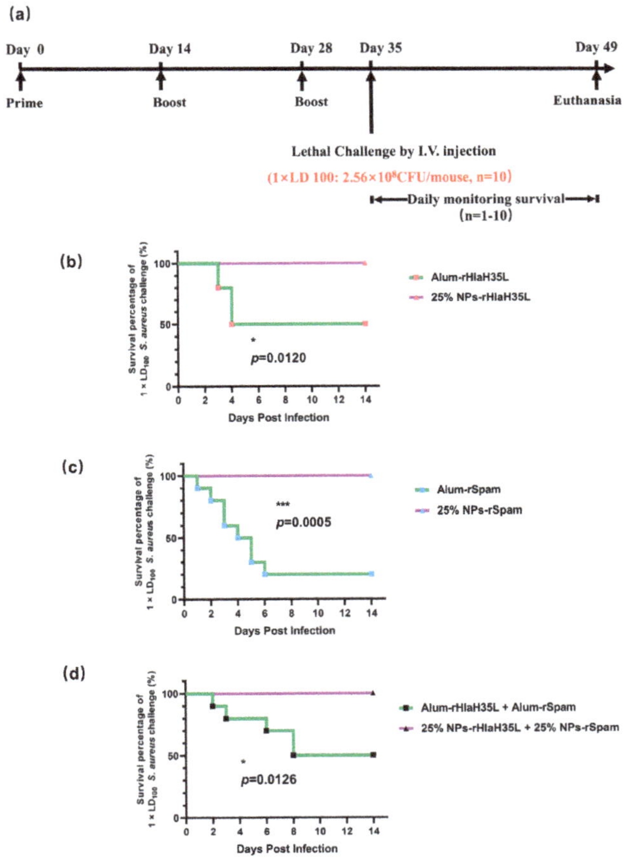

Figure 5. Survival analysis after lethal challenge with *S. aureus* strain ATCC 25923 at concentration of 2.56 × 10^8 CFU per mouse (**a**) Diagram of immunization, lethal challenge, and evaluation of protective efficacy. (**b**) Survival analysis between mice immunized with alum-rHlaH35L and mice immunized with 25% NP-rHlaH35L (n = 10). (**c**) Survival analysis between mice immunized with alum-rSpam and mice immunized with 25% NP-rSpam (n = 10). (**d**) Survival analysis between mice immunized with alum-rHlaH35L + alum-rSpam and mice immunized with 25% NP-rHlaH35L + 25% NP-rSpam (n = 10). Survival rates were analyzed using Log-rank (Mantel–Cox) analysis. Statistical significance is indicated by $p < 0.05$ (* $p < 0.05$; *** $p < 0.001$).

As shown in Figure S5, the same situation happened in the mouse groups treated with pure antigen/the antigen combination or PBS; 100% mortality was observed in all mice after intravenous injection of a 2 × LD_{100} dose of *S. aureus*.". In the alum-adjuvant vaccine groups, all of the mice in the alum-with-a-single-antigen (rHlaH35L or rSpam)-treated groups died within one day after the challenge. Only 20% of the mice survived in the alum-rHlaH35L + alum-rSpam-treated group (Figure S5f). However, the survival rates of the 25% NP vaccine group with pure antigen and the antigen combination were still much higher than those of the alum vaccine groups. A total of 100% of the mice survived in the 25% NP-rHlaH35L-treated group, while 40% and 80% of the mice survived in the 25% NP-rSpam and 25% NP-rHlaH35L + 25% NP-rSpam treated groups (Figure 6b). The addition of rSpam resulted in a decrease in survival percentage in the 25% NP vaccine groups. We further challenged all of the mouse groups with even higher doses.

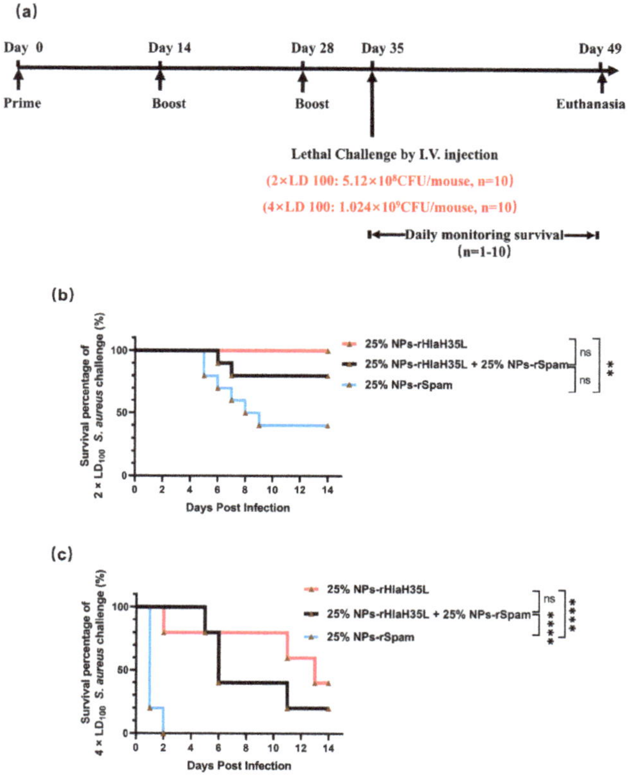

Figure 6. Survival analysis after lethal challenge with *S. aureus* strain ATCC 25923 at different bacteria concentrations. (**a**) Diagram of immunization, lethal challenge, and evaluation of protective efficacy. (**b**) Survival analysis among mice immunized with 25% NP-rHlaH35L, 25% NP-rSpam, and 25% NP-rHlaH35L + 25% NP-rSpam when mice were challenged with *S. aureus* strain ATCC 25923 at concentration of 5.12 × 10^8 CFU per mouse (n = 10). (**c**) Survival analysis among mice immunized with 25% NP-rHlaH35L, 25% NP-rSpam, and 25% NP-rHlaH35L + 25% NP-rSpam when mice were challenged with *S. aureus* strain ATCC 25923 at concentration of 1.024 × 10^9 CFU per mouse (n = 10). Survival rates were analyzed using Log-rank (Mantel–Cox) analysis. Statistical significance is indicated by $p < 0.05$ (** $p < 0.01$, **** $p < 0.0001$).

As expected, all mice in the negative control and alum-adjuvanted groups died following intravenous administration of a 4 × LD_{100} dose of *S. aureus*. (Figure S6). For the

25%-NP-adjuvant-vaccine-treated groups, the survival rate of the mice treated with 25% NP-rSpam is significantly lower than those treated with rHlaH35L (Figure 6c).

The vaccine efficacy and lethal challenge results indicated that for the 25% NP adjuvant, rHlaH35L is a more probable antigen choice, with which 25% NPs could elicit higher neutralizing antibody titers and stronger body protection in mice.

4. Discussion

In recent years, the development of a vaccine for *S. aureus* has been recognized as a key strategy for controlling the prevalence of MRSA. However, to date, no human *S. aureus* vaccine has been successfully tested in clinical trials and commercialized. The critical role of adjuvants in *S. aureus* vaccines has been widely recognized. Due to their excellent biocompatibility and controlled drug release properties, PLGA nanoparticles have been regarded as a promising adjuvant in the development of *S. aureus* vaccines. Cui et al. developed an $IsdB_{137-361}$-encapsulated PLGA-PEI nanoparticle vaccine, which could significantly promote the secretion of anti-infection cytokines and effectively reduce the bacterial loads in the main organs of immunized mice. Compared with the negative control group, the survival rate of the experimental group was significantly improved in the lethal challenge experiment [38]. Based on the foundation of our previous study, a PLGA nanoadjuvant with a specific size, shape, and rigidity (25% NPs) was developed in this study. Compared with the clinically used aluminum adjuvant, 25% NPs significantly elevated neutralizing antibody levels and improved the survival rate of mice in lethal challenge experiments. Vaccines containing 25% NPs as adjuvants have been demonstrated to induce the secretion of high levels of anti-infection cytokines, an effect that cannot be achieved by the alum adjuvant. Meanwhile, 25% NPs have tunable physicochemical properties and freeze-dried characteristics, which expands their potential applications.

The compatibility between antigen and adjuvant may be another vital factor affecting the efficacy of *S. aureus* vaccines, which is less discussed. Mahdavi et al. prepared vaccines using PLGA nanoparticles as adjuvants and PBP-2A protein with MRSA autolysin as antigens. The antigen exerted an effect on the survival rate of mice in the lethal challenge: all of the mice immunized with r-PBP2a-PLGA survived, whereas the survival rate of the mice immunized with r-autolysin-PLGA was 90% [39]. This result suggests that PLGA nanoparticles may have a potential antigen preference.

Based on the principle of the biomimetic design of adjuvants and the pathogenic mechanism of *S. aureus*, the secretion characteristics of *S. aureus* virulence factors may be a pivotal factor influencing adjuvant preference. The majority of exotoxins are soluble proteins secreted extracellularly by *S. aureus*, which can diffuse into the host environment and exert their toxic effects over long distances. Exotoxins are mainly involved in the process of bacterial virulence regulations, whereas cell-wall-anchored proteins are linked to the bacterial cell wall through covalent or non-covalent binding and directly interact with the host cells or immune components. They are involved in the processes of bacterial adhesion and colonization, immune escape, and biofilm formation. The mechanisms of the exotoxin Hla and cell-wall-anchored protein Spa in *S. aureus* infection have been intensively studied and have been widely used in the development of *S. aureus* vaccines. However, there are limitations to these studies. Burns et al. immunized mice with HlaH35L in combination with alum adjuvants and CpG adjuvants. The vaccine significantly reduced the bacterial loads in the kidneys of the mice. However, only one of the two experiments yielded statistically significant results, and the reproducibility of the results needs to be further examined [30]. Schneewind et al. immunized mice with a mixture of HlaH35L and Freund's adjuvant. Lethal challenge experiments were conducted using *S. aureus* strains NEWMAN, LAC, and MW2 [40]. Survival rates were significantly increased in mice. However, this study did not examine the cellular response elicited by vaccines. Schneewind examined the protective ability of the SpA $_{KKAA}$ mutant mixed with Freund's adjuvant [41]. The SpA $_{KKAA}$-immunized mice produced a higher level of specific antibodies and displayed higher survival rates in lethal challenge experiments. Additionally, the bacterial loads and abscess lesions in the kidney tissues of SpA $_{KKAA}$-inoculated mice were substantially reduced. But again, this study did not investigate the effect of the SpA $_{KKAA}$ mutant on cellular immunity.

The present study examined the preference of 25% NPs on rHlaH35L and rSpam from multiple dimensions. As demonstrated in the Results section, 25% NP-rHlaH35L induced higher levels of antibody titers and antimicrobial cytokine secretion levels compared to 25% NP-rSpam and significantly increased the survival rate of mice in high-dose lethal challenge experiments. We speculated that the number of neutralizing antibodies might be one of the key factors leading to the development of antigenic preferences in 25% NPs. Although the lytic capacity of serum antibodies in mice in the 25% NP-rHlaH35L and 25% NP-rSpam groups showed similar trends in the in vitro experiments, the antibody titers of mice in the 25% NP-rHlaH35L group were higher than those of the 25% NP-rSpam group. This indicates that more antibodies with neutralizing activity were produced in the mice in the 25% NP-rHlaH35L group. In addition, mice in the 25% NP-rHlaH35L group exhibited elevated levels of IL-4, IL-17A, and IFN-γ cytokine secretion compared with those in the 25% NP-rSpam group. Elevated levels of IL-4 suggest activation of Th2-type immune responses, while a significant increase in IL-17A reflects activation of Th17 cells, and elevated levels of IFN-γ indicate enhanced Th1-type immune response. This synergistic effect of multidimensional immune response may enhance the body's ability to fight infections by mechanisms such as enhancing antigen presentation, promoting effector T cell differentiation, and strengthening immune memory. The specific mechanisms need to be further investigated.

The present study also has the following limitations. First, despite the use of an intravenous MRSA model to simulate systemic infections, the pathologic process still differs significantly from that of clinical patients. *S. aureus* often achieves persistence in the human body through biofilm-mediated chronic infections (e.g., artificial joint infections) or repeated colonization in immunosuppressed hosts, and it is challenging to reproduce this complex microenvironment of multistage, multiorgan interactions in the acute attack model. Second, there is room for optimization of antigen screening strategies. The current study only focused on the secretion characteristics of bacterial virulence factors, and only two representative antigens were selected. In the future, new antigens need to be introduced and combined with deep-learning-driven epitope prediction to select antigen peptides matching the physicochemical properties of the PLGA adjuvant in order to break through the current bottleneck of antigen selection. Third, the surface functionalization of PLGA adjuvants has not been fully explored. Existing nanoparticles rely on passive targeting strategies, but the introduction of active targeting elements (e.g., CD11c single-chain antibody fragment modification to enhance dendritic cell uptake) or immune signaling synergistic modules (e.g., the covalent coupling of CpG with a TLR9 agonist) can build an integrated delivery/activation smart adjuvant. Finally, the study did not address the potential variability in immune responses between different populations, such as immunocompromised individuals or those with pre-existing conditions. Future studies should include diverse experimental groups to evaluate the broader applicability of the findings.

5. Conclusions

In conclusion, this work initially explores the selective preference of PLGA nanoparticles with a 25% PEG doping ratio to *S. aureus* antigens with different secretion characteristics.

Compared with rSpam, a cell-wall-anchor protein, a nanovaccine composed of 25% NPs and rHlaH35L (an HLA-modified recombinant antigen of a *Staphylococcus aureus* exotoxin) induced higher levels of neutralizing antibody titer and anti-infective cytokine secretion. In addition, the 25% NP-rHlaH35L vaccine provided more efficient immune protection for mice in the lethal challenge experiment with large doses of bacteria, suggesting that in the development of the recombinant *S. aureus* vaccine, in addition to reasonable screening of vaccine components, the selective preference of adjuvants for antigens, which is antigen-dependent adjuvanticity, may be a pivotal factor affecting vaccine efficacy.

Supplementary Materials: The following supporting information can be downloaded at: https://www.mdpi.com/article/10.3390/vaccines13030317/s1, Figure S1: Two antigens were analyzed by SDS-PAGE.; Figure S2: Biocompatibility evaluation of nanovaccines; Figure S3: Representative images of ELISPOT wells are shown for cytokines produced by immunized mice splenocytes.; Figure S4: Survival rate comparisons on day 14 after challenge of *S. aureus* at a concentration of 2.56×10^8 CFU per mouse after immunization (n = 10); Figure S5: Survival rate comparisons on day 14 after challenge of *S. aureus* at a concentration of 5.12×10^8 CFU per mouse after immunization (n = 10); Figure S6: Survival rate comparisons on day 14 after challenge of *S. aureus* at a concentration of 1.024×10^9 CFU per mouse after immunization (n = 10); Table S1: Hydrodynamic diameter of nanoparticles and nano-vaccines; Table S2: Zeta potential of nanoparticles and nano-vaccines.

Author Contributions: Conceptualization, C.S. and L.F.; data curation, Z.M.; formal analysis, L.F.; funding acquisition, L.F.; investigation, M.C., J.X. and Q.Y.; methodology, M.C., Z.S. and C.S.; project administration, L.F.; resources, L.F.; supervision, C.S. and L.F.; validation, M.C., J.X., Y.Z. and M.L.; writing—original draft, M.C.; writing—review and editing, L.F. All authors have read and agreed to the published version of the manuscript.

Funding: This work was financially supported by the National Natural Science Foundation of China (82171962),(Funder: Li, Fan) and the "Clinical medicine + Pharmacy" research center research fund of the Air Force Medical University (LHJJ2023-YX05),(Funder: Li, Fan).

Institutional Review Board Statement: All animal experiments complied with the National Research Council's Guide for the Care and Use of Laboratory Animals and the animal experiments were approved by the Animal Care and Ethics Committee of the Fourth Military Medical University (Approval NO. KY20213144-1).

Informed Consent Statement: Not applicable.

Data Availability Statement: Data will be made available upon request.

Conflicts of Interest: The authors declare that they have no known competing financial interests or personal relationships that could have appeared to influence the work reported in this paper.

References

1. Turner, N.A.; Sharma-Kuinkel, B.K.; Maskarinec, S.A.; Eichenberger, E.M.; Shah, P.P.; Carugati, M.; Holland, T.L.; Fowler, V.G., Jr. Methicillin-resistant *Staphylococcus aureus*: An overview of basic and clinical research. *Nat. Rev. Microbiol.* **2019**, *17*, 203–218. [CrossRef] [PubMed]
2. Maddiboyina, B.; Roy, H.; Ramaiah, M.; Sarvesh, C.N.; Kosuru, S.H.; Nakkala, R.K.; Nayak, B.S. Methicillin-resistant *Staphylococcus aureus*: Novel treatment approach breakthroughs. *Bull. Natl. Res. Cent.* **2023**, *47*, 95. [CrossRef]
3. Nandhini, P.; Kumar, P.; Mickymaray, S.; Alothaim, A.S.; Somasundaram, J.; Rajan, M. Recent Developments in Methicillin-Resistant *Staphylococcus aureus* (MRSA) Treatment: A Review. *Antibiotics* **2022**, *11*, 606. [CrossRef]
4. Karauzum, H.; Datta, S.K. Adaptive Immunity Against *Staphylococcus aureus*. *Curr. Top. Microbiol. Immunol.* **2017**, *409*, 419–439. [CrossRef] [PubMed]
5. Bröker, B.M.; Mrochen, D.; Péton, V. The T Cell Response to *Staphylococcus aureus*. *Pathogens* **2016**, *5*, 31. [CrossRef]
6. Spellberg, B.; Ibrahim Ashraf, S.; Yeaman Michael, R.; Lin, L.; Fu, Y.; Avanesian, V.; Bayer Arnold, S.; Filler Scott, G.; Lipke, P.; Otoo, H.; et al. The Antifungal Vaccine Derived from the Recombinant N Terminus of Als3p Protects Mice against the Bacterium *Staphylococcus aureus*. *Infect. Immun.* **2008**, *76*, 4574–4580. [CrossRef]

7. Yang, E.; Cho, Y.G.; Kim, E.; Chang, E.; Bae, S.; Jung, J.; Kim, M.J.; Chong, Y.P.; Kim, S.-H.; Choi, S.-H.; et al. Clinical and microbiological characteristics of persistent *Staphylococcus aureus* bacteremia, risk factors for mortality, and the role of CD4+ T cells. *Sci. Rep.* **2024**, *14*, 15472. [CrossRef] [PubMed]
8. Jarneborn, A.; Mohammad, M.; Engdahl, C.; Hu, Z.; Na, M.; Ali, A.; Jin, T. Tofacitinib treatment aggravates *Staphylococcus aureus* septic arthritis, but attenuates sepsis and enterotoxin induced shock in mice. *Sci. Rep.* **2020**, *10*, 10891. [CrossRef]
9. Lin, L.; Ibrahim, A.S.; Xu, X.; Farber, J.M.; Avanesian, V.; Baquir, B.; Fu, Y.; French, S.W.; Edwards, J.E., Jr.; Spellberg, B. Th1-Th17 Cells Mediate Protective Adaptive Immunity against *Staphylococcus aureus* and *Candida albicans* Infection in Mice. *PLoS Pathog.* **2009**, *5*, e1000703. [CrossRef]
10. Youn, C.; Pontaza, C.; Wang, Y.; Dikeman, D.A.; Joyce, D.P.; Alphonse, M.P.; Wu, M.J.; Nolan, S.J.; Anany, M.A.; Ahmadi, M.; et al. Neutrophil-intrinsic TNF receptor signaling orchestrates host defense against *Staphylococcus aureus*. *Sci. Adv.* **2023**, *9*, eadf8748. [CrossRef]
11. Archer Nathan, K.; Adappa Nithin, D.; Palmer James, N.; Cohen Noam, A.; Harro Jan, M.; Lee Steven, K.; Miller Lloyd, S.; Shirtliff Mark, E. Interleukin-17A (IL-17A) and IL-17F Are Critical for Antimicrobial Peptide Production and Clearance of *Staphylococcus aureus* Nasal Colonization. *Infect. Immun.* **2016**, *84*, 3575–3583. [CrossRef]
12. Bagnoli, F.; Bertholet, S.; Grandi, G. Inferring Reasons for the Failure of *Staphylococcus aureus* Vaccines in Clinical Trials. *Front. Cell Infect. Microbiol.* **2012**, *2*, 16. [CrossRef] [PubMed]
13. Zhu, Z.; Hu, Z.; Li, S.; Fang, R.; Ono, H.K.; Hu, D.L. Molecular Characteristics and Pathogenicity of *Staphylococcus aureus* Exotoxins. *Int. J. Mol. Sci.* **2023**, *25*, 395. [CrossRef] [PubMed]
14. Foster, T.J.; Geoghegan, J.A.; Ganesh, V.K.; Höök, M. Adhesion, invasion and evasion: The many functions of the surface proteins of *Staphylococcus aureus*. *Nat. Rev. Microbiol.* **2014**, *12*, 49–62. [CrossRef] [PubMed]
15. Turley, J.L.; Lavelle, E.C. Resolving adjuvant mode of action to enhance vaccine efficacy. *Curr. Opin. Immunol.* **2022**, *77*, 102229. [CrossRef] [PubMed]
16. Zhao, T.; Cai, Y.; Jiang, Y.; He, X.; Wei, Y.; Yu, Y.; Tian, X. Vaccine adjuvants: Mechanisms and platforms. *Signal Transduct. Target. Ther.* **2023**, *8*, 283. [CrossRef]
17. Mirshekar, M.; Haghighat, S.; Mousavi, Z.; Abdolghaffari, A.H.; Yazdi, M.H. Monophosphoryl lipid A as a co-adjuvant in methicillin-resistant *Staphylococcus aureus* vaccine development: Improvement of immune responses in a mouse model of infection. *Immunol. Res.* **2024**, *72*, 490–502. [CrossRef]
18. Elhassan Taha, M.M.; Abdelwahab, S.I.; Moni, S.S.; Farasani, A.; Aljahdali, I.A.; Oraibi, B.; Alfaifi, H.A.; Alzahrani, A.H.; Ali Jerah, A. Nano-enhanced immunity: A bibliometric analysis of nanoparticles in vaccine adjuvant research. *Hum. Vaccin. Immunother.* **2024**, *20*, 2427464. [CrossRef]
19. Hassett, K.J.; Higgins, J.; Woods, A.; Levy, B.; Xia, Y.; Hsiao, C.J.; Acosta, E.; Almarsson, Ö.; Moore, M.J.; Brito, L.A. Impact of lipid nanoparticle size on mRNA vaccine immunogenicity. *J. Control Release* **2021**, *335*, 237–246. [CrossRef]
20. Kumar, S.; Anselmo, A.C.; Banerjee, A.; Zakrewsky, M.; Mitragotri, S. Shape and size-dependent immune response to antigen-carrying nanoparticles. *J. Control Release* **2015**, *220*, 141–148. [CrossRef]
21. Song, C.; Hu, J.; Liu, Y.; Tian, Y.; Zhu, Y.; Xi, J.; Cui, M.; Wang, X.; Zhang, B.Z.; Fan, L.; et al. Vaccination-Route-Dependent Adjuvanticity of Antigen-Carrying Nanoparticles for Enhanced Vaccine Efficacy. *Vaccines* **2024**, *12*, 125. [CrossRef]
22. Miao, G.; He, Y.; Shang, Z.; He, P.; Xu, M.; Zhao, X.; Wang, X. Combined Effects of Core Rigidity and Surface Charge of Polymeric Nanomicelles on the Cellular Uptake Efficiency. *Macromolecules* **2023**, *56*, 7663–7674. [CrossRef]
23. Yu, M.; Xu, L.; Tian, F.; Su, Q.; Zheng, N.; Yang, Y.; Wang, J.; Wang, A.; Zhu, C.; Guo, S.; et al. Rapid transport of deformation-tuned nanoparticles across biological hydrogels and cellular barriers. *Nat. Commun.* **2018**, *9*, 2607. [CrossRef] [PubMed]
24. Sarkar, I.; Garg, R.; van Drunen Littel-van den Hurk, S. Selection of adjuvants for vaccines targeting specific pathogens. *Expert. Rev. Vaccines* **2019**, *18*, 505–521. [CrossRef] [PubMed]
25. Zeng, L. Mucosal adjuvants: Opportunities and challenges. *Hum. Vaccin. Immunother.* **2016**, *12*, 2456–2458. [CrossRef]
26. Ferreira, M.U.; da Silva Nunes, M.; Wunderlich, G. Antigenic diversity and immune evasion by malaria parasites. *Clin. Diagn. Lab. Immunol.* **2004**, *11*, 987–995. [CrossRef]
27. Björkman, A.; Benn, C.S.; Aaby, P.; Schapira, A. RTS,S/AS01 malaria vaccine-proven safe and effective? *Lancet. Infect. Dis.* **2023**, *23*, e318–e322. [CrossRef]
28. Didierlaurent, A.M.; Laupèze, B.; Di Pasquale, A.; Hergli, N.; Collignon, C.; Garçon, N. Adjuvant system AS01: Helping to overcome the challenges of modern vaccines. *Expert. Rev. Vaccines* **2017**, *16*, 55–63. [CrossRef]
29. Riley, E.M.; Stewart, V.A. Immune mechanisms in malaria: New insights in vaccine development. *Nat. Med.* **2013**, *19*, 168–178. [CrossRef]
30. Brady, R.A.; Mocca, C.P.; Prabhakara, R.; Plaut, R.D.; Shirtliff, M.E.; Merkel, T.J.; Burns, D.L. Evaluation of genetically inactivated alpha toxin for protection in multiple mouse models of *Staphylococcus aureus* infection. *PLoS ONE* **2013**, *8*, e63040. [CrossRef]

31. Irene, C.; Fantappiè, L.; Caproni, E.; Zerbini, F.; Anesi, A.; Tomasi, M.; Zanella, I.; Stupia, S.; Prete, S.; Valensin, S.; et al. Bacterial outer membrane vesicles engineered with lipidated antigens as a platform for *Staphylococcus aureus* vaccine. *Proc. Natl. Acad. Sci. USA* **2019**, *116*, 21780–21788. [CrossRef] [PubMed]
32. Teymournejad, O.; Li, Z.; Beesetty, P.; Yang, C.; Montgomery, C.P. Toxin expression during *Staphylococcus aureus* infection imprints host immunity to inhibit vaccine efficacy. *NPJ Vaccines* **2023**, *8*, 3. [CrossRef] [PubMed]
33. Zeng, H.; Yang, F.; Feng, Q.; Zhang, J.; Gu, J.; Jing, H.; Cai, C.; Xu, L.; Yang, X.; Xia, X.; et al. Rapid and Broad Immune Efficacy of a Recombinant Five-Antigen Vaccine against *Staphylococcus aureus* Infection in Animal Models. *Vaccines* **2020**, *8*, 134. [CrossRef]
34. Thammavongsa, V.; Rauch, S.; Kim, H.K.; Missiakas, D.M.; Schneewind, O. Protein A-neutralizing monoclonal antibody protects neonatal mice against *Staphylococcus aureus*. *Vaccine* **2015**, *33*, 523–526. [CrossRef] [PubMed]
35. Briand, L.; Marcion, G.; Kriznik, A.; Heydel, J.M.; Artur, Y.; Garrido, C.; Seigneuric, R.; Neiers, F. A self-inducible heterologous protein expression system in Escherichia coli. *Sci. Rep.* **2016**, *6*, 33037. [CrossRef]
36. Fan, L.; Zhang, S.; Zhang, C.; Yin, C.; Chu, Z.; Song, C.; Lin, G.; Li, Q. Multidrug Resistance in Cancer Circumvented Using a Cytosolic Drug Reservoir. *Adv. Sci.* **2018**, *5*, 1700289. [CrossRef]
37. Zhong, L.; Zhang, W.; Liu, H.; Zhang, X.; Yang, Z.; Wen, Z.; Chen, L.; Chen, H.; Luo, Y.; Chen, Y.; et al. A cocktail nanovaccine targeting key entry glycoproteins elicits high neutralizing antibody levels against EBV infection. *Nat. Commun.* **2024**, *15*, 5310. [CrossRef] [PubMed]
38. Wang, B.; Dong, Y.; Cen, Y.; Chen, S.; Wen, X.; Liu, K.; Wu, S.; Yu, L.; Yu, Y.; Zhu, Z.; et al. PEI-PLGA nanoparticles significantly enhanced the immunogenicity of IsdB137-361 proteins from *Staphylococcus aureus*. *Immun. Inflamm. Dis.* **2023**, *11*, e928. [CrossRef]
39. Haghighat, S.; Siadat, S.D.; Akhavan Sepahi, A.; Mahdavi, M. Recombinant PBP2a/autolysin conjugate as PLGA-based nanovaccine induced humoral responses with opsonophagocytosis activity, and protection versus methicillin-resistant *Staphylococcus aureus* infection. *Iran. J. Basic. Med. Sci.* **2022**, *25*, 442–450. [CrossRef]
40. Bubeck Wardenburg, J.; Schneewind, O. Vaccine protection against *Staphylococcus aureus* pneumonia. *J. Exp. Med.* **2008**, *205*, 287–294. [CrossRef]
41. Kim, H.K.; Cheng, A.G.; Kim, H.Y.; Missiakas, D.M.; Schneewind, O. Nontoxigenic protein A vaccine for methicillin-resistant *Staphylococcus aureus* infections in mice. *J. Exp. Med.* **2010**, *207*, 1863–1870. [CrossRef] [PubMed]

vaccines

MDPI

Review

Revolutionizing Nanovaccines: A New Era of Immunization

Mohammed Saleh [1], Ahmed El-Moghazy [2], Adel H. Elgohary [3], WesamEldin I. A. Saber [4] and Yosra A. Helmy [1,*]

1 Department of Veterinary Science, Martin-Gatton College of Agriculture, Food, and Environment, University of Kentucky, Lexington, KY 40546, USA
2 Department of Microbiology and Plant Pathology, University of California, Riverside, CA 92521, USA
3 Department of Hygiene and Zoonoses, Faculty of Veterinary Medicine, Mansoura University, Mansoura 35516, Egypt
4 Microbial Activity Unit, Department of Microbiology, Soils, Water and Environment Research Institute, Agricultural Research Center, Giza 12619, Egypt
* Correspondence: yosra.helmy@uky.edu

Abstract: Infectious diseases continue to pose a significant global health threat. To combat these challenges, innovative vaccine technologies are urgently needed. Nanoparticles (NPs) have unique properties and have emerged as a promising platform for developing next-generation vaccines. Nanoparticles are revolutionizing the field of vaccine development, offering a new era of immunization. They allow the creation of more effective, stable, and easily deliverable vaccines. Various types of NPs, including lipid, polymeric, metal, and virus-like particles, can be employed to encapsulate and deliver vaccine components, such as mRNA or protein antigens. These NPs protect antigens from degradation, target them to specific immune cells, and enhance antigen presentation, leading to robust and durable immune responses. Additionally, NPs can simultaneously deliver multiple vaccine components, including antigens, and adjuvants, in a single formulation, simplifying vaccine production and administration. Nanovaccines offer a promising approach to combat food- and water-borne bacterial diseases, surpassing traditional formulations. Further research is needed to address the global burden of these infections. This review highlights the potential of NPs to revolutionize vaccine platforms. We explore their mechanisms of action, current applications, and emerging trends. The review discusses the limitations of nanovaccines, innovative solutions and the potential role of artificial intelligence in developing more effective and accessible nanovaccines to combat infectious diseases.

Keywords: nanoparticles (NPs); nanoparticle-based vaccines; preclinical and clinical trials; regulatory challenges; future innovations; food safety

1. Introduction

The world is currently grappling with a series of critical public health challenges, including the emergence of multidrug-resistant bacteria, rapidly mutating viruses, increasing populations of superbugs, anthelmintic-resistant parasites, zoonotic pathogens, and secondary tumors. These challenges, compounded by the recent COVID-19 pandemic, pose significant threats to human health and livestock [1]. These epidemic and pandemic outbreaks have far-reaching consequences, contributing to global mortality rates and hindering socioeconomic development. Additionally, they necessitate robust biodefense strategies to protect populations from future health crises [2–5]. The overuse and misuse of antibiotics and other drugs have significantly contributed to the rise of antimicrobial resistance, a major global health crisis [6,7]. Additionally, climate change and environmental pollution

https://doi.org/10.3390/vaccines13020126

have accelerated the emergence, transmission, and spread of infectious diseases, further exacerbating the challenges faced by public health systems worldwide [8,9]. Global cooperation has highlighted the need to tackle these rising resistance and infection levels. The One Health Trust has declared that vaccination is considered an integral strategy to combat antimicrobial resistance, slow the spread of superbugs, and lower the overall burden of infection [10–12].

The immune system is highly sophisticated and is considered the first line of defense against foreign infectious and non-infectious signals. The innate immune cells are composed of a variety of circulating cells, such as natural killer cells, granulocytes (neutrophils, basophils, eosinophils, and mast cells), and antigen-presenting cells (macrophages and dendritic cells). The immune system is responsible for the processes driving the uptake and ingestion of phagocytosed particles as well as the process associated with breaking them down [13]. The five pattern-recognition receptors that can recognize pathogens play important roles in the initiation and activation of the immune response [13,14]. The intended output of vaccination is to stimulate the innate and adaptive immune systems to induce B- or T-cell responses and provide prolonged protection [15]. When a vaccine introduces an antigen into the body, B-lymphocytes produce antibodies that fight infection, while T-cells recognize and kill cells infected with a virus or other foreign cells, thus controlling the infection transmission and spread [16].

Recently, nanotechnology has provided significant advancements in the medical sector, including the production of nanovaccines, nanotherapeutics, diagnostics, and the improvement of drug screening and gene delivery approaches [17–19]. Nanovaccine platforms represent a transformative avenue for vaccine delivery, offering potential solutions to challenges in traditional vaccine development [20]. These platforms use nanoparticles (NPs), such as lipids, polymers, proteins, etc. to improve the delivery of antigens to target cells, resulting in broader and more specific protection against cancers, infectious diseases, inflammatory diseases, atopic diseases, autoimmune diseases, and others [20]. Nanoparticles can be designed to increase the stability and immunogenicity of antigens. They not only serve as antigens carriers but can also function as antigens themselves [21]. NPs can efficiently target vaccine molecules to the desired cell and its receptors, thereby minimizing side effects [22]. Additionally, they have been found to facilitate antigen distribution to APCs and protect antigens and adjuvants from proteolytic and enzymatic degradation [23]. Nanovaccines can evoke both humoral and cell-mediated immune responses [22]. Nanovaccines can persist for a longer time without degradation or alteration, thereby providing enough opportunity for antigen presentation and dendritic cell (DC)-mediated antigen uptake, resulting in mature DC–T-cell interaction, which is necessary to influence the release of pro- and anti-inflammatory cytokines that promote cell-mediated immunity [24]. The NP-mediated vaccine can enhance B-cell antibody production by activating lymphocytes and monocytes. Furthermore, they can enable a targeted release of cargo into draining lymph nodes after passing through the lymphatic drainage system, generating a potent, long-lasting, and antigen-specific immune response [25,26]. These characteristics suggest that NPs might be crucial to vaccines as immune cell stimulators.

In this review, we highlight the transformative impact of NPs on vaccine platforms, focusing on their ability to enhance antigen delivery, improve immune system activation, and address the limitations of traditional vaccines. Specifically, we explore the types of NPs utilized, their mechanisms of action, current applications in vaccine development, and the potential they hold for creating more effective, stable, and accessible immunization strategies in the future. Additionally, we discuss the challenges and opportunities that lie ahead in the development and deployment of nanoparticle-based vaccines.

2. Conventional vs. Nanoparticle-Based Vaccines

2.1. Conventional Vaccines

Conventional vaccines are considered one of the greatest advances in medical history against various diseases. They were discovered in the 1700s when Edward Jenner, a British physician, demonstrated that infection with mild cowpox virus can confer immunity against smallpox virus. This discovery led to the development of the smallpox vaccine, which was the first vaccine to be developed against a contagious disease. Consequently, scientists and vaccinologists began occasional and inconclusive experiments for immunization against specific pathogens, as illustrated in Figure 1 [27–31]. The traditional vaccine development pipeline consists of exploratory research, pre-clinical studies, and clinical trials (Phases I, II, and III), followed by regulatory evaluation and approval. However, there are many types of conventional vaccines, which are classified based on their mechanisms of action.

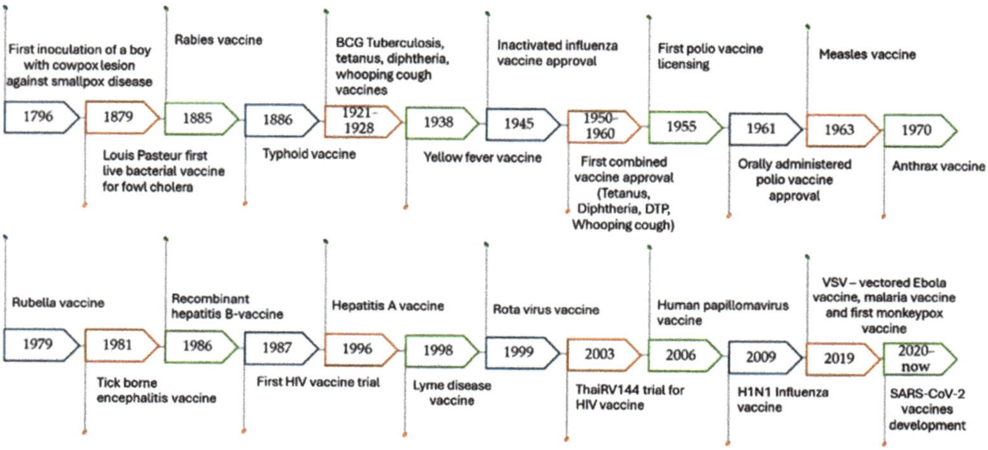

Figure 1. A brief history of conventional vaccine development.

2.1.1. Inactivated Vaccines

This type contains killed pathogens such as influenza, coryza, typhoid, cholera, mycoplasma, hepatitis A, rabies, *Brucella abortus*, *Mannheimia haemolytica*, and *Erysipelothrix rhusipathiae* vaccines [19,32,33]. These vaccines do not offer as much protection as live vaccines do, so booster shots are often required to maintain immunity.

2.1.2. Live Vaccines

Live vaccines contain attenuated forms of pathogens such as measles, rotavirus, avian infectious bronchitis, Newcastle disease, bovine viral diarrhea, canine distemper, and fowl pox [34]. The immune system responds to this type of vaccine by producing strong and long-lasting protection. However, they may not be suitable for hosts with weakened immune systems, due to the possibility of reversion to virulence, interference with maternal immunity, and tissue damage, which may result in pathological disorders or secondary bacterial infection.

2.1.3. Subunit Vaccines

Subunit, recombinant, polysaccharide, and conjugate vaccines utilize specific microbial components, such as proteins, sugars, or capsules, to induce a targeted immune response. Examples include vaccines for hepatitis B, human papillomavirus, and shingles. While these vaccines offer several advantages, including suitability for immunocompromised individuals, they may require booster doses to maintain long-lasting immunity. Additionally, the absence of pathogen-associated molecular patterns in subunit vaccines can limit their ability to stimulate a robust immune response [35,36].

2.1.4. Toxoid Vaccines

Toxoid vaccines, such as tetanus and diphtheria vaccines [37], contain attenuated bacterial toxins, and have a more effective anti-toxin immune response [38,39]. These conventional vaccines have different administration routes, including intramuscular, subcutaneous, ocular, oral, and intranasal routes [13].

Traditional vaccines offer several advantages, including safety, proven efficacy in disease prevention, and the establishment of strong and long-lasting immunity [40,41]. However, their main limitations include a lack of genetic stability, the potential of mutations in live vaccines, and a limited ability to induce a robust immune response for long-lasting immunity for inactivated vaccines [42–44]. This has led to the emergence of nanovaccines as a next-generation approach.

2.2. Nanoparticle-Based Vaccines

NPs can be defined as ultrafine particles with definite dimensions that exhibit unique physicochemical characteristics, supporting their application in various biomedical and food quality fields [45–47]. They have emerged as promising vectors in vaccine technology development and are considered a promising alternative to conventional vaccines (Table 1) [48,49]. Studies have revealed that the physicochemical properties of NPs can influence the immune response in terms of factors such as particle size, hydrophilicity/hydrophobicity ratio, the number of antigens associated with each particle, and the ability of an antigen to raise a B- or T-cell response. Smaller-sized particles can pass biological barriers and epithelia more easily when compared with particles ranging from 20 to 50 nm in diameter [46]. The size of nanoparticle-based vaccines is similar, to some extent, to the size of intracellular components, making it easy for them to enter cells and induce robust and long-lasting immune responses against intracellular pathogens [50]. Spherical NPs induce the most effective immune responses when compared to other nanoparticle shapes [51]. Hydrophobic NPs are more efficiently taken up by specific immune cells and potentially enhance the immune response when compared with hydrophilic NPs that produce a weakened immune response [52,53]. Negatively charged NPs interact more efficiently with macrophages, while positively charged NPs interact more efficiently with dendritic cells [54]. NPs can be composed of polymers, lipids, proteins, emulsions, nanobeads, inorganic nanomaterials, and virus-like particles and have been approved for both animal and human use to prevent infectious and non-infectious diseases [55].

The major types of NPs used for vaccinology against infectious and non-infectious diseases are illustrated in Table 2. They exhibit a wide range of properties that make them valuable humoral and cellular immune stimulators [13]. They can enhance immune responses by effectively presenting or transporting antigens and allowing precise targeting of immune cells [56,57]. They protect antigens from degradation before reaching their target, thus improving vaccine efficacy and shelf life [58]. Some NPs can act as adjuvants themselves, thus stimulating the immune response without the need for additional substances to the vaccine formula [59]. NPs can be designed to control antigen release, extending immune response duration [60]. Additionally, NPs can be tailored for specific pathogens, making them versatile tools for preventing and controlling infectious diseases [49,61–64]. The nanoparticle-based vaccines have different administration routes, including ocular, oral, nasal, sublingual, topical, intradermal, intramuscular, and subcutaneous routes. The subcutaneous route is considered the most advantageous route in stimulating the immune response, as vaccines are injected directly into a site that is rich in immune cells, resulting in the production of antibodies at high concentrations [65].

Table 1. Comparison between conventional and nanoparticle-based vaccines.

Criterion	Conventional Vaccines	Nanoparticle-Based Vaccines
Types	Live attenuated, inactivated, subunit and toxoid vaccines	Protein NPs, lipid NPs, polymer-based NPs, inorganic NPs, and virus-like particles
Advantages	Proven track record, strong and long-lasting immunity	Can carry multiple antigens, enhanced stability, targeted delivery, potentially stronger response
Mechanism	Uses attenuated/killed pathogens or pathogen components to stimulate immune response	Delivers antigens in a targeted manner for better recognition and immune response
Immune response	Immunogenicity can vary and often requires adjuvants to elicit a robust response	Designed for a more specific, potent immune response with built-in adjuvants
Antigen delivery	Direct administration, often less targeted	Targeted delivery of specific cells or tissues
Antigen stability	Require strict temperature control and maintenance	Nanocarriers extend the shelf life of antigens and protect them from degradation
Production Complexity	Long-established manufacturing processes	Advanced bioengineering techniques, more complex production
Stability	Generally unstable, requires cold chain storage	Often more stable, some formulations may require less stringent storage conditions
Customization	Limited adaptability to rapidly mutated pathogens	Highly customizable for emerging pathogens
Dosage	Multiple administrations may be required for optimal immunogenicity, depending on the vaccine and target population	Targeted delivery has minimized the administration of multiple doses
Safety	Almost safe with few risks of reversion to virulent form regarding live attenuated vaccines	Potential toxicity of some NPs
Examples	Polio, hepatitis B, IB, NDV, tetanus, fowl pox	Novavax (protein NPs) and COVID-19 mRNA vaccines (using lipid NPs)

The increasing diversity of nanoparticle-based vaccines has prompted ongoing research toward the optimization of delivery methods and the maximization of immune responses against specific antigens. As depicted in Figure 2, nanoparticle-based vaccines have gained significant prominence in the pharmaceutical and biological industries over the past century. The continued development and implementation of nanoparticle-based vaccines hold immense potential to address a wide range of infectious diseases (viral, bacterial, fungal) and non-infectious diseases (autoimmune diseases, cancer) that pose substantial global health challenges [48,49].

Table 2. List of nanoparticle-based vaccines used for treating infectious and non-infectious diseases.

Nanoparticle Type	Targeted Infectious Pathogen/Disease	Pathogen Classification	Reference
Lipid-based NPs	SARS-CoV-2, herpes simplex virus, leishmania and *Schistosoma*	Virus/parasite	[66–69]
Gold NPs (AuNPs)	SARS-CoV-2, Hepatitis C Virus, Foot and Mouth Disease (FMD), *Clostridia tetani, Burkholderia mallei, Salmonella* Typhi and *Vibrio vulnificus*	Virus/bacteria	[70–76]
Mesoporous Silica NPs	Tumor cells	Tumor cells	[77,78]
Virus-like particles	Hepatitis B virus, malaria, *Theileria parva, Respiratory Syncytial virus* (RSV), influenza, circoviruses and SARS-CoV-2	virus/parasite	[79–84]
Protein-based NPs	SARS-CoV-2	Virus	[85,86]
Self-adjuvanting polyguanidine nanovaccines	Tumor cells	Tumor cells	[87]
Liposomes	Influenza virus, hepatitis A, Human immunosuppressive Virus (HIV), *Pseudomonas aeruginosa*, and methicillin-resistant *Staphylococcus aureus*	Virus/bacteria	[76,88–92]
Silicon scalable lipid NPs	SARS-CoV-2	Virus	[93]
Dendrimer-RNA NPs	H1N1 influenza, Ebola, and *Toxoplasma gondii*	Virus /parasite	[94]
Multiplexed dendrimer with lipid-based NPs	Tumor cells	Tumor cells	[95]
Fluoropolymer-based nanovaccines	Tumor cells	Tumor cells	[96]
Extracellular vesicle-coated multi-antigenic nanovaccines	Multidrug resistant *S. aureus*	Bacteria	[97]
OMV-based nanovaccines	*Escherichia coli* serotype O_2	Bacteria	[98]
Carbon nanotubes and poly-anhydrous NPs	*Anaplasma marginale* and *Mycobacterium paratuberculosis*	Protozoa/bacteria	[99]
Chitosan	*Salmonella* Enteritidis, *Streptococcus*, avian pathogenic *E. coli*, NDV, tuberculosis, *Flavobacterium columnare, S. aureus, E. coli K1* and *Toxoplasma*	Bacteria/virus/parasite	[100–110]
Polyanhydride NPs	*Bacillus anthracis* and *Yersinia pestis*	Bacteria	[111]
Nanoplasmids	*Trypanosoma cruzi*	Parasite	[112]
Antigen-absorbing silica vesicles	Tick-borne diseases	Tick-borne diseases	[113]
Zinc oxide NPs	*E. coli, Salmonella* spp. and H1N1	Bacteria/Virus	[114,115]
Selenium NPs	*S. aureus* and *E. coli*	Bacteria	[116]
Poly (lactic acid-coglycolic acid) polymeric NPs	*C. trachomatis, C. pneumoniae, Helicobacter pylori, Listeria monocytogenes, Salmonella* Typhimurium and *Brucella melitensis*	Bacteria	[117–121]

Table 2. *Cont.*

Nanoparticle Type	Targeted Infectious Pathogen/Disease	Pathogen Classification	Reference
Chitosan and glutamic acid polymeric NPs	*H. pylori*	Bacteria	[122]
Silver NPs (AgNPs)	H3N2, feline calicivirus, infectious bursal disease virus (IBDV), respiratory syncytial virus (RSV), Rift Valley Fever virus (RVF), HIV, adenovirus, poliovirus, norovirus, *Acinetobacter baumannii*, feline coronavirus, *E. coli*, *S. aureus* and *P. aeruginosa*	Virus/bacteria	[123–136]
Dendrimers	*Schistosoma*	Parasite	[137]
Self-assembling polypeptide NPs	Toxoplasmosis	Parasite	[138]
Porous silicon NPs	*P. aeruginosa*	Bacteria	[139]
Calcium-alginate NPs	Methicillin-resistant *S. aureus*	Bacteria	[140]
Ferritin NPs	*Neisseria gonorrhoeae*	Bacteria	[141]
Nanosuspensions	*Cryptosporidium parvum*	Parasite	[142]
Selenium nanoparticle functionalized with oseltamivir	Enterovirus	Virus	[143]
LNP- siVP35-3	EBOV	Virus	[144]
Multivalent peptide–polymer NPs	Influenza	Virus	[145]
Nano-vesicles	*A. marginale* and *shiga toxin by E. coli*	Parasite/bacteria	[146,147]
TMC NPs	Influenza	Virus	[148]

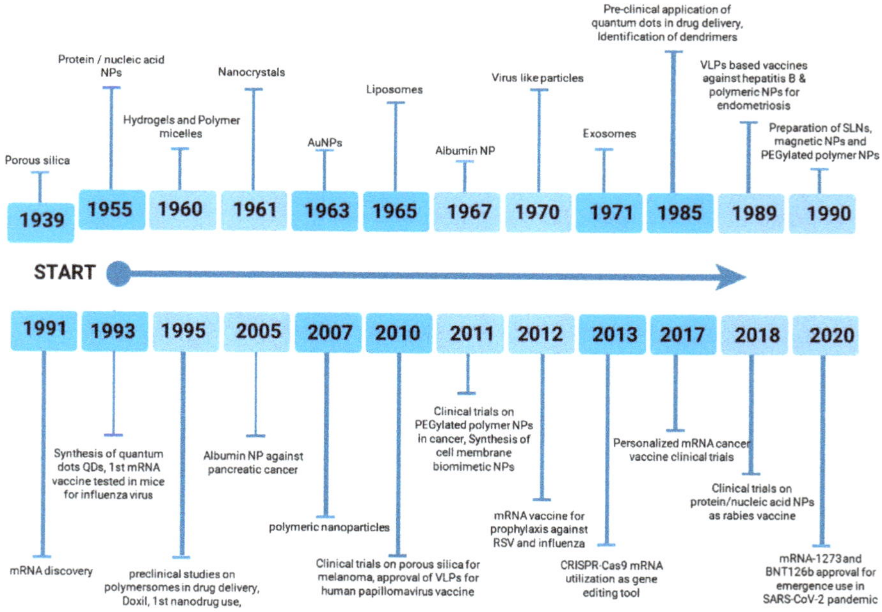

Figure 2. Timeline of nanoparticle-based vaccine development.

Despite significant advancements in nanovaccine development for combating infectious diseases, inflammatory conditions, allergic diseases, and cancers, several challenges persist. Firstly, ensuring consistent size, shape, and composition of NPs remains a complex task, as these properties are crucial for achieving optimal pharmacological, biological, and physicochemical characteristics [149,150]. Secondly, unintended immune reactions or toxicity can arise from nanoparticle accumulation in tissues and organs [20]. Thirdly, stability and shelf-life, under various environmental conditions, pose challenges, particularly for lipid NPs that are sensitive to temperature fluctuations [151,152]. Fourthly, regulatory hurdles, production costs, and safe waste disposal are additional considerations [153]. Lastly, mRNA-based nanovaccines face specific challenges, including the selection of appropriate antigens, the synthesis and purification of mRNA, and ensuring efficient encapsulation within lipid NPs [154].

3. Types of NPs Used in Vaccines

Nanoparticles exhibit specific characteristics that render them more effective for enhancing the immune system and delivering antigens in a controlled and targeted manner [57,155,156]. Several types of NPs are being explored for vaccine development, including lipid-based NPs, polymeric NPs, self-assembled protein or peptide NPs, biomimetic NPs, and inorganic NPs (Figure 3).

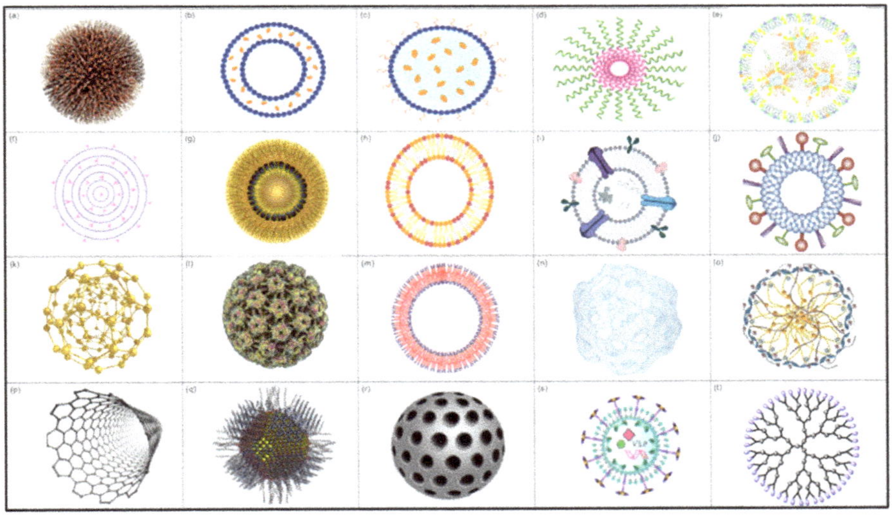

Figure 3. Types of NPs. (**a**) Polymeric nanoparticle, (**b**) liposome, (**c**) solid lipid NP, (**d**) polymeric micelle, (**e**) lipid NP, (**f**) multilamellar lipid vesicles, (**g**) lipid nanoemulsion, (**h**) lipoprotein, (**i**) exosomes, (**j**) virosomes, (**k**) gold NP, (**l**) protein-based NP, (**m**) polymersomes, (**n**) polymeric nanosphere, (**o**) chitosan NP, (**p**) carbon nanotubes, (**q**) quantum dots, (**r**) mesorphous silica NP, (**s**) virus-like particle (VLP), and (**t**) dendrimer.

3.1. Lipid-Based NPs

Lipid-based nanoparticles (LNPs) are also known as lipid-based nanocarriers. They are spherical NPs that use ionizable lipids as their main structural component. These NPs can be categorized based on their structure and composition. In particular, LNPs offer advantages such as biodegradability, biocompatibility, and safety, which are ideal characteristics for a vaccine delivery system [157]. LPNs can facilitate the delivery of mRNA and protect it from rapid degradation by RNases [68]. Additionally, LNPs can interact with the immune system, enhancing cellular and humoral immunity. They have been pivotal in the development of vaccines and other nanomedicines [68,158]. Furthermore, there are many examples of lipid-based NP vaccines that have been widely used in preclinical and clinical studies (Table 3).

Table 3. Examples of lipid-based NP vaccines widely used in preclinical and clinical studies.

Trade Name	Composition	Indication	Ref.
Vaxfectin®	Liposomal vaccine	Herpes simplex virus type 2 and influenza virus	[91]
Epaxal®	Liposomal vaccine	Hepatitis A infection	[92,159]
Doxil® and Abraxane®	Liposomal doxorubicin and albumin-bound paclitaxel	Approved by the FDA and are used in treating Various cancers	[160]
Shingrix®	Recombinant VZV glycoprotein E on liposomes carrier	Approved by FDA against herpes virus	[161]
Onivyde®	Liposomal irinotecan	Pancreatic cancer	[162]
W_ova1	Liposome-formulated mRNAs	Ovarian cancer	[163]
DPX-0907	Liposomes containing a polynucleotide adjuvant	Ovarian, breast, and prostate cancer	[164]
Lipovaxin-MM	Tumor antigen-containing multicomponent liposomes with the DC-targeting molecule DMS-5000	Metastatic melanoma	[165]
Vyxeos®	Liposomal cytarabine and daunorubicin	Acute myeloid leukemia	[166,167]
VaxiSome™	Liposomal adjuvant	Influenza	[168]
MPER-656	HIV-1 gp41 membrane-proximal external region (MPER) on liposome carrier	HIV	[169]
LNP CL-0059/CL-0137	mRNA LNP	RSV	[170]
BNT165b1	mRNA LNP	Malaria	[171]
MRT5413	LNP-formulated, modified mRNA	Influenza	[172]
ARCT-021, MRT5500, Pfizer-BioNTech and Moderna COVID-19, ChulaCov19, mRNA-1273, CVnCoV, LNP-nCoV-saRNA	mRNA LNP	COVID-19	[68,173–180]

3.1.1. Interbilayer-Crosslinked Multilamellar Vesicles (ICMVs)

ICMVs have an aqueous core surrounded by multilayers of lipids. Their robust lipid wall protects the particles from serum-mediated degradation, enabling stable delivery of vaccine components to lymphoid tissues and thus generating potent antibody responses. ICMVs can serve as a delivery system for both hydrophilic and hydrophobic antigens and adjuvants, showing promise as a vaccination strategy against cancers and intracellular pathogens [181]. Studies have demonstrated that ICMVs with both surface-conjugated and encapsulated malarial antigens offer superior vaccination protection compared to soluble antigens due to the prolonged persistence of antigens in the draining lymph nodes [182]. Despite these values, ICMVs have some disadvantages, such as the complexity of production, the risk of triggering an unintended immune response, the unpredictable release of encapsulated substances from ICMVs, and the fact that techniques used to produce ICMVs may not be easily scaled up for industrial production, limiting their widespread application [181].

3.1.2. Lipoproteins

Lipoproteins are complexes of proteins and lipids. They have been reported to transfer and target various lipids, hormones, proteins, vitamins, and endogenous microRNA to recipient cells [183]. They have a hydrophobic core and a hydrophilic surface, that can facilitate the delivery of a broad range of molecules and act as vaccine-delivery carriers [184]. Recombinant lipoproteins with built-in immune stimulators have been established for novel subunit vaccine development. This core platform technology has demonstrated safety in meningococcal group B subunit vaccine, dengue subunit vaccine, novel subunit vaccine against *Clostridium difficile* and HPV-based immunotherapeutic vaccines in animal model studies [185]. The main disadvantage of lipoprotein NPs is the stability issues as they can be unstable in the bloodstream, leading to premature release of their cargo [186]. Additionally, the manufacturing process can be complex and costly, requiring precise control over conditions to ensure consistency and efficacy [187]. They are potentially toxic and some lipoprotein NPs can trigger immune responses and cause adverse reactions [188,189].

3.1.3. Liposomes

Liposomes are composed of phospholipid bilayers and range in size from 50 to 500 nm in diameter. They are amphiphilic in nature, with hydrophobic surfaces and a hydrophilic core. Liposomes have gained significant attention in delivering immunostimulatory molecules, drugs, DNA, and RNA [190]. They are highly versatile, biocompatible, safe, and capable of creating different structures by modulating components of the lipid matrix. They act as carriers for vaccines and facilitate the targeted delivery of viral proteins, triggering antibody production. Liposome carriers influence antigen internalization by immunocompetent cells and immune response induction [191]. They can act as adjuvants employed in nanovaccines formulation, so prolong stability and trigger antigen-presenting cells to initiate an immune response. The impact of liposomes on antigen-presenting cells depends on many factors, such as size, charge, and lipid composition [192]. Immunostimulant components (ligands of pathogen-associated molecular pattern receptors) can be added to liposomal vaccine complexes to modulate the strength and type of immune response [192]. Furthermore, liposomes can induce immunologic reactions to antigens adsorbed on the surface or encapsulated internally [193]. Several studies have shown that liposome-based vaccines have been widely used in preclinical and clinical studies [194]. However, liposomes have some drawbacks as they can be prone to leakage and fusion with the encapsulated drug, which can reduce their effectiveness [195]. Moreover, the manufacturing process for liposomes is often expensive and complex and they may need to be stored at specific temperatures to maintain stability. They also have been reported to have a short half-life in the bloodstream, which can limit their effectiveness [196].

3.1.4. Solid Lipid Nanoparticles (SLNs)

SLNs are particulate carrier systems that are composed of a solid lipid core and a cationic lipid surface [197]. These SLNs can be used to bind DNA, forming an SLN/DNA complex or lipoplex, which shows promise as a potential vaccine [198]. SLNs offer several advantages, including the ability to encapsulate both hydrophilic and hydrophobic drugs, protecting them from degradation in the body and destruction by enzymes. These properties make SLNs suitable for delivering various active ingredients, including vaccines, thereby enhancing therapeutic efficacy [199,200]. SLNs have some limitations, as some biomolecules can be expelled from these NPs during storage. Additionally, the high water content in lipid dispersions can affect stability and drug concentration. SLNs also have a limited ability to deliver drugs through the skin [201].

3.1.5. Exosomes

Exosomes are small vesicles composed of lipophilic bilayers ranging from 50 to 100 nm in size and are used in the formation of nanovaccines. They can cross biological barriers, including the blood-brain barrier, and deliver biological and therapeutic antigens to specific tissues [202,203]. Exosomes are biocompatible and biodegradable particles, so they are safe for clinical use. Additionally, they are immunogenic and can induce the immune response to the vaccine antigen [202]. It has been reported that exosome-based vaccines can improve the therapeutic index against cancer and immunosuppressive conditions [204]. Ongoing research is evaluating the immune response of exosome-based vaccines against Ebola virus, influenza and COVID-19 [107,205].

3.1.6. Virosomes

Virosomes are structures composed of viral proteins and liposomes, ranging from 20 to 50 nm in diameter and with hydrophobic surfaces that facilitate antigen loading. The viral protein components could be glycoproteins of different viruses, such as herpes and influenza viruses [206]. Inflexal®V, an approved FDA nanovaccine, has been used as a subunit influenza vaccine that uses virosomes as carriers to enhance the immune responses against influenza H1N1, H3N2, and B [207].

3.1.7. Emulsions

Emulsions, typically consisting of oil-in-water dispersions, are widely used as delivery systems for vaccines. By incorporating antigens, surfactants, and adjuvants, these emulsions can enhance the immune response to the vaccine. Emulsion-based vaccines have shown significant potential for combating various infectious diseases, including human papillomavirus, hepatitis B virus, and influenza viruses. One notable example of an emulsion-based vaccine adjuvant is MF59, which is licensed for use in certain influenza vaccines [208,209].

3.2. Polymeric NPs

Polymeric nanoparticles (PNPs) have served as versatile carriers for active compounds, either absorbed onto their polymeric core or entrapped within the structure itself. PNPs can be synthetics that are chemically synthesized in the laboratory or natural, originating from biological sources such as polysaccharides, proteins, and lipids. Natural PNPs have more advantages over synthetic PNPs in terms of biodegradability, biocompatibility, and low toxicity [210]. There are different PNP design structures such as nanospheres, micelles, and polymerases. Poly (lactic-co-glycolic acid) (PLGA)-based NPs, chitosan, hyaluronic acid, silk fibroin, alginate, polyethyleneimine, polypropylene sulfide, and polystyrene NPs are the major types of polymeric NPs that are commonly used in vaccine adjuvant delivery systems [211,212]. The production of polymeric NPs generally follows two main strategies, the polymerization of monomers or the dispersion of preformed polymers [213].

PNPs can encapsulate multiple adjuvant materials or antigens within the polymer matrix or adsorb it onto the surface, this can protect these biomedical molecules from degradation and enhance their stability. The release of biomedical molecules can be controlled through diffusion, degradation of the polymer, or a combination of both, this allows sustained and targeted delivery. PNPs can be modified with ligands, antibodies, or other molecules that target specific cells or tissues to ensure that the biomedical molecules are delivered precisely to the desired site, reducing side effects and improving efficacy. Some PNPs are designed to release their payload in response to specific stimuli such as pH, temperature or enzymes present in the target tissue [214]. In addition, PNPs are often taken up by cells through endocytosis, a process where the cell membrane engulfs the NPs. Once inside the cell, NPs can release their biomedical molecules payload in tumors and inflamed tissues, meaning that PNP-based vaccines show significant potential as anti-tumor agents [215]. PNPs can also protect labile antigenic particles from enzymatic degradation in vivo [216,217]. Other biodegradable polymers, such as poly-lactic acid, polyethylene oxide, polyethylene glycol, and poly-ε-caprolactone are also used as adjuvants to boost vaccine immunogenicity [218,219]. Polymeric micelles are self-assembled amphiphilic structures that can efficiently deliver viral proteins. The main advantage of micelle NPs is their ability to present antigens in native-like conformations; however, their poor stability in vivo is a major drawback [220,221]. Chitosan NPs have also been used to improve pulmonary immunity against TB by increasing immune responses at mucosal sites [222]. Overall, the main advantages of PNPs include precise control of particle characteristics, payload flexibility, and ease of surface modification. However, there are potential drawbacks, including the possibility of aggregation and toxicity [223], the complexity of their production, their stability under certain conditions, and the fact that some PNPs could be cytotoxic [156,224,225]. There are different applications of PNPs, either in cancer therapy for the targeted delivery of chemotherapeutic agents or in gene therapy applications through the delivery of genetic materials such as DNA or RNA [215,226]. They have also been investigated against malaria, HIV, COVID-19, and tuberculosis diseases [227]. The main examples of polymeric-based nanovaccines that have been widely used in pre-clinical and clinical trials include NVX-CoV2373, a vaccine produced by Novavax which delivers the spike protein from the original Wuhan strain of SARS-CoV-2. The spike proteins self-assemble into 3D structures called "trimers," which are purified and mixed with an immune-stimulating substance (adjuvant) to form NPs [86]. Clinical trials in adults given two shots of this vaccine 21 days apart demonstrated about 90% efficacy in preventing infection after seven days from the second shot [85,228]. Another example of a PNP-based vaccine is GBP510, a candidate COVID-19 vaccine that has recently undergone a combined Phase 1/2 trial. GBP510 leverages nanoparticle bioengineering technology to stimulate immune responses against COVID-19 [229,230]. Additionally, the TAK-101 vaccine, which incorporates gliadin encapsulated within PLGA nanoparticles (NPs), has been developed for celiac disease and simian immunodeficiency virus (SIV). These PLGA NPs are co-formulated with TLR4 and TLR7/8 agonists, and have demonstrated the ability to induce a prolonged, robust, and high level of protection against SIV in macaque models [231].

3.3. Inorganic NPs

Inorganic NPs are biocompatible, non-toxic, hydrophilic and highly stable particles, presenting significant advantages over their organic counterparts. They can be classified into solids (hydroxyapatite and calcium phosphate) or porous (porous silicon and mesoporous silica) particles. Inorganic NPs exhibit unique properties that hold great promise for enhancing immune responses, improving antigen delivery efficiency, and enabling applications in tumor imaging, drug delivery, vaccine development, and infectious disease prevention. These NPs offer advantages, such as enhanced stability, controlled release kinetics, and targeted delivery, which can significantly improve the efficacy of therapeutic interventions [232]. Gold nanoparticles (AuNPs), carbon nanotubes, quantum dots, and silica NPs are among the inorganic NPs.

AuNPs exhibit unique physical, chemical, and optical properties that differ significantly from bulk gold, making them valuable for a wide range of high-technological applications, such as organic photovoltaics, drug delivery, vaccine development, and catalysis [233,234]. AuNPs are commonly referred to as colloidal gold when dispersed in water.

Their most pronounced properties are the intense absorbance and scattering of incident light at their surface plasmon resonance wavelength [235]. AuNPs have garnered attention in vaccine development due to their ability to enhance immune response, making them potential adjuvants for vaccines. AuNPs can be functionalized with drugs to enhance targeted delivery and reduce side effects [236–238]. Additionally, they provide stability to vaccine formulations during storage [239,240]. Several studies have demonstrated the potential of AuNPs in vaccine development; for example, conjugated antigenic peptides from the SARS-CoV-2 spike protein to AuNPs have shown no inherent toxicity and have been shown to induce significant IgG antibody levels in tested mice [75]. Another example is the increase in the efficacy of the influenza vaccine by over 25% in specific strains through chiral AuNPs [241]. Moreover, AuNPs can attach to and disrupt the cell membranes of bacteria and viruses, leading to cell lysis and death. Additionally, they can induce the production of reactive oxygen species, which cause oxidative stress and damage DNA, proteins, lipids, and other cellular components. They can also trigger apoptosis (programmed cell death) by activating various cellular pathways and can disrupt the metabolic processes of microbial cells through the inhibition of the activity of essential enzymes. AuNPs can induce cellular and humoral immune responses against enterohemorrhagic *E. coli* through increasing antigen uptake and processing [242]. Additionally, they can be loaded with platinum on copper nanosheets, and can promote the glucose oxidation process in tumor tissues, so are considered an optional therapeutic for cancer [243].

The primary limitations associated with AuNPs include (a) potential cytotoxicity at high concentrations, despite their inherent biocompatibility; (b) the requirement for specialized, often complex, and costly techniques for their synthesis and functionalization [244]; (c) their susceptibility to aggregation over time, which can compromise their stability and functionality [245]; and (d) the possibility that they might induce adverse effects like inflammation and immune responses [246].

Silica nanoparticles (SiO_2 NPs), also known as nanosilica are considered another type of inorganic NP. They are typically amorphous substances with a spherical morphology, although they can be synthesized in various shapes and sizes [247]. Their surfaces can be easily modified for specific purposes, making them suitable nanocarriers for drug delivery due to their high surface area and their ability to adsorb active components [248]. SiO_2 NPs possess several advantages including the following: (a) they exhibit low toxicity and are considered safe for biomedical applications; (b) they are thermally stable and can withstand high temperatures without significant degradation; (c) silica is abundant and cost-effective; and (d) their surfaces can be modified and functionalized for specific applications and have gained attention as potential adjuvants for vaccine development. For instance, mesoporous silica nanoparticles (MSNs) are nanomaterials with tunable particle size, high porosity, and versatile surface chemistry. They are commonly used for controlled drug release and delivery due to their low toxicity and biocompatibility [249]. Traditional MSNs have small pores (typically 2–6 nm), which limit their utility for immunotherapy since many antigens and immunostimulatory factors are macromolecules (e.g., proteins, nucleic acids). Recently, MSNs have been developed with extra-large pores (20–30 nm) to incorporate antigens for more effective cancer vaccines. Large-pore MSNs can efficiently load macromolecules, making them suitable for immunotherapy [77]. They target dendritic cells in lymph nodes, delivering tumor antigens and activating DCs, which leads to a cytotoxic T lymphocyte (CTL) antitumor response. Overall, silica-based nanomaterials are biodegradable, biocompatible, and easily conjugated with antigens and adjuvants [250]. They exert immunostimulatory effects, enhancing the immunogenicity of subunit vaccines and offering promise in cancer immunotherapy [78]. There are ongoing clinical trials for various cancers, including ovarian cancer, prostate cancer, and glioblastoma [251]. The primary drawbacks of SiO_2 NPs include the complexity of their synthesis processes and the potential long-term environmental impacts and health risks associated with exposure. These risks may include inflammation, oxidative stress, and cytotoxicity [252–254].

3.4. Virus-like Particles (VLPs)

VLPs are non-infectious, non-replicating viral structural proteins that lack viral genetic material. They mimic the structure of real viruses and do not cause disease. VLPs are safe and effective immune stimulators [255], highly immunogenic, and capable of inducing both cellular and humoral immune responses. They serve as an alternative platform for developing vaccines against infectious diseases, including hepatitis B, malaria, and SARS-CoV-2. VLP-based vaccines have the advantages of efficacy, safety, stability, and diversity [82,83]. Most licensed VLP-based vaccines use classic aluminum adjuvants, which enhance antibody responses through complex cellular reactions and local antigen trapping at the injection site, ensuring slow release and prolonged stimulation of the immune system [256]. The main drawbacks of VLPs include the following: (1) they can be less stable compared to other NPs; (2) they degrade or lose their structural integrity under certain conditions, which can affect their efficacy; (3) the removal of impurities and the achievement of high purity levels of VLPs is a labor-intensive and time-consuming process; (4) their sensitivity to environmental conditions such as temperature and pH can limit their storage and handling options; and (5) VLPs produced in bacterial systems may contain endotoxins, which can cause severe immune reactions [82,255,257,258].

There are numerous examples of VLP-based vaccines in various stages of clinical and pre-clinical development. For instance, VLPs derived from the hepatitis B virus and human papillomavirus (HPV) have been successfully commercialized. Recently, VLP-based vaccines targeting emerging pathogens such as SARS-CoV-2 have shown significant promise. These include VLPs displaying the SARS-CoV-2 spike protein, which have entered clinical trials [259,260]. Additionally, VLP-based vaccines are being explored for other diseases such as influenza, HIV, and malaria. By mimicking the structure of native viruses, VLPs can induce strong immune responses, making them a versatile platform for vaccine development [261–264].

3.5. Self-Assembling Peptide/Protein Nanovaccines

Self-assembling peptides and proteins can form diverse nanostructures, including nanofibers, NPs, and hydrogels. These structures offer unique advantages for vaccine delivery, such as controlled antigen release, enhanced immunogenicity, and targeted delivery to immune cells [265]. Recent advancements in peptide and protein engineering have enabled the development of self-assembling nanovaccines for various diseases. For example, ferritin-based NPs have been explored as a platform for delivering multiple antigens, including those targeting EBV, SARS-CoV-2, and cancer [266,267]. Additionally, peptide-based vaccines have been designed to induce specific immune responses against various pathogens [267,268].

4. Mechanism of Action of Nanovaccines

NPs can be functionalized as antigen carriers through encapsulation or surface conjugation. In encapsulation, antigens are enclosed within the NP matrix, providing a controlled release mechanism. In surface conjugation, antigens are attached to the NP surface, facilitating direct interaction with the immune system [59,269]. The mechanism of action of NPs is governed by their structure, size, surface charge, concentration, dosage, uptake, and processing by the immune system cells [270]. The size of NPs enhances tissue penetration and enables immune cell interaction and activation at the infected site [46]. They can protect antigens from premature enzymatic degradation and facilitate a sustained release, thereby ensuring prolonged antigen exposure and enhancing the immune response [55,271]. NPs can mimic the invasion process of pathogens, enhancing the immune response by presenting antigens more naturally and effectively [87]. NPs can also be functionalized with ligands that attach to certain receptors on antigen-presenting cells, including dendritic cells or macrophages. This targeted approach increases the uptake of antigens by the key immune cells. Once inside the APCs, the NPs degrade, releasing the antigens that are processed and presented on the surface of APCs by major histocompatibility complex molecules [272]. Potentially, the immune response to nanovaccines involves antibodies generated for both the NP itself and the incorporated antigen [273].

Adjuvants can be defined as substances added to vaccines to enhance the immune response, improve antigen presentation, activate the immune cells, and prolong antigen exposure. NPs can act as effective adjuvants by gradually releasing antigens and being engineered to specifically reach antigen-presenting cells. This targeted delivery optimizes immunization strategies, allowing effective immune activation with lower antigen doses [227,274].

Various types of NPs, including gold, dendrimers, carbon, polymers and liposomes, can induce cytokine and antibody responses. LNPs are particularly effective in encapsulating mRNA and delivering antigens to immune cells, facilitating CD8+ T cell activation. These LNPs have demonstrated success in clinical applications, such as in mRNA vaccines for SARS-CoV-2, where they enable efficient delivery of spike proteins to antigen-presenting cells, thereby triggering a strong immune response [189,275]. Additionally, NPs allow for the delivery of multiple antigens within a single vaccine. For example, a single NP could encapsulate spike proteins from various coronaviruses, potentially offering broader protection against related pathogens [276].

5. Nanovaccines for Food Safety

Foodborne and waterborne bacterial pathogens continue to pose significant global health challenges, causing substantial morbidity and mortality worldwide. Recent advances in NP-based vaccines have also emerged to form a promising strategy to combat these pathogens through immunization.

Recent studies have demonstrated significant progress in developing nanovaccines against *Salmonella* spp. Chitosan NPs conjugated with *Salmonella* outer membrane proteins have been shown to induce robust immune responses in broilers, protecting against lethal challenges. The nanoformulation showed superior stability when compared with conventional vaccines and also increased mRNA expression of toll-like receptors [277]. Additionally, oral vaccine administration (*S.* Enteritidis immunogenic outer membrane proteins and flagellin entrapped in mannose chitosan NPs) elicited cross-protective mucosal immune responses against *S.* Typhimurium colonization in broilers [278]

Innovative approaches using gold NPs conjugated with *E. coli* O157:H7 antigens have shown promising results. Recent studies have shown that vaccination with gold NPs conjugated to *E. coli* O157:H7 antigens elicits strong systemic and mucosal immune responses, leading to significant protection against *E. coli* O157:H7 colonization [279]. Furthermore, nanoformulations, combining antibiotics, NPs, and polymeric carriers, can improve drug delivery and efficacy in the treatment of urinary tract infection (UTI) treatment, especially UTIs caused by *E. coli*. These nanoformulations can enhance drug solubility, target specific sites of infection, and control drug release, leading to improved therapeutic outcomes and reduced side effects. These advancements in nanotechnology hold the potential to revolutionize UTI treatment by providing more effective and targeted therapeutic options [280].

A novel lipid nanoparticle (NP)-based vaccine incorporating the cholera toxin B subunit has been developed to combat *V. cholerae*. This formulation has demonstrated enhanced mucosal immunity compared to traditional oral cholera vaccines. Additionally, it has shown improved stability under gastric conditions and increased bioavailability, making it a promising alternative for cholera prevention [281]. Complementary research has shown that nanoparticle-based vaccines combining *V. cholerae* outer membrane vesicles induced long-lasting immunity [282].

Engineered polymer-based NPs incorporating flagellin and outer membrane proteins of *Campylobacter jejuni* have shown a significant reduction in bacterial colonization, achieving protection in poultry models [109]. Additionally, hybrid NPs combining chitosan and alginate have shown enhanced mucosal adherence and improved immune responses against *C. jejuni* [283,284].

Advanced NP formulations targeting *L. monocytogenes* have emerged as potential preventive strategies, where developed NPs conjugated with listeriolysin O have demonstrated enhanced cellular immunity and effective cross-presentation to CD8+ T-cells [285]. In addition, NPs loaded with multiple listeria antigens achieved broad-spectrum protection against different serotypes [285].

Innovative approaches in *Shigella* nanovaccine development have shown significant progress using NPs encapsulating *Shigella* invasion plasmid antigens, demonstrating robust protection against multiple *Shigella* species [286]. Furthermore, success has been reported with pH-responsive NPs that specifically release antigens in the intestinal environment, enhancing the efficiency of immune response induction [287].

6. Nanoparticle-Based Vaccines in Veterinary Medicine

The advancements in nanotechnology have revolutionized the field of vaccination and drug delivery, significantly enhancing the diagnosis, prevention, and treatment of both infectious and non-infectious diseases in humans and animals [269,288]. NPs, when used in vaccines, enable targeted delivery, enhance immune response, and potentially offer prolonged immunity, making them particularly advantageous in veterinary medicine applications. By reducing the reliance on antimicrobials, these nanovaccines play an essential role in promoting animal welfare and addressing antimicrobial resistance, a growing concern in veterinary and public health [289,290]. In veterinary medicine, nanoparticle-based vaccines have demonstrated effectiveness in conferring specific and durable immunity against a range of clinically significant bacterial pathogens, including *Salmonella* spp., *E. coli*, *Staphylococcus* spp., *Helicobacter* spp., *Mycobacterium* spp., *Clostridium* spp., and *Pseudomonas* spp. For example, studies have demonstrated that nanoformulations can enhance antigen stability and allow for controlled release, which can improve vaccine efficacy and immune memory [288–290]. This technology is also being explored to develop novel vaccine platforms that may improve immune responses against difficult-to-treat pathogens and reduce disease burden in livestock industries [290].

Secreting membrane vesicles are nanostructures that can stimulate T-cell responses against different antibiotic-resistant bacteria such as *S. aureus* and avian pathogenic *E. coli*. These extracellular vesicles produced by resistant bacterial strain surfaces and coated with indocyanine-loaded magnetic silica NPs could significantly reduce surface infection, systemic invasiveness, and complications of *S. aureus* in experimental models [97]. The outer membrane vesicles of avian pathogenic *E. coli* serotype O2 have experimentally promoted strong specific and non-specific protective immune responses against homologous infection with *E. coli* in broilers, which was achieved via the reduction of bacterial loads, reduction of proinflammatory cytokine production and activation of T-cell responses [98].

Carbon nanotubes and poly-anhydrous NPs have been tested against anaplasmosis or inactivated *M. paratuberculosis* in murine experimental models. The carbon nanotubes absorbed and displayed the *A. marginale* MSP1 protein on both their outer and inner surfaces. They then traversed biological membranes without compromising their integrity. In contrast, the polyanhydrides encapsulated whole-cell lysates and *M. paratuberculosis* culture filtrates through nanoprecipitation. [99].

Furthermore, nanovaccines based on the oligo-polysaccharide antigen and poly (lactic-co-glycolic acid) NPs have shown promise against *B. melitensis* through efficient engulfment by macrophages and other phagocytic cells [117]. Polyanhydride NPs encapsulating pathogenic fusion proteins co-adjuvanted with noncanonical CDGs are among the latest innovations that rapidly induce specific protective immune responses against highly virulent *B. anthracis* and *Y. pestis* [111]. *F. columnare*, which is a highly contagious bacteria affecting tilapia, can be controlled and suppressed by the immersion of mucoadhesive polymer chitosan-complexed nanovaccine, which modulates the mucosal immune response of tilapia against columnaris disease [103]. Avian pathogenic *E. coli*, which is a highly infectious pathogen impacting poultry production worldwide, has been minimized in chickens by using chitosan-based nanovaccines. These chitosan nanovaccines can encapsulate and load the outer membrane protein and flagellar antigen, resulting in improved vaccine efficacy and protection against avian pathogenic *E. coli* [108].

Additionally, chitosan nanovaccines containing *S. Enteritidis* that are surface-coated with F-protein and O-F antigen have successfully reduced the load of *S.* Enteritidis in chicken and pig intestines by stimulating specific memory B and T cell responses [109,110,291]. Moreover, nanoparticle-based vaccines have been used for the prevention and treatment of different parasitic infections and infestations such as malaria, leishmaniasis, toxoplasmosis, schistosomiasis, anaplasmosis, theileriosis, trypanosomiasis, and coccidiosis. They have also been applied to parasitic vectors like fleas, mosquitoes, and ticks [147,292]. Inoculation of nanovaccines in experimental animals against *T. cruzi* has been developed and optimized for both the protection and control of parasite replication and dissemination [112]. Moreover, new nanovaccine platforms, such as antigen-absorbing silica vesicles, have been evaluated against tick-borne diseases in laboratory models and have shown very promising results [113]. Another novel nano vaccine has shown synergistic efficacy and balanced immune response against *T. parva* [81]. Furthermore, precisely engineered nanovaccines have provided stable, long-lasting immunological and less toxigenic responses against highly pathogenic and mutated viruses, including avian influenza, RSV, and RVF virus [104,147,293,294].

7. Current Research on Nanoparticle-Based Vaccines

The use of NPs in vaccines can improve immunogenicity and offer a promising platform for future vaccine development. Researchers are utilizing microfluidic platforms to evaluate the performance of NPs in preclinical studies. These platforms can mimic in vivo conditions, allowing a more accurate assessment of NPs' efficacy and safety [295]. In addition, membrane-based nanovaccines are under investigation, with several candidates in preclinical and clinical phases for use as effective cancer nanovaccines to inhibit tumor growth and metastasis [296–298]. NPs are being explored to improve the delivery and efficacy of cancer treatments. For instance, in lung cancer therapy, NPs can deliver drugs directly to tumor cells, enhancing the precision and effectiveness of the treatment while minimizing side effects [299]. Different types of NPs, such as polymer NPs, liposomes, quantum dots, dendrimers, and gold NPs, are being studied for their unique properties and applications in targeting tumor tissues [299–301]. Animal models remain the appropriate tool in demonstrating the efficacy of nanoparticle-based vaccines. Rats, guinea pigs, mice, sheep, canines, rabbits, swine, and non-human primates like African green monkeys and Asian macaques are among the most commonly used animal models. Small animal models offer advantages in handling, housing, and maintenance costs; however, selecting a suitable animal model is crucial for translating experimental findings to clinical applications and overcoming developmental barriers. For example, mice are more resistant to classic TB disease which minimizes their utilization for evaluating TB vaccines. The encapsulation of antigens in suitable animal models using microparticles/NPs enhances the immune responses, resulting in better immunity and broader protection compared with conventional vaccines [302]. Self-assembled peptide nanovaccines (SAPNs) can act as innovative nanovaccines that enhance antigen-presenting cell uptake, with B-cell activation and lymph node trafficking, resulting in a robust immune response and protection against various animal infectious diseases [303]. Additionally, VLPs have shown promise in vaccine development and have effectively protected animal models against SARS-CoV-2 infection [304].

8. Hurdles for Nanoparticle-Based Vaccines

Regulatory approval pathways for NPs can be extremely complex due to their unique features and potential applications across various fields, including medicine, cosmetics, and food packaging. The U.S. Food and Drug Administration (FDA) regulates NPs under existing statutory authorities, tailored to the specific product type (e.g., drugs, biologics, devices). For nanomedicine, the FDA requires thorough evaluations of safety, quality, and efficacy and often involves laboratory and clinical studies due to their novel properties. The regulatory bodies perform a risk-benefit analysis tailored to the intended use of the nanoparticle product with continuous monitoring of nanoparticle products post-approval to ensure ongoing safety and effectiveness [305–307]. Regarding NP-based vaccines, key considerations include validation methods for antigen selection, formulation stability, antigen release rates, the characterization of carrier pharmacokinetics and biodistribution, and rigorous clinical trials and safety assessments that include Phase I, II, and III trials which assess safety, efficacy, and immunogenicity [308–310].

Several critical considerations and challenges pertain to NPs, including safety concerns related to adverse reactions, as NPs may accumulate in tissues and organs causing harm. Some NPs may trigger allergic reactions, inflammations, and oxidative stress and there is potential for NPs to induce DNA damage, which could result in mutations and genotoxicity [311]. Regulatory bodies like the FDA, European Medicines Agency (EMA), and National Medical Products Administration (NMPA) have set guidelines and standards to ensure the safety and efficacy of these nanoparticle-based vaccines. Another challenge is the cold chain requirements of LNPs, as they are sensitive to temperature fluctuations and thus require stringent cold chain maintenance during storage and distribution [152]. Global harmonization is another challenge, ensuring consistent standards and social considerations across different countries. The U.S. FDA has set specific guidelines, including preclinical studies, clinical trials, and post-marketing surveillance to ensure the quality, safety, and efficacy of nanoparticle-based vaccines. The European Medicines Agency (EMA) also has stringent regulations focusing on safety, efficacy, and quality control for the approval of NP-based vaccines. In China, the NMPA focuses on clinical trial data and post-marketing monitoring for the regulation and approval of NP-based vaccines. Furthermore, regarding social considerations, there may be different levels of vaccine hesitancy due to mistrust or due to misinformation from the healthcare system, because some communities may have philosophical or religious objections to certain vaccines, and also because there may be disparities in healthcare access and distribution, particularly in low-income countries. Global organizations, like the World Health Organization (WHO), are working to harmonize vaccine standards and guidelines to ensure consistency and safety across countries [312].

Ongoing efforts to overcome cultural barriers and educate and engage communities about the importance of vaccination and increase acceptance. The complexity of nanoparticle design is another challenge and requires careful manufacturing to achieve an ideal product with the appropriate pharmacological, biological, and physicochemical characteristics [149]. The ability to overcome biological barriers to deliver therapeutics effectively to the target sites is a further challenge. There are additional challenges for mRNA-based nanovaccines, which include selecting the appropriate target antigen, synthesizing, and purifying mRNA, and ensuring effective encapsulation of mRNA within lipid NPs. There are several manufacturing challenges for NPs, including top-down manufacturing approaches for breaking down bulk materials into NPs. These techniques, such as milling, lithography, and laser ablation, can be energy-intensive and may lead to issues like contamination and non-uniform particle sizes [313]. In contrast, bottom-up, such as chemical vapor deposition, sol-gel processes, and self-assembly, offer greater control over particle size and composition [313].

There are also scalability challenges for NPs, including the need for uniformity and to achieve consistent size, shape, and distribution of NPs on a large scale, which can affect the performance and safety of the final product [313]. Even small amounts of impurities can significantly alter the properties of NPs, making it crucial to maintain purity and prevent contamination during large-scale production [314]. Additionally, scaling up nanoparticle production can be expensive due to the need for specialized equipment and materials.

Current research focuses on optimizing the manufacturing processes for nanovaccines to enhance their scalability, effectiveness, and cost-efficiency. mRNA vaccines have gained attention due to their rapid design, potent immune responses, and safety. Additionally, they can be designed within days, allowing fast production and scalability [315]. LNPs acting as effective delivery platforms for mRNA vaccines can enhance immunogenicity, target delivery, and improve stability [274]. Interestingly, another platform, called the Silicon Scalable Lipid NP Generation platform, has been developed with a scalable and efficient solution for LNP production and for addressing the challenges exposed during the COVID-19 pandemic [93].

Different measures can help in achieving high-quality and consistent NPs, which are essential for their reliable performance in various applications. Advanced techniques like electron microscopy (scanning, and transmission), and dynamic light scattering are used to analyze the size, shape, and distribution of NPs. Developing standardized protocols for synthesis and characterization helps in maintaining consistency across different batches. Implementing strict control over the production process, including temperature, pH, and reactant concentrations, ensures reproducibility. Regular testing and validation of NPs against predefined criteria help in identifying and mitigating any deviations. Using automated systems for production and quality control can reduce human errors and improve consistency. Adhering to regulatory standards and guidelines ensures that the NPs meet safety and efficacy requirements. Furthermore, rigorous quality control and adherence to regulatory guidelines are essential for successful nanoparticle-based vaccine safety and efficacy. The US Pharmacopeia has developed analytical methods by which to assess attributes such as identity verification, impurity control, and content measurement for mRNA vaccine and these quality standards contribute to enhanced regulatory confidence [316–318]. Preclinical assessment of nanovaccines involves measuring physical, chemical, and stability properties, as well as assessing immunogenicity and toxicity in vitro and in vivo, which are essential to ensure comprehensive characterization throughout vaccine development [319].

9. Innovations in NP Development

9.1. Nanocages, Dendrimers, and Other Novel Structures

Various innovative structures, including nanocages and dendrimers, have emerged in biomedical research and vaccine development. Nanocages are nanoscale, hollow structures with porous walls. Nanocages offer unique properties, including porosity, high loading capacity, controlled release mechanisms, and customizable surface properties. These characteristics make nanocages highly effective for delivering therapeutic and biological molecules. Additionally, nanocages are not recognized as exogenous particles and can be biologically or chemically engineered. Nanocages can be classified into two groups: organic nanocages (e.g., protein nanocages, DNA nanocages, etc.) and inorganic nanocages (e.g., gold nanocages, silica nanocages, carbon-based nanocages, etc.) [320]. Organic nanocages, particularly protein nanocages derived from endogenous proteins, exhibit minimal interactions with living cells. In contrast, inorganic nanocages offer high antigen-loading capacity, controllable drug loading and release kinetics, flexible surface chemical modification, and in vivo safety. They can also incorporate luminescent, radioactive, or magnetic reporter molecules for tracing purposes.

Dendrimers are highly branched nanosized polymers with a central core and multiple layers of branching units. Their characteristics, such as chemical stability, polyvalency, electrostatic interactions, self-assembling, solubility, and low cytotoxicity, make them suitable candidates for functionalization with antigens, adjuvants, and other nucleic acid-based vaccines [321,322]. This technology facilitates a rapid response to vaccines, with broader efficacy and reduced vaccination frequency. Notably, dendrimer-RNA nanoparticles have demonstrated significant protective effects against lethal pathogens such as H1N1 influenza, Ebola, and *T. gondii*, with a single dose protecting challenged mouse models [94].

Outer membrane vesicles (OMVs) are spherical double-layered nanostructures (20 to 250 nm) derived from the surface of gram-negative bacteria. OMVs do not cause diseases because they mimic the structures found on the bacterial cell surface and cannot be replicated. They have a promising role in the creation of nanoparticle vaccines and can be engineered to display multiple antigens against different bacterial and viral infections and malignant tumors. OMVs can strongly stimulate the innate and adaptive immune responses [323]. However, there remain some challenges that are being addressed, including possible degradation by proteases, protein misfolding, and inefficient translocation of complex antigens [324].

Metal–organic frameworks (MOFs) are examples of NPs that can be used in nanovaccines to enhance the immune response to antigens. MOFs can encapsulate and deliver viral proteins, such as the SARS-CoV-2 spike protein. Once inside the cells, MOFs break down and activate the innate immune system through toll-like receptors that lead to increased production of cytokines and other inflammatory molecules [325,326].

Nanogels show great potential as novel antigen-delivery vehicles. They are submicron structures with a network of cross-linked polymer chains. These nanogels can encapsulate biomolecules and target specific immune cells [327,328]. They also can be designed to release antigens in a controlled manner to prolong the immune response. In addition, nanogels can be designed to display multiple antigens to protect against multiple infectious agents [329].

Combinational nanovaccines can combine multiple antigens in a single formulation to induce a broader and stronger immune response. Despite the potential benefits of combinational nanovaccines, several challenges should be considered, as the multiple antigens and adjuvants in the combined nanovaccines can interfere with each other, resulting in adverse effects, while their design can be complex and requiring increased testing and evaluation. The other challenge is regarding combinational nanovaccines' safety. There are different examples of combinational nanovaccines in the preclinical and clinical phases, including an influenza vaccine that can protect against multiple strains of influenza viruses [330], a tuberculosis vaccine that targets different TB antigens [331], a cancer vaccine that combines multiple cancer antigens [332] and an HIV combined vaccine that targets multiple strains of HIV [333].

Most traditional vaccines are administered through injection, which could be painful for some individuals. Needle-free nano vaccine administration (oral delivery, inhalation delivery, and transdermal delivery) offers several advantages, including simple administration, proven safety, reduced pain, and robust immunogenicity. For oral nanovaccine delivery to be suitable for GIT diseases, such as cholera and rotavirus infections, NPs must be designed to withstand the acidic pH of the stomach and to release the antigen in the lumen of the gut to initiate the immune response. Transdermal nanovaccine delivery is another delivery approach through which nanoparticle-based vaccines can penetrate the skin and reach the underlying tissues' immune cells. The inhalation delivery of nanovaccine also ensures that nanoparticle-based vaccines can reach the immune cells in the lung and induce the desirable immune response [334–338]. Overall, needle-free nanovaccine administration offer many advantages, including reducing the risk of blood-borne disease transmission; reducing pain, especially in individuals with needle phobia; eliminating the need for specific equipment and trained personnel due to the ease of administration; and enhancing immunogenicity due to direct delivery of the vaccine to the immune cells.

Interestingly, there have been several suggestions regarding the utilization of spores of *Bacillus* spp. as a mucosal vaccine delivery carrier. These spores have the benefits of dormancy and extreme resistance properties. Many antigens have been presented on these spores through recombinant and non-recombinant methods and have exhibited robust antigen-specific immune responses. Additionally, probiotic-based spores have shown promise in enhancing immune responses by presenting antigens on probiotic strains [339,340].

9.2. Using NPs with Other Advanced Technologies

Other important advances in gene therapy in which NPs play a crucial role include the advancement of cutting-edge technologies like Clustered Regularly Interspaced Short Palindromic Repeats (CRISPR)/associated protein 9 (Cas9) and mRNA-based therapies. CRISPR/Cas9 technology is renowned for its precise gene-editing capabilities and is considered one of the most promising technologies for genome manipulation. However, CRISPR/Cas9 has low intracellular delivery efficiency, which severely affects its potency and clinical application, meaning that the development of an effective delivery system for CRISPR/Cas9 system is crucial. LNPs and other nanocarriers like liposomes, cationic polymers, and gold NPs, have emerged as efficient carriers for CRISPR/Cas9 components, overcoming challenges such as inadequate cellular entry, off-target effects, and nuclease degradation [341,342]. A notable advancement in tumor targeting involves the use of a multiplexed dendrimer with an LNP [95]. This system involves the co-delivery of Cas9 mRNA, focal adhesion kinase siRNA, and sgRNA (siFAK + CRISPR-LNPs) to enable tumor delivery and enhance gene-editing efficacy. The study demonstrated that gene editing was enhanced and increased by more than 10-fold in tumor spheroids, due to increased cellular uptake and NP penetration mediated by FAK-knockdown in tumor tissues.

9.3. Personalized Vaccines

Personalized nanovaccines represent an exciting frontier in vaccine development, tailored to the genetic variations, existing immunity, age, sex, and disease history of individuals. Unlike traditional vaccines, which have a one-size-fits-all approach (standard antigen formulation and dose), personalized nanovaccines are specifically designed for each host's unique immune system. Researchers have been able to develop personalized vaccines after analyzing biomarkers that can predict the host immune response to vaccines. The main advantage of personalized vaccines lies in their ability to target the individual's unique immune system, thereby maximizing efficacy and minimizing adverse reactions [343]. However, their widespread adoption faces challenges including the need for biomarkers development, high costs, and logistical barriers [181,276,344]. Recent examples of engineered personalized nanovaccines include personalized flu vaccines and cancer immunotherapy. In cancer immunotherapy, the tailored nanovaccines not only enhance antigen presentation but also effectively modulate immunosuppression within the tumor microenvironment. A notable example is the mRNA-based personalized cancer vaccine mRNA-4157, which targets twenty tumor-associated antigens specifically expressed by cancer cells, demonstrating the potential for both immunostimulatory and antitumor activities [345]. Fluoropolymer-based nanovaccines represent another class of personalized nanovaccines. When used in combination with immune checkpoint blockade therapy, these tailored nanovaccines have exhibited prevention of post-surgical metastasis and tumor recurrence in breast cancer models [96]. Additionally, neoantigen nanovaccines, which target unique tumor-specific antigens arising from mutations in cancer cells, have shown promise as personalized nanovaccines. For example, a study has demonstrated that delivering neoantigen peptides along with a sting agonist through a nanoparticle-based system significantly improved the immune response and survival rates in a melanoma mouse model [346]. Moreover, the influenza virus is characterized by its rapid mutation, requiring traditional influenza vaccines to be updated periodically according to the circulating strains. However, all individuals cannot equally respond and be protected using these influenza vaccines. These challenges could be controlled with the development of personalized influenza vaccines that optimize the immune response of each individual.

10. Artificial Intelligence in Nanoparticle-Based Vaccines

Artificial intelligence (AI) has emerged as a transformative force in the development of NP-based vaccines, revolutionizing multiple aspects of the development pipeline from initial design to clinical implementation. The integration of AI technologies, particularly machine learning, and deep learning approaches, has significantly accelerated the vaccine development process while improving efficiency and predictive accuracy [347]. These advanced computational tools have enabled researchers to optimize antigen design, predict immune responses, and enhance the overall efficacy of NP-based vaccine platforms.

In the field of antigen design and NPs optimization, AI algorithms have demonstrated remarkable capabilities in predicting optimal epitope sequences and enhancing antigen stability. These systems can analyze vast datasets of protein structures and immunological responses to identify the most promising vaccine candidates, significantly reducing the time and resources required for traditional experimental approaches [348]. AI-driven platforms have revolutionized the characterization of nanoparticle properties, enabling precise control over particle size, shape, and surface chemistry, which are crucial parameters for vaccine efficacy [349].

The application of machine learning in formulation development has particularly enhanced our ability to optimize nanoparticle composition and predict stability profiles. These AI systems can analyze complex relationships between formulation parameters and vaccine performance, leading to more stable and effective vaccine candidates [350]. Additionally, AI models have proven invaluable in predicting immunological responses and potential adverse reactions, allowing researchers to optimize vaccine formulations before clinical trials begin [351].

Looking toward the future, the integration of multiple AI technologies presents exciting opportunities for nanovaccine development. The combination of big data analytics with sophisticated AI models is expected to enable more personalized vaccine approaches and improve prediction accuracy. However, challenges remain, including the need for larger, more diverse datasets and the validation of AI predictions in biological systems. As these technologies continue to evolve, they are expected to play an increasingly crucial role in developing next-generation nanovaccines that are more effective, safer, and more readily available to global populations.

11. Conclusions

Nanoparticle-based vaccines hold immense promise for revolutionizing immunization strategies. These vaccine platforms offer several advantages over traditional approaches, including enhanced immunogenicity, targeted delivery, and sustained antigen release. The ability to incorporate multiple antigens and adjuvants into a single nanoparticle formulation further simplifies vaccine administration and improves efficacy. However, significant challenges remain in translating these promising technologies into practical applications. Overcoming manufacturing hurdles, ensuring long-term stability, and conducting rigorous clinical trials are essential to bringing NP-based vaccines to market. Addressing potential toxicity and immune side effects is crucial for ensuring their safety and efficacy. Researchers can develop highly effective and personalized vaccines by combining nanotechnology with advancements in artificial intelligence and bioengineering. This interdisciplinary approach will facilitate the creation of vaccines capable of combating emerging infectious diseases, reducing the burden of chronic illnesses, and improving global health outcomes.

Author Contributions: Conceptualization, supervision, and funding, Y.A.H.; data curation, M.S. and Y.A.H.; original draft preparation, M.S., A.E.-M., A.H.E., W.I.A.S. and Y.A.H.; Review and editing, M.S., A.E.-M., A.H.E., W.I.A.S. and Y.A.H. All authors have read and agreed to the published version of the manuscript.

Funding: This research was supported by the Center of Biomedical Research Excellence (COBRE) for Translational Chemical Biology (CTCB, NIH P20 GM130456), and the National Center for Advancing Translational Sciences, National Institutes of Health (Grant number KL2TR001996).

Institutional Review Board Statement: Not applicable.

Informed Consent Statement: Not applicable.

Data Availability Statement: All data were included in the published paper.

Conflicts of Interest: The authors declare no conflicts of interest.

References

1. Hedman, H.D.; Krawczyk, E.; Helmy, Y.A.; Zhang, L.; Varga, C. Host Diversity and Potential Transmission Pathways of SARS-CoV-2 at the Human-Animal Interface. *Pathogens* **2021**, *10*, 180. [CrossRef] [PubMed]
2. Hemelaar, J.; Elangovan, R.; Yun, J.; Dickson-Tetteh, L.; Fleminger, I.; Kirtley, S.; Williams, B.; Gouws-Williams, E.; Ghys, P.D.; Alash'le, G.A. Global and regional molecular epidemiology of HIV-1, 1990–2015: A systematic review, global survey, and trend analysis. *Lancet Infect. Dis.* **2019**, *19*, 143–155. [CrossRef] [PubMed]
3. Ksiazek, T.G.; Erdman, D.; Goldsmith, C.S.; Zaki, S.R.; Peret, T.; Emery, S.; Tong, S.; Urbani, C.; Comer, J.A.; Lim, W. A novel coronavirus associated with severe acute respiratory syndrome. *N. Engl. J. Med.* **2003**, *348*, 1953–1966. [CrossRef]
4. Mehendale, R.; Joshi, M.; Patravale, V.B. Nanomedicines for treatment of viral diseases. *Crit. Rev. ™ Ther. Drug Carr. Syst.* **2013**, *30*, 1–49. [CrossRef] [PubMed]
5. Salem, M.; El-Metwally, M.; Saber, W.; Negm, S.; El-Kott, A.; Mazroua, Y.; Makhlouf, A.; Moustafa, M. Secondary antiviral metabolites from fungi with special reference to coronaviruses. *Biocell* **2022**, *46*, 1979–1988. [CrossRef]

6. Helmy, Y.A.; Taha-Abdelaziz, K.; Hawwas, H.A.E.-H.; Ghosh, S.; AlKafaas, S.S.; Moawad, M.M.M.; Saied, E.M.; Kassem, I.I.; Mawad, A.M.M. Antimicrobial Resistance and Recent Alternatives to Antibiotics for the Control of Bacterial Pathogens with an Emphasis on Foodborne Pathogens. *Antibiotics* **2023**, *12*, 274. [CrossRef] [PubMed]
7. Kabir, A.; Lamichhane, B.; Habib, T.; Adams, A.; El-Sheikh Ali, H.; Slovis, N.M.; Troedsson, M.H.T.; Helmy, Y.A. Antimicrobial Resistance in Equines: A Growing Threat to Horse Health and Beyond—A Comprehensive Review. *Antibiotics* **2024**, *13*, 713. [CrossRef]
8. Roope, L.S.; Smith, R.D.; Pouwels, K.B.; Buchanan, J.; Abel, L.; Eibich, P.; Butler, C.C.; Tan, P.S.; Walker, A.S.; Robotham, J.V. The challenge of antimicrobial resistance: What economics can contribute. *Science* **2019**, *364*, eaau4679. [CrossRef]
9. Lamichhane, B.; Mawad, A.M.M.; Saleh, M.; Kelley, W.G.; Harrington, P.J.; Lovestad, C.W.; Amezcua, J.; Sarhan, M.M.; El Zowalaty, M.E.; Ramadan, H.; et al. Salmonellosis: An Overview of Epidemiology, Pathogenesis, and Innovative Approaches to Mitigate the Antimicrobial Resistant Infections. *Antibiotics* **2024**, *13*, 76. [CrossRef]
10. Mullins, L.P.; Mason, E.; Winter, K.; Sadarangani, M. Vaccination is an integral strategy to combat antimicrobial resistance. *PLoS Pathog.* **2023**, *19*, e1011379. [CrossRef]
11. Fawzy, M.; Helmy, Y.A. The One Health Approach is Necessary for the Control of Rift Valley Fever Infections in Egypt: A Comprehensive Review. *Viruses* **2019**, *11*, 139. [CrossRef]
12. Helmy, Y.A.; El-Adawy, H.; Abdelwhab, E.M. A Comprehensive Review of Common Bacterial, Parasitic and Viral Zoonoses at the Human-Animal Interface in Egypt. *Pathogens* **2017**, *6*, 33. [CrossRef] [PubMed]
13. Heng, W.T.; Yew, J.S.; Poh, C.L. Nanovaccines against viral infectious diseases. *Pharmaceutics* **2022**, *14*, 2554. [CrossRef] [PubMed]
14. Helmy, Y.A.; Kassem, I.I.; Rajashekara, G. Immuno-modulatory effect of probiotic *E. coli* Nissle 1917 in polarized human colonic cells against *Campylobacter jejuni* infection. *Gut Microbes* **2021**, *13*, 1857514. [CrossRef] [PubMed]
15. Zepp, F. Principles of vaccine design—Lessons from nature. *Vaccine* **2010**, *28*, C14–C24. [CrossRef] [PubMed]
16. Kiboneka, A.N. Basic concepts in clinical immunology: A review. *World J. Adv. Res. Rev.* **2021**, *12*, 490–496. [CrossRef]
17. Sim, S.; Wong, N.K. Nanotechnology and its use in imaging and drug delivery. *Biomed. Rep.* **2021**, *14*, 42. [CrossRef] [PubMed]
18. Abdelaziz, K.; Helmy, Y.A.; Yitbarek, A.; Hodgins, D.C.; Sharafeldin, T.A.; Selim, M.S.H. Advances in Poultry Vaccines: Leveraging Biotechnology for Improving Vaccine Development, Stability, and Delivery. *Vaccines* **2024**, *12*, 134. [CrossRef] [PubMed]
19. Elaish, M.; Ngunjiri, J.M.; Ali, A.; Xia, M.; Ibrahim, M.; Jang, H.; Hiremath, J.; Dhakal, S.; Helmy, Y.A.; Jiang, X. Supplementation of inactivated influenza vaccine with norovirus P particle-M2e chimeric vaccine enhances protection against heterologous virus challenge in chickens. *PLoS ONE* **2017**, *12*, e0171174. [CrossRef] [PubMed]
20. Murugan, B.; Sagadevan, S. Nano-Vaccines: Opportunities and Challenges in Biomaterial-Based Vaccine Delivery. *Biomater. -Inspired Nanomed. Target. Ther.* **2024**, 101–116. [CrossRef]
21. Peek, L.J.; Middaugh, C.R.; Berkland, C. Nanotechnology in vaccine delivery. *Adv. Drug Deliv. Rev.* **2008**, *60*, 915–928. [CrossRef]
22. Azharuddin, M.; Zhu, G.H.; Sengupta, A.; Hinkula, J.; Slater, N.K.; Patra, H.K. Nano toolbox in immune modulation and nanovaccines. *Trends Biotechnol.* **2022**, *40*, 1195–1212. [CrossRef]
23. Machhi, J.; Shahjin, F.; Das, S.; Patel, M.; Abdelmoaty, M.M.; Cohen, J.D.; Singh, P.A.; Baldi, A.; Bajwa, N.; Kumar, R. Nanocarrier vaccines for SARS-CoV-2. *Adv. Drug Deliv. Rev.* **2021**, *171*, 215–239. [CrossRef] [PubMed]
24. Lu, W.; Cui, C.; Wang, Y.; Sun, X.; Wang, S.; Yang, M.; Yu, Y.; Wang, L. CpG ODN as an adjuvant arouses the vigor of B cells by relieving the negative regulation of surface TLR9 to enhance the antibody response to vaccine. *Appl. Microbiol. Biotechnol.* **2021**, *105*, 4213–4224. [CrossRef] [PubMed]
25. Lozano, D.; Larraga, V.; Vallet-Regí, M.; Manzano, M. An overview of the use of nanoparticles in vaccine development. *Nanomaterials* **2023**, *13*, 1828. [CrossRef]
26. Fredriksen, B.N.; Grip, J. PLGA/PLA micro-and nanoparticle formulations serve as antigen depots and induce elevated humoral responses after immunization of Atlantic salmon (*Salmo salar* L.). *Vaccine* **2012**, *30*, 656–667. [CrossRef]
27. Saleh, A.; Qamar, S.; Tekin, A.; Singh, R.; Kashyap, R. Vaccine development throughout history. *Cureus* **2021**, *13*, e16635. [CrossRef] [PubMed]
28. Taylor, M.W.; Taylor, M.W. *Smallpox*; Springer: Berlin/Heidelberg, Germany, 2014.
29. Davidson, T. *Vaccines: History, Science, and Issues*; Bloomsbury Publishing USA: New York, NY, USA, 2017.
30. Mamelund, S.-E. Influenza, historical. *Medicine* **2008**, *54*, 361–371.
31. Lombard, M.; Pastoret, P.-P.; Moulin, A. A brief history of vaccines and vaccination. *Rev. Sci. Tech. Off. Int. Epizoot.* **2007**, *26*, 29–48. [CrossRef]
32. Meeusen, E.N.; Walker, J.; Peters, A.; Pastoret, P.-P.; Jungersen, G. Current status of veterinary vaccines. *Clin. Microbiol. Rev.* **2007**, *20*, 489–510. [CrossRef] [PubMed]
33. Karin, H.; Lisa, B.; Damer, P. Vaccines as alternatives to antibiotics for food producing animals. Part 1: Challenges and needs. *Vet. Res.* **2018**, *49*, 64.
34. Palomino-Tapia, V. Autogenous Vaccines in the Poultry Industry: A Field Perspective. In *Poultry Farming-New Perspectives and Applications*; IntechOpen: Rijeka, Croatia, 2023.

35. Nagpal, G.; Usmani, S.S.; Raghava, G.P. A web resource for designing subunit vaccine against major pathogenic species of bacteria. *Front. Immunol.* **2018**, *9*, 2280. [CrossRef]
36. Malik, H.; Khan, F.H.; Ahsan, H. Human papillomavirus: Current status and issues of vaccination. *Arch. Virol.* **2014**, *159*, 199–205. [CrossRef] [PubMed]
37. Dai, X.; Xiong, Y.; Li, N.; Jian, C. Vaccine types. In *Vaccines—The History and Future*; IntechOpen: Rijeka, Croatia, 2019.
38. Liljeqvist, S.; Ståhl, S. Production of recombinant subunit vaccines: Protein immunogens, live delivery systems and nucleic acid vaccines. *J. Biotechnol.* **1999**, *73*, 1–33. [CrossRef] [PubMed]
39. Jones, R.G.; Liu, Y.; Rigsby, P.; Sesardic, D. An improved method for development of toxoid vaccines and antitoxins. *J. Immunol. Methods* **2008**, *337*, 42–48. [CrossRef]
40. Andey, T.; Soni, S.; Modi, S. Conventional vaccination methods: Inactivated and live attenuated vaccines. *Adv. Vaccin. Technol. Infect. Chronic Dis.* **2024**, 37–50. [CrossRef]
41. Lee, S.; Nguyen, M.T. Recent advances of vaccine adjuvants for infectious diseases. *Immune Netw.* **2015**, *15*, 51–57. [CrossRef] [PubMed]
42. Minor, P.D. Live attenuated vaccines: Historical successes and current challenges. *Virology* **2015**, *479*, 379–392. [CrossRef]
43. Lopez, S.M.; Sato, A.I.; Chatterjee, A. Vaccines: An overview. *Viral Parasit. Bact. Fungal Infect.* **2023**, *56*, 699–717.
44. Taha-Abdelaziz, K.; Singh, M.; Sharif, S.; Sharma, S.; Kulkarni, R.R.; Alizadeh, M.; Yitbarek, A.; Helmy, Y.A. Intervention Strategies to Control *Campylobacter* at Different Stages of the Food Chain. *Microorganisms* **2023**, *11*, 113. [CrossRef]
45. Teulon, J.-M.; Godon, C.; Chantalat, L.; Moriscot, C.; Cambedouzou, J.; Odorico, M.; Ravaux, J.; Podor, R.; Gerdil, A.; Habert, A. On the operational aspects of measuring nanoparticle sizes. *Nanomaterials* **2018**, *9*, 18. [CrossRef] [PubMed]
46. Cai, T.; Liu, H.; Zhang, S.; Hu, J.; Zhang, L. Delivery of nanovaccine towards lymphoid organs: Recent strategies in enhancing cancer immunotherapy. *J. Nanobiotechnol.* **2021**, *19*, 1–23. [CrossRef] [PubMed]
47. Elattar, K.M.; Al-Otibi, F.O.; El-Hersh, M.S.; Attia, A.A.; Eldadamony, N.M.; Elsayed, A.; Menaa, F.; Saber, W.I. Multifaceted chemical and bioactive features of Ag@ TiO_2 and Ag@ SeO_2 core/shell nanoparticles biosynthesized using *Beta vulgaris* L. extract. *Heliyon* **2024**, *10*, e28359. [CrossRef] [PubMed]
48. Chattopadhyay, S.; Chen, J.-Y.; Chen, H.-W.; Hu, C.-M.J. Nanoparticle vaccines adopting virus-like features for enhanced immune potentiation. *Nanotheranostics* **2017**, *1*, 244. [CrossRef]
49. Zhang, W.; Wang, L.; Liu, Y.; Chen, X.; Liu, Q.; Jia, J.; Yang, T.; Qiu, S.; Ma, G. Immune responses to vaccines involving a combined antigen–nanoparticle mixture and nanoparticle-encapsulated antigen formulation. *Biomaterials* **2014**, *35*, 6086–6097. [CrossRef] [PubMed]
50. Bhardwaj, P.; Bhatia, E.; Sharma, S.; Ahamad, N.; Banerjee, R. Advancements in prophylactic and therapeutic nanovaccines. *Acta Biomater.* **2020**, *108*, 1–21. [CrossRef]
51. Yin, Q.; Wang, Y.; Xiang, Y.; Xu, F. Nanovaccines: Merits, and diverse roles in boosting antitumor immune responses. *Hum. Vaccines Immunother.* **2022**, *18*, 2119020. [CrossRef]
52. Li, M.; Kaminskas, L.M.; Marasini, N. Recent advances in nano/microparticle-based oral vaccines. *J. Pharm. Investig.* **2021**, *51*, 425–438. [CrossRef] [PubMed]
53. Sabourian, P.; Yazdani, G.; Ashraf, S.S.; Frounchi, M.; Mashayekhan, S.; Kiani, S.; Kakkar, A. Effect of physico-chemical properties of nanoparticles on their intracellular uptake. *Int. J. Mol. Sci.* **2020**, *21*, 8019. [CrossRef] [PubMed]
54. González-García, L.E.; MacGregor, M.N.; Visalakshan, R.M.; Lazarian, A.; Cavallaro, A.A.; Morsbach, S.; Mierczynska-Vasilev, A.; Mailänder, V.; Landfester, K.; Vasilev, K. Nanoparticles surface chemistry influence on protein corona composition and inflammatory responses. *Nanomaterials* **2022**, *12*, 682. [CrossRef]
55. Diaz-Arévalo, D.; Zeng, M. Nanoparticle-based vaccines: Opportunities and limitations. In *Nanopharmaceuticals*; Elsevier: Amsterdam, The Netherlands, 2020; pp. 135–150.
56. Kirtane, A.R.; Verma, M.; Karandikar, P.; Furin, J.; Langer, R.; Traverso, G. Nanotechnology approaches for global infectious diseases. *Nat. Nanotechnol.* **2021**, *16*, 369–384. [CrossRef] [PubMed]
57. Gregory, A.E.; Titball, R.; Williamson, D. Vaccine delivery using nanoparticles. *Front. Cell. Infect. Microbiol.* **2013**, *3*, 13. [CrossRef] [PubMed]
58. Manju, K.; Raj, S.N.; Ranjini, H.; Nayaka, S.C.; Ashwini, P.; Satish, S.; Prasad, M.N.; Chouhan, R.S.; Baker, S. Nanovaccines to combat drug resistance: The next-generation immunisation. *Future J. Pharm. Sci.* **2023**, *9*, 64. [CrossRef]
59. Kheirollahpour, M.; Mehrabi, M.; Dounighi, N.M.; Mohammadi, M.; Masoudi, A. Nanoparticles and vaccine development. *Pharm. Nanotechnol.* **2020**, *8*, 6–21. [CrossRef]
60. Liu, J.; Zhang, R.; Xu, Z.P. Nanoparticle-based nanomedicines to promote cancer immunotherapy: Recent advances and future directions. *Small* **2019**, *15*, 1900262. [CrossRef]
61. Kumru, O.S.; Joshi, S.B.; Smith, D.E.; Middaugh, C.R.; Prusik, T.; Volkin, D.B. Vaccine instability in the cold chain: Mechanisms, analysis and formulation strategies. *Biologicals* **2014**, *42*, 237–259. [CrossRef] [PubMed]

62. Torres-Sangiao, E.; Holban, A.M.; Gestal, M.C. Advanced nanobiomaterials: Vaccines, diagnosis and treatment of infectious diseases. *Molecules* **2016**, *21*, 867. [CrossRef] [PubMed]
63. Abusalah, M.A.H.; Chopra, H.; Sharma, A.; Mustafa, S.A.; Choudhary, O.P.; Sharma, M.; Dhawan, M.; Khosla, R.; Loshali, A.; Sundriyal, A. Nanovaccines: A game changing approach in the fight against infectious diseases. *Biomed. Pharmacother.* **2023**, *167*, 115597.
64. Jazayeri, S.D.; Lim, H.X.; Shameli, K.; Yeap, S.K.; Poh, C.L. Nano and microparticles as potential oral vaccine carriers and adjuvants against infectious diseases. *Front. Pharmacol.* **2021**, *12*, 682286. [CrossRef]
65. Rosenbaum, P.; Tchitchek, N.; Joly, C.; Rodriguez Pozo, A.; Stimmer, L.; Langlois, S.; Hocini, H.; Gosse, L.; Pejoski, D.; Cosma, A. Vaccine inoculation route modulates early immunity and consequently antigen-specific immune response. *Front. Immunol.* **2021**, *12*, 645210. [CrossRef]
66. Xie, S.; Pan, B.; Wang, M.; Zhu, L.; Wang, F.; Dong, Z.; Wang, X.; Zhou, W. Formulation, characterization and pharmacokinetics of praziquantel-loaded hydrogenated castor oil solid lipid nanoparticles. *Nanomedicine* **2010**, *5*, 693–701. [CrossRef]
67. Heidari-Kharaji, M.; Taheri, T.; Doroud, D.; Habibzadeh, S.; Badirzadeh, A.; Rafati, S. Enhanced paromomycin efficacy by solid lipid nanoparticle formulation against Leishmania in mice model. *Parasite Immunol.* **2016**, *38*, 599–608. [CrossRef] [PubMed]
68. Thi, T.T.H.; Suys, E.J.; Lee, J.S.; Nguyen, D.H.; Park, K.D.; Truong, N.P. Lipid-based nanoparticles in the clinic and clinical trials: From cancer nanomedicine to COVID-19 vaccines. *Vaccines* **2021**, *9*, 359. [CrossRef] [PubMed]
69. Kondel, R.; Shafiq, N.; Kaur, I.P.; Singh, M.P.; Pandey, A.K.; Ratho, R.K.; Malhotra, S. Effect of acyclovir solid lipid nanoparticles for the treatment of herpes simplex virus (HSV) infection in an animal model of HSV-1 infection. *Pharm. Nanotechnol.* **2019**, *7*, 389–403. [CrossRef] [PubMed]
70. Lee, B.; Park, J.; Ryu, M.; Kim, S.; Joo, M.; Yeom, J.-H.; Kim, S.; Park, Y.; Lee, K.; Bae, J. Antimicrobial peptide-loaded gold nanoparticle-DNA aptamer conjugates as highly effective antibacterial therapeutics against *Vibrio vulnificus*. *Sci. Rep.* **2017**, *7*, 13572. [CrossRef]
71. Chowdhury, R.; Ilyas, H.; Ghosh, A.; Ali, H.; Ghorai, A.; Midya, A.; Jana, N.R.; Das, S.; Bhunia, A. Multivalent gold nanoparticle–peptide conjugates for targeting intracellular bacterial infections. *Nanoscale* **2017**, *9*, 14074–14093. [CrossRef] [PubMed]
72. Gregory, A.E.; Judy, B.M.; Qazi, O.; Blumentritt, C.A.; Brown, K.A.; Shaw, A.M.; Torres, A.G.; Titball, R.W. A gold nanoparticle-linked glycoconjugate vaccine against *Burkholderia mallei*. *Nanomed. Nanotechnol. Biol. Med.* **2015**, *11*, 447–456. [CrossRef]
73. Chen, Y.-S.; Hung, Y.-C.; Lin, W.-H.; Huang, G.S. Assessment of gold nanoparticles as a size-dependent vaccine carrier for enhancing the antibody response against synthetic foot-and-mouth disease virus peptide. *Nanotechnology* **2010**, *21*, 195101. [CrossRef]
74. Lee, M.-Y.; Yang, J.-A.; Jung, H.S.; Beack, S.; Choi, J.E.; Hur, W.; Koo, H.; Kim, K.; Yoon, S.K.; Hahn, S.K. Hyaluronic acid–gold nanoparticle/interferon α complex for targeted treatment of hepatitis C virus infection. *ACS Nano* **2012**, *6*, 9522–9531. [CrossRef]
75. Farfán-Castro, S.; García-Soto, M.J.; Betancourt-Mendiola, L.; Cervantes, J.; Segura, R.; González-Ortega, O.; Rosales-Mendoza, S. Synthesis and evaluation of gold nanoparticles conjugated with five antigenic peptides derived from the spike protein of SARS-CoV-2 for vaccine development. *Front. Nanotechnol.* **2024**, *6*, 1335346. [CrossRef]
76. Halwani, M.; Yebio, B.; Suntres, Z.; Alipour, M.; Azghani, A.; Omri, A. Co-encapsulation of gallium with gentamicin in liposomes enhances antimicrobial activity of gentamicin against *Pseudomonas aeruginosa*. *J. Antimicrob. Chemother.* **2008**, *62*, 1291–1297. [CrossRef] [PubMed]
77. Theivendran, S.; Lazarev, S.; Yu, C. Mesoporous silica/organosilica nanoparticles for cancer immunotherapy. *Exploration* **2023**, *3*, 20220086. [CrossRef]
78. Song, C.; Li, F.; Wang, S.; Wang, J.; Wei, W.; Ma, G. Recent advances in particulate adjuvants for cancer vaccination. *Adv. Ther.* **2020**, *3*, 1900115. [CrossRef]
79. Madapong, A.; Petro-Turnquist, E.M.; Webby, R.J.; McCormick, A.A.; Weaver, E.A. Immunity and Protective Efficacy of a Plant-Based Tobacco Mosaic Virus-like Nanoparticle Vaccine against Influenza a Virus in Mice. *Vaccines* **2024**, *12*, 1100. [CrossRef] [PubMed]
80. Huertas-Díaz, M.C.; Phan, S.; Elson, A.; Nuñez, I.; Wei, H.; Sakamoto, K.; He, B. Parainfluenza virus 5 (PIV5) amplifying virus-like particles expressing respiratory syncytial virus (RSV) antigens protect mice against RSV infection. *Vaccine* **2019**, *37*, 2925–2934. [CrossRef]
81. Lacasta, A.; Mody, K.T.; De Goeyse, I.; Yu, C.; Zhang, J.; Nyagwange, J.; Mwalimu, S.; Awino, E.; Saya, R.; Njoroge, T. Synergistic effect of two nanotechnologies enhances the protective capacity of the *Theileria parva* sporozoite p67C antigen in cattle. *J. Immunol.* **2021**, *206*, 686–699. [CrossRef]
82. Tariq, H.; Batool, S.; Asif, S.; Ali, M.; Abbasi, B.H. Virus-like particles: Revolutionary platforms for developing vaccines against emerging infectious diseases. *Front. Microbiol.* **2022**, *12*, 790121. [CrossRef]
83. Dai, S.; Wang, H.; Deng, F. Advances and challenges in enveloped virus-like particle (VLP)-based vaccines. *J. Immunol. Sci.* **2018**, *2*, 36–41.

84. Michel, M.-L.; Tiollais, P. Hepatitis B vaccines: Protective efficacy and therapeutic potential. *Pathol. Biol.* **2010**, *58*, 288–295. [CrossRef]
85. Keech, C.; Albert, G.; Cho, I.; Robertson, A.; Reed, P.; Neal, S.; Plested, J.S.; Zhu, M.; Cloney-Clark, S.; Zhou, H. Phase 1–2 trial of a SARS-CoV-2 recombinant spike protein nanoparticle vaccine. *N. Engl. J. Med.* **2020**, *383*, 2320–2332. [CrossRef] [PubMed]
86. Vu, M.N.; Kelly, H.G.; Kent, S.J.; Wheatley, A.K. Current and future nanoparticle vaccines for COVID-19. *EBioMedicine* **2021**, *74*, 103699. [CrossRef]
87. Liao, Z.; Huang, J.; Lo, P.-C.; Lovell, J.F.; Jin, H.; Yang, K. Self-adjuvanting cancer nanovaccines. *J. Nanobiotechnol.* **2022**, *20*, 345. [CrossRef]
88. Li, Y.; Su, T.; Zhang, Y.; Huang, X.; Li, J.; Li, C. Liposomal co-delivery of daptomycin and clarithromycin at an optimized ratio for treatment of methicillin-resistant *Staphylococcus aureus* infection. *Drug Deliv.* **2015**, *22*, 627–637. [CrossRef] [PubMed]
89. Wang, J.; Li, P.; Yu, Y.; Fu, Y.; Jiang, H.; Lu, M.; Sun, Z.; Jiang, S.; Lu, L.; Wu, M.X. Pulmonary surfactant–biomimetic nanoparticles potentiate heterosubtypic influenza immunity. *Science* **2020**, *367*, eaau0810. [CrossRef]
90. Hanson, M.C.; Abraham, W.; Crespo, M.P.; Chen, S.H.; Liu, H.; Szeto, G.L.; Kim, M.; Reinherz, E.L.; Irvine, D.J. Liposomal vaccines incorporating molecular adjuvants and intrastructural T-cell help promote the immunogenicity of HIV membrane-proximal external region peptides. *Vaccine* **2015**, *33*, 861–868. [CrossRef]
91. Smith, L.R.; Wloch, M.K.; Ye, M.; Reyes, L.R.; Boutsaboualoy, S.; Dunne, C.E.; Chaplin, J.A.; Rusalov, D.; Rolland, A.P.; Fisher, C.L. Phase 1 clinical trials of the safety and immunogenicity of adjuvanted plasmid DNA vaccines encoding influenza A virus H5 hemagglutinin. *Vaccine* **2010**, *28*, 2565–2572. [CrossRef] [PubMed]
92. Bovier, P.A. Epaxal®: A virosomal vaccine to prevent hepatitis A infection. *Expert Rev. Vaccines* **2008**, *7*, 1141–1150. [CrossRef]
93. Huang, X.; Ma, Y.; Ma, G.; Xia, Y. Unlocking the therapeutic applicability of LNP-mRNA: Chemistry, formulation, and clinical strategies. *Research* **2024**, *7*, 0370. [CrossRef]
94. Chahal, J.S.; Khan, O.F.; Cooper, C.L.; McPartlan, J.S.; Tsosie, J.K.; Tilley, L.D.; Sidik, S.M.; Lourido, S.; Langer, R.; Bavari, S. Dendrimer-RNA nanoparticles generate protective immunity against lethal Ebola, H1N1 influenza, and *Toxoplasma gondii* challenges with a single dose. *Proc. Natl. Acad. Sci. USA* **2016**, *113*, E4133–E4142. [CrossRef]
95. Zhang, D.; Wang, G.; Yu, X.; Wei, T.; Farbiak, L.; Johnson, L.T.; Taylor, A.M.; Xu, J.; Hong, Y.; Zhu, H. Enhancing CRISPR/Cas gene editing through modulating cellular mechanical properties for cancer therapy. *Nat. Nanotechnol.* **2022**, *17*, 777–787. [CrossRef] [PubMed]
96. Xu, J.; Lv, J.; Zhuang, Q.; Yang, Z.; Cao, Z.; Xu, L.; Pei, P.; Wang, C.; Wu, H.; Dong, Z. A general strategy towards personalized nanovaccines based on fluoropolymers for post-surgical cancer immunotherapy. *Nat. Nanotechnol.* **2020**, *15*, 1043–1052. [CrossRef]
97. Chen, G.; Bai, Y.; Li, Z.; Wang, F.; Fan, X.; Zhou, X. Bacterial extracellular vesicle-coated multi-antigenic nanovaccines protect against drug-resistant *Staphylococcus aureus* infection by modulating antigen processing and presentation pathways. *Theranostics* **2020**, *10*, 7131. [CrossRef] [PubMed]
98. Hu, R.; Liu, H.; Wang, M.; Li, J.; Lin, H.; Liang, M.; Gao, Y.; Yang, M. An OMV-based nanovaccine confers safety and protection against pathogenic *Escherichia coli* via both humoral and predominantly Th1 immune responses in poultry. *Nanomaterials* **2020**, *10*, 2293. [CrossRef]
99. Thukral, A.; Ross, K.; Hansen, C.; Phanse, Y.; Narasimhan, B.; Steinberg, H.; Talaat, A.M. A single dose polyanhydride-based nanovaccine against paratuberculosis infection. *npj Vaccines* **2020**, *5*, 15. [CrossRef] [PubMed]
100. Etewa, S.E.; El-Maaty, D.A.A.; Hamza, R.S.; Metwaly, A.S.; Sarhan, M.H.; Abdel-Rahman, S.A.; Fathy, G.M.; El-Shafey, M.A. Assessment of spiramycin-loaded chitosan nanoparticles treatment on acute and chronic toxoplasmosis in mice. *J. Parasit. Dis.* **2018**, *42*, 102–113. [CrossRef]
101. Zhang, J.; Sun, H.; Gao, C.; Wang, Y.; Cheng, X.; Yang, Y.; Gou, Q.; Lei, L.; Chen, Y.; Wang, X. Development of a chitosan--modified PLGA nanoparticle vaccine for protection against *Escherichia coli* K1 caused meningitis in mice. *J. Nanobiotechnol.* **2021**, *19*, 1–15. [CrossRef] [PubMed]
102. Tao, J.; Zhang, Y.; Shen, A.; Yang, Y.; Diao, L.; Wang, L.; Cai, D.; Hu, Y. Injectable chitosan-based thermosensitive hydrogel/nanoparticle-loaded system for local delivery of vancomycin in the treatment of osteomyelitis. *Int. J. Nanomed.* **2020**, *15*, 5855–5871. [CrossRef] [PubMed]
103. Kitiyodom, S.; Trullàs, C.; Rodkhum, C.; Thompson, K.D.; Katagiri, T.; Temisak, S.; Namdee, K.; Yata, T.; Pirarat, N. Modulation of the mucosal immune response of red tilapia (*Oreochromis* sp.) against columnaris disease using a biomimetic-mucoadhesive nanovaccine. *Fish Shellfish. Immunol.* **2021**, *112*, 81–91. [CrossRef]
104. El-Sissi, A.F.; Mohamed, F.H.; Danial, N.M.; Gaballah, A.Q.; Ali, K.A. Chitosan and chitosan nanoparticles as adjuvant in local Rift Valley Fever inactivated vaccine. *3 Biotech* **2020**, *10*, 88. [CrossRef]
105. Nevagi, R.J.; Khalil, Z.G.; Hussein, W.M.; Powell, J.; Batzloff, M.R.; Capon, R.J.; Good, M.F.; Skwarczynski, M.; Toth, I. Polyglutamic acid-trimethyl chitosan-based intranasal peptide nano-vaccine induces potent immune responses against group A streptococcus. *Acta Biomater.* **2018**, *80*, 278–287. [CrossRef] [PubMed]

106. Zhao, K.; Chen, G.; Shi, X.-m.; Gao, T.-t.; Li, W.; Zhao, Y.; Zhang, F.-q.; Wu, J.; Cui, X.; Wang, Y.-F. Preparation and efficacy of a live newcastle disease virus vaccine encapsulated in chitosan nanoparticles. *PLoS ONE* **2012**, *7*, e53314. [CrossRef] [PubMed]
107. Das, I.; Padhi, A.; Mukherjee, S.; Dash, D.P.; Kar, S.; Sonawane, A. Biocompatible chitosan nanoparticles as an efficient delivery vehicle for *Mycobacterium tuberculosis* lipids to induce potent cytokines and antibody response through activation of $\gamma\delta$ T cells in mice. *Nanotechnology* **2017**, *28*, 165101. [CrossRef]
108. Mohammed, G.M.; ElZorkany, H.E.; Farroh, K.Y.; Abd El-Aziz, W.R.; Elshoky, H.A. Potential improvement of the immune response of chickens against *E. coli* vaccine by using two forms of chitosan nanoparticles. *Int. J. Biol. Macromol.* **2021**, *167*, 395–404. [CrossRef] [PubMed]
109. Renu, S.; Renukaradhya, G.J. Chitosan nanoparticle based mucosal vaccines delivered against infectious diseases of poultry and pigs. *Front. Bioeng. Biotechnol.* **2020**, *8*, 558349. [CrossRef] [PubMed]
110. Acevedo-Villanueva, K.; Renu, S.; Gourapura, R.; Selvaraj, R. Efficacy of a nanoparticle vaccine administered in-ovo against *Salmonella* in broilers. *PLoS ONE* **2021**, *16*, e0247938. [CrossRef]
111. Kelly, S.M.; Larsen, K.R.; Darling, R.; Petersen, A.C.; Bellaire, B.H.; Wannemuehler, M.J.; Narasimhan, B. Single-dose combination nanovaccine induces both rapid and durable humoral immunity and toxin neutralizing antibody responses against *Bacillus anthracis*. *Vaccine* **2021**, *39*, 3862–3870. [CrossRef] [PubMed]
112. Chowdhury, I.H.; Lokugamage, N.; Garg, N.J. Experimental nanovaccine offers protection against repeat exposures *to Trypanosoma cruzi* through activation of polyfunctional T cell response. *Front. Immunol.* **2020**, *11*, 595039. [CrossRef]
113. Mody, K.T.; Zhang, B.; Li, X.; Fletcher, N.L.; Akhter, D.T.; Jarrett, S.; Zhang, J.; Yu, C.; Thurecht, K.J.; Mahony, T.J. Characterization of the biodistribution of a silica vesicle nanovaccine carrying a *Rhipicephalus* (*Boophilus*) microplus protective antigen with in vivo live animal imaging. *Front. Bioeng. Biotechnol.* **2021**, *8*, 606652. [CrossRef] [PubMed]
114. Ghaffari, H.; Tavakoli, A.; Moradi, A.; Tabarraei, A.; Bokharaei-Salim, F.; Zahmatkeshan, M.; Farahmand, M.; Javanmard, D.; Kiani, S.J.; Esghaei, M. Inhibition of H1N1 influenza virus infection by zinc oxide nanoparticles: Another emerging application of nanomedicine. *J. Biomed. Sci.* **2019**, *26*, 1–10. [CrossRef] [PubMed]
115. Rajeshkumar, S.; Bharath, L. Controlling of food borne pathogens by nanoparticles. *Bioorganic Phase Nat. Food Overv.* **2018**, *15*, 293–322.
116. Vinayamohan, P.G.; Joseph, D.; Viju, L.S.; Venkitanarayanan, K. Efficacy of selenium for controlling infectious diseases. In *Selenium and Human Health*; IntechOpen: Rijeka, Croatia, 2023.
117. Maleki, M.; Salouti, M.; Shafiee Ardestani, M.; Talebzadeh, A. Preparation of a nanovaccine against Brucella melitensis M16 based on PLGA nanoparticles and oligopolysaccharide antigen. *Artif. Cells Nanomed. Biotechnol.* **2019**, *47*, 4248–4256. [CrossRef]
118. Lee, J.-A.; Jung, B.-G.; Kim, T.-H.; Kim, Y.-M.; Park, M.-H.; Hyun, P.-m.; Jeon, J.-w.; Park, J.-k.; Cho, C.-W.; Suh, G.-H. Poly d, l-lactide-co-glycolide (PLGA) nanoparticle-encapsulated honeybee (*Apis melifera*) venom promotes clearance of *Salmonella enterica serovar Typhimurium* infection in experimentally challenged pigs through the up-regulation of T helper type 1 specific immune responses. *Vet. Immunol. Immunopathol.* **2014**, *161*, 193–204. [PubMed]
119. Demento, S.L.; Cui, W.; Criscione, J.M.; Stern, E.; Tulipan, J.; Kaech, S.M.; Fahmy, T.M. Role of sustained antigen release from nanoparticle vaccines in shaping the T cell memory phenotype. *Biomaterials* **2012**, *33*, 4957–4964. [CrossRef] [PubMed]
120. Tan, Z.; Liu, W.; Liu, H.; Li, C.; Zhang, Y.; Meng, X.; Tang, T.; Xi, T.; Xing, Y. Oral Helicobacter pylori vaccine-encapsulated acid-resistant HP55/PLGA nanoparticles promote immune protection. *Eur. J. Pharm. Biopharm.* **2017**, *111*, 33–43. [CrossRef]
121. Toti, U.S.; Guru, B.R.; Hali, M.; McPharlin, C.M.; Wykes, S.M.; Panyam, J.; Whittum-Hudson, J.A. Targeted delivery of antibiotics to intracellular chlamydial infections using PLGA nanoparticles. *Biomaterials* **2011**, *32*, 6606–6613. [CrossRef] [PubMed]
122. Ramteke, S.; Ganesh, N.; Bhattacharya, S.; Jain, N.K. Amoxicillin, clarithromycin, and omeprazole based targeted nanoparticles for the treatment of *H. pylori*. *J. Drug Target.* **2009**, *17*, 225–234. [CrossRef] [PubMed]
123. Ardestani, M.S.; Fordoei, A.S.; Abdoli, A.; Ahangari Cohan, R.; Bahramali, G.; Sadat, S.M.; Siadat, S.D.; Moloudian, H.; Nassiri Koopaei, N.; Bolhasani, A. Nanosilver based anionic linear globular dendrimer with a special significant antiretroviral activity. *J. Mater. Sci. Mater. Med.* **2015**, *26*, 1–8. [CrossRef]
124. Park, S.; Park, H.H.; Kim, S.Y.; Kim, S.J.; Woo, K.; Ko, G. Antiviral properties of silver nanoparticles on a magnetic hybrid colloid. *Appl. Environ. Microbiol.* **2014**, *80*, 2343–2350. [CrossRef] [PubMed]
125. Huy, T.Q.; Thanh, N.T.H.; Thuy, N.T.; Van Chung, P.; Hung, P.N.; Le, A.-T.; Hanh, N.T.H. Cytotoxicity and antiviral activity of electrochemical–synthesized silver nanoparticles against poliovirus. *J. Virol. Methods* **2017**, *241*, 52–57. [CrossRef]
126. Chen, N.; Zheng, Y.; Yin, J.; Li, X.; Zheng, C. Inhibitory effects of silver nanoparticles against adenovirus type 3 in vitro. *J. Virol. Methods* **2013**, *193*, 470–477. [CrossRef] [PubMed]
127. Sanchez-Villamil, J.I.; Tapia, D.; Torres, A.G. Development of a gold nanoparticle vaccine against enterohemorrhagic *Escherichia coli* O157: H7. *mBio* **2019**, *10*. [CrossRef] [PubMed]
128. Chen, Y.-N.; Hsueh, Y.-H.; Hsieh, C.-T.; Tzou, D.-Y.; Chang, P.-L. Antiviral activity of graphene–silver nanocomposites against non-enveloped and enveloped viruses. *Int. J. Environ. Res. Public Health* **2016**, *13*, 430. [CrossRef] [PubMed]

129. Chen, P.; Wu, Z.; Leung, A.; Chen, X.; Landao-Bassonga, E.; Gao, J.; Chen, L.; Zheng, M.; Yao, F.; Yang, H. Fabrication of a silver nanoparticle-coated collagen membrane with anti-bacterial and anti-inflammatory activities for guided bone regeneration. *Biomed. Mater.* **2018**, *13*, 065014. [CrossRef] [PubMed]
130. Baram-Pinto, D.; Shukla, S.; Perkas, N.; Gedanken, A.; Sarid, R. Inhibition of herpes simplex virus type 1 infection by silver nanoparticles capped with mercaptoethane sulfonate. *Bioconjugate Chem.* **2009**, *20*, 1497–1502. [CrossRef]
131. Tiwari, V.; Tiwari, M.; Solanki, V. Polyvinylpyrrolidone-capped silver nanoparticle inhibits infection of carbapenem-resistant strain of *Acinetobacter baumannii* in the human pulmonary epithelial cell. *Front. Immunol.* **2017**, *8*, 973. [CrossRef]
132. Borrego, B.; Lorenzo, G.; Mota-Morales, J.D.; Almanza-Reyes, H.; Mateos, F.; López-Gil, E.; de la Losa, N.; Burmistrov, V.A.; Pestryakov, A.N.; Brun, A. Potential application of silver nanoparticles to control the infectivity of Rift Valley fever virus in vitro and in vivo. *Nanomed. Nanotechnol. Biol. Med.* **2016**, *12*, 1185–1192. [CrossRef] [PubMed]
133. Yang, X.X.; Li, C.M.; Huang, C.Z. Curcumin modified silver nanoparticles for highly efficient inhibition of respiratory syncytial virus infection. *Nanoscale* **2016**, *8*, 3040–3048. [CrossRef]
134. Pangestika, R.; Ernawati, R. Antiviral activity effect of silver nanoparticles (Agnps) solution against the growth of infectious bursal disease virus on embryonated chicken eggs with Elisa test. *KnE Life Sci.* **2017**, *3*, 536–548. [CrossRef]
135. Kuppan, G.; Sangeeta, K. Dose and Size-Dependent Antiviral Effects of Silver Nanoparticles on Feline Calicivirus, a Human Norovirus Surrogate. *Foodborne Pathog Dis.* **2016**, *13*, 239–244.
136. Xiang, D.; Zheng, Y.; Duan, W.; Li, X.; Yin, J.; Shigdar, S.; O'Connor, M.L.; Marappan, M.; Zhao, X.; Miao, Y. Inhibition of A/Human/Hubei/3/2005 (H3N2) influenza virus infection by silver nanoparticles in vitro and in vivo. *Int. J. Nanomed.* **2013**, *8*, 4103–4114. [CrossRef]
137. Folliero, V.; Zannella, C.; Chianese, A.; Stelitano, D.; Ambrosino, A.; De Filippis, A.; Galdiero, M.; Franci, G.; Galdiero, M. Application of dendrimers for treating parasitic diseases. *Pharmaceutics* **2021**, *13*, 343. [CrossRef] [PubMed]
138. El Bissati, K.; Zhou, Y.; Paulillo, S.M.; Raman, S.K.; Karch, C.P.; Roberts, C.W.; Lanar, D.E.; Reed, S.; Fox, C.; Carter, D. Protein nanovaccine confers robust immunity against Toxoplasma. *npj Vaccines* **2017**, *2*, 24. [CrossRef]
139. Kwon, E.J.; Skalak, M.; Bertucci, A.; Braun, G.; Ricci, F.; Ruoslahti, E.; Sailor, M.J.; Bhatia, S.N. Porous silicon nanoparticle delivery of tandem peptide anti-infectives for the treatment of Pseudomonas aeruginosa lung infections. *Adv. Mater.* **2017**, *29*, 1701527. [CrossRef]
140. Gowri, M.; Latha, N.; Suganya, K.; Murugan, M.; Rajan, M. Calcium alginate nanoparticle crosslinked phosphorylated polyallylamine to the controlled release of clindamycin for osteomyelitis treatment. *Drug Dev. Ind. Pharm.* **2021**, *47*, 280–291. [CrossRef]
141. Wang, L.; Xing, D.; Le Van, A.; Jerse, A.E.; Wang, S. Structure-based design of ferritin nanoparticle immunogens displaying antigenic loops of *Neisseria gonorrhoeae*. *FEBS Open Bio* **2017**, *7*, 1196–1207. [CrossRef] [PubMed]
142. Lemke, A.; Kiderlen, A.F.; Petri, B.; Kayser, O. Delivery of amphotericin B nanosuspensions to the brain and determination of activity against *Balamuthia mandrillaris amebas*. *Nanomed. Nanotechnol. Biol. Med.* **2010**, *6*, 597–603. [CrossRef]
143. Zhong, J.; Xia, Y.; Hua, L.; Liu, X.; Xiao, M.; Xu, T.; Zhu, B.; Cao, H. Functionalized selenium nanoparticles enhance the anti-EV71 activity of oseltamivir in human astrocytoma cell model. *Artif. Cells Nanomed. Biotechnol.* **2019**, *47*, 3485–3491. [CrossRef]
144. Thi, E.P.; Mire, C.E.; Lee, A.C.; Geisbert, J.B.; Zhou, J.Z.; Agans, K.N.; Snead, N.M.; Deer, D.J.; Barnard, T.R.; Fenton, K.A. Lipid nanoparticle siRNA treatment of Ebola-virus-Makona-infected nonhuman primates. *Nature* **2015**, *521*, 362–365. [CrossRef]
145. Lauster, D.; Glanz, M.; Bardua, M.; Ludwig, K.; Hellmund, M.; Hoffmann, U.; Hamann, A.; Böttcher, C.; Haag, R.; Hackenberger, C.P. Multivalent peptide–nanoparticle conjugates for influenza-virus inhibition. *Angew. Chem. Int. Ed.* **2017**, *56*, 5931–5936. [CrossRef]
146. Rodrigues-Jesus, M.; Fotoran, W.L.; Cardoso, R.M.; Araki, K.; Wunderlich, G.; Ferreira, L.C. Nano-multilamellar lipid vesicles (NMVs) enhance protective antibody responses against Shiga toxin (Stx2a) produced by enterohemorrhagic *Escherichia coli* strains (EHEC). *Braz. J. Microbiol.* **2019**, *50*, 67–77. [CrossRef] [PubMed]
147. Zhang, B.; Cavallaro, A.S.; Mody, K.T.; Zhang, J.; Deringer, J.R.; Brown, W.C.; Mahony, T.J.; Yu, C.; Mitter, N. Nanoparticle-based delivery of *Anaplasma marginale* membrane proteins; virb9-1 and virb10 produced in the *Pichia pastoris* expression system. *Nanomaterials* **2016**, *6*, 201. [CrossRef]
148. Rungrojcharoenkit, K.; Sunintaboon, P.; Ellison, D.; Macareo, L.; Midoeng, P.; Chaisuwirat, P.; Fernandez, S.; Ubol, S. Development of an adjuvanted nanoparticle vaccine against influenza virus, an in vitro study. *PLoS ONE* **2020**, *15*, e0237218. [CrossRef] [PubMed]
149. Riehemann, K.; Schneider, S.W.; Luger, T.A.; Godin, B.; Ferrari, M.; Fuchs, H. Nanomedicine—Challenge and perspectives. *Angew. Chem. Int. Ed.* **2009**, *48*, 872–897. [CrossRef] [PubMed]
150. Lung, P.; Yang, J.; Li, Q. Nanoparticle formulated vaccines: Opportunities and challenges. *Nanoscale* **2020**, *12*, 5746–5763. [CrossRef]
151. Verma, M.; Ozer, I.; Xie, W.; Gallagher, R.; Teixeira, A.; Choy, M. The landscape for lipid-nanoparticle-based genomic medicines. *Nat. Rev. Drug Discov.* **2023**, *22*, 349–350. [CrossRef] [PubMed]

152. Uddin, M.N.; Roni, M.A. Challenges of storage and stability of mRNA-based COVID-19 vaccines. *Vaccines* **2021**, *9*, 1033. [CrossRef] [PubMed]
153. Pal, K.; Chakroborty, S.; Nath, N. Limitations of nanomaterials insights in green chemistry sustainable route: Review on novel applications. *Green Process. Synth.* **2022**, *11*, 951–964. [CrossRef]
154. Chen, S.; Huang, X.; Xue, Y.; Álvarez-Benedicto, E.; Shi, Y.; Chen, W.; Koo, S.; Siegwart, D.J.; Dong, Y.; Tao, W. Nanotechnology-based mRNA vaccines. *Nat. Rev. Methods Primers* **2023**, *3*, 63. [CrossRef]
155. Look, M.; Bandyopadhyay, A.; Blum, J.S.; Fahmy, T.M. Application of nanotechnologies for improved immune response against infectious diseases in the developing world. *Adv. Drug Deliv. Rev.* **2010**, *62*, 378–393. [CrossRef]
156. Khan, I.; Saeed, K.; Khan, I. Nanoparticles: Properties, applications and toxicities. *Arab. J. Chem.* **2019**, *12*, 908–931. [CrossRef]
157. Torchilin, V.P. Recent advances with liposomes as pharmaceutical carriers. *Nat. Rev. Drug Discov.* **2005**, *4*, 145–160. [CrossRef] [PubMed]
158. Helmy, Y.A.; Fawzy, M.; Elaswad, A.; Sobieh, A.; Kenney, S.P.; Shehata, A.A. The COVID-19 Pandemic: A Comprehensive Review of Taxonomy, Genetics, Epidemiology, Diagnosis, Treatment, and Control. *J. Clin. Med.* **2020**, *9*, 1225. [CrossRef]
159. Heath, P.T.; Galiza, E.P.; Baxter, D.N.; Boffito, M.; Browne, D.; Burns, F.; Chadwick, D.R.; Clark, R.; Cosgrove, C.; Galloway, J. Safety and efficacy of NVX-CoV2373 COVID-19 vaccine. *N. Engl. J. Med.* **2021**, *385*, 1172–1183. [CrossRef]
160. Wang, F.; Porter, M.; Konstantopoulos, A.; Zhang, P.; Cui, H. Preclinical development of drug delivery systems for paclitaxel-based cancer chemotherapy. *J. Control. Release* **2017**, *267*, 100–118. [CrossRef]
161. Smith, J.F.; Brownlow, M.; Brown, M.; Kowalski, R.; Esser, M.T.; Ruiz, W.; Barr, E.; Brown, D.R.; Bryan, J.T. Antibodies from women immunized with Gardasil® cross-neutralize HPV 45 pseudovirions. *Hum. Vaccines* **2007**, *3*, 109–115. [CrossRef]
162. Passero Jr, F.C.; Grapsa, D.; Syrigos, K.N.; Saif, M.W. The safety and efficacy of Onivyde (irinotecan liposome injection) for the treatment of metastatic pancreatic cancer following gemcitabine-based therapy. *Expert Rev. Anticancer. Ther.* **2016**, *16*, 697–703. [CrossRef] [PubMed]
163. Kantoff, P.W.; Higano, C.S.; Shore, N.D.; Berger, E.R.; Small, E.J.; Penson, D.F.; Redfern, C.H.; Ferrari, A.C.; Dreicer, R.; Sims, R.B. Sipuleucel-T immunotherapy for castration-resistant prostate cancer. *N. Engl. J. Med.* **2010**, *363*, 411–422. [CrossRef]
164. Nordquist, L.T.; Shore, N.D.; Elist, J.J.; Oliver, J.C.; Gannon, W.; Shahlaee, A.H.; Fuller, S.A.; Ghanbari, H.A. Phase 1 open-label trial to evaluate the safety and immunogenicity of PAN-301-1, a novel nanoparticle therapeutic vaccine, in patients with biochemically relapsed prostate cancer. *J. Clin. Oncol.* **2018**, *29* (Suppl. S15), e15166. [CrossRef]
165. Lee, D.-H.; Choi, S.; Park, Y.; Jin, H.-s. Mucin1 and Mucin16: Therapeutic targets for cancer therapy. *Pharmaceuticals* **2021**, *14*, 1053. [CrossRef] [PubMed]
166. Wicki, A.; Witzigmann, D.; Balasubramanian, V.; Huwyler, J. Nanomedicine in cancer therapy: Challenges, opportunities, and clinical applications. *J. Control. Release* **2015**, *200*, 138–157. [CrossRef]
167. Rodríguez, F.; Caruana, P.; De la Fuente, N.; Español, P.; Gámez, M.; Balart, J.; Llurba, E.; Rovira, R.; Ruiz, R.; Martín-Lorente, C. Nano-based approved pharmaceuticals for cancer treatment: Present and future challenges. *Biomolecules* **2022**, *12*, 784. [CrossRef] [PubMed]
168. Cech, P.G.; Aebi, T.; Abdallah, M.S.; Mpina, M.; Machunda, E.B.; Westerfeld, N.; Stoffel, S.A.; Zurbriggen, R.; Pluschke, G.; Tanner, M. Virosome-formulated *Plasmodium falciparum* AMA-1 & CSP derived peptides as malaria vaccine: Randomized phase 1b trial in semi-immune adults & children. *PLoS ONE* **2011**, *6*, e22273.
169. Porras, C.; Sampson, J.N.; Herrero, R.; Gail, M.H.; Cortés, B.; Hildesheim, A.; Cyr, J.; Romero, B.; Schiller, J.T.; Montero, C. Rationale and design of a double-blind randomized non-inferiority clinical trial to evaluate one or two doses of vaccine against human papillomavirus including an epidemiologic survey to estimate vaccine efficacy: The Costa Rica ESCUDDO trial. *Vaccine* **2022**, *40*, 76–88. [CrossRef] [PubMed]
170. Topalidou, X.; Kalergis, A.M.; Papazisis, G. Respiratory syncytial virus vaccines: A review of the candidates and the approved vaccines. *Pathogens* **2023**, *12*, 1259. [CrossRef] [PubMed]
171. Lamb, Y.N. BNT162b2 mRNA COVID-19 vaccine: First approval. *Drugs* **2021**, *81*, 495–501. [CrossRef] [PubMed]
172. Bhangde, S.; Lodaya, R.N.; Amiji, M.M. Nanoscale Vaccines for Influenza. In *Nanomedicines for the Prevention and Treatment of Infectious Diseases*; Springer: Berlin/Heidelberg, Germany, 2023; pp. 331–347.
173. Low, J.G.; De Alwis, R.; Chen, S.; Kalimuddin, S.; Leong, Y.S.; Mah, T.K.L.; Yuen, N.; Tan, H.C.; Zhang, S.L.; Sim, J.X. A phase I/II randomized, double-blinded, placebo-controlled trial of a self-amplifying COVID-19 mRNA vaccine. *npj Vaccines* **2022**, *7*, 161. [CrossRef]
174. Al, L.I.R.; Sönmezer, M.Ç.; Ünal, S. RNA-Based COVID-19 vaccine candidates with clinical phase trials in progress. *Turk. J. Med. Sci.* **2021**, *51*, 3246–3252.
175. Kalnin, K.V.; Plitnik, T.; Kishko, M.; Zhang, J.; Zhang, D.; Beauvais, A.; Anosova, N.G.; Tibbitts, T.; DiNapoli, J.; Ulinski, G. Immunogenicity and efficacy of mRNA COVID-19 vaccine MRT5500 in preclinical animal models. *NPJ Vaccines* **2021**, *6*, 61. [CrossRef]

176. Prompetchara, E.; Ketloy, C.; Alameh, M.-G.; Tharakhet, K.; Kaewpang, P.; Yostrerat, N.; Pitakpolrat, P.; Buranapraditkun, S.; Manopwisedjaroen, S.; Thitithanyanont, A. Immunogenicity and protective efficacy of SARS-CoV-2 mRNA vaccine encoding secreted non-stabilized spike in female mice. *Nat. Commun.* **2023**, *14*, 2309. [CrossRef]
177. World Health Organization. Emergency use designation of COVID-19 candidate vaccines: Ethical considerations for current and future COVID-19 placebo-controlled vaccine trials and trial unblinding: Policy brief, 18 December 2020. In *Emergency Use Designation of COVID-19 Candidate Vaccines: Ethical Considerations for Current and Future COVID-19 Placebo-Controlled Vaccine Trials and Trial Unblinding: Policy Brief, 18 December 2020*; WHO: Geneva, Switzerland, 2020.
178. Shin, M.D.; Shukla, S.; Chung, Y.H.; Beiss, V.; Chan, S.K.; Ortega-Rivera, O.A.; Wirth, D.M.; Chen, A.; Sack, M.; Pokorski, J.K. COVID-19 vaccine development and a potential nanomaterial path forward. *Nat. Nanotechnol.* **2020**, *15*, 646–655. [CrossRef]
179. Silva-Pilipich, N.; Beloki, U.; Salaberry, L.; Smerdou, C. Self-Amplifying RNA: A Second Revolution of mRNA Vaccines against COVID-19. *Vaccines* **2024**, *12*, 318. [CrossRef]
180. Rauch, S.; Roth, N.; Schwendt, K.; Fotin-Mleczek, M.; Mueller, S.O.; Petsch, B. mRNA-based SARS-CoV-2 vaccine candidate CVnCoV induces high levels of virus-neutralising antibodies and mediates protection in rodents. *npj Vaccines* **2021**, *6*, 57. [CrossRef] [PubMed]
181. Moon, J.J.; Suh, H.; Bershteyn, A.; Stephan, M.T.; Liu, H.; Huang, B.; Sohail, M.; Luo, S.; Ho Um, S.; Khant, H. Interbilayer-crosslinked multilamellar vesicles as synthetic vaccines for potent humoral and cellular immune responses. *Nat. Mater.* **2011**, *10*, 243–251. [CrossRef]
182. Moon, J.J.; Suh, H.; Li, A.V.; Ockenhouse, C.F.; Yadava, A.; Irvine, D.J. Enhancing humoral responses to a malaria antigen with nanoparticle vaccines that expand Tfh cells and promote germinal center induction. *Proc. Natl. Acad. Sci. USA* **2012**, *109*, 1080–1085. [CrossRef] [PubMed]
183. Vickers, K.C.; Palmisano, B.T.; Shoucri, B.M.; Shamburek, R.D.; Remaley, A.T. MicroRNAs are transported in plasma and delivered to recipient cells by high-density lipoproteins. *Nat. Cell Biol.* **2011**, *13*, 423–433. [CrossRef] [PubMed]
184. Moon, J.J.; Huang, B.; Irvine, D.J. Engineering nano- and microparticles to tune immunity. *Adv. Mater.* **2012**, *24*, 3724–3746. [CrossRef]
185. Chong, P.; Huang, J.-H.; Leng, C.-H.; Liu, S.-J.; Chen, H.-W. Recombinant lipoproteins as novel vaccines with intrinsic adjuvant. *Adv. Protein Chem. Struct. Biol.* **2015**, *99*, 55–74.
186. Waheed, I.; Ali, A.; Tabassum, H.; Khatoon, N.; Lai, W.-F.; Zhou, X. Lipid-based nanoparticles as drug delivery carriers for cancer therapy. *Front. Oncol.* **2024**, *14*, 1296091. [CrossRef]
187. John, R.; Monpara, J.; Swaminathan, S.; Kalhapure, R. Chemistry and art of developing lipid nanoparticles for biologics delivery: Focus on development and scale-up. *Pharmaceutics* **2024**, *16*, 131. [CrossRef] [PubMed]
188. Ndeupen, S.; Qin, Z.; Jacobsen, S.; Bouteau, A.; Estanbouli, H.; Igyártó, B.Z. The mRNA-LNP platform's lipid nanoparticle component used in preclinical vaccine studies is highly inflammatory. *iScience* **2021**, *24*, 103479. [CrossRef]
189. Hou, X.; Zaks, T.; Langer, R.; Dong, Y. Lipid nanoparticles for mRNA delivery. *Nat. Rev. Mater.* **2021**, *6*, 1078–1094. [CrossRef] [PubMed]
190. Xu, F.; Yuan, Y.; Wang, Y.; Yin, Q. Emerging peptide-based nanovaccines: From design synthesis to defense against cancer and infection. *Biomed. Pharmacother.* **2023**, *158*, 114117. [CrossRef]
191. Abdul Ghaffar, K.; Kumar Giddam, A.; Zaman, M.; Skwarczynski, M.; Toth, I. Liposomes as nanovaccine delivery systems. *Curr. Top. Med. Chem.* **2014**, *14*, 1194–1208. [CrossRef]
192. Tretiakova, D.; Vodovozova, E. Liposomes as adjuvants and vaccine delivery systems. *Biochem. (Mosc.) Suppl. Ser. A Membr. Cell Biol.* **2022**, *16*, 1–20. [CrossRef] [PubMed]
193. Allison, A.; Gregoriadis, G. Liposomes as immunological adjuvants. *Nature* **1974**, *252*, 252. [CrossRef] [PubMed]
194. Yanasarn, N.; Sloat, B.R.; Cui, Z. Negatively charged liposomes show potent adjuvant activity when simply admixed with protein antigens. *Mol. Pharm.* **2011**, *8*, 1174–1185. [CrossRef] [PubMed]
195. Hatakeyama, H.; Akita, H.; Harashima, H. The polyethyleneglycol dilemma: Advantage and disadvantage of PEGylation of liposomes for systemic genes and nucleic acids delivery to tumors. *Biol. Pharm. Bull.* **2013**, *36*, 892–899. [CrossRef]
196. Shetye, L.; Sherlekar, A.; Mendhulkar, V. Liposome-Based Drug Delivery—A New Therapeutic Paradigm. In *Advanced Drug Delivery: Methods and Applications*; Springer: Berlin/Heidelberg, Germany, 2023; pp. 21–48.
197. Gordillo-Galeano, A.; Mora-Huertas, C.E. Solid lipid nanoparticles and nanostructured lipid carriers: A review emphasizing on particle structure and drug release. *Eur. J. Pharm. Biopharm.* **2018**, *133*, 285–308. [CrossRef] [PubMed]
198. Pensado, A.; Seijo, B.; Sanchez, A. Current strategies for DNA therapy based on lipid nanocarriers. *Expert Opin. Drug Deliv.* **2014**, *11*, 1721–1731. [CrossRef]
199. Peres, L.B.; Peres, L.B.; de Araújo, P.H.H.; Sayer, C. Solid lipid nanoparticles for encapsulation of hydrophilic drugs by an organic solvent free double emulsion technique. *Colloids Surf. B Biointerfaces* **2016**, *140*, 317–323. [CrossRef] [PubMed]
200. Pandey, S.; Shaikh, F.; Gupta, A.; Tripathi, P.; Yadav, J.S. A recent update: Solid lipid nanoparticles for effective drug delivery. *Adv. Pharm. Bull.* **2022**, *12*, 17. [CrossRef]

201. Ghasemiyeh, P.; Mohammadi-Samani, S. Solid lipid nanoparticles and nanostructured lipid carriers as novel drug delivery systems: Applications, advantages and disadvantages. *Res. Pharm. Sci.* **2018**, *13*, 288–303. [PubMed]
202. Santos, P.; Almeida, F. Exosome-based vaccines: History, current state, and clinical trials. *Front. Immunol.* **2021**, *12*, 711565. [CrossRef] [PubMed]
203. Hazrati, A.; Soudi, S.; Malekpour, K.; Mahmoudi, M.; Rahimi, A.; Hashemi, S.M.; Varma, R.S. Immune cells-derived exosomes function as a double-edged sword: Role in disease progression and their therapeutic applications. *Biomark. Res.* **2022**, *10*, 30. [CrossRef]
204. Dai, J.; Su, Y.; Zhong, S.; Cong, L.; Liu, B.; Yang, J.; Tao, Y.; He, Z.; Chen, C.; Jiang, Y. Exosomes: Key players in cancer and potential therapeutic strategy. *Signal Transduct. Target. Ther.* **2020**, *5*, 145. [CrossRef]
205. Rezabakhsh, A.; Mahdipour, M.; Nourazarian, A.; Habibollahi, P.; Sokullu, E.; Avci, Ç.B.; Rahbarghazi, R. Application of exosomes for the alleviation of COVID-19-related pathologies. *Cell Biochem. Funct.* **2022**, *40*, 430–438. [CrossRef]
206. Wallis, J.; Shenton, D.; Carlisle, R. Novel approaches for the design, delivery and administration of vaccine technologies. *Clin. Exp. Immunol.* **2019**, *196*, 189–204. [CrossRef] [PubMed]
207. Glück, R.; Metcalfe, I. New technology platforms in the development of vaccines for the future. *Vaccine* **2002**, *20*, B10–B16. [CrossRef] [PubMed]
208. O'Hagan, D.T. MF59 is a safe and potent vaccine adjuvant that enhances protection against influenza virus infection. *Expert Rev. Vaccines* **2007**, *6*, 699–710. [CrossRef]
209. Tang, J.-l.; Sun, J.; He, Z.-G. Self-emulsifying drug delivery systems: Strategy for improving oral delivery of poorly soluble drugs. *Curr. Drug Ther.* **2007**, *2*, 85–93. [CrossRef]
210. Akagi, T.; Baba, M.; Akashi, M. Biodegradable nanoparticles as vaccine adjuvants and delivery systems: Regulation of immune responses by nanoparticle-based vaccine. *Polym. Nanomed.* **2012**, *247*, 31–64.
211. Liu, B.; Wu, Z.; Liu, T.; Qian, R.; Wu, T.; Liu, Q.; Shen, A. Polymeric nanoparticles engineered as a vaccine adjuvant-delivery system. In *Immunization-Vaccine Adjuvant Delivery System and Strategies*; IntechOpen: London, UK, 2018.
212. Danhier, F.; Ansorena, E.; Silva, J.M.; Coco, R.; Le Breton, A.; Préat, V. PLGA-based nanoparticles: An overview of biomedical applications. *J. Control. Release* **2012**, *161*, 505–522. [CrossRef] [PubMed]
213. Reis, C.P.; Neufeld, R.J.; Veiga, F.; Ribeiro, A.J. Preparation of drug-loaded polymeric nanoparticles. In *Nanomedicine in Cancer*; Jenny Stanford Publishing: Oxon, UK, 2017; pp. 171–214.
214. Maji, I.; Mahajan, S.; Sriram, A.; Mehra, N.K.; Singh, P.K. Polymeric nanomaterials: Fundamentals and therapeutic applications. In *Nanomaterial-Based Drug Delivery Systems: Therapeutic and Theranostic Applications*; Springer: Berlin/Heidelberg, Germany, 2023; pp. 33–64.
215. Cano, A.; Ettcheto, M.; Espina, M.; López-Machado, A.; Cajal, Y.; Rabanal, F.; Sánchez-López, E.; Camins, A.; García, M.L.; Souto, E.B. State-of-the-art polymeric nanoparticles as promising therapeutic tools against human bacterial infections. *J. Nanobiotechnology* **2020**, *18*, 1–24. [CrossRef] [PubMed]
216. Jin, Z.; Gao, S.; Cui, X.; Sun, D.; Zhao, K. Adjuvants and delivery systems based on polymeric nanoparticles for mucosal vaccines. *Int. J. Pharm.* **2019**, *572*, 118731. [CrossRef]
217. Elsabahy, M.; Wooley, K.L. Design of polymeric nanoparticles for biomedical delivery applications. *Chem. Soc. Rev.* **2012**, *41*, 2545–2561. [CrossRef] [PubMed]
218. Cohn, D.; Stern, T.; González, M.F.; Epstein, J. Biodegradable poly(ethylene oxide)/poly(ϵ-caprolactone) multiblock copolymers. *J. Biomed. Mater. Res.* **2002**, *59*, 273–281. [CrossRef] [PubMed]
219. Khodaverdi, E.; Tayarani-Najaran, Z.; Minbashi, E.; Alibolandi, M.; Hosseini, J.; Sepahi, S.; Kamali, H.; Hadizadeh, F. Docetaxel-loaded mixed micelles and polymersomes composed of poly (caprolactone)-poly(ethylene glycol)(PEG-PCL) and poly (lactic acid)-poly (ethylene glycol)(PEG-PLA): Preparation and in-vitro characterization. *Iran. J. Pharm. Res. IJPR* **2019**, *18*, 142.
220. Tian, J.-H.; Patel, N.; Haupt, R.; Zhou, H.; Weston, S.; Hammond, H.; Logue, J.; Portnoff, A.D.; Norton, J.; Guebre-Xabier, M. SARS-CoV-2 spike glycoprotein vaccine candidate NVX-CoV2373 immunogenicity in baboons and protection in mice. *Nat. Commun.* **2021**, *12*, 372. [CrossRef] [PubMed]
221. Bangaru, S.; Ozorowski, G.; Turner, H.L.; Antanasijevic, A.; Huang, D.; Wang, X.; Torres, J.L.; Diedrich, J.K.; Tian, J.-H.; Portnoff, A.D. Structural analysis of full-length SARS-CoV-2 spike protein from an advanced vaccine candidate. *Science* **2020**, *370*, 1089–1094. [CrossRef]
222. Bivas-Benita, M.; van Meijgaarden, K.E.; Franken, K.L.; Junginger, H.E.; Borchard, G.; Ottenhoff, T.H.; Geluk, A. Pulmonary delivery of chitosan-DNA nanoparticles enhances the immunogenicity of a DNA vaccine encoding HLA-A* 0201-restricted T-cell epitopes of *Mycobacterium tuberculosis*. *Vaccine* **2004**, *22*, 1609–1615. [CrossRef]
223. Mitchell, M.J.; Billingsley, M.M.; Haley, R.M.; Wechsler, M.E.; Peppas, N.A.; Langer, R. Engineering precision nanoparticles for drug delivery. *Nat. Rev. Drug Discov.* **2021**, *20*, 101–124. [CrossRef] [PubMed]
224. Rezigue, M. *Lipid and Polymeric Nanoparticles: Drug Delivery Applications*; Springer: Cham, Switzerland, 2020; pp. 167–230.

225. Chenthamara, D.; Subramaniam, S.; Ramakrishnan, S.G.; Krishnaswamy, S.; Essa, M.M.; Lin, F.-H.; Qoronfleh, M.W. Therapeutic efficacy of nanoparticles and routes of administration. *Biomater. Res.* **2019**, *23*, 20. [CrossRef]
226. Elzoghby, A.; M Abd-Elwakil, M.; Abd-Elsalam, K.; T Elsayed, M.; Hashem, Y.; Mohamed, O. Natural polymeric nanoparticles for brain-targeting: Implications on drug and gene delivery. *Curr. Pharm. Des.* **2016**, *22*, 3305–3323. [CrossRef] [PubMed]
227. Sun, Z.; Zhao, H.; Ma, L.; Shi, Y.; Ji, M.; Sun, X.; Ma, D.; Zhou, W.; Huang, T.; Zhang, D. The quest for nanoparticle-powered vaccines in cancer immunotherapy. *J. Nanobiotechnol.* **2024**, *22*, 61. [CrossRef] [PubMed]
228. Dunkle, L.M.; Kotloff, K.L.; Gay, C.L.; Áñez, G.; Adelglass, J.M.; Barrat Hernández, A.Q.; Harper, W.L.; Duncanson, D.M.; McArthur, M.A.; Florescu, D.F. Efficacy and safety of NVX-CoV2373 in adults in the United States and Mexico. *N. Engl. J. Med.* **2022**, *386*, 531–543. [CrossRef]
229. Vasudevan, S.S.; Kandrikar, T.Y.; Sayyed, A.A.; Sarker, P.; Nasir, N.S.; Venugopalan, S.; Mariajohn, R.; Chavda, V.P.; Gondaliya, P. Nanoparticle-based vaccines and future vaccine technologies. In *Advanced Vaccination Technologies for Infectious and Chronic Diseases*; Elsevier: Amsterdam, The Netherlands, 2024; pp. 477–495.
230. Cable, J.; Graham, B.S.; Koup, R.A.; Seder, R.A.; Karikó, K.; Pardi, N.; Barouch, D.H.; Sharma, B.; Rauch, S.; Nachbagauer, R. Progress in vaccine development for infectious diseases—A Keystone Symposia report. *Ann. N. Y. Acad. Sci.* **2023**, *1524*, 65–86. [CrossRef]
231. Kasturi, S.P.; Skountzou, I.; Albrecht, R.A.; Koutsonanos, D.; Hua, T.; Nakaya, H.I.; Ravindran, R.; Stewart, S.; Alam, M.; Kwissa, M. Programming the magnitude and persistence of antibody responses with innate immunity. *Nature* **2011**, *470*, 543–547. [CrossRef]
232. Hess, K.L.; Medintz, I.L.; Jewell, C.M. Designing inorganic nanomaterials for vaccines and immunotherapies. *Nano Today* **2019**, *27*, 73–98. [CrossRef]
233. Joseph, T.M.; Kar Mahapatra, D.; Esmaeili, A.; Piszczyk, Ł.; Hasanin, M.S.; Kattali, M.; Haponiuk, J.; Thomas, S. Nanoparticles: Taking a unique position in medicine. *Nanomaterials* **2023**, *13*, 574. [CrossRef]
234. Bansal, S.A.; Kumar, V.; Karimi, J.; Singh, A.P.; Kumar, S. Role of gold nanoparticles in advanced biomedical applications. *Nanoscale Adv.* **2020**, *2*, 3764–3787. [CrossRef]
235. Amendola, V.; Pilot, R.; Frasconi, M.; Maragò, O.M.; Iatì, M.A. Surface plasmon resonance in gold nanoparticles: A review. *J. Phys. Condens. Matter* **2017**, *29*, 203002. [CrossRef] [PubMed]
236. Gnanaraj, M.; Sisubalan, N.; Jebastin, T.; Vijayan, A.; Muneeshwaran, T.; Manikandan, R. Gold Nanoparticles as Antibacterial and Antiviral Agents: Biomedical Applications and Theranostic Potential. In *Nanoparticles in Modern Antimicrobial and Antiviral Applications*; Springer: Berlin/Heidelberg, Germany, 2024; pp. 19–45.
237. Chaudhary, K.; Masram, D.T. Biological activities of nanoparticles and mechanism of action. In *Model Organisms to Study Biological Activities and Toxicity of Nanoparticles*; Springer: Berlin/Heidelberg, Germany, 2020; pp. 19–34.
238. Chernykh, I.; Kopanitsa, M.; Shchul'kin, A.; Yakusheva, E.; Frolova, M. Gold nanoparticles as potential antitumor agents. *Pharm. Chem. J.* **2021**, *55*, 934–941. [CrossRef]
239. He, J.-s.; Liu, S.-j.; Zhang, Y.-r.; Chu, X.-d.; Lin, Z.-b.; Zhao, Z.; Qiu, S.-h.; Guo, Y.-g.; Ding, H.; Pan, Y.-l. The application of and strategy for gold nanoparticles in cancer immunotherapy. *Front. Pharmacol.* **2021**, *12*, 687399. [CrossRef] [PubMed]
240. Farfán-Castro, S.; García-Soto, M.J.; Aguilar-Aguilar, A.; González-Ortega, O.; Rosales-Mendoza, S. Application of gold nanoparticles in vaccine development. In *Gold Nanoparticles for Drug Delivery*; Elsevier: Amsterdam, The Netherlands, 2024; pp. 445–493.
241. Xu, L.; Wang, X.; Wang, W.; Sun, M.; Choi, W.J.; Kim, J.-Y.; Hao, C.; Li, S.; Qu, A.; Lu, M. Enantiomer-dependent immunological response to chiral nanoparticles. *Nature* **2022**, *601*, 366–373. [CrossRef]
242. Tapia, D.; Sanchez-Villamil, J.I.; Torres, A.G. Multicomponent gold nano-glycoconjugate as a highly immunogenic and protective platform against *Burkholderia mallei*. *npj Vaccines* **2020**, *5*, 82. [CrossRef]
243. Gao, L.; Song, Y.; Zhong, J.; Lin, X.; Zhou, S.-F.; Zhan, G. Biocompatible 2D Cu-TCPP nanosheets derived from Cu_2O nanocubes as multifunctional nanoplatforms for combined anticancer therapy. *ACS Biomater. Sci. Eng.* **2022**, *8*, 1074–1086. [CrossRef] [PubMed]
244. Sztandera, K.; Gorzkiewicz, M.; Klajnert-Maculewicz, B. Gold nanoparticles in cancer treatment. *Mol. Pharm.* **2018**, *16*, 1–23. [CrossRef]
245. Vines, J.B.; Yoon, J.-H.; Ryu, N.-E.; Lim, D.-J.; Park, H. Gold nanoparticles for photothermal cancer therapy. *Front. Chem.* **2019**, *7*, 167. [CrossRef]
246. Singh, P.; Mijakovic, I. Advances in gold nanoparticle technology as a tool for diagnostics and treatment of cancer. *Expert Rev. Mol. Diagn.* **2021**, *21*, 627–630. [CrossRef]
247. Nandanwar, R.; Singh, P.; Haque, F.Z. Synthesis and properties of silica nanoparticles by sol-gel method for the application in green chemistry. *Mater. Sci. Res. India* **2013**, *10*, 85–92.
248. Bharti, C.; Nagaich, U.; Pal, A.K.; Gulati, N. Mesoporous silica nanoparticles in target drug delivery system: A review. *Int. J. Pharm. Investig.* **2015**, *5*, 124. [CrossRef] [PubMed]
249. Tang, F.; Li, L.; Chen, D. Mesoporous silica nanoparticles: Synthesis, biocompatibility and drug delivery. *Adv. Mater.* **2012**, *24*, 1504–1534. [CrossRef]

250. Scheiblhofer, S.; Machado, Y.; Feinle, A.; Thalhamer, J.; Hüsing, N.; Weiss, R. Potential of nanoparticles for allergen-specific immunotherapy–use of silica nanoparticles as vaccination platform. *Expert Opin. Drug Deliv.* **2016**, *13*, 1777–1788. [CrossRef] [PubMed]
251. Seré, S.; Vounckx, U.; Seo, J.W.; Lenaerts, I.; Van Gool, S.; Locquet, J.-P. Proof of concept study: Mesoporous silica nanoparticles, from synthesis to active specific immunotherapy. *Front. Nanotechnol.* **2020**, *2*, 584233. [CrossRef]
252. Liu, J.Y.; Sayes, C.M. A toxicological profile of silica nanoparticles. *Toxicol. Res.* **2022**, *11*, 565–582. [CrossRef]
253. Pallavi, P.; Harini, K.; Alshehri, S.; Ghoneim, M.M.; Alshlowi, A.; Gowtham, P.; Girigoswami, K.; Shakeel, F.; Girigoswami, A. From synthetic route of silica nanoparticles to theranostic applications. *Processes* **2022**, *10*, 2595. [CrossRef]
254. Xiao, D.; Qi, H.; Teng, Y.; Pierre, D.; Kutoka, P.T.; Liu, D. Advances and challenges of fluorescent nanomaterials for synthesis and biomedical applications. *Nanoscale Res. Lett.* **2021**, *16*, 1–23. [CrossRef]
255. Nooraei, S.; Bahrulolum, H.; Hoseini, Z.S.; Katalani, C.; Hajizade, A.; Easton, A.J.; Ahmadian, G. Virus-like particles: Preparation, immunogenicity and their roles as nanovaccines and drug nanocarriers. *J. Nanobiotechnol.* **2021**, *19*, 1–27. [CrossRef]
256. Liu, X.; Liu, Y.; Yang, X.; Lu, X.; Xu, X.-N.; Zhang, J.; Chen, R. Potentiating the Immune Responses of HBsAg-VLP Vaccine Using a Polyphosphoester-Based Cationic Polymer Adjuvant. *ACS Appl. Mater. Interfaces* **2023**, *15*, 48871–48881. [CrossRef] [PubMed]
257. Gupta, R.; Arora, K.; Roy, S.S.; Joseph, A.; Rastogi, R.; Arora, N.M.; Kundu, P.K. Platforms, advances, and technical challenges in virus-like particles-based vaccines. *Front. Immunol.* **2023**, *14*, 1123805. [CrossRef] [PubMed]
258. Chu, K.-B.; Quan, F.-S. *Virus-Like Particle Vaccines Against Respiratory Viruses and Protozoan Parasites*; Springer: Berlin/Heidelberg, Germany, 2021.
259. Patil, S. A review of virus-like particle-based SARS-CoV-2 vaccines in clinical trial phases. *Iran. J. Pharm. Res. IJPR* **2021**, *13*, 2901–2905.
260. Aboshi, M.; Matsuda, K.; Kawakami, D.; Kono, K.; Kazami, Y.; Sekida, T.; Komori, M.; Morey, A.L.; Suga, S.; Smith, J.F. Safety and immunogenicity of VLPCOV-02, a SARS-CoV-2 self-amplifying RNA vaccine with a modified base, 5-methylcytosine. *iScience* **2024**, *27*, 108964. [CrossRef] [PubMed]
261. Polack, F.P.; Thomas, S.J.; Kitchin, N.; Absalon, J.; Gurtman, A.; Lockhart, S.; Perez, J.L.; Pérez Marc, G.; Moreira, E.D.; Zerbini, C. Safety and efficacy of the BNT162b2 mRNA COVID-19 vaccine. *N. Engl. J. Med.* **2020**, *383*, 2603–2615. [CrossRef]
262. Vandoolaeghe, P.; Schuerman, L. The RTS, S/AS01 malaria vaccine in children 5 to 17 months of age at first vaccination. *Expert Rev. Vaccines* **2016**, *15*, 1481–1493. [CrossRef] [PubMed]
263. Syomin, B.; Ilyin, Y. Virus-like particles as an instrument of vaccine production. *Mol. Biol.* **2019**, *53*, 323–334. [CrossRef]
264. Erdmann, N.B.; Williams, W.B.; Walsh, S.R.; Grunenberg, N.; Edlefsen, P.T.; Goepfert, P.A.; Cain, D.W.; Cohen, K.W.; Maenza, J.; Mayer, K.H. A HIV-1 Gp41 peptide-liposome vaccine elicits neutralizing epitope-targeted antibody responses in healthy individuals. *medRxiv* **2024**. [CrossRef]
265. Zeigler, D.F.; Gage, E.; Roque, R.; Clegg, C.H. Epitope targeting with self-assembled peptide vaccines. *NPJ Vaccines* **2019**, *4*, 30. [CrossRef]
266. Walls, A.C.; Fiala, B.; Schäfer, A.; Wrenn, S.; Pham, M.N.; Murphy, M.; Longping, V.T.; Shehata, L.; O'Connor, M.A.; Chen, C. Elicitation of potent neutralizing antibody responses by designed protein nanoparticle vaccines for SARS-CoV-2. *Cell* **2020**, *183*, 1367–1382.e17. [CrossRef] [PubMed]
267. Yang, J.; Wang, W.; Chen, Z.; Lu, S.; Yang, F.; Bi, Z.; Bao, L.; Mo, F.; Li, X.; Huang, Y. A vaccine targeting the RBD of the S protein of SARS-CoV-2 induces protective immunity. *Nature* **2020**, *586*, 572–577. [CrossRef]
268. López-Sagaseta, J.; Malito, E.; Rappuoli, R.; Bottomley, M.J. Self-assembling protein nanoparticles in the design of vaccines. *Comput. Struct. Biotechnol. J.* **2016**, *14*, 58–68. [CrossRef]
269. Zhao, L.; Seth, A.; Wibowo, N.; Zhao, C.-X.; Mitter, N.; Yu, C.; Middelberg, A.P. Nanoparticle vaccines. *Vaccine* **2014**, *32*, 327–337. [CrossRef] [PubMed]
270. Bagheri-Josheghani, S.; Bakhshi, B.; Najar-peerayeh, S. The influence of nanoparticle on vaccine responses against bacterial infection. *J. Nanotechnol.* **2022**, *2022*, 6856982. [CrossRef]
271. Chang, H.C.; Zou, Z.Z.; Wang, Q.H.; Li, J.; Jin, H.; Yin, Q.X.; Xing, D. Targeting and specific activation of antigen--presenting cells by endogenous antigen-loaded nanoparticles elicits tumor-specific immunity. *Adv. Sci.* **2020**, *7*, 1900069. [CrossRef] [PubMed]
272. Gamucci, O.; Bertero, A.; Gagliardi, M.; Bardi, G. Biomedical nanoparticles: Overview of their surface immune-compatibility. *Coatings* **2014**, *4*, 139–159. [CrossRef]
273. Zaman, M.; Good, M.F.; Toth, I. Nanovaccines and their mode of action. *Methods* **2013**, *60*, 226–231. [CrossRef]
274. Lin, Y.; Chen, X.; Wang, K.; Liang, L.; Zhang, H. An Overview of Nanoparticle-Based Delivery Platforms for mRNA Vaccines for Treating Cancer. *Vaccines* **2024**, *12*, 727. [CrossRef] [PubMed]
275. Oberli, M.A.; Reichmuth, A.M.; Dorkin, J.R.; Mitchell, M.J.; Fenton, O.S.; Jaklenec, A.; Anderson, D.G.; Langer, R.; Blankschtein, D. Lipid nanoparticle assisted mRNA delivery for potent cancer immunotherapy. *Nano Lett.* **2017**, *17*, 1326–1335. [CrossRef] [PubMed]

276. Pardi, N.; Hogan, M.J.; Porter, F.W.; Weissman, D. mRNA vaccines—A new era in vaccinology. *Nat. Rev. Drug Discov.* **2018**, *17*, 261–279. [CrossRef]
277. Han, Y.; Renu, S.; Patil, V.; Schrock, J.; Feliciano-Ruiz, N.; Selvaraj, R.; Renukaradhya, G.J. Mannose-modified chitosan-nanoparticle-based *Salmonella* subunit oralvaccine-induced immune response and efficacy in a challenge trial in broilers. *Vaccines* **2020**, *8*, 299. [CrossRef]
278. Dolatyabi, S.; Renu, S.; Schrock, J.; Renukaradhya, G.J. Chitosan-nanoparticle-based oral *Salmonella* enteritidis subunit vaccine elicits cross-protection against *Salmonella* typhimurium in broilers. *Poult. Sci.* **2024**, *103*, 103569. [CrossRef] [PubMed]
279. Tapia, D. Evaluation of a Gold Nanoparticle Platform as Highly Immunogenic and Protective Therapy Against *Burkholderia Mallei, B. pseudomallei,* and Enterohemorrhagic *Escherichia coli* O157: H7. Ph.D. Published ETD Collection 2022. Available online: https://hdl.handle.net/2152.3/11436 (accessed on 13 January 2025).
280. Zagaglia, C.; Ammendolia, M.G.; Maurizi, L.; Nicoletti, M.; Longhi, C. Urinary tract infections caused by uropathogenic *Escherichia coli* strains—New strategies for an old pathogen. *Microorganisms* **2022**, *10*, 1425. [CrossRef] [PubMed]
281. Wang, X.; Gao, X.; Wang, L.; Lin, J.; Liu, Y. Advances of Nanotechnology Toward Vaccine Development Against Animal Infectious Diseases. *Adv. Funct. Mater.* **2023**, *33*, 2305061. [CrossRef]
282. Mat Rani, N.N.I.; Alzubaidi, Z.M.; Butt, A.M.; Mohammad Faizal, N.D.F.; Sekar, M.; Azhari, H.; Mohd Amin, M.C.I. Outer membrane vesicles as biomimetic vaccine carriers against infections and cancers. *Wiley Interdiscip. Rev. Nanomed. Nanobiotechnol.* **2022**, *14*, e1784. [CrossRef] [PubMed]
283. Mawad, A.; Helmy, Y.A.; Shalkami, A.-G.; Kathayat, D.; Rajashekara, G.E. *coli* Nissle microencapsulation in alginate-chitosan nanoparticles and its effect on *Campylobacter jejuni* in vitro. *Appl. Microbiol. Biotechnol.* **2018**, *102*, 10675–10690. [CrossRef]
284. Helmy Yosra, A.; Closs, G.; Jung, K.; Kathayat, D.; Vlasova, A.; Rajashekara, G. Effect of Probiotic *E. coli* Nissle 1917 Supplementation on the Growth Performance, Immune Responses, Intestinal Morphology, and Gut Microbes of Campylobacter jejuni Infected Chickens. *Infect. Immun.* **2022**, *90*, e0033722. [CrossRef] [PubMed]
285. Oladejo, M.; Paterson, Y.; Wood, L.M. Clinical experience and recent advances in the development of listeria-based tumor immunotherapies. *Front. Immunol.* **2021**, *12*, 642316. [CrossRef] [PubMed]
286. Datta, M.; Rajeev, A.; Chattopadhyay, I. Application of antimicrobial peptides as next-generation therapeutics in the biomedical world. *Biotechnol. Genet. Eng. Rev.* **2023**, *40*, 2458–2496. [CrossRef]
287. Yao, J.; Chen, Y.; Zhang, L.; Cheng, Y.; Chen, Z.; Zhang, Y.; Zheng, X.; Lv, Y.; Wang, S.; Li, Z. pH-responsive CuS/DSF/EL/PVP nanoplatform alleviates inflammatory bowel disease in mice via regulating gut immunity and microbiota. *Acta Biomater.* **2024**, *178*, 265–286. [CrossRef]
288. Patra, J.K.; Das, G.; Fraceto, L.F.; Campos, E.V.R.; Rodriguez-Torres, M.d.P.; Acosta-Torres, L.S.; Diaz-Torres, L.A.; Grillo, R.; Swamy, M.K.; Sharma, S. Nano based drug delivery systems: Recent developments and future prospects. *J. Nanobiotechnol.* **2018**, *16*, 1–33. [CrossRef] [PubMed]
289. Reddy, P.R.K.; Yasaswini, D.; Reddy, P.P.R.; Kumar, D.S.; Elghandour, M.M.; Salem, A. Nanotechnology in Veterinary Sector: Current Applications, Limitations and Future Perspective. In *Handbook of Green and Sustainable Nanotechnology: Fundamentals, Developments and Applications*; Springer: Berlin/Heidelberg, Germany, 2023; pp. 1541–1567.
290. Celis-Giraldo, C.T.; López-Abán, J.; Muro, A.; Patarroyo, M.A.; Manzano-Román, R. Nanovaccines against animal pathogens: The latest findings. *Vaccines* **2021**, *9*, 988. [CrossRef] [PubMed]
291. Renu, S.; Han, Y.; Dhakal, S.; Lakshmanappa, Y.S.; Ghimire, S.; Feliciano-Ruiz, N.; Senapati, S.; Narasimhan, B.; Selvaraj, R.; Renukaradhya, G.J. Chitosan-adjuvanted *Salmonella* subunit nanoparticle vaccine for poultry delivered through drinking water and feed. *Carbohydr. Polym.* **2020**, *243*, 116434. [CrossRef] [PubMed]
292. Wang, X.; Dai, Y.; Zhao, S.; Tang, J.; Li, H.; Xing, Y.; Qu, G.; Li, X.; Dai, J.; Zhu, Y. PAMAM-Lys, a novel vaccine delivery vector, enhances the protective effects of the SjC23 DNA vaccine against *Schistosoma japonicum* infection. *PLoS ONE* **2014**, *9*, e86578. [CrossRef] [PubMed]
293. Coleman, C.M.; Liu, Y.V.; Mu, H.; Taylor, J.K.; Massare, M.; Flyer, D.C.; Glenn, G.M.; Smith, G.E.; Frieman, M.B. Purified coronavirus spike protein nanoparticles induce coronavirus neutralizing antibodies in mice. *Vaccine* **2014**, *32*, 3169–3174. [CrossRef] [PubMed]
294. Al-Halifa, S.; Gauthier, L.; Arpin, D.; Bourgault, S.; Archambault, D. Nanoparticle-based vaccines against respiratory viruses. *Front. Immunol.* **2019**, *10*, 22. [CrossRef] [PubMed]
295. Zhu, D.; Long, Q.; Xu, Y.; Xing, J. Evaluating nanoparticles in preclinical research using microfluidic systems. *Micromachines* **2019**, *10*, 414. [CrossRef]
296. Desai, N.; Chavda, V.; Singh, T.R.R.; Thorat, N.D.; Vora, L.K. Cancer nanovaccines: Nanomaterials and clinical perspectives. *Small* **2024**, *20*, 2401631. [CrossRef] [PubMed]
297. Zhao, G.; Jiang, Y.; Ma, P.; Wang, S.; Nie, G.; Li, N. Membrane-based cancer nanovaccines: The time is now. *QJM Int. J. Med.* **2023**, *116*, 621–624. [CrossRef]

298. Sharma, H.; Thakur, S. 2 Membrane-Based Nano. In *Nanoparticles in Cancer Therapy: Innovations and Clinical Applications*; CRC Press: Boca Raton, FL, USA, 2024; p. 19.
299. Liu, Y.; Cheng, W.; Xin, H.; Liu, R.; Wang, Q.; Cai, W.; Peng, X.; Yang, F.; Xin, H. Nanoparticles advanced from preclinical studies to clinical trials for lung cancer therapy. *Cancer Nanotechnol.* **2023**, *14*, 28. [CrossRef] [PubMed]
300. Al-Thani, A.N.; Jan, A.G.; Abbas, M.; Geetha, M.; Sadasivuni, K.K. Nanoparticles in cancer theragnostic and drug delivery: A comprehensive review. *Life Sci.* **2024**, *352*, 122899. [CrossRef]
301. Sun, L.; Liu, H.; Ye, Y.; Lei, Y.; Islam, R.; Tan, S.; Tong, R.; Miao, Y.-B.; Cai, L. Smart nanoparticles for cancer therapy. *Signal Transduct. Target. Ther.* **2023**, *8*, 418. [CrossRef] [PubMed]
302. Selvaraj, J.; Rajendran, V.; Ramalingam, B. Nanovaccine: A Modern Approach to Vaccinology. *Nanotechnol. Med.* **2021**, *4*, 57–74.
303. Sun, M.; Pratama, A.C.; Qiu, H.; Liu, Z.; He, F. Toward innovative veterinary nanoparticle vaccines. *Anim. Dis.* **2024**, *4*, 14. [CrossRef]
304. Geng, Q.; Tai, W.; Baxter, V.K.; Shi, J.; Wan, Y.; Zhang, X.; Montgomery, S.A.; Taft-Benz, S.A.; Anderson, E.J.; Knight, A.C. Novel virus-like nanoparticle vaccine effectively protects animal model from SARS-CoV-2 infection. *PLoS Pathog.* **2021**, *17*, e1009897. [CrossRef] [PubMed]
305. Duvall, M.N.; Knight, K. *FDA Regulation of Nanotechnology*; Beveridge and Diamond, PG: Washington, DC, USA, 2012.
306. Csóka, I.; Ismail, R.; Jójárt-Laczkovich, O.; Pallagi, E. Regulatory considerations, challenges and risk-based approach in nanomedicine development. *Curr. Med. Chem.* **2021**, *28*, 7461–7476. [CrossRef] [PubMed]
307. Paliwal, R.; Kumar, P.; Chaurasiya, A.; Kenwat, R.; Katke, S.; Paliwal, S.R. Development of nanomedicines and nano-similars: Recent advances in regulatory landscape. *Curr. Pharm. Des.* **2022**, *28*, 165–177. [CrossRef]
308. Souto, E.B.; Blanco-Llamero, C.; Krambeck, K.; Kiran, N.S.; Yashaswini, C.; Postwala, H.; Severino, P.; Priefer, R.; Prajapati, B.G.; Maheshwari, R. Regulatory Insights into Nanomedicine and Gene Vaccine Innovation: Safety Assessment, Challenges, and Regulatory Perspectives. *Acta Biomater.* **2024**, *180*, 1–17. [CrossRef] [PubMed]
309. Gogoi, N.R.; Bezbaruah, R.; Patel, V.; Satasia, R.; Bhattacharjee, B.; Mazumder, B. Regulatory Pathways for Nanocarrier Vaccine. In *Nanocarrier Vaccines: Biopharmaceutics-Based Fast Track Development*; Wiley Online Library: Hoboken, NJ, USA, 2024; pp. 465–485.
310. Florindo, H.; Lopes, J.; Silva, L.; Corvo, M.; Martins, M.; Gaspar, R. Regulatory development of nanotechnology-based vaccines. In *Micro and Nanotechnology in Vaccine Development*; Elsevier: Amsterdam, The Netherlands, 2017; pp. 393–410.
311. Araste, F.; Bakker, A.D.; Zandieh-Doulabi, B. Potential and risks of nanotechnology applications in COVID-19-related strategies for pandemic control. *J. Nanoparticle Res.* **2023**, *25*, 229. [CrossRef]
312. World Health Organization. *State of the World's Vaccines and Immunization*; World Health Organization: Geneva, Switzerland, 2009.
313. Fernandes, C.; Jathar, M.; Sawant, B.K.S.; Warde, T. Scale-up of nanoparticle manufacturing process. In *Pharmaceutical Process Engineering and Scale-Up Principles*; Springer: Berlin/Heidelberg, Germany, 2023; pp. 173–203.
314. Liu, X.; Meng, H. Consideration for the scale-up manufacture of nanotherapeutics—A critical step for technology transfer. *View* **2021**, *2*, 20200190. [CrossRef]
315. Borrajo, M.L.; Lou, G.; Anthiya, S.; Lapuhs, P.; Álvarez, D.M.; Tobío, A.; Loza, M.I.; Vidal, A.; Alonso, M.J. Nanoemulsions and nanocapsules as carriers for the development of intranasal mRNA vaccines. In *Drug Delivery and Translational Research*; Springer: Berlin/Heidelberg, Germany, 2024; pp. 1–16.
316. Hu, C.; Bai, Y.; Liu, J.; Wang, Y.; He, Q.; Zhang, X.; Cheng, F.; Xu, M.; Mao, Q.; Liang, Z. Research progress on the quality control of mRNA vaccines. *Expert Rev. Vaccines* **2024**, *23*, 570–583. [CrossRef]
317. Whitley, J.; Zwolinski, C.; Denis, C.; Maughan, M.; Hayles, L.; Clarke, D.; Snare, M.; Liao, H.; Chiou, S.; Marmura, T. Development of mRNA manufacturing for vaccines and therapeutics: mRNA platform requirements and development of a scalable production process to support early phase clinical trials. *Transl. Res.* **2022**, *242*, 38–55. [CrossRef] [PubMed]
318. Rosa, S.S.; Prazeres, D.M.; Azevedo, A.M.; Marques, M.P. mRNA vaccines manufacturing: Challenges and bottlenecks. *Vaccine* **2021**, *39*, 2190–2200. [CrossRef]
319. Guerrini, G.; Magrì, D.; Gioria, S.; Medaglini, D.; Calzolai, L. Characterization of nanoparticles-based vaccines for COVID-19. *Nat. Nanotechnol.* **2022**, *17*, 570–576. [CrossRef] [PubMed]
320. Lee, J.H.; Yeo, Y. Controlled drug release from pharmaceutical nanocarriers. *Chem. Eng. Sci.* **2015**, *125*, 75–84. [CrossRef] [PubMed]
321. Svenson, S.; Tomalia, D.A. Dendrimers in biomedical applications—Reflections on the field. *Adv. Drug Deliv. Rev.* **2012**, *64*, 102–115. [CrossRef]
322. Jain, K.; Kesharwani, P.; Gupta, U.; Jain, N. Dendrimer toxicity: Let's meet the challenge. *Int. J. Pharm.* **2010**, *394*, 122–142. [CrossRef]
323. George, E.; Goswami, A.; Lodhiya, T.; Padwal, P.; Iyer, S.; Gauttam, I.; Sethi, L.; Jeyasankar, S.; Sharma, P.R.; Dravid, A.A. Immunomodulatory effect of mycobacterial outer membrane vesicles coated nanoparticles. *Biomater. Adv.* **2022**, *139*, 213003. [CrossRef] [PubMed]

324. Wang, S.; Gao, J.; Wang, Z. Outer membrane vesicles for vaccination and targeted drug delivery. *Wiley Interdiscip. Rev. Nanomed. Nanobiotechnol.* **2019**, *11*, e1523. [CrossRef] [PubMed]
325. Zheng, Y.; Zhao, Y.; Bai, M.; Gu, H.; Li, X. Metal–organic frameworks as a therapeutic strategy for lung diseases. *J. Mater. Chem. B* **2022**, *10*, 5666–5695. [CrossRef]
326. Tousian, B.; Khosravi, A.R.; Ghasemi, M.H.; Kadkhodaie, M. Biomimetic Functionalized Metal Organic Frameworks as Multifunctional Agents: Paving the Way for Cancer Vaccine Advances. *Mater. Today Bio* **2024**, *27*, 101134. [CrossRef] [PubMed]
327. Manimaran, V.; Nivetha, R.; Tamilanban, T.; Narayanan, J.; Vetriselvan, S.; Fuloria, N.K.; Chinni, S.V.; Sekar, M.; Fuloria, S.; Wong, L.S. Nanogels as novel drug nanocarriers for CNS drug delivery. *Front. Mol. Biosci.* **2023**, *10*, 1232109. [CrossRef] [PubMed]
328. Nakahashi-Ouchida, R.; Yuki, Y.; Kiyono, H. Development of a nanogel-based nasal vaccine as a novel antigen delivery system. *Expert Rev. Vaccines* **2017**, *16*, 1231–1240. [CrossRef]
329. Kabanov, A.V.; Vinogradov, S.V. Nanogels as pharmaceutical carriers: Finite networks of infinite capabilities. *Angew. Chem. Int. Ed.* **2009**, *48*, 5418–5429. [CrossRef] [PubMed]
330. Shinde, V.; Fries, L.; Wu, Y.; Agrawal, S.; Cho, I.; Thomas, D.N.; Spindler, M.; Lindner, E.; Hahn, T.; Plested, J. Improved titers against influenza drift variants with a nanoparticle vaccine. *N. Engl. J. Med.* **2018**, *378*, 2346–2348. [CrossRef]
331. Brazzoli, M.; Piccioli, D.; Marchetti, F. Challenges in development of vaccines directed toward antimicrobial resistant bacterial species. *Hum. Vaccines Immunother.* **2023**, *19*, 2228669. [CrossRef] [PubMed]
332. Weber, J.S.; Carlino, M.S.; Khattak, A.; Meniawy, T.; Ansstas, G.; Taylor, M.H.; Kim, K.B.; McKean, M.; Long, G.V.; Sullivan, R.J. Individualised neoantigen therapy mRNA-4157 (V940) plus pembrolizumab versus pembrolizumab monotherapy in resected melanoma (KEYNOTE-942): A randomised, phase 2b study. *Lancet* **2024**, *403*, 632–644. [CrossRef] [PubMed]
333. Saied, A.A. mRNA vaccines and clinical research in Africa-From hope to reality. *Int. J. Surg.* **2022**, *105*, 106833. [CrossRef] [PubMed]
334. Sekiya, T.; Ohno, M.; Nomura, N.; Handabile, C.; Shingai, M.; Jackson, D.C.; Brown, L.E.; Kida, H. Selecting and using the appropriate influenza vaccine for each individual. *Viruses* **2021**, *13*, 971. [CrossRef] [PubMed]
335. Chu, Y.; Liu, Q.; Wei, J.; Liu, B. Personalized cancer neoantigen vaccines come of age. *Theranostics* **2018**, *8*, 4238. [CrossRef]
336. Zhao, W.; Wu, W.; Xu, X. Oral vaccination with liposome-encapsulated recombinant fusion peptide of urease B epitope and cholera toxin B subunit affords prophylactic and therapeutic effects against *H. pylori* infection in BALB/c mice. *Vaccine* **2007**, *25*, 7664–7673. [CrossRef] [PubMed]
337. Kim, B.; Bowersock, T.; Griebel, P.; Kidane, A.; Babiuk, L.; Sanchez, M.; Attah-Poku, S.; Kaushik, R.; Mutwiri, G. Mucosal immune responses following oral immunization with rotavirus antigens encapsulated in alginate microspheres. *J. Control. Release* **2002**, *85*, 191–202. [CrossRef] [PubMed]
338. Riccardi, D.; Baldino, L.; Reverchon, E. Liposomes, transfersomes and niosomes: Production methods and their applications in the vaccinal field. *J. Transl. Med.* **2024**, *22*, 339. [CrossRef]
339. Saggese, A.; Baccigalupi, L.; Donadio, G.; Ricca, E.; Isticato, R. The bacterial spore as a mucosal vaccine delivery system. *Int. J. Mol. Sci.* **2023**, *24*, 10880. [CrossRef] [PubMed]
340. Abouelela, M.E.; Helmy, Y.A. Next-Generation Probiotics as Novel Therapeutics for Improving Human Health: Current Trends and Future Perspectives. *Microorganisms* **2024**, *12*, 430. [CrossRef]
341. Chin, J.S.; Chooi, W.H.; Wang, H.; Ong, W.; Leong, K.W.; Chew, S.Y. Scaffold-mediated non-viral delivery platform for CRISPR/Cas9-based genome editing. *Acta Biomater.* **2019**, *90*, 60–70. [CrossRef]
342. Hii, A.R.K.; Qi, X.; Wu, Z. Advanced strategies for CRISPR/Cas9 delivery and applications in gene editing, therapy, and cancer detection using nanoparticles and nanocarriers. *J. Mater. Chem. B* **2024**, *12*, 1467–1489. [CrossRef] [PubMed]
343. Li, J.; Li, J.; Peng, Y.; Du, Y.; Yang, Z.; Qi, X. Dendritic cell derived exosomes loaded neoantigens for personalized cancer immunotherapies. *J. Control. Release* **2023**, *353*, 423–433. [CrossRef]
344. Ott, P.A.; Hu, Z.; Keskin, D.B.; Shukla, S.A.; Sun, J.; Bozym, D.J.; Zhang, W.; Luoma, A.; Giobbie-Hurder, A.; Peter, L. An immunogenic personal neoantigen vaccine for patients with melanoma. *Nature* **2017**, *547*, 217–221. [CrossRef]
345. Yao, R.; Xie, C.; Xia, X. Recent progress in mRNA cancer vaccines. *Hum. Vaccines Immunother.* **2024**, *20*, 2307187. [CrossRef] [PubMed]
346. Mi, Y.; Smith, C.C.; Serody, J.S.; Vincent, B.G.; Wang, A.Z.; Hyun, H. Neoantigen nanovaccine improves personalized cancer immunotherapy. *Cancer Res.* **2020**, *80*, 2866. [CrossRef]
347. Feng, R.; Yu, F.; Xu, J.; Hu, X. Knowledge gaps in immune response and immunotherapy involving nanomaterials: Databases and artificial intelligence for material design. *Biomaterials* **2021**, *266*, 120469. [CrossRef]
348. Zohuri, B.; Behgounia, F. Application of artificial intelligence driving nano-based drug delivery system. In *A Handbook of Artificial Intelligence in Drug Delivery*; Elsevier: Amsterdam, The Netherlands, 2023; pp. 145–212.
349. Hasanzadeh, A.; Hamblin, M.R.; Kiani, J.; Noori, H.; Hardie, J.M.; Karimi, M.; Shafiee, H. Could artificial intelligence revolutionize the development of nanovectors for gene therapy and mRNA vaccines? *Nano Today* **2022**, *47*, 101665. [CrossRef]

350. Hamilton, S.; Kingston, B.R. Applying artificial intelligence and computational modeling to nanomedicine. *Curr. Opin. Biotechnol.* **2024**, *85*, 103043. [CrossRef]
351. Shi, Y.; Chen, L.; Zhu, M.; Zhao, Y. The Future of Nanomedicine. In *Nanomedicine*; Springer: Berlin/Heidelberg, Germany, 2022; pp. 1–28.

vaccines

MDPI

Article

A Nanoparticle Comprising the Receptor-Binding Domains of Norovirus and *Plasmodium* as a Combination Vaccine Candidate

Ming Xia [1], Pengwei Huang [1], Frank S. Vago [2], Wen Jiang [2], Xi Jiang [1,3,*] and Ming Tan [1,3,*]

1 Division of Infectious Diseases, Cincinnati Children's Hospital Medical Center, 3333 Burnet Avenue, Cincinnati, OH 45229, USA; ming.xia@cchmc.org (M.X.); pengwei.huang@cchmc.org (P.H.)
2 Department of Biological Sciences, Purdue University, West Lafayette, IN 47907, USA; fvago@purdue.edu (F.S.V.); jiang12@purdue.edu (W.J.)
3 Department of Pediatrics, University of Cincinnati College of Medicine, Cincinnati, OH 45229, USA
* Correspondence: jasonjiang@gmail.com (X.J.); ming.tan@cchmc.org (M.T.); Tel.: +1-513-636-0510 (M.T.); Fax: +1-513-636-7655 (M.T.)

Abstract: Background: Noroviruses, which cause epidemic acute gastroenteritis, and *Plasmodium* parasites, which lead to malaria, are two infectious pathogens that pose threats to public health. The protruding (P) domain of norovirus VP1 and the αTSR domain of the circumsporozoite protein (CSP) of *Plasmodium* sporozoite are the glycan receptor-binding domains of the two pathogens for host cell attachment, making them excellent targets for vaccine development. Modified norovirus P domains self-assemble into a 24-meric octahedral P nanoparticle (P_{24} NP). Methods: We generated a unique P_{24}-αTSR NP by inserting the αTSR domain into a surface loop of the P domain. The P-αTSR fusion proteins were produced in the *Escherichia coli* expression system and the fusion protein self-assembled into the P_{24}-αTSR NP. Results: The formation of the P_{24}-αTSR NP was demonstrated through gel filtration, electron microscopy, and dynamic light scattering. A 3D structural model of the P_{24}-αTSR NP was constructed, using the known cryo-EM structure of the previously developed P_{24} NP and P_{24}-VP8* NP as templates. Each P_{24}-αTSR NP consists of a P_{24} NP core, with 24 surface-exposed αTSR domains that have retained their general conformations and binding function to heparan sulfate proteoglycans. The P_{24}-αTSR NP is immunogenic, eliciting strong antibody responses in mice toward both the norovirus P domain and the αTSR domain of *Plasmodium* CSP. Notably, sera from mice immunized with the P_{24}-αTSR NP bound strongly to *Plasmodium* sporozoites and blocked norovirus VLP attachment to their glycan receptors. Conclusion: These data suggest that the P_{24}-αTSR NP may serve as a combination vaccine against both norovirus and *Plasmodium* parasites.

Keywords: norovirus; *Plasmodium*; norovirus P particle; nanoparticle; αTSR domain; nanoparticle vaccine

1. Introduction

Noroviruses belong to the genus *Norovirus*, in the family *Caliciviridae*. They are non-enveloped, single-stranded, positive-sense RNA viruses and are a leading cause of acute gastroenteritis (AGE), affecting individuals of all ages, in both developed and developing countries. These viruses spread rapidly, causing large AGE outbreaks in settings such as schools, cruise ships, military installations, and hospitals [1]. The global burden of norovirus-associated diarrhea is particularly severe in low- and middle-income countries, where it accounts for over 200,000 deaths annually, with USD 4.2 billion in direct health system costs and USD 60.3 billion in socioeconomic losses [2,3]. In the United States alone, norovirus infections cause an estimated 20 million cases of AGE each year, resulting in approximately 70,000 hospitalizations and up to 800 deaths [4,5]. Consequently, noroviruses remain a significant threat to global public health.

Each norovirus virion is encapsulated by an icosahedral capsid composed of capsid proteins (VP1s). The capsid comprises two major parts: an inner shell formed by shell (S) domains and multiple surface protrusions formed by the protruding (P) domains of VP1 [6]. The icosahedral inner shell is responsible for the structural integrity of the virus [7], while the surface protrusions play critical roles in the host interaction and immune response [8]. Accordingly, the S domains are genetically conserved, whereas the P domain sequences vary among different norovirus strains. However, receptor-binding sites within the P domain remain conserved within a genogroup, as these are crucial for receptor binding and viral survival [9,10]. Therefore, the P domain represents an excellent target for vaccine development. Previous studies have demonstrated that in vitro expression of norovirus P domains resulted in the self-assembly of dimers [11–16] and 24-meric P_{24} nanoparticles (P_{24}-NPs) [17,18] through homotypic interactions. The P_{24} NP has been proposed as a norovirus vaccine candidate and as a platform for displaying antigens from other pathogens to develop combination vaccine candidates (reviewed in [19]).

Plasmodium parasites, a group of unicellular protozoans, are causative agents, causing malaria, a serious and sometimes fatal illness [20]. According to the World Health Organization (WHO), there was an estimated 241 million clinical cases of malaria globally in 2020, where approximately 627,000 of those cases resulted in death [21], underscoring malaria as a significant public health threat. Malaria is a mosquito-borne disease, transmitted when infected mosquitoes inject *Plasmodium* sporozoites into human skin through their bites [22]. The sporozoites migrate to the liver, where they replicate in hepatocytes and form merozoites [23,24]. These merozoites subsequently infect erythrocytes, proliferating further and causing the symptoms of malaria sickness [25,26]. Some merozoites also develop into gametocytes [27,28].

The sporozoites introduced by mosquito bites are exposed to host antibodies during their migration to the liver [20,29], making them a prime target for prophylactic and therapeutic strategies to prevent malaria. These sporozoites are enveloped by a multifunctional protein known as the circumsporozoite protein (CSP) [30]. Importantly, the CSP enables sporozoites to recognize heparan sulfate proteoglycans (HSPGs) on liver cells, facilitating their invasion of the liver [30–33]. Our prior study identified the αTSR domain (region II plus) [34] of the CSP as the HSPG binding domain [35], suggesting that the αTSR domain is an ideal target for inhibiting CSP function.

Vaccination is a highly effective prophylactic strategy for preventing infectious diseases. Currently, there are no commercially available vaccines for norovirus, although several candidates have progressed to clinical trials [36–38]. These include recombinant virus-like particle (VLP)-based vaccines administered via intramuscular injection [39–43] and an adenovirus-vectored VP1 protein-based vaccine for oral delivery [44]. In contrast, two malaria vaccines have been licensed for public use. The first one is Mosquirix™, also known as RTS,S/AS01 [45–49], which incorporates the C-terminus and central repeats of the CSP from *Plasmodium falciparum*, displayed on recombinant hepatitis B virus core particles at a 20% ratio. The second is R21/Matrix-MTM [50–55], which also targets the CSP, but is displayed on recombinant hepatitis B virus core particles at a 100% ratio. Both vaccines require three to four doses to achieve a modest protective efficacy, providing approximately 75% protection for up to three years. Given these limitations, there is an urgent need for an effective norovirus vaccine and next-generation malaria vaccines with enhanced efficacy.

In this study, we developed a unique P_{24}-αTSR NP incorporating the receptor-binding domains from both norovirus and *Plasmodium* parasites. Evidence supporting its potential as a combination vaccine candidate against these two pathogens includes the robust antibody responses elicited by the P_{24}-αTSR NP in mice against both the norovirus P domain and the αTSR domain of *Plasmodium* CSP. Additionally, sera from mice immunized with the P_{24}-αTSR NP specifically bound to CSPs on *Plasmodium* sporozoites and blocked norovirus VLP attachment to glycan receptors.

2. Materials and Methods

2.1. DNA Constructs for the Expression of P-αTSR Fusion Proteins

The DNA fragment encoding the αTSR domain of the *P. falciparum* 3D7 strain (GenBank AC#: CAB38998.2, from E309 to C375, 67 amino acids) was amplified using PCR from a previously constructed plasmid containing the αTSR-encoding sequence [35]. This fragment was then subcloned into surface loop 2 of the norovirus P domain in a separate, previously created pGEX-4T-1-based plasmid encoding the glutathione S-transferase (GST)-P-VP8* fusion protein [56], replacing the VP8*-encoding sequence. The resulting plasmid was used to produce GST-tagged P-αTSR fusion protein (Figure 1A), with a thrombin cleavage site between the GST and the P-αTSR protein. An additional plasmid was constructed by subcloning the P-αTSR-encoding DNA sequence into the pET-24b vector (Figure 2A) to produce the tag-free P-αTSR protein, with a stop codon introduced before the C-terminal Hisx6-encoding sequence in the vector.

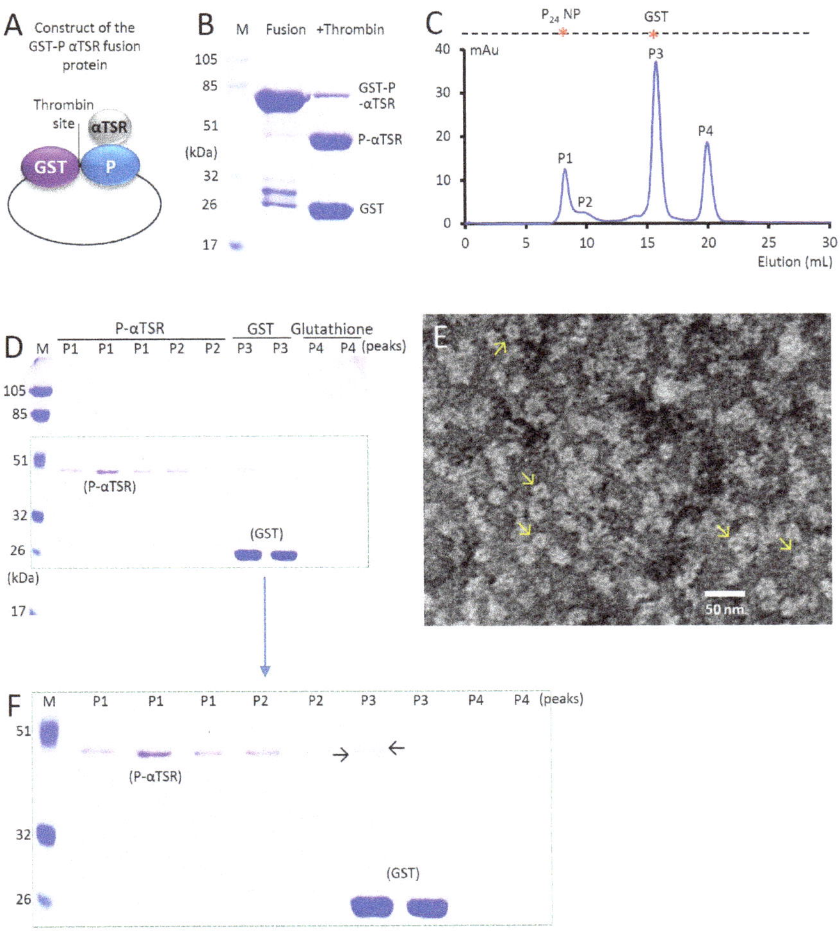

Figure 1. Production of the GST-tagged P-αTSR protein and self-formation of the P_{24}-αTSR NPs. (**A**) Schematic representation of the GST-tagged P-αTSR fusion protein construct. The αTSR domain is inserted in loop 2 of the norovirus P domain, with a thrombin cleavage site present between the GST and P-αTSR protein. (**B**) SDS-PAGE analysis of the GST-P-αTSR fusion protein and its bipartite components after thrombin cleavage. Left lane (M): prestained protein standards, with molecular weights (MWs) indicated in kDa. Middle lane (Fusion): resin-purified GST-P-αTSR fusion protein (~68 kDa) with minor co-purified proteins. Right lane (+Thrombin): two separate components, the P-αTSR protein (~42 kDa) and the GST (~26 kDa), along with minor amounts of uncleaved GST-P-αTSR fusion protein, after thrombin cleavage. (**C**) Elution curve from gel-filtration chromatography of the thrombin-cleaved GST-P-αTSR protein, showing four distinct elution peaks (P1, P2, P3, and P4). The *Y*-axis represents relative protein amounts measured by ultraviolet absorbance at 280 nm (mAU) and the *X*-axis indicates the elution volume (mL). The dashed line at the top, marked with two red stars, indicates the elution positions of the P_{24} NP (MW ~830 kDa) and GST dimer (MW~52 kDa). (**D**) SDS-PAGE analysis of the four peaks from the gel-filtration chromatography, showing the presence of P-αTSR and/or GST protein in each peak. (**E**) A representative transmission electron microscopy (TEM) image of the P-αTSR protein from P1 in (**C**,**D**), showing the formation of the P_{24}-αTSR NPs. Arrows point to typical P_{24}-αTSR NPs. (**F**) Enlarged SDS-PAGE image in (**D**), highlighting the double bands of the P-αTSR protein from P3 in (**C**), indicated by arrows.

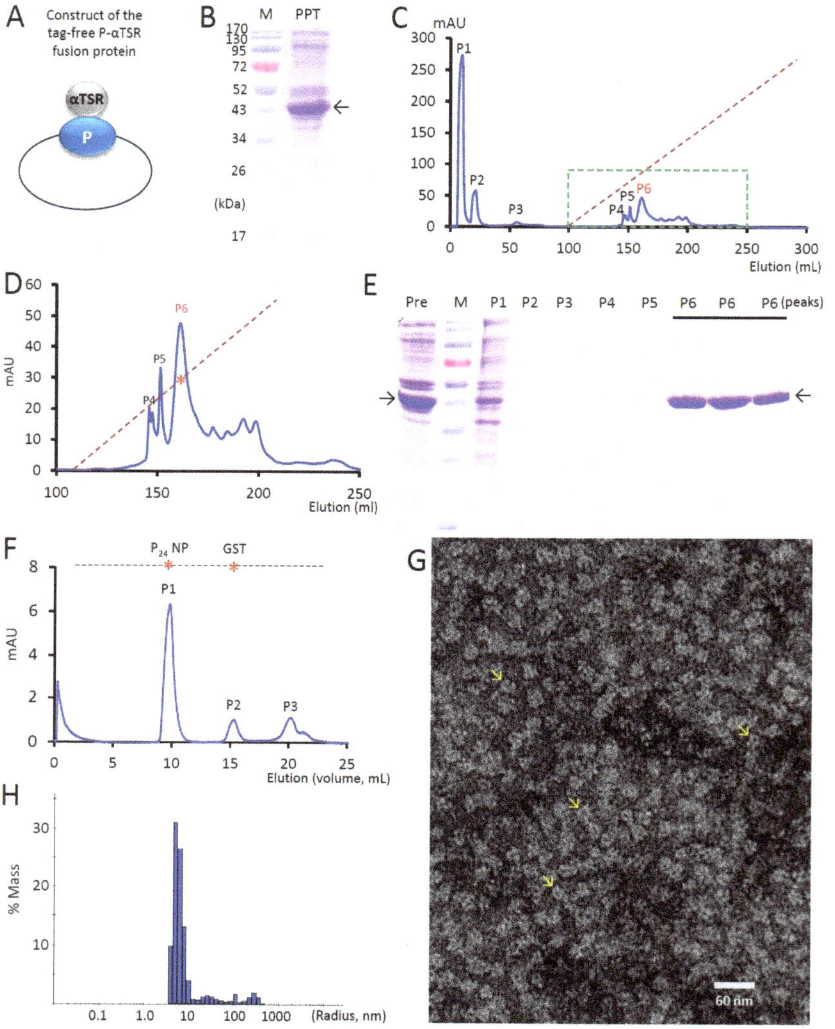

Figure 2. Generation and characterization of tag-free P-αTSR protein and P_{24}-αTSR NPs. (**A**) Schematic representation of the tag-free P-αTSR protein construct. (**B**) SDS-PAGE analysis of the P-αTSR protein precipitated with ammonium sulfate [$(NH4)_2SO_4$]. The left lane (M) shows prestained protein standards, with molecular weights (MWs) labeled in kDa. The right lane (PPT) displays the P-αTSR protein (denoted by an arrow) precipitated from bacterial lysate using 1.2 M $(NH4)_2SO_4$. (**C**,**D**) Representative elution profile from anion exchange chromatography of the $(NH4)_2SO_4$-precipitated P-αTSR protein. (**D**) Is an enlarged view of the framed region in (**C**). The *Y*-axis represents relative protein amounts in the eluents, measured by ultraviolet absorbances at 280 nm (A280, mAU), while the *X*-axis indicates the accumulated elusion volume (mL). The dashed red line shows the linear gradient of buffer B (0–100%), with a red star marking the percentages of buffer B at the elution peak of the P-αTSR protein (33.9%, P6). Six major elution peaks (P1 to P6), analyzed by SDS-PAGE in (**E**), are indicated. (**E**) SDS-PAGE analysis of the six major elution peaks from the anion exchange chromatography in (**C**,**D**). Lane "Pre": redissolved ammonium sulfate-precipitated protein before loading. Lane M: prestained protein standards with MWs indicated in kDa, as in (**B**). The tag-free P-αTSR protein (~68 kDa) is present in P6, as indicated by an arrow. (**F**) Elution curve from gel filtration of the purified tag-free P-αTSR protein. The curve shows a single major peak in the void volume (P1), corresponding to the P_{24}-αTSR NP (MW > 800 kDa). Two minor peaks (P2 and P3)

appeared. P2 likely represents dimeric P-αTSR protein with an MW similar to the GST dimer (~52 kDa), while P3 may represent degraded protein fragments with MWs < 10 kDa. The Y-axis indicates relative protein amounts, while the X-axis represents the elution volume, as in (**C**,**D**). The dashed line at the top, marked with two stars, indicates the elution positions of the P_{24} NP (~830 kDa) and GST dimer (~52 kDa). (**G**) Representative transmission electron microscopy (TEM) image of the P-αTSR protein from P1 in (**F**), showing the formation of the P_{24}-αTSR NPs. Arrows point to typical P_{24}-αTSR NPs. (**H**) The particle size distribution of the tag-free P_{24}-αTSR NPs, determined by dynamic light scattering (DLS).

2.2. Generation of the P-αTSR Proteins

The P-αTSR proteins were produced in *Escherichia coli* BL21 Arctic strain, through induction with 0.25 mM isopropyl-β-D-thiogalactopyranoside (IPTG) at 13 °C for 16 h, as described elsewhere [13,17]. Soluble GST-tagged P-αTSR protein was isolated from the bacterial lysate using GST-binding resin (Glutathione Sepharose 4 Fast Flow, GE Healthcare). Following thrombin cleavage, the GST tag was removed from the P-αTSR protein via gel-filtration chromatography (see below). Additionally, soluble tag-free P-αTSR protein, expressed via the same *E. coli* system, was precipitated from the bacterial lysate with 1.2 M ammonium sulfate [$(NH_4)_2SO_4$] and further purified using anion exchange chromatography (see below).

2.3. SDS-PAGE Analysis and Protein Quantitation

Recombinant proteins produced in this study were analyzed using sodium dodecyl sulfate polyacrylamide gel electrophoresis (SDS-PAGE), with 12% separating gels. The protein concentrations were quantified either by comparison to known concentrations of bovine serum albumin (BSA, Bio-Rad, Hercules, CA, USA) on the same gels [56] and/or by using a NanoDrop spectrophotometer.

2.4. Gel-Filtration Chromatography

This method was carried out using an ÄKTA Fast Performance Liquid Chromatography system (FPLC, ÄKTA pure^{TM} 25 L, GE Healthcare Life Sciences, San Francisco, CA, USA), with a size-exclusion column (Superdex 200, 10/300 GL, 25 mL bed volume, GE Healthcare Life Sciences), as described previously [13]. The elution peaks corresponding to the P_{24}-αTSR NPs and GST were identified using the previously prepared norovirus P_{24} NPs (~830 kDa) [18] and GST dimers (~52 kDa) [35]. The relative protein amount in the eluent was monitored by its absorbance at 280 nm (A_{280}).

2.5. Anion Exchange Chromatography

This procedure was performed using the same ÄKTA FPLC system (see above), equipped with a HiPrep Q HP 16/10 column (GE Healthcare Life Sciences), as previously described [57,58]. Briefly, the column was equilibrated with seven column volumes (CVs) of 20 mM tris buffer (pH 8.0, buffer A). After the protein sample was loaded, unbound proteins were washed out using seven CVs of buffer A. The bound proteins, including the target protein, were eluted using a linear gradient (0 to 100% B) over eight CVs of 1 M NaCl in buffer A (buffer B). The column was subsequently washed using seven CVs of buffer B, followed by re-equilibration with seven CVs of buffer A. The relative protein concentrations in the eluent were monitored by UV absorbances at 280 nm (A_{280}, mAU). The elution positions of the target proteins are indicated as percentages of buffer B.

2.6. Cesium Chloride (CsCl) Density Gradient Ultracentrifugation

This approach was used to assess the density of the P_{24}-αTSR NPs, following a previously established protocol [59]. A 0.5 mL sample of purified tag-free P-αTSR protein was mixed with 10 mL of a CsCl solution, at a density of 1.3630 g/mL. The mixture was centrifuged at 288,000× *g* for 45 h, using an Optima L-90K ultracentrifuge (Beckman Coulter, Brea, CA, USA). The resulting gradient was fractionated into 23 fractions using a bottom puncture. Each fraction was diluted 100-fold with PBS, coated onto 96-well microtiter plates, and analyzed for the presence of the P-αTSR protein using a guinea pig hyperimmune serum against the norovirus VLP [60] and mouse hyperimmune serum against the αTSR protein. The CsCl densities of the fractions containing the P-αTSR protein were determined using the refractive index method.

2.7. Transmission Electron Microscopy (TEM)

Negative stain TEM was used to visually examine the morphology of the P-αTSR NPs, following a previously described protocol [59]. Briefly, purified P-αTSR fusion proteins were absorbed onto grids (FCF200-CV-50, Electron Microscopy Sciences, Hatfield, PA, USA) and stained with 1% ammonium molybdate. The grids were air dried and examined using a Hitachi electron microscope (model H-7650) at 80 kV, with magnifications ranging from 15,000× to 40,000×.

2.8. P_{24}-αTSR NP Glycan Binding Assay

This procedure was carried out as previously described [35]. Briefly, heparin sulfate (HS) in the form of heparin sodium salt purified from porcine intestinal mucosa (Sigma-Aldrich, St. Louis, MO, USA) at a concentration of 25 μg/mL was coated onto 96-well microtiter plates (Thermo Fisher Scientific, Waltham, MA, USA). The P_{24}-αTSR NP, at different concentrations, was incubated with the coated heparin, with the GST-tagged αTSR [35] serving as a positive control and the P_{24} NP [18] and GST serving as negative controls. Bound P_{24}-αTSR NP/P_{24} NP proteins were detected using guinea pig hyperimmune serum against norovirus VLP [60] at 1:3000 dilution, while bound GST-αTSR/GST proteins were detected using mouse hyperimmune serum against GST-αTSR [35] at 1:2000 dilution. Horseradish peroxidase (HRP)-conjugated goat anti-guinea pig IgG or mouse IgG (Thermo Fisher Scientific) at a 1:5000 dilution was used to measure the bound antibodies. Finally, HRP substrates were added to determine the signal intensity, which was defined by the optical density (OD) at 450 nm.

2.9. Dynamic Light Scattering (DLS)

A total of 200 μL of P-αTSR proteins was placed into a well of a clear flat-bottom 96-well microplate (Greiner Bio-One, Monroe, NC, USA) and analyzed with the DynaPro Plate Reader III DLS instrument (Wyatt Technology, Goleta, CA, USA). The size distribution of the P-αTSR NPs was determined using DYNAMICS software (Wyatt Technology).

2.10. Structural Modeling of the P_{24}-αTSR NP

A 3D structural model of the P_{24}-αTSR NP was constructed using UCSF ChimeraX software (version 1.4) [61], based on cryo-EM (cryogenic electron microscopy) density maps of the norovirus P_{24} [18] and the P_{24}-VP8* NPs [56] as templates. In this model, the VP8* domains were replaced with the crystal structures of the αTSR domains of *P. falciparum* (PDB code: 3VDJ) [34]. UCSF ChimeraX software was also utilized for the structural analysis of the P_{24}-αTSR NP model and for generating visualizations.

2.11. Immunization of Mice

A total of 24 pathogen-free BALB/c mice, at an age of ~8 weeks, were randomly divided into three groups of 8 mice each (n = 8). Each group was immunized intramuscularly (IM) three times at 2-week intervals with one of following three immunogens at a dose of 10 μg/mouse/dose: (1) P_{24}-αTSR NP; (2) Hisx6-tagged αTSR protein [35]; and

(3) P_{24} NP [18]. All immunogens were administered with aluminum salt adjuvant (Imject Alum, Thermo Fisher Scientific), at a dose of 25 μL/mouse/dose (20 μg/mouse/dose). Sera samples were prepared from blood specimens that were collected two weeks after the third immunization via cardiac puncture.

2.12. Specific IgG Titer Determination

Serum IgG antibody titers specific to *Plasmodium* αTSR and the norovirus P domain were determined using enzyme immunoassays (EIAs). Briefly, purified GST-αTSR [35] or P_{24} protein at 5 μg/mL were used as capture antigens and coated onto 96-well microtiter plates. The coated antigens were blocked with 5% non-fat milk and then hatched with diluted mouse sera. IgG that bound the coated antigens was detected by HRP-conjugated goat-anti-mouse IgG (1:5000, MP Biomedicals, Irvine, CA, USA). The αTSR and P domain-specific IgG titers were defined as the highest serum dilutions yielding positive signals (OD450 $\geq$ 0.2).

2.13. Blocking of Norovirus VLP–Glycan Receptor Interaction

The norovirus P domain binds to host cell surface histo-blood group antigens (HBGAs) to initiate infection. This binding has been mimicked in an established EIA-based binding assay, with serum antibodies that block this interaction considered surrogate neutralizing antibodies against norovirus [60]. The blocking assay begins by coating HBGAs, in the form of a well-defined type A saliva sample, onto 96-well microtiter plates. Norovirus GII.4 VLPs [60] were pre-incubated with serially diluted sera obtained from mice immunized with the P_{24}-αTSR NPs or controls and then added to the wells with coated HBGAs. A reduction in the binding signal, compared to wells without serum blocking, indicates the blocking effect.

2.14. Immunofluorescence Assays (IFAs)

Sera from mice immunized with the P_{24}-αTSR NP or the P_{24} NP control were used to stain the circumsporozoite proteins (CSPs) on the surface of *P. falciparum* sporozoites, following a procedure reported previously [62,63]. Briefly, slides with air-dried sporozoites of *P. falciparum*, kindly provided by Dr. Photini Sinnis at Johns Hopkins University, were brought to room temperature and blocked with 1% BSA in 1× Tris-Buffered Saline (TBS, pH 7.4). In a humidity chamber, mouse sera diluted at 1:8000 were incubated with the sporozoites. After washing, the sporozoites were incubated with a fluorophore-conjugated secondary antibody (Millipore-Sigma, Norwood, OH, USA), mounted with Citifluor Mountant Media, and sealed with a cover glass using nail polish. The sporozoites were observed with a fluorescence microscope at 20× to 40× magnifications.

2.15. Ethics Statement

The animal study in this project was carried out in accordance with the recommendations outlined in the Guide for the Care and Use of Laboratory Animals (23a) by the National Institute of Health (NIH). The study procedures were reviewed and approved by the Institutional Animal Care and Use Committee (IACUC) at Cincinnati Children's Hospital Research Foundation (Animal Welfare Assurance No. A3108-01).

2.16. Statistical Evaluation

Statistical comparisons between the data groups were conducted using GraphPad Prism 9.0 (GraphPad Software, Inc. Boston, MA, USA) and an unpaired *t* test. Differences were deemed not significant at p-values > 0.05, significant at p-values < 0.05 (marked as "*"), significant at p-values < 0.01 (marked as "**"), and highly significant at p-values < 0.001 (marked as "***").

3. Results

3.1. Production of the P-αTSR Fusion Protein

The P-αTSR protein was initially generated as a soluble GST-tagged protein (Figure 1A) via an *E. coli* expression system, yielding approximately 20 mg of target protein per liter of bacterial culture. This was achieved by isolating the target protein from the bacterial lysate using GST-binding resin (Figure 1B, middle lane, ~68 kDa). The purified GST-P-αTSR fusion protein was then cleaved using thrombin, resulting in two proteins: a ~42 kDa P-αTSR protein and a ~26 kDa GST (Figure 1B, right lane). The two proteins were separated by gel-filtration chromatography, followed by SDS-PAGE analysis (Figure 1C,D). Gel filtration revealed three major peaks, corresponding to the P-αTSR protein in the form of P_{24}-αTSR NPs (P1, see below), GST (P3), and glutathione (P4) from the elution buffer, respectively.

The P-αTSR protein was also produced using a tag-free approach (Figure 2). After expression in the *E. coli* system, the P-αTSR protein was precipitated from the bacterial lysate using 1.2 M ammonium sulfate (Figure 2B). The precipitated P-αTSR protein, along with co-precipitated bacterial proteins, was dissolved in 20 mM tris buffer (pH 8.0) and subjected to anion exchange chromatography, resulting in multiple peaks (Figure 2C–E). The P-αTSR protein was eluted in P6, corresponding to 33.9% of buffer B (339 mM NaCl, Figure 2C–E), with a yield of approximately 15 mg of target protein per liter of bacterial culture.

3.2. Self-Assembly of the P_{24}-αTSR NPs

During the gel-filtration chromatography used to separate the GST tag from the P-αTSR protein (Figure 1C), the majority of intact P-αTSR protein eluted as a peak (P1) in the void volume, indicating a molecular weight (MW) greater than the maximum mass (800 kDa) of the Superdex 200 size-exclusion column. This aligns with the calculated MW of 1008 kDa for the P_{24}-αTSR NP (24 × 42 kDa), suggesting that the P-αTSR protein self-assembled into P_{24}-αTSR NP, consistent with the known self-assembly propensity of the P domain protein into the P_{24} NP [17]. TEM analysis of the P-αTSR protein from the P1 elution revealed a particle morphology characteristic of P_{24} NPs, with a size of approximately 15 nm, showing some variations (Figure 1E), further confirming the formation of the P_{24}-αTSR NPs. While the majority of the intact P-αTSR protein self-assembled into the P_{24}-αTSR NPs, SDS-PAGE analysis (Figure 1D) identified some degraded P-αTSR proteins in P3, co-eluting with the GST dimers. The degraded P-αTSR proteins appeared as two bands, with the lower band indicating partial degradation, which may have impaired their ability to form P_{24}-αTSR NPs (see Section 4).

Additional evidence supporting the self-formation of P_{24}-αTSR NPs was obtained from the study of the P-αTSR protein purified using the tag-free approach. First, gel-filtration analysis of the purified P-αTSR protein from the P6 elution using the same Superdex 200 column revealed a major peak at the void volume (Figure 2F). Second, TEM inspection of the same protein sample showed particles with typical P_{24} NP morphology [18], with an approximate diameter of 15 nm (Figure 2G). Third, DSL analysis revealed that the purified P-αTSR protein formed NPs of diverse sizes, with the major population ranging in diameter from 10 to 20 nm (Figure 2H). Moreover, three minor peaks were observed, corresponding to particle sizes of 50, 200, and 500 nm. Collectively, these data confirmed that the purified P-αTSR protein self-assembled into the P_{24}-αTSR NPs.

3.3. D Structural Modeling of the P_{24}-αTSR NP

TEM micrographs of the P_{24}-αTSR NPs revealed the characteristic morphologies of the octahedral P_{24} NP [18], with recognizable extended protrusions corresponding to the αTSR domains (Figure 3A–D). Additional shapes were also observed in the micrographs. To better understand these morphologies, a 3D model of the P_{24}-αTSR NP was constructed, using the cryo-EM structures of the P_{24} NP [18] (Figure 3E–G) and the P_{24}-VP8* [56] NP as templates. This was achieved by fitting the known crystal structures of the αTSR domains of *P. falciparum* (PDB code: 3VDJ) [34] into the VP8* density maps of the P_{24}-VP8* NP, using UCSF ChimeraX software. Based on this octahedral model, we generated representative images of the P_{24}-αTSR NP from various viewing angles (Figure 3H–P), including views along the three-fold (Figure 3H,K,N) and four-fold (Figure 3I,L,O) symmetry axes, as well as intermediate angles between the two (Figure 3J,M,P). In summary, the octahedral P_{24}-αTSR NPs do not display typical spherical shapes; instead, they may appear as rectangles, pentagons, and/or hexagons, depending on the viewing angle. These visualizations provide valuable insights into the diverse morphologies observed in the TEM micrographs.

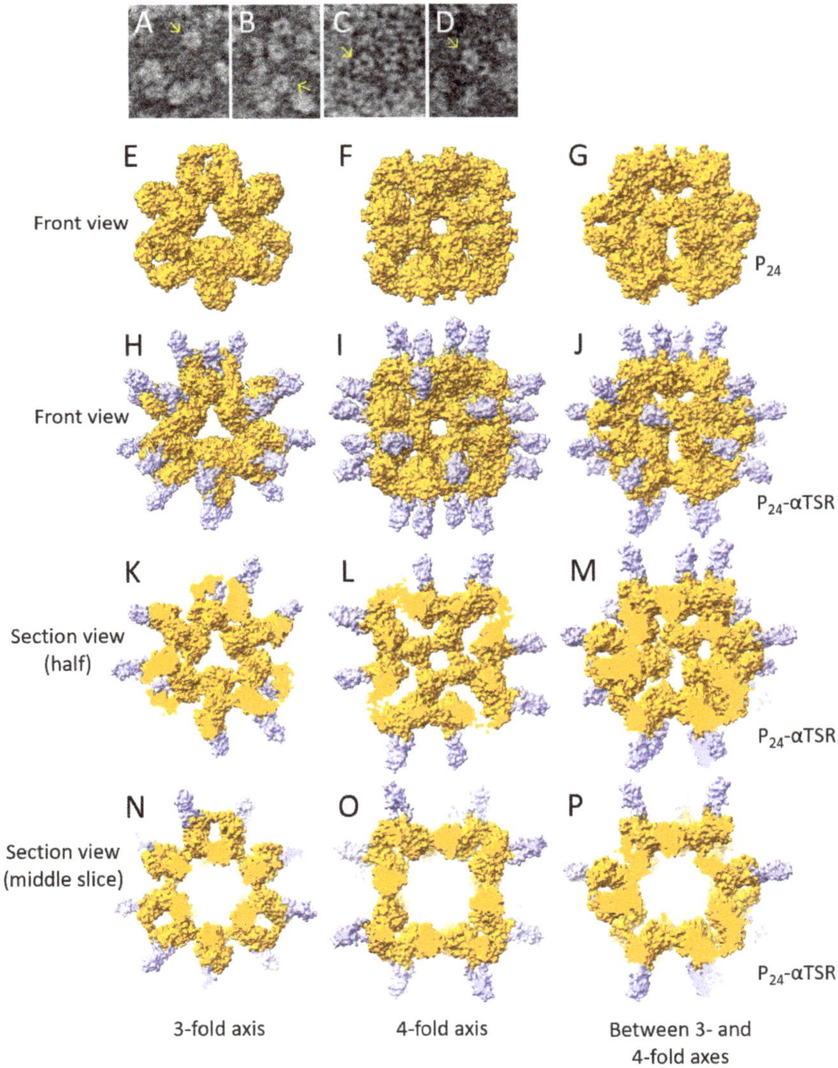

Figure 3. The 3D structural models of the P_{24}-αTSR NP. The model was constructed using UCSF ChimeraX software version 1.4, based on the previously elucidated cryo-EM maps of the norovirus

P_{24} NP and the P_{24}-VP8* chimeric NP. In this model, the rotavirus VP8* domain was replaced by the known crystal structure of the αTSR domain (PDB code: 3VDJ). (**A–D**) Transmission electron microscopy (TEM) images showing the typical morphology of the P_{24}-αTSR NP (indicated by arrows), consisting of a P_{24} NP core with multiple extended protrusions formed by the αTSR domains. (**E–G**) Surface representations of the norovirus P_{24} NP, viewed along the three-fold axis (**E**), the four-fold axis (**F**), and at an intermediate angle between these two axes (**G**). (**H–J**) Surface representations of the P_{24}-αTSR NP, viewed along the three-fold axis (**H**), the four-fold axis (**I**), and at an intermediate angle between these two axes (**J**). (**K–M**) Cross-sectional views of the three P_{24}-αTSR NP structures shown in (**H–J**), cut in half and viewed from the cutting planes along the same three symmetry axes as in (**H–J**). (**N–P**) Middle-section cross-sectional views of the three P_{24}-αTSR NP structures shown in (**H–J**), viewed from the same three symmetry axes as in (**H–J**). In all images from (**H**) to (**P**), the P_{24} NP core is depicted in orange, while the αTSR domains are shown in purple.

3.4. Further Characteristics of the P_{24}-αTSR NP

The P_{24}-αTSR NP was analyzed using a CsCl density gradient. Following ultracentrifugation, the gradient was fractionated, followed by detection of the P_{24}-αTSR NPs in the fractions through EIAs using hyperimmune sera against norovirus VLP [60] and αTSR [35]. This revealed an overlapped peak in the upper half of the gradient, centered at fraction 19 (Figure 4A,B), with a density of 1.2995 g/cm^3. These results confirmed the bipartite composition of the P_{24}-αTSR NP, consisting of the norovirus P_{24} NP core and the *P. falciparum* αTSR domains, which form the surface protrusions (Figure 3). Furthermore, the specific recognition of the P_{24}-αTSR NP by targeted antibodies validated the preservation of the native conformations of both the norovirus P domain and the αTSR domain within the NP.

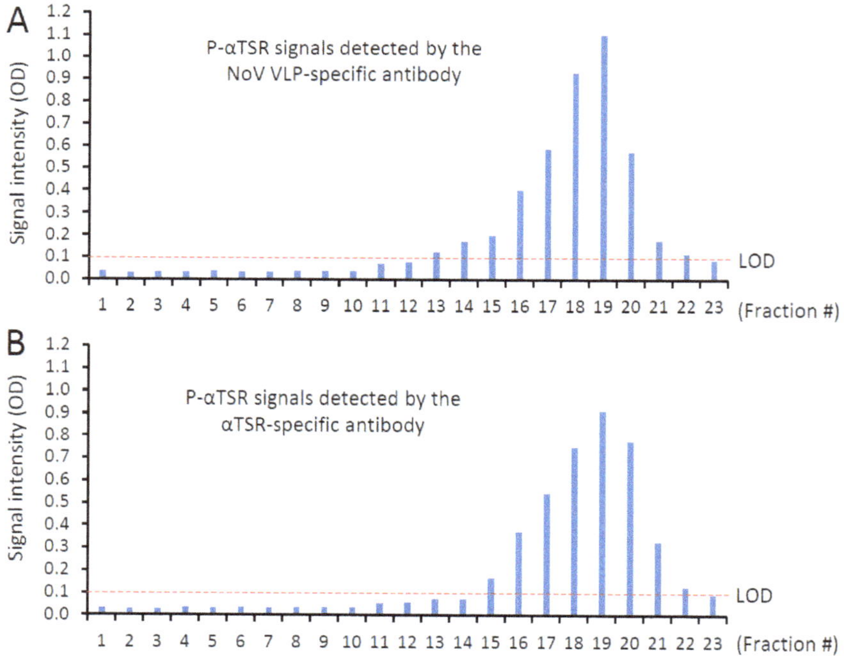

Figure 4. *Cont.*

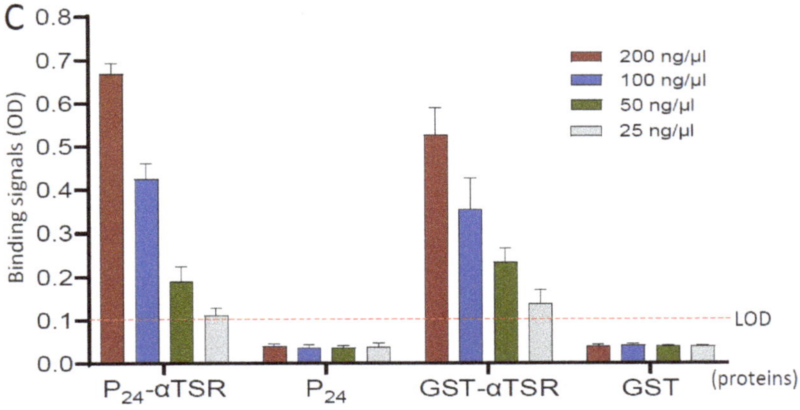

Figure 4. CsCl density gradient ultracentrifugation, specific antibody reactivity, and heparin sulfate glycan binding of the P_{24}-αTSR NP. (**A**,**B**) Each of the 23 fractions from the CsCl density gradient was analyzed for the presence of the P_{24}-αTSR NP using EIA assays. Hyperimmune serum from a guinea pig after immunization with norovirus VLP (**A**) and hyperimmune serum from a mouse after immunization with the *Plasmodium* αTSR domain (**B**) were used as detection antibodies, respectively. The *Y*-axis represents the optical density (OD) as a measure of signal intensity, while the *X*-axis corresponds to the gradient fractions arranged from the bottom (fraction 1) to the top (fraction 23). (**C**) EIA-based binding assay showing the interaction of the P_{24}-αTSR NP with heparin sulfate glycans. The GST-αTSR protein served as a positive control, while the P24 NP and GST were used as negative controls. The *Y*-axis indicates the binding signal intensity of the OD, while the *X*-axis shows the different proteins at the indicated concentrations. The limit of detection (LOD), indicated by a red dashed line in each panel, was set at OD = 0.1.

3.5. Interaction of the P_{24}-αTSR NP with Heparin Glycans

The αTSR domain has been shown to interact with HSPGs [35]. We demonstrated that the P_{24}-αTSR NP, with its surface-exposed αTSR domains, interacted with heparin sulfate in a dose-dependent manner, similar to the GST-αTSR protein (Figure 4C). In contrast, the P_{24} NP and GST, used as negative controls, did not exhibit such interactions. These findings validated that the αTSR domains displayed on the P_{24} NP retain their glycan receptor-binding function, further supporting the conclusion that αTSR domains maintain their native conformations when displayed on the P_{24} NP.

3.6. Robust Antibody Response to the Bipartite Composition of the P_{24}-αTSR NP

Serum IgG and IgA responses specific to the norovirus P domain and *P. falciparum* αTSR domain were evaluated in mice following immunization with the P_{24}-αTSR NP (Figure 5A–D). The norovirus P_{24} NP and free αTSR (with His tag), prepared previously [18,35], were included as controls. After three immunizations, the P_{24}-αTSR NP induced a significantly higher αTSR-specific IgG titer of 1:226,575, compared to the titer of 1:12,200 elicited by the free αTSR protein ($p = 0.0053$, Figure 5A). Conversely, the P_{24}-αTSR NP elicited a high P domain-specific IgG titer of 1:106,057, comparable to the titer of 1:149,943 induced by the P_{24} NP ($p = 0.4402$, Figure 5C).

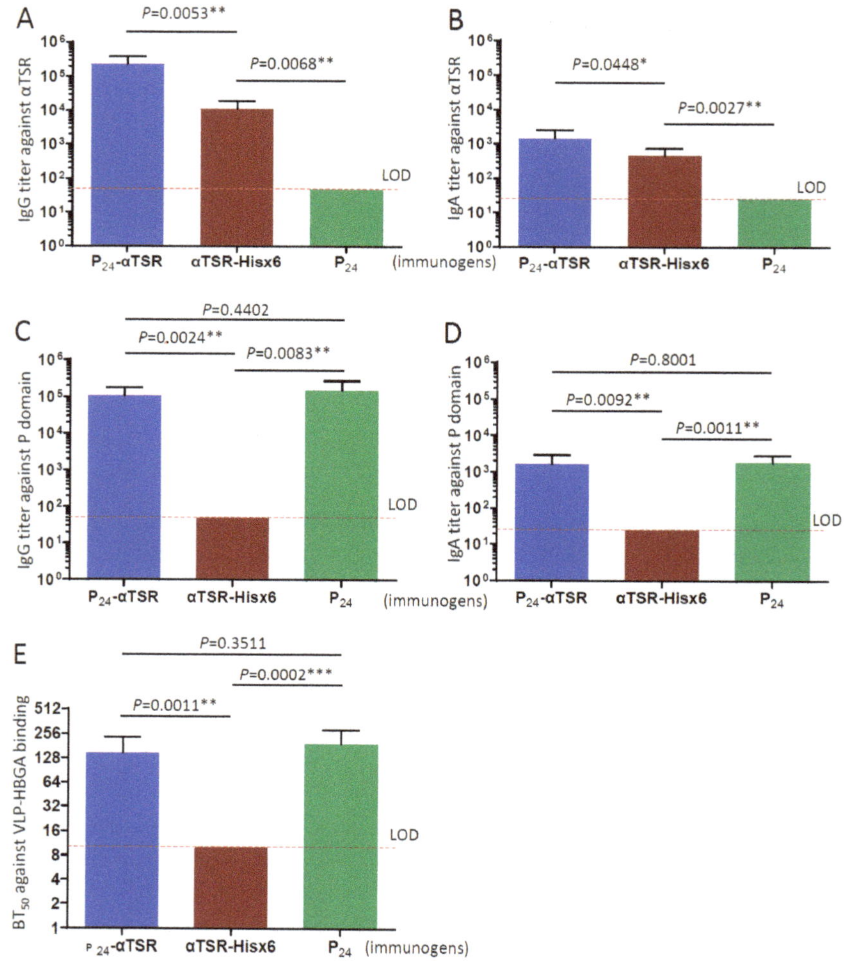

Figure 5. Immune responses to the P_{24}-αTSR NP toward its bipartite components. (**A**,**B**) *Plasmodium* αTSR-specific IgG (**A**) and IgA (**B**) titers in mice immunized with the P_{24}-αTSR NP (blue column) compared to titers induced by free αTSR (red column) and the P_{24} NP (green column) after three intramuscular immunizations. (**C**,**D**) Norovirus P domain-specific IgG (**C**) and IgA (**D**) titers in mice immunized with the P_{24}-αTSR NP (blue column) compared to titers induced by free αTSR (red column) and the P_{24} NP (green column) after three intramuscular immunizations. In (**A**,**C**), the Y-axes represent αTSR-specific (**A**) or P domain-specific (**C**) IgG titers. In (**B**,**D**), the Y-axes represent αTSR-specific (**B**) or P domain-specific (**D**) IgA titers. In (**A–D**), the X-axes indicate various immunogens. (**E**) The 50% blocking titers (BT50) of sera from mice immunized with the P_{24}-αTSR NP (blue column), P_{24} NP (green column), and free αTSR (red column) against norovirus VLP-glycan receptor attachment. The *Y*-axis represents BT50 values, while the *X*-axis indicates various immunogens. Corresponding statistical *p* values between the data groups are displayed above the columns. The limit of detection (LOD) is indicated by a red dashed line in each panel. Differences were deemed not significant at *p*-values > 0.05, significant at *p*-values < 0.05 (marked as "*"), significant at *p*-values < 0.01 (marked as "**"), and highly significant at *p*-values < 0.001 (marked as "***").

IgA responses to the two components of the P_{24}-αTSR NP followed similar trends, but were lower in magnitude than the IgG responses (Figure 5B,D). Specifically, the P_{24}-αTSR NP elicited a significantly higher αTSR-specific IgA titer of 1:1466, compared to 1:467 induced by the free αTSR protein ($p = 0.0448$, Figure 5B). However, the P_{24}-αTSR NP induced a P domain-specific IgA titer of 1:1667, which was comparable to the titer of 1:1828 elicited by the P_{24} NP ($p = 0.8001$, Figure 5D). As expected, the norovirus P_{24} NP did not induce *Plasmodium* αTSR-specific IgG or IgA responses (Figure 5A,B) and the free αTSR protein did not elicit norovirus P domain-specific antibodies (Figure 5C,D).

3.7. Inhibition Against Norovirus VLP–Glycan Receptor Interactions

Norovirus–HBGA attachment is a critical initial step in norovirus infection [64], which can be assessed using an EIA-based binding assay [60]. Studies using clinical trial samples have established a correlation between serum blockade of the norovirus VLP–HBGA interaction and protection against norovirus infection and disease [65–67]. Consequently, blockade of the norovirus VLP–HBGA interaction is regarded as a surrogate marker for neutralization, with the serum functional antibody response assessed through the norovirus VLP–HBGA attachment blocking assay. We observed that sera from mice immunized with the P_{24}-αTSR NP effectively blocked norovirus VLP–HBGA attachment, with a 50% blocking titer of 1:145.6, comparable to the titer of 1:194.3 observed in sera from mice immunized with P_{24} NP ($p = 0.3511$, Figure 5E). In contrast, sera from mice immunized with the αTSR protein did not exhibit such blocking effect. These findings suggest that the sera from mice immunized with the P_{24}-αTSR NP possess the potential to neutralize norovirus infection and/or disease.

3.8. Staining of Plasmodium Sporozoites Using Mouse Hyperimmune Sera

The sera from mice immunized with the P_{24}-αTSR NP were evaluated for their binding capability in regard to surface-exposed CSPs on *Plasmodium* sporozoites, using an IFA method. Two mouse sera with average αTSR-specific IgG titers were tested, and representative micrographs are shown in Figure 6. The air-dried *P. falciparum* sporozoites on slides were specifically stained by the sera from mice immunized with the P_{24}-αTSR NP at 8000-fold dilution (Figure 6A–D), whereas the control sera from mice immunized with the P_{24} NP at the same dilution did not produce staining (Figure 6E,F). The staining was highly specific, as non-sporozoite materials surrounding the sporozoites (Figure 6B,D) remained unstained, whereas the sporozoite surface was evenly stained by the sera. Further dilutions of the sera were not performed due to the limited availability of air-dried *P. falciparum* sporozoites on the slides. The observed specific binding of the hyperimmune sera to CSPs on the sporozoites strongly suggests that the αTSR-specific antibodies may inhibit the function of the αTSR domain of CSPs, potentially blocking the binding of *Plasmodium* sporozoites to their glycan receptors. Thus, the P_{24}-αTSR NP shows potential as a vaccine candidate against *Plasmodium* sporozoite infection and malaria.

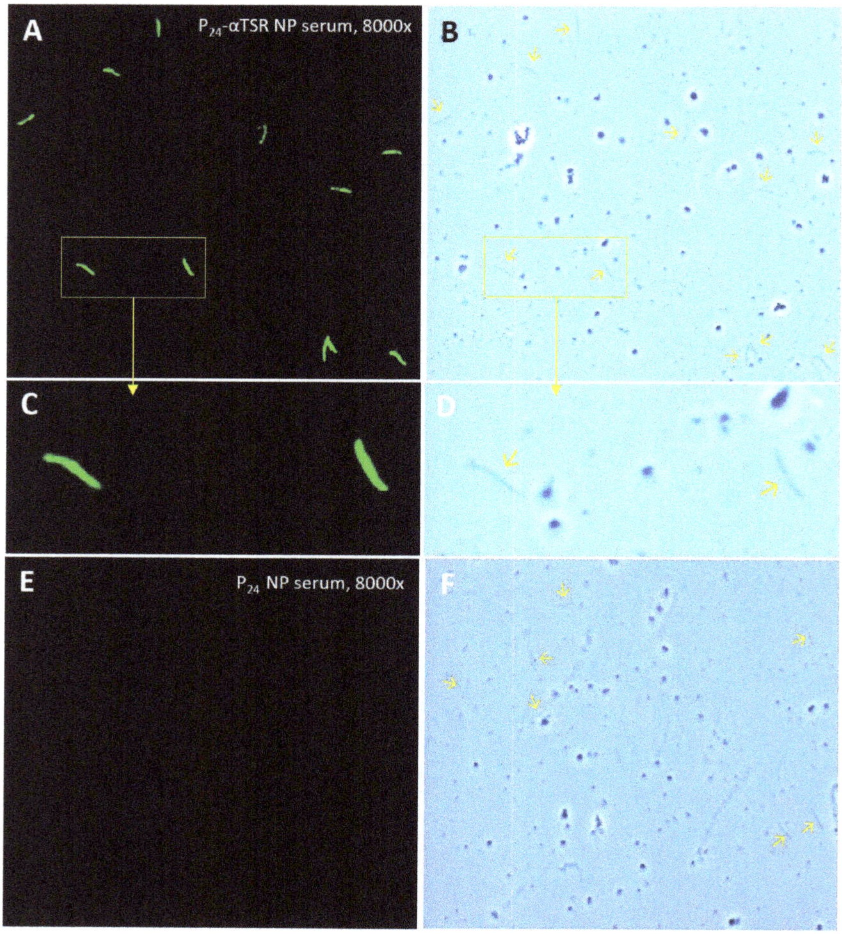

Figure 6. Specific staining of *P. falciparum* sporozoites by sera from mice immunized with the P_{24}-αTSR NP, in immunofluorescence assays (IFAs). (**A**) Representative IFA micrograph showing *Plasmodium* sporozoites stained by mouse sera (1:8000 dilution) after three immunizations with the P_{24}-αTSR NP. (**C**) Enlarged view of the marked regions in (**A**). (**E**) Sera (1:8000 dilution) from mice immunized with the P_{24} NP showed no staining of sporozoites (negative control). (**B**,**D**,**F**) Optical views of the same fields correspond to the IFA images in the left panels. Arrows indicate the visualized sporozoites.

4. Discussion

In this study, we leveraged the self-assembly properties of the P_{24} NP, derived from the modified norovirus protruding (P) domain through homotypic interactions, to design and evaluate a novel chimeric NP. This NP, named P_{24}-αTSR NP, displays the receptor biding αTSR domain of the *Plasmodium* parasite, with the ultimate goal of developing a combination vaccine targeting both pathogens and their associated diseases. The P_{24}-αTSR NP was produced by inserting the *Plasmodium* αTSR domain within an exposed surface loop of the norovirus P domain, followed by the generation of the P-αTSR fusion protein through an *E. coli* expression system, using scalable methods. A series of experiments validated the self-formation of the P_{24}-αTSR NP, consisting of a P_{24} NP core, with multiple αTSR domains forming the extended protrusions on its surface. Supporting evidence includes data from gel-filtration chromatography, a TEM inspection, DLS analysis, and structural modeling of the P_{24}-αTSR NP.

Since both the norovirus P domain and the *Plasmodium* αTSR region are responsible for the attachment to the host glycan receptor, a critical step in initiating pathogenic infection, we further investigated the potential of the P_{24}-αTSR NP as a dual vaccine candidate. Our results demonstrated that the P_{24}-αTSR NP is highly immunogenic in mice, eliciting robust antibody responses against both the norovirus P domain and *Plasmodium* αTSR. Moreover, we showed that the serum antibodies obtained after immunization with the P_{24}-αTSR NP blocked norovirus VLPs from attaching to their histo-blood group antigen (HBGA) receptors, an assay widely regarded as a surrogate for norovirus neutralization. Additionally, the serum antibodies bound specifically to circumsporozoite proteins (CSPs) on the surface of *Plasmodium* sporozoites. We hypothesize that this antibody–αTSR interaction may block the glycan receptor binding site of the αTSR in CSPs, thereby preventing the parasite from attaching to host receptors and, ultimately, inhibiting infection. Collectively, these findings support the potential of the P_{24}-αTSR NP as a promising vaccine candidate against both norovirus and *Plasmodium* parasites, as well as the diseases they cause.

We observed some variability in the size, structure, and morphology of the P_{24}-αTSR NPs in this study compared to similar NPs reported previously. For example, the morphologies of the P_{24}-αTSR NPs observed through TEM appeared less homogeneous than those of the P_{24} NP described earlier [17]. This suggests that the insertion of *Plasmodium* αTSR domains into the surface loops on the protrusions of the P_{24} NPs may influence the global structures and morphologies of the resulting P_{24}-αTSR NPs to some extent. One possibility is the formation of smaller NPs, referred to as P_{12}-αTSR NPs, which comprise 12 norovirus P domains with a calculated MW of 504 kDa, similar to the P_{12} NP reported previously [68].

This hypothesis is supported by the presence of a minor elution peak (P2) in the gel-filtration chromatography, corresponding to a MW smaller than the major P_{24}-αTSR NP elution peak (P1), observed during the analysis of the thrombin-cleaved GST-P-αTSR protein (Figure 1C,D). Additional evidence includes the observation of smaller NPs in the TEM micrographs that differ from the typical P_{24}-αTSR NPs. Notably, the minor peak corresponding to the P_{12}-αTSR NPs did not appear in the P-αTSR protein generated using the tag-free approach (compare Figure 1C with Figure 2F). Consistently, TEM inspection revealed better homogeneity in regard to the size and morphologies of the NPs produced via the tag-free method (compare Figure 1E with Figure 2G). These findings suggest that the tag-free procedure is a superior approach for producing P_{24}-αTSR NPs compared to the GST fusion method. Moreover, the tag-free procedure is easier to scale up as it eliminates the protease cleavage step, which may degrade the target protein to some extent. In fact, such degradation was observed, as evidenced by the appearance of double protein bands representing the P-αTSR protein after thrombin cleavage (Figure 1C,D). These degraded P-αTSR proteins seem to hinder their assembly into P_{24}-αTSR NPs.

The diverse morphologies of the octahedral P_{24}-αTSR NPs observed in the TEM micrographs in this study may be attributed to variations in viewing angles or symmetry axes, as illustrated by the 3D structural model (Figure 3). It is important to emphasize that, although some P_{12}-αTSR NPs may be present among the predominant P_{24}-αTSR NPs, their negative impact on the immune response outcomes is likely minimal. This is because both NP forms share the primary positive factors influencing the immune response of an antigen: (1) Both NP forms appear to preserve authentic pathogen-associated molecular patterns (PAMPs) in the bipartite components, as demonstrated by their strong reactivity to antibodies against norovirus VLP and the *Plasmodium* αTSR domain. (2) Both NPs have repetitive antigen patterns, promoting multiple stimulations of the host immune system.

(3) Both NPs have a large molecular size, serving as intrinsic adjuvants to enhance the immune response. The observed robust antibody responses, high amount of blocking antibody titers against norovirus VLP–glycan receptor attachment, and strong binding ability to *Plasmodium* sporozoites support this hypothesis. Finally, DLS showed that the majority the P-αTSR NP sizes ranged from 10 to 20 nm, consistent with the sizes observed in the TEM. However, DLS also revealed three minor NP populations, corresponding to particle sizes of 50, 200, and 500 nm. These much larger NP sizes may be attributed to aggregation of the P-αTSR protein or the P-αTSR NPs.

A limitation of this study is our inability to evaluate the cell culture-based neutralization or animal model-based protective efficacy of our vaccine candidate against both pathogens. This is due to the lack of a conventional cell culture system for human norovirus and the absence of a small animal model capable of mimicking norovirus infection and disease. However, since human norovirus can replicate to a certain extent in an enteroid culture system, we will collaborate with partners in the future to further assess the neutralization of our vaccine candidate using this platform. Furthermore, our laboratory currently lacks the capability to evaluate the protective efficacy of malaria vaccines. To address this, we plan to engage with potential collaborators who have the necessary expertise to assess the protective efficacy of our P_{24}-αTSR NP vaccine candidate against *Plasmodium* infection.

Although *Plasmodium* parasite-associated malaria and norovirus-induced diarrhea are very different diseases, they are likely to affect similar populations. For example, malaria is known to be prevalent in tropical and subtropical areas, particularly in Africa [69]. Similarly, global surveillance data shown that Africa has the highest prevalence of norovirus infection, with a rate of 15% [70]. Therefore, populations in Africa may be susceptible to both pathogens, justifying the usefulness of a combination vaccine against the two diseases.

5. Conclusions

In this study, we developed and evaluated a novel P_{24}-αTSR NP, which incorporates the receptor-binding domains of both norovirus and *Plasmodium* parasites. Our findings demonstrate its potential as a dual vaccine candidate against these two pathogens. This is evidenced by the robust antibody responses it elicited against the norovirus P domain and the αTSR domain of *Plasmodium* CSP. Moreover, sera from immunized mice bound specifically to CSPs on *Plasmodium* sporozoites and inhibited the attachment of norovirus VLPs to glycan receptors.

Author Contributions: Conceptualization, X.J. and M.T.; data curation, M.X., P.H. and M.T.; formal analysis, M.X. and M.T.; funding acquisition, X.J. and M.T.; investigation, M.X., P.H. and M.T.; methodology, M.X., P.H., F.S.V., W.J., X.J. and M.T.; project administration, M.T.; resources, W.J., X.J. and M.T.; software, F.S.V. and W.J.; supervision, W.J., X.J. and M.T.; validation, M.X.; visualization, M.X., P.H., F.S.V., W.J. and M.T.; writing—original draft, M.T.; writing—review and editing, M.X., F.S.V., W.J., X.J. and M.T. All authors have read and agreed to the published version of the manuscript.

Funding: The research described in this study was supported by funding from Cincinnati Children's Hospital Medical Center (CCHMC, Innovation Funds 2018–2020, GAP Fund 2020–2021, and Research Innovation and Pilot Grant 2020–2021 to M.T.).

Institutional Review Board Statement: All animal studies were conducted in compliance with the recommendations in the Guide for the Care and Use of Laboratory Animals (23a) from the National Institute of Health (NIH). The protocols used were approved by the Institutional Animal Care and Use Committee (IACUC) at Cincinnati Children's Hospital Research Foundation (animal welfare assurance no. A3108-01).

Informed Consent Statement: Not applicable.

Data Availability Statement: The data presented in this study are available in this article.

Acknowledgments: We thank David Haslam at Cincinnati Children's Hospital Medical Center for his assistance in analyzing the particle size distribution of the P_{24}-αTSR NP using dynamic light scattering. We also thank Photini Sinnis at Johns Hopkins University for kindly providing air-dried sporozoites of *Plasmodium falciparum* on slides. The research described in this study was supported by funding from Cincinnati Children's Hospital Medical Center (CCHMC, Innovation Funds 2018–2020, GAP Fund 2020–2021, and Research Innovation and Pilot Grant 2020–2021 to M.T.).

Conflicts of Interest: The authors declare that there are no conflicts of interest.

References

1. Glass, R.I.; Parashar, U.D.; Estes, M.K. Norovirus gastroenteritis. *N. Engl. J. Med.* **2009**, *361*, 1776–1785. [CrossRef] [PubMed]
2. Naghavi, M.; Wang, H.; Lozano, R.; Davis, A.; Liang, X.; Zhou, M.; Vollset, S.E.; Abbasoglu Ozgoren, A.; Abdalla, S.; Abd-Allah, F.; et al. Global, regional, and national age-sex specific all-cause and cause-specific mortality for 240 causes of death, 1990–2013: A systematic analysis for the Global Burden of Disease Study 2013. *Lancet* **2015**, *385*, 117–171. [CrossRef] [PubMed] [PubMed Central]
3. Bartsch, S.M.; Lopman, B.A.; Ozawa, S.; Hall, A.J.; Lee, B.Y. Global Economic Burden of Norovirus Gastroenteritis. *PLoS ONE* **2016**, *11*, e0151219. [CrossRef] [PubMed] [PubMed Central]
4. Hall, A.J.; Lopman, B.A.; Payne, D.C.; Patel, M.M.; Gastanaduy, P.A.; Vinje, J.; Parashar, U.D. Norovirus disease in the United States. *Emerg. Infect. Dis.* **2013**, *19*, 1198–1205. [CrossRef] [PubMed] [PubMed Central]
5. Mullaney, S.B.; Hyatt, D.R.; Salman, M.D.; Rao, S.; McCluskey, B.J. Estimate of the annual burden of foodborne illness in nondeployed active duty US Army Service Members: Five major pathogens, 2010–2015. *Epidemiol. Infect.* **2019**, *147*, e161. [CrossRef]
6. Prasad, B.V.; Hardy, M.E.; Dokland, T.; Bella, J.; Rossmann, M.G.; Estes, M.K. X-ray crystallographic structure of the Norwalk virus capsid. *Science* **1999**, *286*, 287–290. [CrossRef] [PubMed]
7. Hu, L.; Salmen, W.; Chen, R.; Zhou, Y.; Neill, F.; Crowe, J.E., Jr.; Atmar, R.L.; Estes, M.K.; Prasad, B.V.V. Atomic structure of the predominant GII.4 human norovirus capsid reveals novel stability and plasticity. *Nat. Commun.* **2022**, *13*, 1241. [CrossRef] [PubMed] [PubMed Central]
8. Shanker, S.; Choi, J.M.; Sankaran, B.; Atmar, R.L.; Estes, M.K.; Prasad, B.V. Structural analysis of histo-blood group antigen binding specificity in a norovirus GII.4 epidemic variant: Implications for epochal evolution. *J. Virol.* **2011**, *85*, 8635–8645. [CrossRef] [PubMed] [PubMed Central]
9. Tan, M.; Jiang, X. Norovirus-host interaction: Multi-selections by human histo-blood group antigens. *Trends Microbiol.* **2011**, *19*, 382–388. [CrossRef] [PubMed] [PubMed Central]
10. Tan, M.; Jiang, X. Norovirus and its histo-blood group antigen receptors: An answer to a historical puzzle. *Trends Microbiol.* **2005**, *13*, 285–293. [CrossRef] [PubMed]
11. Cao, S.; Lou, Z.; Tan, M.; Chen, Y.; Liu, Y.; Zhang, Z.; Zhang, X.C.; Jiang, X.; Li, X.; Rao, Z. Structural basis for the recognition of blood group trisaccharides by norovirus. *J. Virol.* **2007**, *81*, 5949–5957. [CrossRef] [PubMed] [PubMed Central]
12. Chen, Y.; Tan, M.; Xia, M.; Hao, N.; Zhang, X.C.; Huang, P.; Jiang, X.; Li, X.; Rao, Z. Crystallography of a Lewis-binding norovirus, elucidation of strain-specificity to the polymorphic human histo-blood group antigens. *PLoS Pathog.* **2011**, *7*, e1002152. [CrossRef] [PubMed] [PubMed Central]
13. Tan, M.; Hegde, R.S.; Jiang, X. The P domain of norovirus capsid protein forms dimer and binds to histo-blood group antigen receptors. *J. Virol.* **2004**, *78*, 6233–6242. [CrossRef] [PubMed] [PubMed Central]
14. Singh, B.K.; Leuthold, M.M.; Hansman, G.S. Structural Constraints on Human Norovirus Binding to Histo-Blood Group Antigens. *mSphere* **2016**, *1*, e00049-16. [CrossRef] [PubMed] [PubMed Central]
15. Singh, B.K.; Leuthold, M.M.; Hansman, G.S. Human noroviruses' fondness for histo-blood group antigens. *J. Virol.* **2015**, *89*, 2024–2040. [CrossRef] [PubMed] [PubMed Central]
16. Singh, B.K.; Koromyslova, A.; Hefele, L.; Gurth, C.; Hansman, G.S. Structural Evolution of the Emerging 2014-2015 GII.17 Noroviruses. *J. Virol.* **2016**, *90*, 2710–2715. [CrossRef] [PubMed] [PubMed Central]
17. Tan, M.; Jiang, X. The p domain of norovirus capsid protein forms a subviral particle that binds to histo-blood group antigen receptors. *J. Virol.* **2005**, *79*, 14017–14030. [CrossRef] [PubMed] [PubMed Central]
18. Tan, M.; Fang, P.; Chachiyo, T.; Xia, M.; Huang, P.; Fang, Z.; Jiang, W.; Jiang, X. Noroviral P particle: Structure, function and applications in virus-host interaction. *Virology* **2008**, *382*, 115–123. [CrossRef] [PubMed] [PubMed Central]
19. Tan, M.; Jiang, X. Norovirus Capsid Protein-Derived Nanoparticles and Polymers as Versatile Platforms for Antigen Presentation and Vaccine Development. *Pharmaceutics* **2019**, *11*, 472. [CrossRef] [PubMed] [PubMed Central]
20. Sinnis, P.; Zavala, F. The skin: Where malaria infection and the host immune response begin. *Semin. Immunopathol.* **2012**, *34*, 787–792. [CrossRef] [PubMed] [PubMed Central]

21. World Health Organization. *World Malaria Report 2022*; World Health Organization: Geneva, Switzerland, 2022.
22. Menard, R.; Tavares, J.; Cockburn, I.; Markus, M.; Zavala, F.; Amino, R. Looking under the skin: The first steps in malarial infection and immunity. *Nat. Rev. Microbiol.* **2013**, *11*, 701–712. [CrossRef] [PubMed]
23. Kappe, S.H.; Buscaglia, C.A.; Nussenzweig, V. *Plasmodium* sporozoite molecular cell biology. *Annu. Rev. Cell Dev. Biol.* **2004**, *20*, 29–59. [CrossRef] [PubMed]
24. Prudencio, M.; Rodriguez, A.; Mota, M.M. The silent path to thousands of merozoites: The *Plasmodium* liver stage. *Nat. Rev. Microbiol.* **2006**, *4*, 849–856. [CrossRef] [PubMed]
25. Sturm, A.; Amino, R.; van de Sand, C.; Regen, T.; Retzlaff, S.; Rennenberg, A.; Krueger, A.; Pollok, J.M.; Menard, R.; Heussler, V.T. Manipulation of host hepatocytes by the malaria parasite for delivery into liver sinusoids. *Science* **2006**, *313*, 1287–1290. [CrossRef] [PubMed]
26. Gerald, N.; Mahajan, B.; Kumar, S. Mitosis in the human malaria parasite *Plasmodium falciparum*. *Eukaryot. Cell* **2011**, *10*, 474–482. [CrossRef] [PubMed] [PubMed Central]
27. Josling, G.A.; Llinas, M. Sexual development in *Plasmodium* parasites: Knowing when it's time to commit. *Nat. Rev. Microbiol.* **2015**, *13*, 573–587. [CrossRef] [PubMed]
28. Bancells, C.; Llora-Batlle, O.; Poran, A.; Notzel, C.; Rovira-Graells, N.; Elemento, O.; Kafsack, B.F.C.; Cortes, A. Revisiting the initial steps of sexual development in the malaria parasite *Plasmodium falciparum*. *Nat. Microbiol.* **2019**, *4*, 144–154. [CrossRef] [PubMed] [PubMed Central]
29. Medica, D.L.; Sinnis, P. Quantitative dynamics of *Plasmodium yoelii* sporozoite transmission by infected anopheline mosquitoes. *Infect. Immun.* **2005**, *73*, 4363–4369. [CrossRef] [PubMed] [PubMed Central]
30. Dundas, K.; Shears, M.J.; Sinnis, P.; Wright, G.J. Important Extracellular Interactions between *Plasmodium* Sporozoites and Host Cells Required for Infection. *Trends Parasitol.* **2019**, *35*, 129–139. [CrossRef] [PubMed] [PubMed Central]
31. Pinzon-Ortiz, C.; Friedman, J.; Esko, J.; Sinnis, P. The binding of the circumsporozoite protein to cell surface heparan sulfate proteoglycans is required for *Plasmodium* sporozoite attachment to target cells. *J. Biol. Chem.* **2001**, *276*, 26784–26791. [CrossRef] [PubMed] [PubMed Central]
32. Coppi, A.; Tewari, R.; Bishop, J.R.; Bennett, B.L.; Lawrence, R.; Esko, J.D.; Billker, O.; Sinnis, P. Heparan sulfate proteoglycans provide a signal to *Plasmodium* sporozoites to stop migrating and productively invade host cells. *Cell Host Microbe* **2007**, *2*, 316–327. [CrossRef] [PubMed] [PubMed Central]
33. Frevert, U.; Sinnis, P.; Cerami, C.; Shreffler, W.; Takacs, B.; Nussenzweig, V. Malaria circumsporozoite protein binds to heparan sulfate proteoglycans associated with the surface membrane of hepatocytes. *J. Exp. Med.* **1993**, *177*, 1287–1298. [CrossRef] [PubMed] [PubMed Central]
34. Doud, M.B.; Koksal, A.C.; Mi, L.Z.; Song, G.; Lu, C.; Springer, T.A. Unexpected fold in the circumsporozoite protein target of malaria vaccines. *Proc. Natl. Acad. Sci. USA* **2012**, *109*, 7817–7822. [CrossRef] [PubMed] [PubMed Central]
35. Xia, M.; Vago, F.; Han, L.; Huang, P.; Nguyen, L.; Boons, G.; Klassen, J.; Jiang, W.; Tan, M. The αTSR Domain of *Plasmodium* Circumsporozoite Protein Bound Heparan Sulfates and Elicited High Titers of Sporozoite Binding Antibody After Displayed by Nanoparticles. *Int. J. Nanomed.* **2023**, *2023*, 3087–3107. [CrossRef] [PubMed]
36. Tan, M. Norovirus Vaccines: Current Clinical Development and Challenges. *Pathogens* **2021**, *10*, 1641. [CrossRef] [PubMed]
37. Mattison, C.P.; Cardemil, C.V.; Hall, A.J. Progress on norovirus vaccine research: Public health considerations and future directions. *Expert Rev. Vaccines* **2018**, *17*, 773–784. [CrossRef] [PubMed] [PubMed Central]
38. Lucero, Y.; Vidal, R.; O'Ryan, G.M. Norovirus vaccines under development. *Vaccine* **2018**, *36*, 5435–5441. [CrossRef] [PubMed]
39. Leroux-Roels, G.; Cramer, J.P.; Mendelman, P.M.; Sherwood, J.; Clemens, R.; Aerssens, A.; De Coster, I.; Borkowski, A.; Baehner, F.; Van Damme, P. Safety and Immunogenicity of Different Formulations of Norovirus Vaccine Candidate in Healthy Adults: A Randomized, Controlled, Double-Blind Clinical Trial. *J. Infect. Dis.* **2018**, *217*, 597–607. [CrossRef] [PubMed] [PubMed Central]
40. Bernstein, D.I.; Atmar, R.L.; Lyon, G.M.; Treanor, J.J.; Chen, W.H.; Jiang, X.; Vinje, J.; Gregoricus, N.; Frenck, R.W., Jr.; Moe, C.L.; et al. Norovirus vaccine against experimental human GII.4 virus illness: A challenge study in healthy adults. *J. Infect. Dis.* **2015**, *211*, 870–878. [CrossRef] [PubMed] [PubMed Central]
41. Atmar, R.L.; Bernstein, D.I.; Harro, C.D.; Al-Ibrahim, M.S.; Chen, W.H.; Ferreira, J.; Estes, M.K.; Graham, D.Y.; Opekun, A.R.; Richardson, C.; et al. Norovirus vaccine against experimental human Norwalk Virus illness. *N. Engl. J. Med.* **2011**, *365*, 2178–2187. [CrossRef] [PubMed]
42. Atmar, R.L.; Baehner, F.; Cramer, J.P.; Song, E.; Borkowski, A.; Mendelman, P.M.; Group NORS. Rapid Responses to 2 Virus-Like Particle Norovirus Vaccine Candidate Formulations in Healthy Adults: A Randomized Controlled Trial. *J. Infect. Dis.* **2016**, *214*, 845–853. [CrossRef] [PubMed] [PubMed Central]
43. Atmar, R.L.; Baehner, F.; Cramer, J.P.; Lloyd, E.; Sherwood, J.; Borkowski, A.; Mendelman, P.M.; Group NORS. Persistence of Antibodies to 2 Virus-Like Particle Norovirus Vaccine Candidate Formulations in Healthy Adults: 1-Year Follow-up with Memory Probe Vaccination. *J. Infect. Dis.* **2019**, *220*, 603–614. [CrossRef] [PubMed]

44. Kim, L.; Liebowitz, D.; Lin, K.; Kasparek, K.; Pasetti, M.F.; Garg, S.J.; Gottlieb, K.; Trager, G.; Tucker, S.N. Safety and immunogenicity of an oral tablet norovirus vaccine, a phase I randomized, placebo-controlled trial. *J. Clin. Investig.* **2018**, *3*, e121077. [CrossRef] [PubMed] [PubMed Central]
45. Targett, G. Phase 3 trial with the RTS,S/AS01 malaria vaccine shows protection against clinical and severe malaria in infants and children in Africa. *Evid. Based Med.* **2015**, *20*, 9. [CrossRef] [PubMed]
46. Agnandji, S.T.; Lell, B.; Fernandes, J.F.; Abossolo, B.P.; Methogo, B.G.; Kabwende, A.L.; Adegnika, A.A.; Mordmuller, B.; Issifou, S.; Kremsner, P.G.; et al. A phase 3 trial of RTS,S/AS01 malaria vaccine in African infants. *N. Engl. J. Med.* **2012**, *367*, 2284–2295. [CrossRef] [PubMed]
47. RTS,S Clinical Trials Partnership. Efficacy and safety of RTS,S/AS01 malaria vaccine with or without a booster dose in infants and children in Africa: Final results of a phase 3, individually randomised, controlled trial. *Lancet* **2015**, *386*, 31–45. [CrossRef] [PubMed]
48. RTS,S Clinical Trials Partnership. Efficacy and safety of the RTS,S/AS01 malaria vaccine during 18 months after vaccination: A phase 3 randomized, controlled trial in children and young infants at 11 African sites. *PLoS Med.* **2014**, *11*, e1001685. [CrossRef] [PubMed] [PubMed Central]
49. Nardin, E.H.; Oliveira, G.A.; Calvo-Calle, J.M.; Wetzel, K.; Maier, C.; Birkett, A.J.; Sarpotdar, P.; Corado, M.L.; Thornton, G.B.; Schmidt, A. Phase I testing of a malaria vaccine composed of hepatitis B virus core particles expressing *Plasmodium falciparum* circumsporozoite epitopes. *Infect. Immun.* **2004**, *72*, 6519–6527. [CrossRef] [PubMed] [PubMed Central]
50. Datoo, M.S.; Dicko, A.; Tinto, H.; Ouedraogo, J.B.; Hamaluba, M.; Olotu, A.; Beaumont, E.; Ramos Lopez, F.; Natama, H.M.; Weston, S.; et al. Safety and efficacy of malaria vaccine candidate R21/Matrix-M in African children: A multicentre, double-blind, randomised, phase 3 trial. *Lancet* **2024**, *403*, 533–544. [CrossRef] [PubMed]
51. Datoo, M.S.; Natama, H.M.; Some, A.; Bellamy, D.; Traore, O.; Rouamba, T.; Tahita, M.C.; Ido, N.F.A.; Yameogo, P.; Valia, D.; et al. Efficacy and immunogenicity of R21/Matrix-M vaccine against clinical malaria after 2 years' follow-up in children in Burkina Faso: A phase 1/2b randomised controlled trial. *Lancet Infect. Dis.* **2022**, *22*, 1728–1736. [CrossRef] [PubMed]
52. Genton, B. R21/Matrix-M malaria vaccine: A new tool to achieve WHO's goal to eliminate malaria in 30 countries by 2030? *J. Travel Med.* **2023**, *30*, taad140. [CrossRef] [PubMed] [PubMed Central]
53. Datoo, M.S.; Natama, M.H.; Some, A.; Traore, O.; Rouamba, T.; Bellamy, D.; Yameogo, P.; Valia, D.; Tegneri, M.; Ouedraogo, F.; et al. Efficacy of a low-dose candidate malaria vaccine, R21 in adjuvant Matrix-M, with seasonal administration to children in Burkina Faso: A randomised controlled trial. *Lancet.* **2021**, *397*, 1809–1818. [CrossRef] [PubMed] [PubMed Central]
54. Aderinto, N.; Olatunji, G.; Kokori, E.; Sikirullahi, S.; Aboje, J.E.; Ojabo, R.E. A perspective on Oxford's R21/Matrix-M malaria vaccine and the future of global eradication efforts. *Malar. J.* **2024**, *23*, 16. [CrossRef] [PubMed] [PubMed Central]
55. Hanboonkunupakarn, B.; Mukaka, M.; Jittamala, P.; Poovorawan, K.; Pongsuwan, P.; Stockdale, L.; Provstgaard-Morys, S.; Chotivanich, K.; Tarning, J.; Hoglund, R.M.; et al. A randomised trial of malaria vaccine R21/Matrix-M with and without antimalarial drugs in Thai adults. *NPJ Vaccines* **2024**, *9*, 124. [CrossRef] [PubMed] [PubMed Central]
56. Tan, M.; Huang, P.; Xia, M.; Fang, P.A.; Zhong, W.; McNeal, M.; Wei, C.; Jiang, W.; Jiang, X. Norovirus P particle, a novel platform for vaccine development and antibody production. *J. Virol.* **2011**, *85*, 753–764. [CrossRef] [PubMed] [PubMed Central]
57. Xia, M.; Hoq, M.R.; Huang, P.; Jiang, W.; Jiang, X.; Tan, M. Bioengineered pseudovirus nanoparticles displaying the HA1 antigens of influenza viruses for enhanced immunogenicity. *Nano Res.* **2022**, *15*, 4181–4190. [CrossRef] [PubMed] [PubMed Central]
58. Xia, M.; Huang, P.; Jiang, X.; Tan, M. A Nanoparticle-Based Trivalent Vaccine Targeting the Glycan Binding VP8* Domains of Rotaviruses. *Viruses* **2021**, *13*, 72. [CrossRef] [PubMed] [PubMed Central]
59. Xia, M.; Huang, P.; Sun, C.; Han, L.; Vago, F.S.; Li, K.; Zhong, W.; Jiang, W.; Klassen, J.S.; Jiang, X.; et al. Bioengineered Norovirus S60 Nanoparticles as a Multifunctional Vaccine Platform. *ACS Nano* **2018**, *12*, 10665–10682. [CrossRef] [PubMed] [PubMed Central]
60. Huang, P.; Farkas, T.; Zhong, W.; Tan, M.; Thornton, S.; Morrow, A.L.; Jiang, X. Norovirus and histo-blood group antigens: Demonstration of a wide spectrum of strain specificities and classification of two major binding groups among multiple binding patterns. *J. Virol.* **2005**, *79*, 6714–6722. [CrossRef] [PubMed] [PubMed Central]
61. Pettersen, E.F.; Goddard, T.D.; Huang, C.C.; Meng, E.C.; Couch, G.S.; Croll, T.I.; Morris, J.H.; Ferrin, T.E. UCSF ChimeraX: Structure visualization for researchers, educators, and developers. *Protein Sci.* **2021**, *30*, 70–82. [CrossRef] [PubMed] [PubMed Central]
62. Zhang, M.; Mandraju, R.; Rai, U.; Shiratsuchi, T.; Tsuji, M. Monoclonal Antibodies against *Plasmodium falciparum* Circumsporozoite Protein. *Antibodies* **2017**, *6*, 11. [CrossRef] [PubMed] [PubMed Central]
63. Swearingen, K.E.; Lindner, S.E.; Shi, L.; Shears, M.J.; Harupa, A.; Hopp, C.S.; Vaughan, A.M.; Springer, T.A.; Moritz, R.L.; Kappe, S.H.; et al. Interrogating the *Plasmodium* Sporozoite Surface: Identification of Surface-Exposed Proteins and Demonstration of Glycosylation on CSP and TRAP by Mass Spectrometry-Based Proteomics. *PLoS Pathog.* **2016**, *12*, e1005606. [CrossRef] [PubMed] [PubMed Central]

64. Tenge, V.R.; Hu, L.; Prasad, B.V.V.; Larson, G.; Atmar, R.L.; Estes, M.K.; Ramani, S. Glycan Recognition in Human Norovirus Infections. *Viruses* **2021**, *13*, 2066. [CrossRef] [PubMed] [PubMed Central]
65. Reeck, A.; Kavanagh, O.; Estes, M.K.; Opekun, A.R.; Gilger, M.A.; Graham, D.Y.; Atmar, R.L. Serological correlate of protection against norovirus-induced gastroenteritis. *J. Infect. Dis.* **2010**, *202*, 1212–1218. [CrossRef] [PubMed] [PubMed Central]
66. Atmar, R.L.; Bernstein, D.I.; Lyon, G.M.; Treanor, J.J.; Al-Ibrahim, M.S.; Graham, D.Y.; Vinje, J.; Jiang, X.; Gregoricus, N.; Frenck, R.W.; et al. Serological Correlates of Protection against a GII.4 Norovirus. *Clin. Vaccine Immunol.* **2015**, *22*, 923–929. [CrossRef] [PubMed]
67. Ramani, S.; Estes, M.K.; Atmar, R.L. Correlates of Protection against Norovirus Infection and Disease-Where Are We Now, Where Do We Go? *PLoS Pathog.* **2016**, *12*, e1005334. [CrossRef] [PubMed] [PubMed Central]
68. Tan, M.; Fang, P.A.; Xia, M.; Chachiyo, T.; Jiang, W.; Jiang, X. Terminal modifications of norovirus P domain resulted in a new type of subviral particles, the small P particles. *Virology* **2011**, *410*, 345–352. [CrossRef] [PubMed] [PubMed Central]
69. CDC. Where Malaria Occurs. 2024. Available online: https://www.cdc.gov/malaria/data-research/index.html#:~:text=control/elimination%20programs.-,Generally,%25in%25warmer%25regions%25closer%25to%25the%25equator,%25malaria%25transmission,the%25disease%25remains%25a%25risk (accessed on 28 December 2024).
70. Qi, R.; Huang, Y.T.; Liu, J.W.; Sun, Y.; Sun, X.F.; Han, H.J.; Qin, X.R.; Zhao, M.; Wang, L.J.; Li, W.; et al. Global Prevalence of Asymptomatic Norovirus Infection: A Meta-analysis. *eClinicalMedicine* **2018**, *2–3*, 50–58. [CrossRef] [PubMed] [PubMed Central]

vaccines

MDPI

Review

Regulating Immune Responses Induced by PEGylated Messenger RNA–Lipid Nanoparticle Vaccine

Hyein Jo, Jaewhoon Jeoung, Wonho Kim and Dooil Jeoung *

Department of Biochemistry, College of Natural Sciences, Kangwon National University, Chuncheon 24341, Republic of Korea; qnfdudn1212@gmail.com (H.J.); heyjhw@kangwon.ac.kr (J.J.); kimwonho99@kangwon.ac.kr (W.K.)

* Correspondence: jeoungd@kangwon.ac.kr

Abstract: Messenger RNA (mRNA)-based therapeutics have shown remarkable progress in the treatment and prevention of diseases. Lipid nanoparticles (LNPs) have shown great successes in delivering mRNAs. After an mRNA-LNP vaccine enters a cell via an endosome, mRNA is translated into an antigen, which can activate adaptive immunity. mRNAs can bind to various pattern recognition receptors (PRRs), including toll-like receptors (TLRs), and increase the production of inflammatory cytokines. This review summarizes mechanisms of innate immunity induced by mRNAs. Polyethylene glycol (PEG) has been employed as a component of the mRNA-LNP vaccine. PEGylated nanoparticles display enhanced stability by preventing aggregation of particles. However, PEGylation can cause adverse reactions, including blood clearance (ABC) of nanoparticles via complement activation and anaphylaxis. Mechanisms of PEG-induced ABC phenomenon and anaphylaxis are presented and discussed. There have been studies aimed at reducing immune responses associated with PEG to make safe and effective vaccines. Effects of modifying or replacing PEG in reducing immune responses associated with PEGylated nanoparticles are also discussed. Modifying mRNA can induce immune tolerance, which can prevent hypersensitivity reactions induced by PEGylated mRNA-LNP vaccines. Current progress of immune tolerance induction in association with mRNA-LNP is also summarized. This review might be helpful for developing safe and effective PEGylated mRNA-LNP vaccines.

Keywords: ABC phenomenon; allergies; complement; immune tolerance; innate immunity; mRNA-LNPs; PEG

1. Immune Response Induced by mRNA Vaccines: Antigen Translation

Messenger RNA (mRNA) has been widely applied in therapies, including protein replacement therapies and cancer immunotherapies [1–3]. Successful development of COVID-19 vaccines has greatly contributed to the advancement of mRNA technology.

mRNA can be produced by in vitro transcription (IVT). It can then be translated into functional proteins. mRNA vaccines make it possible to mount an immune response without ever exposing to pathogens. mRNA vaccines show various advantages, including rapid preparation, reduced contamination, no induction of insertional mutagenesis, and high biodegradability [4]. These advantages make it possible to develop mRNA-based personalized vaccines. mRNAs have shown translational efficiency, reliable stability, and controlled immunogenicity [5]. It has been shown that mRNAs can bind to receptors and signaling pathways such as toll-like receptors (TLRs) and the JAK-STAT pathway and promote adaptive immune response [6]. Induction of adaptive immunity involves activation of T cells and the generation of specific antibodies.

Non-replicating mRNA (NRM) and virally generated, self-amplifying RNA (SAM) are now being investigated as vaccines. SAMs encoding a viral replication machinery allow for abundant production of antigen of interest [7]. mRNA structures include 5′ and 3′ untranslated structure regions (UTRs) [8]. 5′ UTR can be optimized to enhance translational efficiency [9]. Replacing ORF coding for viral structure proteins with viral RNA-dependent RNA polymerase can result in cytoplasmic expansion of the replicon structure. Refining 5′ cap structure, 3′ poly (A) tail, and codons can enhance the stability of mRNA [10,11].

mRNA vaccines encoding proteins of interest can be introduced into the cytoplasm of host cells, where they can be expressed into antigens [12]. Internalization of mRNA by antigen-presenting cells (APCs) can be made easier by performing nucleotide modifications and codon optimizations [5]. A single vaccination of mRNA can produce many antigens and induce major histocompatibility complex (MHC)-mediated T cell responses and production of neutralizing antibodies [13]. APCs can internalize the mRNA vaccine via endocytosis. mRNAs can escape endosomes and access the cytoplasm for translation into proteins of interest. The proteasome complex can break down intracellular proteins into antigenic peptides. Antigenic peptides are presented to CD4$^+$ T cells or CD8$^+$ T cells via MHC molecules on surfaces of APCs. Antigenic peptides can induce differentiation of CD4$^+$ T cells into effector T cells, including T helper (TH) cells and T follicular helper (TFH) cells. TFH cells can promote germinal center reactions to induce the production of neutralizing antibodies. Activated CD8$^+$ T cells by MHC-I/Peptide can exert cytotoxic effects. mRNAs can produce extracellular or cell surface proteins that can be recognized by cognate B cells and induce the production of neutralizing antibodies. Targeted delivery and endosomal escape remain challenging issues for developing mRNA-based therapy. Figure 1 shows that proteins encoded by mRNA can induce both innate immunity and adaptive immunity.

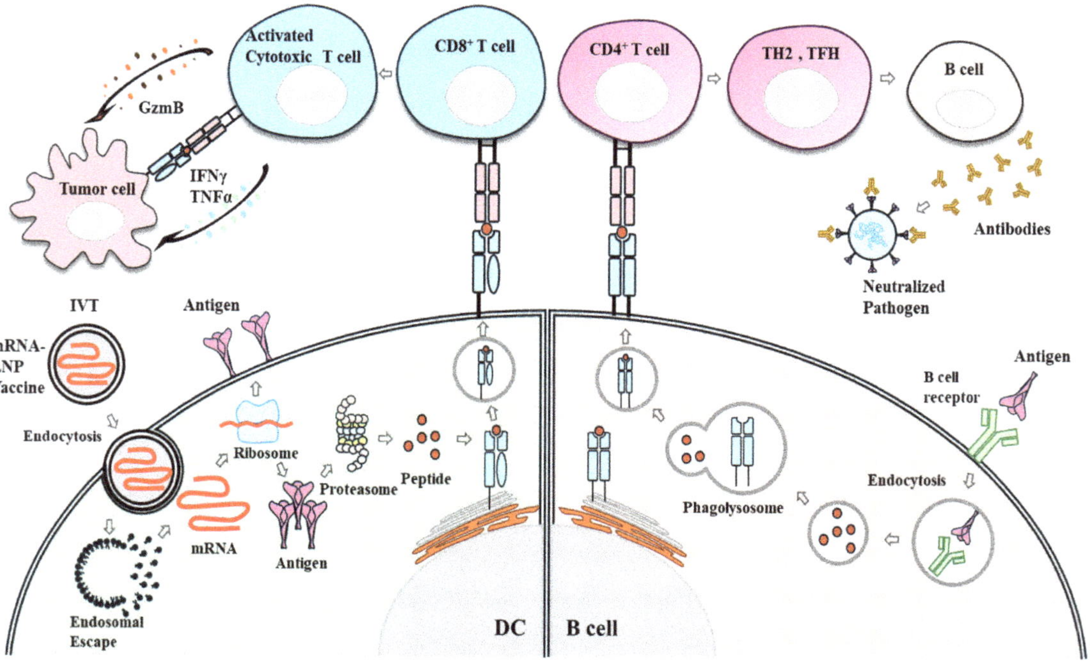

Figure 1. Immune responses induced by mRNA vaccines. mRNA-LNP vaccine is endocytosed by antigen presenting cells. Following endosomal escape, mRNAs are then translated into corresponding

proteins. Proteins undergo proteasomal degradation. Peptides are then presented on MHC-I to induce activation of CD8^{+} T cells. Activated CD8^{+} T cells can secrete IFNγ and TNF-α. Activated CD8^{+} T cells can kill tumor cells by granzyme B (GzmB). Secreted proteins are recognized and engulfed by antigen presenting cells such as B cells. Antigenic peptides are presented on the MHC-II of B cells to induce activation of CD4^{+} T cells. Activated CD4^{+} T cells, such as TH2 cells and TFH cells, can activate B cells to induce production of antigen-specific antibodies. Arrows denote the direction of reaction. TFH denotes follicular T helper cells. IVT denotes in vitro transcription.

2. Structure of mRNA-LNPs

Lipid nanoparticles (LNPs), a leading non-viral delivery system, can efficiently co-deliver vaccines and immune adjuvants to lymphoid organs [14]. LNPs have shown successes as delivery vehicles for small molecules such as mRNAs and siRNAs [15,16]. Negative charges and toxicity of RNAs hinder their efficient uptake by host cells. LNP-encapsulation has been employed to generate safe and effective vaccines. LNPs can enhance the stability of messenger RNA and facilitate endosomal escape [17]. Enhanced half-life stability of LNPs can promote an enhanced permeation and retention (EPR) effect that can result in accumulation of LNPs in cancer tissues [18].

LNPs mostly contain four kinds of lipids: (1) cationic or ionizable cationic lipids such as N-[1-(2,3-dioleyloxy)propyl]-N,N,N-trimethylammonium chloride (DOTMA); (2) phospholipids such as dipalmitoylphosphatidylcholine (DPPC); (3) lipid-anchored polyethylene glycol (PEG) such as DMG-PEG2000; and (4) cholesterol [19]. Ionizable liposomes are neutral in the bloodstream. They display minimal off-target interactions with anionic cell membranes of blood cells [20]. In an endosomal environment, ionizable lipids undergo protonation on the liposome surface, which can promote membrane disruption and endosomal escape of mRNA. PEG-lipids can enhance the stability of mRNA-LNP by decreasing particle aggregation without causing changes in mitochondrial metabolism in neuroblastoma cells [21]. Decreasing PEG molarity or replacing PEG structures can increase protein expression by increasing the size of mRNA-LNP [22]. Large nanoparticles that contain more lipids and mRNAs can promote high endosomal escape and provide more mRNA for protein production. PEG molarity might affect innate and/or adaptive immunity by regulating the production of antigenic epitopes after vaccination with an mRNA-LNP vaccine. It is also probable that decreasing PEG molarity can change the physicochemical properties of nanoparticles to affect protein expression.

3. Induction of Immunity by Lipid Nanoparticles

mRNA-LNPs can act as an immune adjuvant based on the following findings: (1) mRNAs can be sensed by Toll-like receptor (TLR), melanoma differentiation-associated *protein* 5 (MDA5), and nucleotide-binding domain, leucine-rich–containing family, pyrin domain–containing-3 (NLRP3) [23–25]; (2) mRNA-LNPs can increase the production of inflammatory cytokines including IL-1α, IL-1β, and IFN-β, and IFN-γ through RNA sensing receptors [26–29]; and (3) mRNA-LNPs can promote adaptive immunity involving activation of T cells and germinal center (GC) B-cell responses [30–32]. BNT162b2 mRNA-LNP vaccine can increase levels of proinflammatory cytokines including IL-2, CCL2, CCL4, and CCL5 [33]. The BNT162b2 mRNA-LNP vaccine can also induce CD8^{+} T cell responses via type I interferon (IFN)-dependent MDA5 signaling [33]. In addition, ionizable lipids can activate the TLR2, TLR4, and NLRP3 inflammasome [34,35]. Cationic or ionizable cationic lipids and PEGylated lipids of the mRNA-LNP-COVID19 vaccine can also induce acute inflammatory syndrome by increasing the secretion of cytokines, including IL-1α, IFN-γ, IL-1β, and IL-8 by activating the complement system [34,36,37].

4. Innate RNA Sensing

PRRs are known to mediate innate immune responses [38]. PRRs are expressed on APCs such as monocytes and dendritic cells (DCs) [39]. PRRs include TLRs, leucine-rich repeat-containing receptors (NLRs), and retinoic acid-inducible gene 1 (RIG-I) like receptors [38,40]. TLRs are crucial mediators and regulators of host immunity [34].

Nucleic acids can act as molecular patterns [38]. Sensing IVT mRNA by endsomal TLR7/8 [41] can activate the myeloid differentiation factor (MyD) 88 pathway to initiate type1 IFN pathways [41,42]. TLR7/8 agonist can induce production of type1 IFN and IL-6 and adaptive immune responses [41]. Anticancer mRNA-LNP vaccines can trigger TLR4 [43], TLR7 [44], and can be stimulators of IFN gene 16 (STING16) signaling pathways [45]. RNA sensing by PRRs can also activate the IFN pathway, which in turn can increase the production of proinflammatory cytokines, resulting in the activation of APCs [46]. mRNA sensing by TLRs (TLR3, TLR7, and TLR8) can activate the innate immune system to increase production of proinflammatory cytokines [47]. Endosomal TLRs, specifically TLR3, TLR7, and TLR8, can act as viral RNA PRRs. Single-stranded RNA (ssRNA) can bind to TLR-7 and TLR-8 to activate the MyD88 pathway and induce production of the germinal center B cell-dependent IgG antibody [48]. TLR7 can activate NF-κB, MyD88, IRF7 pathways and increase secretion of CCL2, CXCL10, IL-1β, IL-6, IL-8, and type I IFNs [49–51]. TLR-3 can sense double-stranded RNA (dsRNA) and activate the TRIF (TIR-domain containing protein inducing type 1 IFN) pathway while decreasing the expression of adenosine deaminase RNA specific 1 (ADAR1) [52]. TLR-3 can also activate IRF3 and NF-kB to increase production of proinflammatory cytokines and type I IFN [53]. LNP-poly(I:C) can activate lysosomal TLR3 in human and mouse fibroblasts [54].

Viral infection can activate RNA sensors, including RIG-I-like receptors (RLRs), TLRs, protein kinase R (PKR), and ADAR1 [55]. RLRs can sense cytosolic RNA and act as sensors of the BNT162b2 vaccine [33]. Cytosolic RLRs and MDA5 along with mitochondrial antiviral signaling protein (MAVS) can increase production of proinflammatory cytokine [56,57]. DsRNA can activate MAVS and STING-mediated signaling pathways in response to ionizing radiation [56]. MDA5 and RLR can increase the secretion of proinflammatory cytokines by activating IFN response factor 3 (IRF3) and NF-kB [57]. Understanding mechanisms of innate immunity mediated by the IFN pathway is needed to develop safe and effective mRNA-LNP vaccines. Figure 2 shows that RNA sensing by innate PRRs can induce production of proinflammatory cytokines.

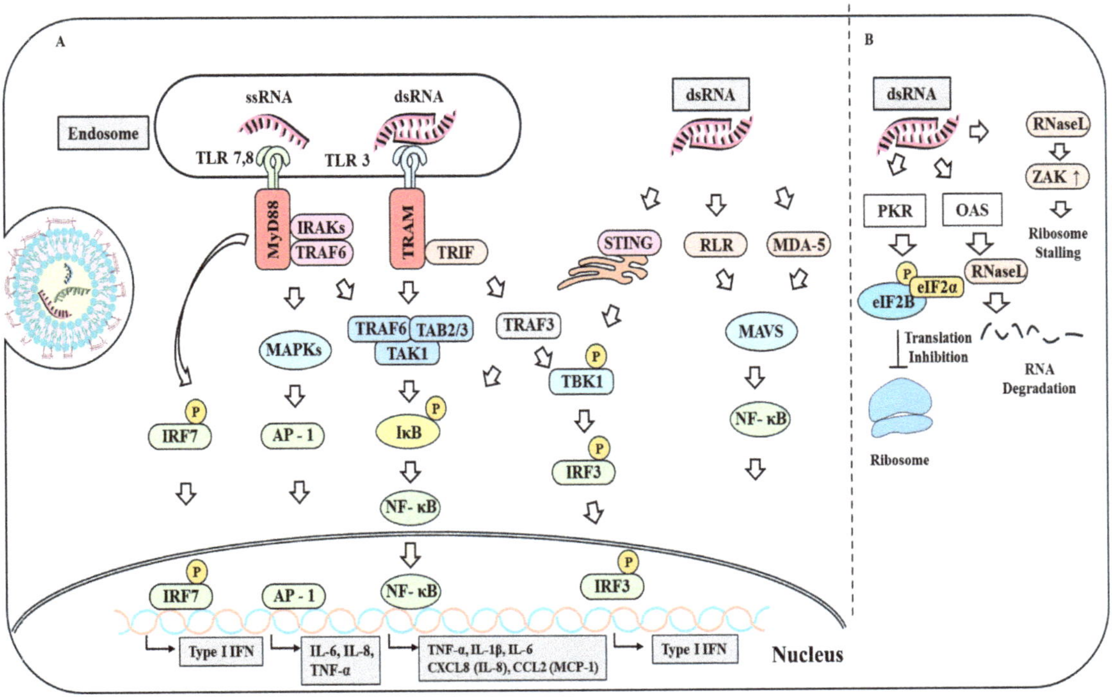

Figure 2. Innate immunity and translational inhibition induced by mRNA. (**A**) Endosomal TLR (TLR3, 7, or 8) can bind to ssRNA or dsRNA. Endosomal TLR7/8 can increase the production of type I IFNs and various proinflammatory cytokines including IL-6, CCL2, and CXCL10 by activating MYD88-NF-kB signaling. DsRNA produced during IVT can activate PKR or OAS pathway, resulting in degradation of mRNA. Endosomal TLR3 can bind to dsRNA and activate TRIF pathway. RLRs and MDA5 can bind to cytosolic RNAs and increase production of proinflammatory cytokines by activating IRF3 and N-kB. (**B**) DsRNA can bind to PKR and OAS. PKR can suppress mRNA translation by inducing phosphorylation of eIF2. OAS can activate RNase L, which in turn can induce translational arrest. Hollow arrows denote the direction of reaction. IRAK, interleukin-1-receptor-associated kinase; MAVS, mitochondrial antiviral-signaling protein; TAB, Mitogen-activated *protein* kinase kinase kinase 7-interacting *protein* 1; TAK, mitogen-activated *protein* kinase kinase kinase 7; TRAF6, tumor necrosis factor receptor-associated factor 6; TRAM, translocating- chain-associated membrane protein; ZAK, mixed lineage kinase. Hollow arrows denote the direction of reaction. T bars denote inhibition of reaction.

5. Translational Inhibition of mRNA by dsRNA

Innate RNA sensors can decrease the expression of antigen from mRNA by activating ribonuclease L (RNase L) that cleaves single stranded RNA (ssRNA) via JNK and p38 MAPK pathways [58]. RNase L can decrease the translation of stress-responsive genes by inducing ribosome stalling through activation of human leucine zipper and sterile alpha motif-containing kinase (ZAK) [58]. Double stranded RNA (dsRNA) produced during IVT can restrict mRNA translation by activating PKR and NF-kB [59,60]. PKR and RIG-1 are known to recognize different types of RNA to interfere with translation [61–63]. PKR can induce phosphorylation of eukaryotic initiation factor 2 (eIF2), which in turn can suppress mRNA translation [63,64].

DsRNA can also trigger innate immunity via oligoadenylate synthetase 3 (OAS) [65]. Upon detecting viral dsRNA, OAS3 can activate RNase L [66], which cleaves both cellular and viral RNAs, further triggering RLR signaling to increase the production of proinflammatory cytokines [67]. RNA-binding protein E3 (E3L) can bind to PKR and inhibit activation of PKR [68]. Thus, E3L can be employed to prevent translational inhibition by dsRNA. This can be used to develop effective mRNA-LNP vaccine. Figure 2 describes the mechanism of RNA degradation by PKR and OAS. To make effective mRNA-LNP vaccines, it is necessary to purify IVT mRNA without dsRNA.

6. Effects of Modifications of mRNAs on Immunogenicity and Translation

mRNA transfection can stimulate TLR3 and TLR7 and induce IFN-β expression in human and mouse fibroblasts [69]. Unmodified mRNA-LNP vaccines can activate innate immune responses that often lead to cytokine storms. Ss RNA 40, a synthetic mimic of SARS-CoV-2 RNA, can increase the expression of proinflammatory cytokines in monocyte-derived dendritic cells (MDDCs) via TLR8 [70]. Inositol requiring enzyme 1 α (IRE1α) inhibitor MKC8866 can prevent ssRNA40 from increasing the production of proinflammatory cytokines [70]. Thus, the TLR8-IRE1α system might serve as a target to control the cytokine storm associated with COVID-19 mRNA-LNP vaccines. TLR signaling can activate IRF3 to induce production of proinflammatory cytokines including IFN [71]. DsRNA can increase the expression of IFN stimulated genes (ISGs) by activating IRF3 and NF-κB [57].

Nucleoside-modified mRNA-LNP vaccines might abrogate innate immunity critical for inducing adaptive immunity. Therefore, it is advisable to modify components of mRNA-LNP vaccines. Post-transcriptional modifications such as RNA pseudouridylation can affect gene expression and regulate various biological processes [72]. Replacing uridine with pseudouridine (Ψ) or other derivatives such as N1-methylpseudouridine (m1ψ), 2-thiouridine, 5-methyluridine, and 5-methylcytidine can prevent endosomal TLR3 and TLR7/8 and cytosolic RIG-1 from recognizing mRNAs and increasing the production of IFN-β [69,73]. Thus, modifications of mRNAs can reduce innate immunity induced by mRNA-LNP vaccines. RNA-editing enzyme ADAR1 (adenosine deaminase acting on RNA) can inhibit immune checkpoint blocker (ICB) responsiveness by suppressing immunogenic dsRNAs arising from dysregulated expression of endogenous retroviral elements (EREs) [74].

Pseudouridylation of alpha ketoglutarate-dependent dioxygenase (ALKBH3), a tumor suppressor gene, can enhance the translation of ALKBH3 and reduce tumor growth [72]. Chemical modification of nucleotides can enhance translation levels by preventing innate sensors from recognizing IVT mRNA [75,76]. Enhanced protein translation efficiency can be attributed to stability of pseudouridine-modified IVT mRNAs resulting from a reduced activity of RNA-dependent PKR [75,76].

7. PEGylated mRNA-LNPs Cause Hypersensitivity Reactions, Allergies, and Complement Activation-Related Pseudoallergy (CARPA)

Safe and effective vaccines can stimulate the innate immune system and prime adaptive immune responses. However, immune-related adverse effects can arise because of immunological actions induced by these vaccines. Stimulation of RNA sensors such as TLRs can induce cytokine storm, airway infiltration of immune cells, and activation of mast cells [77]. It has been shown that mRNA COVID-19 vaccines can cause allergies [78,79]. COVID-19 can activate mast cell-derived proteases and increase levels of eosinophil-associated mediators based on assays employing sera and lung tissues of

COVID-19 patients [77]. The BNT162b2 COVID-19 vaccine can exacerbate asthma by enhancing sensitivity to histamines in a human ex vivo model [80]. Phospholipase A2IIA activity is correlated with severity of COVID-19 vaccine [81]. The COVID-19 vaccine can increase levels of lysophosphatidic acid (LPA) and platelet activating factor (PAF) by phospholipase A2IIA [81]. It is well known that PAF mediates allergic reactions via the PAF receptor [82].

PEGylated proteins along with complete Freund's adjuvant can induce production of PEG-specific antibodies [83]. PEG displays immunogenicity when it is used as an excipient in LNPs [84]. PEG can cause anaphylaxis in some drug reactions [85]. High molecular weight PEGs are known to cause allergic reactions. In animals, anti-PEG immunity consists of mostly the anti-PEG IgM response. PEGylated mRNA-LNP vaccines can induce the production of anti-PEG IgG, anti-PEG IgM, and anti-PEG IgE [86–88]. COVID-19 mRNA-LNP vaccines (Comirnaty and Spikevax) can induce hypersensitivity reactions (HSRs) or anaphylaxis by increasing levels of anti-PEG IgG/IgM [86]. Anti-PEG antibodies can be produced without prior exposure to PEGylated nanoparticles [89]. Anti-PEG Abs have been shown to be correlated with a reduced clinical efficacy of PEGylated therapeutics due to the presence of anti-PEG antibodies.

Anti-PEG antibodies on the surface of PEGylated nanoparticles can induce HSRs such as hypothermia and hypotension by activating Fcγ receptors on innate immune cells [90]. TLRs play a key role in allergic airway inflammatory responses, including airway infiltration of immune cells, increased levels of Th2 cytokines, and metaplasia of lung epithelial cells [91]. The expression levels of inflammation-related genes are decreased in TLR2 KO cells compared to those in wild type cells [91]. The TLR2-ERK signaling pathway can mediate allergic airway inflammation by regulating expression of Gal-3 [92]. The PEGylated siRNA-lipoplex can increase production of anti-PEG IgM by activating the TLR7 signaling pathway [93]. PEGylated liposomes containing TLR agonists can cause acute HSRs by inducing high levels of the anti-PEG antibody [94]. Thus, TLR signaling pathways might be responsible for the induction of allergic symptoms by PEG-mRNA-LNP vaccines.

People who have high levels of anti-PEG Ab in their blood can develop HSRs/anaphylaxis in response to PEGylated vaccines [95]. PEGylated proteins can induce production of anti-PEG antibody by activating T cells responses, while PEGylated liposomes function in a T cell-independent manner [96]. PEG has been implicated in pseudoallergic or anaphylactoid reactions via complement activation-related pseudoallergy (CARPA) [97]. Several types of anti-PEG antibodies are known to contribute to HSRs and premature drug release from PEGylated carriers [95,98,99]. Comirnaty, a PEGylated mRNA vaccine, can induce production of anti-PEG IgG and anti-PEG IgM antibodies, which in turn can activate the complement system to cause pseudoanaphylaxis [95]. Anti-PEG IgG, but not anti-PEG IgM, can induce symptoms of HSRs including hypothermia, altered lung function, and hypotension after administration of PEGylated liposomal doxorubicin (PLD) in C57BL/6 and non-obese diabetic/severe combined immunodeficiency (NOD/SCID) mice [90]. Anti-PEG IgG can cause HSRs through Fcγ receptors independent of complements [90]. Immune cells including basophils, monocytes, neutrophils, and mast cells can mediate anti-PEG IgG-mediated HSRs by recognizing mRNA-LNPs through PRRs such as nucleotide-binding oligomerization domain (NOD)-like receptors (NLR) [100]. This suggests that pathogenesis of allergic diseases involves both innate and adaptive immune responses. Anti-PEG IgE can also mediate HSRs induced by PEGylated nanoparticles [101].

Immunization with the mRNA-LNP COVID-19 vaccine can induce anaphylaxis, like CARPA, following their first or subsequent vaccinations [102,103]. CARPA results from elevated levels of complement C3a and sC5b-C9 [102]. Unlike other IgE-mediated reactions, CARPA can cause anaphylaxis upon first exposure. Allergic inflammations such as asthma involve lectin-dependent activation of the complement system via Ficoli-1 in response to allergens [104,105]. The mRNA-LNP COVID-19 vaccine can activate complements and increase the production of proinflammatory cytokines including IL-1α, IL-1β, IL-6, IL-8, IFN-γ, and TNF-α in PBMC cultures [36]. The COVID-19 vaccine can activate complements in an anti-PEG IgG-dependent manner and induce activation of basophils and M1 polarization of monocytes [78]. Thus, activation of complements caused by the COVID-19 vaccine can increase production of proinflammatory cytokines. PEGylation can reduce clearance of LNPs by mononuclear phagocytic cells by affecting binding of opsonin protein to liposomes [106–109]. In the case of the COVID-19 vaccine, interaction with immune cells can activate the complement system [110]. This implies that the stability of COVID-19 vaccine can be reduced by the complement system. Anti-PEG antibodies can reduce stability of PEGylated nanoparticles [111]. Anti-PEG antibodies can activate complement system by binding to PEG on the surface of PEGylated liposomal doxorubicin (PLD) [112]. Repeated injections of PEG-LNPs can decrease stability of PEG-LNPs [113]. Induction of accelerated blood clearance (ABC) phenomenon by PEG results from anti-PEG IgM produced after the initial injection [113]. Injections of siRNA complexed with PEGylated cationic liposomes (PLpx) can lead to rapid clearance of subsequent doses of PLpx by inducing production of anti-PEG IgM from peritoneal PEG-specific B cells [114]. Anti-PEG IgG and IgM are responsible for loss of efficacy of mRNA-LNP vaccines by inducing ABC phenomenon [115]. LNPs can activate splenic marginal zone B (MZB) cells to induce production of anti-PEG IgM, which in turn can activate the complement system and lead to the ABC phenomenon. Anti-PEG IgM can destabilize mRNA-LNP by releasing mRNA in a complement-dependent way [116]. Complement fragment C3 produced after activation of complement proteases can bind to complement receptors (CRs) expressed on phagocytes, which in turn can mediate the ABC phenomenon [110,117]. C3 convertase (C4b2a) can promote the cleavage of C3 into C3a and C3b, which can promote phagocytosis [118]. C5 convertase activated by C3 can promote the cleavage of C5 into C5a and C5b, which can promote formation of a membrane attack complex (MAC) to cause phagocytosis [118]. Figure 3 shows that PEGylated mRNA-LNP vaccine can induce production of anti-PEG antibodies, which in turn can promote complement activation to cause the ABC phenomenon.

An activated complement system can induce release of histamine and opsonization of LNPs [110]. This implies that activation of complement may lead to allergic inflammation. Mast cells, eosinophils, basophils, macrophages, plasmacytoid dendritic cells, and neutrophils all express receptors for C3a and C5a [119,120]. Anti-PEG IgM and IgG can induce production of anaphylatoxins C3a and C5a via the classical pathway, which in turn can induce degranulation of mast cells [121,122], smooth muscle contraction, and release of histamine, serotonin, PAF, and cysteinyl leukotrienes (CysLTs) [123,124]. C5a contributes to inflammation through activation of the NLRP3 inflammasome and induction of IL-1β [125]. C5a-C5aR signaling can increase numbers of Th1 and Th2 cells while decreasing the number of Treg cells in a mouse model of asthma [126]. C5 can prevent the development of experimental allergic asthma [127], meaning that complement activation can mediate allergic responses induced by mRNA-LNP vaccines. Figure 3 shows that anaphylatoxins can induce degranulation of mast cells to cause anaphylaxis. A better understanding of interactions between PEGylated nanoparticles and the blood immune system is necessary for developing effective and safe PEG-LNP vaccines to reduce ABC phenomenon and allergic responses.

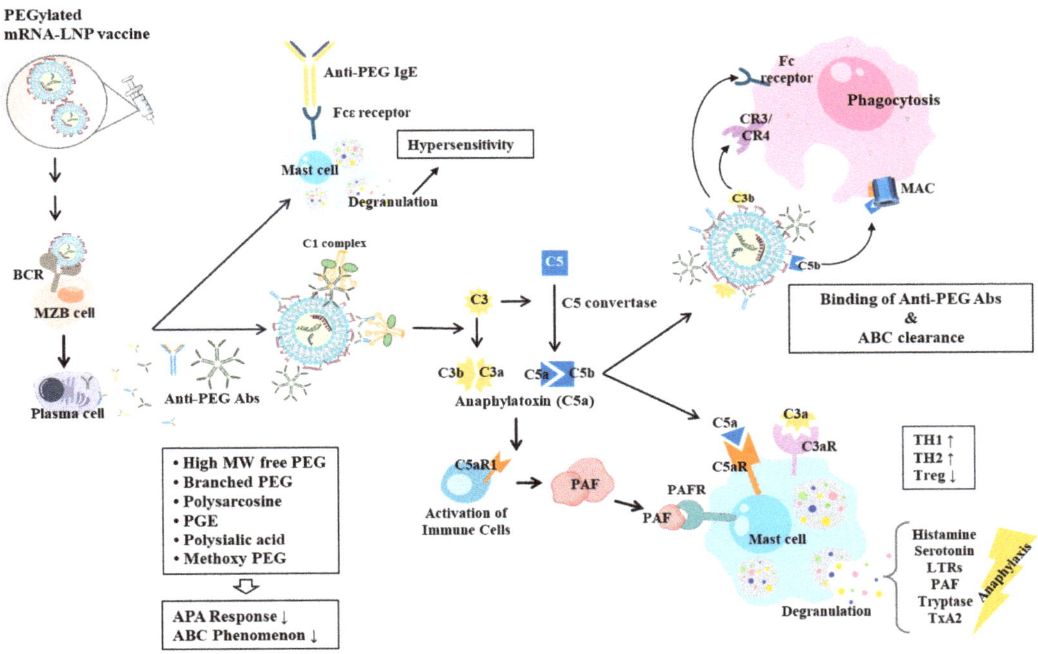

Figure 3. PEG induces complement activation-associated anaphylaxis and ABC phenomenon. PEGylated mRNA-LNP vaccine induces cross linking of B cell receptors. B cells are then differentiated into plasma cells. Plasma cells produce anti-PEG antibodies. Anti-PEG IgE activates immune cells such as mast cells and basophils, resulting in anaphylaxis through Fcε receptor in a complement-independent manner. PEG-anti-PEG-IgM complex can activate complement via classical pathway. Activation of complement produces anaphylatoxins such as C3a and C5a. These anaphylatoxins can stimulate mast cells to release various mediators and induce degranulation of mast cells. Anaphylatoxins can also induce immune cells to release PAF. PAF can bind to mast cells to cause degranulation of mast cells. PEG-mRNA-LNP vaccine can induce the production of anti-PEG IgM, which in turn can bind PEG-LNP. This binding can induce complement activation to produce complement fragments such as C3b and C5b. C3 can activate C5 convertase to produce C5a and C5b. These fragments can bind to complement receptors (CR3 and CR4) to induce ABC phenomenon by macrophages. Arrows denote the direction of reaction. ↓ denotes decreased expression/activity and ↑ denotes increased expression/activity. LTR, leukotriene receptor; TxA2, thromboxane A2; PGE, poly glutamic acid ethylene oxide.

8. How to Reduce Immune Responses Associated with PEG

Many polymers, including polyvinylpyrrolidone and poly(2-methyl-2-oxazoline), can induce polymer-specific antibodies [128]. These antibodies contribute to the pathogenesis of HSRs and the ABC phenomenon [129,130]. mRNA-LNP complexes can cause allergies and autoimmune diseases [90,131,132]. Thus, it is necessary to minimize these immune-related adverse reactions resulting from repeated delivery of mRNA-LNPs. Each component of LNP formulations can elicit an immune response. Replacing each component might reduce immune-related adverse reactions induced by mRNA-LNP vaccines. Since PEG is considered an epitope, PEG density on the surface of LNP can influence the production of anti-PEG IgM. High molecular weight PEGs can cause HSRs more easily than low molecular weight PEGs [133] owing to high level of anti-PEG IgM.

However, 1,2-distearoyl-sn-glycero-3-phosphoethanolamine-poly(ethylene glycol) (DSPE)-PE$_{2,40k}$ (branched PEG lipid derivative) does not trigger the ABC phenomenon owing to the fact that 40 kDa DSPE-$PE_{2,40k}$ can induce lower anti-PEG IgM levels than linear PEG-modified nanocarriers (DSPE-PE_{40k}) [134]. Cleavable-branched and branched PEG-lipid derivatives can reduce the ABC phenomenon by inducing lower levels of anti-PEG antibodies [135].

High molecular weight (MW) free PEG can enhance the stability of PEG liposomes in animals with high levels of pre-existing anti-PEG antibody (APA) by suppressing APA production [136,137]. Free PEG can cause durable suppression of the APA response [136]. Free PEG might also be able to inhibit anaphylaxis by saturating B cell receptors (BCRs) to prevent the infused PEG-liposome from activating BCRs and inducing the production of APA. A lack of APA response might exert a negative effect on the ABC phenomenon.

Polysarcosine (PSar)-liposomes display fewer systemic and off-target interactions and higher stability than PEGylated liposomes [113]. PSar-liposomes can induce low levels of antibodies for PSar-liposomes [113]. Multiple injections of PSar-liposomes can reduce the ABC phenomenon by inhibiting accumulation of liposomes in the liver and spleen [113,138]. Liposomes with the longest PSar chain (68 mers) show enhanced stability and reduced ABC phenomenon compared with PEG-liposomes [139]. Poly (2-methyacryloyloxyethyl phosphorylcholine) (PMPC)-LNPs formulated from 1,2-dipalmitoyl-*sn*-glycero-3-phosphoethanolamine (DPPE)-LNPs display no cytotoxic effects or induce inflammatory responses. They also exhibit higher mRNA transfection efficiency than PEG-LNPs [108]. Poly-glutamic acid-ethylene oxide (PGE) graft copolymers can replace PEG on LNPs [83]. Unlike PEG-LNPs, PGE-LNP does not induce production of the anti-PEG antibody [83]. Injection of Rg3-liposomes does not induce production of IgM or activation of the complement system in blood circulation [140]. Thus, Rg3 can enhance the stability of liposomes without inducing the ABC phenomenon or immune responses. These reports indicate that replacing PEG with other polymers can decrease immune-related adverse reactions induced by PEG-mRNA-LNP vaccines.

Structural modification of the PEG moiety might reduce immunogenicity associated with PEG. Methoxy PEG grafting to allogeneic splenocytes can increase the number of Treg cells while decreasing Th17 lymphocytes [141]. This indicates that modification of PEG can induce immune tolerance. For example, polysialic acid-modified liposomes do not induce immune responses typical of PEGylated liposomes [142]. Inserting gangliosides into the bilayer beside PEG can also suppress the production of anti-PEG IgM [143]. A negative charge on PEG at the liposomal surface can induce HSRs by activating the complement system [144]. However, conjugating PEG200 to cholesterol (CHOL-PEG) can reduce complement activation to attenuate HSRs [144]. Thus, structural modifications of PEG can regulate immune-related adverse reactions by promoting immune tolerance.

9. Tolerance-Inducing mRNAs Can Reduce Allergies and Autoimmunity

Vaccine technology using mRNA-LNP can be employed to suppress antigen-specific immune responses [145,146]. Since mRNA-LNP vaccines can induce allergies and autoimmune response, it is critical to develop tolerogenic vaccines that are safe and effective. PEGylated nanoparticles can create a tolerogenic immune microenvironment by inducing lower complement activation than non-PEGylated PLGA nanoparticles around the injection site [147]. mRNA-LNP encoding non-allergenic MHC-II-binding epitopes has shown tolerogenic effects in a mouse model of peanut anaphylaxis [148]. This tolerogenic effect is mediated by tolerogenic Treg cells and accompanied by suppression of the production of Th2 cytokines (IL-4, IL-5, and IL-13), IgE synthesis, and increased expression of TGF-β and IL-10 [148]. PEG-mRNA-LNP enclosing antigenic T cell epitope can increase frequencies of $Foxp3^{+}$Tregs while decreasing antigen-specific T cells producing TNF [149].

The PEGylated mRNA-LNP vaccine can be considered as a self-antigen that can cause autoimmune diseases. Modification of mRNA can treat allergies and autoimmune diseases associated with mRNA-LNP vaccines [148,150]. Modified mRNA-LNP can enter cells via endocytosis. After endocytosis, mRNA undergoes endosomal escape. mRNA is then translated into antigens. After proteasomal breakdown into peptides, these peptides are then presented on the surface of APCs to induce the generation of Foxp3$^+$ regulatory T cells (Treg), which can induce tolerance to allergens or self-antigens [148].

Modifying nucleosides can suppress inflammation induced by exogenous mRNAs [151]. For example, uridine can be replaced with pseudouridine (Ψ) or m1Ψ, and cytosine can be replaced with 5-methyl cytosine. Unmodified mRNA can induce the production of type I IFN by stimulating innate immune sensors [152]. Unmodified ovalbumin (OVA)-LNP shows better anti-tumor effects than OVA-LNP with m1Ψ modification [153]. Unmodified OVA-LNP can increase the number of CD40$^+$ DCs and the frequency of granzyme B$^+$/IFN-γ^+/TNF-α^+ OVA peptide-specific CD8$^+$ T cells [153]. Unmodified mRNA can also induce the production of proinflammatory cytokines including IFN-γ, TNF-α, and IL-2. However, high doses of Ψ-modified mRNA do not induce the production of proinflammatory cytokines [153,154]. Thus, Ψ-modified mRNA might induce a tolerogenic effect. It has been shown that m1Ψ-modified mRNA displays high translational efficiency without inducing innate immune responses [155,156]. Such m1Ψ-modified mRNA does not induce activation of TLR7 [157]. It has been found that m1Ψ mRNA can induce proliferation of immune suppressive FoxP3$^+$ T reg cells to attenuate peanut-induced anaphylaxis. This effect is accompanied by decreased expression of Th2 cytokines and IgE synthesis [148]. However, m1Ψ mRNA does not lead to generalized immune suppression due to tolerogenic effects on APCs such as DCs [148]. In other words, DCs remain functional after internalization of m1Ψ mRNA.

Unmodified mRNA and small dsRNA can induce innate immunity and immune-related adverse reactions [158]. Modification and codon optimization of mRNA may decrease the risk of unwanted immune responses associated with mRNA-LNP vaccines. There are several ways to achieve immune tolerance: (1) nucleoside modifications, which may suppress activation of innate immune sensors and production of inflammatory cytokines; (2) optimization of UTR, which may increase translation efficiency and reduce immunogenicity of mRNA-LNP; (3) removal of dsRNA during IVT, which may reduce the production of inflammatory cytokines induced by mRNA-LNP vaccine; (4) reducing short RNA production during IVT, which may reduce the induction of innate immunity; and (5) introduction of conformational changes into T7 RNA polymerase, which may reduce production of dsRNA during IVT. Figure 4 shows that mRNA modification can induce immune tolerance to treat allergies, RA, and type I diabetes resulting from injection of PEGylated mRNA-LNP vaccines.

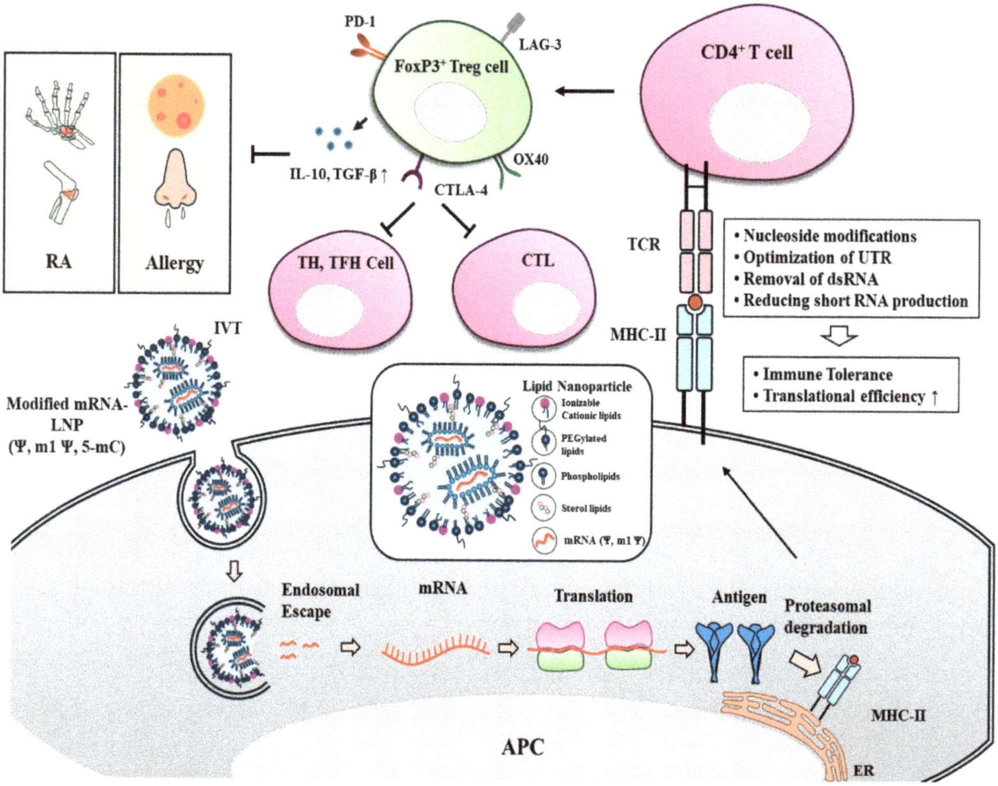

Figure 4. Induction of immune tolerance by mRNA-LNP. After endocytosis of modified PEG-mRNA-LNP vaccine by APC, mRNA undergoes endosomal escape. mRNA is then released and translated into antigen. Antigenic epitopes can be presented on MHC-II and bind to TCR on CD4+ T cells. CD4+ T cells are then differentiated into FoxP3+ Treg cells. Treg cells can secrete immune suppressive cytokines such as IL-10 and TGF-β. These Treg cells can induce immune tolerance. This tolerogenic effect is accompanied by decreased expression TH2 cytokines and IgE. CTLA-4, cytotoxic T-lymphocyte antigen 4; IVT, in vitro transcription; LAG3, lymphocyte activation gene-3; PD-1, programmed death-1; RA, rheumatoid. Arrows and hollow arrows denote the direction of reaction. T bars denote inhibition of reaction. ↑ denotes increased expression/activity.

10. Discussion and Perspectives

Current clinical trials have shown that mRNA-LNP vaccines will continue to be treatment paradigms. However, these mRNA-LNP vaccines are known to cause unwanted immune responses. Vaccination with any COVID-19 vaccine (BNT162b2, Ad26.COV2.S, and mRNA-1273) can reduce risk of developing long COVID [159]. Individuals who have booster vaccination of mRNA-based COVID-19 vaccines (BNT162b2, mRNA-1273) display a higher risk of developing autoimmune connective diseases including rheumatoid arthritis and alopecia areata [160]. Concerns regarding the development of autoimmunity may exert a negative impact on uptake of COVID-19 vaccines. COVID-19 vaccines (BNT162b2 and CoronaVac) can cause autoimmune conditions requiring hospital care [161]. It would be necessary to develop tolerogenic mRNA-LNP vaccines to reduce the risk of developing autoimmune diseases.

The occurrence of PEG antibodies in the general population is increasing every year due to wide use of PEG in daily life and the development of sensitive assay systems for detecting anti-PEG antibodies. The ABC phenomenon induced by anti-PEG antibody could diminish the therapeutic efficacy of PEGylated products and increase the risk of HSR. It will be necessary to better understand the interplay between nanostructures and immune cells to reduce the ABC phenomenon. The contents of LNPs, dosing schedules, and routes of administration may reduce the ABC phenomenon. The role of PEGs in the induction of HSRs, ABC phenomenon, anaphylaxis, and mechanisms associated with these immunological reactions should be further investigated to make safe and effective mRNA-LNP vaccines. It is also necessary to test immunogenicity of PEGylated mRNA-LNP vaccines in animal models in association with the ABC phenomenon, CARPA, and HSRs.

To develop effective mRNA-LNP vaccines, it will be necessary to reduce unwanted immune responses associated with mRNA-LNPs. The following approaches may improve the efficacy of mRNA-LNPs: (1) optimizing mRNA sequence and molecular modification for inducing tolerogenic effects; (2) identifying route of administration for reducing immune-related adverse events; and (3) identifying PEG alternatives to reduce ABC phenomenon and HSRs. The induction of antigen-specific immune tolerance by modifying the mRNA-LNP vaccine has shown clinical benefits for treating allergic diseases. Circular IVT-mRNA produced by permuted intron-exon splicing circularization has shown low immunogenicity without altering protein production [162].

Since PEGylated LNP-mRNA vaccines can cause the ABC phenomenon and allergies, it will be necessary to identify PEG alternatives. Polypropylene glycol (PPG), polytetramethylene ether glycol (PTMEG), and poly-1,4-butylene adipate (PBA) can replace PEG in mRNA-LNPs [160]. However, these polymers may react with the anti-PEG antibody [163] and cause unwanted immune responses, including the ABC phenomenon, CARPA, and various other immune-related adverse reactions. The clinical skin prick test employing a panel consisting of PEG and PEG alternatives can enhance the chance to identify patients who might develop allergic reactions to mRNA-LNP vaccines containing PEG or PEG alternatives. To reduce the risk of the ABC phenomenon and CARPA associated with mRNA-LNP vaccines, it is necessary to identify individuals with natural anti-PEG antibody [78,98].

To develop mRNA-LNP vaccines that would not cause the ABC phenomenon or immune-related adverse reactions resulting from immunogenicity of PEG, it is necessary to understand interactions of PEG and PEG alternatives with immune cells. Studies on the immunogenicity of PEG and PEG alternatives should be carried out in the context of the whole PEG-mRNA-LNP vaccine. Public databases summarizing immunological and physicochemical properties of PEG and PEG alternatives using various excipients and whole products could provide valuable information on the selection of PEG and PEG alternatives for mRNA-LNP vaccines.

Since anti-PEG antibodies can be detected in individuals who have never received PEGylated vaccines, it is necessary to develop vaccines based on PEG-free delivery systems. Since PEG cannot be easily modified, it is reasonable to devise vaccines based on PEG-free systems. There have been reports concerning the effectiveness of PEG-lipid-free vaccines. It has been shown that a PEG-free two-component mRNA vaccine (PFTCmvac) containing a receptor binding domain of SARS-CoV2 can induce strong adaptive and innate immunity without causing immune-related adverse reactions [164]. Furthermore, PFTCmvac does not induce activation of the complement system [164]. Thus, PFTCmvac can be an alternative delivery vehicle for COVID-19 vaccines. mRNA-LNPs based on monoacyl POx/POz-lipids have displayed transfection efficiencies, cytocompatibility, and biophysical properties comparable to PEG-lipid equivalents [165]. PEG-lipid alternatives based on reversible addition-fragmentation chain transfer (RAFT) have displayed higher gene expression and antigen-specific antibody production than conventional PEGylated LNPs [166].

11. Conclusions

COVID-19 has accelerated the acceptance of mRNA-LNPs. mRNA-LNPs have shown clinical benefits. Transcriptomic analysis of individual patients will give valuable information for designing safe and effective COVID-19 vaccines. It has been shown that mRNA-LNPs can elicit unwanted immune reactions including anaphylaxis, hypersensitivity, and autoimmunity. To make more effective and safe vaccines, it will be necessary to develop tolerogenic mRNA-LNPs to prevent the induction of unwanted immune reactions. mRNA-LNPs have a tendency to become trapped in endosomes, which can exert toxic effects. It is necessary to improve endosomal escape of mRNA-LNPs to reduce toxicity and improve efficiency. Overcoming problems associated with mRNA-LNPs will make mRNA-LNPs a cornerstone of personalized therapy.

LNPs can carry Cas-9 mRNA or guide mRNA, suggesting that LNPs can be employed as a delivery vehicle for gene editing. In other words, mRNA-LNPs can be employed to treat various genetic diseases. To broaden the use of mRNA-LNPs, the targeting of mRNA-LNPs to various organs is needed. Based on recent technological progresses, the future of mRNA-LNPs vaccines beyond the pandemic is optimistic.

Author Contributions: H.J., J.J., W.K. and D.J. contributed to the original draft preparation, review, and editing. All authors have read and agreed to the published version of the manuscript.

Funding: This research was funded by National Research Foundation grants (2020R1A2C1006996, 2022R1F1A1060031, and 2017M3A9G7072417) and a grant from the BK21 Four Program.

Institutional Review Board Statement: Not applicable.

Informed Consent Statement: Not applicable.

Data Availability Statement: No new data were created in this study.

Conflicts of Interest: The authors declare no conflicts of interest.

References

1. Berraondo, P.; Cuesta, R.; Sanmamed, M.F.; Melero, I. Immunogenicity and Efficacy of Personalized Adjuvant mRNA Cancer Vaccines. *Cancer Discov.* **2024**, *14*, 2021–2024. [CrossRef] [PubMed]
2. Maher, S.; Assaly, N.M.E.; Aly, D.M.; Atta, S.; Fteah, A.M.; Badawi, H.; Zahran, M.Y.; Kamel, M. Comparative study of neutralizing antibodies titers in response to different types of COVID-19 vaccines among a group of egyptian healthcare workers. *Virol. J.* **2024**, *21*, 277. [CrossRef]
3. Kumar, S.; Hsiao, Y.W.; Wong, V.H.Y.; Aubin, D.; Wang, J.H.; Lisowski, L.; Rakoczy, E.P.; Li, F.; Alarcon-Martinez, L.; Gonzalez-Cordero, A.; et al. Characterization of RNA editing and gene therapy with a compact CRISPR-Cas13 in the retina. *Proc. Natl. Acad. Sci. USA* **2024**, *121*, e2408345121. [CrossRef] [PubMed]
4. Gao, Y.; Yang, L.; Li, Z.; Peng, X.; Li, H. mRNA vaccines in tumor targeted therapy: Mechanism, clinical application, and development trends. *Biomark. Res.* **2024**, *12*, 93. [CrossRef]
5. Yang, L.; Gong, L.; Wang, P.; Zhao, X.; Zhao, F.; Zhang, Z.; Li, Y.; Huang, W. Recent Advances in Lipid Nanoparticles for Delivery of mRNA. *Pharmaceutics* **2022**, *14*, 2682. [CrossRef]
6. Park, S.; Kim, G.; Choi, A.; Kim, S.; Yum, J.S.; Chun, E.; Shin, H. Comparative network-based analysis of toll-like receptor agonist, L-pampo signaling pathways in immune and cancer cells. *Sci. Rep.* **2024**, *14*, 17173. [CrossRef] [PubMed]
7. Presnyak, V.; Alhusaini, N.; Chen, Y.H.; Martin, S.; Morris, N.; Kline, N.; Olson, S.; Weinberg, D.; Baker, K.E.; Graveley, B.R.; et al. Codon optimality is a major determinant of mRNA stability. *Cell* **2015**, *160*, 1111–1124. [CrossRef] [PubMed]
8. Mamaghani, S.; Penna, R.R.; Frei, J.; Wyss, C.; Mellett, M.; Look, T.; Weiss, T.; Guenova, E.; Kündig, T.M.; Lauchli, S.; et al. Synthetic mRNAs Containing Minimalistic Untranslated Regions Are Highly Functional In Vitro and In Vivo. *Cells* **2024**, *13*, 1242. [CrossRef]

9. Cheng, Z.; Islam, S.; Kanlong, J.G.; Sheppard, M.; Seo, H.; Nikolaitchik, O.A.; Kearse, M.G.; Pathak, V.K.; Musier-Forsyth, K.; Hu, W.S. Translation of HIV-1 unspliced RNA is regulated by 5′ untranslated region structure. *J. Virol.* **2024**, *98*, e0116024. [CrossRef]
10. Rohner, E.; Yang, R.; Foo, K.S.; Goedel, A.; Chien, K.R. Unlocking the promise of mRNA therapeutics. *Nat. Biotechnol.* **2022**, *40*, 1586–1600. [CrossRef] [PubMed]
11. Miao, L.; Zhang, Y.; Huang, L. mRNA vaccine for cancer immunotherapy. *Mol. Cancer* **2021**, *20*, 41. [CrossRef] [PubMed]
12. Peng, L.; Renauer, P.A.; Ökten, A.; Fang, Z.; Park, J.J.; Zhou, X.; Lin, Q.; Dong, M.B.; Filler, R.; Xiong, Q.; et al. Variant-specific vaccination induces systems immune responses and potent in vivo protection against SARS-CoV-2. *Cell Rep. Med.* **2022**, *3*, 100634. [CrossRef] [PubMed]
13. Ge, N.; Sun, J.; Liu, Z.; Shu, J.; Yan, H.; Kou, Z.; Wei, Y.; Jin, X. An mRNA vaccine encoding Chikungunya virus E2-E1 protein elicits robust neutralizing antibody responses and CTL immune responses. *Virol. Sin.* **2022**, *37*, 266–276. [CrossRef]
14. Hou, X.; Zaks, T.; Langer, R.; Dong, Y. Lipid nanoparticles for mRNA delivery. *Nat. Rev. Mater.* **2021**, *6*, 1078–1094. [CrossRef] [PubMed]
15. Rivero Berti, I.; Gambaro, R.C.; Limeres, M.J.; Huck-Iriart, C.; Svensson, M.; Fraude-El Ghazi, S.; Pretsch, L.; Si, S.; Lieberwirth, I.; Landfester, K.; et al. Encapsulation of Dexamethasone into mRNA-Lipid Nanoparticles Is a Promising Approach for the Development of Liver-Targeted Anti-Inflammatory Therapies. *Int. J. Mol. Sci.* **2024**, *25*, 11254. [CrossRef]
16. Ding, Y.; Chen, Q.B.; Xu, H.; Adi, D.; Ding, Y.W.; Luo, W.J.; Zhu, W.Z.; Xu, J.C.; Zhao, X.; Shi, X.J.; et al. siRNA nanoparticle targeting Usp20 lowers lipid levels and ameliorates metabolic syndrome in mice. *J. Lipid Res.* **2024**, *65*, 100626. [CrossRef] [PubMed]
17. Tenchov, R.; Sasso, J.M.; Zhou, Q.A. PEGylated Lipid Nanoparticle Formulations: Immunological Safety and Efficiency Perspective. *Bioconjug. Chem.* **2023**, *34*, 941–960. [CrossRef]
18. Paun, R.A.; Jurchuk, S.; Tabrizian, M. A landscape of recent advances in lipid nanoparticles and their translational potential for the treatment of solid tumors. *Bioeng. Transl. Med.* **2023**, *9*, e10601. [CrossRef] [PubMed]
19. Hald Albertsen, C.; Kulkarni, J.A.; Witzigmann, D.; Lind, M.; Petersson, K.; Simonsen, J.B. The role of lipid components in lipid nanoparticles for vaccines and gene therapy. *Adv. Drug Deliv. Rev.* **2022**, *188*, 114416. [CrossRef] [PubMed]
20. Kowalski, P.S.; Rudra, A.; Miao, L.; Anderson, D.G. Delivering the Messenger: Advances in Technologies for Therapeutic mRNA Delivery. *Mol. Ther.* **2019**, *27*, 710–728. [CrossRef]
21. Fuentes, S.; Arancibia, D.; Rojas, M.; Carmona, F.; Ortega, A.; Valenzuela, J.; Hernández-Álvarez, C.; Martín, I.R. Simultaneous Second Harmonic Generation and Multiphoton Excited Photoluminescence in Samarium-Doped $BaTiO_3$ Nanoparticles Functionalized with Poly (ethylene glycol). *ACS Omega* **2024**, *9*, 28061–28071. [CrossRef] [PubMed]
22. Geng, L.; Kato, N.; Kodama, Y.; Mukai, H.; Kawakami, S. Influence of lipid composition of messenger RNA-loaded lipid nanoparticles on the protein expression via intratracheal administration in mice. *Int. J. Pharm.* **2023**, *637*, 122896. [CrossRef] [PubMed]
23. Yang, M.Y.; Zheng, M.H.; Meng, X.T.; Ma, L.W.; Liang, H.Y.; Fan, H.Y. Role of toll-like receptors in the pathogenesis of COVID-19: Current and future perspectives. *Scand. J. Immunol.* **2023**, *98*, e13275. [CrossRef]
24. Klein, C.R.; Heine, A.; Brossart, P.; Karakostas, P.; Schäfer, V.S. Anti-MDA5 autoantibodies predict clinical dynamics of dermatomyositis following SARS-CoV-2 mRNA vaccination: A retrospective statistical analysis of case reports. *Rheumatol. Int.* **2024**, *44*, 2185–2196. [CrossRef] [PubMed]
25. Villacampa, A.; Alfaro, E.; Morales, C.; Díaz-García, E.; López-Fernández, C.; Bartha, J.L.; López-Sánchez, F.; Lorenzo, Ó.; Moncada, S.; Sánchez-Ferrer, C.F.; et al. SARS-CoV-2 S protein activates NLRP3 inflammasome and deregulates coagulation factors in endothelial and immune cells. *Cell Commun. Signal* **2024**, *22*, 38. [CrossRef]
26. Jeon, H.E.; Lee, S.; Lee, J.; Roh, G.; Park, H.J.; Lee, Y.S.; Kim, Y.J.; Kim, H.K.; Shin, J.H.; Lee, Y.J.; et al. SARS-CoV-2 mRNA vaccine intravenous administration induces myocarditis in chronic inflammation. *PLoS ONE* **2024**, *19*, e0311726. [CrossRef] [PubMed]
27. Bredholt, G.; Sævik, M.; Søyland, H.; Ueland, T.; Zhou, F.; Pathirana, R.; Madsen, A.; Vahokoski, J.; Lartey, S.; Halvorsen, B.E.; et al. Three doses of SARS-CoV-2 mRNA vaccine in older adults result in similar antibody responses but reduced cellular cytokine responses relative to younger adults. *Vaccine X* **2024**, *20*, 100564. [CrossRef]
28. Tahtinen, S.; Tong, A.J.; Himmels, P.; Oh, J.; Paler-Martinez, A.; Kim, L.; Wichner, S.; Oei, Y.; McCarron, M.J.; Freund, E.C.; et al. IL-1 and IL-1ra are key regulators of the inflammatory response to RNA vaccines. *Nat. Immunol.* **2022**, *23*, 532–542. [CrossRef] [PubMed]
29. Kim, S.; Jeon, J.H.; Kim, M.; Lee, Y.; Hwang, Y.H.; Park, M.; Li, C.H.; Lee, T.; Lee, J.A.; Kim, Y.M.; et al. Innate immune responses against mRNA vaccine promote cellular immunity through IFN-beta at the injection site. *Nat. Commun.* **2024**, *15*, 7226. [CrossRef] [PubMed]
30. Gangaev, A.; van Sleen, Y.; Brandhorst, N.; Hoefakker, K.; Prajapati, B.; Singh, A.; Boerma, A.; van der Heiden, M.; Oosting, S.F.; van der Veldt, A.A.M.; et al. mRNA-1273 vaccination induces polyfunctional memory CD4 and CD8 T cell responses in patients with solid cancers undergoing immunotherapy or/and chemotherapy. *Front. Immunol.* **2024**, *15*, 1447555. [CrossRef]

31. Lederer, K.; Castaño, D.; Gómez Atria, D.; Oguin, T.H., 3rd; Wang, S.; Manzoni, T.B.; Muramatsu, H.; Hogan, M.J.; Amanat, F.; Cherubin, P.; et al. SARS-CoV-2 mRNA Vaccines Foster Potent Antigen-Specific Germinal Center Responses Associated with Neutralizing Antibody Generation. *Immunity* **2020**, *53*, 1281–1295.e5. [CrossRef]
32. Alameh, M.G.; Tombácz, I.; Bettini, E.; Lederer, K.; Sittplangkoon, C.; Wilmore, J.R.; Gaudette, B.T.; Soliman, O.Y.; Pine, M.; Hicks, P.; et al. Lipid nanoparticles enhance the efficacy of mRNA and protein subunit vaccines by inducing robust T follicular helper cell and humoral responses. *Immunity* **2021**, *54*, 2877–2892. [CrossRef] [PubMed]
33. Li, C.; Lee, A.; Grigoryan, L.; Arunachalam, P.S.; Scott, M.K.D.; Trisal, M.; Wimmers, F.; Sanyal, M.; Weidenbacher, P.A.; Feng, Y.; et al. Mechanisms of innate and adaptive immunity to the Pfizer-BioNTech BNT162b2 vaccine. *Nat. Immunol.* **2022**, *23*, 543–555. [CrossRef]
34. Korzun, T.; Moses, A.S.; Jozic, A.; Grigoriev, V.; Newton, S.; Kim, J.; Diba, P.; Sattler, A.; Levasseur, P.R.; Le, N.; et al. Lipid Nanoparticles Elicit Reactogenicity and Sickness Behavior in Mice Via Toll-Like Receptor 4 and Myeloid Differentiation Protein 88 Axis. *ACS Nano* **2024**, *18*, 24842–24859. [CrossRef] [PubMed]
35. Lonez, C.; Bessodes, M.; Scherman, D.; Vandenbranden, M.; Escriou, V.; Ruysschaert, J.M. Cationic lipid nanocarriers activate toll-like receptor 2 and NLRP3 inflammasome pathways. *Nanomedicine* **2014**, *10*, 775–782. [CrossRef]
36. Bakos, T.; Mészáros, T.; Kozma, G.T.; Berényi, P.; Facskó, R.; Farkas, H.; Dézsi, L.; Heirman, C.; de Koker, S.; Schiffelers, R.; et al. mRNA-LNP COVID-19 Vaccine Lipids Induce Complement Activation and Production of Proinflammatory Cytokines: Mechanisms, Effects of Complement Inhibitors, and Relevance to Adverse Reactions. *Int. J. Mol. Sci.* **2024**, *25*, 3595. [CrossRef] [PubMed]
37. Liu, Y.; Suzuoki, M.; Tanaka, H.; Sakurai, Y.; Hatakeyama, H.; Akita, H. Lymphatic Endothelial Cells Produce Chemokines in Response to the Lipid Nanoparticles Used in RNA Vaccines. *Biol. Pharm. Bull.* **2024**, *47*, 698–707. [CrossRef] [PubMed]
38. Zheng, C.; Zhang, L. Identifying RNA Sensors in Antiviral Innate Immunity. *Methods Mol. Biol.* **2025**, *2854*, 107–115. [PubMed]
39. Kos, M.; Bojarski, K.; Mertowska, P.; Mertowski, S.; Tomaka, P.; Dziki, Ł.; Grywalska, E. Immunological Strategies in Gastric Cancer: How Toll-like Receptors 2, -3, -4, and -9 on Monocytes and Dendritic Cells Depend on Patient Factors? *Cells* **2024**, *13*, 1708. [CrossRef] [PubMed]
40. Domínguez-López, A.; Garfias, Y. Cytokine profile of human limbal myofibroblasts: Key players in corneal antiviral response. *Cytokine* **2022**, *160*, 156047. [CrossRef] [PubMed]
41. Roßmann, L.; Bagola, K.; Stephen, T.; Gerards, A.L.; Walber, B.; Ullrich, A.; Schülke, S.; Kamp, C.; Spreitzer, I.; Hasan, M.; et al. Distinct single-component adjuvants steer human DC-mediated T-cell polarization via Toll-like receptor signaling toward a potent antiviral immune response. *Proc. Natl. Acad. Sci. USA* **2021**, *118*, e2103651118. [CrossRef] [PubMed]
42. King, H.A.D.; Pokkali, S.; Kim, D.; Brammer, D.; Song, K.; McCarthy, E.; Lehman, C.; Todd, J.P.; Foulds, K.E.; Darrah, P.A.; et al. Immune Activation Profiles Elicited by Distinct, Repeated TLR Agonist Infusions in Rhesus Macaques. *J. Immunol.* **2023**, *211*, 1643–1655. [CrossRef] [PubMed]
43. Zhang, H.; You, X.; Wang, X.; Cui, L.; Wang, Z.; Xu, F.; Li, M.; Yang, Z.; Liu, J.; Huang, P.; et al. Delivery of mRNA vaccine with a lipid-like material potentiates antitumor efficacy through Toll-like receptor 4 signaling. *Proc. Natl. Acad. Sci. USA* **2021**, *118*, e2005191118. [CrossRef]
44. Usero, L.; Leal, L.; Gómez, C.E.; Miralles, L.; Aurrecoechea, E.; Esteban, I.; Torres, B.; Inciarte, A.; Perdiguero, B.; Esteban, M.; et al. The Combination of an mRNA Immunogen, a TLR7 Agonist and a PD1 Blocking Agent Enhances In-Vitro HIV T-Cell Immune Responses. *Vaccines* **2023**, *11*, 286. [CrossRef]
45. Zhang, Y.; Yan, J.; Hou, X.; Wang, C.; Kang, D.D.; Xue, Y.; Du, S.; Deng, B.; McComb, D.W.; Liu, S.L.; et al. STING Agonist-Derived LNP-mRNA Vaccine Enhances Protective Immunity Against SARS-CoV-2. *Nano Lett.* **2023**, *23*, 2593–2600. [CrossRef]
46. Abo-Samaha, M.I.; Sharaf, M.M.; El-Nahas, A.F.; Odemuyiwa, S.O. Length-Dependent Modulation of B Cell Activating Factor Transcripts in Chicken Macrophage by Viral Double-Stranded RNA. *Vaccines* **2023**, *11*, 1561. [CrossRef]
47. Choudhury, A.; Das, N.C.; Patra, R.; Mukherjee, S. In silico analyses on the comparative sensing of SARS-CoV-2 mRNA by the intracellular TLRs of humans. *J. Med. Virol.* **2021**, *93*, 2476–2486. [CrossRef]
48. Miquel, C.H.; Abbas, F.; Cenac, C.; Foret-Lucas, C.; Guo, C.; Ducatez, M.; Joly, E.; Hou, B.; Guéry, J.C. B cell-intrinsic TLR7 signaling is required for neutralizing antibody responses to SARS-CoV-2 and pathogen-like COVID-19 vaccines. *Eur. J. Immunol.* **2023**, *53*, e2350437. [CrossRef] [PubMed]
49. Xie, Y.; Chi, Y.; Tao, X.; Yu, P.; Liu, Q.; Zhang, M.; Yang, N.; Liu, S.; Zhu, W. Rabies Virus Regulates Inflammatory Response in BV-2 Cells through Activation of Myd88 and NF-kappaB Signaling Pathways via TLR7. *Int. J. Mol. Sci.* **2024**, *25*, 9144. [CrossRef] [PubMed]
50. Liu, T.; Zhang, L.; Joo, D.; Sun, S.C. NF-κB signaling in inflammation. *Signal Transduct. Target. Ther.* **2017**, *2*, 17023. [CrossRef] [PubMed]
51. Allard, R.L.; Mayfield, J.; Barchiesi, R.; Salem, N.A.; Mayfield, R.D. Toll-like receptor 7: A novel neuroimmune target to reduce excessive alcohol consumption. *Neurobiol. Stress* **2024**, *31*, 100639. [CrossRef] [PubMed]

52. Wang, B.; Le, D.S.; Liu, L.; Zhang, X.X.; Yang, F.; Lai, G.R.; Zhang, C.; Zhao, M.L.; Shen, Y.P.; Liao, P.S.; et al. Targeting exosomal double-stranded RNA-TLR3 signaling pathway attenuates morphine tolerance and hyperalgesia. *Cell Rep. Med.* **2024**, *5*, 101782. [CrossRef] [PubMed]
53. Zhi, Y.; Zhao, X.; Liu, Z.; Shen, G.; Zhang, T.; Zhang, T.; Hu, G. Oxymatrine Modulation of TLR3 Signaling: A Dual-Action Mechanism for H9N2 Avian Influenza Virus Defense and Immune Regulation. *Molecules* **2024**, *29*, 1945. [CrossRef] [PubMed]
54. Lamoot, A.; Jangra, S.; Laghlali, G.; Warang, P.; Singh, G.; Chang, L.A.; Park, S.C.; Singh, G.; De Swarte, K.; Zhong, Z.; et al. Lipid Nanoparticle Encapsulation Empowers Poly(I:C) to Activate Cytoplasmic RLRs and Thereby Increases Its Adjuvanticity. *Small* **2024**, *20*, e2306892. [CrossRef]
55. Zhang, H.; Sandhu, P.K.; Damania, B. The Role of RNA Sensors in Regulating Innate Immunity to Gammaherpesviral Infections. *Cells* **2023**, *12*, 1650. [CrossRef] [PubMed]
56. Mikhalkevich, N.; Russ, E.; Iordanskiy, S. Cellular RNA and DNA sensing pathways are essential for the dose-dependent response of human monocytes to ionizing radiation. *Front. Immunol.* **2023**, *14*, 1235936. [CrossRef] [PubMed]
57. Madaan, V.; Kollara, A.; Spaner, D.; Brown, T.J. ISGylation enhances dsRNA-induced interferon response and NFκB signaling in fallopian tube epithelial cells. *J. Biol. Chem.* **2024**, *300*, 107686. [CrossRef] [PubMed]
58. Karasik, A.; Lorenzi, H.A.; DePass, A.V.; Guydosh, N.R. Endonucleolytic RNA cleavage drives changes in gene expression during the innate immune response. *Cell Rep.* **2024**, *43*, 114287. [CrossRef] [PubMed]
59. Chaumont, L.; Peruzzi, M.; Huetz, F.; Raffy, C.; Le Hir, J.; Minke, J.; Boudinot, P.; Collet, B. Salmonid Double-stranded RNA-Dependent Protein Kinase Activates Apoptosis and Inhibits Protein Synthesis. *J. Immunol.* **2024**, *213*, 700–717. [CrossRef]
60. Yu, H.; Megawati, D.; Zheng, C.; Rothenberg, S. Protein Kinase R (PKR) as a Novel dsRNA Sensor in Antiviral Innate Immunity. *Methods Mol. Biol.* **2025**, *2854*, 265–282.
61. Hu, J.; Hodgkinson, C.P.; Pratt, R.E.; Lee, J.; Sullenger, B.A.; Dzau, V.J. Enhancing cardiac reprogramming via synthetic RNA oligonucleotides. *Mol. Ther. Nucleic Acids* **2020**, *23*, 55–62. [CrossRef] [PubMed]
62. Zhao, M.; Wan, B.; Li, H.; He, J.; Chen, X.; Wang, L.; Wang, Y.; Xie, S.; Qiao, S.; Zhang, G. Porcine 2′,5′-oligoadenylate synthetase 2 inhibits porcine reproductive and respiratory syndrome virus replication in vitro. *Microb. Pathog.* **2017**, *111*, 14–21. [CrossRef]
63. Aloise, C.; Schipper, J.G.; van Vliet, A.; Oymans, J.; Donselaar, T.; Hurdiss, D.L.; de Groot, R.J.; van Kuppeveld, F.J.M. SARS-CoV-2 nucleocapsid protein inhibits the PKR-mediated integrated stress response through RNA-binding domain N2b. *PLoS Pathog.* **2023**, *19*, e1011582. [CrossRef]
64. Tahsin, A.; Bhattacharjee, P.; Al Saba, A.; Yasmin, T.; Nabi, A.H.M.N. Genetic and epigenetic analyses of IFN-γ gene proximal promoter region underlying positive correlation between persistently high anti-SARS-CoV-2 IgG and IFN-γ among COVID-19 vaccinated Bangladeshi adults. *Vaccine* **2024**, *42*, 126157. [CrossRef]
65. Otter, C.J.; Bracci, N.; Parenti, N.A.; Ye, C.; Asthana, A.; Blomqvist, E.K.; Tan, L.H.; Pfannenstiel, J.J.; Jackson, N.; Fehr, A.R.; et al. SARS-CoV-2 nsp15 endoribonuclease antagonizes dsRNA-induced antiviral signaling. *Proc. Natl. Acad. Sci. USA* **2024**, *121*, e2320194121. [CrossRef] [PubMed]
66. Cusic, R.; Burke, J.M. Condensation of RNase L promotes its rapid activation in response to viral infection in mammalian cells. *Sci. Signal* **2024**, *17*, eadi9844. [CrossRef] [PubMed]
67. Yang, K.; Dong, B.; Asthana, A.; Silverman, R.H.; Yan, N. RNA helicase SKIV2L limits antiviral defense and autoinflammation elicited by the OAS-RNase L pathway. *EMBO J.* **2024**, *43*, 3876–3894. [CrossRef] [PubMed]
68. Kim, H.J.; Han, C.W.; Jeong, M.S.; Jang, S.B. Structural study of novel vaccinia virus E3L and dsRNA-dependent protein kinase complex. *Biochem. Biophys. Res. Commun.* **2023**, *665*, 1–9. [CrossRef] [PubMed]
69. Lee, J.; Xu, L.; Gibson, T.M.; Gersbach, C.A.; Sullenger, B.A. Differential effects of toll-like receptor stimulation on mRNA-driven myogenic conversion of human and mouse fibroblasts. *Biochem. Biophys. Res. Commun.* **2016**, *478*, 1484–1490. [CrossRef] [PubMed]
70. Fernández, J.J.; Mancebo, C.; Garcinuño, S.; March, G.; Alvarez, Y.; Alonso, S.; Inglada, L.; Blanco, J.; Orduña, A.; Montero, O.; et al. Innate IRE1α-XBP1 activation by viral single-stranded RNA and its influence on lung cytokine production during SARS-CoV-2 pneumonia. *Genes Immun.* **2024**, *25*, 43–54. [CrossRef] [PubMed]
71. Kodigepalli, K.M.; Nanjundan, M. Induction of PLSCR1 in a STING/IRF3-dependent manner upon vector transfection in ovarian epithelial cells. *PLoS ONE* **2015**, *10*, e0117464. [CrossRef] [PubMed]
72. Chang, Y.; Jin, H.; Cui, Y.; Yang, F.; Chen, K.; Kuang, W.; Huo, C.; Xu, Z.; Li, Y.; Lin, A.; et al. PUS7-dependent pseudouridylation of ALKBH3 mRNA inhibits gastric cancer progression. *Clin. Transl. Med.* **2024**, *14*, e1811. [CrossRef]
73. Andries, O.; Mc Cafferty, S.; De Smedt, S.C.; Weiss, R.; Sanders, N.N.; Kitada, T. N(1)-methylpseudouridine-incorporated mRNA outperforms pseudouridine-incorporated mRNA by providing enhanced protein expression and reduced immunogenicity in mammalian cell lines and mice. *J. Control. Release* **2015**, *217*, 337–344. [CrossRef] [PubMed]
74. Zhang, T.; Yin, C.; Fedorov, A.; Qiao, L.; Bao, H.; Beknazarov, N.; Wang, S.; Gautam, A.; Williams, R.M.; Crawford, J.C.; et al. ADAR1 masks the cancer immunotherapeutic promise of ZBP1-driven necroptosis. *Nature* **2022**, *606*, 594–602. [CrossRef]

75. Anderson, B.R.; Muramatsu, H.; Nallagatla, S.R.; Bevilacqua, P.C.; Sansing, L.H.; Weissman, D.; Kariko, K. Incorporation of pseudouridine into mRNA enhances translation by diminishing PKR activation. *Nucleic Acids Res.* **2010**, *38*, 5884–5892. [CrossRef]
76. Anderson, B.R.; Muramatsu, H.; Jha, B.K.; Silverman, R.H.; Weissman, D.; Kariko, K. Nucleoside modifications in RNA limit activation of 2′-5′-oligoadenylate synthetase and increase resistance to cleavage by RNase L. *Nucleic Acids Res.* **2011**, *39*, 9329–9338. [CrossRef] [PubMed]
77. Gebremeskel, S.; Schanin, J.; Coyle, K.M.; Butuci, M.; Luu, T.; Brock, E.C.; Xu, A.; Wong, A.; Leung, J.; Korver, W.; et al. Mast Cell and Eosinophil Activation Are Associated With COVID-19 and TLR-Mediated Viral Inflammation: Implications for an Anti-Siglec-8 Antibody. *Front. Immunol.* **2021**, *12*, 650331. [CrossRef] [PubMed]
78. Shah, M.M.; Layhadi, J.A.; Hourcade, D.E.; Fulton, W.T.; Tan, T.J.; Dunham, D.; Chang, I.; Vel, M.S.; Fernandes, A.; Lee, A.S.; et al. Elucidating allergic reaction mechanisms in response to SARS-CoV-2 mRNA vaccination in adults. *Allergy* **2024**, *79*, 2502–2523. [CrossRef] [PubMed]
79. Awaya, T.; Hara, H.; Moroi, M. Cytokine Storms and Anaphylaxis Following COVID-19 mRNA-LNP Vaccination: Mechanisms and Therapeutic Approaches. *Diseases* **2024**, *12*, 231. [CrossRef]
80. Calzetta, L.; Chetta, A.; Aiello, M.; Frizzelli, A.; Ora, J.; Melis, E.; Facciolo, F.; Ippoliti, L.; Magrini, A.; Rogliani, P. The BNT162b2 mRNA COVID-19 Vaccine Increases the Contractile Sensitivity to Histamine and Parasympathetic Activation in a Human Ex Vivo Model of Severe Eosinophilic Asthma. *Vaccines* **2023**, *11*, 282. [CrossRef] [PubMed]
81. Farooqui, A.A.; Farooqui, T.; Sun, G.Y.; Lin, T.N.; The, D.B.L.; Ong, W.Y. COVID-19, Blood Lipid Changes, and Thrombosis. *Biomedicines* **2023**, *11*, 1181. [CrossRef]
82. Suzuki, T.; Taketomi, Y.; Yanagida, K.; Yoshida-Hashidate, T.; Nagase, T.; Murakami, M.; Shimizu, T.; Shindou, H. Re-evaluation of the canonical PAF pathway in cutaneous anaphylaxis. *Biochim. Biophys. Acta Mol. Cell Biol. Lipids* **2024**, *1870*, 159563. [CrossRef]
83. Richter, A.W.; Akerblom, E. Antibodies against polyethylene glycol produced in animals by immunization with monomethoxy polyethylene glycol modified proteins. *Int. Arch. Allergy Appl. Immunol.* **1983**, *70*, 124–131. [CrossRef] [PubMed]
84. Qi, Y.; Han, H.; Liu, A.; Zhao, S.; Lawanprasert, A.; Nielsen, J.E.; Choudhary, H.; Liang, D.; Barron, A.E.; Murthy, N. Ethylene oxide graft copolymers reduce the immunogenicity of lipid nanoparticles. *RSC Adv.* **2024**, *14*, 30071–30076. [CrossRef] [PubMed]
85. Perkins, G.B.; Tunbridge, M.J.; Hurtado, P.R.; Zuiani, J.; Mhatre, S.; Yip, K.H.; Le, T.A.; Yuson, C.; Kette, F.; Hissaria, P. PEGylated liposomes for diagnosis of polyethylene glycol allergy. *J. Allergy Clin. Immunol.* **2024**, *154*, 503–507.e1. [CrossRef] [PubMed]
86. Kozma, G.T.; Mészáros, T.; Berényi, P.; Facskó, R.; Patkó, Z.; Oláh, C.Z.; Nagy, A.; Fülöp, T.G.; Glatter, K.A.; Radovits, T.; et al. Role of anti-polyethylene glycol (PEG) antibodies in the allergic reactions to PEG-containing COVID-19 vaccines: Evidence for immunogenicity of PEG. *Vaccine* **2023**, *41*, 4561–4570. [CrossRef]
87. Miao, G.; He, Y.; Lai, K.; Zhao, Y.; He, P.; Tan, G.; Wang, X. Accelerated blood clearance of PEGylated nanoparticles induced by PEG-based pharmaceutical excipients. *J. Control. Release* **2023**, *363*, 12–26. [CrossRef]
88. Carreño, J.M.; Singh, G.; Tcheou, J.; Srivastava, K.; Gleason, C.; Muramatsu, H.; Desai, P.; Aberg, J.A.; Miller, R.L.; Paris Study Group; et al. mRNA-1273 but not BNT162b2 induces antibodies against polyethylene glycol (PEG) contained in mRNA-based vaccine formulations. *Vaccine* **2022**, *40*, 6114–6124. [CrossRef] [PubMed]
89. Li, Y.; Saba, L.; Scheinman, R.I.; Banda, N.K.; Holers, M.; Monte, A.; Dylla, L.; Moghimi, S.M.; Simberg, D. Nanoparticle-Binding Immunoglobulins Predict Variable Complement Responses in Healthy and Diseased Cohorts. *ACS Nano* **2024**, *18*, 28649–28658. [CrossRef]
90. Chen, W.A.; Chang, D.Y.; Chen, B.M.; Lin, Y.C.; Barenholz, Y.; Roffler, S.R. Antibodies against Poly (ethylene glycol) Activate Innate Immune Cells and Induce Hypersensitivity Reactions to PEGylated Nanomedicines. *ACS Nano* **2023**, *17*, 5757–5772. [CrossRef]
91. Park, M.K.; Park, H.K.; Yu, H.S. Toll-like receptor 2 mediates Acanthamoeba-induced allergic airway inflammatory response in mice. *PLoS Negl. Trop. Dis.* **2023**, *17*, e0011085. [CrossRef]
92. Lv, Y.; Jiang, G.; Jiang, Y.; Peng, C.; Li, W. TLR2-ERK signaling pathway regulates expression of galectin-3 in a murine model of OVA-induced allergic airway inflammation. *Toxicol. Lett.* **2024**, *397*, 55–66. [CrossRef] [PubMed]
93. Hashimoto, Y.; Abu Lila, A.S.; Shimizu, T.; Ishida, T.; Kiwada, H. B cell-intrinsic toll-like receptor 7 is responsible for the enhanced anti-PEG IgM production following injection of siRNA-containing PEGylated lipoplex in mice. *J. Control. Release* **2014**, *184*, 1–8. [CrossRef]
94. Stavnsbjerg, C.; Christensen, E.; Münter, R.; Henriksen, J.R.; Fach, M.; Parhamifar, L.; Christensen, C.; Kjaer, A.; Hansen, A.E.; Andresen, T.L. Accelerated blood clearance and hypersensitivity by PEGylated liposomes containing TLR agonists. *J. Control. Release* **2022**, *342*, 337–344. [CrossRef] [PubMed]
95. Barta, B.A.; Radovits, T.; Dobos, A.B.; Tibor Kozma, G.; Mészáros, T.; Berényi, P.; Facskó, R.; Fülöp, T.; Merkely, B.; Szebeni, J. Comirnaty-induced cardiopulmonary distress and other symptoms of complement-mediated pseudo-anaphylaxis in a hyperimmune pig model: Causal role of anti-PEG antibodies. *Vaccine X* **2024**, *19*, 100497. [CrossRef] [PubMed]
96. Yang, Q.; Lai, S.K. Anti-PEG immunity: Emergence, characteristics, and unaddressed questions. *Wiley Interdiscip. Rev. Nanomed. Nanobiotechnol.* **2015**, *7*, 655–677. [CrossRef] [PubMed]

97. Klimek, L.; Novak, N.; Cabanillas, B.; Jutel, M.; Bousquet, J.; Akdis, C.A. Allergenic components of the mRNA-1273 vaccine for COVID-19: Possible involvement of polyethylene glycol and IgG-mediated complement activation. *Allergy* **2021**, *76*, 3307–3313. [CrossRef] [PubMed]
98. Zhu, X.; Luo, W.; Zhang, D.; Liu, R. An Assay for Immunogenic Detection of Anti-PEG Antibody. *ChemBioChem* **2024**, *25*, e202400316. [CrossRef]
99. Davis, E.; Caparco, A.A.; Jones, E.; Steinmetz, N.F.; Pokorski, J.K. Study of uricase-polynorbornene conjugates derived from grafting-from ring-opening metathesis polymerization. *J. Mater. Chem. B* **2024**, *12*, 2197–2206. [CrossRef]
100. Tsang, M.S.; Hou, T.; Chan, B.C.; Wong, C.K. Immunological Roles of NLR in Allergic Diseases and Its Underlying Mechanisms. *Int. J. Mol. Sci.* **2021**, *22*, 1507. [CrossRef]
101. Leven, T.; Coorevits, L.; Vandebotermet, M.; Tuyls, S.; Vanneste, H.; Santy, L.; Wets, D.; Proost, P.; Frans, G.; Devolder, D.; et al. Endotyping of IgE-Mediated Polyethylene Glycol and/or Polysorbate 80 Allergy. *J. Allergy Clin. Immunol. Pract.* **2023**, *11*, 3146–3160.
102. Dézsi, L.; Mészáros, T.; Kozma, G.; H-Velkei, M.; Oláh, C.Z.; Szabó, M.; Patkó, Z.; Fülöp, T.; Hennies, M.; Szebeni, M.; et al. A naturally hypersensitive porcine model may help understand the mechanism of COVID-19 mRNA vaccine-induced rare (pseudo) allergic reactions: Complement activation as a possible contributing factor. *Geroscience* **2022**, *44*, 597–618. [CrossRef] [PubMed]
103. Ribak, Y.; Rubin, L.; Talmon, A.; Dranitzki, Z.; Shamriz, O.; Hershkowitz, I.; Tal, Y.; Hershko, A.Y. Administration of BNT162b2 mRNA COVID-19 vaccine to subjects with various allergic backgrounds. *Front. Immunol.* **2023**, *14*, 1172896. [CrossRef]
104. Gao, P.; Tang, K.; Lu, Y.; Wang, M.; Wang, W.; Wang, T.; Sun, Y.; Zhao, J.; Mao, Y. Increased expression of ficolin-1 is associated with airway obstruction in asthma. *BMC Pulm. Med.* **2023**, *23*, 470. [CrossRef]
105. Kokelj, S.; Östling, J.; Fromell, K.; Vanfleteren, L.E.G.W.; Olsson, H.K.; Nilsson Ekdahl, K.; Nilsson, B.; Olin, A.C. Activation of the Complement and Coagulation Systems in the Small Airways in Asthma. *Respiration* **2023**, *102*, 621–631. [CrossRef]
106. Khan, A.A.; Allemailem, K.S.; Almatroodi, S.A.; Almatroudi, A.; Rahmani, A.H. Recent strategies towards the surface modification of liposomes: An innovative approach for different clinical applications. *3 Biotech* **2020**, *10*, 163. [CrossRef]
107. Münter, R.; Stavnsbjerg, C.; Christensen, E.; Thomsen, M.E.; Stensballe, A.; Hansen, A.E.; Parhamifar, L.; Kristensen, K.; Simonsen, J.B.; Larsen, J.B.; et al. Unravelling Heterogeneities in Complement and Antibody Opsonization of Individual Liposomes as a Function of Surface Architecture. *Small* **2022**, *18*, e2106529. [CrossRef]
108. Khunsuk, P.O.; Pongma, C.; Palaga, T.; Hoven, V.P. Zwitterionic Polymer-Decorated Lipid Nanoparticles for mRNA Delivery in Mammalian Cells. *Biomacromolecules* **2023**, *24*, 5654–5665. [CrossRef] [PubMed]
109. Gabrielaitis, D.; Zitkute, V.; Saveikyte, L.; Labutyte, G.; Skapas, M.; Meskys, R.; Casaite, V.; Sasnauskiene, A.; Neniskyte, U. Nanotubes from bacteriophage tail sheath proteins: Internalisation by cancer cells and macrophages. *Nanoscale Adv.* **2023**, *5*, 3705–3716. [CrossRef] [PubMed]
110. Barbey, C.; Wolf, H.; Wagner, R.; Pauly, D.; Breunig, M. A shift of paradigm: From avoiding nanoparticular complement activation in the field of nanomedicines to its exploitation in the context of vaccine development. *Eur. J. Pharm. Biopharm.* **2023**, *193*, 119–128. [CrossRef]
111. Shi, D.; Beasock, D.; Fessler, A.; Szebeni, J.; Ljubimova, J.Y.; Afonin, K.A.; Dobrovolskaia, M.A. To PEGylate or not to PEGylate: Immunological properties of nanomedicine's most popular component, polyethylene glycol and its alternatives. *Adv. Drug Deliv. Rev.* **2022**, *180*, 114079. [CrossRef] [PubMed]
112. Chen, B.M.; Chen, E.; Lin, Y.C.; Tran, T.T.M.; Turjeman, K.; Yang, S.H.; Cheng, T.L.; Barenholz, Y.; Rofflerm, S.R. Liposomes with Low Levels of Grafted Poly (ethylene glycol) Remain Susceptible to Destabilization by Anti-Poly (ethylene glycol) Antibodies. *ACS Nano* **2024**, *18*, 22122–22138. [CrossRef]
113. Son, K.; Ueda, M.; Taguchi, K.; Maruyama, T.; Takeoka, S.; Ito, Y. Evasion of the accelerated blood clearance phenomenon by polysarcosine coating of liposomes. *J. Control. Release* **2020**, *322*, 209–216. [CrossRef] [PubMed]
114. Shimizu, T.; Lila, A.S.A.; Kitayama, Y.; Abe, R.; Takata, H.; Ando, H.; Ishima, Y.; Ishida, T. Peritoneal B Cells Play a Role in the Production of Anti-polyethylene Glycol (PEG) IgM against Intravenously Injected siRNA-PEGylated Liposome Complexes. *Biol. Pharm. Bull.* **2024**, *47*, 469–477. [CrossRef]
115. Zhang, X.; Pan, J.; Ye, X.; Chen, Y.; Wang, L.; Meng, X.; Chen, W.; Wang, F. Activation of CYP3A by Accelerated Blood Clearance Phenomenon Potentiates the Hepatocellular Carcinoma-Targeting Therapeutic Effects of PEGylated Anticancer Prodrug Liposomes. *Drug Metab. Dispos.* **2023**, *51*, 1651–1662. [CrossRef] [PubMed]
116. Ju, Y.; Lee, W.S.; Pilkington, E.H.; Kelly, H.G.; Li, S.; Selva, K.J.; Wragg, K.M.; Subbarao, K.; Nguyen, T.H.O.; Rowntree, L.C.; et al. Anti-PEG Antibodies Boosted in Humans by SARS-CoV-2 Lipid Nanoparticle mRNA Vaccine. *ACS Nano* **2022**, *16*, 11769–11780. [CrossRef] [PubMed]
117. Smirnov, A.; Daily, K.P.; Gray, M.C.; Ragland, S.; Werner, L.M.; Brittany Johnson, M.; Eby, J.C.; Hewlett, E.L.; Taylor, R.P.; Criss, A.K. Phagocytosis via complement receptor 3 enables microbes to evade killing by neutrophils. *J. Leukoc. Biol.* **2023**, *114*, 1–20. [CrossRef]

118. Zwarthoff, S.A.; Berends, E.T.M.; Mol, S.; Ruyken, M.; Aerts, P.C.; Józsi, M.; de Haas, C.J.C.; Rooijakkers, S.H.M.; Gorham, R.D., Jr. Functional Characterization of Alternative and Classical Pathway C3/C5 Convertase Activity and Inhibition Using Purified Models. *Front. Immunol.* **2018**, *9*, 1691. [CrossRef]
119. Laumonnier, Y.; Korkmaz, R.Ü.; Nowacka, A.A.; Köhl, J. Complement-mediated immune mechanisms in allergy. *Eur. J. Immunol.* **2023**, *53*, e2249979. [CrossRef] [PubMed]
120. Gutzmer, R.; Köther, B.; Zwirner, J.; Dijkstra, D.; Purwar, R.; Wittmann, M.; Werfel, T. Human plasmacytoid dendritic cells express receptors for anaphylatoxins C3a and C5a and are chemoattracted to C3a and C5a. *J. Investig. Dermatol.* **2006**, *126*, 2422–2429. [CrossRef] [PubMed]
121. Schäfer, B.; Piliponsky, A.M.; Oka, T.; Song, C.H.; Gerard, N.P.; Gerard, C.; Tsai, M.; Kalesnikoff, J.; Galli, S.J. Mast cell anaphylatoxin receptor expression can enhance IgE-dependent skin inflammation in mice. *J. Allergy Clin. Immunol.* **2013**, *131*, 541–548. [CrossRef]
122. Kammala, A.K.; Syed, M.; Yang, C.; Occhiuto, C.J.; Subramanian, H. A Critical Role for Na(+)/H(+) Exchanger Regulatory Factor 1 in Modulating FcepsilonRI-Mediated Mast Cell Activation. *J. Immunol.* **2021**, *206*, 471–480. [CrossRef]
123. West, P.W.; Bahri, R.; Garcia-Rodriguez, K.M.; Sweetland, G.; Wileman, G.; Shah, R.; Montero, A.; Rapley, L.; Bulfone-Paus, S. Interleukin-33 Amplifies Human Mast Cell Activities Induced by Complement Anaphylatoxins. *Front. Immunol.* **2021**, *11*, 615236. [CrossRef] [PubMed]
124. Yasuda, M.; Tanaka, Y.; Bando, K.; Sugawara, S.; Mizuta, K. Lipopolysaccharide Priming Exacerbates Anaphylatoxin C5a-Induced Anaphylaxis in Mice. *Biol. Pharm. Bull.* **2023**, *46*, 432–439. [CrossRef] [PubMed]
125. Zhang, T.; Wu, K.Y.; Ma, N.; Wei, L.L.; Garstka, M.; Zhou, W.; Li, K. The C5a/C5aR2 axis promotes renal inflammation and tissue damage. *JCI Insight* **2020**, *5*, e134081. [CrossRef] [PubMed]
126. Hu, X.; Li, X.; Hu, C.; Qin, L.; He, R.; Luo, L.; Tang, W.; Feng, J. Respiratory Syncytial Virus Exacerbates OVA-mediated asthma in mice through C5a-C5aR regulating CD4(+)T cells Immune Responses. *Sci. Rep.* **2017**, *7*, 15207. [CrossRef]
127. Karp, C.L.; Grupe, A.; Schadt, E.; Ewart, S.L.; Keane-Moore, M.; Cuomo, P.J.; Köhl, J.; Wahl, L.; Kuperman, D.; Germer, S.; et al. Identification of complement factor 5 as a susceptibility locus for experimental allergic asthma. *Nat. Immunol.* **2000**, *1*, 221–226. [CrossRef] [PubMed]
128. Rönnau, A.C.; Wulferink, M.; Gleichmann, E.; Unver, E.; Ruzicka, T.; Krutmann, J.; Grewe, M. Anaphylaxis to polyvinylpyrrolidone in an analgesic preparation. *Br. J. Dermatol.* **2000**, *143*, 1055–1058. [CrossRef] [PubMed]
129. Liccioli, G.; Mori, F.; Barni, S.; Pucci, N.; Novembre, E. Anaphylaxis to Polyvinylpyrrolidone in Eye Drops Administered to an Adolescent. *J. Investig. Allergol. Clin. Immunol.* **2018**, *28*, 263–265. [CrossRef]
130. Baysal, S.; Anil, H.; Harmanci, K. A Case Report and Pediatric Literature Review: Povidone as a Rare Cause of Anaphylaxis in Children. *Pediatr. Allergy Immunol. Pulmonol.* **2024**, *37*, 56–59. [CrossRef] [PubMed]
131. Bakhsh, R.; Dairi, K.; Almadabgy, E.; Albiladi, A.; Gamal, L.; Almatrafi, D.; AlShariff, F.; Alsefri, A. New Onset of Neuro-Sjogren's Syndrome Nine Months After the Third COVID-19 Vaccine Dose: A Case Report. *Cureus* **2024**, *16*, e69562. [CrossRef]
132. Cahuapaza-Gutierrez, N.L. Systemic lupus erythematosus following COVID-19 vaccination. A systematic review of case reports and case series. *Lupus* **2024**, *33*, 375–386. [CrossRef] [PubMed]
133. Bruusgaard-Mouritsen, M.A.; Jensen, B.M.; Poulsen, L.K.; Duus Johansen, J.; Garvey, L.H. Optimizing investigation of suspected allergy to polyethylene glycols. *J. Allergy Clin. Immunol.* **2022**, *149*, 168–175. [CrossRef] [PubMed]
134. Liu, M.; Zhao, D.; Yan, N.; Li, J.; Zhang, H.; Liu, M.; Tang, X.; Liu, X.; Deng, Y.; Song, Y.; et al. Evasion of the accelerated blood clearance phenomenon by branched PEG lipid derivative coating of nanoemulsions. *Int. J. Pharm.* **2022**, *612*, 121365. [CrossRef]
135. Sui, D.; Wang, Y.; Sun, W.; Wei, L.; Li, C.; Gui, Y.; Qi, Z.; Liu, X.; Song, Y.; Deng, Y. Cleavable-Branched Polymer-Modified Liposomes Reduce Accelerated Blood Clearance and Enhance Photothermal Therapy. *ACS Appl. Mater. Interfaces* **2023**, *15*, 32110–32120. [CrossRef] [PubMed]
136. McSweeney, M.D.; Shen, L.; DeWalle, A.C.; Joiner, J.B.; Ciociola, E.C.; Raghuwanshi, D.; Macauley, M.S.; Lai, S.K. Pre-treatment with high molecular weight free PEG effectively suppresses anti-PEG antibody induction by PEG-liposomes in mice. *J. Control. Release* **2021**, *329*, 774–781. [CrossRef] [PubMed]
137. Shen, L.; Li, Z.; Ma, A.; Cruz-Teran, C.; Talkington, A.; Shipley, S.T.; Lai, S.K. Free PEG Suppresses Anaphylaxis to PEGylated Nanomedicine in Swine. *ACS Nano* **2024**, *18*, 8733–8744. [CrossRef]
138. Hu, Y.; Hou, Y.; Wang, H.; Lu, H. Polysarcosine as an alternative to PEG for therapeutic protein conjugation. *Bioconjug. Chem.* **2018**, *29*, 2232–2238. [CrossRef] [PubMed]
139. Hu, M.; Taguchi, K.; Matsumoto, K.; Kobatake, E.; Ito, Y.; Ueda, M. Polysarcosine-Coated liposomes attenuating immune response induction and prolonging blood circulation. *J. Colloid Interface Sci.* **2023**, *651*, 273–283. [CrossRef]
140. Xia, J.; Chen, C.; Dong, M.; Zhu, Y.; Wang, A.; Li, S.; Zhang, R.; Feng, C.; Jiang, X.; Xu, X.; et al. Ginsenoside Rg3 endows liposomes with prolonged blood circulation and reduced accelerated blood clearance. *J. Control. Release* **2023**, *364*, 23–36. [CrossRef] [PubMed]

141. Wang, D.; Toyofuku, W.M.; Chen, A.M.; Scott, M.D. Induction of immunotolerance via mPEG grafting to allogeneic leukocytes. *Biomaterials* **2011**, *32*, 9494–9503. [CrossRef] [PubMed]
142. Wu, J.; Lu, S.; Zheng, Z.; Zhu, L.; Zhan, X. Modification with polysialic acid-PEG copolymer as a new method for improving the therapeutic efficacy of proteins. *Prep. Biochem. Biotechnol.* **2016**, *46*, 788–797. [CrossRef] [PubMed]
143. Qelliny, M.R.; Shimizu, T.; Elsadek, N.E.; Emam, S.E.; Takata, H.; Fathalla, Z.M.A.; Hussein, A.K.; Khaled, K.A.; Ando, H.; Ishima, Y.; et al. Incorporating Gangliosides into PEGylated Cationic Liposomes that Complexed DNA Attenuates Anti-PEG Antibody Production but Not Anti-DNA Antibody Production in Mice. *Mol. Pharm.* **2021**, *18*, 2406–2415. [CrossRef] [PubMed]
144. van den Hoven, J.M.; Nemes, R.; Metselaar, J.M.; Nuijen, B.; Beijnen, J.H.; Storm, G.; Szebeni, J. Complement activation by PEGylated liposomes containing prednisolone. *Eur. J. Pharm. Sci.* **2013**, *49*, 265–271. [CrossRef]
145. Polack, F.P.; Thomas, S.J.; Kitchin, N.; Absalon, J.; Gurtman, A.; Lockhart, S.; Perez, J.L.; Pérez Marc, G.; Moreira, E.D.; Zerbini, C.; et al. Safety and Efficacy of the BNT162b2 mRNA COVID-19 Vaccine. *N. Engl. J. Med.* **2020**, *383*, 2603–2615. [CrossRef]
146. La Gualana, F.; Maiorca, F.; Marrapodi, R.; Villani, F.; Miglionico, M.; Santini, S.A.; Pulcinelli, F.; Gragnani, L.; Piconese, S.; Fiorilli, M.; et al. Opposite Effects of mRNA-Based and Adenovirus-Vectored SARS-CoV-2 Vaccines on Regulatory T Cells: A Pilot Study. *Biomedicines* **2023**, *11*, 511. [CrossRef]
147. Li, P.Y.; Bearoff, F.; Zhu, P.; Fan, Z.; Zhu, Y.; Fan, M.; Cort, L.; Kambayashi, T.; Blankenhorn, E.P.; Cheng, H. PEGylation enables subcutaneously administered nanoparticles to induce antigen-specific immune tolerance. *J. Control. Release* **2021**, *331*, 164–175. [CrossRef] [PubMed]
148. Xu, X.; Wang, X.; Liao, Y.P.; Luo, L.; Xia, T.; Nel, A.E. Use of a Liver-Targeting Immune-Tolerogenic mRNA Lipid Nanoparticle Platform to Treat Peanut-Induced Anaphylaxis by Single- and Multiple-Epitope Nucleotide Sequence Delivery. *ACS Nano* **2023**, *17*, 4942–4957. [CrossRef] [PubMed]
149. Pfeil, J.; Simonetti, M.; Lauer, U.; Volkmer, R.; von Thülen, B.; Durek, P.; Krähmer, R.; Leenders, F.; Hamann, A.; Hoffmann, U. Tolerogenic Immunomodulation by PEGylated Antigenic Peptides. *Front. Immunol.* **2020**, *11*, 529035. [CrossRef]
150. Krienke, C.; Kolb, L.; Diken, E.; Streuber, M.; Kirchhoff, S.; Bukur, T.; Akilli-Öztürk, Ö.; Kranz, L.M.; Berger, H.; Petschenka, J.; et al. A noninflammatory mRNA vaccine for treatment of experimental autoimmune encephalomyelitis. *Science* **2021**, *371*, 145–153. [CrossRef]
151. Pardi, N.; Krammer, F. mRNA vaccines for infectious diseases—Advances, challenges and opportunities. *Nat. Rev. Drug Discov.* **2024**, *23*, 838–861. [CrossRef] [PubMed]
152. Al Rahbani, G.K.; Woopen, C.; Dunsche, M.; Proschmann, U.; Ziemssen, T.; Akgün, K. SARS-CoV-2-Specific Immune Cytokine Profiles to mRNA, Viral Vector and Protein-Based Vaccines in Patients with Multiple Sclerosis: Beyond Interferon Gamma. *Vaccines* **2024**, *12*, 684. [CrossRef]
153. Sittplangkoon, C.; Alameh, M.G.; Weissman, D.; Lin, P.J.C.; Tam, Y.K.; Prompetchara, E.; Palaga, T. mRNA vaccine with unmodified uridine induces robust type I interferon-dependent anti-tumor immunity in a melanoma model. *Front. Immunol.* **2022**, *13*, 983000. [CrossRef] [PubMed]
154. Cao, W.; Xia, T. RNA lipid nanoparticles induce immune tolerance to treat human diseases. *Med. Rev.* **2023**, *3*, 180–183. [CrossRef]
155. Wang, Z.; Jacobus, E.J.; Stirling, D.C.; Krumm, S.; Flight, K.E.; Cunliffe, R.F.; Mottl, J.; Singh, C.; Mosscrop, L.G.; Santiago, L.A.; et al. Reducing cell intrinsic immunity to mRNA vaccine alters adaptive immune responses in mice. *Mol. Ther. Nucleic Acids* **2023**, *34*, 102045. [CrossRef] [PubMed]
156. Karikó, K.; Muramatsu, H.; Ludwig, J.; Weissman, D. Generating the optimal mRNA for therapy: HPLC purification eliminates immune activation and improves translation of nucleoside-modified, protein-encoding mRNA. *Nucleic Acids Res.* **2011**, *39*, e142. [CrossRef] [PubMed]
157. Loomis, K.H.; Lindsay, K.E.; Zurla, C.; Bhosle, S.M.; Vanover, D.A.; Blanchard, E.L.; Kirschman, J.L.; Bellamkonda, R.V.; Santangelo, P.J. In Vitro Transcribed mRNA Vaccines with Programmable Stimulation of Innate Immunity. *Bioconjug. Chem.* **2018**, *29*, 3072–3083. [CrossRef]
158. Yuan, X.; Wu, Z.; Guo, J.; Luo, D.; Li, T.; Cao, Q.; Ren, X.; Fang, H.; Xu, D.; Cao, Y. Natural Wood-Derived Macroporous Cellulose for Highly Efficient and Ultrafast Elimination of Double-Stranded RNA from In Vitro-Transcribed mRNA. *Adv. Mater.* **2024**, *36*, e2303321. [CrossRef] [PubMed]
159. Català, M.; Mercadé-Besora, N.; Kolde, R.; Trinh, N.T.H.; Roel, E.; Burn, E.; Rathod-Mistry, T.; Kostka, K.; Man, W.Y.; Delmestri, A.; et al. The effectiveness of COVID-19 vaccines to prevent long COVID symptoms: Staggered cohort study of data from the UK, Spain, and Estonia. *Lancet Respir. Med.* **2024**, *12*, 225–236. [CrossRef] [PubMed]
160. Jung, S.W.; Jeon, J.J.; Kim, Y.H.; Choe, S.J.; Lee, S. Long-Term risk of autoimmune diseases after mRNA-based SARS-CoV-2 vaccination in a Korean, nationwide, population-based cohort study. *Nat. Commun.* **2024**, *15*, 6181. [CrossRef] [PubMed]
161. Li, X.; Gao, L.; Tong, X.; Chan, V.K.Y.; Chui, C.S.L.; Lai, F.T.T.; Wong, C.K.H.; Wan, E.Y.F.; Chan, E.W.Y.; Lau, K.K.; et al. Autoimmune conditions following mRNA (BNT162b2) and inactivated (CoronaVac) COVID-19 vaccination: A descriptive cohort study among 1.1 million vaccinated people in Hong Kong. *J. Autoimmun.* **2022**, *130*, 102830. [CrossRef] [PubMed]

162. Feng, Z.; Zhang, X.; Zhou, J.; Li, Q.; Chu, L.; Di, G.; Xu, Z.; Chen, Q.; Wang, M.; Jiang, X.; et al. An in vitro-transcribed circular RNA targets the mitochondrial inner membrane cardiolipin to ablate $EIF4G2^{+}/PTBP1^{+}$ pan-adenocarcinoma. *Nat. Cancer* **2024**, *5*, 30–46. [CrossRef] [PubMed]
163. McCallen, J.; Prybylski, J.; Yang, Q.; Lai, S.K. Cross-Reactivity of Select PEG-Binding Antibodies to Other Polymers Containing a C-C-O Backbone. *ACS Biomater. Sci. Eng.* **2017**, *3*, 1605–1615. [CrossRef]
164. Li, M.; Huang, Y.; Wu, J.; Li, S.; Mei, M.; Chen, H.; Wang, N.; Wu, W.; Zhou, B.; Tan, X.; et al. A PEG-lipid-free COVID-19 mRNA vaccine triggers robust immune responses in mice. *Mater. Horiz.* **2023**, *10*, 466–472. [CrossRef]
165. He, X.; Payne, T.J.; Takanashi, A.; Fang, Y.; Kerai, S.D.; Morrow, J.P.; Al-Wassiti, H.; Pouton, C.W.; Kempe, K. Tailored Monoacyl Poly (2-oxazoline)- and Poly (2-oxazine)-Lipids as PEG-Lipid Alternatives for Stabilization and Delivery of mRNA-Lipid Nanoparticles. *Biomacromolecules* **2024**, *25*, 4591–4603. [CrossRef] [PubMed]
166. Hassanel, D.N.B.P.; Pilkington, E.H.; Ju, Y.; Kent, S.J.; Pouton, C.W.; Truong, N.P. Replacing poly (ethylene glycol) with RAFT lipopolymers in mRNA lipid nanoparticle systems for effective gene delivery. *Int. J. Pharm.* **2024**, *665*, 124695. [CrossRef] [PubMed]

vaccines

MDPI

Article

Pichia pastoris-Derived β-Glucan Capsules as a Delivery System for DNA Vaccines

Samara Sousa de Pinho †, Maria da Conceição Viana Invenção †, Anna Jéssica Duarte Silva, Larissa Silva de Macêdo, Benigno Cristofer Flores Espinoza, Lígia Rosa Sales Leal, Marco Antonio Turiah Machado da Gama, Ingrid Andrêssa de Moura, Micaela Evellin dos Santos Silva, Débora Vitória Santos de Souza, Marina Linhares Lara, Julia Nayane Soares Azevedo Alves and Antonio Carlos de Freitas *

Laboratory of Molecular Studies and Experimental Therapy—LEMTE, Department of Genetics, Federal University of Pernambuco, Recife 50670-901, Brazil; samara.pinho@ufpe.br (S.S.d.P.); maria.conceicao@ufpe.br (M.d.C.V.I.); anna.jessica@ufpe.br (A.J.D.S.); larissa.smacedo@ufpe.br (L.S.d.M.); benigno.cristofer@ufpe.br (B.C.F.E.); ligia.leal@ufpe.br (L.R.S.L.); marco.turiah@ufpe.br (M.A.T.M.d.G.); ingrid.andressa@ufpe.br (I.A.d.M.); micaela.evellin@ufpe.br (M.E.d.S.S.); debora.vsouza@ufpe.br (D.V.S.d.S.); marina.linharesl@ufpe.br (M.L.L.); julia.nayane@ufpe.br (J.N.S.A.A.)

* Correspondence: antonio.cfreitas@ufpe.br; Tel.: +55-(81)-2126-8569

† These authors contributed equally to this work.

Abstract: Background/Objectives: DNA vaccines are rapidly produced and adaptable to different pathogens, but they face considerable challenges regarding stability and delivery to the cellular target. Thus, effective delivery methods are essential for the success of these vaccines. Here, we evaluated the efficacy of capsules derived from the cell wall of the yeast *Pichia pastoris* as a delivery system for DNA vaccines. Methods: The capsules were extracted from the yeast *Pichia pastoris* strain GS115, previously grown in a YPD medium. pVAX1 expression vector was adopted to evaluate the DNA vaccine insertion and delivery. Three encapsulation protocols were tested to identify the most effective in internalizing the plasmid. The presence of plasmids inside the capsules was confirmed by fluorescence microscopy, and the encapsulation efficiency was calculated by the difference between the initial concentration of DNA used for insertion and the concentration of unencapsulated DNA contained in the supernatant. The capsules were subjected to different temperatures to evaluate their thermostability and were co-cultured with macrophages for phagocytosis analysis. HEK-293T cells were adopted to assess the cytotoxicity levels by MTT assay. Results: The microscopy results indicated that the macrophages successfully phagocytosed the capsules. Among the protocols tested for encapsulation, the one with 2% polyethylenimine for internalization showed the highest efficiency, with an encapsulation rate above 80%. However, the vaccine capsules obtained with the protocol that used 5% NaCl showed better thermal stability and encapsulation efficiency above 63% without induction of cell viability loss in HEK 293T. Conclusions: We successfully described a vaccine delivery system using yeast capsules derived from *Pichia pastoris*, demonstrating its potential for DNA vaccine delivery for the first time. Additional studies will be needed to characterize and improve this delivery strategy.

Keywords: glucan particle; yeast cell wall; nucleic acids

1. Introduction

Vaccination is one of the most effective health interventions for preventing infectious diseases. However, the emergence of pathogens and viral mutations has highlighted the need for new vaccine platforms [1,2]. Although effective, conventional vaccines, such as attenuated, inactivated, and protein subunit vaccines, face limitations regarding scope, development time, and immune response. In this context, innovation in vaccine platforms is essential to meet health demands, especially in outbreak and pandemic situations [3]. Nucleic acid vaccines, such as those based on DNA and RNA, have emerged as a promising alternative being approved for use in humans since the COVID-19 pandemic with three vaccines already licensed: Spikevax (Moderna), Comirnaty (Pfizer-BioNTech), and ZyCoV-D (Cadila). These licenses open doors to new possibilities of approved DNA-based vaccines, as they reinforce the efficacy of this vaccine platform that offers rapid development and flexibility in adapting to different pathogens. However, these vaccines still face challenges, such as the instability of the genetic material and the efficiency of delivery to the cellular target [4]. Overcoming these limitations is crucial to maximize the potential of these new approaches.

A key aspect of the success of nucleic acid vaccines lies in the choice of appropriate delivery systems that can guarantee antigen stability and promote efficient capture by antigen-presenting cells (APCs) [5,6]. In this sense, several delivery systems have been developed to maximize the efficacy of these vaccines, such as delivery systems based on lipids, polymeric structures, and recombinant microorganisms [7–9].

Yeasts have gained recognition as an innovative and promising delivery system for nucleic acid vaccines [10]. Species such as *Pichia pastoris* and *Saccharomyces cerevisiae* are widely employed in biotechnology due to their safety, rapid growth, and ease of genetic manipulation [11]. Yeast-based technology offers a potentially safer and more stable alternative, with a lower risk of toxicity and adverse reactions than lipid systems and viral vectors [12]. These organisms can be explored in another presentation form, based on the composition of the yeast cell wall to generate capsules, also named yeast shells (YS) [13,14]. This system takes advantage of the natural robustness of the cell wall to encapsulate and protect the genetic material or antigens, facilitating controlled delivery and effective interaction with the immune system [15].

Yeast capsules have advantages in storage and transportation. Unlike many lipid systems or viral vectors that require extremely low-temperature storage conditions, yeast capsules are stable over a wider range of conditions, making them ideal for distribution in areas with limited infrastructure. This stability, combined with the ease of large-scale production, makes this technology highly attractive for mass vaccination campaigns, especially in outbreak or pandemic scenarios [15–17].

Additionally, yeast capsules demonstrate superior delivery efficiency, allowing for the controlled release of antigens and enhancing the immune response. This delivery capability is crucial to maximizing the efficacy of vaccines by ensuring the suitable antigen presentation [18]. Additionally, yeast capsules have the potential for an oral route of administration that can increase public acceptance and improve treatment adherence owing to their non-invasive nature and ease of use [18–20]. This method not only simplifies the delivery process but also provides notable immunological advantages [21]. Oral administration effectively mimics the natural route of infection for various pathogens, thereby stimulating a robust mucosal immune response [21,22]. This response includes the production of secretory IgA antibodies and the activation of gut-associated lymphoid tissues (GALT), which are crucial for combating pathogens and promoting both local and systemic immunity [21,23]. The possibility of using them to carry antigens also opens up new perspectives for immunization against infections and the development of therapeutic vaccines and immunotherapies against cancer-targeting components of the tumor microenvironment as tumor-associated macrophages [4,24–26]. Furthermore, the B-glucans present in yeast derivatives interact with Dectin-1 receptors of antigen-presenting cells (APCs), which enables the phagocytosis of this vaccine antigen carried by yeast capsules, overcoming the limitation of naked DNA vaccines, which usually present difficulties in this transport and presentation to immune system cells [27,28].

To date, most studies have focused on the use of *S. cerevisiae*, whose structural and physiological characteristics are well understood, paving the way for significant advances in vaccine application and also in exploring the potential of other yeast species [18]. In this sense, we herein present the development of capsules derived from the cell wall of the yeast *P. pastoris* as a model antigen delivery system. We aimed to present the characterization of this delivery platform as the primary objective of the present study, focusing on its thermostability, phagocytic capacity, cytotoxicity, surface charges, and capacity to internalize DNA without the possibility of degradation, thus demonstrating its potential application as a delivery vehicle for DNA-based vaccines. Thus, this capsule-based transport system has the potential to be improved as a new type of vaccine vehicle, expanding the options available for vaccination strategies.

2. Materials and Methods

2.1. Plasmid DNA

The pVAX1 expression vector (Invitrogen, Carlsbad, CA, USA) was employed to evaluate the insertion and delivery of DNA vaccines. Plasmid DNA was propagated in *Escherichia coli* Top10 [F- *mcrA* Δ (*mrr-hsd*RMS-*mrc*BC) Ø80*lac*ZΔM15 Δ*lac*X74 *rec*A1 *deo*R *ara*D139 Δ(*ara-leu*)7697 *gal*U *gal*K *rsp*L (StrR) *end*A1 *nup*G] (Invitrogen, Carlsbad, CA, USA). Transformed cells were cultured and selected in LB medium supplemented with 50 μg/mL kanamycin.

2.2. Yeast Cultivation

P. pastoris GS115 yeast cells (Genotype: *His4* and Phenotype: Mut^{+} and His-) were cultured in 1 L of liquid YPD medium (1% yeast extract, 2% tryptone, and 2% glucose) in an aerated flask in a shaking incubator at 150 rpm at 30 °C for 48 h. The cells were collected by centrifugation at 2000× *g* for 5 min, washed three times in sterile distilled water, and had their weight measured.

2.3. Obtaining the Capsules

The protocol by Zhu et al. (2021) was adapted to obtain capsules composed of the cell wall of *P. pastoris* GS115 [29]. The yeast biomass obtained after cultivation was suspended in 160 mL of NaOH (1.0 M) and incubated for 1 h at 60 °C under agitation. Then, the mass was centrifuged at 2000× *g* for 10 min. The precipitate was then suspended in 160 mL of water (pH 4.5, adjusted with HCl) and incubated at 55 °C for 1 h. The capsules present in the precipitate were centrifuged and washed once with 70 mL of water. Then, the capsules were washed with 32 mL of isopropanol (4 times) and with 32 mL of acetone (twice) (Figure 1). The precipitate was dried at room temperature and stored at −20 °C until use. The formation of these particles was monitored using optical microscopy, as well as the seeding of the capsules on YPD plates to verify reproductive inactivity, and incubated for 48 h at 30 °C. To evaluate morphological differences or changes in *P. pastoris* yeasts before and after capsule preparation, cells and capsules were stained following the protocol proposed by Kwolek-Mirek e Zadrag-Tecza (2014) [30]. Briefly, a drop of biological material (in PBS 1x) was placed on the slide, followed by a drop of methylene blue. Then, the preparation was covered with a coverslip for observation under an ® K55 OIT optical microscope (Kasvi, Pinhais, PR, Brazil), using magnifications of 40× and 100×. The obtained capsules, also named yeast shells (YS), were quantified using the method described by Huang et al. (2010), where the particles were suspended in 0.9% saline solution and counted in a hemocytometer (5 quadrants) [31]. The particles were kept at −20 °C until use.

2.4. Internalization of Plasmids into Capsules

Three protocols were performed to insert plasmid DNA into the capsules aiming to choose the one with the best encapsulation efficiency (EE%). For all protocols, 1 × 107 capsules were resuspended in 1 mL of PBS for further DNA encapsulation, and three DNA concentrations were tested (50, 250, and 500 ng/μL). In all experiments, the quantification of the capsules was determined using the Countess 3 Automated Cell Counter (Invitrogen, Carlsbad, CA, USA), used to count particles with a diameter of ~4–60 μm. The insertion protocol 1 (PI-1) was based on a method described by Ostroff et al. (2005), with modifications [13]. An aliquot of 25 μL of capsule preparation was mixed with 200 μL of 2% branched polyethylenimine (PEI) solution (Mw ~25,000 by LS; Sigma-Aldrich, St. Louis, MI, USA) and 25 μL of plasmid DNA (Figure 2A). The samples remained for 30 min at room temperature (RT). Then, 75 μL of 95% ethanol was added and incubated at 20 °C for 16 h. The solution was centrifuged at 2000× g for 10 min, washed with 70% ethanol, and centrifuged in the same conditions. Finally, the supernatant was discarded, and the capsules with internalized DNA were stored at −20 °C until use. For the insertion protocol 2 (PI-2), the method described by Paramera et al. (2011) was adopted, with some adaptations [32]. Then, 25 µL of capsules were mixed with 25 µL of 5% NaCl and 25 µL of DNA (Figure 2B) and incubated at 55 °C, under agitation of 150 rpm for 48 h, with subsequent centrifugation at 2000× g for 10 min, washing with 70% ethanol and centrifuged. The resulting capsules carrying the DNA were stored at −20 °C until use. In insertion protocol 3 (PI-3), the capsules were prepared without anionic solutions. Thus, 25 µL of capsules were added to 25 µL of DNA. The capsules were incubated at 55 °C, under 150 rpm agitation, for 48 h. Subsequently, the solution was centrifuged at 2000× g for 10 min and washed with 70% ethanol. The resulting capsules were stored at −20 °C until use.

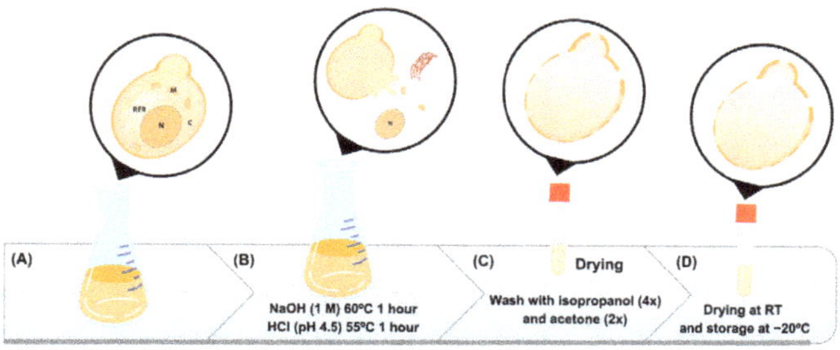

Figure 1. Schematic representation of the methodology used for capsule preparation. This figure outlines the process of yeast cell treatment and preparation. In (**A**), the whole yeast cells are cultured, providing the starting material for the yeast capsule procedure; after that, the cells undergo a chemical treatment (**B**), first with NaOH (1 M) at 60 °C for 1 h to disrupt cell components, followed by HCl (pH 4.5) at 55 °C for 1 h to further process and clean the cellular material. The treated material is subjected to multiple washes (**C**), including four rounds with isopropanol and two with acetone, to remove impurities and ensure purity of the capsules. The final capsules are dried at room temperature (RT) and stored at −20 °C to maintain their integrity and stability for further applications (**D**).

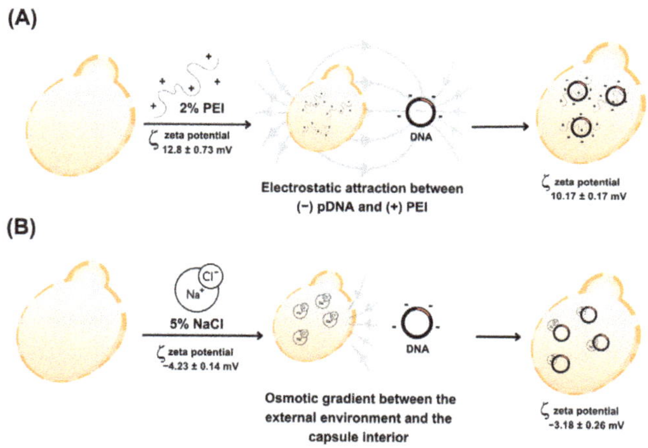

Figure 2. Methodologies for inserting plasmid DNA into the capsules. (**A**) Insertion methodology 1 (PI-1) using 2% PEI; (**B**) insertion methodology 2 (PI-2) using 5% NaCl. Grey arrows represents eletrostatic interactions between DNA and PEI/NaCl.

2.5. Fluorescence Microscopy

The presence of plasmid DNA inside the capsules was verified by fluorescence microscopy following the method described by Paramera et al. (2010), with modifications [32]. The 1 × 10^7 capsules, containing approximately 250 ng/μL DNA, were homogenized in 10 μL of distilled water and placed on slides. Afterward, 5 μL of 4′,6-diamidino-2-phenylindole (DAPI) at a final concentration of 150 ng/mL were added to the slides to allow blue fluorescent labeling when bound to the DNA and washed to remove dye excess. They were then covered with a coverslip. A negative control group (empty capsules) and samples of capsules containing the construction with the pVAX1 vector (2% PEI and 5% NaCl) were used.

2.6. Encapsulation Efficiency (EE%)

To verify how much DNA was internalized in the capsules, the encapsulation efficiency was calculated. This index reflects the difference between the initial concentration of inserted DNA (IC) and the final concentration measured from the supernatant corresponding to the unencapsulated DNA (UED). DNA measurement was performed by spectrophotometry using NanoDrop (Thermo Scientific, Waltham, MA, USA). The encapsulation efficiency calculation, in percentage, was defined from the following equation: $(\%\ \mathrm{EE}) = [(\mathrm{IC}) - (\mathrm{UED})/(\mathrm{IC})] \times 100$.

2.7. Thermal Stability Test

The thermostability of the capsules containing the DNA was evaluated following the protocol proposed by Shi et al. (2007), with modifications [33]. The samples were incubated for 11 days at different temperatures: 8 °C, 26 °C (room temperature), and 65 °C. After this period, the encapsulation efficiency (EE%) and DNA retention after heat treatment were analyzed. The samples were subjected to these temperature variations to evaluate the ability of the capsules to retain the DNA, as well as to determine the best storage conditions. The lost DNA was measured from the supernatant of the preparations using NanoDrop (Thermo Scientific, Waltham, MA, USA), in triplicate. The retained DNA is the result of subtracting the incorporated and lost DNA.

2.8. Phagocytosis Analysis

To visualize internalization in macrophages, the capsules were stained with 5′-DTAF (3-5-(4,6-dichlorotriazinyl)aminofluorescein) (Invitrogen, Carlsbad, CA, USA), according to the method described by Aouadi et al. (2009), with modifications [34]. Thus, 1 × 10^7 capsules were washed with 0.1 M sodium carbonate buffer (pH 9.2) and resuspended in 0.1 L of the buffer. 5′-DTAF was added to the suspension of buffered capsules (10% *v*/*v*), and the mixture was incubated at 20 °C in the dark for 16 h. Then, the capsules were incubated in a 2 mM Tris buffer for 15 min and washed with distilled water to remove excess dye. Finally, the labeled capsules were dehydrated with absolute ethanol and acetone, dried in the dark at 20 °C, and stored at −20 °C until use. The macrophages used for phagocytosis were derived from THP-1 monocytes. For this purpose, the cells were cultured in DMEM medium (Sigma-Aldrich, St. Louis, MI, USA) supplemented with 10% (*v*/*v*) inactivated Fetal Bovine Serum (Sigma-Aldrich, St. Louis, MI, USA) and 1% (*v*/*v*) Penicillin–Streptomycin (Thermo Scientific, Waltham, MA, USA) and maintained in a humid incubator with 5% CO_2 at 37 °C. For differentiation into macrophages, THP-1 cells were seeded at a density of 1 × 10^5/well in an 8-well Nunc® Lab-Tek® II Chamber Slide™ system (Sigma-Aldrich, St. Louis, MI, USA) and incubated in supplemented DMEM medium containing 150 nM PMA (Phorbol 12-Myristate 13-Acetate; Sigma-Aldrich, St. Louis, MI, USA) for 24 h. After this period, the culture medium was exchanged for supplemented DMEM medium, and the cells were incubated for an additional 24 h. After this period, the culture medium was removed to wash the cells with PBS pH 7.4 twice, and fresh supplemented DMEM medium was added. The MØ macrophages were incubated with 1 × 10^7 stained capsules for 4 h. Prior to visualization by microscopy, macrophages were stained with DAPI, and the cells were washed to remove any free YS, ensuring that the observed fluorescence corresponds to phagocytosed YS rather than extracellular particles. This experiment was performed in triplicate. Phagocytosis was evaluated by fluorescence microscopy using a Motic AE31E microscope, and the images were captured by a Moticam S6 camera (Universal City, TX, USA) using Motic Images Plus 3.0 software.

2.9. Cytotoxicity Assessment

For this assay, HEK-293T cells were used, following the protocol described by Monsmann (1983) with adaptations [35]. Briefly, HEK-293T cells were centrifuged, counted, and seeded (1×10^5 cells/well) in 96-well plates containing 100 μL of DMEM medium supplemented with 10% heat-inactivated FBS. The cells were adhered overnight at 37 °C and 5% CO_2. Then, the culture supernatant was removed, and 100 μL of the samples, diluted in DMEM, were added to the cells. The cells were incubated with empty capsules and capsules with DNA (approximately 250 ng/μL) incorporated with 2% PEI and 5% NaCl, constituting three experimental groups. The concentration of yeast shells was 5×10^4/well. The cells were maintained for 24 h. After the incubation period, the supernatant was removed, and the cells adhered to the well were incubated with 5 mg/mL of 3-(4,5-dimethylthiazol-2-yl)-2,5-diphenyltetrazolium bromide (MTT, Sigma-Aldrich) diluted in DMEM, maintained for 4 h at 37 °C and 5% CO_2. After incubation, the supernatant was discarded, and the cells were solubilized in 100 μL of DMSO to resuspend the formazan crystals. Absorbance reading was performed using a GloMax® Discover Microplate Reader (Promega, Madison, WI, USA) at a wavelength of 570 nm. Absorbance values were transformed into percentage of cell viability, based on the mean absorbance of the negative control (untreated HEK 293-T cells) using the GraphPad Prism® version 9.0 program. Tests were performed in 6 biological replicates.

2.10. Zeta Potential (ζ) Measurements

The ζ analysis was performed on empty capsules treated with PEI 2% and NaCl 5% (controls), as well as DNA-loaded capsules treated under the same conditions. Briefly, 1×10^7 capsules from each sample were prepared and adjusted to a final dilution of 1:10 in PBS 1x. The ζ measurements were conducted in triplicate using the Zetasizer Nano ZS90 instrument (Malvern, WR, United Kingdom), employing the dynamic light scattering technique. The data obtained were processed using Technology Software 7.01. Prior to the measurements, all samples were washed with ultra-pure water. Analyses were assessed at room temperature and a pH of approximately 7.0.

2.11. DNAse Assay

The capsules with internalized plasmid DNA were tested for their protective capacity in the presence of DNase. For this purpose, a DNase I assay (Thermo Scientific, Waltham, MA, USA) was performed. The reactions were assembled following the manufacturer's instructions. To visualize the effect of the enzyme, the capsules were labeled with DAPI and visualized by fluorescence microscopy. The experimental groups were divided into (I) empty capsules (control), (II) capsules with PEI + DNA, and (III) capsules with NaCl + DNA. The capsules evaluated were standardized at a concentration of 10^7, with 250 ng/μL of internalized DNA (in the case of groups II and III). For analysis, microscopy was performed before and after enzymatic treatment.

2.12. Statistical Analysis

Graphs and statistical analyses were generated by GraphPad Prism version 9.0. For cytotoxicity assay and encapsulation efficiency analysis, One-way and Two-way ANOVA were applied, respectively, followed by Tukey's post hoc tests. For thermostability evaluation, was employed the One-way ANOVA test followed by Šídák multiple comparison. Results with a p-value < 0.05 were considered statistically significant.

3. Results

3.1. Obtaining β-Glucan Capsules

Growth in culture medium for 48 h resulted in 18 g of wet yeast biomass obtained from 1 L of culture in liquid medium. After the YS preparation steps, the morphological changes between the whole wild yeasts and the capsules were analyzed by optical microscopy (Figure 3). The morphologies did not present relevant changes, demonstrating that the washes did not affect the cell wall architecture. After the treatments, the capsules formed clusters not observed in the whole yeast cells.

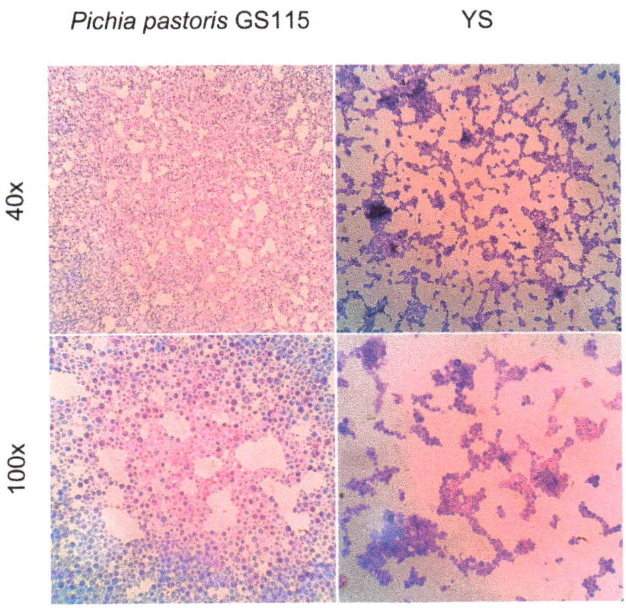

Figure 3. Optical microscopy of *Pichia pastoris* GS115 and yeast shells. Samples were stained with methylene blue and observed at 40× and 100× magnification. Microscopy with K55 OIT optical microscope (Kasvi, Pinhais, PR, Brazil) and image manual capture system.

To prove that the capsules constitute a non-replicative delivery system, they were seeded on plates with a culture medium. It was possible to observe that there was no contamination in obtaining the capsules, and there was no reproductive activity. In contrast, the untreated *P. pastoris* GS115 exhibited the expected reproductive capacity, forming visible colonies in the culture medium (Supplementary Materials, Figure S1).

3.2. DNA Encapsulation Capacity by Yeast Capsules

The capacity to incorporate plasmid DNA was measured using a calculation to verify encapsulation efficiency (%EE). Capsules in which the genetic material was incorporated with PEI 2% showed a higher level of efficiency, regardless of the concentration of plasmid DNA initially used (50, 250, or 500 ng/μL), followed by the protocol using NaCl 5% and the formulation without ionic compounds. There was no significant difference between the use of PEI and NaCl, except for a concentration of 500 ng/μL, where DNA incorporation with the aid of PEI demonstrated greater efficiency ($p < 0.0050$). The absence of 2% PEI and 5% NaCl compounds decreased the encapsulation efficiency rate, regardless of the DNA concentration that was incorporated ($p < 0.0001$) (Figure 4).

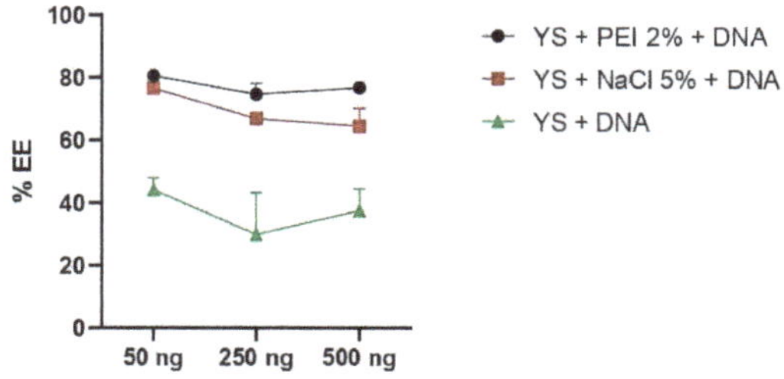

Figure 4. Comparison of the plasmid DNA incorporation efficiency between the three protocols adopted. %EE = encapsulation efficiency percentage. Different DNA concentrations were adopted to verify the loading capacity using 50, 250, and 500 ng/µL (x-axis). Values correspond to mean ± standard deviation.

3.3. Internalization of Plasmid DNA

In Figure 5, we observe the control sample, corresponding to the empty YS, characterized by the absence of fluorescence due to the lack of genetic material. In contrast, in the YS prepared with the insertion of plasmid DNA, fluorescence was detected from the labeling with DAPI, evidencing the presence of genetic material and demonstrating the efficiency of the capsules in incorporating exogenous DNA.

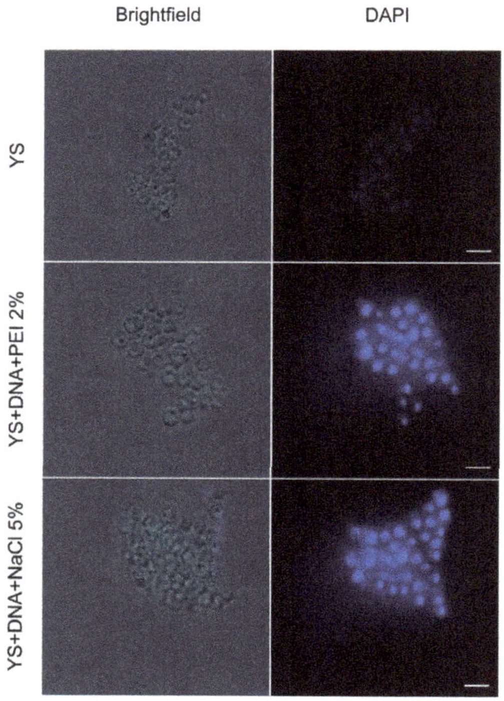

Figure 5. Fluorescence microscopy of capsules yeasts. YS: Empty capsules; YS + DNA + PEI 2%; and YS + DNA + NaCl 5%. Observation of the labeling of YS with DAPI (blue), indicating the presence of nucleic acid in the capsules. YS was used as a negative expression control and displayed only fluorescence background. Images captured through a Leica DMLB epi-fluorescence microscope (Leica, Wetzlar, Alemanha) at 100× magnification. Images were captured by the Leica DFC345 FX. Scale bar: 5 µm.

3.4. Thermostability of Capsules and Retention of Plasmid DNA

The ability to retain the incorporated DNA was evaluated using an 11-day interval between insertion and verification of loss. DNA retention at different temperatures was also evaluated in this study (Figure 6). Although it can be observed that not all of the DNA used for insertion is actually encapsulated, once inside the capsule there is retention of the material to be carried. Capsules whose plasmids were incorporated with the aid of NaCl 5% demonstrated stability in retention despite temperature variations. On the other hand, the material incorporated with PEI 2% is lost when incubating the capsules at 65 °C.

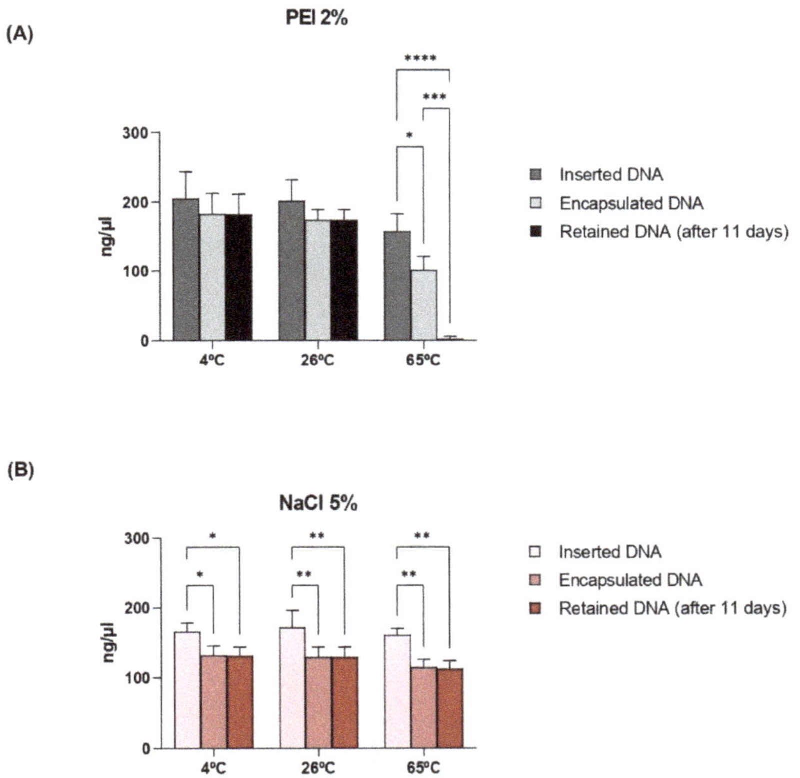

Figure 6. Evaluation of the maintenance of plasmid DNA inside the capsules under different heat treatments. (**A**) PEI 2% and (**B**) NaCl 5% were the compounds used to incorporate the DNA inside the capsules. The y-axis corresponds to the DNA concentration, and the x-axis corresponds to the temperatures that were tested. Bar = means ± SD. Asterisks represent statistical significance (* $p < 0.05$; ** $p < 0.01$; *** $p < 0.001$; **** $p < 0.0001$).

3.5. Phagocytosis of Capsules by Macrophages

The results of the phagocytosis assay are presented in Figure 7. In this figure, we observe macrophages (THP-1) without the addition of YS (negative control) and cells containing phagocytosed YS. The YS, labeled with a cell wall dye with affinity for β-glucans, were visualized by fluorescence microscopy. These results demonstrate the recognition and phagocytosis of YS by the macrophages.

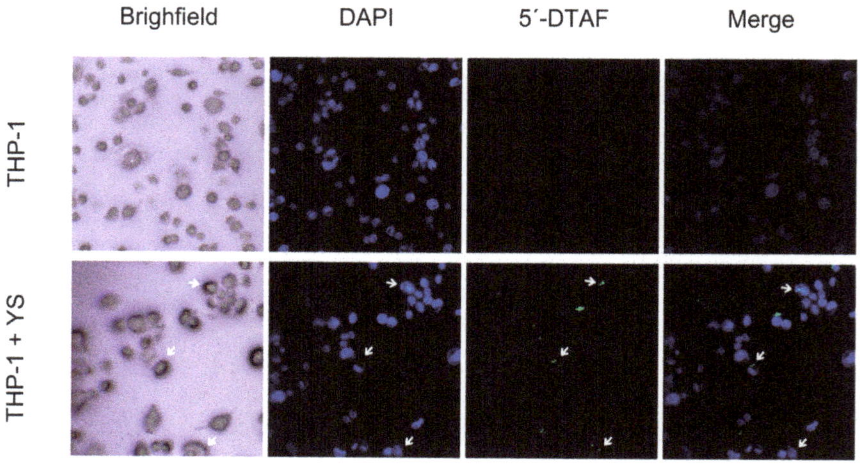

Figure 7. Phagocytosis verification by fluorescence microscopy. Groups: THP-1 cells only and THP-1 + YS (yeast shell), observed under bright-field and fluorescence microscopy. YS were labeled with 5′DTAF (green), and THP-1 cells were labeled with DAPI (blue). The white arrows highlight the YS phagocytosed by the macrophages. Images were captured through a fluorescence microscope (Motic AE31E) at 40× magnification. Images were captured by the Moticam S6 camera using Motic Images Plus 3.0 software. Scale bar: 5 μm.

3.6. Induction of Cytotoxicity

The toxicity potential was assessed using the MTT assay after 24 h of exposure of HEK 293-T cells to the capsules, in a ratio of 2:1 (cells/capsules). It was noted that the preparations with 2% PEI in their composition led to a reduction in cell viability at the concentration tested, in contrast to the other experimental groups (Figure 8).

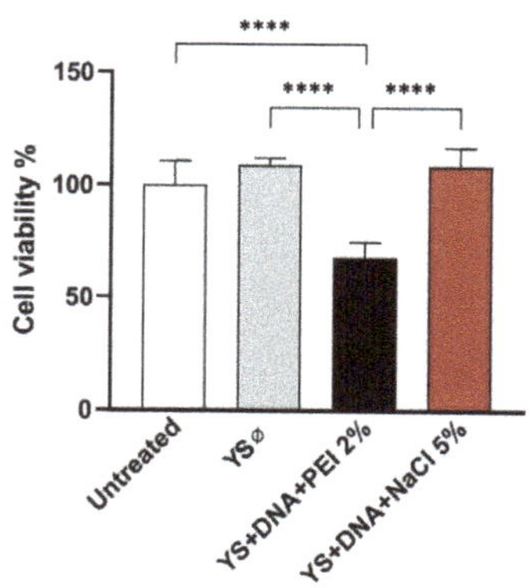

Figure 8. Cell viability evaluation by MTT assay. HEK 293-T cells were treated by incubation with yeast shells (capsules). Asterisks represent statistical significance (**** $p < 0.0001$), determined by the variance test (ANOVA). Values correspond to mean ± standard deviation.

3.7. Zeta Potential (ζ) Analyses

According to the ζ results for empty capsules and DNA-loaded capsules, functionalization using PEI 2% and NaCl 5% proved effective, as evidenced by the variation in ζ between the control and the DNA-containing samples (Table 1).

Table 1. Zeta potential (ζ) values (mean ± SD [1]) of empty capsules and capsules with DNA after treatment with PEI 2% and NaCl 5%.

Sample	ζ (mV)
Empty capsule PEI 2%	12.8 ± 0.73
Capsule pVAX1 PEI 2%	10.17 ± 0.17
Empty capsule NaCl 5%	−4.23 ± 0.14
Capsule pVAX1 NaCl 5%	−3.18 ± 0.26

[1] SD: standard deviation. The ζ measurements were conducted in triplicate.

The empty capsules treated with 2% PEI exhibited a slightly positive zeta potential (ζ = 2.8 ± 0.73 mV), indicating the interaction between PEI and the capsules, resulting in a slightly positive surface charge. This behavior is expected due to the cationic properties of PEI. When the capsules were loaded with pVAX through treatment with PEI 2%, a minimal variation in zeta potential was observed (ζ = 10.17 ± 0.17 mV). This result suggests that the DNA was efficiently incorporated and did not remain adsorbed on the surface, as the change in zeta potential remained consistent with values similar to those observed in the control.

On the other hand, the empty capsules treated with NaCl 5% exhibited a negative zeta potential (ζ = −4.23 ± 0.14 mV), indicating a predominantly negative surface charge, possibly due to the functional groups present on the yeast cell wall, such as β-glucans, without interference from the neutral charge of NaCl. The hypertonic NaCl solution, being electrolytic, may have partially neutralized these charges, reducing the magnitude of the negative potential. The capsules loaded with pVAX and treated with NaCl 5% showed a slightly less negative zeta potential (ζ = −3.18 ± 0.26 mV) compared to the empty capsules. This minimal variation suggests that the DNA was not exclusively retained on the surface.

3.8. Evaluation of Enzymatic Protection

To verify the protection of the internalized plasmid against enzymatic degradation, a DNase assay was performed with microscopy analysis. As shown in Figure 9, there was no visible degradation of the internalized DNA, suggesting that the capsules protect the cargo from degradation and that this cargo is possibly inside the capsular structure and not on its surface since the fluorescence signal is similar in the pre- and post-treatment samples.

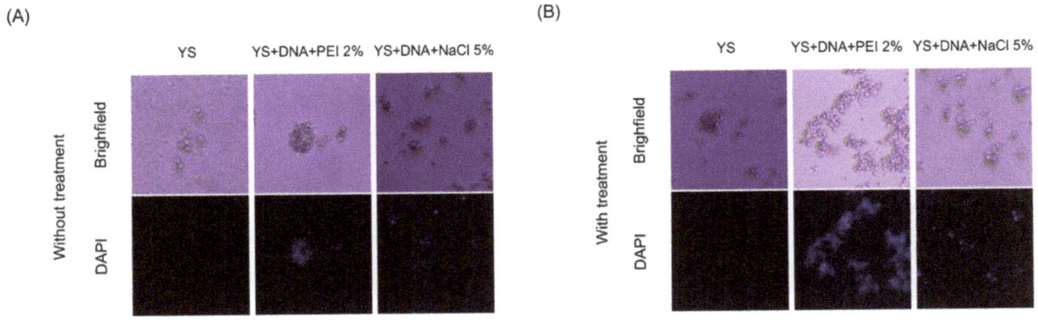

Figure 9. Comparison of DNAse assay results. (**A**) Empty yeast capsules and capsules with two insertion protocols (PEI 2% and NaCl 5%) whose inserted DNA (pVAX1) was stained with DAPI (blue color) without being subjected to DNAse treatment. (**B**) Empty yeast capsules and capsules with two insertion protocols with DNA stained with DAPI but subjected to DNAse treatment. It was possible to observe that there was no degradation of the DNA inserted in the capsules, even in medium containing DNAse. Images captured through a fluorescence microscope (Motic AE31E) at 40× magnification. Images were captured by the Moticam S6 camera using Motic Images Plus 3.0 software.

4. Discussion

Although several yeast species have been proposed as vaccine carriers and adjuvants, only *Saccharomyces cerevisiae* yeast has been applied to produce capsules derived from its cell wall. Despite the morphological similarities between the cell walls of *P. pastoris* and *S. cerevisiae*, there are some differences in the distribution of *P. pastoris* cell wall components, such as higher amounts of β-1,3-glucans, mannans, and chitin, which may increase the immunostimulatory potential concerning T lymphocyte activation [36]. Therefore, we propose a new vaccine delivery platform using β-glucan capsules derived from the *P. pastoris* cell wall.

Yeast cell wall extraction processes included washing steps with isopropyl alcohol [37]. These washes are necessary to remove lipids present in the cell wall, as well as to remove residual moisture and improve YS porosity [20]. However, it is also reported that alcohol can denature proteins, altering their three-dimensional structure [38]. Analyses by optical microscopy after these treatments showed similarities between the capsules and the intact yeasts, with oval structures and apparent preservation in the cell wall architecture (Figure 2). One characteristic that drew attention was the formation of clusters between the capsules that were not observed in the intact cells. This finding may be associated with the increased exposure of β-glucans caused by removing the mannoprotein layer from the cell wall of these yeasts and because these glucans have hydrophobic characteristics [39]. Increasing the hydrophobicity levels of polymers used as delivery systems decreases the release rate and degradation of the polymer matrix due to the increased affinity between polymer and payload [40,41].

In this study, the pVAX1 vector was used as a model plasmid to evaluate the capsules for incorporation and retention of a DNA vaccine. This vector has 3 kb and eukaryotic sequences necessary for expression in mammalian cells to minimize the possibility of chromosomal integration [42]. For the insertion of plasmid DNA into the capsules, three protocols were tested. The first employed a polymer that interacts electrostatically with the capsules and the inserted DNA (PEI 2%), the second using an electrolytic solution (NaCl 5%), and the third without the addition of solutions that have affinity to charges and that interact with the capsules and DNA (PBS). It has been proposed that the porosity of YS can cause leakage of the payload. Therefore, it may be interesting to use polymers that interact electrostatically, such as PEI, forming a protective matrix bypassing these limitations [43].

On the other hand, NaCl, as an electrolyte, provides Na^+ and Cl^- ions that can influence the charge distribution on the surface of yeast capsules and DNA. These ions may reduce the intensity of the negative charges present on these molecules, altering electrostatic interactions and potentially contributing to the partial neutralization of these charges. This neutralization can reduce the electrostatic repulsion between the DNA molecules and the capsule surface, facilitating the approach and internalization of the DNA [44]. In this way, NaCl contributes to creating an ionically balanced environment, which favors the formation of interactions between the DNA and the capsules. However, complete charge neutralization may also be influenced by other factors, such as pH or the presence of additional ions, which play a role in modulating these interactions. Thus, further studies are required to provide a more comprehensive understanding of the mechanism.

The use of NaCl facilitates capsule permeabilization through plasmolysis, creating a significant osmotic gradient between the external environment and the capsule interior. This process enables the diffusion of compounds into the capsule [15]. Several compounds have been incorporated into the capsules following this methodology, such as curcumin, Antarctic krill oil, and chlorogenic acid [32,33,45]. Ours is one of the pioneering works in employing NaCl for plasmid DNA encapsulation, demonstrating its potential use.

The DNA encapsulation efficiency (EE%) was calculated according to the model described by Yu et al. (2015), relating the payload to the difference between the load inserted and the load not absorbed by the capsules [46]. As observed in Figure 4, the capsules formulated with PEI 2% polymer reached above 80% efficiency at different concentrations of inserted DNA, similar to Soto et al. (2008) [14]. Among the advantages of these polymers,

PEI is one of the best cationic polymers for forming nano-complexes with DNA. However, as previously described [47], this compound induces significant cytotoxicity, as illustrated in Figure 7. In contrast, there was no significant change in the viability of eukaryotic cells (HEK 293-T) incubated with capsules with DNA incorporated with NaCl 5%.

Capsules formulated with NaCl 5% showed an EE higher than 60% (Figure 4). These findings demonstrate greater efficiency compared to other studies that used similar methodologies. Paramera et al. (2011) reported that microcapsules containing curcumin achieved an EE% of around 35% [32]. However, the difference in encapsulation efficiency may be related to the nature of the material inserted into the capsules.

In the incorporation protocol that did not employ electrostatic interaction agents, the efficiency was less than 50% at the different DNA concentrations, especially regarding the insertion concentration of 500 ng/μL, which presented an EE% below 3%. Negatively charged materials, such as DNA, tend to adhere more easily to positively charged surfaces compared to uncharged or negatively charged materials [48–51]. This finding highlights the importance of stabilizing agents, such as PEI or NaCl, for effective DNA retention. The absence of these polymers results in rapid release of the payload, indicating that this protocol is not suitable for applications requiring prolonged DNA retention.

The selected formulations, composed of NaCl 5% and PEI 2%, were freeze-dried and verified for their thermal capacity to understand the behavior of YS in retaining this DNA at extreme temperatures. After 11 days, there were minimal losses of inserted DNA at 8 °C, 26 °C, and 65 °C for the protocol with NaCl 5%. On the other hand, the protocol with 2% PEI could not retain the DNA efficiently, resulting in a loss of over 93% of the material when subjected to 65 °C. Although the PEI protocol demonstrated a higher EE%, the protocol with NaCl stood out for its better payload retention, especially under high-temperature conditions. According to Paramera et al. (2011), YS associated with NaCl 5% favors their encapsulation at temperatures above 35 °C, increasing the EE% values by at least two-fold [32].

Fluorescence microscopy was used to visualize the DNA inside the capsules. After the insertion of the plasmid constructs, the fluorescence signal was detected inside the YS, demonstrating the internalization of the genetic material. The DNA delivery system by yeast capsules had already been demonstrated, but only for structures derived from the wall of *S. cerevisiae* [14,52]. It is also worth noting that the internalized material, by both protocols, is protected from enzymatic degradation, as observed in the test performed with DNase. These data are fundamental and should be further investigated because they indicate the functionality of this delivery system in protecting the encapsulated cargo.

The phagocytic process of yeast involves interaction with receptors such as the mannose receptor in macrophages, the β-glucan receptor, and the complement receptor type 3, responsible for recognizing non-opsonized yeast particles [53]. Therefore, an in vitro assay was performed to verify whether macrophages can phagocytose *P. pastoris* capsules, as already described for capsules obtained from *S. cerevisiae* [12,18].

According to Carneiro et al. (2014), most studies focus on the interaction between phagocytes and live yeast cells [54]. However, inactivated cells can also stimulate immune responses since the cell wall PAMPs can be identified in both live and inactivated cells [54]. Thus, despite the structural modifications resulting from the treatments applied to produce yeast capsules, they can be internalized by APCs. However, the mechanisms and parameters of yeast capsule phagocytosis still need to be better elucidated.

The zeta potential of the capsules was analyzed under different conditions to evaluate the effect of PEI or NaCl treatment on the modification of the surface charge and DNA incorporation (pVAX). The minimal variations between the control and the samples containing plasmid indicated that the DNA was not adsorbed on the surface, thus being incorporated inside the capsules. Regarding the final charges of the YS, the capsules treated with PEI presented a positive surface charge, while those treated with NaCl maintained a negative charge, although close to zero.

Previous studies have reported that cationic polymers and DNA complexes with positive surface charge are suitable for efficient endocytosis and gene delivery [55–57]. This behavior can be attributed to the favorable electrostatic interaction between the positive charge of the capsules and the cell membrane, which is negatively charged. This principle is consistent with the results obtained for PEI-functionalized capsules, suggesting that these capsules are more likely to facilitate cell DNA internalization.

On the other hand, NaCl treatment, which resulted in negatively charged capsules, may reduce the interaction with cell membranes, decreasing the internalization efficiency. It has been shown that particles with negative zeta potential exhibit lower interaction with negatively charged cell membranes [58]. In contrast, as observed by Nguyen et al. (2009) studying the CMC compound (P(68)-10 CMC 2) in MTT assays, approximately 80% of the cells remained intact at the highest polymer concentration. In this study, it was suggested that this aspect occurred due to the negative zeta potential, which showed less interaction with the negatively charged cell membrane. Furthermore, the other tested nanocomposite formulations with positive surface charges exhibited higher toxicity profiles [59]. These results reinforce the importance of considering not only zeta potential but also the balance between surface charge and cytotoxicity when designing efficient and safe delivery systems.

Thus, this study demonstrates that YS in different formulations using ionic polymers internalize a large concentration of DNA and contain this material at distinct temperatures. In addition, the *P. pastoris* capsules are recognized and phagocytosed by macrophages, becoming a promising vaccine delivery system. Therefore, more comprehensive studies are needed to make this methodology viable.

5. Conclusions

The results obtained in this study innovatively reveal the development of a vaccine delivery platform based on the use of capsules derived from the *Pichia pastoris* cell wall. The YS showed efficiency in encapsulating DNA through different protocols applied by electrostatic interactions with 5% NaCl and 2% PEI, resulting in encapsulation efficiency (EE%) above 63% and 80%, respectively. Although the PEI protocol demonstrated better encapsulation efficiency, it reduced cell viability at the concentration tested. Furthermore, the protocol using 5% NaCl showed higher thermostability, retaining the inserted DNA even at extreme temperatures such as 65 °C. These findings suggest it is the more suitable methodology for DNA incorporation into the capsules. The retention of DNA for more than ten days at room temperature also reveals an advantage of yeast capsules as a delivery system, considering a storage that does not require refrigeration chains. Phagocytosis assays demonstrated recognition and internalization of YS by macrophages, representing a promising vaccine delivery for antigen-presenting cells. Further studies are needed on the characterization of this approach, especially regarding this system as an immunomodulator. Furthermore, as DNA vaccines are carried within these β-glucan capsules, it will be needed to evaluate the assessment of DNA release in target cells and subsequent expression of the gene inserted in this molecule, which is considered the vaccine antigen.

Supplementary Materials: The following supporting information can be downloaded at: https://www.mdpi.com/article/10.3390/vaccines12121428/s1, Figure S1: Verification of YS metabolic activity.

Author Contributions: Conceptualization, S.S.d.P., M.d.C.V.I., A.J.D.S., L.S.d.M., B.C.F.E., L.R.S.L., M.A.T.M.d.G., I.A.d.M., M.E.d.S.S., D.V.S.d.S., M.L.L., J.N.S.A.A. and A.C.d.F.; investigation, S.S.d.P., M.d.C.V.I., A.J.D.S., L.S.d.M., B.C.F.E., L.R.S.L., M.A.T.M.d.G., I.A.d.M., M.E.d.S.S., D.V.S.d.S., M.L.L., J.N.S.A.A. and A.C.d.F.; methodology, S.S.d.P., M.d.C.V.I., A.J.D.S., L.S.d.M., B.C.F.E., L.R.S.L., M.A.T.M.d.G., I.A.d.M., M.E.d.S.S., D.V.S.d.S., M.L.L., J.N.S.A.A. and A.C.d.F.; validation, S.S.d.P., M.d.C.V.I., A.J.D.S., L.S.d.M., B.C.F.E., L.R.S.L., M.A.T.M.d.G., I.A.d.M., M.E.d.S.S., D.V.S.d.S., M.L.L., J.N.S.A.A. and A.C.d.F.; formal analysis, S.S.d.P., M.d.C.V.I., A.J.D.S., L.S.d.M., B.C.F.E., L.R.S.L., M.A.T.M.d.G., I.A.d.M., M.E.d.S.S., D.V.S.d.S., M.L.L., J.N.S.A.A. and A.C.d.F.; data curation, S.S.d.P., M.d.C.V.I., A.J.D.S., L.S.d.M., B.C.F.E., L.R.S.L., M.A.T.M.d.G., I.A.d.M., M.E.d.S.S., D.V.S.d.S., M.L.L., J.N.S.A.A. and A.C.d.F.; writing—original draft, S.S.d.P., M.d.C.V.I., A.J.D.S., L.S.d.M., B.C.F.E., L.R.S.L., M.A.T.M.d.G., I.A.d.M., M.E.d.S.S., D.V.S.d.S., M.L.L., J.N.S.A.A. and A.C.d.F.; writing—

review and editing, S.S.d.P., M.d.C.V.I., A.J.D.S., L.S.d.M., B.C.F.E., L.R.S.L., M.A.T.M.d.G., I.A.d.M., M.E.d.S.S., D.V.S.d.S., M.L.L., J.N.S.A.A. and A.C.d.F.; visualization, S.S.d.P., M.d.C.V.I., A.J.D.S., L.S.d.M., B.C.F.E., L.R.S.L., M.A.T.M.d.G., I.A.d.M., M.E.d.S.S., D.V.S.d.S., M.L.L., J.N.S.A.A. and A.C.d.F.; supervision, A.J.D.S., L.S.d.M. and A.C.d.F.; project administration, A.J.D.S., L.S.d.M. and A.C.d.F.; resources, A.C.d.F.; funding acquisition, A.C.d.F. All authors have read and agreed to the published version of the manuscript.

Funding: This research was funded by CNPq/MCTI/CT-BIOTEC N° 31/2022, grant number 440315/2022-0 and the Fundação de Amparo à Ciência e Tecnologia de Pernambuco (FACEPE) for grant numbers APQ-1482-2.02/22 (Programa de Apoio à Fixação de Jovens Doutores em Pernambuco), IBPG-0954-2.00/22, and IBPG-1652-2.00/21.

Institutional Review Board Statement: Not applicable.

Informed Consent Statement: Not applicable.

Data Availability Statement: All relevant data from this study are available from the corresponding author.

Acknowledgments: We would like to thank the entire team of the Laboratory of Molecular Studies and Experimental Therapy (LEMTE) for the technical and infrastructure support. We are thankful to Andrea Pedrosa-Harand from the Laboratory of Plant Cytogenetics and Evolution (Genetics Department) at the Federal University of Pernambuco for granting and assisting in the use of the fluorescence microscope. We are thankful to Lucas Matheus Barreto Santana and Tiago Henrique dos Santos Souza from the Instituto Aggeu Magalhães (IAM)—Fundação Oswaldo Cruz—Pernambuco for donating reagents for cytotoxicity assays. We are thankful to Adriana Fontes, Paulo E. Cabral Filho, and João V.A. Lima from the Biomedical Nanotechnology Research Group at the Federal University of Pernambuco (NanoBio/UFPE) for providing fluorescence microscopy and zeta potential analyses.

Conflicts of Interest: The authors declare no conflicts of interest.

References

1. Plotkin, S.A. Vaccines: Past, Present and Future. *Nat. Med.* **2005**, *11*, S5–S11. [CrossRef] [PubMed]
2. Pollard, A.J.; Bijker, E.M. Publisher Correction: A Guide to Vaccinology: From Basic Principles to New Developments. *Nat. Rev. Immunol.* **2021**, *21*, 129. [CrossRef] [PubMed]
3. Gebre, M.S.; Brito, L.A.; Tostanoski, L.H.; Edwards, D.K.; Carfi, A.; Barouch, D.H. Novel Approaches for Vaccine Development. *Cell* **2021**, *184*, 1589–1603. [CrossRef] [PubMed]
4. Wang, C.; Yuan, F. A Comprehensive Comparison of DNA and RNA Vaccines. *Adv. Drug Deliv. Rev.* **2024**, *210*, 115340. [CrossRef]
5. Yang, J.; Li, Y.; Jin, S.; Xu, J.; Wang, P.C.; Liang, X.-J.; Zhang, X. Engineered Biomaterials for Development of Nucleic Acid Vaccines. *Biomater. Res.* **2015**, *19*, 5. [CrossRef]
6. Shah, M.A.A.; Ali, Z.; Ahmad, R.; Qadri, I.; Fatima, K.; He, N. DNA Mediated Vaccines Delivery Through Nanoparticles. *J. Nanosci. Nanotechnol.* **2015**, *15*, 41–53. [CrossRef]
7. Kisakova, L.A.; Apartsin, E.K.; Nizolenko, L.F.; Karpenko, L.I. Dendrimer-Mediated Delivery of DNA and RNA Vaccines. *Pharmaceutics* **2023**, *15*, 1106. [CrossRef]
8. Yang, L.; Gong, L.; Wang, P.; Zhao, X.; Zhao, F.; Zhang, Z.; Li, Y.; Huang, W. Recent Advances in Lipid Nanoparticles for Delivery of mRNA. *Pharmaceutics* **2022**, *14*, 2682. [CrossRef]
9. Chen, G.; Zhao, B.; Ruiz, E.F.; Zhang, F. Advances in the Polymeric Delivery of Nucleic Acid Vaccines. *Theranostics* **2022**, *12*, 4081–4109. [CrossRef]
10. Silva, A.J.D.; de Macêdo, L.S.; Leal, L.R.S.; de Jesus, A.L.S.; Freitas, A.C. Yeasts as a Promising Delivery Platform for DNA and RNA Vaccines. *FEMS Yeast Res.* **2021**, *21*, foab018. [CrossRef]
11. Kumar, R.; Kumar, P. Yeast-Based Vaccines: New Perspective in Vaccine Development and Application. *FEMS Yeast Res.* **2019**, *19*, foz007. [CrossRef]
12. Tan, Y.; Chen, L.; Li, K.; Lou, B.; Liu, Y.; Liu, Z. Yeast as Carrier for Drug Delivery and Vaccine Construction. *J. Control. Release* **2022**, *346*, 358–379. [CrossRef] [PubMed]
13. Ostroff, G. Drug Delivery Product and Methods. U.S. Patent 7,740,861, 22 June 2010.
14. Soto, E.R.; Ostroff, G.R. Characterization of Multilayered Nanoparticles Encapsulated in Yeast Cell Wall Particles for DNA Delivery. *Bioconjugate Chem.* **2008**, *19*, 840–848. [CrossRef] [PubMed]
15. Coradello, G.; Tirelli, N. Yeast Cells in Microencapsulation. General Features and Controlling Factors of the Encapsulation Process. *Molecules* **2021**, *26*, 3123. [CrossRef]
16. Liu, Y.; Wu, Q.; Wu, X.; Algharib, S.A.; Gong, F.; Hu, J.; Luo, W.; Zhou, M.; Pan, Y.; Yan, Y.; et al. Structure, Preparation, Modification, and Bioactivities of β-Glucan and Mannan from Yeast Cell Wall: A Review. *Int. J. Biol. Macromol.* **2021**, *173*, 445–456. [CrossRef] [PubMed]

17. Vetvicka, V.; Vannucci, L.; Sima, P. B-glucan as a New Tool in Vaccine Development. *Scand. J. Immunol.* **2020**, *91*, e12833. [CrossRef] [PubMed]
18. Silva De Macêdo, L.; Sousa De Pinho, S.; Duarte Silva, A.J.; De Moura, I.A.; Flores Espinoza, B.C.; Viana Invenção, M.D.C.; Silva Novis, P.V.; Turiah Machado Da Gama, M.A.; Do Nascimento Carvalho, M.; Rosa Sales Leal, L.; et al. Understanding Yeast Shells: Structure, Properties and Applications. *ADMET DMPK* **2024**, *12*, 299–317. [CrossRef]
19. Hu, X.; Zhang, J. Yeast Capsules for Targeted Delivery: The Future of Nanotherapy? *Nanomedicine* **2017**, *12*, 955–957. [CrossRef]
20. Zhou, X.; Ling, K.; Liu, M.; Zhang, X.; Ding, J.; Dong, Y.; Liang, Z.; Li, J.; Zhang, J. Targeted Delivery of Cisplatin-Derived Nanoprecursors via a Biomimetic Yeast Microcapsule for Tumor Therapy by the Oral Route. *Theranostics* **2019**, *9*, 6568–6586. [CrossRef]
21. Austriaco, N. Yeast Oral Vaccines against Infectious Diseases. *Front. Microbiol.* **2023**, *14*, 1150412. [CrossRef]
22. Alexander, E. Yeasts in Nanotechnology-Enabled Oral Vaccine and Gene Delivery. *Bioengineered* **2021**, *12*, 8325–8335. [CrossRef] [PubMed]
23. De Smet, R.; Allais, L.; Cuvelier, C.A. Recent Advances in Oral Vaccine Development: Yeast-Derived β-Glucan Particles. *Hum. Vaccines Immunother.* **2014**, *10*, 1309–1318. [CrossRef] [PubMed]
24. Yang, F.; Shang, S.; Qi, M.; Xiang, Y.; Wang, L.; Wang, X.; Lin, T.; Hao, D.; Chen, J.; Liu, J.; et al. Yeast Glucan Particles: An Express Train for Oral Targeted Drug Delivery Systems. *Int. J. Biol. Macromol.* **2023**, *253*, 127131. [CrossRef] [PubMed]
25. Geller, A.; Shrestha, R.; Yan, J. Yeast-Derived β-Glucan in Cancer: Novel Uses of a Traditional Therapeutic. *Int. J. Mol. Sci.* **2019**, *20*, 3618. [CrossRef] [PubMed]
26. Fan, T.W.-M.; Daneshmandi, S.; Cassel, T.A.; Uddin, M.B.; Sledziona, J.; Thompson, P.T.; Lin, P.; Higashi, R.M.; Lane, A.N. Polarization and β-Glucan Reprogram Immunomodulatory Metabolism in Human Macrophages and Ex Vivo in Human Lung Cancer Tissues. *J. Immunol.* **2022**, *209*, 1674–1690. [CrossRef]
27. Bazan, S.B.; Geginat, G.; Breinig, T.; Schmitt, M.J.; Breinig, F. Uptake of Various Yeast Genera by Antigen-Presenting Cells and Influence of Subcellular Antigen Localization on the Activation of Ovalbumin-Specific CD8 T Lymphocytes. *Vaccine* **2011**, *29*, 8165–8173. [CrossRef]
28. Brown, G.D. Innate Antifungal Immunity: The Key Role of Phagocytes. *Annu. Rev. Immunol.* **2011**, *29*, 1–21. [CrossRef]
29. Zhu, L.; Lei, Z.; Xia, X.; Zhang, Y.; Chen, Y.; Wang, B.; Li, J.; Li, G.; Yang, G.; Cao, G.; et al. Yeast Shells Encapsulating Adjuvant AS04 as an Antigen Delivery System for a Novel Vaccine against *Toxoplasma gondii*. *ACS Appl. Mater. Interfaces* **2021**, *13*, 40415–40428. [CrossRef]
30. Kwolek-Mirek, M.; Zadrag-Tecza, R. Comparison of Methods Used for Assessing the Viability and Vitality of Yeast Cells. *FEMS Yeast Res.* **2014**, *14*, 1068–1079. [CrossRef]
31. Huang, H.; Ostroff, G.R.; Lee, C.K.; Specht, C.A.; Levitz, S.M. Robust Stimulation of Humoral and Cellular Immune Responses Following Vaccination with Antigen-Loaded β-Glucan Particles. *mBio* **2010**, *1*, e00164-10. [CrossRef]
32. Paramera, E.I.; Konteles, S.J.; Karathanos, V.T. Microencapsulation of Curcumin in Cells of Saccharomyces Cerevisiae. *Food Chem.* **2011**, *125*, 892–902. [CrossRef]
33. Shi, G.; Rao, L.; Yu, H.; Xiang, H.; Pen, G.; Long, S.; Yang, C. Yeast-Cell-Based Microencapsulation of Chlorogenic Acid as a Water-Soluble Antioxidant. *J. Food Eng.* **2007**, *80*, 1060–1067. [CrossRef]
34. Aouadi, M.; Tesz, G.J.; Nicoloro, S.M.; Wang, M.; Chouinard, M.; Soto, E.; Ostroff, G.R.; Czech, M.P. Orally Delivered siRNA Targeting Macrophage Map4k4 Suppresses Systemic Inflammation. *Nature* **2009**, *458*, 1180–1184. [CrossRef] [PubMed]
35. Mosmann, T. Rapid Colorimetric Assay for Cellular Growth and Survival: Application to Proliferation and Cytotoxicity Assays. *J. Immunol. Methods* **1983**, *65*, 55–63. [CrossRef]
36. Bazan, S.B.; Breinig, T.; Schmitt, M.J.; Breinig, F. Heat Treatment Improves Antigen-Specific T Cell Activation after Protein Delivery by Several but Not All Yeast Genera. *Vaccine* **2014**, *32*, 2591–2598. [CrossRef]
37. Jing, Z.; Wang, S.; Xu, K.; Tang, Q.; Li, W.; Zheng, W.; Shi, H.; Su, K.; Liu, Y.; Hong, Z. A Potent Micron Neoantigen Tumor Vaccine GP-Neoantigen Induces Robust Antitumor Activity in Multiple Tumor Models. *Adv. Sci.* **2022**, *9*, 2201496. [CrossRef]
38. Shao, Q.; Fan, Y.; Yang, L.; Qin Gao, Y. From Protein Denaturant to Protectant: Comparative Molecular Dynamics Study of Alcohol/Protein Interactions. *J. Chem. Phys.* **2012**, *136*, 115101. [CrossRef]
39. Kang, X.; Kirui, A.; Muszyński, A.; Widanage, M.C.D.; Chen, A.; Azadi, P.; Wang, P.; Mentink-Vigier, F.; Wang, T. Molecular Architecture of Fungal Cell Walls Revealed by Solid-State NMR. *Nat. Commun.* **2018**, *9*, 2747. [CrossRef]
40. Lee, W.; Park, J.; Yang, E.H.; Suh, H.; Kim, S.H.; Chung, D.S.; Choi, K.; Yang, C.W.; Park, J. Investigation of the Factors Influencing the Release Rates of Cyclosporin A-Loaded Micro- and Nanoparticles Prepared by High-Pressure Homogenizer. *J. Control. Release* **2002**, *84*, 115–123. [CrossRef]
41. Park, T.G. Degradation of Poly(Lactic-Co-Glycolic Acid) Microspheres: Effect of Copolymer Composition. *Biomaterials* **1995**, *16*, 1123–1130. [CrossRef]
42. Leventhal, S.S.; Clancy, C.; Erasmus, J.; Feldmann, H.; Hawman, D.W. An Intramuscular DNA Vaccine for SARS-CoV-2 Decreases Viral Lung Load but Not Lung Pathology in Syrian Hamsters. *Microorganisms* **2021**, *9*, 1040. [CrossRef] [PubMed]
43. Tan, C.; Wang, J.; Sun, B. Polysaccharide Dual Coating of Yeast Capsules for Stabilization of Anthocyanins. *Food Chem.* **2021**, *357*, 129652. [CrossRef] [PubMed]
44. Gavryushov, S. Electrostatics of B-DNA in NaCl and $CaCl_2$ Solutions: Ion Size, Interionic Correlation, and Solvent Dielectric Saturation Effects. *J. Phys. Chem. B* **2008**, *112*, 8955–8965. [CrossRef] [PubMed]

45. Fu, J.; Song, L.; Guan, J.; Sun, C.; Zhou, D.; Zhu, B. Encapsulation of Antarctic Krill Oil in Yeast Cell Microcarriers: Evaluation of Oxidative Stability and in Vitro Release. *Food Chem.* **2021**, *338*, 128089. [CrossRef]
46. Hong, Z.; Yu, M.; Chen, Z.; Guo, W.; Wang, J.; Feng, Y.; Kong, X. Specifically Targeted Delivery of Protein To Phagocytic Macrophages. *Int. J. Nanomed.* **2015**, *10*, 1743–1757. [CrossRef]
47. Pack, D.W.; Hoffman, A.S.; Pun, S.; Stayton, P.S. Design and Development of Polymers for Gene Delivery. *Nat. Rev. Drug Discov.* **2005**, *4*, 581–593. [CrossRef]
48. Silveira, N.; Longuinho, M.M.; Leitão, S.G.; Silva, R.S.F.; Lourenço, M.C.; Silva, P.E.A.; Pinto, M.D.C.F.R.; Abraçado, L.G.; Finotelli, P.V. Synthesis and Characterization of the Antitubercular Phenazine Lapazine and Development of PLGA and PCL Nanoparticles for Its Entrapment. *Mater. Sci. Eng. C* **2016**, *58*, 458–466. [CrossRef]
49. Booysen, L.L.I.J.; Kalombo, L.; Brooks, E.; Hansen, R.; Gilliland, J.; Gruppo, V.; Lungenhofer, P.; Semete-Makokotlela, B.; Swai, H.S.; Kotze, A.F.; et al. In Vivo/in Vitro Pharmacokinetic and Pharmacodynamic Study of Spray-Dried Poly-(Dl-Lactic-Co-Glycolic) Acid Nanoparticles Encapsulating Rifampicin and Isoniazid. *Int. J. Pharm.* **2013**, *444*, 10–17. [CrossRef]
50. Do, D.P.; Pai, S.B.; Rizvi, S.A.A.; D'Souza, M.J. Development of Sulforaphane-Encapsulated Microspheres for Cancer Epigenetic Therapy. *Int. J. Pharm.* **2010**, *386*, 114–121. [CrossRef]
51. Doostmohammadi, A.; Monshi, A.; Salehi, R.; Fathi, M.H.; Golniya, Z.; Daniels, A.U. Bioactive Glass Nanoparticles with Negative Zeta Potential. *Ceram. Int.* **2011**, *37*, 2311–2316. [CrossRef]
52. Pan, Y.; Li, X.; Kang, T.; Meng, H.; Chen, Z.; Yang, L.; Wu, Y.; Wei, Y.; Gou, M. Efficient Delivery of Antigen to DCs Using Yeast-Derived Microparticles. *Sci. Rep.* **2015**, *5*, 10687. [CrossRef] [PubMed]
53. Bos, H.; De Souza, W. Phagocytosis of Yeast: A Method for Concurrent Quantification of Binding and Internalization Using Differential Interference Contrast Microscopy. *J. Immunol. Methods* **2000**, *238*, 29–43. [CrossRef] [PubMed]
54. Carneiro, C.; Vaz, C.; Carvalho-Pereira, J.; Pais, C.; Sampaio, P. A New Method for Yeast Phagocytosis Analysis by Flow Cytometry. *J. Microbiol. Methods* **2014**, *101*, 56–62. [CrossRef] [PubMed]
55. Yang, L.; Liang, X.; Ren, X.; Liu, Z.; Liu, Y.; Wang, J.; Wang, J.; Zhang, L.-M.; Deng, D.Y.B.; Quan, D. An Efficient Nonviral Gene-Delivery Vector Based on Hyperbranched Cationic Glycogen Derivatives. *Int. J. Nanomed.* **2014**, *9*, 419–435. [CrossRef]
56. Deng, J.; Zhou, Y.; Xu, B.; Mai, K.; Deng, Y.; Zhang, L.-M. Dendronized Chitosan Derivative as a Biocompatible Gene Delivery Carrier. *Biomacromolecules* **2011**, *12*, 642–649. [CrossRef]
57. Farahpour, A.; Ramezanian, N.; Gholami, L.; Askarian, S.; Banisadr, A.; Kazemi Oskuee, R. Synthesis and Characterization of Polyethyleneimine-Terminated Poly(β-Amino Esters) Conjugated with Pullulan for Gene Delivery. *Pharm. Dev. Technol.* **2022**, *27*, 606–614. [CrossRef]
58. Jin, J.W.; Rong, M.Z.; Zhang, M.Q.; Wong, W. Preparation of a Water Soluble Aminated B-1, 3-D -glucan for Gene Carrier: The in Vitro Study of the Anti-inflammatory Activity and Transfection Efficiency. *J. Biomed. Mater. Res.* **2021**, *109*, 2506–2515. [CrossRef]
59. Nguyen, J.; Reul, R.; Betz, T.; Dayyoub, E.; Schmehl, T.; Gessler, T.; Bakowsky, U.; Seeger, W.; Kissel, T. Nanocomposites of Lung Surfactant and Biodegradable Cationic Nanoparticles Improve Transfection Efficiency to Lung Cells. *J. Control. Release* **2009**, *140*, 47–54. [CrossRef]

vaccines

Review

Advances in Vaccine Adjuvants for Teleost Fish: Implications for Aquatic Welfare and the Potential of Nanoparticle-Based Formulations

Iosif Tammas [1], Konstantina Bitchava [1,*] and Athanasios I. Gelasakis [2,*]

[1] Laboratory of Applied Hydrobiology, Department of Animal Science, Agricultural University of Athens, 11855 Athens, Greece; stud217095@aua.gr

[2] Laboratory of Anatomy & Physiology of Farm Animals, Agricultural University of Athens, 11855 Athens, Greece

* Correspondence: bitchava@aua.gr (K.B.); gelasakis@aua.gr (A.I.G.); Tel.: +30-2105294387 (A.I.G.)

Abstract: Vaccine adjuvants are crucial for reinforcing the immunogenicity of vaccines. Therefore, they are widely used in the aquaculture sector as vaccine components, facilitating the efficient prevention of infectious diseases and promoting sustainable teleost fish growth. Despite their benefits, there has been a growing concern about the potential adverse effects of vaccine adjuvants in teleost fish, connoting a valid impact on their overall health and welfare. Among the adjuvants used in aquaculture vaccinology, nanoparticle-based formulations have given rise to a promising new alternative to traditional options, such as oil-based emulsions and aluminum compounds, offering the benefit of minimizing relevant side effects. The aim of this paper was to review the current status of the adjuvants used in aquaculture, provide a description and an evaluation of their mode of action and side effects, and explore the potential use of nanoparticle formulations as adjuvants to improve the efficacy of aquaculture vaccines. By demonstrating and assessing the equilibrium between teleost fish welfare and immunological efficacy, this review presents a collective perspective that will assist in establishing a framework for the utilization of effective species-specific practices around adjuvant use in aquaculture, while also addressing the challenges of welfare-friendly immunization.

Keywords: aquaculture; vaccines; adjuvants; nanoparticles; immunization; inflammation; granulomas; lesions; sustainability; welfare; health; side effects; teleost fish

1. Introduction

In the dynamic landscape of aquaculture, the use of vaccines has emerged as a cornerstone for disease prevention, marking an essential checkpoint in sustainable industry growth [1]. Vaccines constitute the most efficient solution to combat impeding pathogen threats, especially in the context of modern aquaculture, where increased rearing densities have caused a spike in microbial agent transmission [2,3]. The ecological pressure exerted by aquaculture practices in the global transmission of pathogens in wild fish populations has added an additional layer to the issue, proving that the efficient and timely development of vaccines is paramount in safeguarding the health of teleost fish species [4–6].

To further boost the immunogenicity and the elicitation of robust immune responses, auxiliary substances called adjuvants have long been used in aquaculture vaccine formulations to maximize the effectiveness of the administered vaccine preparations. Traditionally, these substances were abstractly described as agents that either improved vaccine potency or efficiency, two terms that were used to signify boosts to adaptive immune responses and the prevention of infection, respectively [7]. Considerable advancements in the field of immunology, however, led to a novel, clearer, and more refined term for adjuvants, defining them as structurally heterogenous compounds that regulate the inherent immunogenicity of vaccine antigens [7]. This new definition acknowledges the role of adjuvants in boosting immunizations through specific and intricate mechanisms of action, thereby consolidating a promising field for scientific exploration in the field of immunology.

The use of adjuvants in aquaculture is compatible with every method of vaccine administration, including injection, immersion, and oral vaccination. The vast majority of vaccines utilized in aquaculture, however, are administered via intraperitoneal (IP) injection, since this method of vaccine delivery provides the most adequate and long-lasting immunization in teleost fish [7–9]. Nevertheless, the use of specific adjuvants has enabled other vaccine delivery methods, such as immersion vaccination, to be efficiently utilized in aquaculture, thus opening the gateway for alternative methods of vaccine administration. By enhancing the effectiveness of alternative vaccination methods, the industry can potentially shift from the exclusive use of injection vaccination to other forms of vaccine delivery that will offer benefits not only in logistical and cost-related aspects, but also in animal welfare, by reducing the stress induced during the immunization process.

In this quest for innovative and sustainable immunization strategies, nanoparticles have paved the way for a promising alternative to traditional adjuvant formulations in the aquaculture sector. Thanks to their small size and their ability to provide tailored platforms for controlled release, these formulations offer enhanced antigen delivery and improved immune response, while minimizing the required dosages and mitigating any relevant side effects associated with traditional adjuvants [10–12]. Recent data suggest that nanoparticles not only boost the efficacy of conventional injectable vaccines, but also improve immunization methods that involve mucosal delivery, such as immersion and oral vaccination [13]. By promoting enhanced antigen uptake and by stimulating both innate and adaptive immunity, nanoparticle-based adjuvants can revolutionize vaccine protocols in aquaculture, potentially reducing the reliance on invasive methods in favor of more cost-efficient and welfare-friendly routes. Furthermore, their capacity to encapsulate and protect antigens improves vaccine stability, prospectively broadening their use across various aquatic species and environmental conditions [14].

This review article aims to synthesize a comprehensive perspective on the most popular adjuvants currently used in aquaculture vaccinology, drawing information from literature not only about their exact modes of action, but also about advances made in understanding and mitigating their side effects. Additionally, the potential of utilizing nanoparticle formulations as an alternative to traditional vaccine adjuvants in the industry of aquaculture is discussed, offering a complete overview of the future directions for optimizing vaccine safety and enhancing immune responses. Spreading research findings regarding adjuvant side effects across different and diverse combinations of adjuvants, vaccines, fish species, and rearing conditions is paramount in broadening the knowledge on the assessment of these adverse effects, thereby safeguarding aquatic organism health and sustainability in practice. Delving deeper into the matter, understanding exactly how adjuvants work is essential for unraveling their importance and integration in vaccination strategies and their implications for teleost fish welfare.

2. Mode of Action: The Two-Signal Model

The precise mechanisms by which adjuvants enhance immune responses have long been, and continue to be, a subject of significant debate and investigation. While adjuvants are known to enhance the body's immune response to antigens, the detailed processes by which they achieve this remain elusive [15,16]. Researchers have proposed many assumptions, including the stimulation of innate immune receptors, the creation of a "depot effect" that prolongs antigen exposure, and the modulation of the immune environment; yet, a complete understanding is still lacking, especially in the context of aquatic organisms [7]. In this quest for understanding how adjuvants work, Schijns' model, also known as the two-signal model [17], stands out as the most comprehensive model for explaining and categorizing their mechanisms of action, providing a structured approach to these complex interactions, while highlighting the need for continued research to fully unravel the intricacies of adjuvant function to improve vaccine development. According to this proposed model, vaccine adjuvants can be categorized based on their capacity to facilitate two types of signals: type 1 and type 2 signals.

2.1. Type 1 Signals

Type 1 signals typically refer to the process of antigen presentation by specific types of immune cells, called professional antigen-presenting cells (pAPCs) [7]. These cells can bind to antigens through highly polymorphic MHC class II proteins expressed on their surface and subsequently display them to T-cells to facilitate their differentiation into CD4+ T-cells, known as helper T-cells or T_h cells. This sequence of events is essential not only to the production of antibodies, but also to the induction of innate immune responses, since T_h cells can aid by providing necessary co-stimulatory signals that are unique to the type of antigen that is recognized [18–21]. Several immune cells are known to possess antigen-presenting functions, including B-cells, macrophages, and dendritic cells (DCs), and this is a pattern that is conserved in teleost fish, albeit with some differences [22].

Adjuvants that influence type 1 signals and antigen presentation thereof can be categorized in a distinct group that encompasses type 1 signal facilitators that can essentially catalyze the antigen presentation process. The exact way in which type 1 adjuvants influence can vary; however, it all boils down to improving the immunoavailability of the vaccine's antigens to antigen-presenting cells [16]. Based on that, and by taking the geographical concept of immune reactivity into consideration [23], adjuvants that have a significant effect on an antigen's presentability in the context of space, time, and availability can collectively be categorized as type 1 adjuvants according to the two-signal model.

Although type 1 adjuvants are widely implemented in vaccinology and are considered crucial for eliciting robust immune responses, the precise mechanisms behind their effects remain largely elusive at present [16]. Since type 1 adjuvants are used extensively in injectable vaccines, the hypothesis for the most important mechanism of action thus far proposes that these adjuvants facilitate a "depot effect", in which a localized reservoir of the antigen is formed at the site of injection [7,17]. This allows for a controlled, gradual, and sustained release of the antigen over an extended period of time, therefore affecting its overall immunoavailability to the host's immune system cells [16]. By maintaining the antigen's presence in the body for a longer period, the depot effect maximizes the immune system's exposure to the vaccine's antigen, thereby promoting a more robust and prolonged immune response.

The evidence on whether the "depot effect" is solely responsible for the immunostimulating effects of type 1 adjuvants or whether it simply affects it remains debatable, keeping many researchers conflicted on the topic [24]. Many studies imply that the tissue microenvironment of the injection site plays an important role in immunostimulation, as many antigen-presenting cells are attracted to chemotactic factors produced locally at this site [16,24]. Many key players of the innate wing of the immune system enter the injured tissue and influence the function of APCs. Based on that, it appears that type 1 adjuvants can potentially promote inflammatory signals, adding an additional layer of complexity to the modes of action supporting their immune-enhancing effects.

2.2. Type 2 Signals

On the other hand, type 2 signals, according to the two-signal model, are regarded as secondary, ancillary signals that facilitate the activation of adaptive immune cells like B-cells and T-cells [7,16,17]. These signals provide necessary co-stimulation during the process of antigen presentation, which is fulfilled through the equilibrated and combinatory effects of co-stimulatory or co-inhibitory signals [16]. These signals can thus tailor the scale and the effect of adaptive immune responses, resulting in varying outcomes of immunostimulation that aid in enhancing the process of immunization. As such, adjuvants that facilitate type 2 signals can collectively be grouped into type 2 adjuvants.

The activity of adjuvants that facilitate type 2 signals has been well described, even down to the molecular level, as opposed to those that facilitate type 1 signals [16]. The cells of the innate immune system express pattern-recognition receptors (PRRs) that are used to recognize pathogens by binding to specific pathogen-associated molecular patterns (PAMPs) [16]. These molecular patterns are typically structural elements that are unique to different classes of pathogens and are recognized by different kinds of PRRs [25,26]. The recognition of pathogen-associated molecular patterns by the receptors of the innate immune cells orchestrates signaling cascades that facilitate the induction of both humoral and cellular immune responses. Some of the most important examples of pattern-recognition receptors' subfamilies and their associated ligands identified so far in teleost fish are summarized in Table 1.

Table 1. A cumulative summary of the three major pattern-recognition receptor (PRR) types identified in teleost fish according to the current and up-to-date literature. It is evident that the teleost fish species is an important parameter in the presence or absence of certain PRRs, as well as in their associated ligands. Homologues of mammalian PRRs have been characterized in a plethora of teleost fish species. This table includes some indicative ones for reference [26–41].

Toll-like Receptors (TLRs)		
Receptor	**Associated Ligands**	**Indicative Teleost Fish Species**
TLR1	Lipopeptides	Rainbow Trout, Large Yellow Croaker, Carp, Pufferfish, Orange-Spotted Grouper, European Sea Bass, Turbot
TLR2	Lipopeptides, PGN, LTA, Pam_3CSK_4	Carp, Catfish, Orange-Spotted Grouper, European Sea Bass, Turbot, Gilthead Seabream
TLR3	dsRNA, poly(I:C)	Carp, Pufferfish, Zebrafish
TLR4	Unknown	Carp, Catfish, Rare Minnow, Zebrafish
TLR5	Flagellin	Atlantic Salmon, Japanese Flounder, Channel Catfish, Gilthead Seabream, Rainbow Trout, Pufferfish, Zebrafish, Turbot
TLR7	ssRNA	Channel Catfish, Grass Carp, Pufferfish, Rainbow Trout, Zebrafish, Turbot
TLR8	ssRNA	Atlantic Salmon, Channel Catfish, Pufferfish, Rainbow Trout, Turbot
TLR9	CpG motifs	Atlantic Salmon, Cobia, Japanese Flounder, Rainbow Trout, Zebrafish, European Seabass, Gilthead Seabream, Turbot
TLR13	rRNA	Atlantic Salmon, Channel Catfish, Orange-Spotted Grouper
TLR14	Unknown	Japanese Flounder, Orange-Spotted Grouper, Pufferfish
TLR18	Unknown	Channel Catfish, Grass Carp, Zebrafish
TLR19	dsRNA	Channel Catfish, Grass Carp, Zebrafish
TLR20	Unknown	Carp, Channel Catfish, Zebrafish
TLR21	CpG motifs	Channel Catfish, Grass Carp, Orange-Spotted Grouper, Zebrafish, Turbot
TLR22	dsRNA, poly(I:C)	Atlantic Cod, Channel Catfish, Grass Carp, Pufferfish, Zebrafish, European Seabass, Turbot, Gilthead Seabream
TLR23	Unknown	Atlantic Cod, Pufferfish
TLR25	Unknown	Channel Catfish, Fathead Minnow, Nile Tilapia
TLR26	LPS, poly(I:C)	Channel Catfish, Yellow Catfish
TLR28	LPS, poly(I:C)	Brown Croaker

Table 1. *Cont.*

NOD-like Receptor (NLR) Superfamilies		
Receptor	**Associated Ligands**	**Indicative Teleost Fish Species**
NLR-A	LPS, PGN (iE-DAP, MDP), Poly (I:C)	Grass Carp, Nile Tilapia, Rainbow Trout, Channel Catfish, Zebrafish
NLR-B	Unknown	Grass Carp
NLR-C	Unknown	Grass Carp, Brown Croaker
RIG-1-like Receptors (RLRs)		
RIG-1	dsRNA	Zebrafish
MDA5	dsRNA	Grass Carp
LGP2	dsRNA	Grass Carp

Type 2 adjuvants facilitate the stimulation of the immune system by simulating the recognition of PAMPs by specific pattern-recognition receptors. However, another way by which type 2 adjuvants can stimulate the innate immune system lies in the induction of the so-called danger signals that are associated with tissue injury and damage. According to the model proposed by Matzinger, the immune system does not necessarily focus on the recognition of microbial agents by themselves, but rather on the co-recognition of antigens associated with danger signals [42]. Upon entry and infection, microbial agents cause damage and stress to surrounding tissues, enabling the induction of signals that are materialized through the release of damage-associated molecular patterns, or DAMPs. These patterns can include both segments of dead cells and molecules like heat-shock proteins, uric acid, and cytokines that promote the activation of stimulatory signals for the pAPCs [16]. According to Ribeiro and Schijns, this can elucidate the mechanism of action behind several type 1 adjuvants as well, adding an interesting element of overlapping between the immunostimulating machineries involved in both type 1 and type 2 adjuvants.

Overall, it is accepted that the categorization of adjuvants to type 1 or type 2 facilitators is not necessarily mutually exclusive [16]. The intricate mechanisms behind the activation of signaling cascades that modulate immune responses can be affected in numerous ways, highlighting that the study of different adjuvants is important in the elucidation of the interplay between immune-enhancing compounds and the elements of the immune system. Adjuvants that facilitate type 2 signals can influence a plethora of immunostimulating pathways, and that is evidenced by the fact that one of the most well-studied PRR receptors, the toll-like receptors (TLRs), can aid in the activation of both innate and adaptive immunity [16], bridging the gap between the two immune subsystems. Upon binding to pattern-recognition receptors, different molecules can have different outcomes in terms of immune-enhancing capabilities, thus facilitating the induction of different stimulatory signals that are materialized through the secretion of cytokines [26–28,43].

2.3. *Further Elaborations on the Two-Signal Model*

Since the binding of PRR ligands precedes the secretion of cytokines and co-stimulatory molecules, it has been suggested that the binding of PAMPs to the PRRs constitutes a separate type of signal, often referred to as a type 0 signal [16]. According to this proposal, this type of signal facilitates the initiation of all down-stream pathways involved in the secretion of co-stimulatory signals that make up the type 2 signal category, ultimately making a distinction between the effects of PRR ligand binding and the stimulatory effects of type 2 signals on the activation of naive B-cells and T-cells. This distinction can potentially aid in advancing the field of adjuvant immunology by allowing research to focus on adjuvants that have profound stimulatory effects on each type of signal. This can include studies on PRR agonists co-delivered with antigens, inhibitory molecules that block immune-attenuating signals, and recombinant co-stimulatory molecules that effectively emulate the immune-amplifying capabilities of endogenous cytokines directly [16].

In a recent update of events, further elaborations on Schijns' proposed model appear to have been made, making the case for a fourth type of signal (type 3 signal) that regulates the proliferation of T-cells into different antigen-specific subpopulations, as with regulatory T-cells (Tregs) and different subsets of T_h cells [44]. Additionally, a fifth type of signal called a type 4 signal can be used to describe adjuvants that affect the localization of vaccine antigens by directing them to specific anatomical regions within the host's body [44]. This aspect adds another source of complexity to the activity of adjuvants, as vaccine formulations need to not only activate the immune system, but also ensure the correct direction of immune effector cells and the induced immune response. In aquaculture vaccinology, this can be especially important for vaccines targeting mucosal routes, like the skin, the gills, and the intestines.

The mucosal surfaces are primary entry points for many aquatic pathogens [45–47]. By potentially utilizing adjuvants that enhance the mucosal imprinting of type 4 homing signals, vaccines can be designed to induce robust and localized immune responses that will reduce disease incidence and improve overall teleost health and aquaculture sustainability. This lies in accordance with the current direction of aquaculture vaccinology that aims to develop vaccines that will not only be able to elicit mucosal immunity, but also be administered through routes that are more preferrable to the industry, such as immersion and oral vaccination [1]. Effectively, this can also reduce the stress induced on fish during the vaccination process, therefore promoting their overall welfare. A schematic overview synthesizing the most recent insights incorporated into the signal modeling of vaccine adjuvant activity can be seen in Figure 1.

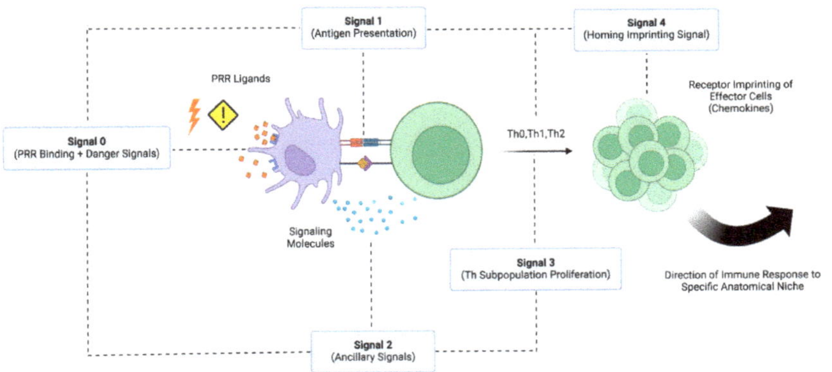

Figure 1. An overview of all elaborations made in the two-signal model to effectively explain the mode of action behind several vaccine adjuvants. An effective immunization is usually the result of a well-orchestrated immune response, and thus, the synergistic effect of antigens and vaccine adjuvants facilitating certain signals can aid in eliciting a robust and long-lasting protective effect.

3. Traditional Adjuvants Used in Aquaculture Vaccinology

In the realm of aquaculture vaccinology, adjuvants are key components that enhance the immunogenicity of vaccines, which makes them indispensable in the development of efficient vaccines for aquatic species. This section will delve into the major adjuvants currently employed in aquaculture vaccines, exploring their modes of action, efficacy, and their influence on immune cell activation and homing. By gaining a deeper grasp on the unique properties and applications of these adjuvants, a better understanding of their potential exploitation can be established, leading to the improved assessment of their side effects and their mitigation thereof in different fish species. An illustration summarizing the main adjuvants discussed below can be seen in Figure 2.

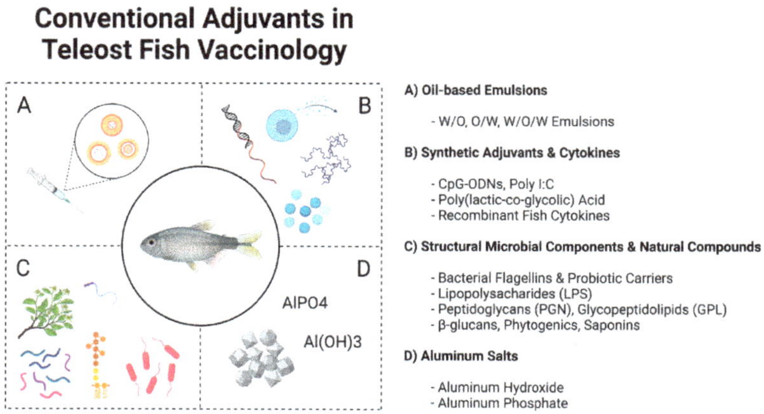

Figure 2. A schematic illustration showcasing the main types of vaccine adjuvants used conventionally so far in teleost fish vaccinology.

3.1. Oil-Based Emulsions

Oil-based emulsions are widely used in aquaculture vaccine formulations, especially due to their proven efficacy in enhancing immunization [48]. Emulsions consist of a dispersed liquid distributed inside a second liquid, forming two phases, the continuous and the dispersed, where typically the two liquids would not be able to mix naturally. In vaccine formulations, these phases typically consist of an aqueous continuous phase, loaded with the vaccine antigen, and an oily phase that is dispersed [7]. The formation of the emulsion is facilitated by the use of emulsifiers that act as surfactants, stabilizing the emulsion by reducing the surface tension between the two liquids thanks to their polar groups, which are hydrophilic, and their non-polar groups, which are hydrophobic [7,48]. Surfactants can be categorized by taking into consideration the value of the analogy of the hydrophilic to the hydrophobic groups, known as hydrophilic–lipophilic balance (HLB), a term first proposed by Griffin in 1949 [49].

The HLB value of surfactants is a determining factor for the production of different types of emulsions [49–51] that have different efficiencies as vaccine adjuvants [7]. Emulsifiers with low HLB values lead to the creation of water-in-oil emulsions (W/O), whereas those with high HLB values tend to form oil-in-water emulsions (O/W) that are typically better tolerated but exhibit weaker immune responses [7]. Certain surfactant systems with an intermediate HLB value allow for the creation of W/O/W emulsions, where the continuous phase is aqueous and the dispersed phase is oil, within which an entrapped aqueous phase containing water-soluble or suspended antigens is found. This third type of emulsion has been shown to elicit robust immune responses with a variety of antigens, as summarized by Dalmo et al. [7].

The majority of aquaculture vaccines rely on adjuvants that are based on mineral oil emulsions [7,52]. The most common example of such adjuvants consists of Freund's Complete Adjuvant (FCA), which is a mineral oil-based formulation used to create a W/O emulsion with the addition of a surfactant and heat-killed *Mycobacteria* [7,52]. Despite being effective, however, this adjuvant has raised concerns due to its association with severe side effects during administration, making its implementation restricted to mainly experimental settings, while gradually being replaced by Freund's Incomplete Adjuvant (FIA), which lacks the mycobacterial elements [7]. Montanide adjuvants are a series of commercially available adjuvants based on various oil emulsions formulated by the company SEPPIC and licensed under the common trade name of Montanide™. These adjuvants are not exclusively mineral oil-based, but their use has been well-established in the realm of veterinary vaccines, constituting the adjuvant of choice for many commercially available vaccines alongside other mineral oil-based adjuvant brand series in the market. A list of the currently commercially available Montanide^{TM} series formulations intended for use in teleost fish can be read below in Table 2.

Table 2. Commercially available Montanide^{TM} adjuvants intended for use in teleost fish according to the manufacturer, at the time of writing. The vast majority are based on mineral oils and are intended for use via injection vaccination.

Currently Available Montanide^{TM} Adjuvants for Use in Aquaculture Vaccinology		
Series Name	**Technology**	**Route**
Montanide^{TM} ISA 61 VG	Mineral Oil base for W/O Emulsions	Injection
Montanide^{TM} ISA 50 V2	Mineral Oil base for W/O Emulsions	Injection
Montanide^{TM} ISA 70 VG	Mineral Oil base for W/O Emulsions	Injection
Montanide^{TM} ISA 71 VG	Mineral Oil base for W/O Emulsions	Injection
Montanide^{TM} ISA 71 R VG	Mineral Oil base for W/O Emulsions	Injection
Montanide^{TM} ISA 761 VG	Mineral Oil base for W/O Emulsions	Injection
Montanide^{TM} ISA 78 VG	Mineral Oil base for W/O Emulsions	Injection
Montanide^{TM} ISA 763B VG	Non-Mineral Oil base for W/O Emulsions	Injection
Montanide^{TM} ISA 660 VG	Non-Mineral Oil base for W/O Emulsions	Injection
Montanide^{TM} GR	Oil base for W/O Emulsions containing a Protective Matrix	Oral
Montanide^{TM} IMS 1312 VG	Combination of Micro-Emulsions with an Immunostimulating Compound	Immersion

Some of the most commonly used Montanide^{TM} adjuvants in aquaculture vaccines include the Montanide^{TM} ISA 763A and 763B VG series, whose efficiency has been highlighted in numerous publications dealing with different species of fish and pathogens [53–56]. However, in accordance with evidence provided by the recent elaborations on the two-signal model, recent studies appear to suggest that the use of different adjuvants, including even commercialized ones, with different vaccine antigens, can result in a totally different type and scale of immunization [57–60]. This further strengthens the notion that the appropriate selection of the adjuvant–antigen combination is paramount in ensuring a robust and long-lasting immune response upon vaccination in the context of aquatic organisms, alluding to the fact that more research should be focused on such combinations to shed light into the optimization of vaccine formulations.

3.2. *Aluminum-Based Compounds*

Another category of popular vaccine adjuvants for aquaculture vaccines are aluminum compounds that contain aluminum salts, such as aluminum hydroxide or aluminum phosphate, collectively referred to as alum [7]. The mechanism of action of these adjuvants remains elusive, though quite a number of mechanisms have been proposed to explain their immune-enhancing capabilities [15,61]. Similarly to oil-based emulsions, it has been suggested that alum provides a "depot" effect by adsorbing the vaccine antigen onto its particles and creating an antigen reservoir at the site of administration [61–64]. Studies, however, have shown that an antigen depot is not necessary for the adjuvanticity of alum [63], alluding to the fact that there might be other mechanisms at play.

Today, it is strongly suggested that alum has a profound effect on phagocytosis, inflammation, dendritic cell maturation, and the secretion of cytokines [62,63]. Several studies have highlighted its role in the activation of the NLRP3 inflammasome, yet the definite mode of action is still under debate, as there is no universally established standard as far as immunization protocol, type of aluminum adjuvant, and animal model are concerned [62]. Alum does not appear to have the ability to influence dendritic cells directly through TLR binding pathways, though some level of interaction has been reported via TLR-independent binding to lipid cues in the dendritic cell membrane [63]. Additionally, alum has been shown to promote the induction of endogenous danger signals and facilitate the activation of the complement system, with the primary characteristic of its adjuvant activity lying in the generation of Th2-mediated humoral immune responses, as summarized by He et al.'s review [63].

The validation of these hypotheses and the elucidation of the intrinsic machinery involved in alum's adjuvant action remains an interesting and elusive topic for researchers so far. In the context of aquatic organism immunology, the scarcity of information makes understanding the action of aluminum-based adjuvants additionally challenging; nevertheless, recent studies have shown that alum is a potent adjuvant in teleost fish, able to enhance immune responses upon vaccination. In Angosto et al.'s recent study, alum was shown to trigger the expression of the gene encoding IL-1β in both the European seabass (*Dicentrarchus labrax*) and gilthead seabream (*Sparus aurata*), matching the hypothesis that aluminum-based adjuvants can promote the secretion of inflammatory cytokines by activating the NLRP3 inflammasome [65]. Surprisingly though, no IL-1β secretion was observed in seabream leukocytes and macrophages, alluding to the fact that alum's adjuvant action might be independent of the inflammasome in teleost fish, as further strengthened by the lack of conservation of caspase-1 processing sites in non-mammalian IL-1β, the inability of caspase-1 to modify IL-1β, and the failure of caspase-1 activation through NLRP3 inflammasome facilitators previously reported in gilthead seabream [66–68].

Despite the forementioned findings, alum remains a strong adjuvant in aquaculture vaccinology, able to facilitate the production of reactive oxygen species through a NADPH oxidase-mediated route [65]. Adding to the fact that alum crystals can potentially induce endogenous signals like uric acid and release DNA from dead cells [69–71], these results tilt towards the validation of the notion that alum's adjuvant capacity is achieved through the induction of danger signals in the host's body. However, the existence of side effects can also be attributed to alum's immunostimulating capacity, as inflammatory reactions, granulomas, melanizations, and toxicity in the splenic melanomacrophage centers have been documented with alum use in teleost fish [72].

3.3. Synthetic Adjuvants and Cytokines

Synthetically derived adjuvants represent a unique category of adjuvants that have been facilitated by the relatively recent advent of scientific and technological innovation. In the realm of aquaculture, one of the most popular adjuvants of this type explored consists of synthetic oligodeoxynucleotides (ODNs) that express unmethylated motifs known as CpG. Such motifs are abundant in microbial DNA and can be recognized by immune cells expressing TLR9 [7], though recent reviews also portray TLR21 as a receptor [27,28]. Synthetic double-stranded polyribonucleotides called Poly I:C (Polyinosinic–polycytidylic acid) have also been used as adjuvants in aquaculture vaccinology, particularly to combat pathogens of viral origins. Their mode of action relies on simulating a viral infection, leading to the production of antiviral cytokines like type 1 interferons (IFNs), through binding to the TLR3 receptor [7]. However, as mentioned previously, recent data have shown that Poly I:C can also act as a ligand to the TLR22, TLR26, and TLR28 receptors in teleost fish, opening the possibilities for additional signaling pathways to be involved in the adjuvanticity of Poly I:C.

The exploration of fish cytokines as potential vaccine adjuvants in aquaculture was recently summarized in our previous paper [1], incorporating insights from Guo and Li's review on the topic [73], along with findings from several recent studies highlighting their effects. Overall, recombinant cytokines and cytokine genes incorporated into vectors have shown promising results when incorporated as adjuvants into vaccine formulations. Although in its infancy, this approach has recently garnered attention, as cytokines can modulate immune responses and provide certain immunopotentiation effects during fish vaccination, without the danger of a strong generalized response and potential side effects [74]. Th0-derived cytokines like IL-1β, IL-8, IL-12, IL-15, IL-17, G-CSF, and TNF-α have been shown to have potential adjuvant activity in various teleost fish species and in combination with various pathogens and vaccine technologies, as is the case with Th1-derived IFN-α, IFN-γ, IFN-c, IL-2, and Th-2 derived IL-6 and β-chemokines [1,73,74].

Lastly, for oral vaccination, synthetic and biodegradable poly (lactic-co-glycolic) acid polymers, known simply as PLGA, have long been prevalent as vaccine carriers and potential adjuvants in teleost fish vaccination. Even though the vast majority of studies have explored PLGA microparticles as a vaccine delivery carrier through the method of antigen encapsulation, there have been a few reported publications that exhibit PLGA's adjuvant activity as per Dalmo et al.'s review [7]. Since then, recent reports have successfully implemented PLGA-mediated delivery of DNA [75] and inactivated vaccines [76] in teleost fish immunization protocols, showcasing PLGA's ability in conferring increased levels of protection against both viral and bacterial pathogens. In an exciting and promising new series of events, recent studies combining PLGA microparticles with cytokines [77] and CpG-ODNs [78] as adjuvants have also surfaced, demonstrating that synthetic adjuvants remain a fertile area of research in the future of aquaculture vaccinology, since these combinations show better immunization efficacy when compared to using each component individually.

3.4. Structural Microbial Components and Natural Compounds

Naturally occurring substances have similarly been explored for their adjuvant and immunostimulating capabilities in teleost fish immunization. Typically, this category encompasses structural components of microorganisms that exhibit promising adjuvant effects, since they constitute distinctive pathogen-associated molecular patterns and can act as PRR ligands for the host's immune cells. Prime examples include lipopolysaccharides (LPSs), flagellins, β-glucans, and peptidoglycans (PGNs), which have been implemented as both feed immunostimulants and as adjuvants in aquaculture. These macromolecules induce the secretion of signaling cytokines and potentiate immune responses, opening the gateway for their application as both functional feed additives and vaccine adjuvants in a wide array of fish species, as evidenced by numerous publications on the topic [7,79–84]. Similarly, lipopeptides have been successfully used as vaccine adjuvants in fish, since they are abundant in many microbial species, including *Mycobacteria* and *Mycoplasma*. Polar glycopeptidolipids (pGPLs) of mycobacterial origin have shown promising immune-enhancing capabilities in fish, without the presence of any adverse side effects [85].

Besides PAMPs, however, other naturally occurring substances have exhibited promising results when used as potential aquaculture vaccine adjuvants. Saponins are a naturally occurring group of steroid or terpenoid glycosides that are typically extracted from plants. They have been shown to enhance innate immune responses in many animal species, including teleost fish and aquatic invertebrates [86]. In aquaculture vaccinology, saponins extracted from the soapbark tree (*Quillaja Saponaria*) have given rise to a promising vaccine adjuvant, known as Quil-A. The implementation of saponins as aquaculture vaccine adjuvants, however, has been met with adversity, as they can be toxic in high doses, and their stability is compromised in aqueous solutions [7,87]. Nevertheless, recent studies show that saponins can effectively augment immune responses, even when administered together with other adjuvants or through mucosal routes [88–90]. Additionally, several other plant-derived saponins, like ginseng stem leaf saponins, have successfully been reported to enhance immune responses in fish upon administration, as seen in recent studies [91,92].

The adjuvant capacity of several phytogenic substances for aquaculture vaccinology has been well described in a review recently published by Soltani et al. [93]. Furthermore, probiotics appear to gain traction as adjuvants in aquaculture vaccines, especially in oral vaccination applications dealing with encapsulation platforms. Coupled with modern genetic bioengineering techniques, probiotic bacteria like Bacilli, Lactobacilli, and Lactococci can have profound effects on fish immunization, opening the gateway for interesting vaccine delivery applications by combining the immune-stimulating effects with improved antigen delivery through the oral route [94–103]. Finally, the use of chitosan, the deacetylation product of the naturally occurring polysaccharide chitin derived from insects, fungi, and aquatic invertebrates, has been met with promise in the realm of aquaculture vaccinology, as it can be similarly used in oral routes as an immune-enhancing vehicle for vaccine delivery [104–106]. The adjuvant capacity of chitosan microparticles can be traced in a plethora of recent publications, both by itself, and in combination with other vaccine adjuvants or polymers [107–115].

4. Assessing the Side Effects of Traditional Adjuvants

Despite vaccine adjuvants being crucial components in enhancing the efficacy of aquaculture vaccines, it is well recognized that their use can be accompanied by several side effects and drawbacks. Oil-based adjuvants are currently dominating formulations used in aquaculture, as they are predominantly implemented in vaccines administered by way of intraperitoneal injection [8,116–118]. However, these vaccines are often associated with adverse side effects, stemming primarily from inflammatory reactions that are triggered during immunization. These reactions can exhibit varying degrees of intensity and severity, and the most typical symptoms include adhesions of tissues and organs to each other or to the peritoneal cavity, granuloma formation, and discoloration at the site of injection [117–125]. A schematic illustration summarizing the adverse effects of currently employed vaccine formulations in teleost fish can be seen below in Figure 3.

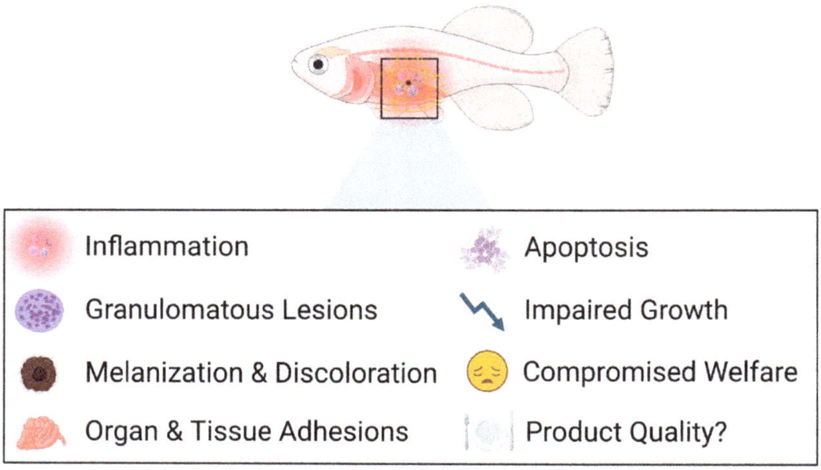

Figure 3. Potential side effects of the current adjuvanted injectable vaccines in teleost fish. In severe cases, the physiological functioning of the digestive system and thus the growth rate can be impaired, potentially undermining animal welfare and lowering the quality of aquatic food products.

The Spielberg scale, developed in 1996 by Midtlyng et al., was designed to assess the severity of intraperitoneal symptoms in Atlantic salmon (*Salmo salar*) following immunization against furunculosis [126]. Remarkably, this scale and its underlying methodology remain valid and therefore prevalent in the international literature, and are particularly referenced in studies examining the side effects of vaccines in teleost fish of similar species. In this framework, individual fish are evaluated based on the severity of their symptoms and the associated pathological changes post-vaccination, by implementing a scale that typically ranges from 0 to 6; a higher score indicates more pronounced post-vaccination pathological findings, offering a tool for establishing a standardized approach in evaluating the impact of adjuvanted vaccines in teleost fish health. Recently, in 2021, Tziouvas and Varvarigos adapted a Spielberg-type scale to the European sea bass (*Dicentrarchus labrax*) [125]; a side-by-side adaptation comparison of the two scales can be seen in Table 3.

Table 3. A side-by-side adaptation comparison between Spielberg's original scale for assessing the Atlantic salmon post-furunculosis vaccination and the newly proposed scale for assessing vaccination side effects in the European seabass by Tziouvas and Varvarigos.

Score (0–6)	Spielberg Scale [126] (Atlantic Salmon)	Tziouvas and Varvarigos Scale [125] (European Sea Bass)
0	• No visible lesions.	• No adhesions or peritoneal lesions. Fish with this score may be unvaccinated by mistake.
1	• Very slight adhesions, usually localized near the injection site, with minimal peritoneal opacity post evisceration. Unlikely to be noticeable by laypeople.	• Soft, localized adhesions around organs, mainly the intestine; the swim bladder and the abdominal wall are unaffected. Organs separate easily, and nodules or lesions are absent. Typically seen soon post vaccination.
2	• Same as the score above, but with minor adhesions connecting the colon, spleen, or caudal pyloric caeca to the abdominal wall. Peritoneum remains slightly opaque after adhesion removal.	• Soft, localized adhesions affecting some organs, but sparing the swim bladder and the abdominal wall. Few small, non-pigmented granulomatous nodules are present.
3	• Moderate adhesions extending to more cranial organs, like the pyloric caeca, liver, or ventricle, attaching them to the abdominal wall. Minor lesions visible post evisceration, removable manually.	• Widespread and soft adhesions involving multiple organs, but not the swim bladder or the abdominal wall. Small, mostly non-pigmented nodules are frequent and under 1 mm in diameter.
4	• Major adhesions with granulomas and extensively connected internal organs appearing as a single unit. Moderate lesions may be difficult to remove manually. Likely noticeable by laypeople during evisceration.	• Soft to moderately hard widespread adhesions, affecting the swim bladder and the abdominal wall. Separation requires moderate force, with small and some larger nodules (up to 2 mm), mostly non-pigmented, but some pigmented.
5	• Extensive lesions affecting nearly all internal organs. Peritoneum is thickened and opaque, with prominent, pigmented lesions on the fillet. Visible damage remains after lesion removal. It is likely to be noticed by laypeople during evisceration.	• Moderately hard, widespread adhesions involving the swim bladder and the abdominal wall. Moderate force is needed to separate organs, risking rupture. Numerous nodules from pin-point to >4 mm, non-pigmented and pigmented.
6	• Very pronounced lesions, often with significant melanization. Viscera cannot be removed without damaging fillet integrity, leaving substantial carcass damage.	• Strong adhesions connecting visceral and parietal peritoneum, with numerous nodules exceeding 8 mm. Organ separation risks rupture; no fillet lesions, but potential growth impacts are possible.

This study involved the collection of data from fish growing in Greek aquaculture sites that were vaccinated through intraperitoneal injection with licensed oil-adjuvanted vaccines over a three-year period (2016–2018) and the meticulous evaluation of the pathological findings found within the peritoneal cavity. As expected, many of the sampled fish exhibited the side effects described in the literature, such as chronic peritonitis, granulomatous lesions, discolorations, and tissue and organ adhesions. Notably, smaller-sized fish appeared to experience more severe side effects following vaccination, potentially due to the challenges associated with the administration method of IP injection, which requires careful handling and execution. Furthermore, vaccines formulated with mineral oil adjuvants that are not easily metabolized by fish, such as paraffin oils, resulted in more pronounced symptoms associated with intense inflammatory responses. The feed conversion rate (FCR) remained largely unaffected in the aquaculture sites, except in cases where there were severe granulomatous lesions and strong tissue adhesions hindering the physiological functioning of the digestive system.

The findings from the previous study, combined with simple macroscopic and histological analyses conducted during the autopsies, led to the development of a scale similar to that of Midtlyng et al. to estimate the severity of symptoms post vaccination. The newly proposed six-point scale can potentially serve as a valuable tool for aquaculture industry professionals and scientist alike to assess the welfare of teleost fish and evaluate relevant side effects of oil-adjuvanted vaccines in aquaculture. This advancement holds particular significance for marine finfish and European aquaculture, where species like the European sea bass (*Dicentrarchus labrax*) and the gilthead seabream (*Sparus aurata*), two of the most economically important species within the Mediterranean region, are routinely vaccinated with commercial vaccines. Simultaneously, this work also lays the groundwork for further research to be conducted into the implications of contemporary vaccines for farmed fish, expanding the scope to include various factors such as teleost fish species, type of vaccine and adjuvant used, growth rates, final body weight, and the quality of the final products. In conclusion, this showcases the necessity for similar studies to be carried out in order to fully understand the impact of current aquaculture vaccines on aquatic welfare and health. As such, the development of species-specific and tailored scales is essential for enhancing sustainability and promoting humane practices in the sector.

Today, it is widely acknowledged that the mitigation of the side effects of adjuvanted injectable vaccines hinges on optimizing vaccination regimens, refining formulations, and reducing dosage volumes. Strategies like lowering vaccine volume, thereby increasing the antigen density, have shown promise [116]. However, the quest for innovative adjuvants to address the enduring challenge of side effects continues, with nanoparticle-based formulations emerging as a promising breakthrough in aquaculture vaccinology.

5. The Promise of Nanoparticle-Based Formulations for the Future

The association between traditional vaccine adjuvants and the adverse post-vaccination side effects on the welfare and health of teleost fish has prompted researchers to explore novel and alternative options, with nanoparticle-based formulations gaining significant traction as the future of this field. As technological expertise progresses, these innovative formulations pave the way for increased vaccination efficacy as novel adjuvants, with the capacity to minimize relevant side effects and adverse reactions. In this section, recent advances in the field of nanoparticle-based adjuvants for aquaculture vaccines will be reviewed, highlighting their position as a potential avenue for ongoing research, where the crossroads of aquatic immunology, aquaculture health management, and technological innovation are met.

5.1. Polymeric Nanoparticle Formulations

As already mentioned, polymers represent a significant portion of current nanoparticle-based research attempts focused on vaccine delivery in aquaculture. The shift in scale from microparticles to nanoparticles can offer advantages in immunization enhancement, as smaller-sized particles can activate the immune system more efficiently, triggering both cellular and humoral immune responses in teleost fish. In addition, nanoformulations seem to improve crucial vaccine factors, such as solubility, stability, targeting, permeability, and biocompatibility [127].

PLGA-based nanoparticle applications in aquaculture vaccinology have maintained a prominent presence in the international literature over the past decade, as seen in some of earlier publications [128–131]. Focusing on more recent studies, PLGA nanoparticles have been utilized for immunization against a diverse range of aquatic pathogens, employing various vaccination technologies and administration methods, while targeting numerous teleost fish species. Dubey et al. have demonstrated that the oral administration of recombinant outer-membrane protein W (OmpW) of *Aeromonas hydrophila* encapsulated in PLGA nanoparticles shows a dose-dependent protective effect in rohu fish (*Labeo rohita*) [132]. Since then, Harshita et al. have also recently shown that the recombinant OmpA can similarly be encapsulated in PLGA nanoparticles and be orally delivered to confer protection in *Aeromonas hydrophila*-challenged zebrafish (*Danio rerio*) [133]. The compatibility of PLGA nanoparticle-based vaccine formulations has also been tested with DNA vaccines against *Aeromonas hydrophila*, both in oral and injectable applications, as seen in the recent works by Alishahi et al. [134,135]. In the case of viral pathogens, a very early first attempt of implementing a PLGA nanoparticle-based oral DNA vaccine against infectious hematopoietic necrosis was made in Adomako's study [130]; however, recent endeavors have expanded towards more viruses, such as the spring viremia of carp virus (SVCV) and the hemorrhagic septicemia virus (HSV). In a study published in 2019, the efficacy of an inactivated viral vaccine against HSV encapsulated in PLGA nanoparticles was demonstrated by Kole et al. [76], whereas in 2022, Zhang et al. successfully developed a dual-targeting polymer containing PLGA and chitosan nanoparticles to encapsulate a DNA vaccine that was able to confer robust protective immunity in common carp (*Cyprinus carpio*) [136].

The combination of polymers with chitosan nanoparticles as vaccine adjuvants is not coincidental, as chitosan is non-toxic, biodegradable, and has potent mucoadhesive properties that can aid in eliciting strong protective effects upon mucosal immunization in fish [104,137,138]. Several studies combining chitosan nanoparticles with other polymers like alginate and PLGA for various vaccination applications in teleost fish have been published in recent years [139–144]. Apart from those, chitosan-complexed nanovaccines have been implemented to counter both bacterial aquatic diseases [145–157] and viral ones [158–162] in a plethora of new studies. Additionally, novel technologies like chitosan nanoparticles complexed with oxygen nanobubbles have emerged, facilitating the enhanced uptake of vaccines and resulting in superior mucosal immune responses [163]. The synergy of chitosan nanoparticles with traditional adjuvants has also been studied, highlighting that the combination of aluminum adjuvants with chitosan-based nanovaccines can trigger both local and systemic immune responses [164], while on the contrary, Poly I:C's adjuvanticity appears to deteriorate, presumably due to the delayed, gradual, and stable antigen release that the chitosan particles provide on the microscale [109].

Nevertheless, Collado-González and Esteban highlighted that the inconsistencies in the characterization of the physicochemical properties of chitosan particles, like the degree of acetylation (DA) and the molecular weight (MW), and the different preparation methods used have made it difficult to determine which specific characteristics are most effective in stimulating the teleost fish immune system [106]. This underscores the need for standardizing chitosan nanoparticle-based treatments and improving reporting practices to enhance data comparability and reproducibility. Overall, studies that properly characterized chitosan suggest that low DA (10–15%) and a MW ranging from 50 to 800 kDa may enhance immune gene expression in teleost fish, though the differing amounts of chitosan used and the different immunization regimens followed in different species make direct comparisons difficult at this stage. Collectively, the implementation of polymeric nanoparticle-based vaccine formulations in teleost fish appears promising so far, and there is a pressing need for more studies to explore and correctly assess the advantages of the shift from micro- to nanoscale polymeric materials in aquaculture vaccinology.

5.2. Lipid-Based Nanoparticle Formulations

Another type of nanoparticle-based formulation that has gained traction for vaccine delivery and the enhancement of immune responses encompasses lipid nanoparticles, either through nanoliposomes or by using cationic lipid surfactants. These substances typically aid in the stabilization of antigens and can serve as interesting tools for the development of nanoformulations in a wide array of vaccine technologies. Liposomes consist of a hydrophilic core surrounded by hydrophobic bilayers and can be used to encapsulate both hydrophilic and lipophilic compounds [165]. This offers advantages in enhanced solubility, bioavailability, and antigen stability, offering a solid delivery platform that could potentially reinforce immune responses. Alternatively, surfactants like cetrimonium bromide, known as CTAB, are currently being explored for the construction of cationic lipid-based vaccines due to their ability to modify nanoparticle size and surfaces. Additionally, they can alter vaccine charges, therefore opening the gateways for improved delivery, especially through mucosal routes [166,167].

The use of such technologies at the nanoscale are not extensively studied in the field of aquaculture vaccinology; however, there are a few studies that have delved into the use of lipid-based nanoparticle vaccine formulations. In 2019, Ji et al. published a study where nanoliposomes loaded with bacterial LPS and Poly I:C were successfully used to enhance innate immune responses in zebrafish larvae, leading to the upregulation of several immune-related genes and an increase in survival post *Aeromonas hydrophila* challenge following bath immersion vaccination [165]. These results have been succeeded by similar studies, where immunostimulant-loaded nanoliposomes were tested on adult zebrafish, resulting in protection against *Pseudomonas aeruginosa* and SVCV infection both in injection and in immersion applications [168,169]. These results also highlight the recent shift in size from liposome to nanoliposome formulations in teleost fish and pave the way for fish fry vaccination applications by lifting size limitations and improving antigen delivery [165]. A recent development was also made in 2024, where an mRNA-lipid nanoparticle vaccine formulation was successfully applied in the Atlantic salmon (*Salmo salar*), demonstrating that the relatively novel mRNA vaccine technology encapsulated in lipid nanoparticles can indeed be used to express antigens in teleost fish, both in vitro and in vivo [170].

Cationic lipid-based nanovaccines represent the other major part of this category as far as nanoparticle-based formulations in aquaculture vaccinology are concerned. Despite the lack of reports on teleost fish applications, recent years have witnessed the emergence of a few studies implementing this technology to develop mucoadhesive immersion nanovaccines against bacterial pathogens. In Thangsunnan et al.'s published study, a mucoadhesive nanovaccine against *Francisella noatunensis* subsp. *orientalis* was successfully constructed using CTAB as a surfactant [167]. This formulation conferred protection comparable to that of a whole-cell inactivated vaccine, inducing strong immune responses after boost immunization. Gene expression analysis revealed a significant upregulation of both IgM and IgT in vaccinated fish, confirming enhanced mucosal and systemic responses in red tilapia (*Oreochromis* sp.). Similarly, Bunnoy et al. developed a vaccine of similar technology to combat columnaris disease [168]. The nanovaccine effectively targeted the mucosal-associated lymphoid tissues (MALT) in the skin and gills of the Asian seabass (*Lates calcarifer*), resulting in a strong immune response, where vaccinated fish showed higher antibody levels and survival rates (65.83–72.50%) compared to the controls. This formulation stimulated both innate and adaptive immunity, resulting in an increased expression of immune-related genes like IgM and MHC-IIα. However, while effective, the vaccine was linked with early mortality, indicating that further optimization is needed in vaccination protocols.

Below, we present a table (Table 4) that collectively summarizes the efficacy of lipid-based nanoparticle vaccine formulations discussed above. Special reference is given to the teleost fish species studied, the type of lipid-based nanoparticle technology implemented, the pathogen species targeted, the vaccine technology used, the method of administration, and the immunological efficacy reached in these studies following vaccination. This summary provides a unified approach, showcasing the potential applicability of different approaches across a diverse range of aquaculture contexts, despite lipid-based nanoparticle technologies being in their early stages as far as fish vaccinology is concerned.

Table 4. Recent studies on teleost fish vaccinations utilizing lipid nanoparticle-based formulations. So far, the two approaches implemented in in vivo vaccination trials mainly include encapsulating nanoliposomes and cationic lipid-based nanoemulsions.

Vaccine Technology	Delivery Route	Pathogen (Antigen)	Teleost Species	Efficacy	Source
Encapsulating Nanoliposomes	Immersion	*Aeromonas hydrophila* (*E. coli* LPS + Poly I:C)	Zebrafish (*Danio rerio*)	Increased survival rate post challenge and upregulation of immune-related genes	[165]
CTAB-Cationic Lipid Nanoemulsion	Immersion	*Flavobacterium oreochromis* (Formalin-Inactivated)	Asian Seabass (*Lates calcalifer*)	Increased survival rate post challenge and upregulation of immune-related genes	[166]
CTAB-Cationic Lipid Nanoemulsion	Immersion	*Francisella noatunensis* subsp. *Orientalis* (Formalin Inactivated)	Red Tilapia (*Oreochromis* sp.)	Reduced bacterial load in tissues, enhanced antibody titers, and upregulation of immune-related genes	[167]
Encapsulating Nanoliposomes	Injection and Immersion	*Pseudomonas aeruginosa* and SVCV (*E. coli* LPS + Poly I:C)	Zebrafish (*Danio rerio*)	Increased survival rates against both bacterial and viral challenges	[169]

5.3. Carbon Nanotubes and Inorganic Nanoparticle Formulations

Lastly, carbon nanotubes (CNTs) represent a cutting-edge paradigm with regard to nanoparticle-based vaccine formulations for teleost fish, offering an alternative tool to enhance the delivery and immunogenicity of aquaculture vaccines, particularly through mucosal routes. Carbon nanotubes comprise hollow, cylindrical structures made from rolled up graphene sheets. These nanotubes can either include a single, or many graphene layers, making the distinction between single-walled carbon nanotubes (SWCNTs) and multi-walled carbon nanotubes (MWCNTs) [171].

Carbon nanotubes are considered inert and not immunogenic by themselves, having relatively low toxicity and high biocompatibility [172]. A recent study conducted by Cimbaluk et al. demonstrated that this nanomaterial does not induce any genotoxic effects like single or double DNA strand breaks in teleost fish erythrocytes. However, CNT-DNA crosslinks were observed, alluding to CNTs preventing the separation of DNA strands and potentially compromising the process of replication [173]. Chronic exposure to this material has indeed been linked to the induction of oxidative stress, pollutant accumulation, neurotoxicity, inflammation, apoptotic signaling, and histological pathologies in fish [173–181]. Nonetheless, the relevance of these studies to vaccination regimens remains to be elucidated and warrants further research to be conducted, especially since the physicochemical properties of the CNTs appear to make a difference in their toxicity according to Jiang et al. [182]. The modification of CNTs with other polymers also appears to lower toxicity claims, as is the case with chitosan grafting in Wisdom et al.'s study [183], which underpins the possibility of an ideal formulation to be discovered soon through the combination of nanopolymeric materials.

The functionalization of CNTs for immunization applications in teleost fish typically shows reduced toxicity and can have profound effects on the stimulation of immune cells [184]. During the past decade, several studies have accumulated in the international literature dealing with CNT-based vaccination in a wide range of species, yet the focus seems to be particularly set in immunization applications against viral aquatic diseases. Nevertheless, there have been a few studies showcasing that CNTs can be used to combat bacterial diseases as well, with *Aeromonas hydrophila* centered as the main pathogen of interest. In 2015, Gong et al. developed a novel formulation using SWCNTs to deliver a recombinant vaccine against the pathogen in juvenile grass carp (*Ctenopharyngodon idellus*) [185]. The vaccine, administered via both injection and immersion, led to a higher antibody production and a significant upregulation of immune markers, achieving similar survival rates to traditional injection methods. A year later, the same group published another study, highlighting the upregulation of immune-related genes, which was induced by their SWCNT-based vaccine compared to a SWCNT-free control [186]. More recent efforts were reported in 2020, with Zhang et al. successfully demonstrating the immunoprotective effect of a whole-cell lysed inactivated vaccine, utilizing SWCNTs as a carrier [187]. In 2024, Cao et al. shifted the focus by publishing an interesting study in which they constructed a CNT nanocarrier immersion vaccine encoding the immunogenic recombinant protein Sip [188]. This vaccine formulation promoted the expression of immune genes, induced relatively high serum antibody production, and enhanced related enzyme activities, ultimately conferring significant cross-protection against *Streptococcus agalactiae* and *Streptococcus iniae* in red tilapia.

As previously mentioned, however, the main body of work revolving around CNT-based vaccination in teleost fish stems primarily from applications dealing with viral diseases. The grass carp reovirus (GCRV) constitutes a prime example, since there are several publications documenting the efficacy of CNT-based nanovaccines in a plethora of different vaccine technologies, such as recombinant subunit vaccines [189–191] and DNA vaccines [192–194]. Another grass carp virus, the spring viraemia carp virus (SVCV), has similarly garnered the attention of CNT-based vaccination applications, as evidenced by recent studies [195–199]. Zhao et al. have successfully constructed and demonstrated the effect of a SWCNT-based subunit vaccine against the infectious spleen and kidney necrosis virus (ISKNV) in mandarin fish (*Siniperca chuatsi*) [200,201], and the same strategy was implemented for a DNA vaccine [202]. The most recent publications from this group are currently showcasing the efficacy of a mannose-modified CNT subunit vaccine designed to target antigen-presenting cells both in injection and in immersion formats [203,204]. Apart from ISKNV, CNT-based vaccination has been reported to be efficient against another iridovirus, named the grouper iridovirus of Taiwan (TGIV) [205,206], as well as other viral families, such as herpesviruses [207,208], rhabdoviruses [209], betanodaviruses [210], and the recently classified largemouth bass ulcerative syndrome ranavirus [211]. An indicative table (Table 5) summarizing the recent studies of CNT-based vaccination studies in teleost fish mentioned can be read below:

Staying within the context of nanotube formulations, it is worth noting that an inorganic alternative of CNTs has recently emerged in the realm of aquaculture vaccines, offering an alternative to its organic relative and addressing the gap in the literature regarding the use of inorganic nanoparticle-based formulations in teleost fish [212]. Halloysite nanotubes (HNTs) are naturally occurring tubular clay structures similar to CNTs, featuring a unique disparity in charge between their inner and outer surfaces [213]. They provide an environmentally safe and cost-efficient platform for vaccine delivery throughout the gastrointestinal tract, allowing for the encapsulation of susceptible antigens and their subsequent release in a pH-dependent manner, as well as for their modification with other materials to enhance their properties [213,214].

Pumchan et al. have recently harnessed the power of HNTs to test out and develop a novel oral nanovaccine delivery system to combat streptococcosis in tilapia (*Oreochromis* sp.) [214]; remarkably, they found that chitosan-modified HNTs demonstrated strong resilience to stomach acidity, bypassing gastric digestion and delivering a whole-cell inactivated vaccine to the intestines that promoted an immune response via the gut-associated lymphoid tissues (GALTs). This formulation effectively induced immunity against *Streptococcus agalactiae* serotype Ia, exhibiting similar disease protection and survival rates when compared to traditional injectable vaccines. The chitosan coating of HNTs enhanced the immune response and adherence to intestinal cells, positioning this nanoparticle-based formulation as a promising candidate for efficacious oral vaccination. Taken together with Zhang et al.'s recent publication on the use of a mesoporous silica nanoparticle-based delivery system for oral vaccination against *Vibrio alginolyticus* [215], these few studies highlight the use of inorganic nanoparticles in teleost fish immunization, ultimately completing the puzzle of a much promising arsenal for the future of adjuvant research in aquaculture vaccinology.

Table 5. A summary of recent carbon nanotube-based applications in teleost fish vaccinology, based on the recent literature. Findings suggest that CNT-based vaccination aligns predominantly with two underutilized vaccine technologies in commercial applications, recombinant subunit, and DNA vaccine platforms, indicating a potential shift in the future of aquaculture vaccines thanks to nanoparticle formulations.

CNT Type	Vaccine Technology	Delivery Route	Pathogen (Antigen)	Teleost Species	Source
Single-Walled	Recombinant Subunit	Intramuscular Injection and Bath Immersion	*Aeromonas hydrophila* (aerA)	Grass Carp	[185]
Single-Walled	DNA Vaccine	Intramuscular Injection	*Aeromonas hydrophila* (aerA)	Grass Carp	[186]
Single-Walled	Whole-cell Inactivated Vaccine	Intraperitoneal Injection and Bath Immersion	*Aeromonas hydrophila* (Bacterial Lysate)	Grass Carp	[187]
Single-Walled	Recombinant Subunit	Bath Immersion	*Streptococcus* sp. (rSip)	Tilapia	[188]
Single-Walled (Mannose Modified)	Recombinant Subunit	Bath Immersion	GCRV (VP7)	Grass Carp	[189]
Single-Walled	Recombinant Subunit	Bath Immersion	GCRV (VP7)	Grass Carp	[190]
Single-Walled	Recombinant Subunit	Bath Immersion	GCRV (VP4-3)	Grass Carp	[191]
Single-Walled	DNA Vaccine	Intramuscular Injection and Bath Immersion	GCRV (VP4-3, VP5, VP7)	Grass Carp	[192–194]
Single-Walled	DNA Vaccine	Intramuscular Injection	SVCV (M)	Common Carp	[195]
Single-Walled	Recombinant Subunit	Intramuscular Injection and Bath Immersion	SVCV (G)	Common Carp	[196]
Single-Walled	DNA Vaccine	Intramuscular Injection and Bath Immersion	SVCV (G)	Common Carp	[197]
Single-Walled	DNA Vaccine	Bath Immersion	SVCV (M)	Common Carp	[198]
Single-Walled (Mannose Modified)	Recombinant Subunit	Bath Immersion	SVCV (G)	Common Carp	[199]
Single-Walled	Recombinant Subunit	Bath Immersion	ISKNV (MCP)	Mandarin Fish	[200,201]
Single-Walled	DNA Vaccine	Bath Immersion	ISKNV (MCP)	Mandarin Fish	[202]
Single-Walled (Mannose Modified)	Recombinant Subunit	Intramuscular Injection and Bath Immersion	ISKNV (MCP)	Mandarin Fish	[203,204]
Single-Walled	Recombinant Subunit	Bath Immersion	TGIV (MCP, P2)	Pearl Gentian Grouper	[205,206]
Single-Walled	DNA Vaccine	Intramuscular Injection and Bath Immersion	KHV (ORF 149)	Koi Fish	[207,208]
Single-Walled	Recombinant Subunit	Bath Immersion	MSRV (G)	Largemouth Bass	[209]
Single-Walled	Recombinant Subunit	Bath Immersion	NNV (MCP)	Pearl Gentian Grouper	[210]
Single-Walled	Recombinant Subunit	Bath Immersion	LBUSV (MCP)	Largemouth Bass	[211]

6. Conclusions

Adjuvants are quintessential elements of modern aquaculture vaccines designed for teleost fish, as they aid in inducing potent and long-lasting immune responses that can protect reared species against aquatic diseases. Progress made in understanding the exact mechanisms of action behind various adjuvants has facilitated the development of many innovative and novel vaccine formulations, designed specifically to induce desired immunization outcomes. Combined with recent insights into aquatic and teleost fish immunology, these advancements have enabled the introduction of progressively more sophisticated vaccine and adjuvant technologies to be implemented in aquaculture settings, indicating that the future of vaccinology research and vaccine development in one of the fastest growing sectors is an excellent field for investment and exploration.

Despite vaccine adjuvants forming an integral part of aquaculture vaccines, there have been growing concerns about their potential adverse effects in teleost fish health, raising questions regarding product quality assurance and associated welfare issues in modern aquaculture. The use of contemporary vaccine adjuvants, such as oil-based emulsions, has been linked with side effects like tissue and organ adhesions, inflammatory reactions, granulomas, and spatial discolorations at the sites of injections, since intraperitoneal injection is currently the most widely used method of vaccine administration in the field. These findings have led researchers and aquaculture professionals alike to explore alternative ways of teleost fish immunization, in an effort to minimize cost, labor, and potential side effects, while maximizing the immunization outcomes. Apart from the shift in the administration route, as current trends dictate, the exploration of novel adjuvants offers another backroad for circumnavigating the limitations of current aquaculture vaccines, providing cutting-edge solutions to persisting issues, such as impotent immunization and compromised fish welfare. Taken together, these two elements can significantly improve the sustainability of the aquaculture industry and facilitate the establishment of more welfare-friendly practices that will ensure the quality of aquatic food products in accordance with the consumer demands.

This is particularly true for mucosal routes of vaccination, such as oral or immersion vaccination, where the "depot" effect observed in injectable vaccines is typically absent or significantly reduced. Unlike traditionally adjuvanted injectable vaccines, which provide a localized, prolonged release of antigens, vaccine administration through mucosal routes often lacks the advantage of concentrated antigen delivery. This can lead to shorter durations of immune responses and reduced vaccination efficacy in teleost fish. Consequently, these methods are presently better suited for booster vaccination applications in today's aquaculture practice. Addressing this issue requires innovative delivery technologies, like encapsulation or slow-releasing technologies, as well as novel adjuvants with potent mucoadhesive or mucosal-homing properties that can simulate depot-like behavior by prolonging antigen exposure and enhancing immune stimulation.

Currently, the spearhead of such attempts, as far as modern innovation is concerned, lies in the utilization of nanoparticle-based formulations as a means of replacing or improving the traditional vaccine adjuvants in aquaculture vaccinology. By harnessing the power of technological advances and biological knowledge, the utilization of nanoparticles can enhance vaccine delivery, potentiate immune responses, and ultimately confer superior protective effects in reared species, improving vaccine formulations and mitigating relative side effects linked to conventional vaccine adjuvants. Polymeric nanoparticles, lipid-based nanoformulations, and nanomaterials of organic or inorganic origin are at the forefront of current research endeavors employing nanoparticle-based formulations for immunization purposes in teleost fish. However, even though they appear promising, these applications are still considered to be in their relative infancy, as most of these technologies have only been tested in fish for no more than a decade. Ultimately, this warrants further research efforts to be put into assessing the application of these materials, so that concrete and relative evidence will be accumulated on their efficacy, optimization, and safety in aquatic species.

Overall, the assessment of animal safety and welfare should be placed as a top priority in trials exploring novel formulations, as this can potentially ensure faster application approval and the establishment of welfare-friendly vaccination practices in the sector. The need for tools and attention to effectively monitor and evaluate teleost fish welfare is pressing, especially in the realm of aquaculture immunization, as many different technologies, employing various administration routes in several species, are currently being utilized globally, in a plethora of aquaculture sites. A series of properly standardized technology- and species-specific guidelines can potentially help in establishing a framework where vaccine adjuvants and their effects can be tracked, compared, and evaluated on an evidential basis, in an effort to accelerate vaccine development and sort out potential candidates for novel vaccine formulations. By striking a balance between vaccination efficacy and teleost fish welfare, the next generation of aquaculture vaccines will probably be soon available in the mass market, ushering in a new era for the future of the aquaculture industry.

Author Contributions: Conceptualization, I.T., K.B. and A.I.G.; methodology, I.T., K.B. and A.I.G.; software, I.T.; validation, I.T., K.B. and A.I.G.; formal analysis, I.T.; investigation, I.T., K.B. and A.I.G.; resources, I.T. and A.I.G.; data curation, I.T., K.B. and A.I.G.; writing—original draft preparation, I.T.; writing—review and editing, K.B. and A.I.G.; visualization, I.T., K.B. and A.I.G.; supervision, K.B. and A.I.G.; project administration, K.B. and A.I.G. All authors have read and agreed to the published version of the manuscript.

Funding: This research received no external funding.

Acknowledgments: All of the figures in this manuscript were created using BioRender.com: (Figure 1: Created in BioRender. Tammas, I. (2024) https://BioRender.com/s47j443 (accessed on 26 November 2024), Figure 2: Created in BioRender. Tammas, I. (2024) https://BioRender.com/n42u649 (accessed on 26 November 2024), Figure 3: Created in BioRender. Tammas, I. (2024) https://BioRender.com/l57y747 (accessed on 26 November 2024).

Conflicts of Interest: The authors declare no conflicts of interest.

References

1. Tammas, I.; Bitchava, K.; Gelasakis, A.I. Transforming Aquaculture through Vaccination: A Review on Recent Developments and Milestones. *Vaccines* **2024**, *12*, 732. [CrossRef] [PubMed]
2. Schar, D.; Klein, E.Y.; Laxminarayan, R.; Gilbert, M.; Van Boeckel, T.P. Global Trends in Antimicrobial Use in Aquaculture. *Sci. Rep.* **2020**, *10*, 21878. [CrossRef]
3. Vincent, A.T.; Gauthier, J.; Derome, N.; Charette, S.J. The Rise and Fall of Antibiotics in Aquaculture. In *Microbial Communities in Aquaculture Ecosystems: Improving Productivity and Sustainability*; Derome, N., Ed.; Springer International Publishing: Cham, Switzerland, 2019; pp. 1–19, ISBN 978-3-030-16190-3.
4. Arechavala-Lopez, P.; Sanchez-Jerez, P.; Bayle-Sempere, J.T.; Uglem, I.; Mladineo, I. Reared Fish, Farmed Escapees and Wild Fish Stocks—A Triangle of Pathogen Transmission of Concern to Mediterranean Aquaculture Management. *Aquac. Environ. Interact.* **2013**, *3*, 153–161. [CrossRef]
5. Mordecai, G.J.; Miller, K.M.; Bass, A.L.; Bateman, A.W.; Teffer, A.K.; Caleta, J.M.; Di Cicco, E.; Schulze, A.D.; Kaukinen, K.H.; Li, S.; et al. Aquaculture Mediates Global Transmission of a Viral Pathogen to Wild Salmon. *Sci. Adv.* **2021**, *7*, eabe2592. [CrossRef]
6. Johansen, L.-H.; Jensen, I.; Mikkelsen, H.; Bjørn, P.-A.; Jansen, P.A.; Bergh, Ø. Disease Interaction and Pathogens Exchange between Wild and Farmed Fish Populations with Special Reference to Norway. *Aquaculture* **2011**, *315*, 167–186. [CrossRef]
7. Dalmo, R.; Bøgwald, J.; Tafalla, C. Adjuvants and Delivery Methods: Current and Novel. In *Fish Vaccines*; Adams, A., Ed.; Springer: Basel, Switzerland, 2016; pp. 75–103, ISBN 978-3-0348-0980-1.
8. Mondal, H.; Thomas, J. A Review on the Recent Advances and Application of Vaccines against Fish Pathogens in Aquaculture. *Aquacult Int.* **2022**, *30*, 1971–2000. [CrossRef]
9. Subramani, P.A.; Michael, R.D. Chapter 4—Prophylactic and Prevention Methods Against Diseases in Aquaculture. In *Fish Diseases*; Jeney, G., Ed.; Academic Press: Cambridge, MA, USA, 2017; pp. 81–117, ISBN 978-0-12-804564-0.
10. Nasr-Eldahan, S.; Nabil-Adam, A.; Shreadah, M.A.; Maher, A.M.; El-Sayed Ali, T. A Review Article on Nanotechnology in Aquaculture Sustainability as a Novel Tool in Fish Disease Control. *Aquacult Int.* **2021**, *29*, 1459–1480. [CrossRef]
11. Shah, B.R.; Mraz, J. Advances in Nanotechnology for Sustainable Aquaculture and Fisheries. *Rev. Aquac.* **2020**, *12*, 925–942. [CrossRef]
12. Sarkar, B.; Mahanty, A.; Gupta, S.K.; Choudhury, A.R.; Daware, A.; Bhattacharjee, S. Nanotechnology: A next-Generation Tool for Sustainable Aquaculture. *Aquaculture* **2022**, *546*, 737330. [CrossRef]

13. Thompson, K.D.; Rodkhum, C.; Bunnoy, A.; Thangsunan, P.; Kitiyodom, S.; Sukkarun, P.; Yostawornkul, J.; Yata, T.; Pirarat, N. Addressing Nanovaccine Strategies for Tilapia. *Vaccines* **2023**, *11*, 1356. [CrossRef]
14. Radhakrishnan, A.; Vaseeharan, B.; Ramasamy, P.; Jeyachandran, S. Oral Vaccination for Sustainable Disease Prevention in Aquaculture—An Encapsulation Approach. *Aquacult. Int.* **2023**, *31*, 867–891. [CrossRef] [PubMed]
15. Turley, J.L.; Lavelle, E.C. Resolving Adjuvant Mode of Action to Enhance Vaccine Efficacy. *Curr. Opin. Immunol.* **2022**, *77*, 102229. [CrossRef] [PubMed]
16. Ribeiro, C.M.S.; Schijns, V.E.J.C. Immunology of Vaccine Adjuvants. In *Vaccine Adjuvants: Methods and Protocols*; Davies, G., Ed.; Humana Press: Totowa, NJ, USA, 2010; pp. 1–14, ISBN 978-1-60761-585-9.
17. Schijns, V.E. Immunological Concepts of Vaccine Adjuvant Activity: Commentary. *Curr. Opin. Immunol.* **2000**, *12*, 456–463. [CrossRef] [PubMed]
18. Wu, L.; Qin, Z.; Liu, H.; Lin, L.; Ye, J.; Li, J. Recent Advances on Phagocytic B Cells in Teleost Fish. *Front. Immunol.* **2020**, *11*, 824. [CrossRef] [PubMed]
19. Secombes, C.J.; Belmonte, R. Overview of the Fish Adaptive Immune System. In *Fish Vaccines*; Adams, A., Ed.; Springer: Basel, Switzerland, 2016; pp. 35–52, ISBN 978-3-0348-0978-8.
20. Kordon, A.O.; Pinchuk, L.; Karsi, A. Adaptive Immune System in Fish. *Turk. J. Fish. Aquat. Sci.* **2021**, *22*, TRJFAS20235. [CrossRef]
21. Díaz-Rosales, P.; Muñoz-Atienza, E.; Tafalla, C. Role of Teleost B Cells in Viral Immunity. *Fish Shellfish Immunol.* **2019**, *86*, 135–142. [CrossRef]
22. Johnstone, C.; Chaves-Pozo, E. Antigen Presentation and Autophagy in Teleost Adaptive Immunity. *Int. J. Mol. Sci.* **2022**, *23*, 4899. [CrossRef]
23. Zinkernagel, R.M.; Ehl, S.; Aichele, P.; Oehen, S.; Kündig, T.; Hengartner, H. Antigen Localisation Regulates Immune Responses in a Dose- and Time-Dependent Fashion: A Geographical View of Immune Reactivity. *Immunol. Rev.* **1997**, *156*, 199–209. [CrossRef]
24. Awate, S.; Babiuk, L.A.; Mutwiri, G. Mechanisms of Action of Adjuvants. *Front. Immunol.* **2013**, *4*, 114. [CrossRef]
25. Kordon, A.O.; Karsi, A.; Pinchuk, L. Innate Immune Responses in Fish: Antigen Presenting Cells and Professional Phagocytes. *Turk. J. Fish. Aquat. Sci.* **2018**, *18*, 1123–1139. [CrossRef]
26. Sahoo, B.R. Structure of Fish Toll-like Receptors (TLR) and NOD-like Receptors (NLR). *Int. J. Biol. Macromol.* **2020**, *161*, 1602–1617. [CrossRef] [PubMed]
27. Liao, Z.; Su, J. Progresses on Three Pattern Recognition Receptor Families (TLRs, RLRs and NLRs) in Teleost. *Dev. Comp. Immunol.* **2021**, *122*, 104131. [CrossRef]
28. Zhang, J.; Kong, X.; Zhou, C.; Li, L.; Nie, G.; Li, X. Toll-like Receptor Recognition of Bacteria in Fish: Ligand Specificity and Signal Pathways. *Fish Shellfish Immunol.* **2014**, *41*, 380–388. [CrossRef] [PubMed]
29. Ferreira, I.A.; Peixoto, D.; Losada, A.P.; Quiroga, M.I.; Vale, A.D.; Costas, B. Early Innate Immune Responses in European Sea Bass (*Dicentrarchus labrax* L.) Following Tenacibaculum Maritimum Infection. *Front. Immunol.* **2023**, *14*, 1254677. [CrossRef]
30. Leiva-Rebollo, R.; Gémez-Mata, J.; Castro, D.; Borrego, J.J.; Labella, A.M. Immune Response of DNA Vaccinated-Gilthead Seabream (*Sparus aurata*) against LCDV-Sa Infection: Relevance of the Inflammatory Process. *Front. Immunol.* **2023**, *14*, 1209926. [CrossRef]
31. Franch, R.; Cardazzo, B.; Antonello, J.; Castagnaro, M.; Patarnello, T.; Bargelloni, L. Full-Length Sequence and Expression Analysis of Toll-like Receptor 9 in the Gilthead Seabream (*Sparus aurata* L.). *Gene* **2006**, *378*, 42–51. [CrossRef]
32. Miccoli, A.; Buonocore, F.; Picchietti, S.; Scapigliati, G. The Sea Bass *Dicentrarchus labrax* as a Marine Model Species in Immunology: Insights from Basic and Applied Research. *Aquac. Fish.* **2024**, *9*, 136–143. [CrossRef]
33. Muñoz, I.; Sepulcre, M.P.; Meseguer, J.; Mulero, V. Toll-like Receptor 22 of Gilthead Seabream, *Sparus aurata*: Molecular Cloning, Expression Profiles and Post-Transcriptional Regulation. *Dev. Comp. Immunol.* **2014**, *44*, 173–179. [CrossRef]
34. Chen, Z.; Ceballos-Francisco, D.; Guardiola, F.A.; Huang, D.; Esteban, M.Á. The Alleviation of Skin Wound-Induced Intestinal Barrier Dysfunction via Modulation of TLR Signalling Using Arginine in Gilthead Seabream (*Sparus aurata* L). *Fish Shellfish Immunol.* **2020**, *107*, 519–528. [CrossRef]
35. Wang, B.; Liu, X.; Zhao, J.; Cao, M.; Yu, Z.; Fu, Q.; Tan, F.; Yang, N.; Li, C. Characterization, Evolution and Expression Analysis of Toll-like Receptor 7 (TLR7) in Turbot (*Scophthalmus maximus* L.). *Fish Shellfish Immunol.* **2022**, *125*, 9–16. [CrossRef]
36. Li, S.; Wang, G.; Liu, D.; Liu, Q.; Hu, G. Cloning and Expression Analysis of a *Toll-like Receptor 21* (*TLR21*) Gene from Turbot, *Scophthalmus maximus*. *Dev. Comp. Immunol.* **2017**, *73*, 163–168. [CrossRef] [PubMed]
37. Zhang, H.; Hu, G.; Liu, Q.; Zhang, S. Cloning and Expression Study of a Toll-like Receptor 2 (*Tlr2*) Gene from Turbot, *Scophthalmus maximus*. *Fish Shellfish Immunol.* **2016**, *59*, 137–148. [CrossRef] [PubMed]
38. Dong, X.; Su, B.; Zhou, S.; Shang, M.; Yan, H.; Liu, F.; Gao, C.; Tan, F.; Li, C. Identification and Expression Analysis of Toll-like Receptor Genes (TLR8 and TLR9) in Mucosal Tissues of Turbot (*Scophthalmus maximus* L.) Following Bacterial Challenge. *Fish Shellfish Immunol.* **2016**, *58*, 309–317. [CrossRef]
39. Yu, J.; Liu, X.; Yang, N.; Wang, B.; Su, B.; Fu, Q.; Zhang, M.; Tan, F.; Li, C. Characterization of Toll-like Receptor 1 (TLR1) in Turbot (*Scophthalmus maximus* L.). *Fish Shellfish Immunol.* **2021**, *115*, 27–34. [CrossRef]
40. Hu, G.-B.; Zhang, S.-F.; Yang, X.; Liu, D.-H.; Liu, Q.-M.; Zhang, S.-C. Cloning and Expression Analysis of a Toll-like Receptor 22 (*Tlr22*) Gene from Turbot, *Scophthalmus maximus*. *Fish Shellfish Immunol.* **2015**, *44*, 399–409. [CrossRef] [PubMed]

41. Liu, F.; Su, B.; Fu, Q.; Shang, M.; Gao, C.; Tan, F.; Li, C. Identification, Characterization and Expression Analysis of TLR5 in the Mucosal Tissues of Turbot (*Scophthalmus maximus* L.) Following Bacterial Challenge. *Fish Shellfish Immunol.* **2017**, *68*, 272–279. [CrossRef]
42. Matzinger, P. Tolerance, Danger, and the Extended Family. *Annu. Rev. Immunol.* **1994**, *12*, 991–1045. [CrossRef]
43. Tanekhy, M. The Role of Toll-like Receptors in Innate Immunity and Infectious Diseases of Teleost. *Aquac. Res.* **2016**, *47*, 1369–1391. [CrossRef]
44. Savelkoul, H.F.J.; Ferro, V.A.; Strioga, M.M.; Schijns, V.E.J.C. Choice and Design of Adjuvants for Parenteral and Mucosal Vaccines. *Vaccines* **2015**, *3*, 148–171. [CrossRef]
45. Salinas, I.; Ding, Y.; Fernández-Montero, Á.; Sunyer, J.O. Mucosal Immunity in Fish. In *Principles of Fish Immunology: From Cells and Molecules to Host Protection*; Buchmann, K., Secombes, C.J., Eds.; Springer International Publishing: Cham, Switzerland, 2022; pp. 387–443, ISBN 978-3-030-85420-1.
46. Gomez, D.; Sunyer, J.O.; Salinas, I. The Mucosal Immune System of Fish: The Evolution of Tolerating Commensals While Fighting Pathogens. *Fish Shellfish Immunol.* **2013**, *35*, 1729–1739. [CrossRef]
47. Yu, Y.; Wang, Q.; Huang, Z.; Ding, L.; Xu, Z. Immunoglobulins, Mucosal Immunity and Vaccination in Teleost Fish. *Front. Immunol.* **2020**, *11*, 567941. [CrossRef] [PubMed]
48. Villumsen, K.R.; Koppang, E.O.; Christensen, D.; Bojesen, A.M. Alternatives to Mineral Oil Adjuvants in Vaccines against *Aeromonas salmonicida* Subsp. *salmonicida* in Rainbow Trout Offer Reductions in Adverse Effects. *Sci. Rep.* **2017**, *7*, 5930. [CrossRef] [PubMed]
49. Pasquali, R.C.; Sacco, N.; Bregni, C. The Studies on Hydrophilic-Lipophilic Balance (HLB): Sixty Years after William C. Griffin's Pioneer Work (1949–2009). *Lat. Am. J. Pharm* **2009**, *28*, 313–317.
50. Hong, I.K.; Kim, S.I.; Lee, S.B. Effects of HLB Value on Oil-in-Water Emulsions: Droplet Size, Rheological Behavior, Zeta-Potential, and Creaming Index. *J. Ind. Eng. Chem.* **2018**, *67*, 123–131. [CrossRef]
51. Schmidts, T.; Dobler, D.; Guldan, A.-C.; Paulus, N.; Runkel, F. Multiple W/O/W Emulsions—Using the Required HLB for Emulsifier Evaluation. *Colloids Surf. A Physicochem. Eng. Asp.* **2010**, *372*, 48–54. [CrossRef]
52. Raman, R.P.; Kumar, S. Adjuvants for Fish Vaccines. In *Fish Immune System and Vaccines*; Makesh, M., Rajendran, K.V., Eds.; Springer Nature: Singapore, 2022; pp. 231–244, ISBN 978-981-19126-8-9.
53. Tepparin, S.; Unajak, S.; Hirono, I.; Kondo, H.; Areechon, N. Efficacy of Adjuvanted *Streptococcus agalactiae* Vaccine by Montanide ISA 763 A VG in Nile Tilapia (*Oreochromis niloticus* Linn.). *J. Fish. Environ.* **2018**, *42*, 26–38.
54. Soltani, M.; Mokhtari, A.; Mirzargar, S.S.; Taherimirghaed, A.; Zargar, A.; Shafiei, S.; Hosseini Shekarabi, S.P. Efficacy and Immune Response of Intraperitoneal Vaccination of Rainbow Trout (*Oncorhynchus mykiss*) with a Yersinia Ruckeri Bacterin Formulated with Montanide (TM) ISA 763 AVG Adjuvant. *Bull. Eur. Assoc. Fish Pathol.* **2016**, *36*, 225–236.
55. Xu, W.; Jiao, C.; Bao, P.; Liu, Q.; Wang, P.; Zhang, R.; Liu, X.; Zhang, Y. Efficacy of Montanide^{TM} ISA 763 A VG as Aquatic Adjuvant Administrated with an Inactivated *Vibrio harveyi* Vaccine in Turbot (*Scophthalmus maximus* L.). *Fish Shellfish Immunol.* **2019**, *84*, 56–61. [CrossRef]
56. Jaafar, R.M.; Chettri, J.K.; Dalsgaard, I.; Al-Jubury, A.; Kania, P.W.; Skov, J.; Buchmann, K. Effects of Adjuvant Montanide^{TM} ISA 763 A VG in Rainbow Trout Injection Vaccinated against *Yersinia Ruckeri*. *Fish Shellfish Immunol.* **2015**, *47*, 797–806. [CrossRef]
57. Wangkahart, E.; Thongsrisuk, A.; Vialle, R.; Pholchamat, S.; Sunthamala, P.; Phudkliang, J.; Srisapoome, P.; Wang, T.; Secombes, C.J. Comparative Study of the Effects of Montanide^{TM} ISA 763A VG and ISA 763B VG Adjuvants on the Immune Response against *Streptococcus agalactiae* in Nile Tilapia (*Oreochromis niloticus*). *Fish Shellfish Immunol.* **2023**, *134*, 108563. [CrossRef]
58. Wangkaghart, E.; Deville, S.; Wang, B.; Srisapoome, P.; Wang, T.; Secombes, C.J. Immune Response and Protective Efficacy of Two New Adjuvants, Montanide^{TM} ISA 763B VG and Montanide^{TM} GEL02, Administered with a *Streptococcus agalactiae* Ghost Vaccine in Nile Tilapia (*Oreochromis niloticus*). *Fish Shellfish Immunol.* **2021**, *116*, 19–29. [CrossRef] [PubMed]
59. Veenstra, K.A.; Wang, T.; Russell, K.S.; Tubbs, L.; Ben Arous, J.; Secombes, C.J. Montanide^{TM} ISA 763A VG and ISA 761 VG Induce Different Immune Pathway Responses in Rainbow Trout (*Oncorhynchus mykiss*) When Used as Adjuvant for an *Aeromonas salmonicida* Bacterin. *Fish Shellfish Immunol.* **2021**, *114*, 171–183. [CrossRef] [PubMed]
60. Thim, H.L.; Villoing, S.; McLoughlin, M.; Christie, K.E.; Grove, S.; Frost, P.; Jørgensen, J.B. Vaccine Adjuvants in Fish Vaccines Make a Difference: Comparing Three Adjuvants (Montanide ISA763A Oil, CpG/Poly I:C Combo and VHSV Glycoprotein) Alone or in Combination Formulated with an Inactivated Whole Salmonid Alphavirus Antigen. *Vaccines* **2014**, *2*, 228–251. [CrossRef]
61. Huang, M.; Wang, W. Factors Affecting Alum–Protein Interactions. *Int. J. Pharm.* **2014**, *466*, 139–146. [CrossRef] [PubMed]
62. HogenEsch, H. Mechanism of Immunopotentiation and Safety of Aluminum Adjuvants. *Front. Immunol.* **2013**, *3*, 406. [CrossRef]
63. He, P.; Zou, Y.; Hu, Z. Advances in Aluminum Hydroxide-Based Adjuvant Research and Its Mechanism. *Hum. Vaccines Immunother.* **2015**, *11*, 477–488. [CrossRef] [PubMed]
64. HogenEsch, H.; O'Hagan, D.T.; Fox, C.B. Optimizing the Utilization of Aluminum Adjuvants in Vaccines: You Might Just Get What You Want. *npj Vaccines* **2018**, *3*, 51. [CrossRef]
65. Angosto, D.; López-Muñoz, A.; García-Alcazar, A.; Meseguer, J.; Sepulcre, M.P.; Mulero, V. Aluminum Is a Powerful Adjuvant in Teleost Fish despite Failing to Induce Interleukin-1β Release. *Dev. Comp. Immunol.* **2018**, *85*, 18–24. [CrossRef]
66. Compan, V.; Baroja-Mazo, A.; López-Castejón, G.; Gomez, A.I.; Martínez, C.M.; Angosto, D.; Montero, M.T.; Herranz, A.S.; Bazán, E.; Reimers, D.; et al. Cell Volume Regulation Modulates NLRP3 Inflammasome Activation. *Immunity* **2012**, *37*, 487–500. [CrossRef]

67. Angosto, D.; López-Castejón, G.; López-Muñoz, A.; Sepulcre, M.P.; Arizcun, M.; Meseguer, J.; Mulero, V. Evolution of Inflammasome Functions in Vertebrates: Inflammasome and Caspase-1 Trigger Fish Macrophage Cell Death but Are Dispensable for the Processing of IL-1β. *Innate Immun.* **2012**, *18*, 815–824. [CrossRef]
68. Bird, S.; Zou, J.; Wang, T.; Munday, B.; Cunningham, C.; Secombes, C.J. Evolution of Interleukin-1β. *Cytokine Growth Factor Rev.* **2002**, *13*, 483–502. [CrossRef] [PubMed]
69. McKee, A.S.; Burchill, M.A.; Munks, M.W.; Jin, L.; Kappler, J.W.; Friedman, R.S.; Jacobelli, J.; Marrack, P. Host DNA Released in Response to Aluminum Adjuvant Enhances MHC Class II-Mediated Antigen Presentation and Prolongs CD4 T-Cell Interactions with Dendritic Cells. *Proc. Natl. Acad. Sci. USA* **2013**, *110*, E1122–E1131. [CrossRef] [PubMed]
70. Marichal, T.; Ohata, K.; Bedoret, D.; Mesnil, C.; Sabatel, C.; Kobiyama, K.; Lekeux, P.; Coban, C.; Akira, S.; Ishii, K.J.; et al. DNA Released from Dying Host Cells Mediates Aluminum Adjuvant Activity. *Nat. Med.* **2011**, *17*, 996–1002. [CrossRef]
71. Kool, M.; Pétrilli, V.; De Smedt, T.; Rolaz, A.; Hammad, H.; van Nimwegen, M.; Bergen, I.M.; Castillo, R.; Lambrecht, B.N.; Tschopp, J. Cutting Edge: Alum Adjuvant Stimulates Inflammatory Dendritic Cells through Activation of the NALP3 Inflammasome. *J. Immunol.* **2008**, *181*, 3755–3759. [CrossRef] [PubMed]
72. Galindo-Villegas, J.; García-Alcazar, A.; Meseguer, J.; Mulero, V. Aluminum Adjuvant Potentiates Gilthead Seabream Immune Responses but Induces Toxicity in Splenic Melanomacrophage Centers. *Fish Shellfish Immunol.* **2019**, *85*, 31–43. [CrossRef]
73. Guo, M.; Li, C. An Overview of Cytokine Used as Adjuvants in Fish: Current State and Future Trends. *Rev. Aquac.* **2021**, *13*, 996–1014. [CrossRef]
74. Wang, Q.; Ji, W.; Xu, Z. Current Use and Development of Fish Vaccines in China. *Fish Shellfish Immunol.* **2020**, *96*, 223–234. [CrossRef] [PubMed]
75. Thirumalaikumar, E.; Vimal, S.; Sathishkumar, R.; Ravi, M.; Karthick, V.; Ramya, S.; Thomas, J.; Kumar, V.; Kamaraj, C.; Citarasu, T. DNA Vaccine Incorporated Poly (Lactic-*Co*-Glycolic) Acid (PLGA) Microspheres Offer Enhanced Protection against *Aeromonas hydrophila* Infection. *Int. J. Biol. Macromol.* **2023**, *253*, 127182. [CrossRef]
76. Kole, S.; Qadiri, S.S.N.; Shin, S.-M.; Kim, W.-S.; Lee, J.; Jung, S.-J. PLGA Encapsulated Inactivated-Viral Vaccine: Formulation and Evaluation of Its Protective Efficacy against Viral *Haemorrhagic septicaemia* Virus (VHSV) Infection in Olive Flounder (*Paralichthys olivaceus*) Vaccinated by Mucosal Delivery Routes. *Vaccine* **2019**, *37*, 973–983. [CrossRef]
77. Jiang, C.; Huo, X.; Tang, L.; Hu, M.; Yang, C.; Luo, D.; Su, J. Oral PLGA-Based DNA Vaccines Using Interferons as Adjuvants Remarkably Promote the Immune Protection of Grass Carp (*Ctenopharyngodon idella*) against GCRV Infection. *Water Biol. Secur.* **2023**, *2*, 100143. [CrossRef]
78. Yogeshwari, G. Poly D, L-Lactide-Co-Glycolic Acid (PLGA)-Encapsulated CpG-Oligonucleotide (ODN) on Immune Response in *Cyprinus carpio* against *Aeromonas hydrophila*. *J. Aquac. Res. Dev.* **2015**, *6*, 327. [CrossRef]
79. Diao, J.; Ye, H.; Yu, X.; Fan, Y.; Xu, L.; Li, T.; Wang, Y. Adjuvant and Immunostimulatory Effects of LPS and β-Glucan on Immune Response in Japanese Flounder, *Paralichthys olivaceus*. *Vet. Immunol. Immunopathol.* **2013**, *156*, 167–175. [CrossRef] [PubMed]
80. Wangkahart, E.; Secombes, C.J.; Wang, T. Studies on the Use of Flagellin as an Immunostimulant and Vaccine Adjuvant in Fish Aquaculture. *Front. Immunol.* **2019**, *9*, 3054. [CrossRef]
81. Sahoo, L.; Parhi, J.; Debnath, C.; Prasad, K.P. Effect of Feeding Lipopolysaccharide as an Immunostimulant on Immune Response and Immune Gene Expression of *Labeo bata*. *Vet. Immunol. Immunopathol.* **2017**, *188*, 48–58. [CrossRef]
82. Munang'andu, H.M.; Salinas, I.; Tafalla, C.; Dalmo, R.A. Editorial: Vaccines and Immunostimulants for Finfish. *Front. Immunol.* **2020**, *11*, 573771. [CrossRef]
83. Vallejos-Vidal, E.; Reyes-López, F.; Teles, M.; MacKenzie, S. The Response of Fish to Immunostimulant Diets. *Fish Shellfish Immunol.* **2016**, *56*, 34–69. [CrossRef] [PubMed]
84. Giri, S.S.; Chi, C.; Jun, J.W.; Park, S.C. Use of Bacterial Subcellular Components as Immunostimulants in Fish Aquaculture. *Rev. Aquac.* **2018**, *10*, 474–492. [CrossRef]
85. Hoel, K.; Lillehaug, A. Adjuvant Activity of Polar Glycopeptidolipids from Mycobacterium Chelonaein Experimental Vaccines against *Aeromonas salmonicida* in Salmonid Fish. *Fish Shellfish Immunol.* **1997**, *7*, 365–376. [CrossRef]
86. Burakova, Y.; Madera, R.; McVey, S.; Schlup, J.R.; Shi, J. Adjuvants for Animal Vaccines. *Viral Immunol.* **2018**, *31*, 11–22. [CrossRef]
87. Vinay, T.-N.; Park, C.-S.; Kim, H.-Y.; Jung, S.-J. Toxicity and Dose Determination of Quillaja Saponin, Aluminum Hydroxide and Squalene in Olive Flounder (*Paralichthys olivaceus*). *Vet. Immunol. Immunopathol.* **2014**, *158*, 73–85. [CrossRef]
88. Wang, Y.; Wang, X.; Huang, J.; Li, J. Adjuvant Effect of Quillaja Saponaria Saponin (QSS) on Protective Efficacy and IgM Generation in Turbot (*Scophthalmus maximus*) upon Immersion Vaccination. *Int. J. Mol. Sci.* **2016**, *17*, 325. [CrossRef] [PubMed]
89. Jung, M.-H.; Jung, S.-J.; Kim, T. Saponin and Chitosan-Based Oral Vaccine against Viral *Haemorrhagic septicaemia* Virus (VHSV) Provides Protective Immunity in Olive Flounder (*Paralichthys olivaceus*). *Fish Shellfish Immunol.* **2022**, *126*, 336–346. [CrossRef] [PubMed]
90. Jung, M.-H.; Kole, S.; Jung, S.-J. Efficacy of Saponin-Based Inactivated Rock Bream Iridovirus (RBIV) Vaccine in Rock Bream (*Oplegnathus fasciatus*). *Fish Shellfish Immunol.* **2022**, *121*, 12–22. [CrossRef] [PubMed]
91. Sun, F.; Zhou, J.; Zhang, Y.; Liu, Q.; Wang, Q.; Liu, X. A Compound Ginseng Stem Leaf Saponins and Aluminium Adjuvant Enhances the Potency of Inactivated *Aeromonas salmonicida* Vaccine in Turbot. *Fish Shellfish Immunol.* **2022**, *128*, 60–66. [CrossRef]
92. Sun, F.; Wu, Y.; Zhang, Y.; Liu, Q.; Wang, Q.; Liu, X. An Aluminium Adjuvant Compound with Ginseng Stem Leaf Saponins Enhances the Potency of Inactivated *Pseudomonas plecoglossicida* Vaccine in Large Yellow Croaker (*Larimichthys crocea*). *Fish Shellfish Immunol.* **2024**, *144*, 109243. [CrossRef]

93. Soltani, M.; Lymbery, A.; Song, S.K.; Hosseini Shekarabi, P. Adjuvant Effects of Medicinal Herbs and Probiotics for Fish Vaccines. *Rev. Aquac.* **2019**, *11*, 1325–1341. [CrossRef]
94. Beck, B.R.; Lee, S.H.; Kim, D.; Park, J.H.; Lee, H.K.; Kwon, S.-S.; Lee, K.H.; Lee, J.I.; Song, S.K. A *Lactococcus lactis* BFE920 Feed Vaccine Expressing a Fusion Protein Composed of the OmpA and FlgD Antigens from *Edwardsiella tarda* Was Significantly Better at Protecting Olive Flounder (*Paralichthys olivaceus*) from Edwardsiellosis than Single Antigen Vaccines. *Fish Shellfish Immunol.* **2017**, *68*, 19–28. [CrossRef]
95. Lee, S.H.; Beck, B.R.; Hwang, S.-H.; Song, S.K. Feeding Olive Flounder (*Paralichthys olivaceus*) with *Lactococcus lactis* BFE920 Expressing the Fusion Antigen of *Vibrio* OmpK and FlaB Provides Protection against Multiple *Vibrio* Pathogens: A Universal Vaccine Effect. *Fish Shellfish Immunol.* **2021**, *114*, 253–262. [CrossRef]
96. Kim, D.; Beck, B.R.; Lee, S.M.; Jeon, J.; Lee, D.W.; Lee, J.I.; Song, S.K. Pellet Feed Adsorbed with the Recombinant *Lactococcus lactis* BFE920 Expressing SiMA Antigen Induced Strong Recall Vaccine Effects against *Streptococcus iniae* Infection in Olive Flounder (*Paralichthys olivaceus*). *Fish Shellfish Immunol.* **2016**, *55*, 374–383. [CrossRef]
97. Anuradha, K.; Foo, H.L.; Mariana, N.S.; Loh, T.C.; Yusoff, K.; Hassan, M.D.; Sasan, H.; Raha, A.R. Live Recombinant *Lactococcus lactis* Vaccine Expressing Aerolysin Genes D1 and D4 for Protection against *Aeromonas hydrophila* in Tilapia (*Oreochromis niloticus*). *J. Appl. Microbiol.* **2010**, *109*, 1632–1642. [CrossRef]
98. Naderi-Samani, M.; Soltani, M.; Dadar, M.; Taheri-Mirghaed, A.; Zargar, A.; Ahmadivand, S.; Hassanzadeh, R.; Goudarzi, L.M. Oral Immunization of Trout Fry with Recombinant *Lactococcus lactis* NZ3900 Expressing G Gene of Viral Hemorrhagic Septicaemia Virus (VHSV). *Fish Shellfish Immunol.* **2020**, *105*, 62–70. [CrossRef] [PubMed]
99. Docando, F.; Nuñez-Ortiz, N.; Gonçalves, G.; Serra, C.R.; Gomez-Casado, E.; Martín, D.; Abós, B.; Oliva-Teles, A.; Tafalla, C.; Díaz-Rosales, P. *Bacillus subtilis* Expressing the Infectious Pancreatic Necrosis Virus VP2 Protein Retains Its Immunostimulatory Properties and Induces a Specific Antibody Response. *Front. Immunol.* **2022**, *13*, 888311. [CrossRef]
100. Chen, D.-D.; Yao, Y.-Y.; Cui, Z.-W.; Zhang, X.-Y.; Guo, X.; Zhou, Y.-Y.; Zhang, Y.-A. Comparative Study of the Immunoprotective Effect of Two Grass Carp-Sourced *Bacillus subtilis* Spore-Based Vaccines against Grass Carp Reovirus. *Aquaculture* **2019**, *504*, 88–95. [CrossRef]
101. Gonçalves, G.; Santos, R.A.; Coutinho, F.; Pedrosa, N.; Curado, M.; Machado, M.; Costas, B.; Bonneville, L.; Serrano, M.; Carvalho, A.P.; et al. Oral Vaccination of Fish against Vibriosis Using Spore-Display Technology. *Front. Immunol.* **2022**, *13*, 1012301. [CrossRef] [PubMed]
102. Wang, Q.; Liang, X.; Ning, Y.; Liu, S.; Liang, Z.; Zhang, Z.; Chen, Y.; Cao, J.; Wang, F.; Lan, L.; et al. Surface Display of Major Capsid Protein on *Bacillus subtilis* Spores against Largemouth Bass Virus (LMBV) for Oral Administration. *Fish Shellfish Immunol.* **2023**, *135*, 108627. [CrossRef]
103. Huang, X.; Ma, Y.; Wang, Y.; Niu, C.; Liu, Z.; Yao, X.; Jiang, X.; Pan, R.; Jia, S.; Li, D.; et al. Oral Probiotic Vaccine Expressing Koi Herpesvirus (KHV) ORF81 Protein Delivered by Chitosan-Alginate Capsules Is a Promising Strategy for Mass Oral Vaccination of Carps against KHV Infection. *J. Virol.* **2021**, *95*, e00415-21. [CrossRef]
104. Wu, Y.; Rashidpour, A.; Almajano, M.P.; Metón, I. Chitosan-Based Drug Delivery System: Applications in Fish Biotechnology. *Polymers* **2020**, *12*, 1177. [CrossRef]
105. Mohan, K.; Rajan, D.K.; Ganesan, A.R.; Divya, D.; Johansen, J.; Zhang, S. Chitin, Chitosan and Chitooligosaccharides as Potential Growth Promoters and Immunostimulants in Aquaculture: A Comprehensive Review. *Int. J. Biol. Macromol.* **2023**, *251*, 126285. [CrossRef]
106. Collado-González, M.; Esteban, M.Á. Chitosan-Nanoparticles Effects on Mucosal Immunity: A Systematic Review. *Fish Shellfish Immunol.* **2022**, *130*, 1–8. [CrossRef]
107. Tian, J.; Yu, J.; Sun, X. Chitosan Microspheres as Candidate Plasmid Vaccine Carrier for Oral Immunisation of Japanese Flounder (*Paralichthys olivaceus*). *Vet. Immunol. Immunopathol.* **2008**, *126*, 220–229. [CrossRef]
108. León-Rodríguez, L.; Luzardo-Álvarez, A.; Blanco-Méndez, J.; Lamas, J.; Leiro, J. A Vaccine Based on Biodegradable Microspheres Induces Protective Immunity against Scuticociliatosis without Producing Side Effects in Turbot. *Fish Shellfish Immunol.* **2012**, *33*, 21–27. [CrossRef] [PubMed]
109. Jin, P.; Sun, F.; Liu, Q.; Wang, Q.; Zhang, Y.; Liu, X. An Oral Vaccine Based on Chitosan/Aluminum Adjuvant Induces Both Local and Systemic Immune Responses in Turbot (*Scophthalmus maximus*). *Vaccine* **2021**, *39*, 7477–7484. [CrossRef]
110. Xu, F.-F.; Jiang, F.-Y.; Zhou, G.-Q.; Xia, J.-Y.; Yang, F.; Zhu, B. The Recombinant Subunit Vaccine Encapsulated by Alginate-Chitosan Microsphere Enhances the Immune Effect against *Micropterus salmoides* Rhabdovirus. *J. Fish Dis.* **2022**, *45*, 1757–1765. [CrossRef]
111. Wang, E.; Wang, X.; Wang, K.; He, J.; Zhu, L.; He, Y.; Chen, D.; Ouyang, P.; Geng, Y.; Huang, X.; et al. Preparation, Characterization and Evaluation of the Immune Effect of Alginate/Chitosan Composite Microspheres Encapsulating Recombinant Protein of *Streptococcus iniae* Designed for Fish Oral Vaccination. *Fish Shellfish Immunol.* **2018**, *73*, 262–271. [CrossRef] [PubMed]
112. Behera, T.; Swain, P. Antigen Encapsulated Alginate-Coated Chitosan Microspheres Stimulate Both Innate and Adaptive Immune Responses in Fish through Oral Immunization. *Aquacult Int* **2014**, *22*, 673–688. [CrossRef]
113. Behera, T.; Swain, P. Alginate–Chitosan–PLGA Composite Microspheres Induce Both Innate and Adaptive Immune Response through Parenteral Immunization in Fish. *Fish Shellfish Immunol.* **2013**, *35*, 785–791. [CrossRef] [PubMed]
114. Dezfuly, Z.T.; Alishahi, M.; Ghorbanpoor, M.; Tabandeh, M.R.; Mesbah, M. Immunogenicity and Protective Efficacy of *Yersinia Ruckeri* Lipopolysaccharide (LPS), Encapsulated by Alginate-Chitosan Micro/Nanoparticles in Rainbow Trout (*Oncorhyncus Mykiss*). *Fish Shellfish Immunol.* **2020**, *104*, 25–35. [CrossRef] [PubMed]

115. González-Chavarría, I.; Roa, F.J.; Sandoval, F.; Muñoz-Flores, C.; Kappes, T.; Acosta, J.; Bertinat, R.; Altamirano, C.; Valenzuela, A.; Sánchez, O.; et al. Chitosan Microparticles Enhance the Intestinal Release and Immune Response of an Immune Stimulant Peptide in *Oncorhynchus mykiss*. *Int. J. Mol. Sci.* **2023**, *24*, 14685. [CrossRef]
116. Brudeseth, B.E.; Wiulsrød, R.; Fredriksen, B.N.; Lindmo, K.; Løkling, K.-E.; Bordevik, M.; Steine, N.; Klevan, A.; Gravningen, K. Status and Future Perspectives of Vaccines for Industrialised Fin-Fish Farming. *Fish Shellfish Immunol.* **2013**, *35*, 1759–1768. [CrossRef]
117. Hoare, R.; Jung, S.-J.; Ngo, T.P.H.; Bartie, K.; Bailey, J.; Thompson, K.D.; Adams, A. Efficacy and Safety of a Non-Mineral Oil Adjuvanted Injectable Vaccine for the Protection of Atlantic Salmon (*Salmo salar* L.) against *Flavobacterium psychrophilum*. *Fish Shellfish Immunol.* **2019**, *85*, 44–51. [CrossRef]
118. Poppe, T.T.; Koppang, E.O. Side-Effects of Vaccination. In *Fish Vaccination*; John Wiley & Sons, Ltd.: Hoboken, NJ, USA, 2014; pp. 153–161, ISBN 978-1-118-80691-3.
119. Mutoloki, S.; Cooper, G.A.; Marjara, I.S.; Koop, B.F.; Evensen, Ø. High Gene Expression of Inflammatory Markers and IL-17A Correlates with Severity of Injection Site Reactions of Atlantic Salmon Vaccinated with Oil-Adjuvanted Vaccines. *BMC Genom.* **2010**, *11*, 336. [CrossRef] [PubMed]
120. Jiao, X.; Cheng, S.; Hu, Y.; Sun, L. Comparative Study of the Effects of Aluminum Adjuvants and Freund's Incomplete Adjuvant on the Immune Response to an *Edwardsiella tarda* Major Antigen. *Vaccine* **2010**, *28*, 1832–1837. [CrossRef] [PubMed]
121. Gjessing, M.C.; Falk, K.; Weli, S.C.; Koppang, E.O.; Kvellestad, A. A Sequential Study of Incomplete Freund's Adjuvant-Induced Peritonitis in Atlantic Cod. *Fish Shellfish Immunol.* **2012**, *32*, 141–150. [CrossRef] [PubMed]
122. Spinos, E.; Kokkoris, G.D.; Bakopoulos, V. Prevention of Sea Bass (*Dicentrarchus labrax*, L. 1758) Photobacteriosis and Vibriosis. Long Term Efficacy Study of Intraperitoneally Administered Bivalent Commercial Vaccines. *Aquaculture* **2017**, *471*, 172–184. [CrossRef]
123. Veenstra, K.A.; Wang, T.; Alnabulsi, A.; Douglas, A.; Russell, K.S.; Tubbs, L.; Arous, J.B.; Secombes, C.J. Analysis of Adipose Tissue Immune Gene Expression after Vaccination of Rainbow Trout with Adjuvanted Bacterins Reveals an Association with Side Effects. *Mol. Immunol.* **2017**, *88*, 89–98. [CrossRef]
124. Li, J.; Tang, L.; Li, S.; Li, G.; Mo, Z. The Efficacy and Side-Effects of Oil-Based Adjuvants Emulsified *Vibrio anguillarum* Bivalent Inactivated Vaccine in Turbot (*Scophthalmus maximus*) under Production Mode. *Aquaculture* **2020**, *524*, 735259. [CrossRef]
125. Tziouvas, H.; Varvarigos, P. Intensity Scale of Side Effects in European Sea Bass (*Dicentrarchus labrax*) Post Intraperitoneal Injection with Commercial Oil-Adjuvanted Vaccines. *Bull. EAFP* **2021**, *41*, 103–110. [CrossRef]
126. Midtlyng, P.J.; Reitan, L.J.; Speilberg, L. Experimental Studies on the Efficacy and Side-Effects of Intraperitoneal Vaccination of Atlantic Salmon (*Salmo salar* L.) against Furunculosis. *Fish Shellfish Immunol.* **1996**, *6*, 335–350. [CrossRef]
127. Harshitha, M.; Nayak, A.; Disha, S.; Akshath, U.S.; Dubey, S.; Munang'andu, H.M.; Chakraborty, A.; Karunasagar, I.; Maiti, B. Nanovaccines to Combat *Aeromonas hydrophila* Infections in Warm-Water Aquaculture: Opportunities and Challenges. *Vaccines* **2023**, *11*, 1555. [CrossRef]
128. Fredriksen, B.N.; Grip, J. PLGA/PLA Micro- and Nanoparticle Formulations Serve as Antigen Depots and Induce Elevated Humoral Responses after Immunization of Atlantic Salmon (*Salmo salar* L.). *Vaccine* **2012**, *30*, 656–667. [CrossRef]
129. Fredriksen, B.N.; Sævareid, K.; McAuley, L.; Lane, M.E.; Bøgwald, J.; Dalmo, R.A. Early Immune Responses in Atlantic Salmon (*Salmo salar* L.) after Immunization with PLGA Nanoparticles Loaded with a Model Antigen and β-Glucan. *Vaccine* **2011**, *29*, 8338–8349. [CrossRef] [PubMed]
130. Adomako, M.; St-Hilaire, S.; Zheng, Y.; Eley, J.; Marcum, R.D.; Sealey, W.; Donahower, B.C.; LaPatra, S.; Sheridan, P.P. Oral DNA Vaccination of Rainbow Trout, *Oncorhynchus mykiss* (Walbaum), against Infectious Haematopoietic Necrosis Virus Using PLGA [Poly(D,L-Lactic-Co-Glycolic Acid)] Nanoparticles. *J. Fish Dis.* **2012**, *35*, 203–214. [CrossRef]
131. Tian, J.; Yu, J. Poly(Lactic-Co-Glycolic Acid) Nanoparticles as Candidate DNA Vaccine Carrier for Oral Immunization of Japanese Flounder (*Paralichthys olivaceus*) against Lymphocystis Disease Virus. *Fish Shellfish Immunol.* **2011**, *30*, 109–117. [CrossRef] [PubMed]
132. Dubey, S.; Avadhani, K.; Mutalik, S.; Sivadasan, S.M.; Maiti, B.; Paul, J.; Girisha, S.K.; Venugopal, M.N.; Mutoloki, S.; Evensen, Ø.; et al. *Aeromonas hydrophila* OmpW PLGA Nanoparticle Oral Vaccine Shows a Dose-Dependent Protective Immunity in Rohu (*Labeo rohita*). *Vaccines* **2016**, *4*, 21. [CrossRef] [PubMed]
133. Harshitha, M.; D'souza, R.; Akshay, S.D.; Nayak, A.; Disha, S.; Aditya, V.; Akshath, U.S.; Dubey, S.; Munang'andu, H.M.; Chakraborty, A.; et al. Oral Administration of Recombinant Outer Membrane Protein A-Based Nanovaccine Affords Protection against *Aeromonas hydrophila* in Zebrafish. *World J. Microbiol. Biotechnol.* **2024**, *40*, 250. [CrossRef]
134. Alishahi, M.; Vaseghi, M.; Tabandeh, M.R.; Khosravi, M. Immunogenic and Protective Effects of an Oral Polylactic-Co-Glycolic Acid Nano Encapsulated DNA Vaccine Encoding aopB Gene of *Aeromonas hydrophila* in Common Carp. *Aquac. Int.* **2024**, *32*, 1169–1190. [CrossRef]
135. Alishahi, M.; Lababian, H.; Heidari, H.; Tabandeh, M.R.; Khosravi, M. Development of an Injectable DNA Vaccine Against *Aeromonas hydrophila* Infection Nanoencapsulated With Poly(Lactic-Co-Glycolic) Acid (PLGA) in Common Carp. *Aquac. Res.* **2024**, *2024*, 7270489. [CrossRef]
136. Zhang, C.; Zhang, P.-Q.; Guo, S.; Zhao, Z.; Wang, G.-X.; Zhu, B. Dual-Targeting Polymer Nanoparticles Efficiently Deliver DNA Vaccine and Induce Robust Prophylactic Immunity against Spring Viremia of Carp Virus Infection. *Microbiol. Spectr.* **2022**, *10*, e03085-22. [CrossRef]

137. Ahmed, F.; Soliman, F.M.; Adly, M.A.; Soliman, H.A.M.; El-Matbouli, M.; Saleh, M. Recent Progress in Biomedical Applications of Chitosan and Its Nanocomposites in Aquaculture: A Review. *Res. Vet. Sci.* **2019**, *126*, 68–82. [CrossRef]
138. El-Naggar, M.; Medhat, F.; Taha, A. Applications of Chitosan and Chitosan Nanoparticles in Fish Aquaculture. *Egypt. J. Aquat. Biol. Fish.* **2022**, *26*, 23–43. [CrossRef]
139. Kole, S.; Dar, S.A.; Shin, S.-M.; Jeong, H.-J.; Jung, S.-J. Potential Efficacy of Chitosan-Poly (Lactide-Co-Glycolide)-Encapsulated Trivalent Immersion Vaccine in Olive Flounder (*Paralichthys olivaceus*) Against Viral Hemorrhagic Septicemia Virus, Streptococcus Parauberis Serotype I, and *Miamiensis avidus* (Scuticociliate). *Front. Immunol.* **2021**, *12*, 761130. [CrossRef]
140. Ahmadivand, S.; Soltani, M.; Behdani, M.; Evensen, Ø.; Alirahimi, E.; Hassanzadeh, R.; Soltani, E. Oral DNA Vaccines Based on CS-TPP Nanoparticles and Alginate Microparticles Confer High Protection against Infectious Pancreatic Necrosis Virus (IPNV) Infection in Trout. *Dev. Comp. Immunol.* **2017**, *74*, 178–189. [CrossRef] [PubMed]
141. Rao, B.M.; Kole, S.; Gireesh-Babu, P.; Sharma, R.; Tripathi, G.; Bedekar, M.K. Evaluation of Persistence, Bio-Distribution and Environmental Transmission of Chitosan/PLGA/pDNA Vaccine Complex against *Edwardsiella tarda* in *Labeo rohita*. *Aquaculture* **2019**, *500*, 385–392. [CrossRef]
142. Halimi, M.; Alishahi, M.; Abbaspour, M.R.; Ghorbanpoor, M.; Tabandeh, M.R. Valuable Method for Production of Oral Vaccine by Using Alginate and Chitosan against *Lactococcus garvieae/Streptococcus iniae* in Rainbow Trout (*Oncorhynchus mykiss*). *Fish Shellfish Immunol.* **2019**, *90*, 431–439. [CrossRef]
143. Huo, X.; Tang, L.; Liu, Q.; Zhu, W.; Zhang, J.; Hu, M.; Zhao, F.; Wang, P.; Yuan, G.; Yang, C.; et al. Oral pcDNA3.1-VP4/VP56-FlaC DNA Vaccine Encapsulated by Chitosan/Sodium Alginate Nanoparticles Confers Remarkable Protection against GCRV Infection in Grass Carp. *Aquaculture* **2023**, *577*, 739996. [CrossRef]
144. Leya, T.; Ahmad, I.; Sharma, R.; Tripathi, G.; Kurcheti, P.P.; Rajendran, K.V.; Bedekar, M.K. Bicistronic DNA Vaccine Macromolecule Complexed with Poly Lactic-*Co*-Glycolic Acid-Chitosan Nanoparticles Enhanced the Mucosal Immunity of *Labeo rohita* against *Edwardsiella tarda* Infection. *Int. J. Biol. Macromol.* **2020**, *156*, 928–937. [CrossRef]
145. Li, L.; Lin, S.-L.; Deng, L.; Liu, Z.-G. Potential Use of Chitosan Nanoparticles for Oral Delivery of DNA Vaccine in Black Seabream *Acanthopagrus schlegelii* Bleeker to Protect from *Vibrio parahaemolyticus*. *J. Fish Dis.* **2013**, *36*, 987–995. [CrossRef] [PubMed]
146. Rajesh Kumar, S.; Ishaq Ahmed, V.P.; Parameswaran, V.; Sudhakaran, R.; Sarath Babu, V.; Sahul Hameed, A.S. Potential Use of Chitosan Nanoparticles for Oral Delivery of DNA Vaccine in Asian Sea Bass (*Lates calcarifer*) to Protect from *Vibrio* (*Listonella*) *Anguillarum*. *Fish Shellfish Immunol.* **2008**, *25*, 47–56. [CrossRef]
147. Sukkarun, P.; Kitiyodom, S.; Yostawornkul, J.; Chaiin, P.; Yata, T.; Rodkhum, C.; Boonrungsiman, S.; Pirarat, N. Chitosan-Polymer Based Nanovaccine as Promising Immersion Vaccine against *Aeromonas veronii* Challenge in Red Tilapia (*Oreochromis* sp.). *Fish Shellfish Immunol.* **2022**, *129*, 30–35. [CrossRef]
148. Kitiyodom, S.; Yata, T.; Yostawornkul, J.; Kaewmalun, S.; Nittayasut, N.; Suktham, K.; Surassmo, S.; Namdee, K.; Rodkhum, C.; Pirarat, N. Enhanced Efficacy of Immersion Vaccination in Tilapia against Columnaris Disease by Chitosan-Coated "Pathogen-like" Mucoadhesive Nanovaccines. *Fish Shellfish Immunol.* **2019**, *95*, 213–219. [CrossRef]
149. Dubey, S.; Avadhani, K.; Mutalik, S.; Sivadasan, S.M.; Maiti, B.; Girisha, S.K.; Venugopal, M.N.; Mutoloki, S.; Evensen, Ø.; Karunasagar, I.; et al. *Edwardsiella tarda* OmpA Encapsulated in Chitosan Nanoparticles Shows Superior Protection over Inactivated Whole Cell Vaccine in Orally Vaccinated Fringed-Lipped Peninsula Carp (*Labeo fimbriatus*). *Vaccines* **2016**, *4*, 40. [CrossRef] [PubMed]
150. Thirumalaikumar, E.; Lelin, C.; Sathishkumar, R.; Vimal, S.; Anand, S.B.; Babu, M.M.; Citarasu, T. Oral Delivery of pVAX-OMP and pVAX-*Hly* DNA Vaccine Using Chitosan-Tripolyphosphate (Cs-TPP) Nanoparticles in Rohu, (*Labeo rohita*) for Protection against *Aeromonas hydrophila* Infection. *Fish Shellfish Immunol.* **2021**, *115*, 189–197. [CrossRef]
151. Kole, S.; Kumari, R.; Anand, D.; Kumar, S.; Sharma, R.; Tripathi, G.; Makesh, M.; Rajendran, K.V.; Bedekar, M.K. Nanoconjugation of Bicistronic DNA Vaccine against *Edwardsiella tarda* Using Chitosan Nanoparticles: Evaluation of Its Protective Efficacy and Immune Modulatory Effects in *Labeo rohita* Vaccinated by Different Delivery Routes. *Vaccine* **2018**, *36*, 2155–2165. [CrossRef]
152. Zhang, Z.; Jin, Z.; Wang, G.; Ling, F. Mannosylated Chitosan Nanoparticles Loaded with ABP Antigen Server as a Novel Nucleic Acid Vaccine against *Nocardia Seriolae* Infection in *Micropterus salmoides*. *Aquaculture* **2023**, *574*, 739635. [CrossRef]
153. Ponce, M.; Zuasti, E.; Anguís, V.; Fernández-Díaz, C. Anti-Bacterial and Immunostimulatory Properties of Ulvan-Loaded Chitosan Nanoparticles for Use in Aquaculture. *Mar. Biotechnol.* **2024**, *26*, 19–27. [CrossRef] [PubMed]
154. Sukkarun, P.; Kitiyodom, S.; Kamble, M.T.; Bunnoy, A.; Boonanuntanasarn, S.; Yata, T.; Boonrungsiman, S.; Thompson, K.D.; Rodkhum, C.; Pirarat, N. Systemic and Mucosal Immune Responses in Red Tilapia (*Oreochromis* sp.) Following Immersion Vaccination with a Chitosan Polymer-Based Nanovaccine against *Aeromonas veronii*. *Fish Shellfish Immunol.* **2024**, *146*, 109383. [CrossRef]
155. Kitiyodom, S.; Trullàs, C.; Rodkhum, C.; Thompson, K.D.; Katagiri, T.; Temisak, S.; Namdee, K.; Yata, T.; Pirarat, N. Modulation of the Mucosal Immune Response of Red Tilapia (*Oreochromis* sp.) against Columnaris Disease Using a Biomimetic-Mucoadhesive Nanovaccine. *Fish Shellfish Immunol.* **2021**, *112*, 81–91. [CrossRef] [PubMed]
156. Kitiyodom, S.; Yata, T.; Thompson, K.D.; Costa, J.; Elumalai, P.; Katagiri, T.; Temisak, S.; Namdee, K.; Rodkhum, C.; Pirarat, N. Immersion Vaccination by a Biomimetic-Mucoadhesive Nanovaccine Induces Humoral Immune Response of Red Tilapia (*Oreochromis* sp.) against Flavobacterium Columnare Challenge. *Vaccines* **2021**, *9*, 1253. [CrossRef]

157. Zhang, W.; Zhao, Z.; Zhou, J.; Wang, W.; Su, J.; Yuan, G. Carboxymethyl Chitosan Nanoparticles Loaded with *Ctenopharyngodon idella* Interferon-Γ2 (CiIFN-Γ2) Enhance Protective Efficacy against Bacterial Infection in Grass Carp. *Aquaculture* **2023**, *572*, 739554. [CrossRef]
158. Tattiyapong, P.; Kitiyodom, S.; Yata, T.; Jantharadej, K.; Adamek, M.; Surachetpong, W. Chitosan Nanoparticle Immersion Vaccine Offers Protection against Tilapia Lake Virus in Laboratory and Field Studies. *Fish Shellfish Immunol.* **2022**, *131*, 972–979. [CrossRef]
159. Zheng, F.; Liu, H.; Sun, X.; Zhang, Y.; Zhang, B.; Teng, Z.; Hou, Y.; Wang, B. Development of Oral DNA Vaccine Based on Chitosan Nanoparticles for the Immunization against Reddish Body Iridovirus in Turbots (*Scophthalmus maximus*). *Aquaculture* **2016**, *452*, 263–271. [CrossRef]
160. Kole, S.; Qadiri, S.S.N.; Shin, S.-M.; Kim, W.-S.; Lee, J.; Jung, S.-J. Nanoencapsulation of Inactivated-Viral Vaccine Using Chitosan Nanoparticles: Evaluation of Its Protective Efficacy and Immune Modulatory Effects in Olive Flounder (*Paralichthys olivaceus*) against Viral *Haemorrhagic septicaemia* Virus (VHSV) Infection. *Fish Shellfish Immunol.* **2019**, *91*, 136–147. [CrossRef] [PubMed]
161. Vimal, S.; Abdul Majeed, S.; Nambi, K.S.N.; Madan, N.; Farook, M.A.; Venkatesan, C.; Taju, G.; Venu, S.; Subburaj, R.; Thirunavukkarasu, A.R.; et al. Delivery of DNA Vaccine Using Chitosan–Tripolyphosphate (CS/TPP) Nanoparticles in Asian Sea Bass, *Lates calcarifer* (Bloch, 1790) for Protection against Nodavirus Infection. *Aquaculture* **2014**, *420–421*, 240–246. [CrossRef]
162. Valero, Y.; Awad, E.; Buonocore, F.; Arizcun, M.; Esteban, M.Á.; Meseguer, J.; Chaves-Pozo, E.; Cuesta, A. An Oral Chitosan DNA Vaccine against Nodavirus Improves Transcription of Cell-Mediated Cytotoxicity and Interferon Genes in the European Sea Bass Juveniles Gut and Survival upon Infection. *Dev. Comp. Immunol.* **2016**, *65*, 64–72. [CrossRef]
163. Thu Lan, N.G.; Dong, H.T.; Vinh, N.T.; Salin, K.R.; Senapin, S.; Pimsannil, K.; St-Hilaire, S.; Shinn, A.P.; Rodkhum, C. A Novel Vaccination Strategy against *Vibrio harveyi* Infection in Asian Seabass (*Lates calcarifer*) with the Aid of Oxygen Nanobubbles and Chitosan. *Fish Shellfish Immunol.* **2024**, *149*, 109557. [CrossRef]
164. Andresen, A.M.S.; Gjøen, T. Chitosan Nanoparticle Formulation Attenuates Poly (I:C) Induced Innate Immune Responses against Inactivated Virus Vaccine in Atlantic Salmon (*Salmo salar*). *Comp. Biochem. Physiol. Part D Genom. Proteom.* **2021**, *40*, 100915. [CrossRef]
165. Ji, J.; Merino, S.; Tomás, J.M.; Roher, N. Nanoliposomes Encapsulating Immunostimulants Modulate the Innate Immune System and Elicit Protection in Zebrafish Larvae. *Fish Shellfish Immunol.* **2019**, *92*, 421–429. [CrossRef]
166. Bunnoy, A.; Thangsunan, P.; Chokmangmeepisarn, P.; Yata, T.; Klongklaew, N.; Pirarat, N.; Kitiyodom, S.; Srisapoome, P.; Rodkhum, C. Mucoadhesive Cationic Lipid-Based *Flavobacterium oreochromis* Nanoencapsulation Enhanced the Efficacy of Mucoadhesive Immersion Vaccination against Columnaris Disease and Strengthened Immunity in Asian Sea Bass (*Lates calcarifer*). *Fish Shellfish Immunol.* **2022**, *127*, 633–646. [CrossRef] [PubMed]
167. Thangsunan, P.; Kitiyodom, S.; Srisapoome, P.; Pirarat, N.; Yata, T.; Thangsunan, P.; Boonrungsiman, S.; Bunnoy, A.; Rodkhum, C. Novel Development of Cationic Surfactant-Based Mucoadhesive Nanovaccine for Direct Immersion Vaccination against *Francisella Noatunensis* Subsp. *Orientalis in Red Tilapia (Oreochromis sp.). Fish Shellfish Immunol.* **2022**, *127*, 1051–1060. [CrossRef]
168. Ruyra, A.; Cano-Sarabia, M.; MacKenzie, S.A.; Maspoch, D.; Roher, N. A Novel Liposome-Based Nanocarrier Loaded with an LPS-dsRNA Cocktail for Fish Innate Immune System Stimulation. *PLoS ONE* **2013**, *8*, e76338. [CrossRef]
169. Ruyra, A.; Cano-Sarabia, M.; García-Valtanen, P.; Yero, D.; Gibert, I.; Mackenzie, S.A.; Estepa, A.; Maspoch, D.; Roher, N. Targeting and Stimulation of the Zebrafish (*Danio rerio*) Innate Immune System with LPS/dsRNA-Loaded Nanoliposomes. *Vaccine* **2014**, *32*, 3955–3962. [CrossRef] [PubMed]
170. Dahl, L.O.S.; Hak, S.; Braaen, S.; Molska, A.; Rodà, F.; Parot, J.; Wessel, Ø.; Fosse, J.H.; Bjørgen, H.; Borgos, S.E.; et al. Implementation of mRNA–Lipid Nanoparticle Technology in Atlantic Salmon (*Salmo salar*). *Vaccines* **2024**, *12*, 788. [CrossRef] [PubMed]
171. Giri, S.S.; Kim, S.G.; Kang, J.W.; Kim, S.W.; Kwon, J.; Lee, S.B.; Jung, W.J.; Park, S.C. Applications of Carbon Nanotubes and Polymeric Micro-/Nanoparticles in Fish Vaccine Delivery: Progress and Future Perspectives. *Rev. Aquac.* **2021**, *13*, 1844–1863. [CrossRef]
172. Giri, S.S.; Park, S.C. Application of Carbon Nanotubes in the Advancement of Fish Vaccine. In *Biotechnological Advances in Aquaculture Health Management*; Gupta, S.K., Giri, S.S., Eds.; Springer Nature: Singapore, 2021; pp. 61–78, ISBN 9789811651953.
173. Cimbaluk, G.V.; Ramsdorf, W.A.; Perussolo, M.C.; Santos, H.K.F.; Da Silva De Assis, H.C.; Schnitzler, M.C.; Schnitzler, D.C.; Carneiro, P.G.; Cestari, M.M. Evaluation of Multiwalled Carbon Nanotubes Toxicity in Two Fish Species. *Ecotoxicol. Environ. Saf.* **2018**, *150*, 215–223. [CrossRef]
174. Chowdhry, A.; Kaur, J.; Khatri, M.; Puri, V.; Tuli, R.; Puri, S. Characterization of Functionalized Multiwalled Carbon Nanotubes and Comparison of Their Cellular Toxicity between HEK 293 Cells and Zebra Fish in Vivo. *Heliyon* **2019**, *5*, e02605. [CrossRef]
175. Su, Y.; Yan, X.; Pu, Y.; Xiao, F.; Wang, D.; Yang, M. Risks of Single-Walled Carbon Nanotubes Acting as Contaminants-Carriers: Potential Release of Phenanthrene in Japanese Medaka (*Oryzias latipes*). *Environ. Sci. Technol.* **2013**, *47*, 4704–4710. [CrossRef]
176. Gao, X.; Zheng, X.; Gao, S.; Huang, Y.; Xiong, J.; Ren, H. Toxicity of Amine-Functionalized Single-Carbon Nanotube (NH2 f-SWCNT) to Channel Catfish (*Ietalurus punetaus*): Organ Pathologies, Oxidative Stress, Inflammation, and Apoptosis. *Chemosphere* **2021**, *282*, 131133. [CrossRef]
177. Smith, C.J.; Shaw, B.J.; Handy, R.D. Toxicity of Single Walled Carbon Nanotubes to Rainbow Trout, (*Oncorhynchus mykiss*): Respiratory Toxicity, Organ Pathologies, and Other Physiological Effects. *Aquat. Toxicol.* **2007**, *82*, 94–109. [CrossRef]
178. Zhu, B.; Liu, G.-L.; Ling, F.; Song, L.-S.; Wang, G.-X. Development Toxicity of Functionalized Single-Walled Carbon Nanotubes on Rare Minnow Embryos and Larvae. *Nanotoxicology* **2015**, *9*, 579–590. [CrossRef]

179. Li, Y.; Men, B.; He, Y.; Xu, H.; Liu, M.; Wang, D. Effect of Single-Wall Carbon Nanotubes on Bioconcentration and Toxicity of Perfluorooctane Sulfonate in Zebrafish (*Danio rerio*). *Sci. Total Environ.* **2017**, *607–608*, 509–518. [CrossRef]
180. Bisesi, J.H.; Ngo, T.; Ponnavolu, S.; Liu, K.; Lavelle, C.M.; Afrooz, A.R.M.N.; Saleh, N.B.; Ferguson, P.L.; Denslow, N.D.; Sabo-Attwood, T. Examination of Single-Walled Carbon Nanotubes Uptake and Toxicity from Dietary Exposure: Tracking Movement and Impacts in the Gastrointestinal System. *Nanomaterials* **2015**, *5*, 1066–1086. [CrossRef] [PubMed]
181. Sohn, E.K.; Chung, Y.S.; Johari, S.A.; Kim, T.G.; Kim, J.K.; Lee, J.H.; Lee, Y.H.; Kang, S.W.; Yu, I.J. Acute Toxicity Comparison of Single-Walled Carbon Nanotubes in Various Freshwater Organisms. *BioMed Res. Int.* **2015**, *2015*, 323090. [CrossRef]
182. Jiang, T.; Amadei, C.A.; Gou, N.; Lin, Y.; Lan, J.; Vecitis, C.D.; Gu, A.Z. Toxicity of Single-Walled Carbon Nanotubes (SWCNTs): Effect of Lengths, Functional Groups and Electronic Structures Revealed by a Quantitative Toxicogenomics Assay. *Environ. Sci. Nano* **2020**, *7*, 1348–1364. [CrossRef] [PubMed]
183. Wisdom, K.S.; Bhat, I.A.; Chanu, T.I.; Kumar, P.; Pathakota, G.-B.; Nayak, S.K.; Walke, P.; Sharma, R. Chitosan Grafting onto Single-Walled Carbon Nanotubes Increased Their Stability and Reduced the Toxicity in Vivo (Catfish) Model. *Int. J. Biol. Macromol.* **2020**, *155*, 697–707. [CrossRef]
184. Rosales-Mendoza, S.; González-Ortega, O. Carbon Nanotubes-Based Mucosal Vaccines. In *Nanovaccines: An Innovative Technology to Fight Human and Animal Diseases*; Rosales-Mendoza, S., González-Ortega, O., Eds.; Springer International Publishing: Cham, Switzerland, 2019; pp. 159–179, ISBN 978-3-030-31668-6.
185. Gong, Y.-X.; Zhu, B.; Liu, G.-L.; Liu, L.; Ling, F.; Wang, G.-X.; Xu, X.-G. Single-Walled Carbon Nanotubes as Delivery Vehicles Enhance the Immunoprotective Effects of a Recombinant Vaccine against *Aeromonas hydrophila*. *Fish Shellfish Immunol.* **2015**, *42*, 213–220. [CrossRef] [PubMed]
186. Liu, L.; Gong, Y.-X.; Liu, G.-L.; Zhu, B.; Wang, G.-X. Protective Immunity of Grass Carp Immunized with DNA Vaccine against *Aeromonas hydrophila* by Using Carbon Nanotubes as a Carrier Molecule. *Fish Shellfish Immunol.* **2016**, *55*, 516–522. [CrossRef]
187. Zhang, Z.; Liu, G.; Ma, R.; Qi, X.; Wang, G.; Zhu, B.; Ling, F. The Immunoprotective Effect of Whole-Cell Lysed Inactivated Vaccine with SWCNT as a Carrier against *Aeromonas hydrophila* Infection in Grass Carp. *Fish Shellfish Immunol.* **2020**, *97*, 336–343. [CrossRef] [PubMed]
188. Cao, Y.; Liu, J.; Liu, G.; Du, H.; Liu, T.; Liu, T.; Li, P.; Yu, Q.; Wang, G.; Wang, E. A Nanocarrier Immersion Vaccine Encoding Surface Immunogenic Protein Confers Cross-Immunoprotection against *Streptococcus agalactiae* and *Streptococcus iniae* Infection in Tilapia. *Fish Shellfish Immunol.* **2024**, *144*, 109267. [CrossRef]
189. Zhu, B.; Zhang, C.; Zhao, Z.; Wang, G.-X. Targeted Delivery of Mannosylated Nanoparticles Improve Prophylactic Efficacy of Immersion Vaccine against Fish Viral Disease. *Vaccines* **2020**, *8*, 87. [CrossRef]
190. Zhu, B.; Liu, G.-L.; Gong, Y.-X.; Ling, F.; Song, L.-S.; Wang, G.-X. Single-Walled Carbon Nanotubes as Candidate Recombinant Subunit Vaccine Carrier for Immunization of Grass Carp against Grass Carp Reovirus. *Fish Shellfish Immunol.* **2014**, *41*, 279–293. [CrossRef]
191. Qiu, D.-K.; Jia, Y.-J.; Gong, Y.-M.; Zheng, Y.-Y.; Wang, G.-X.; Zhu, B. Optimizing the Immunization Procedure of Single-Walled Carbon Nanotubes Based Vaccine against Grass Carp Reovirus for Grass Carp. *Aquaculture* **2021**, *533*, 736152. [CrossRef]
192. Wang, Y.; Liu, G.-L.; Li, D.-L.; Ling, F.; Zhu, B.; Wang, G.-X. The Protective Immunity against Grass Carp Reovirus in Grass Carp Induced by a DNA Vaccination Using Single-Walled Carbon Nanotubes as Delivery Vehicles. *Fish Shellfish Immunol.* **2015**, *47*, 732–742. [CrossRef] [PubMed]
193. Zhu, B.; Liu, G.-L.; Gong, Y.-X.; Ling, F.; Wang, G.-X. Protective Immunity of Grass Carp Immunized with DNA Vaccine Encoding the *Vp7* Gene of Grass Carp Reovirus Using Carbon Nanotubes as a Carrier Molecule. *Fish Shellfish Immunol.* **2015**, *42*, 325–334. [CrossRef] [PubMed]
194. Zheng, Y.-Y.; Qiu, D.-K.; Guo, Z.-R.; Gong, Y.-M.; Wang, G.-X.; Zhu, B. Evaluation of SWCNTs-Loaded DNA Vaccine Encoding Predominant Antigen Epitope VP4-3 against Type II GCRV. *Aquaculture* **2021**, *534*, 736197. [CrossRef]
195. Zhang, C.; Zhao, Z.; Liu, G.-Y.; Li, J.; Wang, G.-X.; Zhu, B. Immune Response and Protective Effect against Spring Viremia of Carp Virus Induced by Intramuscular Vaccination with a SWCNTs-DNA Vaccine Encoding Matrix Protein. *Fish Shellfish Immunol.* **2018**, *79*, 256–264. [CrossRef]
196. Zhang, C.; Li, L.-H.; Wang, J.; Zhao, Z.; Li, J.; Tu, X.; Huang, A.-G.; Wang, G.-X.; Zhu, B. Enhanced Protective Immunity against Spring Viremia of Carp Virus Infection Can Be Induced by Recombinant Subunit Vaccine Conjugated to Single-Walled Carbon Nanotubes. *Vaccine* **2018**, *36*, 6334–6344. [CrossRef] [PubMed]
197. Zhang, C.; Zhao, Z.; Zha, J.-W.; Wang, G.-X.; Zhu, B. Single-Walled Carbon Nanotubes as Delivery Vehicles Enhance the Immunoprotective Effect of a DNA Vaccine against Spring Viremia of Carp Virus in Common Carp. *Fish Shellfish Immunol.* **2017**, *71*, 191–201. [CrossRef]
198. Zhang, C.; Zheng, Y.-Y.; Gong, Y.-M.; Zhao, Z.; Guo, Z.-R.; Jia, Y.-J.; Wang, G.-X.; Zhu, B. Evaluation of Immune Response and Protection against Spring Viremia of Carp Virus Induced by a Single-Walled Carbon Nanotubes-Based Immersion DNA Vaccine. *Virology* **2019**, *537*, 216–225. [CrossRef]
199. Gong, Y.-M.; Zhang, C.; Li, Y.; Chen, G.; Wang, G.-X.; Zhu, B. Optimization of Immunization Procedure for SWCNTs-Based Subunit Vaccine with Mannose Modification against Spring Viraemia of Carp Virus in Common Carp. *J. Fish Dis.* **2021**, *44*, 1925–1936. [CrossRef]

200. Zhao, Z.; Zhang, C.; Jia, Y.-J.; Qiu, D.-K.; Lin, Q.; Li, N.-Q.; Huang, Z.-B.; Fu, X.-Z.; Wang, G.-X.; Zhu, B. Immersion Vaccination of Mandarin Fish *Siniperca chuatsi* against Infectious Spleen and Kidney Necrosis Virus with a SWCNTs-Based Subunit Vaccine. *Fish Shellfish Immunol.* **2019**, *92*, 133–140. [CrossRef]
201. Zhao, Z.; Xiong, Y.; Zhang, C.; Jia, Y.-J.; Qiu, D.-K.; Wang, G.-X.; Zhu, B. Optimization of the Efficacy of a SWCNTs-Based Subunit Vaccine against Infectious Spleen and Kidney Necrosis Virus in Mandarin Fish. *Fish Shellfish Immunol.* **2020**, *106*, 190–196. [CrossRef] [PubMed]
202. Zhao, Z.; Zhang, C.; Lin, Q.; Li, N.; Huang, Z.-B.; Zhao, M.; Fu, X.-Z.; Wang, G.-X.; Zhu, B. Single-Walled Carbon Nanotubes as Delivery Vehicles Enhance the Immunoprotective Effect of an Immersion DNA Vaccine against Infectious Spleen and Kidney Necrosis Virus in Mandarin Fish. *Fish Shellfish Immunol.* **2020**, *97*, 432–439. [CrossRef] [PubMed]
203. Zhao, Z.; Li, Y.; Chen, G.; Zhang, C.; Wang, G.-X.; Zhu, B. Protective Immunity against Infectious Spleen and Kidney Necrosis Virus Induced by Mannose Modified Subunit Vaccine with Carbon Nanotubes in Mandarin Fish. *Aquac. Res.* **2022**, *53*, 2175–2184. [CrossRef]
204. Zhao, Z.; Meng, Q.; Sun, T.-Z.; Zhu, B. Mannose Modified Targeted Immersion Vaccine Delivery System Improves Protective Immunity against Infectious Spleen and Kidney Necrosis Virus in Mandarin Fish (*Siniperca chuatsi*). *Vaccine* **2024**, *42*, 2886–2894. [CrossRef] [PubMed]
205. Liu, G.-Y.; Wang, E.-L.; Qu, X.-Y.; Yang, K.-C.; Zhang, Z.-Y.; Liu, J.-Y.; Zhang, C.; Zhu, B.; Wang, G.-X. Single-Walled Carbon Nanotubes Enhance the Immune Protective Effect of a Bath Subunit Vaccine for Pearl Gentian Grouper against Iridovirus of Taiwan. *Fish Shellfish Immunol.* **2020**, *106*, 510–517. [CrossRef]
206. Zhang, Z.; Liu, G.; Liu, J.; Zhu, B.; Wang, G.; Ling, F. Epitope Screening of the Major Capsid Protein within Grouper Iridovirus of Taiwan and the Immunoprotective Effect with SWCNTs as the Vaccine Carrier. *Fish Shellfish Immunol.* **2021**, *117*, 17–23. [CrossRef]
207. Hu, F.; Li, Y.; Wang, Q.; Wang, G.; Zhu, B.; Wang, Y.; Zeng, W.; Yin, J.; Liu, C.; Bergmann, S.M.; et al. Carbon Nanotube-Based DNA Vaccine against Koi Herpesvirus given by Intramuscular Injection. *Fish Shellfish Immunol.* **2020**, *98*, 810–818. [CrossRef]
208. Hu, F.; Li, Y.; Wang, Q.; Zhu, B.; Wu, S.; Wang, Y.; Zeng, W.; Yin, J.; Liu, C.; Bergmann, S.M.; et al. Immersion Immunization of Koi (*Cyprinus carpio*) against Cyprinid Herpesvirus 3 (CyHV-3) with Carbon Nanotube-Loaded DNA Vaccine. *Aquaculture* **2021**, *539*, 736644. [CrossRef]
209. Guo, Z.-R.; Zhao, Z.; Zhang, C.; Jia, Y.-J.; Qiu, D.-K.; Zhu, B.; Wang, G.-X. Carbon Nanotubes-Loaded Subunit Vaccine Can Increase Protective Immunity against Rhabdovirus Infections of Largemouth Bass (*Micropterus salmoides*). *Fish Shellfish Immunol.* **2020**, *99*, 548–554. [CrossRef]
210. Liu, G.-Y.; Li, J.; Zhang, Z.-Y.; Liu, J.-Y.; Zhang, C.; Zhu, B.; Wang, G.-X. An Immersion Subunit Vaccine Loaded by Single-Walled Carbon Nanotube Protects Pearl Gentian Grouper from Viral Nervous Necrosis Virus. *Aquaculture* **2021**, *541*, 736813. [CrossRef]
211. Jia, Y.-J.; Guo, Z.-R.; Ma, R.; Qiu, D.-K.; Zhao, Z.; Wang, G.-X.; Zhu, B. Immune Efficacy of Carbon Nanotubes Recombinant Subunit Vaccine against Largemouth Bass Ulcerative Syndrome Virus. *Fish Shellfish Immunol.* **2020**, *100*, 317–323. [CrossRef] [PubMed]
212. Vijayaram, S.; Tsigkou, K.; Zuorro, A.; Sun, Y.-Z.; Rabetafika, H.; Razafindralambo, H. Inorganic Nanoparticles for Use in Aquaculture. *Rev. Aquac.* **2023**, *15*, 1600–1617. [CrossRef]
213. Lisuzzo, L.; Cavallaro, G.; Milioto, S.; Lazzara, G. Layered Composite Based on Halloysite and Natural Polymers: A Carrier for the pH Controlled Release of Drugs. *New J. Chem.* **2019**, *43*, 10887–10893. [CrossRef]
214. Pumchan, A.; Sae-Ueng, U.; Prasittichai, C.; Sirisuay, S.; Areechon, N.; Unajak, S. A Novel Efficient Piscine Oral Nano-Vaccine Delivery System: Modified Halloysite Nanotubes (HNTs) Preventing Streptococcosis Disease in Tilapia (*Oreochromis* sp.). *Vaccines* **2022**, *10*, 1180. [CrossRef]
215. Zhang, W.; Zhu, C.; Xiao, F.; Liu, X.; Xie, A.; Chen, F.; Dong, P.; Lin, P.; Zheng, C.; Zhang, H.; et al. pH-Controlled Release of Antigens Using Mesoporous Silica Nanoparticles Delivery System for Developing a Fish Oral Vaccine. *Front. Immunol.* **2021**, *12*, 644396. [CrossRef]

vaccines

MDPI

Review

Advancements in Nanoparticle-Based Adjuvants for Enhanced Tuberculosis Vaccination: A Review

Jiao Wang [1,2], Zian Zhao [1], Quan Wang [1], Jingyu Shi [1], Duo Wai-Chi Wong [1,*] and James Chung-Wai Cheung [1,*]

1 Department of Biomedical Engineering, Faculty of Engineering, The Hong Kong Polytechnic University, Hong Kong 999077, China
2 Department of Clinical Laboratory, Hubei Provincial Hospital of Traditional Chinese Medicine, Wuhan 430073, China
* Correspondence: duo.wong@polyu.edu.hk (D.W.-C.W.); james.chungwai.cheung@polyu.edu.hk (J.C.-W.C.)

Abstract: Tuberculosis (TB) remains a leading cause of morbidity and mortality worldwide, necessitating the development of more effective vaccines. Nanoparticle-based adjuvants represent a promising approach to enhancing tuberculosis vaccine efficacy. This review focuses on the advantages of nanoparticulate-loaded vaccines, emphasizing their ability to improve antigen delivery, safety, and immunogenicity. We discuss the various types of nanoparticles and their unique physicochemical properties that contribute to improved antigen delivery and sustained immune activation. Additionally, we highlight the advantages of nanoparticle-based adjuvants in inducing strong cellular and humoral immunity, enhancing vaccine stability, and reducing adverse effects. Finally, we address current challenges and future perspectives in the application of these novel adjuvants, emphasizing their potential to transform TB vaccine strategies and ultimately contribute to better global health outcomes.

Keywords: tuberculosis; vaccines; nanoparticles; adjuvants

1. Introduction

Tuberculosis (TB), caused by the bacterium *Mycobacterium tuberculosis* (*M. tuberculosis*), remains a significant global health challenge, with an estimated 10 million new cases and 1.5 million deaths annually [1,2], as reported by the World Health Organization (WHO). This infectious disease primarily affects the lungs but can also manifest in other organs, leading to extrapulmonary TB. The burden of TB is particularly pronounced in low- and middle-income countries [3], particularly in Asia and Africa, where inadequate healthcare infrastructure, high rates of comorbidities like HIV [4], and the emergence of drug-resistant strains (multidrug-resistant TB, MDR-TB [5], and extensively drug-resistant TB, XDR-TB [6,7]) complicate control efforts. The only currently approved vaccine, Bacillus Calmette-Guérin (BCG) [8], while providing some protection against severe forms of TB in children [9], exhibits inconsistent efficacy in preventing adult pulmonary TB with highly variable protection ranging from 0% to 80%, which is highly variable and generally inadequate [10–12]. Furthermore, BCG fails to prevent reactivation of latent TB infections, which remains a major source of new cases.

To address these limitations, numerous new vaccine candidates are under development, targeting both pre-exposure and post-exposure scenarios. However, the absence of a highly effective TB vaccine underscores the need for innovative strategies that can enhance vaccine efficacy, safety, and stability. Nanoparticle-based adjuvants represent a promising advancement in the quest for more effective TB vaccines, offering several advantages over adjuvant candidates such as alum, MF59, and AS01 [13,14]. Unlike these conventional adjuvants, which may enhance only specific aspects of the immune response, nanoparticle formulations can be engineered for targeted delivery to antigen-presenting cells, facilitating improved antigen uptake and sustained release and ensuring prolonged exposure to the immune system [15,16].

This targeted approach not only enhances immunogenicity-promoting robust cellular and humoral responses but also allows for effective immune activation with lower doses of antigens, potentially reducing the overall amount needed and potentially decreasing the frequency of vaccinations required [17,18]. Their biocompatibility and safety profile are superior to traditional adjuvants. Furthermore, the modular design of nanoparticles enables the incorporation of multiple antigens or immunostimulatory agents within a single formulation, fostering a comprehensive immune response that is critical for combating TB [14,19]. Earlier research demonstrated that antigens incorporated into nanoparticulate systems can effectively amplify immunogenic responses against various pathogens responsible for infectious diseases. This suggests a similar promising application for tuberculosis vaccines once developed [20,21]. Nanoparticle-based adjuvants offer significant advantages over traditional vaccine formulations by improving antigen delivery, boosting immune responses, and reducing adverse effects [22–25]. All these features position nanoparticle-based adjuvants as a transformative strategy for enhancing TB vaccination efforts and addressing the ongoing global health challenge posed by this disease.

These innovative platforms facilitate targeted delivery to antigen-presenting cells (APCs), enhancing antigen uptake and presentation. Experimental studies have highlighted the potential of biocompatible and nontoxic methods in establishing prolonged immunity [22]. Moreover, incorporating adjuvants either triggers innate immunity or serves as a delivery mechanism to direct antigens specifically to APCs. This process promotes both cell-mediated and humoral immune responses through the activation of major histocompatibility complex (MHC) molecules [26]. The unique physicochemical properties of nanoparticles allow for controlled and sustained release of antigens, promoting long-lasting immune activation. Moreover, nanoparticles can be engineered to co-deliver multiple antigens or adjuvants, tailoring the immune response to effectively combat *M. tuberculosis* [15]. By protecting encapsulated antigens from degradation and improving bioavailability, nanoparticle-based adjuvants ensure that vaccines maintain their potency until reaching target tissues. In addition to enhancing immunogenicity, nanoparticle-based adjuvants can minimize adverse effects through targeted delivery, reducing off-target immune activation and lowering the required doses of antigens. This dual benefit not only improves safety but also optimizes the overall efficacy of TB vaccination strategies.

As research progresses, nanoparticle-based adjuvants hold transformative potential for TB vaccine development. By addressing the shortcomings of current vaccines and enabling new approaches for effective immune responses, these adjuvants could significantly impact global TB control efforts. Ultimately, the integration of nanoparticle-based adjuvants into TB vaccination strategies has the potential to contribute to better health outcomes worldwide, aligning with the WHO's End TB Strategy and Sustainable Development Goals.

2. Types of Nanoparticles Used in TB Vaccines

In TB vaccine research, the size of nanoparticles used can vary depending on the type of material and the specific application. Generally, nanoparticles used as adjuvants in TB vaccines are within the range of 20 nm to 500 nm [16,27–29] (Figure 1). Due to their small size, nanoparticles exhibit unique physical, chemical, and biological properties that differ significantly from their bulk counterparts [15,30]. In the context of vaccinology, nanoparticles can be engineered from various materials, including lipids, polymers, metals, and proteins, to serve as carriers for antigens and adjuvants (Table 1).

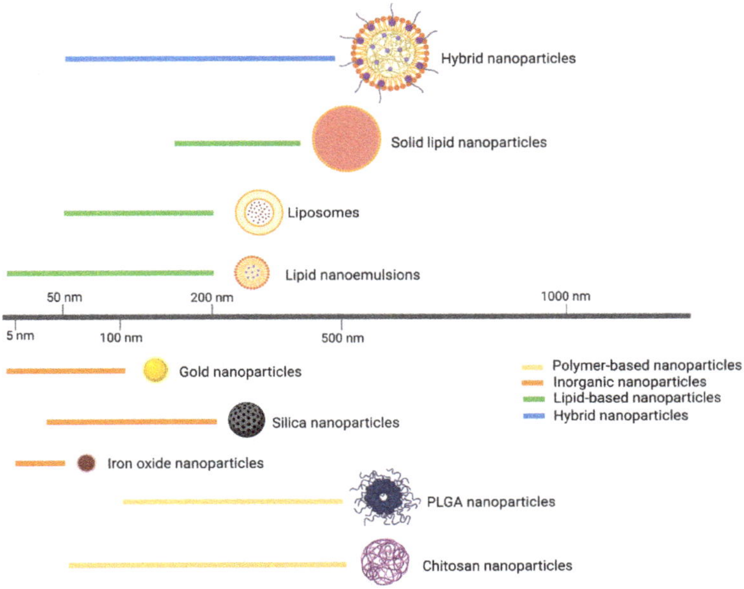

Figure 1. Average size range of different nanoadjuvants: lipid-based nanoparticles (50–500 nm); polymer-based nanoparticles (50–500 nm); inorganic nanoparticles (<200 nm); hybrid nanoparticles (50–500 nm).

Table 1. Structural features and functions of nanoparticles as adjuvants in TB vaccines.

Category	Type	Structure	Function
Lipid-based	Liposomes [31,32]	- Phospholipid bilayer - Encapsulating hydrophilic antigens - Hydrophobic antigens	- Delivers TB antigens like Ag85B, ESAT-6 - Enhances antigen uptake and presentation
	SLNs [33]	- Solid lipid core w/ surfactant coating	- Enhances stability and immune response of TB antigens like Ag85A
	Lipid NEM [34]	- Oil-in-water/water-in-oil droplets - Stabilized by surfactants	- Delivers hydrophobic TB antigens - Promotes antigen uptake
Polymer-based	PLGA NPs [35,36]	- Biodegradable PLGA matrix - Encapsulating antigens	- Sustained release of antigens like HspX, ESAT-6 - Enhances immune response
	Chitosan NPs [37,38]	- Cationic chitosan structure - Interacting w/ negatively charged antigens	- Ideal for mucosal TB vaccine delivery - Enhances local immune responses
Inorganic	AuNPs [39–41]	- Gold core w/ tunable surface functionalization	- Delivers Mtb antigens like CFP-10 - Facilitates uptake by dendritic cells
	Silica NPs [42,43]	- Porous structure w/ high surface area	- Co-delivers antigens and adjuvants - Enhances stability and immune activation
	Iron oxide NPs [18,44]	- Magnetic core w/ modifiable surface	- Magnetic targeting of TB antigens - Imaging and diagnostics
Hybrid	Hybrid NPs [45]	- Combination of materials (e.g., lipid–polymer) - Multilayered	- Co-delivers antigens and adjuvants - Enhances stability and immune response

AuNPs: gold NPs; NEM: nanoemulsions; NPs: nanoparticles; PLGA: polylactic-co-glycolic acid; SLNs: solid lipid NPs; TB: tuberculosis; Mtb: Mycobacterium tuberculosis; w/: with.

2.1. Lipid-Based Nanoparticles

Lipid-based nanoparticles, such as liposomes, solid lipid nanoparticles (SLNs), and nanostructured lipid carriers (NLCs), are widely used in TB vaccine formulations due to their biocompatibility, ability to encapsulate both hydrophilic and hydrophobic antigens, and ease of surface modification. Liposomes are highly versatile and effective delivery systems, offering advantages such as biocompatibility, high loading capacity, and customizable surface modifications [31]. Composed of phospholipid bilayers, they mimic natural cell membranes, facilitating fusion with antigen-presenting cells (APCs) and enhancing antigen uptake. Additionally, liposome-based systems protect DNA from endonuclease degradation, significantly improving plasmid-DNA transfection and antigen presentation compared to naked DNA vaccines [32]. SLNs and NLCs offer advantages in terms of improved stability and controlled release, making them suitable for sustained antigen release and long-term immune stimulation [15]. Additionally, lipid-based nanoparticles can be engineered to target specific tissues or cells by conjugating targeting ligands, further enhancing their effectiveness as vaccine delivery platforms.

2.2. Polymer-Based Nanoparticles

Polymer-based nanoparticles, such as those made from poly (lactic-co-glycolic acid) (PLGA) [46,47], chitosan, and polyethylene glycol (PEG), are another class of nanocarriers used in TB vaccine development [48]. These nanoparticles can be engineered to tailor immune responses, making them suitable for TB vaccination. PLGA nanoparticles are highly valued for their biodegradability, FDA approval, and ability to provide controlled antigen release over extended periods, making them cost-effective and versatile for TB DNA vaccine development [49–52]. These nanoparticles protect encapsulated antigens and adjuvants from enzymatic degradation while enhancing immunogenicity. Chitosan-based nanoparticles, derived from natural polysaccharides, are ideal for mucosal TB vaccines due to their mucoadhesive properties, efficient antigen transport, and ability to open tight cell junctions, enabling effective nasal DNA vaccine delivery [53–55]. PEGylated nanoparticles can evade the mononuclear phagocyte system, prolonging circulation time and improving antigen presentation [56,57]. By adjusting the polymer composition and surface properties, polymer-based nanoparticles can be tailored to achieve desired immune responses and biodistribution profiles.

2.3. Inorganic Nanoparticles

Inorganic nanoparticles, including gold nanoparticles (AuNPs), silica nanoparticles, and metal oxides (e.g., iron oxide), are increasingly being explored for TB vaccine applications because of their unique physicochemical properties like ease of functionalization and stability. AuNPs, for example, are highly versatile and can be functionalized with various biomolecules to enhance antigen stability and uptake by dendritic cells (DCs). Several studies confirm the efficacy of AuNPs for DNA vaccination by injection via T cell activation and protection [41,58–60]. With their capacity for cargo delivery, biocompatibility, and ease of synthesis, AuNPs have been frequently studied as delivery vehicles [61]. Their inherent optical properties also allow for real-time tracking and monitoring of antigen delivery and immune response. Silica nanoparticles, with their high surface area and porous structure, enable efficient antigen loading and controlled release, while metal oxide nanoparticles can act as both delivery vehicles and imaging agents, aiding in the visualization of immune responses [18]. Inorganic nanoparticles can be designed to overcome delivery barriers and shape adaptive immunity, making them promising candidates for future TB vaccine development [23,62].

2.4. Hybrid Nanoparticles

Hybrid nanoparticles combine the properties of two or more materials, such as lipids, polymers, and inorganic components, to achieve synergistic effects. These nanoparticles can integrate the biocompatibility of lipid-based systems, the tunability of polymer-based carriers, and the structural stability of inorganic nanoparticles. For instance, lipid–polymer hybrid nanoparticles can provide the controlled release of polymer-based systems with the biocompatibility and cellular uptake efficiency of lipid-based carriers [19]. The use of lipid–polymer hybrid nanoparticles, such as those combining PLGA with DDA and TDB, has demonstrated the ability to tailor humoral and cellular immunity, which is crucial for protection against intracellular bacteria like *M. tuberculosis* [19]. Another study highlighted the use of biocompatible polymer poly (methyl methacrylate) (PMMA) in combination with cationic lipids to create hybrid nanoparticles that overcame the poor biocompatibility of traditional lipid-based adjuvants, suggesting their potential for a variety of vaccines, including TB [63]. Similarly, inorganic–lipid hybrids can enhance antigen stability and improve delivery efficiency. The versatility of hybrid nanoparticles allows for the incorporation of multiple functionalities, such as targeted delivery, enhanced antigen protection, and multimodal imaging, making them highly adaptable platforms for developing next-generation TB vaccines.

Each type of nanoparticle offers distinct advantages and can be tailored to address specific challenges in TB vaccine development (Table 2). By leveraging the unique properties of these diverse nanocarriers, researchers can design more effective and safer TB vaccines that elicit strong and durable immune responses, ultimately contributing to improved TB prevention and control.

Table 2. Advantages and disadvantages of nanoparticles as adjuvants in TB vaccines.

Category	Type	Advantages	Disadvantages
Lipid-based	Liposomes [32]	- Protects antigens - Increases antigen uptake; - Can be surface-modified for targeted delivery.	- Stability and leakage issues; - Scale-up difficulties.
	SLNs [33,64]	- Biocompatible and biodegradable; - Sustained release; - Good protection of antigens.	- Limited drug-loading capacity; - Lipid crystallization can affect release.
	Lipid NEM [65]	- High surface area; - Enhanced antigen stability; - Modifiable with immune-stimulatory molecules.	- Stability issues like coalescence; - Maintaining homogeneity.
Polymer-based	PLGA NPs [66,67]	- FDA-approved; - Sustained release; - Protection from degradation.	- Burst release; - Acidic degradation products affecting stability.
	Chitosan NPs [68]	- Good mucoadhesive properties; - Enhances antigen uptake; - Biodegradable.	- Limited solubility in physiological conditions; - Variability in properties.
Inorganic	AuNPs [39,40]	- Easy surface modification; - Biocompatible; - Useful for imaging and diagnostics.	- Long-term accumulation; - High production costs.
	Silica NPs [42,69]	- High surface area; - Good stability; - Customizable properties.	- Limited biodegradability; - Potential toxicity.
	Iron oxide NPs [18,44]	- Precise magnetic targeting; - Good biocompatibility.	- Oxidative stress; - Toxicity at high concentrations.
Hybrid	Hybrid NPs [70,71]	- Synergistic effect; - Flexibility in design and functionality.	- Complex formulation; - Stability and scalability issues.

AuNPs: gold NPs; NEM: nanoemulsions; NPs: nanoparticles; PLGA: polylactic-co-glycolic acid; SLNs: solid lipid NPs.

3. Physicochemical Properties and Their Role in Antigen Delivery

3.1. Physicochemical Properties

The physicochemical properties of nanoparticles, such as size, shape, surface charge, hydrophobicity/hydrophilicity, and composition, play a crucial role in determining their behavior in biological systems and their effectiveness as vaccine delivery platforms. These properties influence the nanoparticles' interactions with cells, their biodistribution, the efficiency of antigen delivery, and the type of immune response they elicit.

The size of nanoparticles significantly affects their uptake by APCs and subsequent trafficking within the body. The immunogenicity of smaller particles surpasses that of larger ones, making particle size a critical factor in using nanoparticles as vaccine adjuvants or delivery systems [72]. Nanoparticle-based tuberculosis vaccine delivery systems with sizes ranging from 300 to 600 nm are efficiently taken up by APCs and promote a Th1 immune response, essential for combating TB infection. In contrast, micron-sized particles tend to induce Th2 responses [68]. Nanoparticles with sizes ranging from 20 to 200 nm are generally considered ideal for vaccine delivery because they can efficiently be taken up by APCs such as DCs and macrophages [73]. Smaller nanoparticles (20–50 nm) can penetrate deeper into tissues and facilitate intracellular delivery, whereas larger nanoparticles (100–200 nm) are more suitable for lymph node targeting [16,74–76]. For instance, chitosan nanoparticles with an average diameter of 189 nm showed effective antigen delivery and sustained release, which is crucial for prolonged immune stimulation [77]. Optimal size selection ensures efficient cellular uptake and enhances antigen presentation to T cells, leading to robust immune activation.

The shape of nanoparticles influences their cellular uptake and distribution within the body. Spherical nanoparticles are more readily internalized by cells compared to rod-shaped or elongated particles, which can remain in circulation longer and exhibit different biodistribution patterns. Examples are the lipid nanoparticles (LNPs) which assume a micelle-like structure, encapsulating therapeutic molecules in a nonaqueous core [78]. Elongated or rod-shaped nanoparticles can enhance cellular interactions by increasing the contact area with the cell membrane, which may benefit specific antigen delivery strategies aimed at targeting tissues or organs involved in TB infection, such as lungs. Rod-shaped nanoparticles, such as aluminum oxyhydroxide nanorods, have been shown to induce higher levels of cellular uptake and immune activation compared to spherical particles [79]. Hydroxyapatite nanorods with a high aspect ratio were found to induce stronger pro-inflammatory responses and higher antigen-specific antibody production than their spherical counterparts [80].

The surface charge of nanoparticles determines their interaction with cell membranes, which are typically negatively charged. Positively charged nanoparticles exhibit higher affinity for cell membranes, leading to enhanced uptake by APCs [61,73,81]. PLGA nanoparticles modified with cationic surfactants and immunopotentiators were shown to induce strong Th1/Th17 responses, which are essential for protection against *M. tuberculosis* [19]. Comparative studies by Yan et al. demonstrated that among neutral, negatively, and positively charged liposomes, only cationic liposomes could upregulate the CCL2 chemokine gene in DCs. This induction was mediated through the ERK pathway both in vitro and in vivo [82]. Similarly, NH2-functionalized aluminum oxyhydroxide nanorods exhibited higher cellular uptake and NLRP3 inflammasome activation, leading to stronger immunogenicity [79]. The surface charge of adjuvants also influences the degree of association of antigens and the physical stability of vaccine formulations [73,83]. However, excessive positive charge can cause toxicity and lead to aggregation. Therefore, a balance must be achieved to maximize uptake without compromising safety. Neutral or slightly anionic nanoparticles can also facilitate longer circulation times, which may be beneficial for certain vaccine delivery applications [61,73].

The hydrophobic or hydrophilic nature of nanoparticles affects their stability in biological fluids, interactions with proteins, and cellular uptake. Hydrophobic nanoparticles may exhibit strong interactions with cell membranes, promoting uptake but potentially causing protein aggregation and rapid clearance from the bloodstream [73,83]. Hydrophobicity of microparticles is a key factor for determining the magnitude of immunogenicity stimulated by vaccination containing different particulate systems [84]. Increased surface hydrophobicity on polymeric microparticles, such as PLGA-based microparticles, was found to promote antigen internalization into DCs and enhance cytokine secretion levels, thereby boosting the immune response [84]. Hydrophilic nanoparticles, such as those coated with PEG, are more resistant to protein adsorption and can evade clearance by the mononuclear phagocyte system, resulting in prolonged circulation times [85,86]. Hydrophobic amino acid-based nanoparticles also demonstrated effective immune responses by leveraging their hydrophobic properties to enhance antigen delivery [87]. Modifying the surface hydrophobicity/hydrophilicity balance is key to optimizing nanoparticle behavior in vivo.

The composition of nanoparticles determines their biodegradability, biocompatibility, and ability to deliver antigens effectively. The chemical components used to prepare particulate vaccine adjuvants significantly influence their adjuvant potency [73]. Each nanoparticle material presents unique advantages, allowing for the design of optimized vaccine delivery systems.

3.2. The Role in Antigen Delivery

The surface of nanoparticles can be functionalized with various ligands, such as antibodies, peptides, or carbohydrates, to enhance targeting specificity. For TB vaccines, targeting ligands can be used to direct nanoparticles to specific receptors on APCs, such as mannose receptors, which are highly expressed on DCs [15]. For instance, guanidinylated cationic nanoparticles were shown to enhance antigen uptake and presentation by DCs, leading to a robust immune response [88]. Similarly, the use of Toll-like receptor (TLR) ligands as surface functionalizers can significantly boost the immunogenicity of the encapsulated antigens [89]. This targeted approach not only improves the efficiency of antigen delivery to immune cells but also reduces off-target effects and systemic distribution, minimizing potential side effects.

The ability of nanoparticles to encapsulate or conjugate antigens is another critical factor influencing their performance as vaccine carriers. As shown in Figure 2, nanoparticles offer high encapsulation efficiency, allowing for the presentation of multiple antigens directly to APCs while protecting them from early enzymatic and endosomal degradation, even under high-temperature conditions. This ensures that antigens, along with adjuvants, can reach the cytosol and initiate signaling cascades via TLRs, facilitating immediate immune stimulation, as highlighted in the literature [90]. The high encapsulation efficiency and antigen-loading capacity ensure that an adequate amount of antigen is delivered to target cells while preserving stability and enabling controlled release [15,91]. In one study, PLGA nanoparticles encapsulated with Angelica sinensis polysaccharide and ovalbumin showed an encapsulation efficiency of around 66.28%, which was effective in inducing a strong immune response [92]. PLGA nanoparticles have demonstrated high encapsulation efficiencies, making them suitable for vaccine delivery [92,93]. Nanoparticles with a high surface-to-volume ratio can encapsulate multiple antigens or adjuvants, providing opportunities for multivalent vaccine formulations. For instance, guanidinylated nanoparticles were able to load approximately 200 μg of ovalbumin per mg of polymer, which significantly enhanced the antigen-specific immune response [88]. Similarly, mesoporous silica nanoparticles (MSNs) have shown potential for high antigen loading, making them effective carriers for vaccine delivery [42]. Higher antigen loading can lead to a more robust immune response.

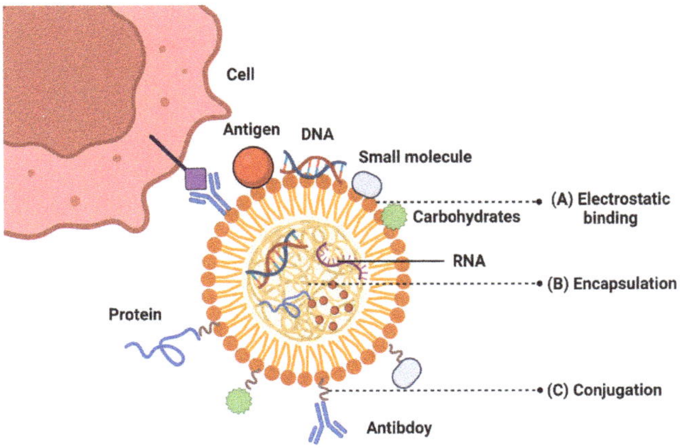

Figure 2. Schematic representation of nanocarriers. Different approaches can be utilized for the incorporation of targeting molecules (e.g., antigen, adjuvants, ligands, etc.) into/onto liposomes.

The degradation rate of nanoparticles and the release kinetics of the encapsulated antigen are crucial determinants of the duration of immune stimulation. Nanoparticles can be engineered to degrade at specific rates, ensuring a sustained release of the antigen. Nanoparticles made from biodegradable polymers like PLGA can provide sustained antigen release over days or weeks, ensuring prolonged immune activation and memory formation [35,94,95]. Similarly, silk fibroin nanoparticles demonstrated a continuous and slow release of the antigen over 56 days, which helped in maintaining a prolonged immune response [96]. By adjusting the polymer composition or incorporating additional stabilizing agents, it is possible to fine-tune the degradation and release profiles to match the desired immune response.

By carefully engineering these physicochemical properties, researchers can optimize nanoparticles to enhance antigen stability, improve cellular uptake, promote targeted delivery, and elicit desired immune responses. This rational design approach holds promise for developing next-generation TB vaccines that are safer, more effective, and capable of providing long-lasting protection against the disease.

4. Mechanisms of Improved Antigen Delivery and Immunogenicity of Nanoparticle-Based TB Vaccines

Currently, the majority of adjuvants effective in TB animal models, including those in clinical development, share a common feature: they combine the dual functionality of an immunomodulator, which activates pathogen-recognition receptors (PRRs), and a delivery system that ensures efficient vaccine targeting [97,98]. Nanoparticle-based adjuvants offer unique mechanisms that enhance the delivery and immunogenicity of TB vaccines, overcoming several limitations of traditional vaccine strategies. These mechanisms include targeted delivery to APCs, controlled and sustained antigen release, improved cross-presentation, and the ability to serve as multifunctional platforms for the co-delivery of antigens and additional adjuvants.

4.1. *Targeted Delivery and Uptake by Antigen-Presenting Cells*

Nanoparticles offer unique capabilities to enhance antigen delivery specifically to APCs, such as DCs and macrophages, thereby optimizing immune activation and antigen presentation. Antigen presentation by APCs is a crucial step that links innate and adaptive immunity and involves distinctive mechanisms [4].

Nanoparticles can be functionalized with surface ligands (e.g., mannose, antibodies, or peptides) that specifically bind to receptors on APCs, such as C-type lectin receptors, scavenger receptors, or DC-SIGN [99,100]. These modifications increase the affinity and specificity of nanoparticles for APCs, ensuring that the encapsulated antigen is preferentially delivered to these immune cells rather than being distributed nonspecifically throughout the body. For example, liposomes modified with appropriate targeting ligands activate cells via pattern-recognition receptors, leading to the maturation of APCs and efficient antigen processing and presentation [15]. Surface modifications of PLGA nanoparticles with biomaterials like mannose or cationic lipids, or the use of adjuvants, have improved antigen presentation to DCs, although this may come at the cost of reduced encapsulation efficiency, morphology, and yield [45,101].

The physicochemical properties of nanoparticles play a critical role in determining their interaction with APCs. Nanoparticles typically range from 20 to 200 nm in size, a range optimal for endocytosis and phagocytosis by APCs. Additionally, positively charged nanoparticles show enhanced uptake by APCs due to electrostatic interactions with the negatively charged cell membranes. For instance, the positively charged surface of cationic liposomes favors interactions with the negatively charged surface of DCs, facilitating antigen delivery and uptake [102]. These nanoparticles stimulate the maturation of DCs and the activation of macrophages, leading to a higher production of cytokines such as TNF-α, IL-6, and IL-10, which are essential for a strong immune response [88]. Additionally, cationic lipid nanoparticles have been formulated with triple adjuvants to enhance immune responses against respiratory pathogens, suggesting their potential for TB vaccination [103]. The surface charge and particle size of liposomes influence vaccine delivery by enhancing their ability to attract, interact with, and activate APCs, such as DCs, macrophages, and B cells [104]. Shape also influences uptake, as rod-shaped or discoidal nanoparticles have shown superior internalization compared to spherical particles in some studies.

Once inside APCs, nanoparticles can facilitate efficient antigen processing and presentation via both MHC class I and class II pathways. This dual presentation is crucial for the activation of both $CD8^+$ cytotoxic T cells and $CD4^+$ helper T cells, leading to a more comprehensive immune response [15,105,106]. Additionally, some nanoparticles can promote the escape of antigens from endosomes into the cytosol, further enhancing cross-presentation and the generation of a robust cellular response. Another study demonstrated that lipid nanoparticles could significantly enhance B cell and T cell responses to subunit vaccine antigens, indicating their potential as adjuvants in TB vaccines [107]. LNPs also enhance the uptake of antigens by APCs, leading to stronger $CD4^+$ and $CD8^+$ T cell responses [107].

By targeting delivery specifically to APCs, nanoparticles prevent premature degradation of the antigen in the extracellular environment. This targeted approach ensures that a higher proportion of the administered antigen reaches its intended cellular target, leading to a more effective and efficient immune activation. For instance, guanidinylated cationic nanoparticles were shown to encapsulate antigens effectively and release them in a controlled manner within the acidic lysosomal compartments of APCs. This targeted delivery not only protects the antigen from degradation but also enhances its uptake and presentation by APCs, leading to a more robust immune response [88]. Lipid-based nanoparticles were demonstrated to enhance the delivery of antigens to lymph nodes, where they are efficiently taken up by APCs [15].

4.2. Controlled and Sustained Release of Antigens

Nanoparticles can be engineered to control the release kinetics of encapsulated antigens, providing sustained exposure to the immune system over time. This prolonged release enhances the magnitude and duration of the immune response, which is critical for effective TB vaccination.

Encapsulation in nanoparticles protects antigens from harsh environmental conditions, such as low pH or enzymatic degradation, during storage or after administration. Polymers like PLGA and chitosan are commonly used to encapsulate antigens in nanoparticles. These polymers degrade slowly over time, releasing the encapsulated antigen in a controlled manner. This gradual degradation ensures a steady supply of antigen to APCs, mimicking the natural pathogen persistence and resulting in prolonged immune stimulation. PLGA possesses immunological properties, such as receptor recognition through cytokine stimulation and APCs, which protect vaccines from enzymatic degradation and offer greater immune safety compared to live bacterial carriers [108,109]. The use of lipid-based nanoparticles also allows for the encapsulation of antigens and adjuvants, providing a controlled release that enhances both humoral and cellular immune responses [15]. This protection ensures that the antigen remains in its native conformation, preserving its immunogenicity and enhancing its recognition by the immune system.

The release rate of antigens can be fine-tuned by altering the properties of the nanoparticles, such as polymer composition, molecular weight, and hydrophobicity. For example, increasing the hydrophobicity of the polymer matrix or cross-linking density can slow down the release rate, leading to prolonged antigen availability. MnO_2 nanoparticles not only deliver antigens but also act as a depot to release Mn^{2+} ions, which enhance the immune response through the stimulator of interferon genes (STING) pathway in DCs [110]. This feature is particularly important for generating memory responses, as it allows for continuous stimulation of the immune system without the need for multiple booster doses.

By providing a steady release of antigens, nanoparticles ensure sustained activation of APCs, which, in turn, continue to present antigens to T and B cells over an extended period. Particle replication in nonwetting template (PRINT) nanoparticles were shown to sustain prolonged antigen presentation to APCs, leading to a stronger and more sustained immune response compared to rapidly clearing soluble antigens [111]. This prolonged presentation increases the likelihood of forming long-lived memory T cells and effector T cells, leading to long-term protection against TB.

4.3. Multivalent and Multifunctional Nanoparticles

Nanoparticles can be designed to incorporate multiple antigens, adjuvants, or immunomodulatory molecules, thereby acting as multifunctional platforms capable of orchestrating complex immune responses [15,23,112]. This versatility enables nanoparticles to enhance the breadth and quality of the immune response.

Nanoparticles can encapsulate or conjugate multiple antigens that represent different stages or proteins of *M. tuberculosis* [45]. This multivalent approach ensures that the immune system is trained to recognize various targets on the pathogen, reducing the likelihood of immune escape and increasing vaccine efficacy [15]. For example, combining early secreted antigens (e.g., ESAT-6) with later-stage antigens (e.g., Ag85 complex) can generate a more comprehensive response capable of targeting both acute and latent TB infections [30].

Nanoparticles can be co-loaded with potent adjuvants, such as TLR agonists, cytokines, or small molecule immunomodulators, to further enhance immune responses [110,113]. For instance, lipid nanoparticles combined with a synthetic TLR9 agonist were shown to elicit stronger Th1-type responses and multifunctional $CD8^+$ T cell responses compared to unadjuvanted vaccines [107]. These adjuvants can activate innate immune pathways, promote APC maturation, and induce the production of pro-inflammatory cytokines, thereby boosting the overall immune response to the antigen. Co-delivery of antigens and adjuvants within the same nanoparticle ensures that these signals are presented simultaneously to APCs, leading to synergistic effects and optimal immune activation [114, 115]. Self-immolative micelles can co-deliver antigens and adjuvants to DCs, leading to the release of pristine antigens and the upregulation of costimulatory molecules, which are essential for a strong cellular immune response [116].

The surface of nanoparticles can be modified with pathogen-associated molecular patterns (PAMPs) or other targeting moieties to mimic natural pathogen structures. PAMPs are critical components in the design of nanoparticle-based adjuvants due to their ability to activate pattern-recognition receptors on immune cells [117,118]. The incorporation of multiple PAMPs into nanoparticles can mimic the natural infection process, leading to a more potent immune response [119,120]. This functionalization enhances the ability of nanoparticles to activate innate immune receptors, such as TLRs or NOD-like receptors (NLRs), resulting in stronger and more rapid immune responses. For instance, incorporating lipopolysaccharide or CpG motifs on the nanoparticle surface can activate TLR4 and TLR9 pathways [121–123], respectively, promoting a Th1-biased response crucial for TB protection. Similarly, nanoparticles co-loaded with CpG oligodeoxynucleotide (a TLR9 agonist) and 3pRNA (a RIG-I agonist) were shown to significantly enhance APC activation and antigen cross-presentation, resulting in strong humoral and cytotoxic T lymphocyte responses [115].

By delivering a combination of antigens and adjuvants that activate different immune pathways, multivalent nanoparticles can elicit both Th1 and Th17 responses, which are essential for protection against TB. Th1 responses promote IFN-γ production and macrophage activation, while Th17 responses enhance mucosal immunity and neutrophil recruitment [16,19,124]. Aluminum hydroxide nanoparticles co-loaded with CpG oligodeoxynucleotide (CpG-ODN) and 3pRNA activate TLR9 and retinoic acid-inducible gene I (RIG-I), respectively. This dual activation leads to robust humoral and cellular immune responses, including strong cytotoxic T lymphocyte (CTL) responses and IFN-γ secretion, which are crucial for effective TB vaccination [115]. This balanced activation is necessary for clearing both intracellular and extracellular *M. tuberculosis* bacilli.

The ability to deliver multiple components in a controlled manner and target specific immune pathways makes nanoparticle-based TB vaccines a versatile and powerful tool for inducing robust and long-lasting immunity against *M. tuberculosis*. Figure 3 shows these features, positioning them as promising candidates to enhance current TB vaccination strategies and improve global health outcomes.

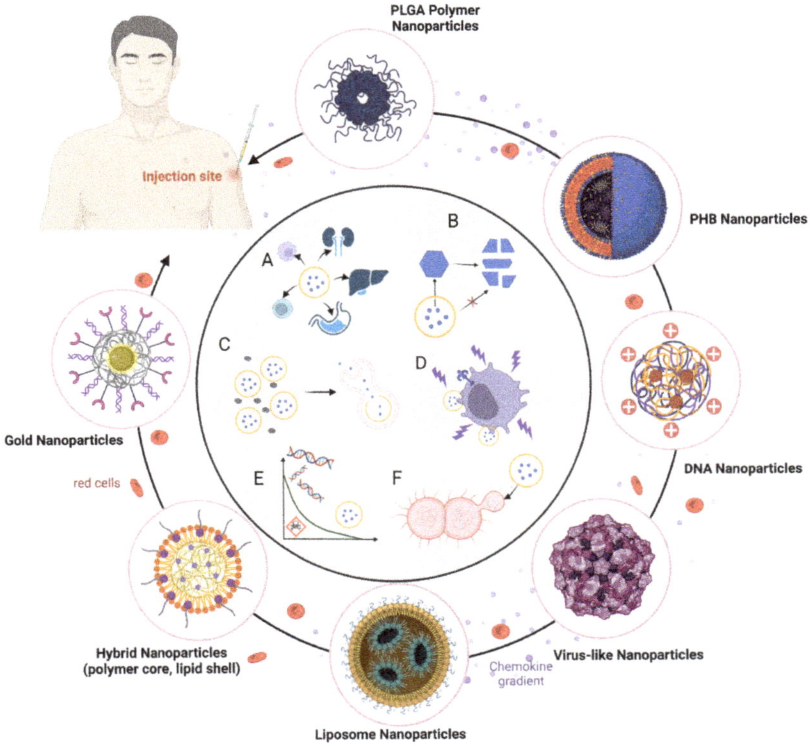

Figure 3. Characteristics of vaccine delivery systems with different nanocarriers in TB vaccines. A: target antigen to specific cells/organs. B: protect the antigen from degradation. C: slow release of antigen. D: enhance antigen uptake/presentation. E: dampen PAMP systemic toxicity. F: direct stimulatory effect of the APC or adjacent cells.

5. Advantages of Nanoparticle-Based Adjuvants in TB Vaccination

Nanoparticle-based adjuvants in TB vaccines enhance immune protection by inducing strong cellular and humoral responses, which are crucial for effective, long-term immunity against tuberculosis. They also improve the stability of vaccines and protect antigens from degradation, ensuring efficient delivery to target cells. Additionally, these adjuvants allow targeted delivery and controlled release, which helps minimize adverse effects and enables reduced antigen doses without sacrificing effectiveness. This combination of benefits makes nanoparticle-based adjuvants a promising approach for advancing TB vaccination strategies.

5.1. Induction of Strong Cellular and Humoral Immunity

Nanoparticle-based adjuvants have the capability to simultaneously stimulate robust T cell-mediated (Th1/Th17) and B cell-mediated responses, which are crucial for effective protection against *M. tuberculosis*. Unlike traditional adjuvants, which may primarily induce one type of immune response, nanoparticles provide a more balanced and comprehensive activation of both cellular and humoral immunity (Figure 4).

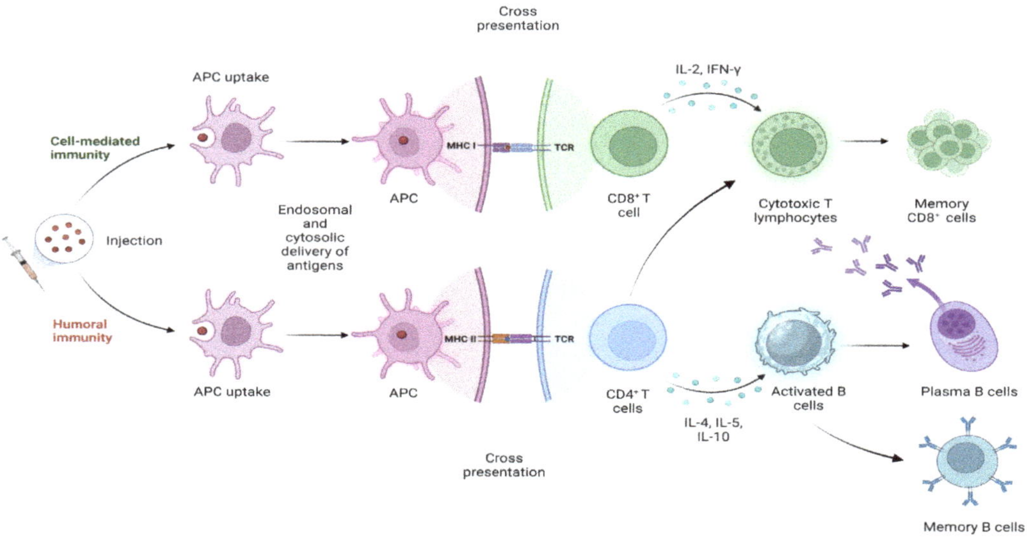

Figure 4. Different nanoparticle-based TB vaccine delivery systems function by activating cellular and humoral immunity to varying degrees.

Nanoparticles enhance antigen uptake and processing by APCs, leading to the efficient presentation of antigens via both MHC class I and II pathways [125–127]. This dual pathway activation is essential for the differentiation of $CD4^+$ T helper cells into Th1 and Th17 subsets and the activation of $CD8^+$ cytotoxic T cells. Th1 cells produce cytokines such as IFN-γ and TNF-α, which activate macrophages to kill intracellular *M. tuberculosis*, while Th17 cells secrete IL-17, promoting mucosal immunity and recruitment of neutrophils to the site of infection [15]. Polypropylene sulfide nanoparticles conjugated with antigens were shown to enhance both MHC I and MHC II presentation pathways, leading to robust activation of both $CD8^+$ and $CD4^+$ T cells [126,128]. This comprehensive activation of T cells provides enhanced control over both acute and latent TB infections. Previous studies showed that TB vaccines encapsulating plasmid DNA and immunostimulants in PLGA nanoparticles induce higher levels of interferon-gamma (IFN-γ) and immunoglobulin (Ig) G2a antibodies in mice via injection compared to naked plasmid-DNA vaccines, indicating a strong Th1 polarization-directed response [129,130]. Furthermore, chitosan nanoparticles co-administered with mucosal adjuvants were found to stimulate strong Th1/Th17 responses and induce $CD8^+$ T cells, which are essential for robust cellular immunity [131].

Although the role of humoral immunity in TB protection is less prominent compared to cellular immunity, antibodies can neutralize extracellular bacilli and prevent their dissemination. Nanoparticles enhance B cell activation by facilitating the delivery of antigens to follicular dendritic cells in lymphoid organs, promoting the formation of germinal centers where B cell maturation occurs [132,133]. This process leads to the generation of high-affinity antibodies and long-lived plasma cells, contributing to a strong and sustained humoral response [134]. The mechanisms by which nanoparticles enhance B cell responses and antibody production are multifaceted. One key mechanism is the ability of nanoparticles to target lymph nodes, where B cells are activated. For instance, PEGylated lipid nanoparticles encapsulating cyclic di-GMP (cdGMP) were shown to redirect the adjuvant to draining lymph nodes (dLNs), leading to increased germinal center B cell differentiation and higher antibody titers [113]. This targeted delivery ensures that the antigens and adjuvants are presented in an optimal environment for B cell activation and differentiation. Additionally, nanoparticles can modulate the immune response by activating specific pathways.

By providing prolonged antigen exposure and enhanced stimulation of APCs, nanoparticle-based adjuvants promote the formation of memory T and B cells [17,95,135]. These memory cells are crucial for long-term protection and rapid immune responses upon re-exposure to the pathogen, ensuring lasting immunity against TB. Nanoparticles can be engineered to target lymph nodes, where immune responses are initiated and memory cells are formed. For example, PEGylated lipid nanoparticles encapsulating cyclic dinucleotides (CDNs) were shown to enhance the accumulation of adjuvants in draining lymph nodes, leading to increased $CD8^{+}$ T cell responses and durable antibody titers [113]. Certain nanoparticle formulations can induce tissue-resident memory T cells (TRM), which provide long-term protection at mucosal and barrier sites such as the lungs [136].

5.2. Enhanced Vaccine Stability and Antigen Protection

Nanoparticles offer significant advantages in terms of antigen stability and protection, ensuring that the vaccine remains potent and effective under various environmental conditions.

Encapsulation of antigens within nanoparticles shields them from enzymatic degradation by proteases and other enzymes present in the body. This protection is particularly important for peptide or protein antigens that are susceptible to rapid breakdown [114,137,138]. By preserving the integrity and conformation of the antigen, nanoparticles ensure that the antigen remains bioactive and capable of eliciting a strong immune response [18]. The encapsulation of antigens in PLGA nanoparticles was shown to increase the in vivo stability and lymph node accumulation of the antigenic peptide, thereby enhancing the overall efficacy of the vaccine [137]. Additionally, encapsulation within nanoparticles prevents antigen aggregation and the formation of inactive multimers, which can occur during storage or under harsh conditions [138]. This reduction in aggregation ensures that the antigen maintains its structural and functional properties, leading to a more consistent and effective immune response.

Nanoparticle-based formulations are more stable under varying pH and temperature conditions compared to traditional adjuvants or free antigen formulations [136,139]. This stability reduces the risk of antigen denaturation during storage and transport, making nanoparticle-based vaccines more suitable for use in diverse geographical regions, including areas with limited cold chain infrastructure [138,140]. One of the significant benefits of nanoparticle-based TB vaccines is their enhanced thermostability. For instance, spray drying has been used to encapsulate a TB subunit vaccine containing a nanoemulsion adjuvant into dry powder microparticles. This method has shown that the powder form of the vaccine remains physically stable for up to 26 months at temperatures as high as 40 °C. The integrity of the adjuvant system is maintained at temperatures up to 25 °C for 26 months and even after one month at 40 °C, demonstrating improved antigen thermostability compared to liquid formulations [138]. The use of cationic solid lipid nanoparticles (cSLNs) was demonstrated to enhance the stability of antigens, as evidenced by an increase in the transition temperature related to antigen dissociation, indicating improved stability [141].

Nanoparticles can be engineered to release their antigen payloads in a controlled and sustained manner, providing continuous stimulation to the immune system. This sustained release mimics the natural course of infection, where the immune system is exposed to antigens over a prolonged period. The result is a more effective induction of both primary and memory immune responses, which are essential for long-term protection against TB [103,112,142]. Additionally, targeted delivery to specific tissues or cells can reduce the exposure of the vaccine components to degrading environmental factors. This approach has been demonstrated in various studies, including the use of PLGA nanoparticles and lipid-based nanoparticles for TB vaccination [16,19].

5.3. Reduction of Adverse Effects

One of the key challenges in vaccine development is minimizing adverse effects while maintaining strong immunogenicity. By targeting antigens specifically to APCs or other immune cells, nanoparticles minimize off-target effects and reduce systemic exposure [17]. This targeted approach reduces the likelihood of nonspecific immune activation and systemic inflammatory responses, which are common with traditional adjuvants. Nanoparticle-based TB vaccines offer the advantage of precise targeting, which significantly reduces systemic distribution and associated adverse effects. Encapsulating adjuvants like cyclic dinucleotides (CDNs) within PEGylated lipid nanoparticles redirects the adjuvant to draining lymph nodes (dLNs) rather than the bloodstream, thereby minimizing systemic inflammation [113]. Similarly, synthetic vaccine particles (SVPs) co-encapsulating antigens and adjuvants result in strong local immune activation without inducing systemic cytokine release, thus reducing potential adverse events [143].

Nanoparticles enhance the potency of encapsulated antigens and adjuvants, allowing for lower doses to achieve the desired immune response. The use of PLGA:DDA hybrid nanoparticles for nasal delivery of TB antigens demonstrated efficient immune responses with lower doses, which can reduce the risk of adverse effects associated with higher doses [45]. This dose-sparing effect reduces the risk of local reactogenicity (e.g., pain, redness, or swelling at the injection site) and systemic side effects (e.g., fever, fatigue) associated with high-dose vaccinations.

Traditional adjuvants such as alum or oil-in-water emulsions can cause strong local inflammatory reactions. Nanoparticles, on the other hand, can be engineered to minimize local reactogenicity by modulating their surface properties, composition, and release kinetics. The co-delivery of antigens with adjuvants in synthetic polymer nanoparticles was shown to result in minimal systemic production of inflammatory cytokines, thereby reducing potential adverse events and safety concerns [143]. Using biocompatible and biodegradable materials like PLGA or chitosan can reduce local irritation and enhance biocompatibility, making nanoparticle-based vaccines more tolerable. The use of biocompatible materials such as chitosan for nanoparticle formulation ensures that the vaccine is nontoxic and well tolerated, further minimizing adverse effects [131].

The ability of nanoparticles to co-deliver antigens and immune-stimulatory molecules in a controlled manner prevents excessive immune activation, which can lead to adverse effects such as cytokine storms or autoimmune reactions. By carefully tuning the composition and release profile of nanoparticles, it is possible to achieve a balanced immune activation that is both potent and safe. The use of cationic lipid nanoparticles with a triple adjuvant system for intranasal vaccination against respiratory pathogens was shown to favor a Th1 type of response, promoting higher IgG2a and IgA serum antibody titers without causing excessive immune activation [103]. Additionally, the modular self-assembling nanoparticle platform used in SVPs enables strong local immune activation and balanced humoral and cellular responses without systemic overstimulation [143].

These advantages make nanoparticle-based adjuvants a promising platform for developing safer and more effective TB vaccines, addressing the limitations of traditional adjuvants and enhancing the overall efficacy and safety profile of TB vaccination strategies.

6. Current Technical and Manufacturing Challenges in the Development of Nanoparticle-Based TB Vaccines

Developing nanoparticle-based TB vaccines at a large scale presents numerous technical and manufacturing challenges that must be addressed to ensure the consistency and reproducibility of formulations [144]. While many nanoparticle formulations have shown promise at the laboratory scale, translating these methods to industrial-scale production is complex. This complexity arises due to the need for precise control over particle size, shape, surface characteristics, and encapsulation efficiency [51,145]. Small changes in manufacturing parameters can result in variations in these properties, which can significantly impact vaccine performance and safety [30]. Scaling up the production while maintaining these critical quality attributes requires sophisticated equipment and process optimization.

Consistency between different batches is crucial for regulatory approval and clinical application. Even minor variations in raw materials, environmental conditions, or processing steps can lead to inconsistencies in nanoparticle size, surface charge, or antigen loading, affecting the vaccine's immunogenicity and safety [146,147]. Developing robust manufacturing protocols and in-process control measures is essential to ensure that every batch of nanoparticle-based vaccines meets stringent quality requirements. For instance, the engineering of lipid–polymer hybrid nanoparticles requires precise control over several critical process parameters (CPPs) such as acetone concentration, stabilizer concentration, and lipid-to-total solid ratio to achieve consistent particle size, polydispersity index (PDI), and zeta potential [19]. Variations in these parameters can significantly affect the immunogenicity and stability of the vaccine.

Developing multifunctional nanoparticles that can co-deliver multiple antigens and adjuvants in a controlled manner increases the complexity of the formulation. These delivery systems often require multistep processes, sophisticated techniques (e.g., emulsion–evaporation or microfluidic methods), and strict environmental controls, all of which add to the cost and difficulty of large-scale production [51]. For example, the development of mesoporous silica nanoparticles as vaccine carriers involves optimizing synthesis, adsorption, and release kinetics to achieve effective antigen delivery and immune response [42]. Nanoparticle-based formulations are often sensitive to environmental conditions such as temperature and humidity. This sensitivity can lead to issues related to the long-term stability of the vaccine, complicating storage and distribution, particularly in resource-limited settings [34,148]. Formulating nanoparticles to enhance their stability under a wide range of conditions or developing appropriate stabilization techniques (e.g., lyophilization) is necessary to ensure that the vaccines remain effective throughout their shelf life.

7. Future Perspectives and Transformative Potential

Nanoparticle-based adjuvants have the potential to transform TB vaccination by enhancing both efficacy and safety, allowing precise antigen delivery and minimizing side effects. Some nanoparticle-based TB vaccines, including RUTI and ID93 + GLA-SE, have entered preclinical and clinical trials [149–153], exhibiting favorable safety profiles and encouraging immunogenicity results (Table 3). These adjuvants also pave the way for effective mucosal TB vaccines, which target the respiratory tract directly, providing localized immunity at the primary infection site. Furthermore, integrating nanoparticle-based adjuvants into global TB control strategies could enhance vaccine accessibility and impact, supporting worldwide TB eradication efforts with a focus on long-lasting, robust immunity.

Table 3. Selected clinical trials of nanoparticle-based TB vaccines.

Vaccine	Nanoparticle Type	Antigen	Trial	Key Findings
RUTI [149,150]	Lipid-based NPs	Heat-inactivated Mtb fragments	Phase II	- Strong immune response; - Good safety profile; - Potential to reduce TB reactivation risk.
ID93 + GLA-SE [149,151,152]	Glucopyranosyl lipid A (GLA-SE)	ID93 (fusion of four Mtb antigens)	Phase I/II	- Robust Th1 response; - Improved antigen stability and immunogenicity; - Safe with manageable adverse effects.
H56:IC31 [153,154]	Peptide NPs with adjuvant	H56 (ESAT-6, Ag85B, Rv2660c)	Phase I/II	- Robust Th1-biased immunity; - Potential as a booster vaccine or adjunct to TB treatment.
M72/AS01E [155–157]	Liposomal NPs	M72 (fusion of Rv1196 and Rv0125)	Phase IIb/III	- ~50% efficacy in preventing active TB (adults); - Strong humoral and cellular immune responses.

Mtb: Mycobacterium tuberculosis; NPs: nanoparticles.

7.1. Innovative Approaches for Enhancing Efficacy and Safety

The future perspectives for nanoparticle-based TB vaccines are highly promising due to ongoing advancements in nanotechnology and vaccine design. Innovations in precision medicine enable the development of personalized nanoparticle vaccines, which can be tailored to an individual's immune profile for a more effective response [158]. Smart-release systems using nanoparticles can be designed to respond to environmental cues like pH, allowing for precise antigen delivery and timing [159,160]. Moreover, emerging nanoparticle platforms can co-deliver multiple antigens or combine them with adjuvants, leading to a more robust immune response [161,162]. Hybrid nanoparticles and biomimetic approaches offer improved antigen delivery by mimicking natural pathogens and overcoming TB's complex life cycle [19]. These advancements have the potential to transform TB vaccine strategies, enhancing both their efficacy and safety.

7.2. Potential for Mucosal TB Vaccines

Mucosal vaccination, particularly targeting the respiratory tract where *M. tuberculosis* initially infects, is a promising area for nanoparticle-based TB vaccine development [163]. The nasal cavity offers several advantages, including better patient compliance, improved epithelium permeability, and the potential for needle-free self-administration [164,165]. While mucosal vaccination holds significant promise for providing physiological and immunological protection against *M. tuberculosis*, safety concerns related to adjuvants and administration routes must be considered [70,166]. Evidence suggests that mucosal vaccination via the respiratory tract offers superior immune protection against pathogenic bacteria compared to other administration sites [167,168]. Additionally, PLGA/PLA nanoparticles administered intranasally or intratracheally have shown effectiveness in mediating protective immune responses against respiratory bacterial infections [169,170].

Mucosal administration of nanoparticle-based vaccines (e.g., intranasal or aerosol delivery) can induce strong local immunity at the primary site of TB infection [171]. Localized immunity at the respiratory mucosa is crucial for preventing the establishment of infection and controlling bacterial replication early in the infection process. Amini and colleagues developed a trimethyl chitosan (TMC)-based vaccine candidate for intranasal administration. These TMC nanoparticles encapsulate the antigen and release it gradually, enhancing uptake by immune cells in the nasal-associated lymphoid tissue and triggering a targeted immune response in the respiratory mucosa [172].

Mucosal vaccines are generally noninvasive and can be self-administered, making them more acceptable to patients and easier to deploy in large-scale vaccination campaigns. Mucosal nanoparticle-based vaccines can be used as booster vaccines in individuals previously vaccinated with BCG or other TB vaccines [18,173,174]. This strategy can enhance local immunity and provide an additional layer of protection without the need for multiple systemic doses.

7.3. Integration with Global TB Control Strategies

Nanoparticle-based TB vaccines hold significant promise in advancing global TB control efforts, aligning with the WHO's End TB Strategy and Sustainable Development Goals (SDGs). By offering improved protection and reduced adverse effects and targeting latent TB infections, these vaccines could play a crucial role in achieving the End TB Strategy's targets of reducing TB incidence by 90% and TB deaths by 95% by 2035 [175]. Additionally, South African researchers have advanced a nanoparticle-based vaccine formulation to clinical trials, utilizing a synthetic nanoemulsion adjuvant, GLA-SE, combined with the TB antigen ID93 [176]. The development of nanoparticle-based TB vaccines provides opportunities for global collaboration in research and development. Such collaborative efforts can accelerate the progress of TB vaccine research, foster innovation, and encourage the sharing of knowledge across borders.

8. Conclusions

In summary, nanoparticle-based adjuvants represent a promising advancement in TB vaccine development by offering improved antigen delivery, enhanced immune activation, and sustained antigen release, leading to potent cellular and humoral responses crucial for TB protection. Despite challenges such as manufacturing scalability, regulatory concerns, and accessibility in low-resource settings, these innovative platforms hold transformative potential. With further research and strategic implementation, nanoparticle-based vaccines could significantly contribute to global TB control efforts, especially through novel approaches like mucosal vaccination, ultimately supporting the WHO's goal of eradicating TB.

Author Contributions: Conceptualization, J.W., J.S. and J.C.-W.C.; methodology, J.W., D.W.-C.W. and J.C.-W.C.; validation, Q.W. and J.S.; formal analysis, J.W.; investigation, J.W., Z.Z. and Q.W.; data curation: J.W., Z.Z. and Q.W.; writing—original draft preparation, J.W.; writing—review and editing, J.S., D.W.-C.W. and J.C.-W.C.; visualization, J.W., Z.Z. and Q.W.; supervision, D.W.-C.W. and J.C.-W.C.; project administration, D.W.-C.W. and J.C.-W.C.; funding acquisition, J.C.-W.C. All authors have read and agreed to the published version of the manuscript.

Funding: This research received no external funding.

Institutional Review Board Statement: Not applicable.

Informed Consent Statement: Not applicable.

Data Availability Statement: No new data were created or analyzed in this study. Data sharing is not applicable to this article.

Conflicts of Interest: The authors declare no conflicts of interest.

References

1. Ghebreyesus, T.A.; Lima, N.T. The TB Vaccine Accelerator Council: Harnessing the power of vaccines to end the tuberculosis epidemic. *Lancet Infect. Dis.* **2023**, *23*, 1222–1223. [CrossRef] [PubMed]
2. Zaman, K. Tuberculosis: A global health problem. *J. Health Popul. Nutr.* **2010**, *28*, 111. [CrossRef] [PubMed]
3. Reid, M.J.; Arinaminpathy, N.; Bloom, A.; Bloom, B.R.; Boehme, C.; Chaisson, R.; Chin, D.P.; Churchyard, G.; Cox, H.; Ditiu, L. Building a tuberculosis-free world: The Lancet Commission on tuberculosis. *Lancet* **2019**, *393*, 1331–1384. [CrossRef] [PubMed]
4. Bekale, R.B.; Du Plessis, S.-M.; Hsu, N.-J.; Sharma, J.R.; Sampson, S.L.; Jacobs, M.; Meyer, M.; Morse, G.D.; Dube, A. Mycobacterium Tuberculosis and Interactions with the Host Immune System: Opportunities for Nanoparticle Based Immunotherapeutics and Vaccines. *Pharm. Res.* **2018**, *36*, 8. [CrossRef] [PubMed]
5. Marks, S.M.; Flood, J.; Seaworth, B.; Hirsch-Moverman, Y.; Armstrong, L.; Mase, S.; Salcedo, K.; Oh, P.; Graviss, E.A.; Colson, P.W. Treatment practices, outcomes, and costs of multidrug-resistant and extensively drug-resistant tuberculosis, United States, 2005–2007. *Emerg. Infect. Dis.* **2014**, *20*, 812. [CrossRef] [PubMed]
6. Mullerpattan, J.B.; Udwadia, Z.Z.; Banka, R.A.; Ganatra, S.R.; Udwadia, Z.F. Catastrophic costs of treating drug resistant TB patients in a tertiary care hospital in India. *Indian J. Tuberc.* **2019**, *66*, 87–91. [CrossRef] [PubMed]
7. Abate, G.; Hoft, D.F. Immunotherapy for tuberculosis: Future prospects. *ImmunoTargets Ther.* **2016**, *5*, 37–45.
8. Puvačić, S.; Dizdarević, J.; Santić, Z.; Mulaomerović, M. Protective effect of neonatal BCG vaccines against tuberculous meningitis. *Biomol. Biomed.* **2004**, *4*, 46–49. [CrossRef] [PubMed]
9. Schrager, L.K.; Vekemens, J.; Drager, N.; Lewinsohn, D.M.; Olesen, O.F. The status of tuberculosis vaccine development. *Lancet Infect. Dis.* **2020**, *20*, e28–e37. [CrossRef]
10. Lowrie, D.B.; Tascon, R.E.; Colston, M.J.; Silva, C.L. Towards a DNA vaccine against tuberculosis. *Vaccine* **1994**, *12*, 1537–1540. [CrossRef]
11. Cable, J.; Srikantiah, P.; Crowe Jr, J.E.; Pulendran, B.; Hill, A.; Ginsberg, A.; Koff, W.; Mathew, A.; Ng, T.; Jansen, K. Vaccine innovations for emerging infectious diseases—A symposium report. *Ann. N. Y. Acad. Sci.* **2020**, *1462*, 14–26. [CrossRef] [PubMed]
12. Bastola, R.; Noh, G.; Keum, T.; Bashyal, S.; Seo, J.-E.; Choi, J.; Oh, Y.; Cho, Y.; Lee, S. Vaccine adjuvants: Smart components to boost the immune system. *Arch. Pharmacal Res.* **2017**, *40*, 1238–1248. [CrossRef] [PubMed]
13. Firdaus, F.Z.; Skwarczynski, M.; Toth, I. Developments in vaccine adjuvants. In *Vaccine Design: Methods and Protocols, Volume 3. Resources for Vaccine Development*; Humana: New York, NY, USA,, 2022; pp. 145–178.
14. Gupta, A.; Das, S.; Schanen, B.; Seal, S. Adjuvants in micro-to nanoscale: Current state and future direction. *Wiley Interdiscip. Rev. Nanomed. Nanobiotechnol.* **2016**, *8*, 61–84. [CrossRef] [PubMed]
15. Chatzikleanthous, D.; O'Hagan, D.T.; Adamo, R. Lipid-Based Nanoparticles for Delivery of Vaccine Adjuvants and Antigens: Toward Multicomponent Vaccines. *Mol. Pharm.* **2021**, *18*, 2867–2888. [CrossRef] [PubMed]
16. Ballester, M.; Nembrini, C.; Dhar, N.; de Titta, A.; de Piano, C.; Pasquier, M.; Simeoni, E.; van der Vlies, A.J.; McKinney, J.D.; Hubbell, J.A.; et al. Nanoparticle conjugation and pulmonary delivery enhance the protective efficacy of Ag85B and CpG against tuberculosis. *Vaccine* **2011**, *29*, 6959–6966. [CrossRef]
17. Hart, P.; Copland, A.; Diogo, G.R.; Harris, S.; Spallek, R.; Oehlmann, W.; Singh, M.; Basile, J.; Rottenberg, M.; Paul, M.J.; et al. Nanoparticle-Fusion Protein Complexes Protect against Infection. *Mol. Ther.* **2018**, *26*, 822–833. [CrossRef]
18. Neto, L.M.M.; Zufelato, N.; de Sousa, A.A.; Trentini, M.M.; da Costa, A.C.; Bakuzis, A.F.; Kipnis, A.; Junqueira-Kipnis, A.P. Specific T cell induction using iron oxide based nanoparticles as subunit vaccine adjuvant. *Hum. Vaccines Immunother.* **2018**, *14*, 2786–2801. [CrossRef]

19. Rose, F.; Wern, J.E.; Ingvarsson, P.T.; van de Weert, M.; Andersen, P.; Follmann, F.; Foged, C. Engineering of a novel adjuvant based on lipid-polymer hybrid nanoparticles: A quality-by-design approach. *J. Control. Release* **2015**, *210*, 48–57. [CrossRef]
20. Collins, K.A.; Snaith, R.; Cottingham, M.G.; Gilbert, S.C.; Hill, A.V. Enhancing protective immunity to malaria with a highly immunogenic virus-like particle vaccine. *Sci. Rep.* **2017**, *7*, 46621. [CrossRef]
21. González-Miró, M.; Rodríguez-Noda, L.M.; Fariñas-Medina, M.; Cedré-Marrero, B.; Madariaga-Zarza, S.; Zayas-Vignier, C.; Hernández-Cedeño, M.; Kleffmann, T.; García-Rivera, D.; Vérez-Bencomo, V. Bioengineered polyester beads co-displaying protein and carbohydrate-based antigens induce protective immunity against bacterial infection. *Sci. Rep.* **2018**, *8*, 1888. [CrossRef]
22. Barik, A.; Dey, S. Nanotechnology-Based Vaccination for TB. In *Infectious Diseases Drug Delivery Systems*; Shegokar, R., Pathak, Y., Eds.; Springer International Publishing: Cham, Switzerland, 2023; pp. 147–168.
23. Lin, L.C.-W.; Chattopadhyay, S.; Lin, J.-C.; Hu, C.-M.J. Advances and Opportunities in Nanoparticle-and Nanomaterial-Based Vaccines against Bacterial Infections. *Adv. Healthc. Mater.* **2018**, *7*, 1701395. [CrossRef]
24. Moon, J.J.; Huang, B.; Irvine, D.J. Engineering nano-and microparticles to tune immunity. *Adv. Mater.* **2012**, *24*, 3724–3746. [CrossRef] [PubMed]
25. Chattopadhyay, S.; Chen, J.-Y.; Chen, H.-W.; Hu, C.-M.J. Nanoparticle vaccines adopting virus-like features for enhanced immune potentiation. *Nanotheranostics* **2017**, *1*, 244. [CrossRef] [PubMed]
26. Ojha, R.; Pandey, R.K.; Prajapati, V.K. Vaccine delivery systems against tuberculosis. In *Nanotechnology Based Approaches for Tuberculosis Treatment*; Elsevier: Amsterdam, The Netherlands, 2020; pp. 75–90.
27. Narisetty, S. Inulin Acetate Nanoparticles as Vaccine Adjuvants. *FASEB J.* **2017**, *31*, 938.5. [CrossRef]
28. Oyewumi, M.O.; Kumar, A.; Cui, Z. Nano-microparticles as immune adjuvants: Correlating particle sizes and the resultant immune responses. *Expert Rev. Vaccines* **2010**, *9*, 1095–1107. [CrossRef] [PubMed]
29. Jia, J.; Zhang, W.; Liu, Q.; Yang, T.; Wang, L.; Ma, G. Adjuvanticity regulation by biodegradable polymeric nano/microparticle size. *Mol. Pharm.* **2017**, *14*, 14–22. [CrossRef] [PubMed]
30. Yu, F.; Wang, J.; Dou, J.; Yang, H.; He, X.; Xu, W.; Zhang, Y.; Hu, K.; Gu, N. Nanoparticle-based adjuvant for enhanced protective efficacy of DNA vaccine Ag85A-ESAT-6-IL-21 against Mycobacterium tuberculosis infection. *Nanomed. Nanotechnol. Biol. Med.* **2012**, *8*, 1337–1344. [CrossRef]
31. Nisini, R.; Poerio, N.; Mariotti, S.; De Santis, F.; Fraziano, M. The Multirole of Liposomes in Therapy and Prevention of Infectious Diseases. *Front. Immunol.* **2018**, *9*, 155. [CrossRef]
32. Schwendener, R.A.; Ludewig, B.; Cerny, A.; Engler, O. Liposome-based vaccines. In *Liposomes: Methods and Protocols, Volume 1: Pharmaceutical Nanocarriers*; Humana Press: Totowa, NJ, USA, 2010; pp. 163–175.
33. Vieira, A.C.C.; Chaves, L.L.; Pinheiro, S.; Pinto, S.; Pinheiro, M.; Lima, S.C.; Ferreira, D.; Sarmento, B.; Reis, S. Mucoadhesive chitosan-coated solid lipid nanoparticles for better management of tuberculosis. *Int. J. Pharm.* **2018**, *536*, 478–485. [CrossRef] [PubMed]
34. Kramer, R.M.; Archer, M.C.; Orr, M.T.; Dubois Cauwelaert, N.; Beebe, E.A.; Huang, P.D.; Dowling, Q.M.; Schwartz, A.M.; Fedor, D.M.; Vedvick, T.S.; et al. Development of a thermostable nanoemulsion adjuvanted vaccine against tuberculosis using a design-of-experiments approach. *Int. J. Nanomed.* **2018**, *13*, 3689–3711. [CrossRef]
35. Khademi, F.; Sahebkar, A.; Fasihi-Ramandi, M.; Taheri, R.A. Induction of strong immune response against a multicomponent antigen of Mycobacterium tuberculosis in BALB/c mice using PLGA and DOTAP adjuvant. *Apmis* **2018**, *126*, 509–514. [CrossRef]
36. Malik, A.; Gupta, M.; Mani, R.; Bhatnagar, R. Single-dose Ag85B-ESAT6–loaded poly (lactic-co-glycolic acid) nanoparticles confer protective immunity against tuberculosis. *Int. J. Nanomed.* **2019**, *14*, 3129–3143. [CrossRef]
37. Karimi, S.M.; Sankian, M.; Khademi, F.; Tafaghodi, M. Chitosan (CHT) and trimethylchitosan (TMC) nanoparticles as adjuvant/delivery system for parenteral and nasal immunization against Mycobacterium tuberculosis (MTb) ESAT-6 antigen. *Nanomed. J.* **2016**, *3*, 223–229.
38. Lin, Y.H.; Sun, B.N.; Jin, Z.; Zhao, K. Enhanced Immune Responses to Mucosa by Functionalized Chitosan-Based Composite Nanoparticles as a Vaccine Adjuvant for Intranasal Delivery. *ACS Appl. Mater. Interfaces* **2022**, *14*, 52691–52701. [CrossRef]
39. Toraskar, S.; Madhukar Chaudhary, P.; Kikkeri, R. The shape of nanostructures encodes immunomodulation of carbohydrate antigen and vaccine development. *ACS Chem. Biol.* **2022**, *17*, 1122–1130. [CrossRef]
40. Li, L.; Yan, X.; Xia, M.; Shen, B.; Cao, Y.T.; Wu, X.Y.; Sun, J.W.; Zhang, Y.; Zhang, M.C. Nanoparticle/Nanocarrier Formulation as an Antigen: The Immunogenicity and Antigenicity of Itself. *Mol. Pharm.* **2022**, *19*, 148–159. [CrossRef]
41. Neto, L.M. Kipnis a, Junqueira-Kipnis AP. Role of metallic nanoparticles in vaccinology: Implications for infectious disease vaccine development. *Front. Immunol.* **2017**, *8*, 239.
42. Mody, K.T.; Popat, A.; Mahony, D.; Cavallaro, A.S.; Yu, C.; Mitter, N. Mesoporous silica nanoparticles as antigen carriers and adjuvants for vaccine delivery. *Nanoscale* **2013**, *5*, 5167–5179. [CrossRef]
43. Skrastina, D.; Petrovskis, I.; Lieknina, I.; Bogans, J.; Renhofa, R.; Ose, V.; Dishlers, A.; Dekhtyar, Y.; Pumpens, P. Silica Nanoparticles as the Adjuvant for the Immunisation of Mice Using Hepatitis B Core Virus-Like Particles. *PLoS ONE* **2014**, *9*, e114006. [CrossRef]
44. Behzadi, M.; Vakili, B.; Ebrahiminezhad, A.; Nezafat, N. Iron nanoparticles as novel vaccine adjuvants. *Eur. J. Pharm. Sci.* **2021**, *159*, 105718. [CrossRef]
45. Khademi, F.; Derakhshan, M.; Yousefi-Avarvand, A.; Najafi, A.; Tafaghodi, M. A novel antigen of *Mycobacterium tuberculosis* and MPLA adjuvant co-entrapped into PLGA:DDA hybrid nanoparticles stimulates mucosal and systemic immunity. *Microb. Pathog.* **2018**, *125*, 507–513. [CrossRef]

46. Chan, J.M.; Valencia, P.M.; Zhang, L.; Langer, R.; Farokhzad, O.C. Polymeric nanoparticles for drug delivery. In *Cancer Nanotechnology: Methods and Protocols*; Humana Press: Totowa, NJ, USA, 2010; pp. 163–175.
47. Anderson, J.M.; Shive, M.S. Biodegradation and biocompatibility of PLA and PLGA microspheres. *Adv. Drug Deliv. Rev.* **1997**, *28*, 5–24. [CrossRef] [PubMed]
48. Hans, M.L.; Lowman, A.M. Biodegradable nanoparticles for drug delivery and targeting. *Curr. Opin. Solid State Mater. Sci.* **2002**, *6*, 319–327. [CrossRef]
49. Moon, J.J.; Suh, H.; Polhemus, M.E.; Ockenhouse, C.F.; Yadava, A.; Irvine, D.J. Antigen-displaying lipid-enveloped PLGA nanoparticles as delivery agents for a Plasmodium vivax malaria vaccine. *PLoS ONE* **2012**, *7*, e31472. [CrossRef]
50. Gutjahr, A.; Phelip, C.; Coolen, A.-L.; Monge, C.; Boisgard, A.-S.; Paul, S.; Verrier, B. Biodegradable polymeric nanoparticles-based vaccine adjuvants for lymph nodes targeting. *Vaccines* **2016**, *4*, 34. [CrossRef]
51. Luo, X.; Zeng, X.; Gong, L.; Ye, Y.; Sun, C.; Chen, T.; Zhang, Z.; Tao, Y.; Zeng, H.; Zou, Q. Nanomaterials in tuberculosis DNA vaccine delivery: Historical perspective and current landscape. *Drug Deliv.* **2022**, *29*, 2912–2924. [CrossRef]
52. Mikušová, V.; Mikuš, P. Advances in chitosan-based nanoparticles for drug delivery. *Int. J. Mol. Sci.* **2021**, *22*, 9652. [CrossRef]
53. Ahmed, M.; Jiao, H.; Domingo-Gonzalez, R.; Das, S.; Griffiths, K.L.; Rangel-Moreno, J.; Nagarajan, U.M.; Khader, S.A. Rationalized design of a mucosal vaccine protects against Mycobacterium tuberculosis challenge in mice. *J. Leucoc. Biol.* **2017**, *101*, 1373–1381. [CrossRef]
54. Wu, M.; Zhao, H.; Li, M.; Yue, Y.; Xiong, S.; Xu, W. Intranasal vaccination with mannosylated chitosan formulated DNA vaccine enables robust IgA and cellular response induction in the lungs of mice and improves protection against pulmonary mycobacterial challenge. *Front. Cell. Infect. Microbiol.* **2017**, *7*, 445. [CrossRef]
55. Csaba, N.; Garcia-Fuentes, M.; Alonso, M.J. Nanoparticles for nasal vaccination. *Adv. Drug Deliv. Rev.* **2009**, *61*, 140–157. [CrossRef] [PubMed]
56. Majzoub, R.N.; Ewert, K.K.; Jacovetty, E.L.; Carragher, B.; Potter, C.S.; Li, Y.; Safinya, C.R. Patterned threadlike micelles and DNA-tethered nanoparticles: A structural study of PEGylated cationic liposome–DNA assemblies. *Langmuir* **2015**, *31*, 7073–7083. [CrossRef] [PubMed]
57. Garbuzenko, O.; Barenholz, Y.; Priev, A. Effect of grafted PEG on liposome size and on compressibility and packing of lipid bilayer. *Chem. Phys. Lipids* **2005**, *135*, 117–129. [CrossRef] [PubMed]
58. Zhang, X.; Divangahi, M.; Ngai, P.; Santosuosso, M.; Millar, J.; Zganiacz, A.; Wang, J.; Bramson, J.; Xing, Z. Intramuscular immunization with a monogenic plasmid DNA tuberculosis vaccine: Enhanced immunogenicity by electroporation and co-expression of GM-CSF transgene. *Vaccine* **2007**, *25*, 1342–1352. [CrossRef]
59. Villarreal, D.O.; Siefert, R.J.; Weiner, D.B. Alarmin IL-33 elicits potent TB-specific cell-mediated responses. *Hum. Vaccines Immunother.* **2015**, *11*, 1954–1960. [CrossRef]
60. Tang, J.; Cai, Y.; Liang, J.; Tan, Z.; Tang, X.; Zhang, C.; Cheng, L.; Zhou, J.; Wang, H.; Yam, W.-C. In vivo electroporation of a codon-optimized BERopt DNA vaccine protects mice from pathogenic Mycobacterium tuberculosis aerosol challenge. *Tuberculosis* **2018**, *113*, 65–75. [CrossRef] [PubMed]
61. Zaman, M.; Good, M.F.; Toth, I. Nanovaccines and their mode of action. *Methods* **2013**, *60*, 226–231. [CrossRef] [PubMed]
62. Amini, Y.; Moradi, B.; Tafaghodi, M.; Meshkat, Z.; Ghazvini, K.; Fasihi-Ramandi, M. TB trifusion antigen adsorbed on calcium phosphate nanoparticles stimulates strong cellular immunity in mice. *Biotechnol. Bioprocess Eng.* **2016**, *21*, 653–658. [CrossRef]
63. Pérez-Betancourt, Y.; Araujo, P.M.; Távora, B.d.C.L.F.; Pereira, D.R.; Faquim-Mauro, E.L.; Carmona-Ribeiro, A.M. Cationic and biocompatible polymer/lipid nanoparticles as immunoadjuvants. *Pharmaceutics* **2021**, *13*, 1859. [CrossRef]
64. Carneiro, S.P.; dos Santos, O.D.H. Nanostructured lipid carrier-based drug delivery systems for tuberculosis treatment. In *Nanotechnology Based Approaches for Tuberculosis Treatment*; Elsevier: Amsterdam, The Netherlands, 2020; pp. 193–205.
65. Ahmed, M.; Smith, D.M.; Hamouda, T.; Rangel-Moreno, J.; Fattom, A.; Khader, S.A. A novel nanoemulsion vaccine induces mucosal Interleukin-17 responses and confers protection upon challenge in mice. *Vaccine* **2017**, *35*, 4983–4989. [CrossRef]
66. Zare, S.; Kabiri, M.; Amini, Y.; Najafi, A.; Mohammadpour, F.; Ayati, S.H.; Nikpoor, A.R.; Tafaghodi, M. Immunological Assessment of Chitosan or Trimethyl Chitosan-Coated PLGA Nanospheres Containing Fusion Antigen as the Novel Vaccine Candidates Against Tuberculosis. *AAPS Pharmscitech* **2021**, *23*, 15. [CrossRef] [PubMed]
67. Malathi, S.; Balasubramanian, S. PLGA Nanoparticles for Anti Tuberculosis Drug Delivery. *Adv. Mater. Res. Switz.* **2012**, *584*, 465–469. [CrossRef]
68. Tafaghodi, M.; Khademi, F.; Firouzi, Z. Polymer-based nanoparticles as delivery systems for treatment and vaccination of tuberculosis. In *Nanotechnology Based Approaches for Tuberculosis Treatment*; Elsevier: Amsterdam, The Netherlands, 2020; pp. 123–142.
69. Mahony, D.; Cavallaro, A.S.; Stahr, F.; Mahony, T.J.; Qiao, S.Z.; Mitter, N. Mesoporous Silica Nanoparticles Act as a Self-Adjuvant for Ovalbumin Model Antigen in Mice. *Small* **2013**, *9*, 3138–3146. [CrossRef] [PubMed]
70. Manjaly Thomas, Z.-R.; McShane, H. Aerosol immunisation for TB: Matching route of vaccination to route of infection. *Trans. R. Soc. Trop. Med. Hyg.* **2015**, *109*, 175–181. [CrossRef]
71. Khademi, F.; Yousefi-Avarvand, A.; Derakhshan, M.; Abbaspour, M.R.; Sadri, K.; Tafaghodi, M. Formulation and Optimization of a New Cationic Lipid-Modified PLGA Nanoparticle as Delivery System for HspX/EsxS Fusion Protein: An Experimental Design. *Iran. J. Pharm. Res.* **2019**, *18*, 446–458.

72. Garg, N.K.; Dwivedi, P.; Jain, A.; Tyagi, S.; Sahu, T.; Tyagi, R.K. Development of novel carrier(s) mediated tuberculosis vaccine: More than a tour de force. *Eur. J. Pharm. Sci.* **2014**, *62*, 227–242. [CrossRef] [PubMed]
73. Bastola, R.; Lee, S. Physicochemical properties of particulate vaccine adjuvants: Their pivotal role in modulating immune responses. *J. Pharm. Investig.* **2019**, *49*, 279–285. [CrossRef]
74. Nuhn, L.; Vanparijs, N.; De Beuckelaer, A.; Lybaert, L.; Verstraete, G.; Deswarte, K.; Lienenklaus, S.; Shukla, N.M.; Salyer, A.C.; Lambrecht, B.N. pH-degradable imidazoquinoline-ligated nanogels for lymph node-focused immune activation. *Proc. Natl. Acad. Sci. USA* **2016**, *113*, 8098–8103. [CrossRef]
75. Qiao, D.; Liu, L.; Chen, Y.; Xue, C.; Gao, Q.; Mao, H.-Q.; Leong, K.W.; Chen, Y. Potency of a scalable nanoparticulate subunit vaccine. *Nano Lett.* **2018**, *18*, 3007–3016. [CrossRef]
76. Nakamura, T.; Harashima, H. Dawn of lipid nanoparticles in lymph node targeting: Potential in cancer immunotherapy. *Adv. Drug Deliv. Rev.* **2020**, *167*, 78–88. [CrossRef]
77. Mehrabi, M.; Dounighi, N.M.; Rezayat Sorkhabadi, S.M.; Doroud, D.; Amani, A.; Khoobi, M.; Ajdary, S.; Pilehvar-Soltanahmadi, Y. Development and physicochemical, toxicity and immunogenicity assessments of recombinant hepatitis B surface antigen (rHBsAg) entrapped in chitosan and mannosylated chitosan nanoparticles: As a novel vaccine delivery system and adjuvant. *Artif. Cells Nanomed. Biotechnol.* **2018**, *46*, 230–240. [CrossRef]
78. Samaridou, E.; Heyes, J.; Lutwyche, P. Lipid nanoparticles for nucleic acid delivery: Current perspectives. *Adv. Drug Deliv. Rev.* **2020**, *154–155*, 37–63. [CrossRef] [PubMed]
79. Bingbing, S.; Zhaoxia, J.; Yu-Pei, L.; Hyun, C.C.; Xiang, W.; Justine, K.; Changying, X.; Vahid, M.; Tian, X. Enhanced Immune Adjuvant Activity of Aluminum Oxyhydroxide Nanorods through Cationic Surface Functionalization. *ACS Appl. Mater. Interfaces* **2017**, *9*, 21697–21705.
80. Zhang, L.; Liang, Z.; Chen, C.; Yang, X.; Fu, D.; Bao, H.; Li, M.; Shi, S.; Yu, G.; Zhang, Y. Engineered hydroxyapatite nanoadjuvants with controlled shape and aspect ratios reveal their immunomodulatory potentials. *ACS Appl. Mater. Interfaces* **2021**, *13*, 59662–59672. [CrossRef] [PubMed]
81. Christensen, D.; Korsholm, K.S.; Andersen, P.; Agger, E.M. Cationic liposomes as vaccine adjuvants. *Expert Rev. Vaccines* **2011**, *10*, 513–521. [CrossRef] [PubMed]
82. Yan, W.; Chen, W.; Huang, L. Mechanism of adjuvant activity of cationic liposome: Phosphorylation of a MAP kinase, ERK and induction of chemokines. *Mol. Immunol.* **2007**, *44*, 3672–3681. [CrossRef]
83. Foged, C. Subunit vaccines of the future: The need for safe, customized and optimized particulate delivery systems. *Ther. Deliv.* **2011**, *2*, 1057–1077. [CrossRef]
84. Liu, Y.; Yin, Y.; Wang, L.; Zhang, W.; Chen, X.; Yang, X.; Xu, J.; Ma, G. Surface hydrophobicity of microparticles modulates adjuvanticity. *J. Mater. Chem. B* **2013**, *1*, 3888–3896. [CrossRef]
85. Allen, T.M.; Cullis, P.R. Liposomal drug delivery systems: From concept to clinical applications. *Adv. Drug Deliv. Rev.* **2013**, *65*, 36–48. [CrossRef]
86. Khan, A.A.; Allemailem, K.S.; Almatroodi, S.A.; Almatroudi, A.; Rahmani, A.H. Recent strategies towards the surface modification of liposomes: An innovative approach for different clinical applications. *3 Biotech* **2020**, *10*, 163. [CrossRef]
87. Azuar, A.; Li, Z.; Shibu, M.A.; Zhao, L.; Luo, Y.; Shalash, A.O.; Khalil, Z.G.; Capon, R.J.; Hussein, W.M.; Toth, I. Poly (hydrophobic amino acid)-based self-adjuvanting nanoparticles for Group A Streptococcus vaccine delivery. *J. Med. Chem.* **2021**, *64*, 2648–2658. [CrossRef]
88. Li, P.; Shi, G.; Zhang, X.; Song, H.; Zhang, C.; Wang, W.; Li, C.; Song, B.; Wang, C.; Kong, D. Guanidinylated cationic nanoparticles as robust protein antigen delivery systems and adjuvants for promoting antigen-specific immune responses in vivo. *J. Mater. Chem. B* **2016**, *4*, 5608–5620. [CrossRef]
89. Bal, S.M.; Slütter, B.; Verheul, R.; Bouwstra, J.A.; Jiskoot, W. Adjuvanted, antigen loaded N-trimethyl chitosan nanoparticles for nasal and intradermal vaccination: Adjuvant-and site-dependent immunogenicity in mice. *Eur. J. Pharm. Sci.* **2012**, *45*, 475–481. [CrossRef] [PubMed]
90. Pati, R.; Shevtsov, M.; Sonawane, A. Nanoparticle vaccines against infectious diseases. *Front. Immunol.* **2018**, *9*, 2224. [CrossRef] [PubMed]
91. Rueda, F.; Eich, C.; Cordobilla, B.; Domingo, P.; Acosta, G.; Albericio, F.; Cruz, L.J.; Domingo, J.C. Effect of TLR ligands co-encapsulated with multiepitopic antigen in nanoliposomes targeted to human DCs via Fc receptor for cancer vaccines. *Immunobiology* **2017**, *222*, 989–997. [CrossRef]
92. Gu, P.; Liu, Z.; Sun, Y.; Ou, N.; Hu, Y.; Liu, J.; Wu, Y.; Wang, D. Angelica sinensis polysaccharide encapsulated into PLGA nanoparticles as a vaccine delivery and adjuvant system for ovalbumin to promote immune responses. *Int. J. Pharm.* **2019**, *554*, 72–80. [CrossRef]
93. Du, G.; Hathout, R.M.; Nasr, M.; Nejadnik, M.R.; Tu, J.; Koning, R.I.; Koster, A.J.; Slütter, B.; Kros, A.; Jiskoot, W. Intradermal vaccination with hollow microneedles: A comparative study of various protein antigen and adjuvant encapsulated nanoparticles. *J. Control. Release* **2017**, *266*, 109–118. [CrossRef]
94. Slütter, B.; Plapied, L.; Fievez, V.; Sande, M.A.; des Rieux, A.; Schneider, Y.-J.; Van Riet, E.; Jiskoot, W.; Préat, V. Mechanistic study of the adjuvant effect of biodegradable nanoparticles in mucosal vaccination. *J. Control. Release* **2009**, *138*, 113–121. [CrossRef]

95. De Titta, A.; Ballester, M.; Julier, Z.; Nembrini, C.; Jeanbart, L.; Van Der Vlies, A.J.; Swartz, M.A.; Hubbell, J.A. Nanoparticle conjugation of CpG enhances adjuvancy for cellular immunity and memory recall at low dose. *Proc. Natl. Acad. Sci. USA* **2013**, *110*, 19902–19907. [CrossRef]
96. Rezaei, F.; Keshvari, H.; Shokrgozar, M.A.; Doroud, D.; Gholami, E.; Khabiri, A.; Farokhi, M. Nano-adjuvant based on silk fibroin for the delivery of recombinant hepatitis B surface antigen. *Biomater. Sci.* **2021**, *9*, 2679–2695. [CrossRef]
97. Agger, E.M. Novel adjuvant formulations for delivery of anti-tuberculosis vaccine candidates. *Adv. Drug Deliv. Rev.* **2016**, *102*, 73–82. [CrossRef]
98. Kaufmann, S.H. Vaccination against tuberculosis: Revamping BCG by molecular genetics guided by immunology. *Front. Immunol.* **2020**, *11*, 316. [CrossRef] [PubMed]
99. Gupta, A.; Gupta, G.S. Applications of mannose-binding lectins and mannan glycoconjugates in nanomedicine. *J. Nanoparticle Res.* **2022**, *24*, 228. [CrossRef] [PubMed]
100. Garcia-Vallejo, J.; Van Kooyk, Y. DC-sign. C-type lectin with prominent role in immune system. In *Glycoscience: Biology and Medicine*; Springer: Tokyo, Japan, 2015; pp. 649–659.
101. Khademi, F.; Yousefi, A.; Derakhshan, M.; Najafi, A.; Tafaghodi, M. Enhancing immunogenicity of novel multistage subunit vaccine of Mycobacterium tuberculosis using PLGA: DDA hybrid nanoparticles and MPLA: Subcutaneous administration. *Iran. J. Basic Med. Sci.* **2019**, *22*, 893. [PubMed]
102. Zahednezhad, F.; Saadat, M.; Valizadeh, H.; Zakeri-Milani, P.; Baradaran, B. Liposome and immune system interplay: Challenges and potentials. *J. Control. Release* **2019**, *305*, 194–209. [CrossRef]
103. Aibani, N.; Patel, P.; Buchanan, R.; Strom, S.; Wasan, K.M.; Hancock, R.E.; Gerdts, V.; Wasan, E.K. Assessing the In Vivo Effectiveness of Cationic Lipid Nanoparticles with a Triple Adjuvant for Intranasal Vaccination against the Respiratory Pathogen Bordetella pertussis. *Mol. Pharm.* **2022**, *19*, 1814–1824. [CrossRef]
104. Song, Y.; Zhou, Y.; Chen, L. Cellulose-based polyelectrolyte complex nanoparticles for DNA vaccine delivery. *Biomater. Sci.* **2014**, *2*, 1440–1449. [CrossRef]
105. De Serrano, L.O.; Burkhart, D.J. Liposomal vaccine formulations as prophylactic agents: Design considerations for modern vaccines. *J. Nanobiotechnol.* **2017**, *15*, 83. [CrossRef]
106. Zhao, L.; Seth, A.; Wibowo, N.; Zhao, C.-X.; Mitter, N.; Yu, C.; Middelberg, A.P. Nanoparticle vaccines. *Vaccine* **2014**, *32*, 327–337. [CrossRef]
107. Swaminathan, G.; Thoryk, E.A.; Cox, K.S.; Meschino, S.; Dubey, S.A.; Vora, K.A.; Celano, R.; Gindy, M.; Casimiro, D.R.; Bett, A.J. A novel lipid nanoparticle adjuvant significantly enhances B cell and T cell responses to sub-unit vaccine antigens. *Vaccine* **2016**, *34*, 110–119. [CrossRef]
108. Saxena, M.; Van, T.T.H.; Baird, F.J.; Coloe, P.J.; Smooker, P.M. Pre-existing immunity against vaccine vectors–friend or foe? *Microbiology* **2013**, *159*, 1–11. [CrossRef] [PubMed]
109. Tajmir-Riahi, H.; Pillai, C.; Thomas, T. Biodegradable Polymers for Gene Delivery. *Molecules* **2019**, *24*, 3744. [CrossRef] [PubMed]
110. Song, T.; Liao, Y.; Zuo, Q.; Liu, N.; Liu, Z. MnO_2 nanoparticles as a minimalist multimode vaccine adjuvant/delivery system to regulate antigen presenting cells for tumor immunotherapy. *J. Mater. Chem. B* **2022**, *10*, 3474–3490. [CrossRef] [PubMed]
111. Mueller, S.N.; Tian, S.; DeSimone, J.M. Rapid and persistent delivery of antigen by lymph node targeting PRINT nanoparticle vaccine carrier to promote humoral immunity. *Mol. Pharm.* **2015**, *12*, 1356–1365. [CrossRef]
112. Desai, D.N.; Mahal, A.; Varshney, R.; Obaidullah, A.J.; Gupta, B.; Mohanty, P.; Pattnaik, P.; Mohapatra, N.C.; Mishra, S.; Kandi, V. Nanoadjuvants: Promising bioinspired and biomimetic approaches in vaccine innovation. *ACS Omega* **2023**, *8*, 27953–27968. [CrossRef] [PubMed]
113. Hanson, M.C.; Crespo, M.P.; Abraham, W.; Moynihan, K.D.; Szeto, G.L.; Chen, S.H.; Melo, M.B.; Mueller, S.; Irvine, D.J. Nanoparticulate STING agonists are potent lymph node–targeted vaccine adjuvants. *J. Clin. Investig.* **2015**, *125*, 2532–2546. [CrossRef]
114. Zhang, S.; Huang, S.; Lu, L.; Song, X.; Li, P.; Wang, F. Curdlan sulfate–O-linked quaternized chitosan nanoparticles: Potential adjuvants to improve the immunogenicity of exogenous antigens via intranasal vaccination. *Int. J. Nanomed.* **2018**, *13*, 2377–2394. [CrossRef]
115. Hou, Y.; Wang, Y.; Tang, Y.; Zhou, Z.; Tan, L.; Gong, T.; Zhang, L.; Sun, X. Co-delivery of antigen and dual adjuvants by aluminum hydroxide nanoparticles for enhanced immune responses. *J. Control. Release* **2020**, *326*, 120–130. [CrossRef]
116. Ren, H.; Li, J.; Liu, G.; Sun, Y.; Yang, X.; Jiang, Z.; Zhang, J.; Lovell, J.F.; Zhang, Y. Anticancer vaccination with immunogenic micelles that capture and release pristine $CD8^+$ T-cell epitopes and adjuvants. *ACS Appl. Mater. Interfaces* **2022**, *14*, 2510–2521. [CrossRef]
117. Akira, S.; Uematsu, S.; Takeuchi, O. Pathogen recognition and innate immunity. *Cell* **2006**, *124*, 783–801. [CrossRef]
118. Mogensen, T.H. Pathogen recognition and inflammatory signaling in innate immune defenses. *Clin. Microbiol. Rev.* **2009**, *22*, 240–273. [CrossRef]
119. Speth, M.T.; Repnik, U.; Müller, E.; Spanier, J.; Kalinke, U.; Corthay, A.; Griffiths, G. Poly (I:C)-encapsulating nanoparticles enhance innate immune responses to the tuberculosis vaccine Bacille Calmette–Guérin (BCG) via synergistic activation of innate immune receptors. *Mol. Pharm.* **2017**, *14*, 4098–4112. [CrossRef] [PubMed]
120. Siefert, A.L.; Caplan, M.J.; Fahmy, T.M. Artificial bacterial biomimetic nanoparticles synergize pathogen-associated molecular patterns for vaccine efficacy. *Biomaterials* **2016**, *97*, 85–96. [CrossRef] [PubMed]

121. Kuramoto, Y.; Kawakami, S.; Zhou, S.; Fukuda, K.; Yamashita, F.; Hashida, M. Efficient peritoneal dissemination treatment obtained by an immunostimulatory phosphorothioate-type CpG DNA/cationic liposome complex in mice. *J. Control. Release* **2008**, *126*, 274–280. [CrossRef] [PubMed]
122. Mansourian, M.; Badiee, A.; Jalali, S.A.; Shariat, S.; Yazdani, M.; Amin, M.; Jaafari, M.R. Effective induction of anti-tumor immunity using p5 HER-2/neu derived peptide encapsulated in fusogenic DOTAP cationic liposomes co-administrated with CpG-ODN. *Immunol. Lett.* **2014**, *162*, 87–93. [CrossRef] [PubMed]
123. Wang, C.; Zhuang, Y.; Zhang, Y.; Luo, Z.; Gao, N.; Li, P.; Pan, H.; Cai, L.; Ma, Y. Toll-like receptor 3 agonist complexed with cationic liposome augments vaccine-elicited antitumor immunity by enhancing TLR3–IRF3 signaling and type I interferons in dendritic cells. *Vaccine* **2012**, *30*, 4790–4799. [CrossRef]
124. Avarvand, A.Y.; Meshkat, Z.; Khademi, F.; Aryan, E.; Sankian, M.; Tafaghodi, M. Enhancement of the immunogenicity of a Mycobacterium tuberculosis fusion protein using ISCOMATRIX and PLUSCOM nano-adjuvants as prophylactic vaccine after nasal administration in mice. *Iran. J. Basic Med. Sci.* **2024**, *27*, 24.
125. Hernandez-Franco, J.F.; Jan, I.M.; Goldsmith, M.; HogenEsch, H. Optimizing vaccine performance through improved cross-presentation with a nanoparticle adjuvant. *J. Immunol.* **2022**, *208*, 123.03. [CrossRef]
126. Hirosue, S.; Kourtis, I.C.; van der Vlies, A.J.; Hubbell, J.A.; Swartz, M.A. Antigen delivery to dendritic cells by poly (propylene sulfide) nanoparticles with disulfide conjugated peptides: Cross-presentation and T cell activation. *Vaccine* **2010**, *28*, 7897–7906. [CrossRef]
127. Lee, W.; Suresh, M. Vaccine adjuvants to engage the cross-presentation pathway. *Front. Immunol.* **2022**, *13*, 940047. [CrossRef]
128. Stano, A.; Nembrini, C.; Swartz, M.A.; Hubbell, J.A.; Simeoni, E. Nanoparticle size influences the magnitude and quality of mucosal immune responses after intranasal immunization. *Vaccine* **2012**, *30*, 7541–7546. [CrossRef]
129. Lima, K.d.M.; Santos, S.A.d.; Lima, V.; Coelho-Castelo, A.; Rodrigues, J.; Silva, C.L. Single dose of a vaccine based on DNA encoding mycobacterial hsp65 protein plus TDM-loaded PLGA microspheres protects mice against a virulent strain of Mycobacterium tuberculosis. *Gene Ther.* **2003**, *10*, 678–685. [CrossRef] [PubMed]
130. Wedlock, D.; Keen, D.; McCarthy, A.; Andersen, P.; Buddle, B. Effect of different adjuvants on the immune responses of cattle vaccinated with Mycobacterium tuberculosis culture filtrate proteins. *Vet. Immunol. Immunopathol.* **2002**, *86*, 79–88. [CrossRef] [PubMed]
131. Ebensen, T.; Arntz, A.; Schulze, K.; Hanefeld, A.; Guzmán, C.A.; Scherließ, R. Pulmonary application of novel antigen-loaded chitosan nano-particles co-administered with the mucosal adjuvant C-Di-AMP resulted in enhanced immune stimulation and dose sparing capacity. *Pharmaceutics* **2023**, *15*, 1238. [CrossRef] [PubMed]
132. Moon, J.J.; Suh, H.; Li, A.V.; Ockenhouse, C.F.; Yadava, A.; Irvine, D.J. Enhancing humoral responses to a malaria antigen with nanoparticle vaccines that expand Tfh cells and promote germinal center induction. *Proc. Natl. Acad. Sci. USA* **2012**, *109*, 1080–1085. [CrossRef] [PubMed]
133. Hernandez-Franco, J.F.; Mosley, Y.-Y.C.; Franco, J.; Ragland, D.; Yao, Y.; HogenEsch, H. Effective and safe stimulation of humoral and cell-mediated immunity by intradermal immunization with a cyclic Dinucleotide/Nanoparticle combination adjuvant. *J. Immunol.* **2021**, *206*, 700–711. [CrossRef]
134. Shute, T.; Amiel, E.; Alam, N.; Yates, J.L.; Mohrs, K.; Dudley, E.; Salas, B.; Mesa, C.; Serrata, A.; Angel, D. Glycolipid-containing nanoparticle vaccine engages invariant NKT cells to enhance humoral protection against systemic bacterial infection but abrogates T-independent vaccine responses. *J. Immunol.* **2021**, *206*, 1806–1816. [CrossRef] [PubMed]
135. Darling, R.; Senapati, S.; Christiansen, J.; Liu, L.; Ramer-Tait, A.E.; Narasimhan, B.; Wannemuehler, M. Polyanhydride nanoparticles induce low inflammatory dendritic cell activation resulting in $CD8^+$ T cell memory and delayed tumor progression. *Int. J. Nanomed.* **2020**, *15*, 6579–6592. [CrossRef]
136. Knight, F.C.; Gilchuk, P.; Kumar, A.; Becker, K.W.; Sevimli, S.; Jacobson, M.E.; Suryadevara, N.; Wang-Bishop, L.; Boyd, K.L.; Crowe Jr, J.E. Mucosal immunization with a pH-responsive nanoparticle vaccine induces protective $CD8^+$ lung-resident memory T cells. *ACS Nano* **2019**, *13*, 10939–10960. [CrossRef]
137. Zhang, Q.; Huang, W.; Yuan, M.; Li, W.; Hua, L.; Yang, Z.; Gao, F.; Li, S.; Ye, C.; Chen, Y. Employing ATP as a new adjuvant promotes the induction of robust antitumor cellular immunity by a PLGA nanoparticle vaccine. *ACS Appl. Mater. Interfaces* **2020**, *12*, 54399–54414. [CrossRef]
138. Gomez, M.; Archer, M.; Barona, D.; Wang, H.; Ordoubadi, M.; Karim, S.B.; Carrigy, N.B.; Wang, Z.; McCollum, J.; Press, C. Microparticle encapsulation of a tuberculosis subunit vaccine candidate containing a nanoemulsion adjuvant via spray drying. *Eur. J. Pharm. Biopharm.* **2021**, *163*, 23–37. [CrossRef]
139. Khandhar, A.P.; Liang, H.; Simpson, A.C.; Reed, S.G.; Carter, D.; Fox, C.B.; Orr, M.T. Physicochemical structure of a polyacrylic acid stabilized nanoparticle alum (nanoalum) adjuvant governs TH1 differentiation of CD4+ T cells. *Nanoscale* **2020**, *12*, 2515–2523. [CrossRef] [PubMed]
140. Orr, M.T.; Kramer, R.M.; Barnes, L.; Dowling, Q.M.; Desbien, A.L.; Beebe, E.A.; Laurance, J.D.; Fox, C.B.; Reed, S.G.; Coler, R.N. Elimination of the cold-chain dependence of a nanoemulsion adjuvanted vaccine against tuberculosis by lyophilization. *J. Control. Release* **2014**, *177*, 20–26. [CrossRef] [PubMed]
141. Li, S.; Yang, Y.; Lin, X.; Li, Z.; Ma, G.; Su, Z.; Zhang, S. Biocompatible cationic solid lipid nanoparticles as adjuvants effectively improve humoral and T cell immune response of foot and mouth disease vaccines. *Vaccine* **2020**, *38*, 2478–2486. [CrossRef] [PubMed]

142. Poecheim, J.; Heuking, S.; Brunner, L.; Barnier-Quer, C.; Collin, N.; Borchard, G. Nanocarriers for DNA vaccines: Co-delivery of TLR-9 and NLR-2 ligands leads to synergistic enhancement of proinflammatory cytokine release. *Nanomaterials* **2015**, *5*, 2317–2334. [CrossRef]
143. Ilyinskii, P.O.; Roy, C.J.; O'Neil, C.P.; Browning, E.A.; Pittet, L.A.; Altreuter, D.H.; Alexis, F.; Tonti, E.; Shi, J.; Basto, P.A. Adjuvant-carrying synthetic vaccine particles augment the immune response to encapsulated antigen and exhibit strong local immune activation without inducing systemic cytokine release. *Vaccine* **2014**, *32*, 2882–2895. [CrossRef]
144. Rodrigues, M.Q.; Alves, P.M.; Roldão, A. Functionalizing Ferritin Nanoparticles for Vaccine Development. *Pharmaceutics* **2021**, *13*, 1621. [CrossRef]
145. Al-Halifa, S.; Gauthier, L.; Arpin, D.; Bourgault, S.; Archambault, D. Nanoparticle-based vaccines against respiratory viruses. *Front. Immunol.* **2019**, *10*, 22. [CrossRef]
146. Dormont, F.; Rouquette, M.; Mahatsekake, C.; Gobeaux, F.; Peramo, A.; Brusini, R.; Calet, S.; Testard, F.; Lepetre-Mouelhi, S.; Desmaële, D. Translation of nanomedicines from lab to industrial scale synthesis: The case of squalene-adenosine nanoparticles. *J. Control. Release* **2019**, *307*, 302–314. [CrossRef] [PubMed]
147. Huynh, M.T.; Veyan, J.F.; Pham, H.; Rahman, R.; Yousuf, S.; Brown, A.; Lin, J.; Balkus Jr, K.J.; Diwakara, S.D.; Smaldone, R.A. The Importance of Evaluating the Lot-to-Lot Batch Consistency of Commercial Multi-Walled Carbon Nanotube Products. *Nanomaterials* **2020**, *10*, 1930. [CrossRef]
148. Qi, Y.; Fox, C.B. Development of thermostable vaccine adjuvants. *Expert Rev. Vaccines* **2021**, *20*, 497–517. [CrossRef]
149. Saramago, S.; Magalhaes, J.; Pinheiro, M. Tuberculosis Vaccines: An Update of Recent and Ongoing Clinical Trials. *Appl. Sci.* **2021**, *11*, 9250. [CrossRef]
150. Nell, A.S.; D'lom, E.; Bouic, P.; Sabaté, M.; Bosser, R.; Picas, J.; Amat, M.; Churchyard, G.; Cardona, P.J. Safety, Tolerability, and Immunogenicity of the Novel Antituberculous Vaccine RUTI: Randomized, Placebo-Controlled Phase II Clinical Trial in Patients with Latent Tuberculosis Infection. *PLoS ONE* **2014**, *9*, e89612. [CrossRef] [PubMed]
151. Day, T.A.; Penn-Nicholson, A.; Luabeya, A.K.K. Safety and immunogenicity of the adjunct therapeutic vaccine ID93 + GLA-SE in adults who have completed treatment for tuberculosis: A randomised, double-blind, placebo-controlled, phase 2a trial. *Lancet Respir. Med.* **2021**, *9*, 373–386. [CrossRef] [PubMed]
152. Sagawa, Z.K.; Goman, C.; Frevol, A.; Blazevic, A.; Tennant, J.; Fisher, B.; Day, T.; Jackson, S.; Lemiale, F.; Toussaint, L.; et al. Safety and immunogenicity of a thermostable ID93 + GLA-SE tuberculosis vaccine candidate in healthy adults. *Nat. Commun.* **2023**, *14*, 1138. [CrossRef] [PubMed]
153. Suliman, S.; Luabeya, A.K.K.; Geldenhuys, H.; Tameris, M.; Hoff, S.T.; Shi, Z.K.; Tait, D.; Kromann, I.; Ruhwald, M.; Rutkowski, K.T.; et al. Dose Optimization of H56:IC31 Vaccine for Tuberculosis-Endemic Populations A Double-Blind, Placebo-controlled, Dose-Selection Trial. *Am. J. Respir. Crit. Care* **2019**, *199*, 220–231. [CrossRef] [PubMed]
154. Luabeya, A.K.K.; Kagina, B.M.N.; Tameris, M.D.; Geldenhuys, H.; Hoff, S.T.; Shi, Z.K.; Kromann, I.; Hatherill, M.; Mahomed, H.; Hanekom, W.A.; et al. First-in-human trial of the post-exposure tuberculosis vaccine H56:IC31 in infected and non-infected healthy adults. *Vaccine* **2015**, *33*, 4130–4140. [CrossRef]
155. Tait, D.R.; Hatherill, M.; Van Der Meeren, O.; Ginsberg, A.M.; Van Brakel, E.; Salaun, B.; Scriba, T.J.; Akite, E.J.; Ayles, H.M.; Bollaerts, A. Final analysis of a trial of M72/AS01E vaccine to prevent tuberculosis. *N. Engl. J. Med.* **2019**, *381*, 2429–2439. [CrossRef]
156. Wilson, L.; Gracie, L.; Kidy, F.; Thomas, G.N.; Nirantharakumar, K.; Greenfield, S.; Manaseki-Holland, S.; Ward, D.J.; Gooden, T.E. Safety and efficacy of tuberculosis vaccine candidates in low-and middle-income countries: A systematic review of randomised controlled clinical trials. *BMC Infect. Dis.* **2023**, *23*, 120. [CrossRef]
157. Kumarasamy, N.; Poongulali, S.; Bollaerts, A.; Moris, P.; Beulah, F.E.; Ayuk, L.N.; Demoitié, M.-A.; Jongert, E.; Ofori-Anyinam, O. A randomized, controlled safety, and immunogenicity trial of the M72/AS01 candidate tuberculosis vaccine in HIV-positive Indian adults. *Medicine* **2016**, *95*, e2459. [CrossRef]
158. Zhao, G.; Sathkumara, H.D.; Miranda-Hernandez, S.; Seifert, J.; Valencia-Hernandez, A.M.; Puri, M.; Huang, W.; Toth, I.; Daly, N.; Skwarczynski, M.; et al. A modular self-assembling and self-adjuvanting multiepitope peptide nanoparticle vaccine platform to improve the efficacy and immunogenicity of BCG. *bioRxiv* **2024**. bioRxiv:2024.2008.2004.606253. [CrossRef]
159. Meher, M.K.; Poluri, K.M. Chapter 2—pH-Sensitive Nanomaterials for Smart Release of Drugs. In *Intelligent Nanomaterials for Drug Delivery Applications*; Ahmad, N., Gopinath, P., Eds.; Elsevier: Amsterdam, The Netherlands, 2020; pp. 17–41.
160. Khan, M.I.; Hossain, M.I.; Hossain, M.K.; Rubel, M.; Hossain, K.; Mahfuz, A.; Anik, M.I. Recent progress in nanostructured smart drug delivery systems for cancer therapy: A review. *ACS Appl. Bio Mater.* **2022**, *5*, 971–1012. [CrossRef]
161. Abdelwahab, W.M.; Riffey, A.; Buhl, C.; Johnson, C.; Ryter, K.; Evans, J.T.; Burkhart, D.J. Co-adsorption of synthetic Mincle agonists and antigen to silica nanoparticles for enhanced vaccine activity: A formulation approach to co-delivery. *Int. J. Pharm.* **2021**, *593*, 120119. [CrossRef] [PubMed]
162. Li, X.; Pan, C.; Li, C.; Wang, K.; Ye, J.; Sun, P.; Guo, Y.; Wu, J.; Wang, H.; Zhu, L. Self-Assembled Proteinaceous Nanoparticles for Co-Delivery of Antigens and Cytosine Phosphoguanine (CpG) Adjuvants: Implications for Nanovaccines. *ACS Appl. Nano Mater.* **2023**, *6*, 7637–7648. [CrossRef]
163. Rhee, J.H. Chapter 19—Current and New Approaches for Mucosal Vaccine Delivery. In *Mucosal Vaccines*, 2nd ed.; Kiyono, H., Pascual, D.W., Eds.; Academic Press: Cambridge, MA, USA, 2020; pp. 325–356.

164. Wang, S.; Liu, H.; Zhang, X.; Qian, F. Intranasal and oral vaccination with protein-based antigens: Advantages, challenges and formulation strategies. *Protein Cell* **2015**, *6*, 480–503. [CrossRef] [PubMed]
165. Tafaghodi, M.; Tabassi, S.A.S.; Jaafari, M.-R.; Zakavi, S.R.; Momen-Nejad, M. Evaluation of the clearance characteristics of various microspheres in the human nose by gamma-scintigraphy. *Int. J. Pharm.* **2004**, *280*, 125–135. [CrossRef] [PubMed]
166. Sarkar, I.; Garg, R.; van Drunen Littel-van den Hurk, S. Selection of adjuvants for vaccines targeting specific pathogens. *Expert Rev. Vaccines* **2019**, *18*, 505–521. [CrossRef] [PubMed]
167. Chen, Y.; Inobe, J.-I.; Marks, R.; Gonnella, P.; Kuchroo, V.K.; Weiner, H.L. Peripheral deletion of antigen-reactive T cells in oral tolerance. *Nature* **1995**, *376*, 177–180. [CrossRef]
168. Giri, P.K.; Verma, I.; Khuller, G.K. Enhanced immunoprotective potential of Mycobacterium tuberculosis Ag85 complex protein based vaccine against airway Mycobacterium tuberculosis challenge following intranasal administration. *FEMS Immunol. Med. Microbiol.* **2006**, *47*, 233–241. [CrossRef]
169. Huang, S.-S.; Li, I.-H.; Hong, P.-D.; Yeh, M.-K. Development of Yersinia pestis F1 antigen-loaded microspheres vaccine against plague. *Int. J. Nanomed.* **2014**, *9*, 813–822.
170. Conway, M.A.; Madrigal-Estebas, L.; McClean, S.; Brayden, D.J.; Mills, K.H. Protection against Bordetella pertussis infection following parenteral or oral immunization with antigens entrapped in biodegradable particles: Effect of formulation and route of immunization on induction of Th1 and Th2 cells. *Vaccine* **2001**, *19*, 1940–1950. [CrossRef]
171. Jin, Z.; Gao, S.; Cui, X.; Sun, D.; Zhao, K. Adjuvants and delivery systems based on polymeric nanoparticles for mucosal vaccines. *Int. J. Pharm.* **2019**, *572*, 118731. [CrossRef]
172. Amini, Y.; Tebianian, M.; Mosavari, N.; Fasihi Ramandi, M.; Ebrahimi, S.M.; Najminejad, H.; Dabaghian, M.; Abdollahpour, M. Development of an effective delivery system for intranasal immunization against Mycobacterium tuberculosis ESAT-6 antigen. *Artif. Cells Nanomed. Biotechnol.* **2017**, *45*, 291–296. [CrossRef] [PubMed]
173. Hoseinpur, R.; Hasani, A.; Baradaran, B.; Abdolalizadeh, J.; Amini, Y.; Salehi, R.; Samadi Kafil, H.; Azizian, K.; Memar, M.Y.; Gholizadeh, P. Chitosan nanoparticles containing fusion protein (Hspx–PPE44–EsxV) and resiquimod adjuvant (HPERC) as a novel booster vaccine for Mycobacterium tuberculosis. *J. Biomater. Appl.* **2022**, *37*, 40–47. [CrossRef] [PubMed]
174. Wang, H.; Wang, S.; Fang, R.; Li, X.; Xing, J.; Li, Z.; Song, N. Enhancing TB Vaccine Efficacy: Current Progress on Vaccines, Adjuvants and Immunization Strategies. *Vaccines* **2023**, *12*, 38. [CrossRef] [PubMed]
175. Kaufmann, S.H.; Weiner, J.; von Reyn, C.F. Novel approaches to tuberculosis vaccine development. *Int. J. Infect. Dis.* **2017**, *56*, 263–267. [CrossRef]
176. Penn-Nicholson, A.; Tameris, M.; Smit, E.; Day, T.A.; Musvosvi, M.; Jayashankar, L.; Vergara, J.; Mabwe, S.; Bilek, N.; Geldenhuys, H. Safety and immunogenicity of the novel tuberculosis vaccine ID93 + GLA-SE in BCG-vaccinated healthy adults in South Africa: A randomised, double-blind, placebo-controlled phase 1 trial. *Lancet Respir. Med.* **2018**, *6*, 287–298. [CrossRef]

vaccines

Brief Report

Demonstration of Tunable Control over a Delayed-Release Vaccine Using Atomic Layer Deposition

Sky W. Brubaker *, Isabella R. Walters, Emily M. Hite, Lorena R. Antunez, Emma L. Palm, Hans H. Funke and Bryan L. Steadman

VitriVax, Inc., 5435 Airport Blvd Suite 106, Boulder, CO 80301, USA; iwalters@vitrivaxbio.com (I.R.W.); ehite@vitrivaxbio.com (E.M.H.); lnapolitano@vitrivaxbio.com (L.R.A.); epalm@vitrivaxbio.com (E.L.P.); hfunke@vitrivaxbio.com (H.H.F.); bsteadman@vitrivaxbio.com (B.L.S.)

* Correspondence: sbrubaker@vitrivaxbio.com

Abstract: Many vaccines require multiple doses for full efficacy, posing a barrier for patient adherence and protection. One solution to achieve full vaccination may be attained with single-administration vaccines containing multiple controlled release doses. In this study, delayed-release vaccines were generated using atomic layer deposition (ALD) to coat antigen-containing powders with alumina. Using in vitro and in vivo methods, we show that increasing the coat thickness controls the kinetics of antigen release and antibody response, ranging from weeks to months. Our results establish an in vitro–in vivo correlation with a level of tunable control over the antigen release and antibody response times with the potential to impact future vaccine design.

Keywords: single-administration vaccines; atomic layer deposition; controlled antigen release; in vitro–in vivo correlation

1. Introduction

There is great promise in developing new vaccine technologies with improved safety, efficacy, and durability. Current vaccines often require multiple doses spaced over time to provide optimal protection [1,2]. Poor compliance with vaccine schedules can reduce efficacy, posing a critical public health challenge from preventable infectious diseases throughout the world [3]. To overcome this hurdle, controlled-release technology is a promising strategy for delivering more than one vaccine dose on a predictable schedule using a single administration [4]. However, precision in tuning the controlled-release technologies to dictate the kinetics of the vaccine release and immune response times remains a challenge.

Several controlled-release vaccine technologies have emerged [5–7], which differ in their release profiles. Sustained release profiles provide continuous delivery of the antigen over a long period of time [8]. Pulsatile release profiles are characterized by a delay period and one or more pulses of antigen. Combining vaccine materials with different delay periods in a single dose may achieve an antigen delivery profile that mimics exposure from temporally spaced, repeated bolus injections (e.g., a standard vaccine schedule), thus approaching 100% efficacy without the need for multiple vaccine administrations.

Spray drying and ALD have previously been combined to achieve thermostability and delayed release in vivo with a single administration of a human papillomavirus (HPV) vaccine [9,10]. The vaccine preparations of HPV L1 capsomers were spray dried to make thermostable powders that were subsequently coated with alumina by ALD. Spray-dried particles have previously been imaged by scanning electron microscopy before and after ALD coating to establish particle size and shape characteristics [9]. In addition, transition electron microscopy was used to confirm the estimates that sequential ALD reactions deposit roughly 2.3-Å-thick layers of alumina on the particle surface [9]. The alumina coating prolonged the antigen retention at the site of injection, suggesting that the antigen release was delayed in vivo [9]. Furthermore, animals vaccinated using ALD-coated powders elicited neutralizing antibody titers that were comparable to the animals receiving a conventional prime and boost administration of the vaccine [9,10]. However, the range of achievable immune response times remains to be determined for ALD-coated powders.

2. Materials and Methods

2.1. Spray Drying, ALD Coating, and Other Materials

Intermediate spray-dried powders were made using a Buchi Mini B-290 spray dryer with a B-296 Dehumidifier (BUCHI, New Castle, DE, USA). The proprietary formulations contained 13% total solids, including 1% ovalbumin (OVA) (InvivoGen, San Diego, CA, USA, #vac-pova-100). Spray-drying setpoints were chosen to target the desired powder properties, including particle size, residual moisture content, and spray dryer yield. ALD-coated powders were generated using the intermediate spray-dried powders in a custom, mechanically agitated fluidized-bed ALD reactor, using trimethylaluminum and water as precursors [9,10]. Purge steps were employed after each precursor exposure to avoid chemical vapor deposition. The coating was performed at 50 °C, using argon as the fluidization gas. The number of ALD cycles used to coat powders in this study were 50, 100, 250, 500, and 1000.

The Alhydrogel® (InvivoGen, San Diego, CA, USA, #vac-alu-50) and OVA used in the animal studies were prepared following the manufacturer's instruction. Briefly, Alhydrogel® adjuvant 2% was combined with the OVA antigen in a 1:1 volume, mixed, and allowed to adsorb for 30 min at room temperature prior to administration. For the animal groups receiving Alhydrogel®, the mass of aluminum administered was approximately 11.6 µg per 500 ng dose of OVA (Figure 1C,D) or 1.6 µg per 250 ng dose of OVA (Figure 1E–G).

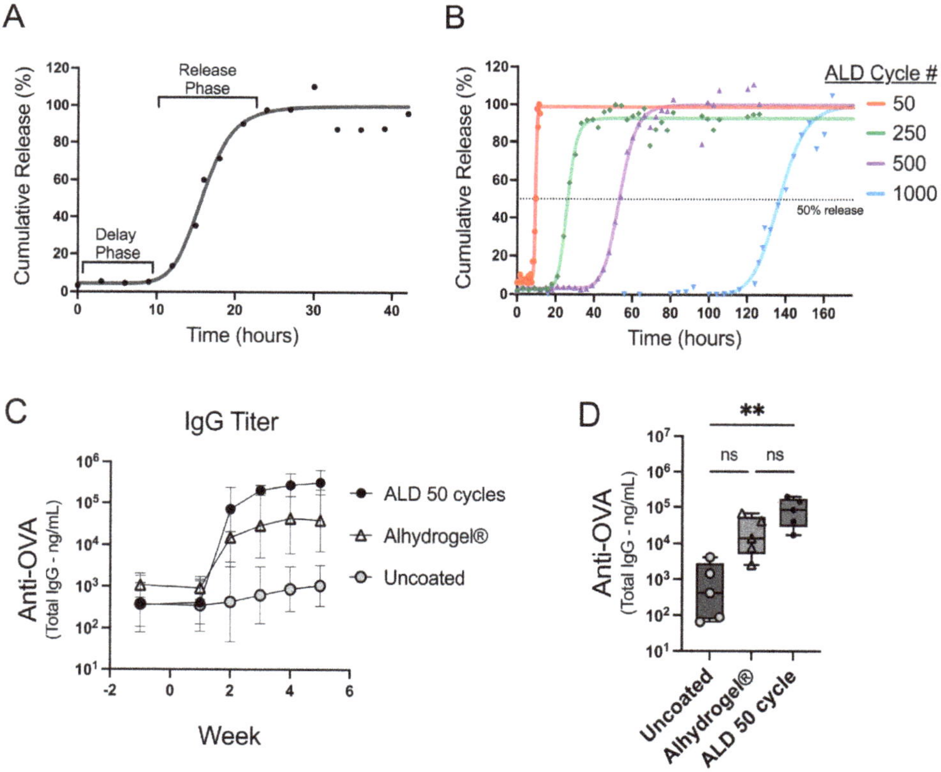

Figure 1. *Cont.*

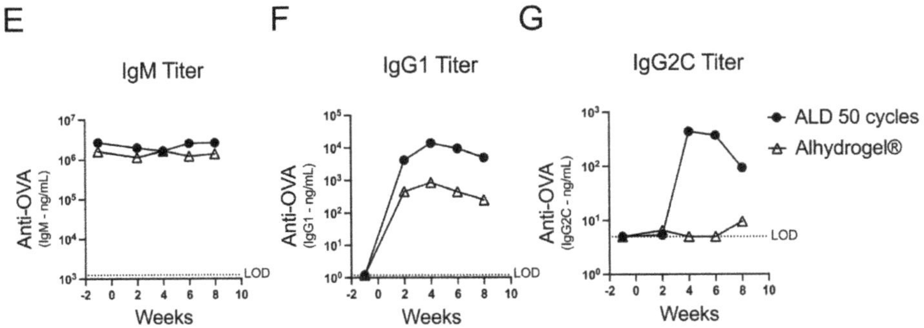

Figure 1. In vitro dissolution and early antibody responses kinetics. (**A**) The in vitro dissolution from powder coated with 100 ALD cycles is presented as a percentage of the total content released over time. Individual data points are depicted (dots) with a four-parameter, nonlinear logistic (4PL) regression (line). (**B**) The in vitro dissolution curves are shown for powders coated with 50, 250, 500, and 1000 ALD cycles. A dotted line at 50% gives an approximation of the inflection point for each 4PL regression (Supplementary Table S2). (**C**,**D**) Mice (n = 5) were vaccinated on day 0 with 500 ng OVA either from uncoated spray-dried powder (gray circles) or a 50-cycle ALD-coated powder (black circles), or with an Alhydrogel® formulation (open triangles). (**C**) OVA-specific total IgG from serum was plotted over time as the GMT for each group with 95% confidence intervals. (**D**) OVA-specific total IgG serum titers at two weeks post-injection are shown as box plots (+/−; min/max), with an overlay of individual data points. (**E**–**G**) Mice (n = 5) were vaccinated on day 0 with 250 ng OVA either from 50-cycle ALD-coated powder (black circles) or formulated with Alhydrogel® (open triangles). The serum from the individual mice was pooled at each timepoint, and titer concentrations for each timepoint are reported for OVA-specific IgM (**E**), IgG1 (**F**), or IgG2C (**G**) antibodies. The statistical analysis for panel D was performed by Kruskal-Wallis ANOVA with Dunn's post-hoc test (a = 0.05) ** indicates p-value = 0.0044, ns indicates not significant, and # indicates number.

2.2. Vaccine Administration and Animal Studies

All animal studies were conducted under approval from the Institutional Animal Care and Use Committee at the University of Colorado, Boulder, CO (protocol #2835). Vaccine administrations were given by intramuscular injection to the right flank of female C57BL/6J mice aged 6–8 weeks (The Jackson Laboratory, Bar Harbor, ME, USA, Strain #000664). The dosing was based on OVA mass and delivered with diluent (saline containing 6% hydroxyethyl starch) in a final volume of 50 µL. The serum was isolated from submandibular blood samples using gel separation tubes (Sarstedt, Nümbrecht, Germany, #41.1378.005) and stored at −80 °C prior to analysis.

2.3. Quantitation of Antibody Titers by ELISA

The immunized mouse serum samples were analyzed by an indirect, enzyme-linked immunosorbent assay (ELISA) to determine anti-OVA total IgG, IgG1, IgG2c, IgA, and IgM antibody titers. Nunc 96-well flat bottom high-binding plates (Thermo Scientific, Norristown, PA, USA, #442404) were coated with 5 µg/well of OVA (Fisher, Hampton, NH, USA, #BP2535-5) in phosphate-buffered saline (PBS), incubated at 4 °C overnight, and then rinsed four times using a wash buffer (0.05% Tween 20 in PBS). The plates were then incubated with an assay-blocking buffer (3% BSA, 0.05% Tween 20 in PBS) at room temperature for two hours. A standard curve was established with a dilution series of calibrated primary control (total IgG: Biolegend, San Diego, CA, USA, #520501, IgG1: Chondrex, Woodinville, WA, USA, #7093, IgG2c: Chondrex, #7109, and IgM: Chondrex, #7097). The calibrated primary control and mouse sera samples were incubated for 1 hour at room temperature, followed by a wash step to remove the unbound antibody. Horseradish peroxide-conjugated secondary antibody (Biolegend, #405306 @ 1:2000 dilution; Southern Biotech, Birmingham, AL, USA, #1071-05 @ 1:4000 dilution; Southern Biotech,

#1078-05 @ 1:4000 dilution; and Southern Biotech, #1021-05 @ 1:1000 dilution) was added (50 µL/well) and incubated for one hour at room temperature. Excess secondary antibody was washed off, and the plates were developed for 20 min using Ultra TMB (Thermo Fisher, #34028) and quenched using sulfuric acid (H_2SO_4), per the manufacturer's protocol. Absorbance was measured at 450 nm and 650 nm on a BioTek Synergy plate reader (Agilent, Santa Clara, CA, USA) five minutes after the addition of H_2SO_4. The standard curves and interpolated data for each plate were independently generated using Gen5 software (Agilent, Santa Clara, CA, USA).

2.4. Accelerated Dissolution of ALD Coated Particles In Vitro

ALD-coated particles were suspended at a concentration of 10 mg/mL buffered solution in Lo-bind microcentrifuge tubes (Eppendorf, Hamburg, Germany, #022431102) and agitated on an Eppendorf Thermomixer C set at 50 °C and 950 rpm. The dissolutions were performed at 50 °C to fully capture the dissolution profiles of the ALD-coated particles in a reasonable time frame while minimizing the risk of evaporation and microbial contamination. The supernatants were sampled at the indicated times after brief centrifuging to pellet the ALD powder. The volume removed was replaced with an equivalent volume of fresh buffer, the pellets were resuspended, and the tubes were placed back in the Thermomixer. The supernatant samples were added to a 96-well assay plate, and content released from the ALD-coated particles was measured using a Fluoraldehyde o-phthaldialdehyde (OPA) Assay Reagent Mix (Thermo Scientific, #26025) per the manufacturer's instructions. Fluorescence was measured after excitation at 360 nm and detected at 450 nm using a BioTek Synergy plate reader (Agilent, Santa Clara, CA, USA). Data are presented as percent recovery and calculated by normalizing the fluorescent signal against the maximum signal from each sample. The data were analyzed using four-parameter logistic (4PL) regressions.

2.5. Statistical Analysis

The Prism (Graph Pad, Boston, MA, USA) software version 10.2.3 was used for the statistical analysis and for the linear as well as nonlinear regression analyses. The nonparametric statistical analysis was conducted using a Kruskal-Wallis ANOVA with Dunn's post-hoc test $\alpha = 0.05$ [11].

3. Results

The tunable release of antigen from ALD-coated powders was shown using an accelerated in vitro dissolution assay. The coated powders were incubated in a dissolution buffer at 50 °C and sampled over time to monitor the release. Powders coated with 100 cycles of ALD demonstrate a delay phase followed by a release phase (Figure 1A). Nonlinear regression (4PL) analysis of the data generated a curve and parameters reporting on the initial and final percent release, the time to the inflection point, and the slope of the curve at the inflection point (Supplementary Table S1). The initial release is indicative of the coat quality or the containment of the antigen after the suspension of the powder. The inflection point represents the length of time it takes to release roughly 50 percent of the total content.

To achieve products with different release times, the number of ALD cycles used for alumina coating can be adjusted to increase or decrease the coat thickness [9]. The in vitro dissolution assay confirmed that the timing of release is controlled by the coat thickness, using coated powders generated with 50–1000 ALD cycles (Figure 1B). The dissolution profiles showed that all the powders had a robust coating (greater than 90% initial containment) and that increasing the number of ALD cycles resulted in increasingly delayed releases (Supplementary Table S2).

To explore the potential for an early immune response from ALD-coated vaccine powders, mice were either administered coated powder that released quickly in vitro (50 ALD cycles; Figure 1B) or uncoated powder. The powders were spray dried in a buffer formulation containing the model antigen ovalbumin (OVA) but without adjuvant. The coated and uncoated powders were then compared against a liquid formulation of

OVA and Alhydrogel®, a conventional aluminum-based adjuvant. The total IgG antibody responses to OVA were similar between the animals receiving 50-cycle ALD powders and Alhydrogel® (Figure 1C). Lower IgG titers were detected in the animals treated with uncoated spray-dried powders, likely due to the absence of the adjuvant or aluminum coating. There was a statistically significant difference in the total IgG titer at two weeks between the animals receiving powders that were coated vs. the animals that received the uncoated powders (**, p 0.0044); however, there was no significant difference when comparing 50-cycle ALD against the Alhydrogel®-formulated OVA (ns, p 0.6093 Figure 1D). A subset of antibody isotypes was measured, and the increase in the total IgG can be predominantly attributed to the increased production of OVA-specific IgG1 (Figure 1E–G). These data confirm that ALD-coated vaccines effectively trigger a robust humoral response with the antibody class switching predominantly toward IgG1 [12]. Furthermore, the data establish that 50-cycle powder induces an early antibody response that mimics a conventionally adjuvanted liquid vaccine formulation (i.e., the kinetic comparison with the Alhydrogel®-formulated OVA in these experiments is similar to the kinetics of ALD 50-cycle powder). In summary, ALD-coated vaccine powders are immunogenic and can achieve similar early response kinetics comparable to a conventionally adjuvanted liquid vaccine formulation.

A challenge in expanding the use of ALD for single-administration vaccines is demonstrating its ability to control payload release for predictable delayed immune responses in vivo. Similar to the delayed in vitro vaccine release with increasing coat thickness (Figure 1B), the in vivo immune response kinetics for the same coated powders confirmed increasingly delayed immune response with increasing coat thickness (Figure 2). The geometric mean titers (GMT) of OVA-specific IgG1 increased rapidly following the administration of 50-cycle ALD powders, whereas the increase in titer was delayed following the administration of 250-, 500-, or 1000-cycle ALD powders (Figure 2A).

There was not a statistically significant difference in the peak antibody titer responses compared between 50 and 250 (ns, $p > 0.9999$)- or 50 and 500 (ns, $p > 0.9999$)-cycle ALD powders (Figure 2B). Thus, similar immunogenicity can be elicited from the delayed-release ALD powders as compared to the early response elicited from 50-cycle powder. In contrast, the median peak titer was significantly lower among the animals that received 1000-cycle ALD powders as compared to the animals that received 50-cycle powders (**, p 0.0024). Two of the animals in this group responded robustly, whereas the rest were classified as non-responders (peak titers < 10^3 ng/mL). It is possible that the experiment was terminated before the vaccine release occurred in the non-responder mice, or the results may indicate that higher doses are required for powders coated with many ALD cycles. The median time to peak titer (using the cutoff > 10^3 ng/mL to exclude non-responders) for 50-, 250-, 500-, and 1000-cycle powders was 4, 15, 30, and 38 weeks, respectively (Figure 2C). This result demonstrates that increasing the number of ALD cycles delays the antibody response in vivo. However, the peak antibody titer in each animal may not reflect the onset of antigen-specific antibody production.

Seroconversion, a detectable increase in titer over pre-vaccination levels, is a better measure for the onset of the antibody response. Thus, seroconversion was deemed a more suitable parameter for establishing an in vitro–in vivo correlation between the time to the antigen release and the time to the antibody response. The threshold determination for seroconversion used in this study was a two-log increase in OVA-specific IgG1 titer. Increasing the number of ALD cycles slowed the rate of seroconversion in vivo (Figure 2D) with median times to seroconversion for 50-, 250-, 500-, and 1000-cycle powders of 2, 4, 13, and 36 weeks, respectively (Figure 2E). The in vitro release kinetics (Figure 1B; Supplementary Table S2) for different coating thicknesses linearly correlates with the time to in vivo seroconversion (Figure 2F) as shown in Figure 2E ($R^2 = 0.9031$). Thus, vaccines which induce tunable, predictably delayed immune responses can be formulated by encapsulating spray-dried powders containing antigen with ALD alumina of varying thicknesses.

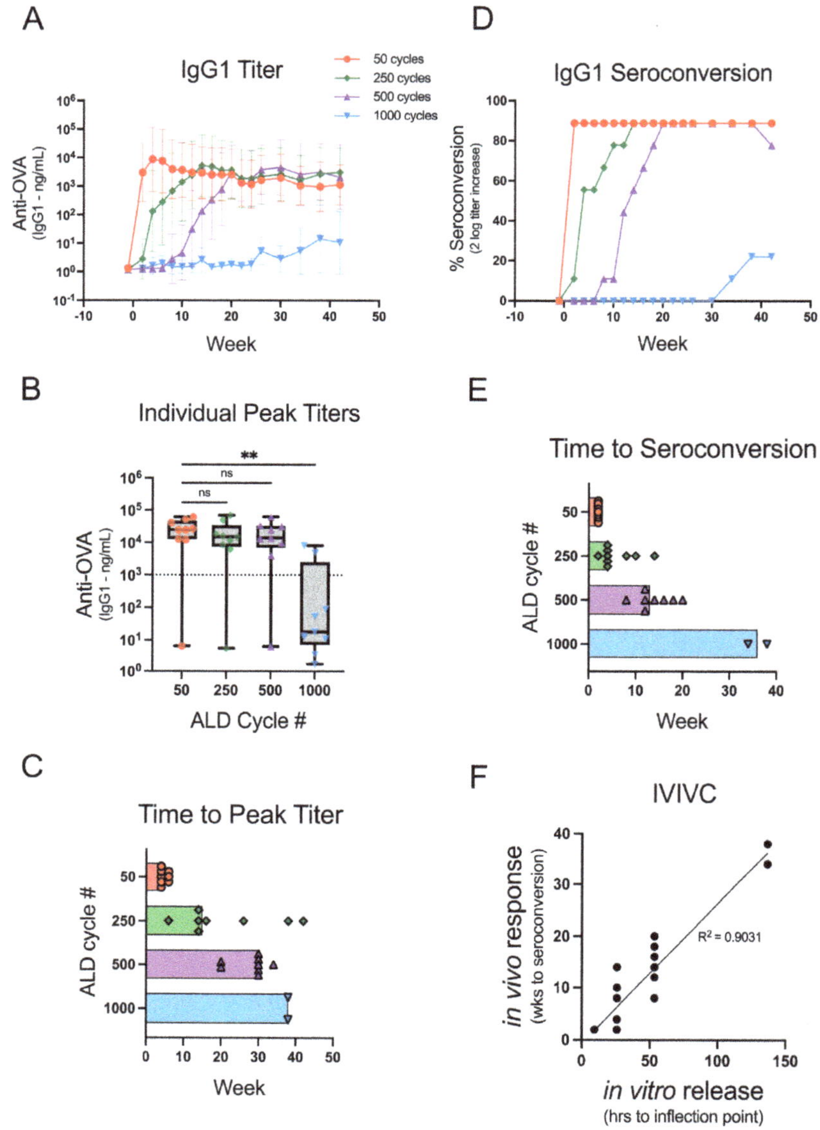

Figure 2. Coat thickness delays the antibody response times in vivo. Mice (n = 9) were administered a single 250 ng dose of OVA on day 0 using reconstituted powders coated with 50 (red circle), 250 (green diamond), 500 (purple triangle), or 1000 (blue triangle) cycles of ALD, which increases the coat thickness. Individual animal sera were analyzed for OVA-specific IgG1 antibody titers over the course of 42 weeks. (**A**) The GMT for each animal group is depicted over time with 95% confidence intervals. (**B**) The peak titers are shown for each group as box plots (+/−; min/max) with an overlay of individual data points. The dotted line on the graph at 10^3 ng/mL represents a cutoff between the responders and non-responders. (**C**) The time to peak titer for each individual mouse (excluding the non-responders) is shown, and the bars extend to the median time to peak titer for the group. (**D**) The percent seroconversion (as determined by a two-log increase in titer over pre-vaccination) is shown for each group over the course of 42 weeks. (**E**) The time to seroconversion for each individual mouse is shown, and the bars extend to the median time to seroconversion for the group. (**F**) An in vitro to in vivo correlation was drawn by plotting the time to the inflection point of the

in vitro dissolution (Figure 1B, Supplementary Table S2) against the individual mouse times to seroconversion. The correlation line was generated using a simple linear regression. Statistical analysis for panel B was performed by Kruskal-Wallis ANOVA with Dunn's post-hoc test ($\alpha = 0.05$) ** indicates p-value = 0.0024, ns indicates not significant, and # indicates number.

4. Discussion

The use of atomic layer deposition as a method to coat a thermostable dried vaccine product is a relatively new. Prior research has demonstrated how the approach maintains thermostability for three clinically relevant antigens, the L1 capsomere proteins from HPV 16, 18, and 31 [10]. In addition, ALD coating has been shown to impart delayed-release capabilities with the potential for creating single-administration vaccines [9].

In this report, we expand on these previous studies to show the temporal control of vaccine-induced immune response using ALD alumina-coated powders. An accelerated in vitro dissolution method was developed to characterize coated powders and was used to demonstrate that increasing ALD cycle numbers delayed the vaccine release (Figure 1A,B). Furthermore, the time to release in vitro was shown to correlate with the antibody response times in vivo (Figure 2F).

Robust and delayed antibody response times between 2 and 13 weeks were achieved by varying the thickness of the ALD coating. In an extreme example using particles coated with 1000 cycles, the data showed that response times 30 weeks after injection may be achievable with ALD technology (Figure 2). While impressive, this may not be necessary or desirable when considering the design of a clinical vaccine product. However, combining ALD-coated vaccines in a single administration, providing two doses (an immediate and delayed dose) separated by eight weeks, may be more desirable. The differently coated products could deliver the same antigen for a classic prime and boost, or they could contain different antigens depending on the goal. Future combination studies testing these concepts are needed to uncover factors to consider when making a combined product.

The performance characteristics of ALD-coated vaccines compared against competing vaccine platforms or adjuvants is another area for future research. Many head-to-head comparisons between technologies using different types of antigens are likely needed. Alhydrogel® and other aluminum-based adjuvants are generally considered to direct strong humoral or Th2 responses. It will be interesting to understand whether ALD-coated vaccines behave similarly or whether they stimulate distinct cellular immune responses.

Other approaches to achieving controlled-release vaccine technology have been reported previously [13–21]. To our knowledge, this is the first example demonstrating controlled-release vaccine technology that stimulates a delayed antibody response in the absence of any priming dose. This is critical for product development to obtain in vivo response times that align with a desired immunization schedule. This report also describes an approach for characterizing ALD-coated vaccine powders that may guide rational vaccine design to elicit immune responses that mimic repeated administrations of a standard vaccine product. ALD-coated vaccines provide a critical step toward generating single-administration products that can have a significant impact against preventable infectious diseases.

5. Conclusions

This study provides an example of how to establish an in vitro to in vivo correlation for the characterization of delayed-release vaccine technologies. Specifically, we show that increasing the thickness of ALD-coated vaccines results in longer delays before release in vitro that correlate with longer delays in antibody response times in vivo. The results demonstrate that ALD coating technology can be used to achieve tunable control over a wide range of immune response times from a vaccine.

Supplementary Materials: The following are available online at https://www.mdpi.com/article/10.3390/vaccines12070761/s1, Table S1: The 4PL terms generated from in vitro dissolution of powder coated with 100 ALD cycles and the 95% confidence intervals; Table S2: The 4PL terms generated from in vitro dissolution of powder coated with 50, 250, 500, and 1000 ALD cycles and the 95% confidence intervals.

Author Contributions: Conceptualization, S.W.B. and B.L.S.; methodology, S.W.B., I.R.W., E.M.H., L.R.A., E.L.P. and H.H.F.; investigation, S.W.B., I.R.W., E.M.H., L.R.A., E.L.P. and H.H.F.; writing—original draft preparation, S.W.B.; writing—review and editing, S.W.B., I.R.W., E.M.H., L.R.A., E.L.P. and H.H.F.; supervision, B.L.S. All authors have read and agreed to the published version of the manuscript.

Funding: This research received no external funding.

Institutional Review Board Statement: Not applicable.

Informed Consent Statement: Not applicable.

Data Availability Statement: The original contributions presented in the study are included in the article/Supplementary Material, further inquiries can be directed to the corresponding author.

Conflicts of Interest: S.W.B., I.R.W., E.M.H., L.R.A., E.L.P., H.H.F. and B.L.S. are employees of VitriVax, Inc. and are entitled to ownership in the form of stocks or shares. The research described in this report was privately funded. However, Vitrivax, Inc. has received support from commercial sources of funding by companies that sell drugs, medical devices, or provide medical services. In addition, VitriVax, Inc. has received research grant funding from the Defense Threat Reduction Agency, Department of Defense (HDTRA123C0009) and the Bill and Melinda Gates Foundation (INV-064824).

References

1. Kroger, A.; Bahta, L.; Long, S.; Sanchez, P. *General Best Practice Guidelines for Immunization*; Best Practice Guidance of the Advisory Committee on Immunization Practices (ACIP): Atlanta, GA, USA, 2023.
2. Plotkin, S.A.; Orenstein, W.A.; Offit, P.A. *Vaccines*, 6th ed.; Elsevier Saunders: Philadelphia, PA, USA, 2013; 1550p.
3. Rainey, J.J.; Watkins, M.; Ryman, T.K.; Sandhu, P.; Bo, A.; Banerjee, K. Reasons related to non-vaccination and under-vaccination of children in low and middle income countries: Findings from a systematic review of the published literature, 1999–2009. *Vaccine* **2011**, *29*, 8215–8221. [CrossRef] [PubMed]
4. Cleland, J.L. Single-administration vaccines: Controlled-release technology to mimic repeated immunizations. *Trends Biotechnol.* **1999**, *17*, 25–29. [CrossRef] [PubMed]
5. McHugh, K.J.; Guarecuco, R.; Langer, R.; Jaklenec, A. Single-injection vaccines: Progress, challenges, and opportunities. *J. Control Release* **2015**, *219*, 596–609. [CrossRef] [PubMed]
6. Li, W.; Meng, J.; Ma, X.; Lin, J.; Lu, X. Advanced materials for the delivery of vaccines for infectious diseases. *Biosaf. Health* **2022**, *4*, 95–104. [CrossRef]
7. Ray, S.; Puente, A.; Steinmetz, N.F.; Pokorski, J.K. Recent advancements in single dose slow-release devices for prophylactic vaccines. *Wiley Interdiscip Rev. Nanomed Nanobiotechnol.* **2023**, *15*, e1832. [CrossRef] [PubMed]
8. Laracuente, M.L.; Yu, M.H.; McHugh, K.J. Zero-order drug delivery: State of the art and future prospects. *J. Control Release* **2020**, *327*, 834–856. [CrossRef] [PubMed]
9. Garcea, R.L.; Meinerz, N.M.; Dong, M.; Funke, H.; Ghazvini, S.; Randolph, T.W. Single-administration, thermostable human papillomavirus vaccines prepared with atomic layer deposition technology. *NPJ Vaccines* **2020**, *5*, 45. [CrossRef] [PubMed]
10. Witeof, A.E.; Meinerz, N.M.; Walker, K.D.; Funke, H.H.; Garcea, R.L.; Randolph, T.W. A Single Dose, Thermostable, Trivalent Human Papillomavirus Vaccine Formulated Using Atomic Layer Deposition. *J. Pharm. Sci.* **2023**, *112*, 2223–2229. [CrossRef] [PubMed]
11. Olsen, C.H. Review of the use of statistics in infection and immunity. *Infect Immun.* **2003**, *71*, 6689–6692. [CrossRef] [PubMed]
12. Murphy, K.; Weaver, C.; Janeway, C. *Janeway's Immunobiology*, 9th ed.; W.W. Norton & Company: New York, NY, USA, 2017.
13. Chua, B.Y.; Sekiya, T.; Al Kobaisi, M.; Short, K.R.; Mainwaring, D.E.; Jackson, D.C. A single dose biodegradable vaccine depot that induces persistently high levels of antibody over a year. *Biomaterials* **2015**, *53*, 50–57. [CrossRef] [PubMed]
14. Guarecuco, R.; Lu, J.; McHugh, K.J.; Norman, J.J.; Thapa, L.S.; Lydon, E.; Langer, R.; Jaklenec, A. Immunogenicity of pulsatile-release PLGA microspheres for single-injection vaccination. *Vaccine* **2018**, *36*, 3161–3168. [CrossRef]
15. Tzeng, S.Y.; McHugh, K.J.; Behrens, A.M.; Rose, S.; Sugarman, J.L.; Ferber, S.; Langer, R.; Jaklenec, A. Stabilized single-injection inactivated polio vaccine elicits a strong neutralizing immune response. *Proc. Natl. Acad. Sci. USA* **2018**, *115*, E5269–E5278. [CrossRef] [PubMed]

16. Tran, K.T.M.; Gavitt, T.D.; Farrell, N.J.; Curry, E.J.; Mara, A.B.; Patel, A.; Brown, L.; Kilpatrick, S.; Piotrowska, R.; Mishra, N.; et al. Transdermal microneedles for the programmable burst release of multiple vaccine payloads. *Nat. Biomed. Eng.* **2021**, *5*, 998–1007. [CrossRef] [PubMed]
17. Tran, K.T.M.; Gavitt, T.D.; Le, T.T.; Graichen, A.; Lin, F.; Liu, Y.; Tulman, E.R.; Szczepanek, S.M.; Nguyen, T.D. A Single-Administration Microneedle Skin Patch for Multi-Burst Release of Vaccine against SARS-CoV-2. *Adv. Mater. Technol.* **2022**, *8*, 2200905. [CrossRef] [PubMed]
18. Wang, H.; Cui, L.; Luo, Y.; Zhou, X.; Liu, R.; Chen, Q.; Guan, Y.; Zhang, Y. Construction of single-injection vaccine using new time-controlled release system. *Biomater. Adv.* **2022**, *137*, 212812. [CrossRef] [PubMed]
19. Zhou, X.; Wang, H.; Zhang, J.; Guan, Y.; Zhang, Y. Single-injection subunit vaccine for rabies prevention using lentinan as adjuvant. *Int. J. Biol. Macromol.* **2024**, *254*, 128118. [CrossRef] [PubMed]
20. McHugh, K.J.; Nguyen, T.D.; Linehan, A.R.; Yang, D.; Behrens, A.M.; Rose, S.; Tochka, Z.L.; Tzeng, S.Y.; Norman, J.J.; Anselmo, A.C.; et al. Fabrication of fillable microparticles and other complex 3D microstructures. *Science* **2017**, *357*, 1138–1142. [CrossRef] [PubMed]
21. Roth, G.A.; Gale, E.C.; Alcantara-Hernandez, M.; Luo, W.; Axpe, E.; Verma, R.; Yin, Q.; Yu, A.C.; Lopez Hernandez, H.; Maikawa, C.L.; et al. Injectable Hydrogels for Sustained Codelivery of Subunit Vaccines Enhance Humoral Immunity. *ACS Cent. Sci.* **2020**, *6*, 1800–1812. [CrossRef] [PubMed]

vaccines

MDPI

Article

Polymeric Caffeic Acid Acts as an Antigen Delivery Carrier for Mucosal Vaccine Formulation by Forming a Complex with an Antigenic Protein

Rui Tada [1,*], Yuzuho Nagai [1], Miki Ogasawara [1], Momoko Saito [1], Akihiro Ohshima [1], Daisuke Yamanaka [2], Jun Kunisawa [3], Yoshiyuki Adachi [2] and Yoichi Negishi [1]

1 Department of Drug Delivery and Molecular Biopharmaceutics, School of Pharmacy, Tokyo University of Pharmacy and Life Sciences, 1432-1 Horinouchi, Hachioji 192-0392, Tokyo, Japan; yuzuho21@gmail.com (Y.N.); miki.ogasawara.0012012@gmail.com (M.O.); momomo3p@hotmail.co.jp (M.S.); ba.aky@docomo.ne.jp (A.O.); negishi@toyaku.ac.jp (Y.N.)

2 Laboratory for Immunopharmacology of Microbial Products, School of Pharmacy, Tokyo University of Pharmacy and Life Sciences, 1432-1 Horinouchi, Hachioji 192-0392, Tokyo, Japan; ymnk@toyaku.ac.jp (D.Y.); adachi@toyaku.ac.jp (Y.A.)

3 Laboratory of Vaccine Materials and Laboratory of Gut Environmental System, Microbial Research Center for Health and Medicine, National Institutes of Biomedical Innovation, Health and Nutrition (NIBIOHN), 7-6-8 Saito-Asagi, Ibaraki 567-0085, Osaka, Japan; kunisawa@nibiohn.go.jp

* Correspondence: ruitada@toyaku.ac.jp; Tel.: +81-42-676-5166

Abstract: The development of mucosal vaccines, which can generate antigen-specific immune responses in both the systemic and mucosal compartments, has been recognized as an effective strategy for combating infectious diseases caused by pathogenic microbes. Our recent research has focused on creating a nasal vaccine system in mice using enzymatically polymerized caffeic acid (pCA). However, we do not yet understand the molecular mechanisms by which pCA stimulates antigen-specific mucosal immune responses. In this study, we hypothesized that pCA might activate mucosal immunity at the site of administration based on our previous findings that pCA possesses immune-activating properties. However, contrary to our initial hypothesis, the intranasal administration of pCA did not enhance the expression of various genes involved in mucosal immune responses, including the enhancement of IgA responses. Therefore, we investigated whether pCA forms a complex with antigenic proteins and enhances antigen delivery to mucosal dendritic cells located in the lamina propria beneath the mucosal epithelial layer. Data from gel filtration chromatography indicated that pCA forms a complex with the antigenic protein ovalbumin (OVA). Furthermore, we examined the promotion of OVA delivery to nasal mucosal dendritic cells (mDCs) after the intranasal administration of pCA in combination with OVA and found that OVA uptake by mDCs was increased. Therefore, the data from gel filtration chromatography and flow cytometry imply that pCA enhances antigen-specific antibody production in both mucosal and systemic compartments by serving as an antigen-delivery vehicle.

Keywords: polyphenol; antigen delivery; mucosal vaccine; mucosal adjuvant

1. Introduction

Infectious diseases pose a significant risk, and despite advances in modern medicine, they remain challenging to eliminate. Currently, infectious diseases rank as the second most common cause of death globally [1,2]. The Coronavirus Disease 2019 (COVID-19) pandemic has highlighted the severe repercussions of highly virulent and contagious pathogenic microorganisms on a global scale. Although the creation of innovative RNA vaccinations and improvements in vaccination initiatives have been made, the number of new cases did not significantly decline worldwide during the pandemic [3,4]. Conventional vaccine techniques that involve introducing substances beneath the skin or within the muscle tissue have not yet been successful in eradicating the pandemic. One possible reason for this is their inability to stimulate immune responses within the mucous membranes, including the generation of immunoglobulin A (IgA), which targets particular pathogens [5]. Given that current vaccines have not been effective in preventing infection, particularly in regard to the initial colonization of mucosal surfaces by pathogens [6], developing a robust vaccine that elicits antigen-specific mucosal immune responses is essential for the preparation of the next pandemic [7].

Mucosal vaccination is widely regarded as the most effective approach for combating infectious diseases caused by pathogens. Recent advancements in mucosal vaccines have demonstrated their ability to provide superior protection against both systemic and mucosal infections, which are the primary entry points for most pathogenic microorganisms [8–10].

Despite the fact that mucosal vaccines have demonstrated greater effectiveness than conventional parenteral vaccines, only a limited number of live-attenuated and inactivated mucosal vaccines have been authorized for clinical use [11]. Considering that both forms of mucosal vaccines contain entire pathogenic microorganisms, there exists a persistent threat of unforeseen adverse consequences attributable to the toxicity and antigenicity of these pathogens. In order to effectively address this issue, the development of subunit mucosal vaccines that exclusively contain a microbial antigen is essential for the widespread implementation of mucosal vaccine systems in clinical settings. Subunit mucosal vaccines offer several advantages over live-attenuated or inactivated vaccines in terms of their ability to elicit a targeted immune response with fewer adverse effects. One of the key benefits of these vaccines is that they can be customized to incorporate specific antigens that trigger a protective immune response while avoiding the need to include the entire pathogen. This customization allows for a more targeted and effective approach to vaccine development [12]. Unfortunately, there are currently no vaccines of this kind that have been approved because of the lack of a proven and dependable method for stimulating mucosal immune responses in hosts using mucosal adjuvants [13,14] or for delivering microbial antigens to mucosal dendritic cells (mDCs) using antigen-delivery vehicles [15,16]. The primary cause of this situation is the insufficient ability of the mucosa to respond to antigenic proteins, which is the result of the immune system's tolerance to foreign antigens [11,17–19]. In addition, mucosal epithelial cells establish tight junctions that hinder the passage of macromolecules, such as antigens, across the physical barrier. This prevents access to mDCs situated in the lamina propria, which is essential for initiating mucosal immune responses [20]. A safer and more efficient method is necessary to stimulate antigen-specific immune responses in mucosal areas in order to develop successful mucosal vaccines.

Polyphenols, also known as phenolic compounds, are substances that occur naturally in plants, such as fruits, vegetables, and coffee. These compounds are consumed in significant amounts in our daily lives, with some people consuming up to 1 g per day [21,22]. In recent years, polyphenols have received considerable attention because of their potential health benefits. Research has demonstrated that the daily consumption of polyphenols can aid in preventing cardiovascular disease, cancer, and infectious diseases [23–25]. There are numerous sources of polyphenols, and their biological activities have been extensively studied. Traditionally, polyphenols are known for their antioxidant properties [26–28], but recent research suggests that they also have immunomodulatory effects on various immune cells. Polyphenols, particularly those with high molecular weights, like lignin, which is produced through the enzymatic polymerization of phenolic compounds, have been discovered to possess antiviral and antibacterial properties that bolster the immune system [29–31].

Our research has focused on the immune-regulating properties of lignin-like polyphenols that were generated in vitro through the use of horseradish peroxidase and phenylpropanoids. We have been exploring these effects for an extended period of time. Recently, it was discovered that intranasal immunization with an antigenic protein and enzymatically polymerized caffeic acid (pCA) induced specific antibody responses in both systemic and mucosal compartments in mice. This makes pCA a promising mucosal adjuvant for nasal vaccines against infectious diseases [32–34]. However, the molecular mechanisms underlying pCA's mucosal adjuvant properties are still unclear. Adjuvants generally enhance the uptake and presentation of antigens to antigen-presenting cells (APCs), particularly dendritic cells (DCs), and/or activate the innate immune system [35]. Our study aimed to uncover the mechanism responsible for the enhanced production of antigen-specific antibodies after the intranasal administration of pCA. Specifically, we investigated whether pCA not only activates innate immune responses at the administration site but also acts as an antigen-delivery vehicle, facilitating antigen delivery to the lamina propria, where mDCs are located.

2. Materials and Methods

2.1. Mice

Six-week-old female BALB/cCrSlc mice were procured from Japan SLC in Shizuoka, Japan, and kept under specific pathogen-free (SPF) conditions throughout the study. All experimental procedures involving animals were conducted using mice aged between 7 and 10 weeks. The animal experimental procedures were approved by the Tokyo University of Pharmacy and Life Sciences Committee for Laboratory Animal Experiments, and the approval numbers for these procedures were P16-12, P17-26, P18-71, and P19-58.

2.2. Materials

Sigma-Aldrich (St. Louis, MO, USA) was the supplier of horseradish peroxidase (HRP), and 3-(3,4-dihydroxyphenyl)-2-propenoic acid (also known as caffeic acid; CA) was obtained from Tokyo Chemical Industry Co., Ltd. (Tokyo, Japan). Low endotoxin egg white ovalbumin (OVA) (less than 1 EU/mg) and cholera toxin were procured from FUJIFILM Wako Pure Chemical Corporation (Osaka, Japan). Additionally, Thermo Fisher Scientific (Waltham, MA, USA) was the source of Alexa Fluor^{TM} 488-conjugated OVA (AF488-OVA).

2.3. Preparation of a Polymeric Caffeic Acid

A lignin-like polymeric caffeic acid was synthesized through oxidative polymerization, as described in our previous study [36]. The synthesis was initiated by neutralizing 200 mg of caffeic acid with sodium hydroxide and diluting the resulting solution to 10 mL with phosphate-buffered saline containing 1 mg of horseradish peroxidase (HRP). Next, we added 1.5 molar equivalents of hydrogen peroxide dropwise to the mixture while stirring for 1 h at 25 °C. The mixture was stirred for an additional two hours at 25 °C and then heated to 100 °C for 20 min to inactivate and precipitate the HRP. After centrifugation (12,000 rpm, 4 °C), the supernatant was collected and dialyzed against deionized water for two days using a regenerated cellulose membrane with a molecular weight cut-off of 50,000 (Repligen, Waltham, MA, USA). This process was followed by lyophilization to obtain the polymeric caffeic acid (pCA). We tested the pCA preparation for endotoxin contamination using an Endospecy ES-50 M kit from Seikagaku Biobusiness Corporation (Tokyo, Japan), and the results showed a very low endotoxin content of 231.5 pg/mg. To prepare a stock solution (10 mg/mL), we dissolved all samples in endotoxin-free phosphate-buffered saline (PBS; pH 7.4) from FUJIFILM Wako Pure Chemical Corporation and filtered the solution through 0.45 μm filter membranes from Osaka Chemical Co., Ltd. (Osaka, Japan) to remove any contaminants. The stock solution was then stored at −20 °C until use.

2.4. Expression of the Genes Induced by pCA at Nasal Mucosa

Mice were administered a mix of medetomidine (0.75 mg/kg), midazolam (4 mg/kg), and butorphanol tartrate (5 mg/kg) via intraperitoneal injection to induce anesthesia. Afterward, they were administered either PBS or pCA (100 μg/mouse) nasally, with a volume of 13 μL (6.5 μL per nostril). The mice were later euthanized by cervical dislocation, and their nasal tissues and spleens were collected for the examination of gene expression changes resulting from intranasal pCA administration. RNA was extracted from the nasal tissue using the FavorPrep Tissue Total RNA Mini Kit (Favorgen Biotech

Corporation, Ping-Tung, Taiwan) and treated with DNase I (Roche Life Science, Penzberg, Germany) to ensure purity. The RNA concentration was measured using spectrophotometry. Complementary DNA (cDNA) was synthesized using the ReverTra Ace qPCR RT Master Mix (Toyobo, Tokyo, Japan) following the manufacturer's instructions. qRT-PCR was performed on a CFX Connect Real-Time PCR Detection System (Bio-Rad, Hercules, CA, USA) using Thunderbird SYBR qPCR Mix (Toyobo) and the primers synthesized by Eurofins Genomics (Tokyo, Japan). For this research, we chose beta-2-microglobulin (B2M) as the housekeeping gene for normalization. B2M is well-established as a dependable reference gene in a variety of tissues and experimental conditions. Previous studies, including our own, have successfully utilized B2M as a housekeeping gene [32]. The expression of genes of interest was normalized to the reference gene B2M, and the relative gene expression among samples was calculated using the comparative Ct (ΔΔCt) method. Melting curve analysis was performed to confirm specific amplification, with the qRT-PCR conditions including an initial step at 95 °C for 1 min, followed by 40 cycles at 95 °C for 15 s and 60 °C for 1 min. The primer pairs used in the experiment are as follows: IL-1α, 5′-ACGTCAAGCAACGGGAAGATTC-3′ (forward) and 5′-ATGGGTTATGGACTGCAGGTC-3′ (reverse); IL-1β, 5′-TACAAGGAGAACCAAGCAACGAC-3′ (forward) and 5′-TGCCGTC TTTCATTACACAGGAC-3′ (reverse); IL-6, 5′-GAAATGATGGATGCTACCAAACTG-3′ (forward) and 5′-CTCTCTGAAGGACTCTGGCTTTG-3′ (reverse); GM-CSF, 5′-TGGGCATT GTGGTCTACAGC-3′ (forward) and 5′-GCGGGTCTGCACACATGTTA-3′ (reverse); TNF-α, 5′-ACCGTCAGCCGATTTGCTATC-3′ (forward) and 5′-ATGGGCTCATACCAGGGTTTG-3′ (reverse); CCL3, 5′-ACCATGACACTCTGCAACCAAG-3′ (forward) and 5′-CAACGATG AATTGGCGTGGAATC-3′ (reverse); TGF-β, 5′-TACGCCTGAGTGGCTGTCTTTTG-3′ (forward) and 5′-CGTGGAGTTTGTTATCTTTGCTGTC-3′ (reverse); IFN-β1, 5′-TTACACTGCC TTTGCCATCC-3′ (forward) and 5′-ACTGTCTGCTGGTGGAGTTCAT-3′ (reverse); IL-12p40, 5′-TGGTTTGCCATCGTTTTGCTG-3′ (forward) and 5′-ACAGGTGAGGTTCACTGT TTCT-3′ (reverse); β2-microglobulin (B2M), 5′-TTCTGGTGCTTGTCTCACTGA-3′ (forward) and 5′-CAGTATGTTCGGCTTCCCATTC-3′ (reverse). The CFX manager software version 3.1 (BIO-RAD) was used to analyze the data, and cycle threshold (Ct) values were calculated. The results are expressed as relative gene expression levels normalized to B2M.

2.5. Assessment of Complex Formation of pCA and Antigen

The formation of complexes between pCA and OVA was determined by conducting an analysis using size-exclusion chromatography (SEC). A mixture of pCA and OVA was prepared in a mass ratio of 1:10 and allowed to stand at 25 °C for 30 min. Subsequently, the mixture was subjected to SEC. After the incubation process, the solution was passed through a filter membrane with a pore size of 0.45 μm to eliminate any particulate matter that was insoluble. High-performance liquid chromatography (HPLC) was used for this analysis with a size-exclusion column (OHpak SB-803 HQ column, 8.0 × 300 mm; Resonac Corporation, Tokyo, Japan). Throughout the experiment, PBS was used as the mobile phase, with a flow rate of 0.5 mL/min. Eluted compounds were detected using a UV detector set at a wavelength of 280 nm.

2.6. Flow Cytometric Analysis of Antigen Uptake

Six hours after intranasally administering 50 μg/mouse AF488-OVA and 100 μg/mouse pCA to mice, single-cell suspensions were obtained from nasal tissues to analyze the uptake of AF488-OVA by dendritic cells (DCs) in the nasal mucosa. The nasal tissues were homogenized in PBS and centrifuged. To obtain single-cell suspensions, red blood cells were lysed with RBC lysis buffer (BioLegend, San Diego, CA, USA). The cells were then washed with staining buffer (PBS containing 2% heat-inactivated fetal bovine serum (FBS; Biowest, Nuaillé, France) and 0.1% sodium azide) and blocked with anti-mouse CD16/CD32 (clone 2.4G2, Tonbo Biosciences, San Diego, CA, USA) for 20 min on ice to block Fc receptors. After being washed with staining buffer, cells were stained with PE/Cy7 anti-mouse CD45 (clone 30-F11, BioLegend) in combination with APC anti-mouse CD11c (clone N418, BioLegend) or an isotype control for 30 min on ice. Finally, the samples were analyzed using a FACSCanto instrument (BD Biosciences, San Jose, CA, USA).

2.7. Statistics

Welch's correction was employed to conduct an unpaired *t*-test and determine statistical differences. The significance level was set at 0.05, and all data were analyzed using GraphPad Prism 9 software (GraphPad Software, La Jolla, CA, USA).

3. Results

3.1. Expression of Genes Related to the Mucosal Immune Responses of Mice That Received Nasal Administration of pCA

Typically, the induction of antigen-specific immune responses involves the activation of the innate immune system and/or enhancement of antigen uptake and presentation by APCs, such as DCs [35]. Soluble factors, such as cytokines, secreted by both immune and non-immune cells typically regulate immune responses to antigens [37–39]. Previous studies have shown that pCA induces the secretion of cytokines by various immune cells [36,40,41]. We aimed to investigate the mechanisms underlying the mucosal adjuvant activity of pCA by identifying genes involved in innate immune responses that have the potential to enhance mucosal IgA responses (interleukin (IL-1)α, IL-1β, IL-6, granulocyte-macrophage colony-stimulating factor (GM-CSF), TNF-α, C-C motif chemokine ligand 3 (CCL3), transforming growth factor-β (TGF-β), IFN-β1, and IL-12p40) following the intranasal administration of pCA. The results shown in Figure 1 demonstrate that the intranasal administration of pCA did not stimulate the expression of genes associated with innate immune responses that could potentially enhance mucosal IgA responses in the nasal region [42–48]. These results clearly indicated that the activation of mucosal immunity through the activation of innate immune responses did not occur as a result of pCA administration.

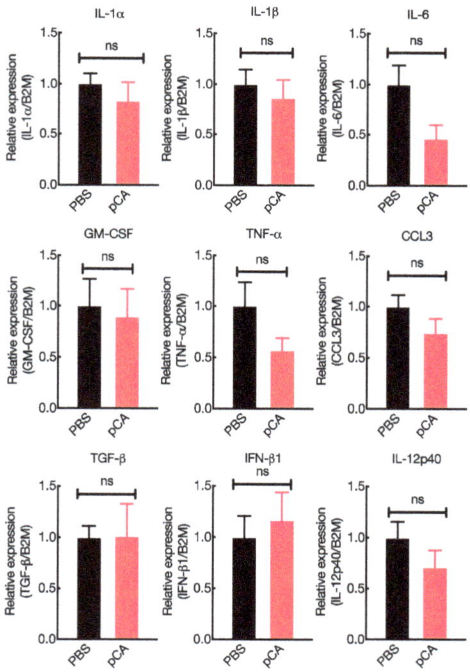

Figure 1. Gene expression associated with the development of antigen-specific IgA responses in mice following the intranasal administration of pCA. Nasal tissues were obtained from mice that received

either PBS or pCA (100 μg/mouse) and harvested 16 h after administration. Total RNA was extracted and cDNA synthesized, and gene expression changes were determined using quantitative reverse transcriptase-polymerase chain reaction (qRT-PCR). The average ± standard deviation (SD) values of three biological replicates were reported. The experiments were independently repeated three times. Statistical differences were assessed using an unpaired t-test with Welch's correction, and a p-value of less than 0.05 was considered statistically significant. ns: not significant ($p > 0.05$), * $p < 0.05$.

3.2. pCA Forms Complexes with Antigenic Proteins

For the induction of antigen-specific mucosal immune responses, it is crucial that the antigen be phagocytosed by antigen-presenting cells, such as mucosal dendritic cells, which are located in the lamina propria underlying the mucosal epithelial layers. Thus, improving the efficiency of antigen delivery to mucosal dendritic cells is critical for the development of mucosal vaccines [49,50]. To achieve this, it is essential that pCA functions as an antigenic protein carrier; that is, pCA interacts with antigenic proteins to form complexes. Therefore, we investigated the potential of pCA to form complexes with antigenic proteins by size-exclusion chromatography (SEC). Figure 2 shows the SEC chromatograms of pCA and OVA separately and their mixtures. The SEC chromatogram showed monomeric peaks eluting at 17.2 and 15.5 min, representing pCA and OVA, respectively. In addition, the SEC chromatogram of the pCA–OVA mixture showed a distinct new elution peak at 12.2 min, indicating the formation of the pCA–OVA complex.

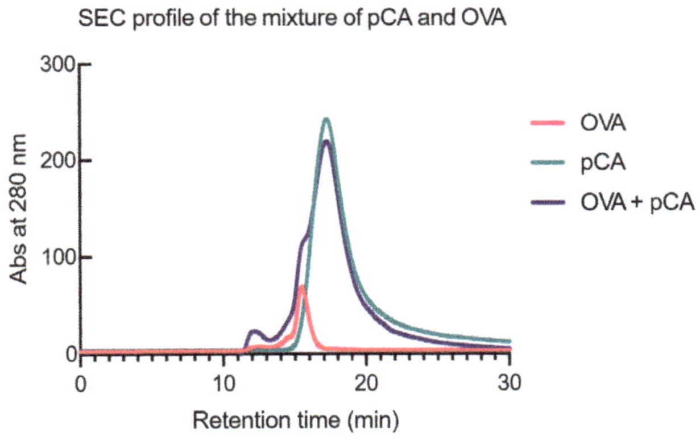

Figure 2. Gel filtration chromatography analysis examining the pCA complex formation with OVA. The OHpak SB-803 HQ column (8.0 × 300 mm) was pre-equilibrated with PBS and eluted with PBS at a flow rate of 0.5 mL/min. The samples, comprising OVA alone, pCA alone, or a mixture of pCA and OVA incubated at room temperature, were injected into the column and monitored using a UV detector set to 280 nm.

3.3. pCA Promotes the Delivery of Antigenic Proteins to Nasal DCs

The successful delivery of antigenic proteins to antigen-presenting cells (APCs) is crucial for inducing antigen-specific immunity in the host. In this study, we examined the effect of pCA on the uptake of AF488-OVA by nasal dendritic cells (DCs). Our flow cytometric analyses, as shown in Figure 3, revealed a significant increase in the proportion of $AF488^+$ cells among the $CD11c^+$ cells in the DC population of mice receiving both AF488-OVA and pCA (2.5 ± 0.30%) compared with those receiving AF488-OVA alone (1.1 ± 0.24%). This suggests that the pCA-enhanced uptake may serve as a mechanism for inducing both mucosal and systemic antigen-specific immune responses.

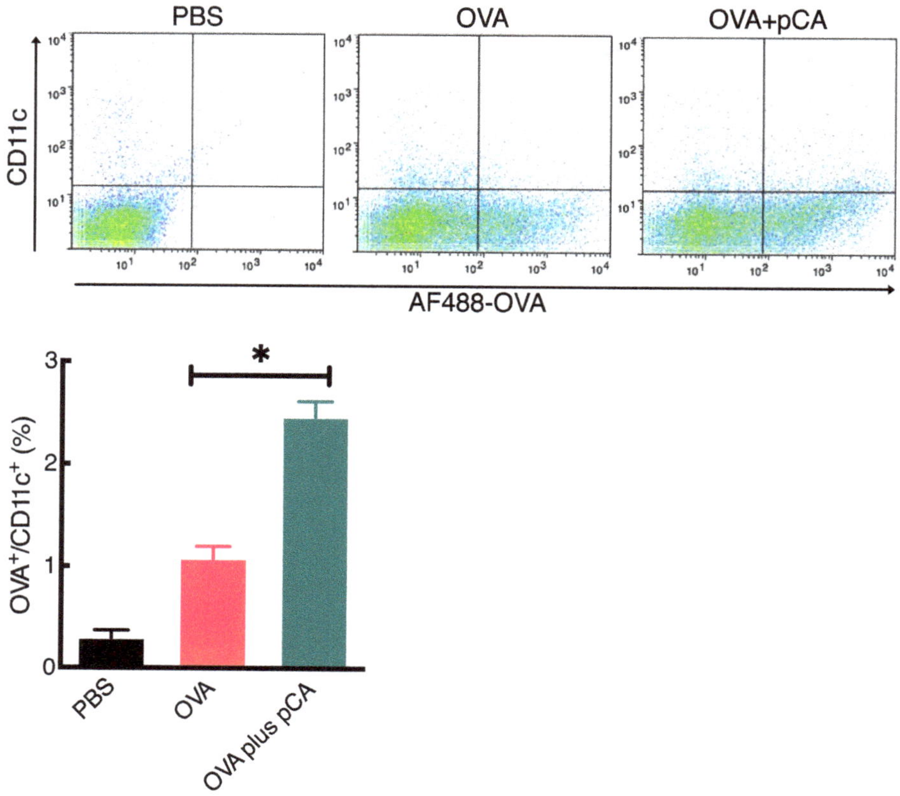

Figure 3. The uptake of OVA by nasal $CD11c^+$ dendritic cells (DCs) following the intranasal administration of OVA with pCA. To investigate this, female BALB/c mice were immunized intranasally with PBS, OVA alone (50 µg/mouse), or OVA with pCA (50 µg/mouse and 100 µg/mouse, respectively). After 6 h, single-cell suspensions were isolated from the nasal tissues and treated with anti-mouse CD11c mAb or isotype control for staining. The uptake of AF488-OVA was evaluated using flow cytometry, with four mice in each group used for analysis. A *t*-test with Welch's correction was used to determine significance, and the results showed a *p*-value of less than 0.05. * $p < 0.05$.

4. Discussion

The scientific community has prioritized the development of safe and efficient vaccines that can protect against infectious diseases, particularly after the global pandemic caused by severe acute respiratory syndrome coronavirus 2 (SARS-CoV-2). This study aimed to explore the potential of enzymatically polymerized caffeic acid (pCA) as a mucosal adjuvant for nasal vaccines against infectious diseases. Prior research has suggested that pCA, along with other enzymatically polymerized polyphenols, can enhance antigen-specific immune responses when administered intranasally with antigenic proteins. However, the precise molecular mechanisms underlying the mucosal adjuvant effects of pCA remain unidentified.

The study findings, which reveal that pCA forms complexes with antigenic proteins and enhances antigen uptake by mucosal dendritic cells, provide insights into the molecular mechanisms underlying pCA's function as a promising adjuvant for mucosal vaccines. The formation of complexes between pCA and antigens may facilitate the antigen passage through the mucosal barrier and promote antigen delivery to mucosal dendritic cells, which is crucial for inducing mucosal immune responses. Additionally, the results suggest that pCA may directly enhance antigen uptake and presentation by mucosal dendritic cells. Future research is anticipated to elucidate the detailed mechanisms by which the pCA–antigen complex formation contributes to the induction of mucosal immune responses.

This study aimed to investigate the mechanisms underlying the mucosal adjuvant effects of pCA. Adjuvants typically work through two mechanisms: activating the innate immune system and improving antigen uptake and presentation to APCs, particularly DCs [35]. Additionally, previous research has shown that pCA has immunomodulatory properties. The oral administration of pCA has been found to activate NK cells and induce cytokine production in mouse splenocytes in vitro [36,40,41,51]. Therefore, it is believed that pCA enhances antigen-specific immune responses by activating innate immunity in the mucosa. However, contrary to our initial expectations, the intranasal administration of pCA did not stimulate the expression of genes associated with innate immune responses that could potentially enhance mucosal IgA responses in the nasal region (Figure 1). Next, we investigated the potential of pCA as a delivery vehicle for antigenic proteins in mucosal dendritic cells. Recent studies have shown that polyphenols can effectively interact with proteins through non-covalent intermolecular forces, such as hydrogen bonding, hydrophobic interactions, van der Waals forces, and electrostatic interactions. These interactions result in the formation of soluble protein–polyphenol complexes, which can affect the bioavailability and properties of both components [52–55]. In our SEC analysis, we observed that the mixture of pCA and OVA eluted faster than the individual components, indicating the formation of a complex between pCA and OVA (Figure 2). Moreover, pCA acted as an antigen-delivery vehicle, facilitating antigen delivery to mDCs (Figure 3). This could potentially lead to an enhanced production of antigen-specific antibodies in both systemic and mucosal compartments. However, the study has some limitations. It is not clear how pCA enhances the delivery of antigenic proteins to mDCs in the lamina propria. While it is known that low-molecular-weight polyphenols can cross the small intestine without mucosal absorption and are transported by active transport or sodium-dependent glucose transport [56–60], high-molecular-weight polyphenols have limited tissue penetration due to their polymeric and hydrophilic nature [61]. Nevertheless, given that pCA has been shown to activate various immune cells in vitro [40,41,51,62], it is possible that there is an unknown transport pathway involving specific receptors. Further research is needed to understand the molecular mechanisms involved. Additionally, the observed activation of splenocytes and dendritic cells by pCA, leading to cytokine production, suggests a potential interaction with dendritic cell membrane receptors [36]. This interaction could facilitate antigen uptake and presentation, a hypothesis supported by our previous research indicating enhanced dendritic cell uptake through the complex formation between pCA and antigens. Identifying the specific receptors and pathways involved in pCA's immunomodulatory effects remains a critical area for future investigation.

This research highlights the potential of pCA as an antigen delivery carrier, but it is important to acknowledge certain limitations. Firstly, the model antigen OVA utilized in this study may not be comparable to actual pathogenic antigens. Future studies must investigate the mucosal adjuvant effects of pCA using disease-specific antigens. Moreover, preclinical trials involving human cells or tissues, along with a comprehensive safety profile assessment, will be necessary for clinical application. Furthermore, to gain a deeper insight into the mechanisms underlying pCA's mucosal adjuvant effects, it is crucial to examine the influence of the pCA–antigen complex formation on mucosal immune responses at the molecular and cellular levels.

The results of this study have potential implications for the development of safe and effective mucosal vaccines against various infectious diseases. The use of pCA as a mucosal adjuvant could greatly enhance the efficacy of mucosal vaccines by facilitating antigen delivery to nasal DCs while minimizing unexpected side effects due to pathogen toxicity and antigenicity. Further preclinical and clinical studies on the safety and efficacy of pCA are necessary to develop novel mucosal vaccines that can effectively prevent infectious diseases.

The mucosal immune system is a vital initial line of defense against pathogens that enter through mucosal surfaces, such as the respiratory and gastrointestinal tracts. The development of effective mucosal vaccines has been a long-standing challenge due to the unique features of the mucosal immune system, including the presence of physical and chemical barriers, the complexity of mucosal antigen presentation, and the induction of tolerance. The use of mucosal adjuvants, such as pCA, represents a promising strategy to overcome these challenges and enhance the efficacy of mucosal vaccines. By facilitating antigen delivery to mucosal dendritic cells and potentially modulating immune responses, mucosal adjuvants can help to break mucosal tolerance and induce robust antigen-specific immunity. The results of this study, demonstrating the capacity of pCA to form complexes with antigenic proteins and enhance the antigen uptake by nasal dendritic cells, provide a strong rationale for further exploring its potential as a mucosal adjuvant. Future studies should investigate the impact of pCA on the induction of mucosal antibody responses, such as secretory IgA, which plays a crucial role in neutralizing pathogens at mucosal surfaces. Additionally, evaluating the ability of pCA to induce long-lasting memory immune responses and its compatibility with different vaccine platforms, such as subunit vaccines or nanoparticle-based vaccines, will be important for its clinical translation.

The importance of this research lies in its showing of pCA's capacity to serve as a promising adjuvant for mucosal vaccines. The creation of effective mucosal adjuvants is a critical challenge in the development of mucosal vaccines for the prevention of infectious diseases. As an antigen-delivery carrier, pCA has the potential to emerge as a novel mucosal adjuvant capable of effectively inducing mucosal immune responses. Furthermore, pCA is a safe and naturally derived substance, which suggests that the barriers to clinical application may be relatively low. According to the results of this study, the development of mucosal vaccines using pCA is expected to make a significant contribution to the prevention of infectious diseases in the future.

5. Conclusions

The findings of this study indicate that enzymatically polymerized caffeic acid (pCA) presents a promising prospect as a mucosal adjuvant for nasal vaccines. The capacity of pCA to form complexes with antigenic proteins and enhance antigen uptake by mucosal dendritic cells suggests its potential to augment the efficacy of mucosal vaccines significantly. In facilitating the delivery of antigens to nasal dendritic cells and potentially improving antigen passage through the mucosal barrier, pCA offers a novel strategy to surmount the limitations of conventional vaccine formulations. As a naturally derived substance, pCA's favorable safety profile further supports its potential for clinical applications. To fully recognize the potential of pCA as a mucosal adjuvant, future investigations should concentrate on exploring its efficacy using disease-specific antigens and evaluating its safety in human cell or tissue models. Moreover, elucidating the molecular mechanisms underpinning pCA's immunomodulatory effects will provide invaluable insights into its role in enhancing mucosal immune responses. The development of pCA as a mucosal adjuvant represents a considerable advancement in the field of vaccine development. By harnessing the potential of this innovative approach, we can create safer and more effective mucosal vaccines to prevent a wide array of infectious diseases. The continued research and clinical development of pCA-based mucosal vaccines hold great promise for addressing the global burden of infectious diseases and enhancing public health outcomes worldwide.

Author Contributions: Conceptualization, R.T.; data curation, R.T., Y.N. (Yuzuho Nagai), M.O., M.S., A.O. and D.Y.; formal analysis, R.T.; funding acquisition, R.T., D.Y., J.K., Y.A. and Y.N. (Yoichi Negishi); investigation, R.T., Y.N. (Yuzuho Nagai), M.O. and D.Y.; methodology, R.T. and D.Y.; project administration, R.T. and Y.N. (Yoichi Negishi); resources, R.T., J.K., Y.A. and Y.N. (Yoichi Negishi); supervision, R.T. and Y.N. (Yoichi Negishi); visualization, R.T. and Y.N. (Yuzuho Nagai); writing—original draft preparation, R.T.; writing—review and editing, D.Y., J.K., Y.A. and Y.N. (Yoichi Negishi). All authors have read and agreed to the published version of the manuscript.

Funding: This study was partly funded by JSPS KAKENHI Grant Number 21K06715 (Grant-in-Aid for Scientific Research C to R.T.) and the Japan Agency for Medical Research and Development (grant numbers: 223fa727001h0001 to J.K.).

Institutional Review Board Statement: Animal use protocol and relevant experiments procedures were approved by the Committee of Animal Use and Welfare of Tokyo University of Pharmacy and Life Sciences (authorization numbers: P16-12, P17-26, P18-71, and P19-58).

Informed Consent Statement: Not applicable.

Data Availability Statement: The data presented in this study are available in the article.

Conflicts of Interest: The authors declare no conflicts of interest.

References

1. Fauci, A.S.; Touchette, N.A.; Folkers, G.K. Emerging Infectious Diseases: A 10-Year Perspective from the National Institute of Allergy and Infectious Diseases. *Emerg. Infect. Dis.* **2005**, *11*, 519–525. [CrossRef] [PubMed]
2. Fauci, A.S. Infectious Diseases: Considerations for the 21st Century. *Clin. Infect. Dis.* **2001**, *32*, 675–685. [CrossRef] [PubMed]
3. Rothan, H.A.; Byrareddy, S.N. The Epidemiology and Pathogenesis of Coronavirus Disease (COVID-19) Outbreak. *J. Autoimmun.* **2020**, *109*, 102433. [CrossRef] [PubMed]
4. Huang, C.; Wang, Y.; Li, X.; Ren, L.; Zhao, J.; Hu, Y.; Zhang, L.; Fan, G.; Xu, J.; Gu, X.; et al. Clinical Features of Patients Infected with 2019 Novel Coronavirus in Wuhan, China. *Lancet* **2020**, *395*, 497–506. [CrossRef] [PubMed]
5. Alturaiki, W. Considerations for Novel COVID-19 Mucosal Vaccine Development. *Vaccines* **2022**, *10*, 1173. [CrossRef] [PubMed]
6. Bleier, B.S.; Ramanathan, M., Jr.; Lane, A.P. COVID-19 Vaccines May Not Prevent Nasal SARS-CoV-2 Infection and Asymptomatic Transmission. *Otolaryngol. Head Neck Surg.* **2021**, *164*, 305–307. [CrossRef] [PubMed]
7. Lapuente, D.; Fuchs, J.; Willar, J.; Vieira Antão, A.; Eberlein, V.; Uhlig, N.; Issmail, L.; Schmidt, A.; Oltmanns, F.; Peter, A.S.; et al. Protective Mucosal Immunity against SARS-CoV-2 after Heterologous Systemic Prime-Mucosal Boost Immunization. *Nat. Commun.* **2021**, *12*, 6871. [CrossRef] [PubMed]
8. Tamura, S.-I.; Kurata, T. Defense Mechanisms against Influenza Virus Infection in the Respiratory Tract Mucosa. *Jpn. J. Infect. Dis.* **2004**, *57*, 236–247. [CrossRef] [PubMed]
9. Moyes, D.L.; Naglik, J.R. Mucosal Immunity and Candida Albicans Infection. *Clin. Dev. Immunol.* **2011**, *2011*, 346307. [CrossRef]
10. Faden, H.; Duffy, L.; Wasielewski, R.; Wolf, J.; Krystofik, D.; Tung, Y.; Pediatrics, T. Relationship between Nasopharyngeal Colonization and the Development of Otitis Media in Children. *J. Infect. Dis.* **1997**, *175*, 1440–1445. [CrossRef] [PubMed]
11. Kim, S.-H.; Jang, Y.-S. The Development of Mucosal Vaccines for Both Mucosal and Systemic Immune Induction and the Roles Played by Adjuvants. *Clin. Exp. Vaccine Res.* **2017**, *6*, 15–21. [CrossRef] [PubMed]
12. Lavelle, E.C.; Ward, R.W. Mucosal Vaccines—Fortifying the Frontiers. *Nat. Rev. Immunol.* **2022**, *22*, 236–250. [CrossRef] [PubMed]
13. Rhee, J.H.; Lee, S.E.; Kim, S.Y. Mucosal Vaccine Adjuvants Update. *Clin. Exp. Vaccine Res.* **2012**, *1*, 50–63. [CrossRef] [PubMed]
14. Aoshi, T. Modes of Action for Mucosal Vaccine Adjuvants. *Viral Immunol.* **2017**, *30*, 463–470. [CrossRef] [PubMed]
15. Nakahashi-Ouchida, R.; Yuki, Y.; Kiyono, H. Development of a Nanogel-Based Nasal Vaccine as a Novel Antigen Delivery System. *Expert Rev. Vaccines* **2017**, *16*, 1231–1240. [CrossRef] [PubMed]
16. Zho, F.; Neutra, M.R. Antigen Delivery to Mucosa-Associated Lymphoid Tissues Using Liposomes as a Carrier. *Biosci. Rep.* **2002**, *22*, 355–369. [CrossRef] [PubMed]
17. Czerkinsky, C.; Anjuere, F.; McGhee, J.R.; George-Chandy, A.; Holmgren, J.; Kieny, M.P.; Fujiyashi, K.; Mestecky, J.F.; Pierrefite-Carle, V.; Rask, C.; et al. Mucosal Immunity and Tolerance: Relevance to Vaccine Development. *Immunol. Rev.* **1999**, *170*, 197–222. [CrossRef] [PubMed]
18. Mestecky, J.; Moldoveanu, Z.; Elson, C.O. Immune Response versus Mucosal Tolerance to Mucosally Administered Antigens. *Vaccine* **2005**, *23*, 1800–1803. [CrossRef]
19. Mestecky, J.; Russell, M.W.; Elson, C.O. Perspectives on Mucosal Vaccines: Is Mucosal Tolerance a Barrier? *J. Immunol.* **2007**, *179*, 5633–5638. [CrossRef]
20. Corthésy, B.; Bioley, G. Lipid-Based Particles: Versatile Delivery Systems for Mucosal Vaccination against Infection. *Front. Immunol.* **2018**, *9*, 431. [CrossRef]
21. Grosso, G.; Stepaniak, U.; Topor-Mądry, R.; Szafraniec, K.; Pająk, A. Estimated Dietary Intake and Major Food Sources of Polyphenols in the Polish Arm of the HAPIEE Study. *Nutrition* **2014**, *30*, 1398–1403. [CrossRef] [PubMed]
22. Scalbert, A.; Williamson, G. Dietary Intake and Bioavailability of Polyphenols. *J. Nutr.* **2000**, *130*, 2073S–2085S. [CrossRef]
23. Corcoran, M.P.; McKay, D.L.; Blumberg, J.B. Flavonoid Basics: Chemistry, Sources, Mechanisms of Action, and Safety. *J. Nutr. Gerontol. Geriatr.* **2012**, *31*, 176–189. [CrossRef] [PubMed]
24. Tomás-Barberán, F.A.; Andrés-Lacueva, C. Polyphenols and Health: Current State and Progress. *J. Agric. Food Chem.* **2012**, *60*, 8773–8775. [CrossRef] [PubMed]
25. Kishimoto, Y.; Tani, M.; Kondo, K. Pleiotropic Preventive Effects of Dietary Polyphenols in Cardiovascular Diseases. *Eur. J. Clin. Nutr.* **2013**, *67*, 532–535. [CrossRef] [PubMed]

26. Stoclet, J.-C.; Chataigneau, T.; Ndiaye, M.; Oak, M.-H.; El Bedoui, J.; Chataigneau, M.; Schini-Kerth, V.B. Vascular Protection by Dietary Polyphenols. *Eur. J. Pharmacol.* **2004**, *500*, 299–313. [CrossRef] [PubMed]
27. Yan, J.; Zhao, Y.; Suo, S.; Liu, Y.; Zhao, B. Green Tea Catechins Ameliorate Adipose Insulin Resistance by Improving Oxidative Stress. *Free Radic. Biol. Med.* **2012**, *52*, 1648–1657. [CrossRef] [PubMed]
28. Meng, Q.; Velalar, C.N.; Ruan, R. Regulating the Age-Related Oxidative Damage, Mitochondrial Integrity, and Antioxidative Enzyme Activity in Fischer 344 Rats by Supplementation of the Antioxidant Epigallocatechin-3-Gallate. *Rejuvenation Res.* **2008**, *11*, 649–660. [CrossRef]
29. Singh, S.; Aggarwal, B.B. Activation of Transcription Factor NF-Kappa B Is Suppressed by Curcumin (Diferuloylmethane) [Corrected]. *J. Biol. Chem.* **1995**, *270*, 24995–25000. [CrossRef]
30. Zhong, Y.; Chiou, Y.-S.; Pan, M.-H.; Shahidi, F. Anti-Inflammatory Activity of Lipophilic Epigallocatechin Gallate (EGCG) Derivatives in LPS-Stimulated Murine Macrophages. *Food Chem.* **2012**, *134*, 742–748. [CrossRef]
31. Pan, W.; Yu, H.; Huang, S.; Zhu, P. Resveratrol Protects against TNF-α-Induced Injury in Human Umbilical Endothelial Cells through Promoting Sirtuin-1-Induced Repression of NF-KB and P38 MAPK. *PLoS ONE* **2016**, *11*, e0147034. [CrossRef]
32. Tada, R.; Yamanaka, D.; Ogasawara, M.; Saito, M.; Ohno, N.; Kiyono, H.; Kunisawa, J.; Aramaki, Y. Polymeric Caffeic Acid Is a Safer Mucosal Adjuvant That Augments Antigen-Specific Mucosal and Systemic Immune Responses in Mice. *Mol. Pharm.* **2018**, *15*, 4226–4234. [CrossRef]
33. Tada, R.; Ogasawara, M.; Yamanaka, D.; Sakurai, Y.; Negishi, Y.; Kiyono, H.; Ohno, N.; Kunisawa, J.; Aramaki, Y. Enzymatically Polymerised Polyphenols Prepared from Various Precursors Potentiate Antigen-Specific Immune Responses in Both Mucosal and Systemic Compartments in Mice. *PLoS ONE* **2021**, *16*, e0246422. [CrossRef]
34. Tada, R.; Suzuki, H.; Ogasawara, M.; Yamanaka, D.; Adachi, Y.; Kunisawa, J.; Negishi, Y. Polymeric Caffeic Acid Acts as a Nasal Vaccine Formulation against Streptococcus Pneumoniae Infections in Mice. *Pharmaceutics* **2021**, *13*, 585. [CrossRef]
35. Awate, S.; Babiuk, L.A.; Mutwiri, G. Mechanisms of Action of Adjuvants. *Front. Immunol.* **2013**, *4*, 114. [CrossRef] [PubMed]
36. Yamanaka, D.; Tamiya, Y.; Motoi, M.; Ishibashi, K.-I.; Miura, N.N.; Adachi, Y.; Ohno, N. The Effect of Enzymatically Polymerised Polyphenols on CD4 Binding and Cytokine Production in Murine Splenocytes. *PLoS ONE* **2012**, *7*, e36025. [CrossRef]
37. Thompson, A.L.; Staats, H.F. Cytokines: The Future of Intranasal Vaccine Adjuvants. *Clin. Dev. Immunol.* **2011**, *2011*, 289597. [CrossRef]
38. Boyaka, P.N.; McGhee, J.R. Cytokines as Adjuvants for the Induction of Mucosal Immunity. *Adv. Drug Deliv. Rev.* **2001**, *51*, 71–79. [CrossRef]
39. Boyaka, P.N.; Lillard, J.W.; McGhee, J. Interleukin 12 and Innate Molecules for Enhanced Mucosal Immunity. *Immunol. Res.* **1999**, *20*, 207–217. [CrossRef]
40. Yamanaka, D.; Motoi, M.; Ishibashi, K.-I.; Miura, N.N.; Adachi, Y.; Ohno, N. Modulation of Interferon-γ Synthesis by the Effects of Lignin-like Enzymatically Polymerized Polyphenols on Antigen-Presenting Cell Activation and the Subsequent Cell-to-Cell Interactions. *Food Chem.* **2013**, *141*, 4073–4080. [CrossRef] [PubMed]
41. Yamanaka, D.; Ishibashi, K.-I.; Adachi, Y.; Ohno, N. Species Difference in Reactivity to Lignin-like Enzymatically Polymerized Polyphenols on Interferon-γ Synthesis and Involvement of Interleukin-2 Production in Mice. *Int. Immunopharmacol.* **2016**, *38*, 443–449. [CrossRef]
42. Kayamuro, H.; Yoshioka, Y.; Abe, Y.; Arita, S.; Katayama, K.; Nomura, T.; Yoshikawa, T.; Kubota-Koketsu, R.; Ikuta, K.; Okamoto, S.; et al. Interleukin-1 Family Cytokines as Mucosal Vaccine Adjuvants for Induction of Protective Immunity against Influenza Virus. *J. Virol.* **2010**, *84*, 12703–12712. [CrossRef]
43. Loudon, P.T.; Yager, E.J.; Lynch, D.T.; Narendran, A.; Stagnar, C.; Franchini, A.M.; Fuller, J.T.; White, P.A.; Nyuandi, J.; Wiley, C.A.; et al. GM-CSF Increases Mucosal and Systemic Immunogenicity of an H1N1 Influenza DNA Vaccine Administered into the Epidermis of Non-Human Primates. *PLoS ONE* **2010**, *5*, e11021. [CrossRef] [PubMed]
44. Kayamuro, H.; Abe, Y.; Yoshioka, Y.; Katayama, K.; Nomura, T.; Yoshida, T.; Yamashita, K.; Yoshikawa, T.; Kawai, Y.; Mayumi, T.; et al. The Use of a Mutant TNF-α as a Vaccine Adjuvant for the Induction of Mucosal Immune Responses. *Biomaterials* **2009**, *30*, 5869–5876. [CrossRef]
45. Lysén, A.; Braathen, R.; Gudjonsson, A.; Tesfaye, D.Y.; Bogen, B.; Fossum, E. Dendritic Cell Targeted Ccl3- and Xcl1-Fusion DNA Vaccines Differ in Induced Immune Responses and Optimal Delivery Site. *Sci. Rep.* **2019**, *9*, 1820. [CrossRef]
46. Borsutzky, S.; Cazac, B.B.; Roes, J.; Guzmán, C.A. TGF-Beta Receptor Signaling Is Critical for Mucosal IgA Responses. *J. Immunol.* **2004**, *173*, 3305–3309. [CrossRef] [PubMed]
47. Bracci, L.; Canini, I.; Puzelli, S.; Sestili, P.; Venditti, M.; Spada, M.; Donatelli, I.; Belardelli, F.; Proietti, E. Type I IFN Is a Powerful Mucosal Adjuvant for a Selective Intranasal Vaccination against Influenza Virus in Mice and Affects Antigen Capture at Mucosal Level. *Vaccine* **2005**, *23*, 2994–3004. [CrossRef]
48. Metzger, D.W. Interleukin-12 as an Adjuvant for Induction of Protective Antibody Responses. *Cytokine* **2010**, *52*, 102–107. [CrossRef] [PubMed]
49. Suzuki, H.; Nagatake, T.; Nasu, A.; Lan, H.; Ikegami, K.; Setou, M.; Hamazaki, Y.; Kiyono, H.; Yagi, K.; Kondoh, M.; et al. Impaired Airway Mucociliary Function Reduces Antigen-Specific IgA Immune Response to Immunization with a Claudin-4-Targeting Nasal Vaccine in Mice. *Sci. Rep.* **2018**, *8*, 2904. [CrossRef]
50. Suzuki, H.; Watari, A.; Hashimoto, E.; Yonemitsu, M.; Kiyono, H.; Yagi, K.; Kondoh, M.; Kunisawa, J. C-Terminal Clostridium Perfringens Enterotoxin-Mediated Antigen Delivery for Nasal Pneumococcal Vaccine. *PLoS ONE* **2015**, *10*, e0126352. [CrossRef]

51. Yamanaka, D.; Tajima, K.; Adachi, Y.; Ishibashi, K.-I.; Miura, N.N.; Motoi, M.; Ohno, N. Effect of Polymeric Caffeic Acid on Antitumour Activity and Natural Killer Cell Activity in Mice. *J. Funct. Foods* **2014**, *6*, 513–522. [CrossRef]
52. Bandyopadhyay, P.; Ghosh, A.K.; Ghosh, C. Recent Developments on Polyphenol–Protein Interactions: Effects on Tea and Coffee Taste, Antioxidant Properties and the Digestive System. *Food Funct.* **2012**, *3*, 592–605. [CrossRef]
53. Siebert, K.J.; Troukhanova, N.V.; Lynn, P.Y. Nature of Polyphenol−Protein Interactions. *J. Agric. Food Chem.* **1996**, *44*, 80–85. [CrossRef]
54. Sun, X.; Sarteshnizi, R.A.; Udenigwe, C.C. Recent Advances in Protein–Polyphenol Interactions Focusing on Structural Properties Related to Antioxidant Activities. *Curr. Opin. Food Sci.* **2022**, *45*, 100840. [CrossRef]
55. Li, M.; Ritzoulis, C.; Du, Q.; Liu, Y.; Ding, Y.; Liu, W.; Liu, J. Recent Progress on Protein-Polyphenol Complexes: Effect on Stability and Nutrients Delivery of Oil-in-Water Emulsion System. *Front. Nutr.* **2021**, *8*, 765589. [CrossRef]
56. Ozdal, T.; Sela, D.A.; Xiao, J.; Boyacioglu, D.; Chen, F.; Capanoglu, E. The Reciprocal Interactions between Polyphenols and Gut Microbiota and Effects on Bioaccessibility. *Nutrients* **2016**, *8*, 78. [CrossRef]
57. Shimizu, M. Multifunctions of Dietary Polyphenols in the Regulation of Intestinal Inflammation. *J. Food Drug Anal.* **2017**, *25*, 93–99. [CrossRef]
58. Bernardi, S.; Del Bo', C.; Marino, M.; Gargari, G.; Cherubini, A.; Andrés-Lacueva, C.; Hidalgo-Liberona, N.; Peron, G.; González-Dominguez, R.; Kroon, P.; et al. Polyphenols and Intestinal Permeability: Rationale and Future Perspectives. *J. Agric. Food Chem.* **2020**, *68*, 1816–1829. [CrossRef]
59. Bohn, T. Dietary Factors Affecting Polyphenol Bioavailability. *Nutr. Rev.* **2014**, *72*, 429–452. [CrossRef]
60. Ed Nignpense, B.; Francis, N.; Blanchard, C.; Santhakumar, A.B. Bioaccessibility and Bioactivity of Cereal Polyphenols: A Review. *Foods* **2021**, *10*, 1595. [CrossRef]
61. Manach, C.; Scalbert, A.; Morand, C.; Rémésy, C.; Jiménez, L. Polyphenols: Food Sources and Bioavailability. *Am. J. Clin. Nutr.* **2004**, *79*, 727–747. [CrossRef] [PubMed]
62. Tajima, K.; Akanuma, S.; Matsumoto-Akanuma, A.; Yamanaka, D.; Ishibashi, K.-I.; Adachi, Y.; Ohno, N. Activation of Macrophages by a Laccase-Polymerized Polyphenol Is Dependent on Phosphorylation of Rac1. *Biochem. Biophys. Res. Commun.* **2018**, *495*, 2209–2213. [CrossRef] [PubMed]

vaccines

MDPI

Article

Vaccination-Route-Dependent Adjuvanticity of Antigen-Carrying Nanoparticles for Enhanced Vaccine Efficacy

Chaojun Song [1,†], Jinwei Hu [2,†], Yutao Liu [2], Yi Tian [3], Yupu Zhu [2], Jiayue Xi [2], Minxuan Cui [2], Xiaolei Wang [4], Bao-Zhong Zhang [5], Li Fan [2,*] and Quan Li [6,*]

1 School of Life Science, Northwestern Polytechnical University, 127th Youyi West Road, Xi'an 710072, China; cj6005@nwpu.edu.cn
2 Department of Pharmaceutical Chemistry and Analysis, School of Pharmacy, Airforce Medical University, 169th Changle West Road, Xi'an 710032, China; hujinwei@fmmu.edu.cn (J.H.); liuyutao@fmmu.edu.cn (Y.L.); zhuyupupapa@fmmu.edu.cn (Y.Z.); 0518xjy@fmmu.edu.cn (J.X.); cuiminxuan2521@fmmu.edu.cn (M.C.)
3 Department of Oncology, Airforce Medical Center of PLA, 30th Fu Cheng Road, Beijing 100142, China; tiannyii@163.com
4 School of Biomedical Sciences, Li Ka Shing Faculty of Medicine, The University of Hong Kong, Pokfulam, Hong Kong SAR, China; xiaoleiw@hku.hk
5 Key Laboratory of Quantitative Engineering Biology, Shenzhen Institute of Synthetic Biology, Shenzhen Institutes of Advanced Technology, Chinese Academy of Sciences, Shenzhen 518055, China; bz.zhang3@siat.ac.cn
6 Department of Physics, The Chinese University of Hong Kong, Shatin, New Territories, Hong Kong SAR, China
* Correspondence: xxfanny@fmmu.edu.cn (L.F.); liquan@cuhk.edu.hk (Q.L.)
† These authors contributed equally to this work.

Abstract: Vaccination-route-dependent adjuvanticity was identified as being associated with the specific features of antigen-carrying nanoparticles (NPs) in the present work. Here, we demonstrated that the mechanical properties and the decomposability of NP adjuvants play key roles in determining the antigen accessibility and thus the overall vaccine efficacy in the immune system when different vaccination routes were employed. We showed that soft nano-vaccines were associated with more efficient antigen uptake when administering subcutaneous (S.C.) vaccination, while the slow decomposition of hard nano-vaccines promoted antigen uptake when intravenous (I.V.) vaccination was employed. In comparison to the clinically used aluminum (Alum) adjuvant, the NP adjuvants were found to stimulate both humoral and cellular immune responses efficiently, irrespective of the vaccination route. For vaccination via S.C. and I.V. alike, the NP-based vaccines show excellent protection for mice from *Staphylococcus aureus* (*S. aureus*) infection, and their survival rates are 100% after lethal challenge, being much superior to the clinically used Alum adjuvant.

Keywords: nanoparticle adjuvant; adjuvanticity; nano-vaccine; vaccination route; mechanical property; decomposability; *Staphylococcus aureus*

1. Introduction

A vaccine requires the proper combination of pathogen-specific antigens and immunostimulating adjuvants to induce a protective adaptive immune response. NPs acting as both antigen carriers and adjuvants have been reported to improve the antigen stability and cellular uptake [1], promote the humoral and/or cell-mediated immune response [2–4], and thus enhance vaccine efficacy with antigen dosage reduction. For example, NPs loaded with detained toxins effectively trigger the formation of germinal centers and induce a highly neutralizing antibody titer [5].

One of the major contributions of NP-associated adjuvanticity is ascribed to the increased availability of antigens when carried by NPs. On the one hand, specific forms of antigen loading (e.g., encapsulation) in the NP carrier was found to be effective in protecting the antigen (forms such as DNA or RNA [6]) against decomposition in the biological environment [7]. On the other hand, the promotion of antigen uptake also results from the rather easy cellular uptake of NPs at both the cellular and system level, especially when a special size range of the NPs is chosen [8,9] and/or they are equipped with surface-targeting functions (e.g., to lymph nodes [10–12] or dendritic cells (DCs) [13,14]). A report is also available on liposome-encapsulated hen egg-white lysozyme (HEL) targeting early endosomes and thus entering the MHC class I and II presentation pathways [15]. Though NPs could increase the availability of antigens by changing the antigen loading patterns and surface targeting modification, the underlying mechanism for how the physicochemical parameters of NPs affect the antigen availability is unclarified. Moreover, different vaccination routes have different requirements regarding the NPs to enhance the antigen availability. For example, ideal vaccines for the intravenous vaccination route require better stability of the NPs in order to avoid antigen release during the circulation, as well as long circulation properties for enhancing the immune cells' opportunities for uptake. On the other hand, in the subcutaneous vaccination route, better decomposability is needed for faster antigen release, leading to faster antigen presentation and immune response.

As a matter of fact, the effects of some of the material parameters on NPs' adjuvanticity have been studied in the literature, but in a rather scattered manner. For example, the geometrical configuration match was identified as an effective cue to enhance the antibody binding affinity and broaden its viral neutralizing spectrum [16]. The degradability of a lipid NP carrier was observed to be a physical cue to direct antigens towards cross presentation [15]. Mechanical forces are also important determinants in immune cell activation, such as regulating cell-surface receptor activation, cell migration, intracellular signaling and intercellular communication [17,18]. More specifically, NPs that retain their force-dependent deformability were found to stimulate both humoral and cellular adaptive responses, and thus increase the survival of mice upon a lethal dose of influenza virus [18]. The stiffness of the NPs was also recognized recently as an important factor in affecting immune cell behaviors via modulating the FAK-NF-κB signaling pathway [19].

In the present work, we took *S. aureus* as a model system, due to its growing impact on public health resulting from the emergence and spread of methicillin-resistant *S. aureus* (MRSA) strains. In fact, the emergence of drug-resistant strains urged the search for alternative treatments such as immunotherapeutic approaches. Here we fabricated both soft and hard PLGA-based NPs as carriers for an *S. aureus* antigen (i.e., recombinant ess extracellular B (rEsxB)) [20] and the medium for adjuvanticity (Figure 1). We discovered that tuning the stiffness of the carrier materials by modulating the carrier composition promoted the cellular uptake of rEsxB-carrying NPs, but at the same time affected the decomposability of the NPs (and thus rEsxB release). In the mouse group receiving a S.C. vaccination, a significantly higher rEsxB-specific antibody titer was found in mice treated with a soft nano-vaccine, while similar results were obtained in mouse groups receiving a I.V. vaccination but with a hard nano-vaccine. For vaccination via S.C. and I.V. alike, the NP-based *S. aureus* vaccines show excellent protection for mice against *S. aureus* lethal challenge, and their survival rates are 100% after the inoculation of ATCC25923 strain, being far superior to the clinically used Alum adjuvant.

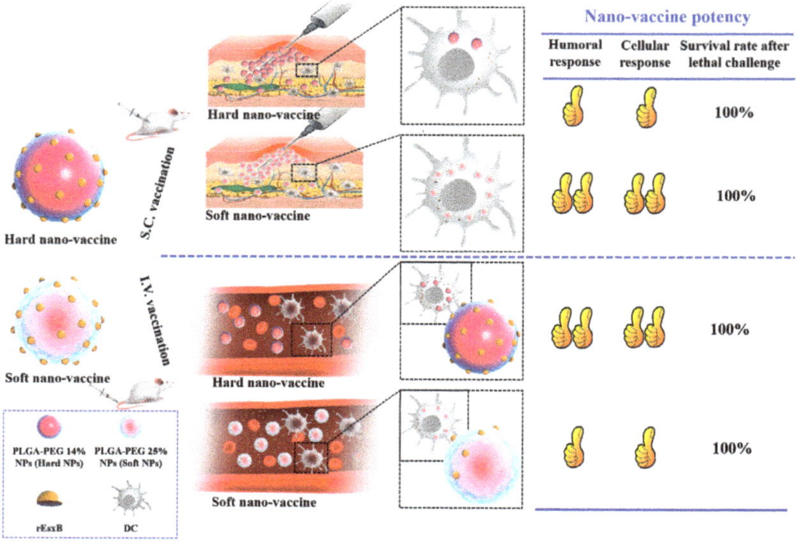

Figure 1. Schematic illustration of working mechanism of the developed soft and hard PLGA-based NPs as carriers for *S. aureus* rEsxB and the medium for adjuvanticity. The hard PLGA-based nano-vaccine could elicit stronger cellular and humoral immune responses through I.V. vaccination, while the soft one elicited stronger cellular and humoral immune responses via S.C. vaccination. Both nano-vaccines (either via I.V. or S.C. immunization) could protect the mice from lethal challenge of *S. aures* with survival rates of 100%.

2. Materials and Methods

2.1. Materials

BL21 (DE3)-competent *Escherichia coli* (*E. coil*) was purchased from TIANGEN BIOTECH (BEIJING) Co., Ltd. (Beijing, China). Ni-Sepharose was purchased from Cytiva (Shanghai, China). A High-Capacity Endotoxin Removal column was purchased from Xiamen Bioendo Technology Co., Ltd. (Xiamen, China). PLGA-PEG X% polymers (X = 0, 14, 20, 25, 33, PLGA with a lactic acid to glycolic acid ratio of 50:50) and fluorescein isothiocyanate (FITC)-PEG_{2000}-NH_2 were purchased from Xi'an Ruixi Biological Technology Co., Ltd. (Xi'an, China). Isopropylbeta-D-thiogalactopyranoside (IPTG), imidazole, polyvinyl alcohol (PVA), dichloromethane (DCM), acetone, 2-(N-morpholino) ethanesulfonic acid hydrate (MES), 1-ethyl-3-(3-dimethylaminopropyl)-carbodiimide (EDC), N-Hydroxysuccinimide (NHS), bovine serum albumin (BSA), Tween 20, 2,2′-azino-bis (3-ethylbenzothiazoline-6-sulfonic acid) (ABTS), horseradish peroxidase (HRP)-labeled goat anti-mouse IgG, and 0.22 μm and 0.45 μm sterile filters were purchased from Merck (Shanghai, China). Imject™ Alum adjuvant was purchased from Thermo Fisher Scientific (Waltham, MA, USA). Ninety-six-well polystyrene plates for the enzyme-linked immunosorbent assay (ELISA) were purchased from Corning (New York, NY, USA). Enzyme-linked immunospot (ELISPOT) kits were purchased from Mabtech (Stockholm, Sweden). All other solvents are of analytical grade and were used without further purification.

2.2. Metheds

2.2.1. Expression and Purification of rEsxB Antigen

rEsxB was chosen as the model antigen to investigate the adjuvanticity of the NP carriers. His-tagged rEsxB was prepared using the standard IPTG induction protocol [21]. Briefly, full-length wild-type EsxB DNA was subcloned into the pET-28a vector and expressed as a 6×His-tag fusion protein in BL21 (DE3) *E. coil*. rEsxB expression was induced by IPTG, purified from bacterial cell lysates by means of Ni-Sepharose chromatography, and was gradient eluted from the Ni-Sepharose with 100 mmol/L, 300 mmol/L and 500 mmol/L imidazole. The sample was then dialyzed with phosphate-buffered saline (PBS) and endotoxin was removed using the High-Capacity Endotoxin Removal column. Finally, the rEsxB solution was filtered through a 0.22 μm sterile filter and the total protein concentration was quantified using a bicinchoninic acid (BCA) protein assay. The purity and specificity of the rEsxB antigen were analyzed by means of SDS-PAGE and Western Blot.

2.2.2. Preparation and Characterization of the PLGA-PEG NPs Conjugated with rEsxB

The apparent Young's modulus of the PLGA-PEG X% (X = 0, 14, 20, 25, 33) polymers was measured using a atomic force microscope (AFM, Bruker BioScope Catalyst). Referring to the stiffness of the cell membrane [22], two sets of PLGA-PEG X% polymers (PEG_{5000}-$PLGA_{30,000}$, PEG_{5000}-$PLGA_{15,000}$) were chosen for the synthesis of PLGA-PEG X% NPs (X = 14, 25). The PLGA-PEG X% NPs were prepared via the precipitation/solvent diffusion method according to our previous work with small modifications [23,24]. Briefly, 100 mg of PLGA-PEG was dissolved in a mixture of 1.5 mL of DCM and 1 mL acetone. The polymer solution was directly added to 10 mL of a 5% PVA (Mw 31,000~50,000 Da, 87~89% hydrolyzed) solution and stirred at 600 rpm to evaporate the DCM and acetone. The mixture was then homogenized for 2 min by using a probe sonicator at 20 W to generate an oil-in-water (O/W) emulsion. The formed emulsion was added to 50 mL of deionized (DI) water and stirred for 4 h at room temperature. After that, the PLGA-PEG X% NPs were collected and filtered using a 0.45 μm filter. The NP solution was centrifuged at 16,500× *g* force at 4 °C for 1 h, washed with DI water 3 times, and lyophilized for storage.

The rEsxB was chemically linked to the PLGA-PEG X% NPs via a condensation reaction between the carboxyl group on the NPs and the amino group of the rEsxB. Then, a 10 mg sample of PLGA-PEG X% NPs was re-suspended in 10 mL of 25 mmol/L MES buffer, and 0.4 mL of 1 mol/L EDC and 0.25 mL of 1 mol/L NHS were added into the mixture and stirred for 4 h at the room temperature to activate the carboxyl groups on the NPs. After that, the NPs were centrifuged at 16,500× *g* force at 4 °C for 1 h and washed with DI water 3 times. The activated NPs were re-suspended in 1 mL of 1 mg/mL rEsxB in PBS (pH 8.0) overnight at 4 °C, then the PLGA-PEG X% NPs-rEsxB were centrifuged at 16,500× *g* force at 4 °C for 1 h, washed with DI water 3 times, and lyophilized for later use.

The hydrodynamic diameter, polydispersity index (PDI), and zeta potential of the PLGA-PEG X% NPs and PLGA-PEG X% NPs-rEsxB were measured via dynamic light scattering (DLS) (Delsa™ Nano, Beckman-Coulter, High Wycombe, UK). The successful synthesis of the PLGA-PEG X% NPs-rEsxB was determined by means of Fourier Transform infrared spectroscopy (FTIR, Thermo Nicolet IS50, Thermo Fisher Scientific, Waltham, MA, USA). The BCA protein quantification assay was used to quantify the total rEsxB concentration loaded. The loading efficiency was calculated using the following formula:

$$\text{Loading efficiency}(\%) = {m_{\text{rEsxB on the NPs (mg)}}}/{m_{\text{NPs (mg)}}} \times 100\%$$

2.2.3. Evaluations of the Cellular Uptake and Load Release of NPs-rEsxB

Quantitative evaluation of the cellular uptake of NPs was carried out by means of the analysis of the intracellular fluorescein intensity using FITC-loaded NPs (NPs-FITC), with FITC as the florescence label. The PLGA-PEG X% NPs (X = 14, 25) were conjugated with FITC-PEG_{2000}-NH_2 via the condensation reaction between the carboxyl groups on NPs and the amino groups on FITC-PEG_{2000}-NH_2 via sulfo-NHS-catalyzed EDC activation of carboxyl groups. To simulate different administration routes of S.C. and I.V. injection, DCs (5×10^4 cells/well) were fed with the 2.5 mL NPs-FITC solution at two different concentrations—2 mg/mL, for simulating a local high concentration after S.C. injection, and 10 μg/mL, for simulating a diluted environment after I.V. injection. At specific time intervals (0.5 h, 1 h, 2 h, 4 h, 6 h and 8 h), DCs were collected by means of centrifugation (250× *g* for 5 min) then re-suspended in 5 mL of RPMI 1640 medium without fetal bovine serum (FBS). The cell suspensions were frozen and thawed 4~6 times repeatedly then centrifuged at 1500× g for 10 min. Emission of the supernatants was measured at 528/20 nm with an excitation wavelength of 485/20 nm (Biotek SYNERGY LX Instruments, Santa Clara, CA, USA).

The load release experiment was carried out using the fluorescence method. FITC was chosen as a fluorescence tag and conjugated to the NPs in a similar manner to that of the rEsxB, then the NPs-FITC was used to examine the payload release profile of all NP samples. For the payload release studies, 50% FBS simulating the humoral environment was used. The experiments were carried out at 37 °C, and data were collected for a 12 h duration. In detail, 10 mg of PLGA-PEG X% NPs-FITC was re-suspended in 10 mL of 50% FBS at 37 °C and 1 mL of each NPs-FITC sample was collected at specific time points (1 h, 2 h, 4 h, 8 h and 12 h after release) then centrifuged at 16,500× g at 4 °C for 1 h. The emission of the supernatants was measured at 528/20 nm with an excitation wavelength at 485/20 nm (Biotek SYNERGY LX Instruments, Santa Clara, CA, USA).

2.2.4. Evaluation of Biocompatibility of PLGA-PEG X% NPs-rEsxB In Vitro and In Vivo

We first examined the cytotoxicity of the NPs-rEsxB conjugates in vitro using fibroblast L929 cell line using the cell counting kit-8 (CCK-8) assay. Cells were seeded in a 96-well plate (5×10^3 cells/well) for overnight attachment. In this process, 100 μL of the NPs-rEsxB medium solution with specific concentrations (10 ng/mL, 100 ng/mL, 1 μg/mL, 5 μg/mL, 10 μg/mL, 50 μg/mL, 100 μg/mL, 200 μg/mL, 500 μg/mL and 1 mg/mL) was added to the cell solution, respectively. Each sample was tested using 6 replicated wells. Then, the cells were incubated for another 24 h. Then,20 μL of the CCK-8 solution was added to each well and incubated at 37 °C for 1~4 h. The absorbance of cells was then measured with a microplate reader at a wavelength of 450 nm (Biotek SYNERGY LX Instruments, Santa Clara, CA, USA). The cell viabilities were calculated using the classic cytotoxicity equation:

$$\text{Cell Viability} = (\text{OD}_{\text{test}} - \text{OD}_{\text{background}}) / (\text{OD}_{\text{control}} - \text{OD}_{\text{background}}) \times 100\%$$

Further evaluation of the biocompatibility of the NPs-rEsxB was carried out in vivo using BALB/c mice. All animals were monitored for their activity, physical condition, and body weight. The body weight of each mouse was measured and recorded every other day until 30 days after the first vaccination.

Hematoxylin and eosin (H&E) staining of major organs by means of the standard method published in our previous paper [25] was then carried out after the animals were sacrificed at day 42 after the first vaccination.

2.2.5. Animal Immunization

All of the BALB/c mice were purchased from the Animal Experiment Center of Air Force Medical University. All animal experiments complied with the National Research Council's Guide for the Care and Use of Laboratory Animals and the animal experiments were approved by the Animal Care and Ethic Committee of Fourth Military Medical University (Approval NO. KY20213144-1). Female 6–8-week-old BALB/c mice were divided randomly into 5 groups (n = 6 or 10 each) and immunized with PBS and rEsxB as negative control groups, clinically used Alum mixed with rEsxB (Alum-rEsxB) as the positive control group and PLGA-PEG X% NPs-rEsxB (X = 14, 25) as the nano-vaccine groups. Four groups (except PBS) were immunized either S.C. or I.V. with the same rEsxB concentration of 50 μg/mouse. To determine humoral and cell mediated response all groups were boosted with the same formulation (rEsxB concentration of 25 μg/mice) 14 and 28 days after the first immunization. Mouse serum samples were collected at specific time points.

2.2.6. ELISA Assay

The rEsxB-specific antibody titer was determined through indirect ELISA on days 35 and 208 after the first immunization. Briefly, a 96-well ELISA plate was coated with rEsxB (500 ng/well) followed by blocking with 3% BSA on the next day. The plate was then washed 5 times with PBS containing 0.1% Tween 20 (PBST) and incubated with serially diluted immunized mouse serum ranging from 1:200 to 1:204,800 (100 μL/well) at 37 °C for 1 h. followed by five washes with PBST. HRP-labeled goat anti-mouse IgG (1:5000, 100 μL/well) was added and incubated at 37 °C for 45 min. This was followed by washing (as above) and development with ABTS and H_2O_2 as substrate chromogens. Finally, the color development for 30 min at room temperature in the dark was read at 405 nm by a microplate reader (Biotek SYNERGY LX Instruments, Santa Clara, CA, USA). The cut-off value was calculated as 2.1 times the absorbance from the negative control serum assayed [26]. In the sample group comparisons of the ELISA titer results, the area under the curve (AUC) method had similar power to the absorbance summation (AS) method, and better power than the endpoint titer (ET) method. The rEsxB-specific antibody titer results were represented by the mean serum ELISA AUC [27,28] with the S.D. of 6 mice in each group.

2.2.7. ELISPOT Assay

The splenocytes of immunized animals 7 days after the third immunization (both S.C. and I.V. vaccination groups) were analyzed for IL-4 (representative of humoral immunity), IFN-γ (representative of cell-mediated immunity), IL-12 (representative of DC activation) and IL-17A (representative of TH17 T cell activation) production by means of ELISPOT assays using standard protocols [29]. The ELISPOT plates (capture antibody precoated) were washed 4 times with sterile PBS and incubated with a complete culture medium (RPMI-1640, 10% FBS) for at least 30 min at room temperature. After removing the complete culture medium, splenocytes from the immunized mice (2.5×10^6 cells/well) and stimuli (4 μg/mL rEsxB) was mixed, added in the plate and incubated at 37 °C for 24 h. Phytohemagglutinin (PHA, 10 μg/mL) and the media alone served as positive and negative controls, respectively. Then, the splenocytes were removed and the plates were washed 5 times with PBS, followed by adding the detection antibodies into PBS containing 0.5% FBS (PBS-0.5% FBS) and incubation for 2 h at room temperature. The plates were washed as above and a TMB substrate solution was added to the plates until distinct spots emerged. The spots' development was stopped by washing the plates extensively in DI water. The plates were dried in the dark at room temperature, and the spots were counted in an ELISPOT reader (Cellular Technology Limited, Cleveland, OH, USA).

2.2.8. Lethal Challenge

On the 7th day after the third immunization, 100 μL of *S. aureus* (ATCC 25923) at a concentration of 1.67×10^8 colony-forming units (CFU)/mL was administered to the mice in each of the groups via I.V. injection. Mice were then monitored daily for mortality and clinical signs. At the end point of the experiment, all remaining animals were euthanized.

2.2.9. Serum Bactericidal Test

Standard Neisser–Wechsberg bacteriolysis evaluation [30] was employed to investigate the bacteriolysis effects of the antibody in all nano-vaccines, namely the Alum-rEsxB, rEsxB and PBS groups, with some modifications. The *S. aureus* ATCC 25923 strain was cultured until reaching the mid-exponential phase (OD_{600} = 0.5), then the *S. aureus* concentrations were adjusted to 3×10^6 CFU/mL with PBS (pH 7.4). The serum of immunized mice and normal guinea pig serum were added to the *S. aureus* at a final concentration of 2% (v/v), respectively. The mixture was incubated at 37 °C for 0.5 h, plated on LB agar plates and incubated overnight at 37 °C. Finally, the number of colony-forming units of *S. aureus* was counted and the percentage of lytic bacteria was calculated using the following formula:

$$\text{Percentage of lytic bacteria}(\%) = \left(1 - {}^{\text{The CFU of treated group}}/_{\text{The CFU of control group}}\right) \times 100\%$$

2.2.10. Statistical Analysis

All results were expressed as mean ± standard deviation (SD) and statistical analysis was performed using the software of GraphPad Prism 9.0. Analysis of variance (students' *t* test) was employed to determine the statistical significance of differences, and log-rank (Mantel–Cox) analysis was employed to determine the statistical significance of the survival rate. The statistical differences were defined as * $p < 0.05$, ** $p < 0.01$, *** $p < 0.001$, and **** $p < 0.0001$, and $p < 0.05$ were considered significant.

3. Results

3.1. General Characterizations of the PLGA-PEG X% NPs Conjugated with rEsxB

By manipulating the composition of the PLGA-PEG composite (i.e., PEG content), the stiffness of the material can be varied. The apparent Young's moduli of the polymer with different PEG contents are compared in Figure S1a. The higher the PEG content, the lower the apparent Young's modulus of the composite sample. This is consistent with the literature [24]. Using the same composition recipe, we fabricated two sets of PLGA-PEG NPs with 14%, and 25% PEG (wt%), corresponding to the stiffness of ~0.18 MPa and ~0.02 MPa, respectively (Figure S1b). The respective samples were named PLGA-PEG 14% NPs and PLGA-PEG 25% NPs. The stiffness of the softer sample (PLGA-PEG 25% NP) is comparable to that of the plasma membrane (~0.01 MPa) [22]. The average sizes of the NPs in both sample groups (of different PEG content) were similar (~170 nm) (Figure S1c). With the PEG content increasing, the zeta potential of the NPs gradually changed from slightly negative to slightly positive (Figure S1d). The purification of the rEsxB protein was confirmed by both SDS-PAGE and Western blot (Figure S2). A small increase of ~20 nm in the average size of the NPs was found after the rEsxB conjugation (Figure S3b). The zeta potentials of PLGA-PEG 14% NPs-rEsxB and PLGA-PEG 25% NPs-rEsxB appeared to be slightly more negative (−2 mV~0 mV) (Figure S3c). The rEsxB loading efficiencies of PLGA-PEG 14% NPs and PLGA-PEG 25% NPs were both ~3.5% (mass ratio) (Figure S3d). Successful covalent binding of the rEsxB to the two groups of selected NPs was verified by the existence of amide bonds in the FTIR taken from the conjugated NPs samples (Figure S3e).

3.2. Biocompatibility of the rEsxB Conjugated NPs

The rEsxB-conjugated NPs showed excellent biocompatibility at both the cell and animal level. Cell viability was always >95% in all the experimental groups, including rEsxB-conjugated NPs of 14 and 25 wt% PEG (Figure S4a). No significant weight loss was identified in all mouse groups receiving either the S.C. (Figure S4b) or the I.V. injection (Figure S4c) for the administration of the NPs-rEsxB. As shown in Figure S4d,e, no obvious lesions were observed in the H&E staining of all major organs (heart, liver, spleen, kidney and lung), irrespective of the NPs-rEsxB administration routes.

3.3. rEsxB-Specific Antibody

In the S.C. vaccination groups, the antibody titer was read at the background level for either the PBS- or rEsxB-treated mouse groups on days 35 and 208 after the first vaccination, respectively (Figure 2b,c). A moderate antibody level (AUC of 3365) was found at day 35 for the mouse group treated with Alum-rEsxB. The rEsxB-specific antibody for the same mouse group dropped to a barely detectable level at day 208 with an AUC of only about 332. A high rEsxB-specific antibody level in both mouse groups treated with PLGA-PEG 14% NPs-rEsxB and PLGA-PEG 25% NPs-rEsxB appeared at day 35, with an AUC of 12,857 and 66,728, respectively. Moreover, on day 208 after vaccination, the rEsxB-specific antibody for these two NPs-rEsxB-treated mouse groups maintained a relative high titer level, with an AUC of 3927 for the PLGA-PEG 14% NPs-rEsxB group and 8496 for the PLGA-PEG 25% NPs-rEsxB group.

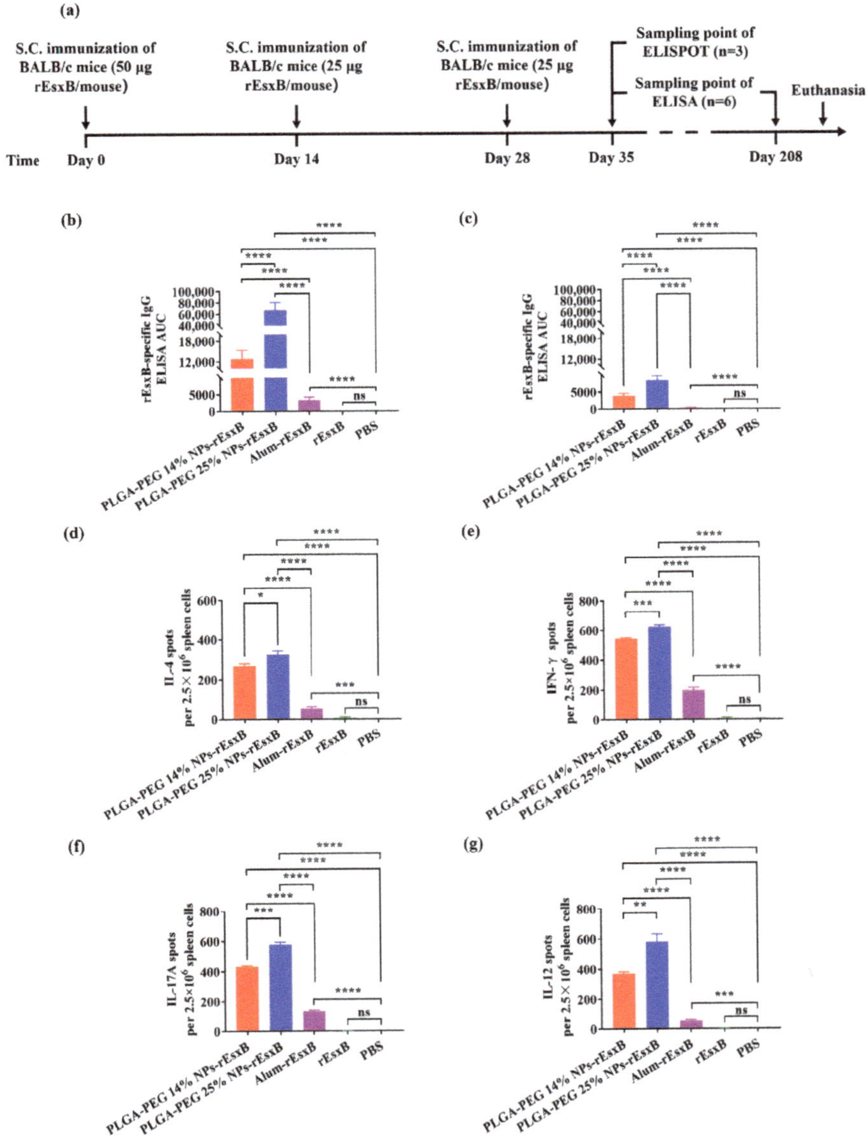

Figure 2. Potent humoral and cellular response to *S. aureus* vaccinations via S.C. (**a**) Illustration of the immunization protocol and the sampling time points. Serum rEsxB-specific IgG titer at (**b**) day 35 and (**c**) day 208 (n = 6). ELISpot analysis of (**d**) IL-4, (**e**) IFN-γ, (**f**) IL-12 and (**g**) IL-17A spot-forming cells among splenocytes (n = 3). All data were expressed as mean ± S.D. A p value < 0.05 was considered as statistically significant (* $p < 0.05$; ** $p < 0.01$; *** $p < 0.001$; **** $p < 0.0001$).

It is worth noting that, on both day 35 and day 208 after the first vaccination, much higher antibody titer levels were found in the mouse groups treated with the softer NPs-rEsxB (PLGA-PEG 25% NPs-rEsxB) when compared to the harder NPs-rEsxB (PLGA-PEG 14% NPs-rEsxB), while both levels of the NPs-rEsxB groups were comparable to the peak level of Alum-rEsxB, even after 6 months (Figure 2b,c), when that of Alum-rEsxB dropped to a very low AUC of 332.

In the I.V. vaccination groups, superior antibody titer results were also found in the NPs-rEsxB in comparison with the Alum-rEsxB (Figure 3b,c) About 13~50 higher levels of antibody titer were found in the NP groups compared with Alum-rEsxB group on day 35 after the first vaccination. Also, longer maintenance of the antibody at a rather high level was observed in mice vaccinated with NPs-rEsxB. On day 208 after the first vaccination, a barely detectable antibody level (AUC of 5.37) could be found in the Alum-rEsxB group. However, in both NP-treated groups, the antibody titer was maintained at moderate levels, with an AUC of 4291 for the harder NPs-rEsxB and an AUC of 1965 for the softer NPs-rEsxB.

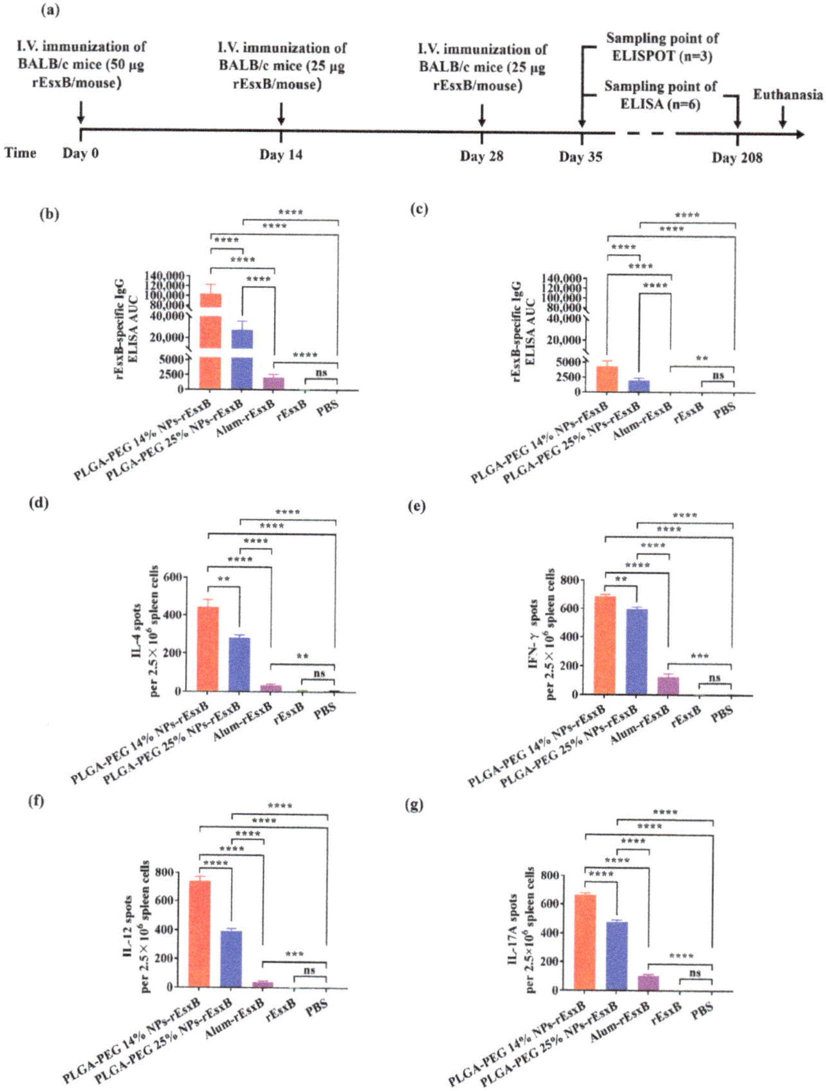

Figure 3. Potent humoral and cellular response to *S. aureus* vaccinations via I.V. (**a**) Illustration of the immunization protocol and the sampling time points. Serum rEsxB-specific IgG titer at (**b**) day 35 and (**c**) day 208 (n = 6). ELISpot analysis of (**d**) IL-4, (**e**) IFN-γ, (**f**) IL-12 and (**g**) IL-17A spot-forming cells among splenocytes (n = 3). All data were expressed as the means ± S.D. A p value < 0.05 was considered as statistically significant (** $p < 0.01$; *** $p < 0.001$; **** $p < 0.0001$).

Nonetheless, an opposite trend was observed (vs. the S.C. set above) when we compared the results from the mouse groups receiving the harder or softer NPs—that is, superior antibody titer levels were found in the harder NP-treated mouse group. In particular, the PLGA-PEG 14% NPs-rEsxB demonstrated the highest level of the antibody titer among all vaccination formulations, and the peak level represented a 50-fold increase compared to that of the Alum-rEsxB on day 35 after the first vaccination. In fact, a very low antibody titer level persisted in the Alum-rEsxB using I.V. vaccination.

3.4. rEsxB-Specific IL-4^{+}, IFN-γ^{+}, IL-12^{+} and IL-17A^{+} T Cell Responses

For both the S.C. and I.V. vaccination routes, the IL-4, IFN-γ, IL-12 and IL-17A levels were always found to be much higher in the splenocyte of NPs-rEsxB-vaccinated mice than those in the splenocytes of Alum-rEsxB-, rEsxB-, or PBS-treated ones. On the other hand, the opposite trend was identified in comparing the results from the softer and the harder NPs-rEsxB groups when employing different vaccination routes—that is, the IL-4, IFN-γ, IL-12 and IL-17A levels were found to be consistently higher in the case of softer NPs (PLGA-PEG 25% NPs-rEsxB) when empoying the S.C. route (Figure 2d–g and Figure S5a), while these cytokines' levels appeared higher in the case of harder NPs (PLGA-PEG 14% NPs-rEsxB) when I.V. was adopted (Figure 3d–g and Figure S5b). The opposite behavior of the softer and the harder NPs-rEsxB in cytokine secretion was consistent with that in antibody titer measurement. As IL-4 and IFN-γ serve as effective indicators for humoral and cellular immunity, respectively, the experiment results suggest that stronger immune responses (both humoral and cellular immune responses) were achieved in the NPs-rEsxB groups than in the Alum-rEsxB group. It is also interesting to point out that stronger immune responses were obtained when using softer NPs in the S.C. vaccination and harder NPs in the I.V. vaccination.

3.5. Lethal Challenge

Before the lethal challenge, all mice were healthy. For mouse groups treated with rEsxB or PBS, all mice died after a lethal dose of *S. aureus* challenged via intravenous injection. Overall, 30% of mice survived in the Alum-rEsxB-treated group. In comparison, the survival rates of the NPs-rEsxB groups using both vaccination routes (S.C. and I.V.) were much higher (100%) than those of the Alum-rEsxB and rEsxB groups in general (Figures 4b–g and 5b–g). No significant difference was identified when using the softer or harder nano-vaccines. Moreover, no obvious body weight changes were observed in both NP-treated groups (Figures 4h and 5h).

3.6. Neutralizing Capability of the Antibody

Bacteriolysis evaluation was employed to investigate the bacteriolysis effects of the antibody produced by means of vaccination with NPs-rEsxB or Alum-rEsxB (with rEsxB and PBS as the control groups). The results (Figure 6) reveal the significantly higher bacteria-neutralizing capability of antibodies produced by mice vaccinated with NPs-rEsxB than with Alum-rEsxB. The results are consistent with the mouse survival rates in the lethal challenge experiment.

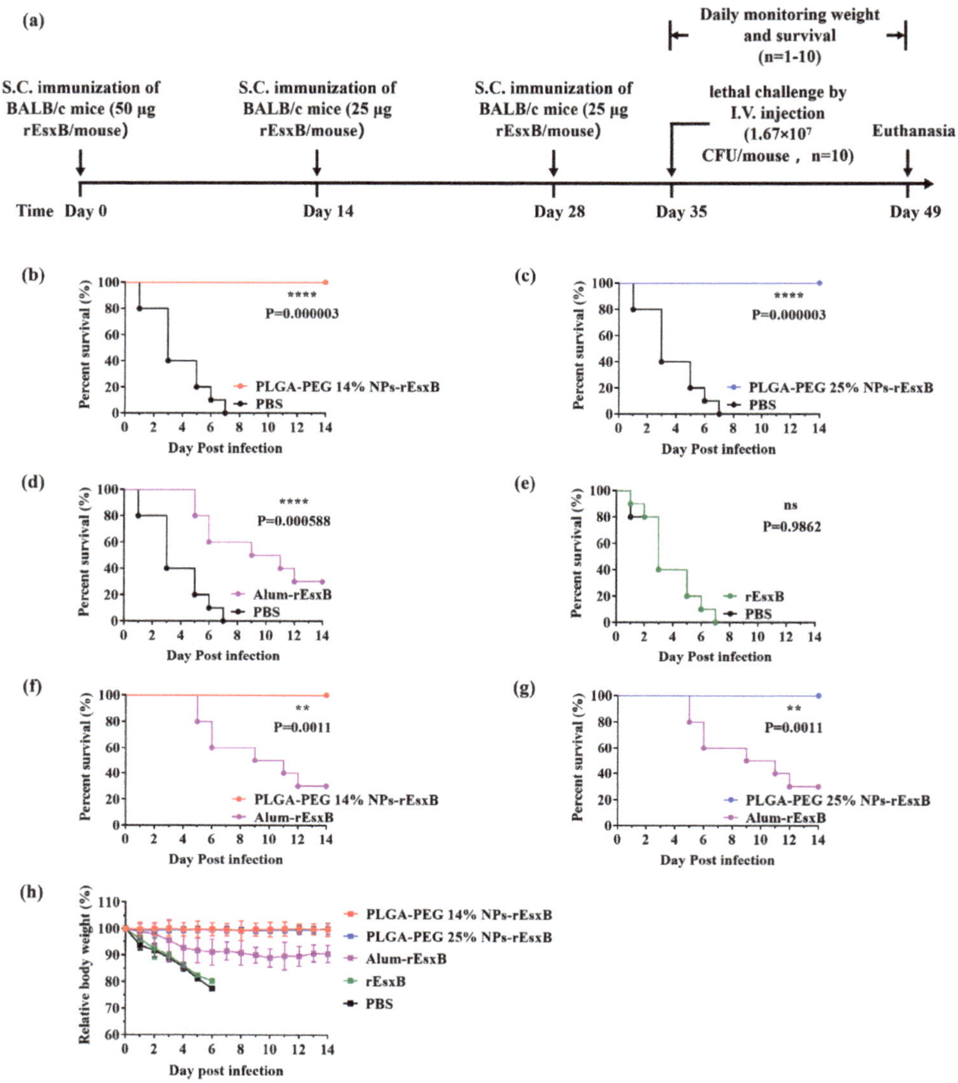

Figure 4. Survival rate comparisons on day 14 after lethal challenge of *S. aureus* in the S.C. vaccination groups. (**a**) Timeline for immunization, challenge and evaluation of protective efficacy. (**b–g**) Survival rate (n = 10) and (**h**) weight loss (n = 1–10) of the immunized mice. Survival rates were analyzed with Log-rank (Mantel–Cox) analysis, and mouse weight is shown as the mean ± S.D. A p value < 0.05 was considered as statistically significant (** $p < 0.01$; **** $p < 0.0001$).

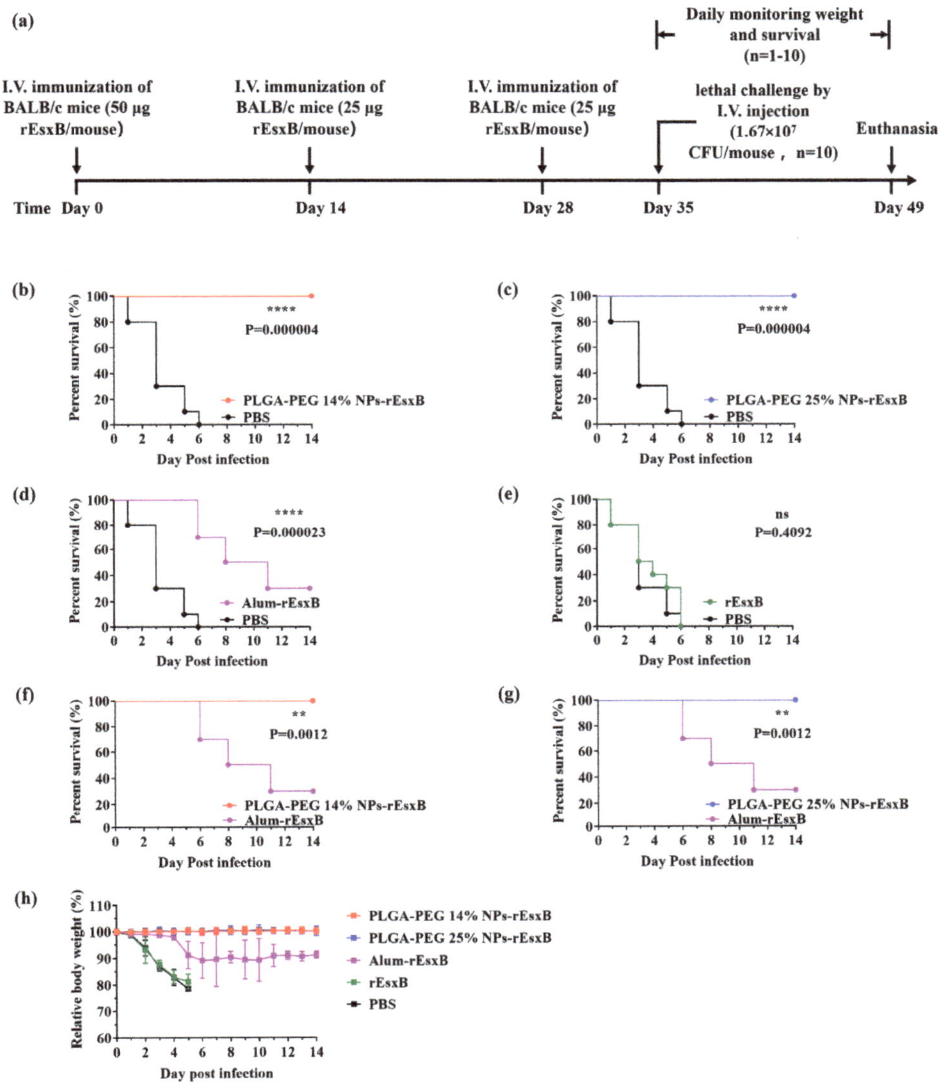

Figure 5. Survival rate comparisons on day 14 after lethal challenge of *S. aureus* in the I.V. vaccination groups. (**a**) Timeline for immunization, challenge and evaluation of protective efficacy. (**b**–**g**) Survival rate (n = 10) and (**h**) weight loss (n = 1–10) of the immunized mice. Survival rates were analysed with Log-rank (Mantel–Cox) analysis, and mouse weight is shown as the mean ± S.D. A p value < 0.05 was considered as statistically significant (** $p < 0.01$; **** $p < 0.0001$).

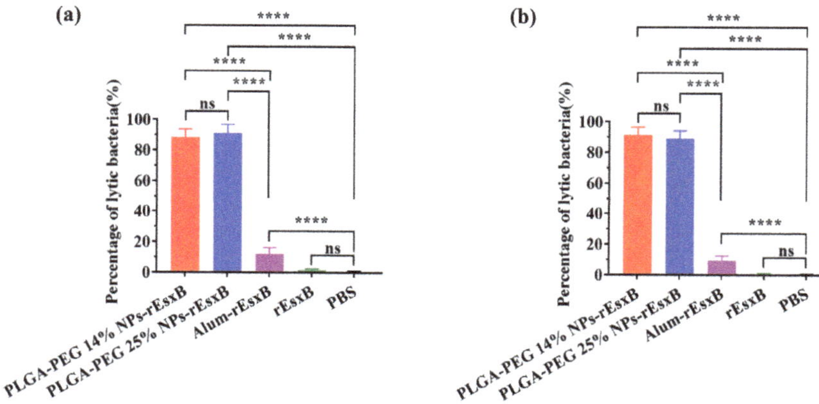

Figure 6. The bacteriolysis of serum antibody by (**a**) S.C. and (**b**) I.V. (n = 6). All data were expressed as the means ± S.D. A p value < 0.05 was considered as statistically significant (**** $p < 0.0001$).

3.7. Mechanism Hypothesis

The difference observed in the immune response of mice vaccinated with the softer or the harder NPs in specific vaccination routes (S.C. and I.V.) may be understood by the difference in the cellular uptake and pay load release pattern.

We first look at the cellular uptake of softer and harder NPs. For mouse groups receiving the S.C. injection, the NPs are expected to be accumulated at the injection site with a rather high concentration. A high NP feeding concentration of 2 mg/mL was chosen as one scenario for the in vitro test of cellular uptake. For mouse groups receiving the I.V. injection, the NPs will be diluted soon after entering the circulation system. When they are caught by the DCs later, the local concentration of NPs is expected to be much lower than that in the S.C. group. A low NP feeding concentration of 10 μg/mL was therefore chosen as the other scenario.

Figure 7a plots the difference in cellular uptake of the harder and the softer NPs (PLGA-PEG 14% NPs-FITC and PLGA-PEG 25% NPs-FITC) at a feeding concentration of 2 mg/mL with an incubation time from 0.5 h to 8 h. A significant difference can be found between the softer and the harder NPs. The endocytosis amount is higher when softer NPs are employed. On the other hand, when DCs were fed with a low concentration of NPs, no significant differences in cellular uptake were found when feeding the two NP samples (Figure 7b). The results reveal that when using the S.C. vaccination route, a much higher uptake amount would be associated with softer NPs when compared to the harder ones. In contrast, a similar cellular uptake amount of the NPs is expected when using the I.V. route. The higher uptake amount is associated with the higher antigen level presented by DC cells, initiating the antigen-specific immune response. This provides an explanation for the stronger immune response observed in the case of softer NPs when the S.C. vaccination route is adopted.

On the other hand, one also needs to consider that when the antigen-loaded NPs were not taken up by antigen-presenting cells in a short period of time, payload (the antigen in the present case) release from the NP carriers would take place. When this happens, physical separation between the antigen and the carrier adjuvant occurs, and the antigen is no longer bound by the carrier adjuvant, but behaves as a free antigen. This can be particularly significant when I.V. vaccination is adopted. In this regard, we observed a significantly faster release in the case of softer NPs in 50% FBS (Figure 7c). This result therefore explains the stronger immune response obtained in harder NPs compared to the softer ones when employing I.V. as the vaccination route.

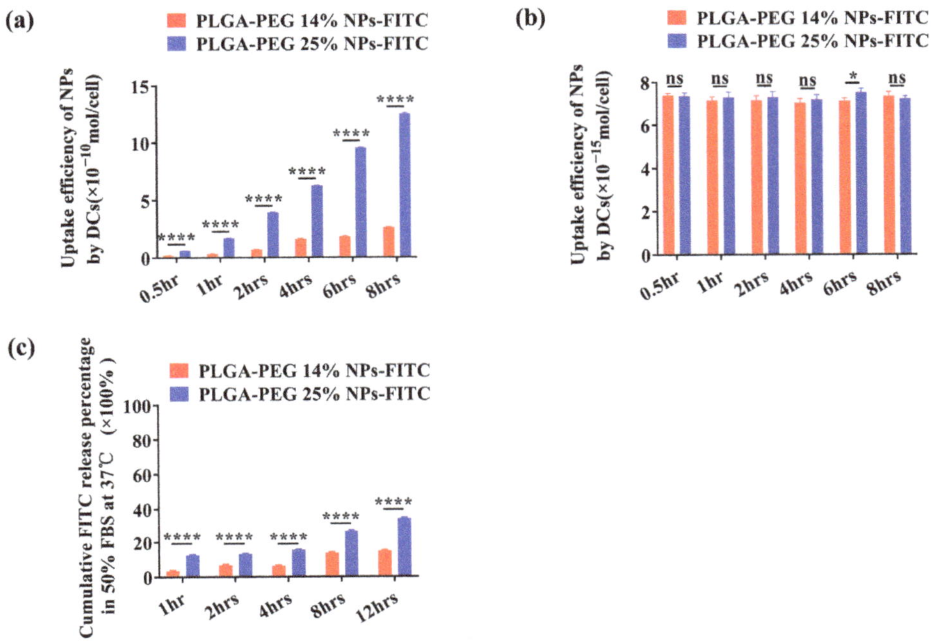

Figure 7. Cellular uptake efficiency of PLGA-PEG X% NPs-FITC at (**a**) a high feeding concentration of 2 mg/mL and (**b**) a low feeding concentration of 10 µg/mL. (**c**) In vitro cumulative FITC release behavior of PLGA-PEG X% NPs-FITC was studied in 50% FBS at 37 °C. The release of FITC was quantified in triplicate wells for each group. All data are expressed as the means ± S.D. p values < 0.05 were considered as statistically significant (* $p < 0.05$; **** $p < 0.0001$).

4. Discussion and Conclusions

In order to enhance the protective efficacy of vaccines, including the induction of rapid and long-lasting immunity, new adjuvants, especially nanoadjuvants, have been studied as a promising platform with rational designs. By rationally selecting/designing polymers based on their physicochemical properties, and considering the antigen and vaccine regimen, it is possible to modulate appropriate immune responses for specific diseases.

In this study, by introducing PEG to the PLGA, we manipulated the mechanical properties (stiffness) of the NP adjuvants, which also served as carriers for the antigen of *S. aureus* (EsxB). Two typical stiffness values were chosen for preparing the NP adjuvants—PLGA-PEG 14% NPs for the hard stiffness value and PLGA-PEG 25% NPs for the soft stiffness value. Two nanovaccine systems, PLGA-PEG 14% NPs-EsxB and PLGA-PEG 25% NPs-EsxB, were then obtained for immune efficacy evaluation using two separate vaccination routes (I.V. and S.C.). As shown in the Results section, a significantly higher level of protection against *S. aureus* infection was always achieved with both hard and soft NP formulations (vs. the clinically used Alum adjuvant), as suggested by the lethal challenge experiments, with all animals being kept alive after the administration of a lethal dose of *S. aureus*. This was mainly attributed to the production of neutralized antibodies induced by NP vaccines. Also, the higher protection efficacy results of NP vaccines were consistent with the antibody titer and the cytokine analysis. Moreover, the enhanced vaccine efficacy results from improved antigen accessibility when delivered by the PLGA-PEG nanocarriers, which were also found to serve as excellent adjuvants to effectively stimulate both humoral and cellular immune responses in mice. On the other hand, a higher PEG content led to softer NPs that promoted cellular uptake, but itself was also prone to faster decomposition (and thus faster release of the antigen payload). The differences in the decomposition speed between PLGA-PEG nanocarriers may be attributed to two reasons—the molecular weight of the polymer and the existence of enzymes in the surrounding environment. The molecular weight of the PLGA chain mostly affects its degradation in indirect ways, e.g., by determining its morphology and porosity [31]. Moreover, the pancreatic lipase, which is the main component of serum lipase and subcutaneous tissue fluid, played a key role in PLGA degradation in the circulation [32,33]. The molecular weight of PLGA in PLGA-PEG 14% NPs is 30 KD, while that in PLGA-PEG 25% NPs is just 15 KD. These differences may affect the degradation of PLGA-based NPs by pancreatic lipase, so that the two PLGA-based NPs are subject to differential degradation. The interplay between these factors then determined the actual antigen uptake efficiency at the system level when different vaccination routes are utilized—that is, S.C. or I.V. injection. This provides a possible mechanism explanation for the observed higher antibody/cytokine levels induced by the rEsxB-loaded soft (hard) NPs in the case of S.C. (I.V.) vaccination. The present work provides a guideline for the vaccination-route-specific design of nanoparticle-based vaccines with maximized potency and minimized side effects.

Supplementary Materials: The following supporting information can be downloaded at: https://www.mdpi.com/article/10.3390/vaccines12020125/s1, Figure S1: Nanoadjuvant via the precipitation/solvent diffusion method; Figure S2: rEsxB antigen was analyzed using (a) SDS-PAGE and (b) Western blot; Figure S3: Nanovaccine via the condensation reaction as an enhanced *S. aureus* vaccine adjuvant; Figure S4: Biocompatibility evaluation of PLGA-PEG X% NPs-rEsxB; Figure S5: Representative images of ELISPOT wells are shown for cytokines produced by splenocytes with (a) S.C. and (b) I.V. References [34,35] are cited in the Supplementary Materials.

Author Contributions: Q.L. conceived the research idea. C.S., L.F. and Q.L. designed the experiments. J.H., Y.L., Y.T., Y.Z., M.C., J.X., X.W. and B.-Z.Z. performed the synthesis and characterization of the nanovaccine and animal experiments. C.S., L.F. and Q.L. analyzed the experimental results. L.F. and Q.L. wrote the manuscript. L.F. and Q.L. supervised the study and provided the infrastructure for the study. All authors discussed the results and commented on the manuscript. All authors have read and agreed to the published version of the manuscript.

Funding: Funding included financial support of the Health and Medical Research Fund, the Food and Health Bureau, the Government of the Hong Kong Special Administrative Region under project No. 18170262, The National Natural Science Foundation of China (No. 82171962), the Key Research and Development Program of Science and Technology Department of Shaanxi Province (No. 2022ZDLSF01-03).

Institutional Review Board Statement: All animal experiments complied with the National Research Council's Guide for the Care and Use of Laboratory Animals and the animal experiments were approved by the Animal Care and Ethic Committee of Fourth Military Medical University (Approval NO. KY20213144-1).

Data Availability Statement: Data will be made available upon request.

Conflicts of Interest: The authors declare that they have no known competing financial interests or personal relationships that could have appeared to influence the work reported in this paper.

References

1. Garg, A.; Dewangan, H.K. Nanoparticles as Adjuvants in Vaccine Delivery. *Crit. Rev. Ther. Drug Carr.* **2020**, *37*, 183–204. [CrossRef] [PubMed]
2. Diaz-Arévalo, D.; Zeng, M. Chapter 7—Nanoparticle-based vaccines: Opportunities and limitations. In *Nanopharmaceuticals*; Shegokar, R., Ed.; Elsevier: Amsterdam, The Netherlands, 2020; pp. 135–150.
3. Chaudhary, N.; Weissman, D.; Whitehead, K.A. mRNA vaccines for infectious diseases: Principles, delivery and clinical translation. *Nat. Rev. Drug Discov.* **2021**, *20*, 817–838. [CrossRef] [PubMed]
4. Mao, L.; Chen, Z.; Wang, Y.; Chen, C. Design and application of nanoparticles as vaccine adjuvants against human corona virus infection. *J. Inorg. Biochem.* **2021**, *219*, 111454. [CrossRef] [PubMed]
5. Hu, C.M.; Fang, R.H.; Wang, K.C.; Luk, B.T.; Thamphiwatana, S.; Dehaini, D.; Nguyen, P.; Angsantikul, P.; Wen, C.H.; Kroll, A.V.; et al. Nanoparticle biointerfacing by platelet membrane cloaking. *Nature* **2015**, *526*, 118–121. [CrossRef]

6. Pardi, N.; Hogan, M.J.; Porter, F.W.; Weissman, D. mRNA vaccines—A new era in vaccinology. *Nat. Rev. Drug Discov.* **2018**, *17*, 261–279. [CrossRef]
7. Moon, J.J.; Suh, H.; Bershteyn, A.; Stephan, M.T.; Liu, H.; Huang, B.; Sohail, M.; Luo, S.; Um, S.H.; Khant, H.; et al. Interbilayer-crosslinked multilamellar vesicles as synthetic vaccines for potent humoral and cellular immune responses. *Nat. Mater.* **2011**, *10*, 243–251. [CrossRef]
8. Roth, G.A.; Picece, V.; Ou, B.S.; Luo, W.; Pulendran, B.; Appel, E.A. Designing spatial and temporal control of vaccine responses. *Nat. Rev. Mater.* **2022**, *7*, 174–195. [CrossRef] [PubMed]
9. Hassett, K.J.; Higgins, J.; Woods, A.; Levy, B.; Xia, Y.; Hsiao, C.J.; Acosta, E.; Almarsson, Ö.; Moore, M.J.; Brito, L.A. Impact of lipid nanoparticle size on mRNA vaccine immunogenicity. *J. Control. Release* **2021**, *335*, 237–246. [CrossRef] [PubMed]
10. Zhao, L.; Seth, A.; Wibowo, N.; Zhao, C.X.; Mitter, N.; Yu, C.; Middelberg, A.P. Nanoparticle vaccines. *Vaccine* **2014**, *32*, 327–337. [CrossRef] [PubMed]
11. Kheirollahpour, M.; Mehrabi, M.; Dounighi, N.M.; Mohammadi, M.; Masoudi, A. Nanoparticles and Vaccine Development. *Pharm. Nanotechnol.* **2020**, *8*, 6–21. [CrossRef] [PubMed]
12. Li, T.; Song, R.; Sun, F.; Saeed, M.; Guo, X.; Ye, J.; Chen, F.; Hou, B.; Zhu, Q.; Wang, Y.; et al. Bioinspired magnetic nanocomplexes amplifying STING activation of tumor-associated macrophages to potentiate cancer immunotherapy. *Nano Today* **2022**, *43*, 101400. [CrossRef]
13. Pati, R.; Shevtsov, M.; Sonawane, A. Nanoparticle Vaccines Against Infectious Diseases. *Front. Immunol.* **2018**, *9*, 2224. [CrossRef]
14. Reddy, S.T.; van der Vlies, A.J.; Simeoni, E.; Angeli, V.; Randolph, G.J.; O'Neil, C.P.; Lee, L.K.; Swartz, M.A.; Hubbell, J.A. Exploiting lymphatic transport and complement activation in nanoparticle vaccines. *Nat. Biotechnol.* **2007**, *25*, 1159–1164. [CrossRef]
15. Belizaire, R.; Unanue, E.R. Targeting proteins to distinct subcellular compartments reveals unique requirements for MHC class I and II presentation. *Proc. Natl. Acad. Sci. USA* **2009**, *106*, 17463–17468. [CrossRef]
16. Kanekiyo, M.; Wei, C.J.; Yassine, H.M.; McTamney, P.M.; Boyington, J.C.; Whittle, J.R.; Rao, S.S.; Kong, W.P.; Wang, L.; Nabel, G.J. Self-assembling influenza nanoparticle vaccines elicit broadly neutralizing H1N1 antibodies. *Nature* **2013**, *499*, 102–106. [CrossRef]
17. Huse, M. Mechanical forces in the immune system. *Nat. Rev. Immunol.* **2017**, *17*, 679–690. [CrossRef] [PubMed]
18. Xia, Y.; Wu, J.; Wei, W.; Du, Y.; Wan, T.; Ma, X.; An, W.; Guo, A.; Miao, C.; Yue, H.; et al. Exploiting the pliability and lateral mobility of Pickering emulsion for enhanced vaccination. *Nat. Mater.* **2018**, *17*, 187–194. [CrossRef] [PubMed]
19. Chen, N.; Wang, Y.; Song, X.; Li, Y.; Deng, Z. Steering DNA Condensation on Engineered Nanointerfaces. *Nano Lett.* **2022**, *22*, 8550–8558. [CrossRef] [PubMed]
20. Eleftherianos, I.; More, K.; Spivack, S.; Paulin, E.; Khojandi, A.; Shukla, S. Nitric oxide levels regulate the immune response of Drosophila melanogaster reference laboratory strains to bacterial infections. *Infect. Immun.* **2014**, *82*, 4169–4181. [CrossRef] [PubMed]
21. Briand, L.; Marcion, G.; Kriznik, A.; Heydel, J.M.; Artur, Y.; Garrido, C.; Seigneuric, R.; Neiers, F. A self-inducible heterologous protein expression system in Escherichia coli. *Sci. Rep.* **2016**, *6*, 33037. [CrossRef] [PubMed]
22. Bálint, Z.; Krizbai, I.A.; Wilhelm, I.; Farkas, A.E.; Párducz, A.; Szegletes, Z.; Váró, G. Changes induced by hyperosmotic mannitol in cerebral endothelial cells: An atomic force microscopic study. *Eur. Biophys. J. EBJ* **2007**, *36*, 113–120. [CrossRef]
23. Xu, P.; Gullotti, E.; Tong, L.; Highley, C.B.; Errabelli, D.R.; Hasan, T.; Cheng, J.X.; Kohane, D.S.; Yeo, Y. Intracellular drug delivery by poly(lactic-co-glycolic acid) nanoparticles, revisited. *Mol. Pharm.* **2009**, *6*, 190–201. [CrossRef]
24. Brennan, M.; Monahan, D.S.; Brulin, B.; Gallinetti, S.; Humbert, P.; Tringides, C.; Canal, C.; Ginebra, M.P.; Layrolle, P. Biomimetic versus sintered macroporous calcium phosphate scaffolds enhanced bone regeneration and human mesenchymal stromal cell engraftment in calvarial defects. *Acta Biomater.* **2021**, *135*, 689–704. [CrossRef]
25. Fan, L.; Zhang, S.; Zhang, C.; Yin, C.; Chu, Z.; Song, C.; Lin, G.; Li, Q. Multidrug Resistance in Cancer Circumvented Using a Cytosolic Drug Reservoir. *Adv. Sci.* **2018**, *5*, 1700289. [CrossRef]
26. Sun, G.G.; Wang, Z.Q.; Liu, C.Y.; Jiang, P.; Liu, R.D.; Wen, H.; Qi, X.; Wang, L.; Cui, J. Early serodiagnosis of trichinellosis by ELISA using excretory-secretory antigens of Trichinella spiralis adult worms. *Parasites Vectors* **2015**, *8*, 484. [CrossRef]
27. Hartman, H.; Wang, Y.; Schroeder, H.W., Jr.; Cui, X. Absorbance summation: A novel approach for analyzing high-throughput ELISA data in the absence of a standard. *PLoS ONE* **2018**, *13*, e0198528. [CrossRef]
28. Yu, X.; Gilbert, P.B.; Hioe, C.E.; Zolla-Pazner, S.; Self, S.G. Statistical approaches to analyzing HIV-1 neutralizing antibody assay data. *Stat. Biopharm. Res.* **2012**, *4*, 1–13. [CrossRef]
29. Ma, Y.; Tang, K.; Zhang, Y.; Zhang, C.; Cheng, L.; Zhang, F.; Zhuang, R.; Jin, B.; Zhang, Y. Protective CD8(+) T-cell response against Hantaan virus infection induced by immunization with designed linear multi-epitope peptides in HLA-A2.1/K(b) transgenic mice. *Virol. J.* **2020**, *17*, 146. [CrossRef]
30. Honda-Ogawa, M.; Sumitomo, T.; Mori, Y.; Hamd, D.T.; Ogawa, T.; Yamaguchi, M.; Nakata, M.; Kawabata, S. Streptococcus pyogenes Endopeptidase O Contributes to Evasion from Complement-mediated Bacteriolysis via Binding to Human Complement Factor C1q. *J. Biol. Chem.* **2017**, *292*, 4244–4254. [CrossRef]
31. Machatschek, R.; Lendlein, A. Fundamental insights in PLGA degradation from thin film studies. *J. Control. Release* **2020**, *319*, 276–284. [CrossRef]

32. Mante, A.; Heider, M.; Zlomke, C.; Mäder, K. PLGA nanoparticles for peroral delivery: How important is pancreatic digestion and can we control it? *Eur. J. Pharm. Biopharm.* **2016**, *108*, 32–40. [CrossRef] [PubMed]
33. Tietz, N.W.; Shuey, D.F. Lipase in serum–the elusive enzyme: An overview. *Clin. Chem.* **1993**, *39*, 746–756. [CrossRef] [PubMed]
34. Deng, J.; Wang, X.; Zhang, B.Z.; Gao, P.; Lin, Q.; Kao, R.Y.; Gustafsson, K.; Yuen, K.Y.; Huang, J.D. Broad and Effective Protection against Staphylococcus aureus Is Elicited by a Multivalent Vaccine Formulated with Novel Antigens. *mSphere* **2019**, *4*. [CrossRef] [PubMed]
35. Silverstein, R.M.; Webster, F.X.; Kiemle, D.J.; Bryce, D.L. *Spectrometric Identification of Organic Compounds*, 8th ed.; Wiley: New York, NY, USA, 2017.

vaccines

MDPI

Review

Efficiency of Chitosan Nanocarriers in Vaccinology for Mucosal Immunization

Salvatore Calogero Gaglio [1], Massimiliano Perduca [1,*], Donato Zipeto [2,*] and Giuseppe Bardi [3,*]

1 Department of Biotechnology, University of Verona, Cà Vignal 1, Strada Le Grazie 15, 37134 Verona, Italy; salvatorecalogero.gaglio@univr.it
2 Department of Neurosciences, Biomedicine and Movement Sciences, University of Verona, Strada Le Grazie 8, 37134 Verona, Italy
3 Nanobiointeractions & Nanodiagnostics, Istituto Italiano di Tecnologia, Via Morego 30, 16163 Genova, Italy
* Correspondence: massimiliano.perduca@univr.it (M.P.); donato.zipeto@univr.it (D.Z.); giuseppe.bardi@iit.it (G.B.)

Abstract: The mucosal barrier constitutes a huge surface area, close to 40 m^2 in humans, located mostly in the respiratory, gastrointestinal and urogenital tracts and ocular cavities. It plays a crucial role in tissue interactions with the microbiome, dietary antigens and other environmental materials. Effective vaccinations to achieve highly protective mucosal immunity are evolving strategies to counteract several serious diseases including tuberculosis, diphtheria, influenzae B, severe acute respiratory syndrome, Human Papilloma Virus infection and Acquired Immune Deficiency Syndrome. Interestingly, one of the reasons behind the rapid spread of severe acute respiratory syndrome coronavirus 2 variants has been the weakness of local immunization at the level of the respiratory mucosa. Mucosal vaccines can outperform parenteral vaccination as they specifically elicit protective mucosal immune responses blocking infection and transmission. In this scenario, chitosan-based nanovaccines are promising adjuvants-carrier systems that rely on the ability of chitosan to cross tight junctions and enhance particle uptake due to chitosan-specific mucoadhesive properties. Indeed, chitosan not only improves the adhesion of antigens to the mucosa promoting their absorption but also shows intrinsic immunostimulant abilities. Furthermore, by finely tuning the colloidal properties of chitosan, it can provide sustained antigen release to strongly activate the humoral defense. In the present review, we agnostically discuss the potential reasons why chitosan-based vaccine carriers, that efficiently elicit strong immune responses in experimental setups and in some pre-clinical/clinical studies, are still poorly considered for therapeutic formulations.

Keywords: mucosal barrier; chitosan nanoparticles; vaccines; mucosal vaccines; local immunization

1. Introduction

The outbreak of the COVID-19 pandemic reminded us of the importance of vaccines in the fight against infectious diseases induced by emerging and re-emerging uncontrolled pathogens continuously worrying the World Heathy Organization (WHO) [1]. It is estimated that one-third of the world's population is infected with *Mycobacterium tuberculosis* (*M. tuberculosis*) and that approximately 9.6 million may develop active tuberculosis each year resulting in approximately 1.5 million deaths [2]. In addition, almost 70% of all deaths in very young children are caused by infectious diseases, with the highest percentages due to infections of the mucosal membranes, such as pneumonia (18%) and diarrhea (15%) [3].

Interestingly, one of the reasons behind the rapid spread of severe acute respiratory syndrome coronavirus 2 (SARS-CoV-2) was the lack of previous immunity and the availability of mucosal immunization [4]. The latter was likely responsible for infections of vaccinated people who did not show disease symptoms. Mucosal membranes constitute a huge surface area of approximately 40 m^2 in humans, predominantly in the respiratory, gastrointestinal and urogenital tracts, as well as eye cavities. They play a key role in the interactions with the microbiome, dietary antigens and several environmental materials [5]. Mucosae are considered hot spots where the chance of pathogen entry is greatest. Therefore, an effective vaccination to elicit highly protective local immunity in the mucosa is an emerging strategy to counteract severe infectious diseases including tuberculosis, diphtheria, influenzae B, severe acute respiratory syndrome (SARS), Human Papilloma Virus infection (HPV) and Acquired Immune Deficiency Syndrome (AIDS) [6–11]. In this regard, mucosal vaccination may outperform the parenteral route in eliciting local and specific immune responses capable of blocking infection or transmission [5]. Among the materials used to develop mucosal vaccines, the natural polymer chitosan has shown promising features as a delivery system, in particular, its mucoadhesive [12] and intrinsic immunostimulant properties [13]. Therefore, chitosan can be used as an advantageous delivery system also acting as an adjuvant to provide sustained vaccine release and strongly activate the local defense system.

In nanomedicine, several well-described facilitations of specific drug delivery through mucosal barriers by chitosan nanoparticles (NPs) have been reported. For instance, intestinal adsorption of doxorubicin and tamoxifen was improved by encapsulating these drugs in chitosan NPs [14,15]. Furthermore, brain targeting and increased bioavailability of carbamazepine were achieved by using chitosan-based NPs as nasal carriers [16]. In addition, the positive charge of the polymer is another feature widely exploited to design chitosan NPs suitable for gene and antiviral-drug delivery [17,18].

In this review, we focus our attention in particular on the efficacy in vaccinology of chitosan-based carriers for the mucosa and its potential and limitations for future vaccine formulations.

2. Mucosal Structure

The mucous membrane, known as mucosa, is the soft tissue covering the body's channels and organs in the digestive, respiratory and reproductive systems. The mucosa constitutes a huge surface area in humans, allowing various interactions with the microbiome, dietary antigens and diverse environmental materials. The structure of the mucosa presents several physiological barriers that interfere with the uptake of vaccine antigens and determine their rapid clearance. These barriers are determined by the cellular structure of epithelial surfaces (i.e., tight junctions), the presence of enzymatic barriers (i.e., nucleases and proteases) and the mucus. The latter creates a gelatinous layer which, due to its sticky nature and rapid turnover, prevents the entry of potentially dangerous substances or infectious agents. At the same time, the mucus inhibits the diffusion of NPs linked to drugs or vaccine antigens, thereby greatly reducing their efficacy. As the first barrier against pathogens, it also exerts critical roles in immunity and several physiological processes, such as the absorption of nutrients in the gastrointestinal tract.

The mucosa is a complex structure consisting of three different overlapping layers (Figure 1) [19]:

i. The epithelial layer is the most superficial. Epithelial cells secrete a thick, gel-like mucus that protects the body from irritants and gives the mucous membrane its name. There may be one or more layers of cells and they may be stacked in columns or arranged like bricks. Epithelial cells also have a high turnover rate, frequently replacing each other to clear out invasive particles. Some cells in this layer have small complex structures called cilia, which help to remove extraneous substances.
ii. The lamina layer is a connective tissue to which the epithelium attaches, and it is considered the middle layer of the mucosa. The lamina is composed of structural protein molecules, nerves and veins. It thus plays a crucial role in blood supply to the epithelium, keeping cells in place and binding them to the underlying smooth muscle. Its nerves respond to muscle fluctuations to change the shape of the epithelium as needed. Several varieties of immune cells are present in this layer seeking out and destroying intrusive microbes.

iii. The deeper layer is named muscularis mucosae, as it is mainly composed of smooth muscles. The muscularis mucosae provides a perpetual motor function that keeps the mucosa moving. This dynamic feature helps the lining to stretch and contract along with the various organs of the digestive system during their activity. In addition, it helps the mucous membrane to perform cleaning functions by keeping the hair-like cilia moving on the surface layer cells.

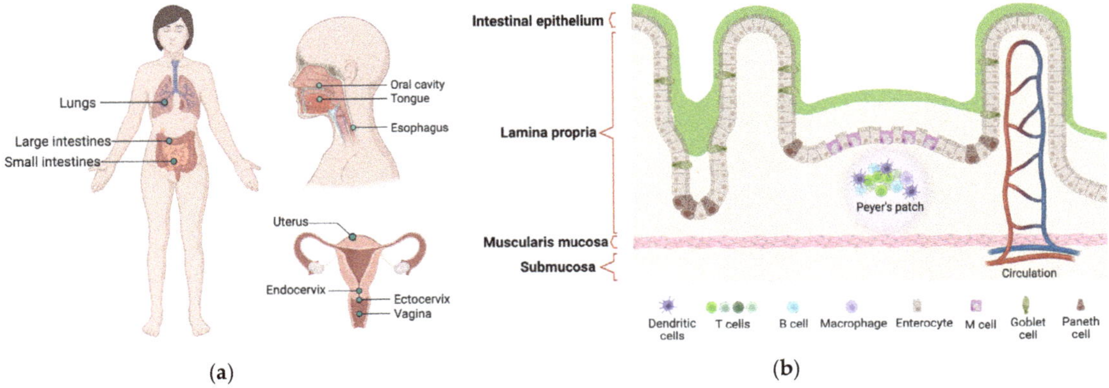

Figure 1. (**a**) Mucosal tissue of the human body. (**b**) The architecture of the small intestinal mucosa as an example of the tissue structure.

The intricate and complex structure of the mucosa poses a challenge for the development of mucosal vaccines: the delivery of antigens must be able to overcome the mucus layer present in the epithelium in order to achieve sufficient quantities to effectively activate adaptive immunity.

3. Needle-Free Vaccines and Mucosal Vaccination

Vaccines were developed to provide direct protection to immunized individuals through B and T cell-dependent mechanisms. Two crucial "players" that define vaccine performance are certainly the type of antigen and the adjuvant. However, the route of administration also has a strong influence on efficacy, along with the ability to be unaffected by storage temperature variations [20]. Depending on the site of entry, the vaccine encounters different types of "antigen presenting cells" (APCs), such as dendritic cells or macrophages [21]. APCs expose antigens on their surface and present them to lymphocytes, the effector cells of the adaptive immune system. The way the antigen is presented also determines the quality of the response in terms of efficacy, cellular memory, prolonged response and the specific cell subset involved.

Therefore, the antigen, adjuvant and route of administration must be chosen for each type of vaccine to optimize the induced immunological response [22,23]. While mandatory vaccines (i.e., anti-polio, anti-diphtheria, anti-tetanus, anti-hepatitis B, etc.) are injected into muscles, different needle-free vaccination systems using alternative routes of administration (i.e., oral, nasal, urogenital) also exist or are being developed [24].

Upon vaccine injection, the immune reaction causes a systemic effect. The antigen encounters the cells of the immune system, and the presence of an adjuvant causes a slow and continuous release depot (depot effect). For vaccines containing adjuvants, intramuscular injection is preferred because it reduces the risk of local skin reactions. Although, paradoxically, in remote areas with limited sterile conditions, injections aimed at inducing an immune response could be risky, providing a route of entry (i.e., the lesion caused by the syringe) for unexpected external pathogens. However, data show that protection against respiratory and sexually transmitted pathogens from locally administered vaccines induces greater protection than systemically/parenterally administered vaccines [25]. In general, injectable vaccines induce high antibody production but a lower T lymphocyte-mediated cellular response [21] and weaker induction of mucosal immunity [26]. For instance, intranasal vaccines are better suited to stimulate mucosal-specific immunity [27] as they are able to elicit the immune defense mechanism in situ.

The advantages and disadvantages of "common" injectable vaccines versus needle-free mucosal ones are summarized in Table 1.

Table 1. Comparison between systemic and mucosal vaccination. [22,26,28] *, [24,29] **, [27] ***.

Administration	Advantages	Disadvantages
Injectable vaccines	i. systemic effect * ii. antibody response * iii. depot effect * iv. simple and rapid * v. intradermal immunogenicity *	i. weak cell-mediated response * ii. weak mucosal immunity *** iii. need for a specialized operator ** iv. maintenance of the cold chain **
Needle-free/mucosal	i. mucosal immunity *** ii. tolerability and safety ** iii. low costs ** iv. ease of production, storage and distribution ** v. antigen stability ** vi. ease of administration **	i. immunogenic tolerance * ii. barriers in the gastrointestinal tract (pH, proteolytic enzymes) * iii. not suitable for all types of antigens * iv. most are preliminary **

Mucous membranes are the hot spots of contact between the organism and the external environment, including potential pathogens facing a tightly regulated and specific mucosal immunity (Figure 2).

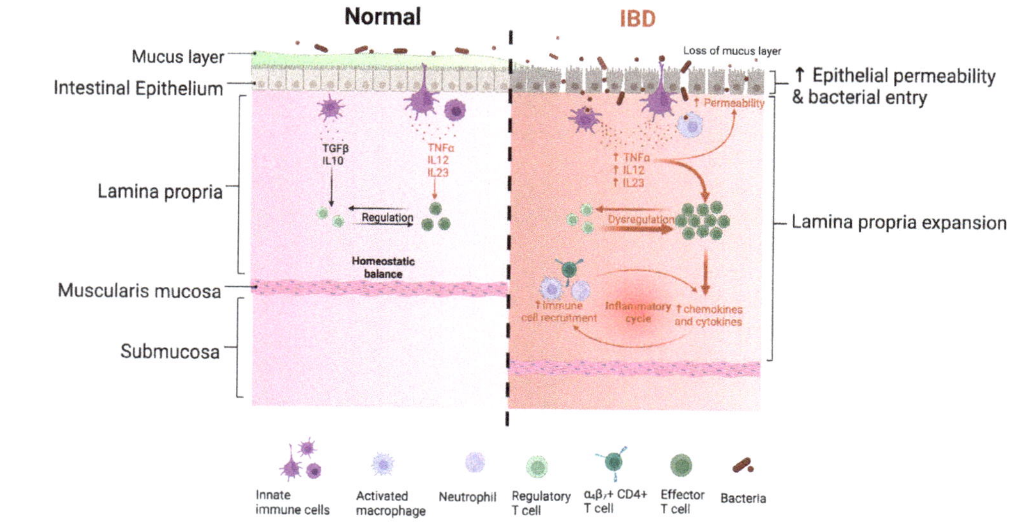

Figure 2. Schematic drawing of the immune response regulation in normal tissue and during inflammatory bowel disease (IBD) as an example of the pathological condition. The arrow (↑) means: "upregulation" of cytokines and "increased" permeability of the epithelium.

Unlike injected vaccines, mucosal vaccines are captured by innate immune cells patrolling the lamina propria close to the epithelium (Figure 2). Mucosal vaccine internalization under physiological conditions allows activation of the adaptive response without any injury to the tissue. Although very limited, the injection induces local damage stimulating the dysregulated activation of effector cells similar to local pathological conditions, as exemplified schematically by the IBD on the right side of Figure 2.

The mucosa is also particularly rich in secondary lymphoid organs harboring different immune cells producing antimicrobial proteins and B lymphocytes producing immunoglobulin A (IgA). These cellular configurations create a protective inner mucus layer (Figure 3).

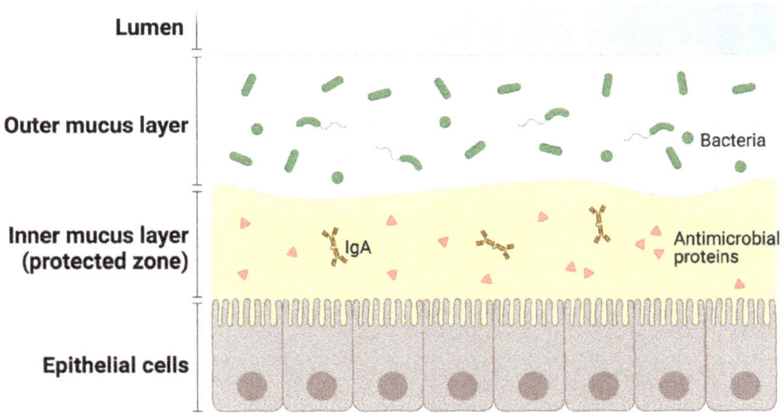

Figure 3. Organization of the mucosal defense barrier: the outer section of the mucus layer is in direct contact with the lumen and is exposed to external pathogens; the inner one, facing the epithelial cells, is rich in IgA and antimicrobial factors.

Vaccine-activated B lymphocytes in the mucosa enrich the antimicrobial protein barrier of the inner mucus layer (Figure 3), increasing the efficiency of the "protected zone". Moreover, IgA antibodies elicited by intramuscular vaccination differ from those induced by the compartmentalized, mucosal immune response to natural infection. Monomeric IgA present in plasma are less potent than the corresponding IgG, while dimeric IgA, the main type of antibody elicited at the mucosal level, are more potent against the same targets [30–32]. For instance, dimeric IgA deployed in mucosal secretions as the most effective antibody class able to neutralize air-borne pathogens, like SARS-CoV-2, in infected patients [27,29,33]. Interestingly, recent studies have shown that parenteral vaccines used against SARS-CoV-2 induce a weak mucosal response, although IgG can be detected in saliva and nasal fluid. Similarly, IgA are elicited in plasma [32], but they were detected at very low levels in saliva. This suggests that systemic vaccination does not stimulate an immunoglobulin response at the mucosal level, leading to reduced protection and a rapid decrease in the defense from infection in vaccinated individuals [34].

It is worth mentioning that in some cases, comparable results have been obtained between systemic and mucosal immunization. Nevertheless, the latter is often a safe, effective and less invasive alternative to parenteral vaccination [25].

4. Mucosal Immunity and Immunization

Mucosal vaccines could outperform parenteral vaccination in blocking infection and transmission by providing robust protective immune responses in the sites of pathogen penetration [7,9,35]. The induction of specific adaptive immunity in the mucosal first line of defense has the ability to prevent the onset of infection, rather than limiting the development of disease symptoms, making mucosal vaccines important tools for epidemic spreading [36].

As we previously mentioned, the predominant Ig isotype in the mucosal epithelium is the secretory IgA (SIgA) [37]. SIgA antibodies are able to neutralize toxins or pathogens in the mucosa through three different pathways: immune exclusion, antigen excretion and intracellular neutralization. In addition, they exert higher neutralizing activity than

IgG due to their extracellular immune exclusion effect. After vaccination, it is desirable to achieve a sterilizing condition to prevent the spread of infection, which can be particularly dangerous in hospital settings during the management of frail patients [37–39].

Tissue-resident memory T ($_{\mathrm{TRM}}$T) cells also participate in the activation of adaptive immunity in the mucosa: indeed, they confer a rapid re-stimulation response that induces the production of inflammatory cytokines mediating tissue resistance and chemokines recruiting immune helper cells [40]. $_{\mathrm{TRM}}$T cells have been found at several mucosal sites exerting a decisive role in rapid responses to infections [41,42]. For example, lung-resident T and memory T cells play an essential role in fighting frontline viral and bacterial pathogens through direct mechanisms and by coordinating the adaptive immune system through the crosstalk pathway [43].

Therefore, mucosal immunization can trigger extensive adaptive IgA- and $_{\mathrm{TRM}}$T-mediated responses, which are, for instance, crucial for the establishment of protective lung immunity against tuberculosis [44]. Furthermore, strong cellular and humoral immune responses in the mucosa have the potential to induce sterilizing immunity by preventing pathogen binding and uptake across epithelial surfaces and, eventually, greatly reducing the risk of asymptomatic or paucisymptomatic infections. As the recent COVID-19 pandemic has taught, asymptomatic individuals are often the most difficult to intercept and heavily contribute to the diffusion of the virus and consequently to the pandemic risk [33].

5. Chitosan in Mucosal Vaccine Design

The adaptive immune response in the mucosa relies on the secretion of IgA and the activation of tissue-resident memory T cells. A mucosal vaccine should be able to efficiently deliver its cargo in order to trigger the secretion of antibodies penetrating the lamina layer and activate $_{\mathrm{TRM}}$T lymphocytes by continuous and prolonged exposure to the antigen. Furthermore, the rapid clearance by cilia and enzymes degradation by nucleases/proteases should be avoided. Nanotechnologies could overcome these obstacles by providing protection to the antigen, enhancing cellular uptake and increasing the retention time.

Often, the vaccine-active molecule is not sufficiently immunogenic to ensure a complete response. This requires stimulators (adjuvants) to successfully promote the immunization process. Chitosan-based nanomaterials have attracted great attention as adjuvants and delivery systems able to effectively boost mucosal immune responses [13,45].

To better understand the advantages of chitosan-based NPs, it is worth first recalling the chemical properties of the chitosan polymer and its valuable manipulation at the nano-scale. Chitosan is a biopolymer consisting of randomly distributed N-acetylated and deacetylated glucosamines [46,47]. It is produced by the deacetylation process of chitin [48–50], which is a component of the exoskeleton of shrimps, crabs and fungi. Interestingly, chitosan effectively fits into the circular economy as a high-added value material obtained from the waste product of chitin. In addition, chitosan is widely regarded as a nontoxic and hypoallergenic material suitable for use in medical and pharmaceutical applications [51].

The ionic cross-linking of chitosan NPs is the most common chitosan-nanomaterial preparation method. It is based on the interaction between a cross-linking agent and the amine or hydroxyl groups of chitosan (Figure 4a,b) [49]. Additionally, NPs are spontaneously formed in the process of polyelectrolyte complexation by adding nuclide acids to a chitosan acetic acid solution.

The reverse-micelle preparation method could also be used to prepare chitosan NPs of small size. This synthesis consists of dropping a polymer aqueous solution into an organic phase containing a surfactant with constant agitation to allow the formation of reverse micelles.

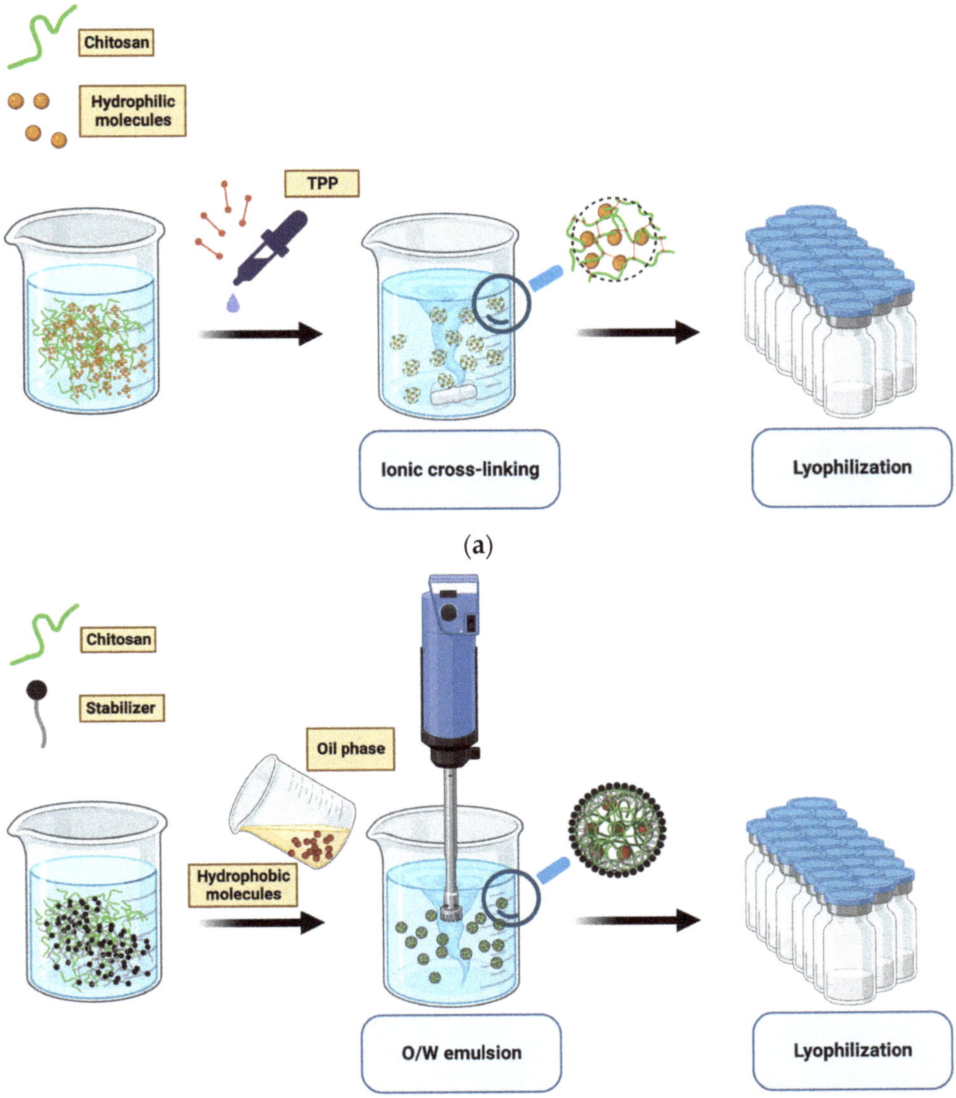

(b)

Figure 4. (**a**) Schematic illustration of chitosan NPs preparation by ionic cross-linking method; (**b**) schematic illustration of chitosan NPs preparation by O/W emulsion method.

The described methods are suitable for loading hydrophilic molecules, while oil-in-water emulsion is used to produce 200–500 nm NPs embedded with hydrophobic molecules. An oil-in-water (O/W) emulsion is prepared by mixing an aqueous solution of stabilizer and chitosan with an organic solvent under mechanical shaking. Precipitation is the driving force for the nanostructure arrangement. Finally, NPs are collected and purified from solvent residues [18,49,52].

In vaccinology, chitosan mucoadhesive properties make this biopolymer suitable for the design of mucosal vaccines. Indeed, several studies have demonstrated the potential use of chitosan NPs and derivatives as promising vehicles for vaccine delivery, due to the presence of numerous positive charges on its surface facilitating electrostatic interactions with the negative charges that characterize the sialic acid present in the mucus lining the epithelium [53]. The increased retention time of the loaded antigen [54–56] also makes antigen uptake possible [57] by promoting permeability relying on its ability to open tight junctions between epithelial cells [58]. It is possible to further increase the permeability and diffusion capacity of chitosan nanoparticles by chemically modifying them by adding short tails of poly(ethylene glycol), poly(2-hydroxyethyl acrylate), poly(2-ethyl-2-oxazoline) or poly(N-vinyl pyrrolidone) [59]. Thus, chitosan NPs can more effectively target the immune cell population in the lamina layer. These NPs can provide sustained antigen release to strongly activate the humoral defense system by adjusting the colloidal properties. The immune triggering mechanism of chitosan NPs is based on enhancing antigen uptake and inducing macrophages to secrete inflammatory factors regulating Th1/Th2 balance tailoring a specific immune response [46].

Nanoscale chitosan intrinsic immunostimulatory properties have been observed in different studies and are reviewed elsewhere [60]. For example, Nantachit et al. [61] demonstrated that in vitro stimulation of human nasal epithelial cells with trimethyl chitosan NPs embedded with EDIII-D3 stimulated the secretion of several proinflammatory, Th1 and Th2 cytokines as well as chemokines. Particles uptake by DCs also upregulated the maturation markers (CD80, CD83, CD86, and HLA-DR) of DCs.

6. Chitosan Outcome in SARS-CoV-2 Immunization

The SARS-CoV-2 pandemic has also become a worldwide problem due to the rapid spread of variants, which limits the effectiveness of current intramuscularly administrated vaccines. The rapid establishment of an immune barrier at the level of the respiratory mucous membranes would be essential to counter the future spread of newly born viral variants. Moreover, the use of nanocarriers with adjuvant properties would certainly be helpful to allow a complete adaptive response in the mucosa.

In this regard, Shao-Hua Zhuo et al. designed a chitosan-based nanovaccine for inhalation. Spike protein-loaded chitosan NPs ((S@CS)NPs) were prepared using polyelectrolyte complexation. sIgA were detected in the bronchoalveolar fluid of mice immunized with (S@CS)NPs, whereas free Spike failed to elicit a sIgA response. In addition, SARS-CoV-2 Spike-specific IgG were also stimulated in bronchoalveolar liquid. Intriguingly, the IgG_1/IgG_{2a} ratio in serum samples suggested that the intranasally administrated nanovaccine enhanced Th2 immunity. The authors highlighted the active role of (S@CS) in enhancing T cell recognition and activation due to the dual role of chitosan as an adjuvant and transporter. Additionally, memory T cells were detected in splenocytes after the third dose. Thus, S@CS challenge enhanced the production of $CD4^+$ T cells and memory effector cells ($CD3^+$ $CD4^+$, $CD44^+$ and $CD62L^-$) [62]. The potential of mucosal chitosan-based nanovaccines to elicited not only a local mucosal immunization but also a systemic response needs to be further explored.

7. Modified Chitosan Nanocarriers and Pre-Clinical Trials

To design efficient carrier-adjuvant material for different administration routes, chitosan is often chemically modified (Figure 5) and prepared in different versions, including microparticles, NPs and hydrogels. Microparticles are suitable for sustained release and oral administration, whereas NPs optimize cellular interaction and can be preferred for nasal administration. Chitosan hydrogel can increase retention time depending on the target tissue and the specific formulation. The biopolymer can be easily prepared under mild conditions (such as ionic gelation) and functionalized by surface chemistry modulation.

In the following subchapters, the different chitosan modifications are also described in terms of employment in pre-clinical tests, and the data are summarized in Table 2.

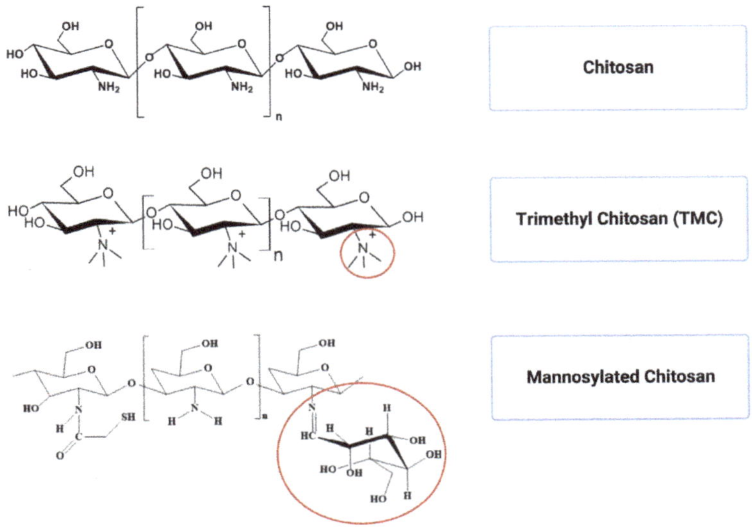

Figure 5. Molecular structure of chitosan, trimethyl chitosan and mannosylated chitosan.

7.1. Trimethyl Chitosan

Trimethyl chitosan (TMC) shows a higher aqueous solubility and stability over a wide range of ionic conditions than the unmodified polymer. Furthermore, the modification does not affect the mucoadhesive properties and it allows improved crossing through the tight junctions of epithelial cells [54]. Because TMC-based NPs are smaller in size than unmodified chitosan NPs, they also show increased cellular uptake. The modified release pattern, higher loading efficiency and stronger positive surface potential make TMC-based NPs very attractive for the design of high-performance mucosal vaccines [63].

The potential of ovalbumin-conjugated N-trimethylaminoethylmethacrylate chitosan NPs (OVA-NP) in eliciting mucosal immune responses after nasal administration has been demonstrated in animal models by in vivo assays. Interestingly, the mice group treated with nasally administered OVA-NPs showed a 5.3-fold higher lymph node targeting index (LNTI) than the group receiving an intramuscular injection. When OVA was administered intranasally alone, it did not induce strong immune responses in the nasal cavity because of its low immunogenicity. Similarly, the injection of alum-precipitated OVA failed to induce mucosal immune responses. On the contrary, TMC-conjugation dramatically elicited sIgA levels in the mucosa [64].

These findings suggest the potential advantages of nasal vaccination in the prevention of respiratory infectious diseases by using a TMC-based carrier to enhance mucosal immune response. Other studies have demonstrated the ability of TMC-based carriers to elicit strong immunization by penetrating the mucosal epithelium and increasing the likelihood of boosting the antigen-presenting cell mechanism [54,65,66].

Upregulation of the B lymphocyte-mediated response has been also demonstrated by TMC NPs loaded with bacterial toxins. Tetanus toxoid-loaded TMC NPs (TMC(TT) NPs) were administered nasally in BALB/c mice. After the administration, antibody production against the tetanus toxoid antigen was detected. Unlike intramuscular tetanus toxoid injection, TMC(TT) NPs enhanced IgA secretion in saliva and IgG in blood serum, which significantly increased over the following 42 days [67].

In search of easy-to-use vaccines for tropical infectious diseases, TMC NPs embedded with domain III of dengue serotype-3 E protein (EDIII-D3) were prepared. The ability to stimulate human nasal epithelial cells by EDIII-D3 TMC NPs was proven, and the potential nanoformulated vaccine increased the secretion of several proinflammatory cytokines and chemokines. NP internalization by dendritic cells also upregulated the maturation markers

(CD80, CD83, CD86 and HLA-DR) of DCs [61]. Antigen presentation-mediated overexpression of inflammatory mediators underlines the high potential of nasally administered TMC-based nanovaccines.

Further confirmation of TMC NPs' enhanced cellular-mediated mucosal adaptive response has been shown by their use as mucosal immunopotentiators against Hepatitis B virus (HBV). TMC NPs embedded with an HBV surface antigen induced T cell proliferation and higher serum and nasal antibody titers than free antigen [68].

Efficient carriers are characterized by efficient protection of drug/nucleic acid load and its effective release on site. It has been proven that TMC encapsulation can also provide gastroprotection to the cargo emphasizing TMC NP suitability for oral administration. Actually, TMC(Omp31) NPs elicited a Th1–Th17 immune response. Moreover, when mice were challenged with *B. melitensis* 16 M, those orally vaccinated with TMC(Omp31) NPs were more protected than those immunized intraperitoneally [69].

Efficacy in mucosal vaccinology was also evaluated on chickens challenged with Newcastle disease virus (NDV). Compared with the commercial live attenuated NDV vaccine, chicken treated with O-2′-HACC/pFDNA NPs induced higher production of anti-NDV IgG and sIgA antibodies and increased lymphocyte proliferation. Higher levels of IL-2, IL-4, IFN-γ, CD4+ and CD8+ T lymphocytes were also recorded exhibiting a complete immune response [70].

7.2. Mannosylated Chitosan

A synergy between specific humoral response and cellular immunity in the lung airways would be crucial for vaccine protection against tuberculosis infection. In fact, IgA could directly block the entry of *Mycobacterial tuberculosis* and modulate pro-inflammatory responses. Conjugation of mannose to chitosan (MCS) is another valuable strategy to enhance chitosan NP uptake by APCs [71]. In fact, mannose receptors are mainly expressed on macrophages and dendritic cell membranes to enable recognition and endocytosis of mannose-enriched pathogens [72].

An interesting hint for a notable application of mannosylated chitosan has been published by Manli Wu et al., who designed an MCS-DNA vaccine that could dramatically increase SIgA production in C57BL mice, contributing to the significant reduction of bacterial CFU in the lung. Adaptive poly-functional $CD4^+/CD8^+$ T responses were also detected [73].

7.3. Chitosan Hydrogel

To overcome the rapid clearance of the antigens in mucosal vaccination through nasal administration, which is mainly due to natural defense systems such as cilia, James G. Bedford et al. developed a chitosan-based hydrogel. The hydrogel structure aims at improving antigen retention time and, therefore, the chances of interacting with the target immune cells [66,74]. Highly protective local immunity was observed after treatment with a chitosan hydrogel vaccine loaded with influenza virus peptides. Immunized mice showed significant protection when infected by the influenza virus. The prolonged antigen retention provided by the hydrogel increased the proliferation of tissue-resident memory $CD8^+$ T cells within the nasal mucosa [74]. Fascinatingly, these types of chitosan-based hydrogel could be explored to elicit nasal ${}_{Trm}CD8^+$ activation towards SARS-CoV-2 and other clinically relevant respiratory viruses.

Table 2. Summary of modified chitosan nanocarriers present in pre-clinical trials.

Chitosan-Based Nanocarriers	Antigen	Application	Activity and Immune Response	Advantages and Disadvantages	Administration Route	Reference
Trimethyl chitosan	Ovalbumin	Prevention of respiratory infectious diseases	Eliciting sIgA levels in the mucosa	High water solubility and increased stability	Intranasal	[48,55–57]
		Penetration of the mucosal epithelium and boosting the antigen presentation	Preserving the mucoadhesive properties and favouring the tight junctions crossing of epithelial cells			
		Upregulation of B lymphocyte				
	Bacterial toxin	Prevention against Tetanus	IgA secretion in saliva and IgG were detected in blood serum	Decrease in nanoparticle size than non-modified chitosan	Intranasal	[58]
			Increasing the secretion of several proinflammatory cytokines and chemokines	Increased cellular uptake		
				Increased loading efficiency and stronger positive surface		
	Domain III of dengue serotype-3 E protein	Prevention of tropical infectious diseases	Increase the secretion of several proinflammatory cytokines and chemokines		Intranasal	[52]
			Up-regulation of maturation markers (CD80, CD83, CD86 and HLA-DR) for dendritic cells			
	Hepatitis B surface antigen	Immunization against Hepatitis B virus	Induced T cell proliferation and increased serum and nasal antibody titer		Intranasal	[59]
	Omp31	Protection against Brucellosis	Eliciting the Th1–Th17 immune response		Oral	[60]

Table 2. *Cont.*

Chitosan-Based Nanocarriers	Antigen	Application	Activity and Immune Response	Advantages and Disadvantages	Administration Route	Reference
	O-2′-HACC/pFDNA	Complete protection to Newcastle disease virus	Massive production of anti-NDV IgG and sIgA antibodies and increased lymphocyte proliferation		Intranasal	[61]
			High levels of IL-2, IL-4, IFN-γ, CD4+ and CD8+ T lymphocytes			
Mannosylated chitosan	MSC-DNA	Protection against Mycobacterium tuberculosis	Increase SIgA production in C57BL mice	Enhance chitosan nanoparticle uptake by APCs	Intranasal	[62,64]
			Trigger a poly-functional CD4+/CD8+ T responses			
Chitosan hydrogel	Influenza virus peptides	Protection against Influenza virus	Increase the proliferation of tissue resident memory CD8+ T cells within the nasal mucosa	Improve the antigen retention time	Intranasal	[65]
			Confers high protective local immunity			

8. Chitosan-Based Vaccines in Clinical Trials

Despite chitosan showing promising evidence in eliciting strong mucosal immunization in experimental setups, clinical efforts to standardize evaluation criteria and protocols to assess its potential use in human mucosal vaccination are lacking. Thus, we refer to efficacy in terms of antibody production and cellular response resulting from different pre-clinical studies that occurred in the last twenty years. All data in the chapter are summarized in Table 3.

Several studies have demonstrated that intranasally administered vaccines provide combined systemic and secretory immunity even in other distant mucosal membranes, such as the lung and the genital tract [75,76]. Intranasal immunization is therefore of great interest for infections caused by pathogens acquired via the mucosa, like *C. diphtheriae* or sexually transmitted microbes. Volunteers treated with intranasal CRM_{197}-chitosan showed an sIgA response, while for those treated with intramuscular injection, no antitoxin sIgA was collected in the lavage fluid. Only intranasal administration activated nose-associated lymphoid tissue (NALT) for the sIgA response. Indeed, after the second immunization, a local antitoxin sIgA response was observed in the nasal lavage fluid of the vaccinated nostril and in the circulating IgA antitoxin antibody secreting cells (ASCs). The immune response induced by CRM_{197}-chitosan was greater than CRM_{197} alone. The authors outlined the role of chitosan in stimulating efficient mucosal activation. The bio-adhesive property of the polymer could counteract the mucociliary action by favoring the localization of the antigen inside the nose, allowing its prolonged absorption. Furthermore, chitosan-based delivery preserved the intact antigen structure without causing histological changes [77].

Other studies demonstrated the potential role of CRM_{197} loaded in chitosan NPs to prepare a mucosal diphtheria vaccine. After nasal administration, a strong increase in Th2-type responses was observed in volunteers and it correlated with protective levels of toxin-neutralizing antibodies [78].

To develop an inexpensive but effective needle-free vaccine against both C *N. meningitidis* and *diphtheria*, healthy volunteers were given either a single intramuscular injection of (MCP)-CRM197 conjugate vaccine in alum or two nasal insufflations of chitosan-based (MCP)-CRM197. Nasal insufflation was generally safe and showed a serum bactericidal antibody (SBA) titer after two nasal immunizations in naïve subjects comparable to parenteral immunization. Furthermore, a concentration of MCP-specific immunoglobulin G close to that obtained with parental administration was recorded in naïve subjects. Interestingly, an almost four-fold higher concentration was collected in previously parenterally immunized subjects, suggesting that nasal administration could also be used as a booster in a second vaccination. Finally, only MCP-specific sIgA were induced at the mucosal site. In this phase I trial, authors highlighted the role of chitosan-based vaccines in eliciting both systemic and mucosal immunization, particularly against C *N. meningitidis* [79].

Table 3. Summary of modified chitosan nanocarriers present in clinical trials.

Chitosan-Based Nanocarriers	Antigen	Application	Activity and Immune Response	Advantages and Disadvantages	Administration Route	Reference
CRM_{197}-Chitosan	CRM_{197}	Protection against Diphtheria	sIgA response	Efficient mucosal activation	Intranasal	[68,69]
			Activation of nose-associated lymphoid tissue	Preserving the antigen structure without histological changes		
			Presence of local antitoxin sIgA in the nasal mucus and in the circulating IgA antitoxin antibody secreting cells	Bio-adhesive property counteracts the mucociliary action, favoring the localization of the antigen inside the nose and allowing for prolonged absorption		
			Strong increase in Th2-type responses, correlated with protective levels of toxin-neutralizing antibodies			
(MCP)-CRM197 Chitosan	(MCP)-CRM197 conjugated	Protection against both C N. meningitidis and diphtheriae	Serum bactericidal antibody titer after two nasal immunizations	Improve the antigen retention time	Intranasal	[70]
			Immunoglobulin G levels close to the ones obtained with parenteral administration			
VLP-adjuvanted chitosan	Norwalk virus-like particle	Protection against Norwalk virus	Elicited mucosal dendritic cells to enhance the immune response locally and systemically	Triggering the mucosal priming phenomenon	Intranasal	[71]
			Immunoglobulin G levels close to the ones obtained with parenteral administration	Inducing potent mucosal and systemic immune responses		

Table 3. *Cont.*

Chitosan-Based Nanocarriers	Antigen	Application	Activity and Immune Response	Advantages and Disadvantages	Administration Route	Reference
VLP-adjuvanted chitosan			High levels of IgA and IgG ASCs in peripheral blood	Inducing mucosal immune responses at distant sites from the administration point, such as the gastrointestinal tract		
			Presence of circulating ASCs of VLP-specific IgA and IgG with different homing potentials			

Samer S. El-Kamary et al. investigated a chitosan-adjuvanted Norwalk virus-like particle (VLP) intranasal vaccine, which elicited mucosal dendritic cells to enhance the immune response locally and systemically. Two phase 1 studies were conducted in healthy subjects (18–49 years) with no vaccine-related serious adverse events. One dose was sufficient to activate antibody secreting cells, while two doses were required to achieve an increased serologic antibody response. The observed mucosal priming phenomenon was supported by the presence of high frequencies of IgA and IgG ASCs in peripheral blood. Authors highlighted the ability of chitosan-adjuvanted VLP vaccines to induce potent mucosal and systemic immune responses including effector cell activity at a distant site of infection such as the gastrointestinal tract. Moreover, intranasal immunization elicited circulating ASCs of VLP-specific IgA and IgG with different homing potentials. IgA-specific ASCs upregulated receptors capable of conferring their homing ability in both intestinal mucosa and peripheral lymphoid tissues, while IgG-ASCs expressed homing receptors that support the localization in peripheral lymphoid tissues. This study provided evidence for the use of a nasal vaccine to combat gastrointestinal infections by triggering mucosal immune responses at distant sites from the administration point, such as the gastrointestinal tract [80].

9. Conclusions

Chitosan features allow for nanoscale manipulation resulting in several chitosan/chitosan-modified NP preparations, and it has been experimentally demonstrated to facilitate a defined mucosal immune response. However, its application in vaccine formulations is still absent as major market production. Different issues could be raised for this limitation. Though chitosan is generally recognized as safe (GRAS) by the FDA, the potential pro-inflammatory chitosan-induced reactions (adjuvant activity) might discourage pharmaceutical companies to face unexpected side effects. Furthermore, despite the easy scalability of chitosan production, the chemical processes employed may be harmful to the environment due to the large amount of alkaline waste and organic material produced [81]. On the other hand, chitosan itself is a natural compound that is easily obtainable from natural bio-sources and wastes from the crustacean food industry as a low-cost biocompatible nanoparticle precursor [82]. The natural bio-source, however, reduces the possibility to obtain chitosan batches with identical chemico-physical characteristics [83], making standardization more complicated. This can affect the final properties of the nanocarrier for large-scale products, such as vaccines, less reproducible [84]. Additionally, the production of chitosan NPs with modified surfaces could be challenging as functionalization processes increase manufacturing costs.

Nevertheless, several experimental results, pre-clinical and clinical trials of effective chitosan-based nanomaterials for vaccine preparation, as well as their economic [85] and ecological advantages, may lead to possible reconsideration of this material for the production of broad-spectrum mucosal vaccines against different infectious diseases.

Author Contributions: Conceptualization, G.B. and D.Z.; methodology, M.P., G.B. and D.Z.; writing—original draft preparation, S.C.G. and M.P.; writing—review and editing, S.C.G., M.P., D.Z. and G.B. All authors have read and agreed to the published version of the manuscript.

Funding: This research received no external funding. The work of M.P. and D.Z. was supported by the Department of Excellence 2023/2027, MUR, Italy.

Institutional Review Board Statement: Not applicable.

Informed Consent Statement: Not applicable.

Data Availability Statement: Not applicable.

Conflicts of Interest: The authors declare no conflict of interest.

Abbreviations

AIDS	Acquired Immune Deficiency Syndrome
ASCs	Antibody secreting cells
APCs	Antigen presenting cells
AUC0→4 h	Area under the concentration time-curve
DC	Dendritic cells
EDIII-D3	Domain III of dengue serotype-3 E protein
FDA	Food and Drug Administration
GRAS	Generally recognized as safe
HBV	Hepatitis B virus
HLA-DR	Human Leukocyte Antigen–DR isotype
HPV	Human Papilloma Virus infection
IgA	Immunoglobulin A
IBD	Inflammatory bowel disease
LNTI	Lymph node targeting index
MCS	Mannosylated chitosan
NPs	Nanoparticles
NDV	Newcastle disease virus
NALT	Nose-associated lymphoid tissue
O/W	Oil in water
OVA-NP	Ovalbumin-conjugated N-trimethylaminoethylmethacrylate chitosan NPs
SIgA	Secretory IgA
SBA	Serum bactericidal antibody
SARS	Severe acute respiratory syndrome
SARS-CoV-2	Severe acute respiratory syndrome coronavirus 2
(S@CS)NPs	Spike protein-loaded chitosan NPs
TMC(TT) NPs	Tetanus toxoid loaded TMC NPs
TRMT	Tissue-resident memory T
TMC	Trimethyl chitosan
VLP	Virus-like particle
WHO	World Health Organization

References

1. Nabel, G.J. Designing Tomorrow's Vaccines. *N. Engl. J. Med.* **2013**, *368*, 551–560. [CrossRef]
2. Lin, C.-H.; Lin, C.-J.; Kuo, Y.-W.; Wang, J.-Y.; Hsu, C.-L.; Chen, J.-M.; Cheng, W.-C.; Lee, L.-N. Tuberculosis mortality: Patient characteristics and causes. *BMC Infect. Dis.* **2014**, *14*, 5. [CrossRef]
3. Black, R.E.; Cousens, S.; Johnson, H.L.; Lawn, J.E.; Rudan, I.; Bassani, D.G.; Jha, P.; Campbell, H.; Walker, C.F.; Cibulskis, R.; et al. Global, regional, and national causes of child mortality in 2008: A systematic analysis. *Lancet* **2010**, *375*, 1969–1987. [CrossRef]
4. Miteva, D.; Peshevska-Sekulovska, M.; Snegarova, V.; Batselova, H.; Alexandrova, R.; Velikova, T. Mucosal COVID-19 vaccines: Risks, benefits and control of the pandemic. *World J. Virol.* **2022**, *11*, 221–236. [CrossRef]
5. Lavelle, E.C.; Ward, R.W. Mucosal vaccines—Fortifying the frontiers. *Nat. Rev. Immunol.* **2022**, *22*, 236–250. [CrossRef]
6. Pollard, A.J.; Bijker, E.M. A guide to vaccinology: From basic principles to new developments. *Nat. Rev. Immunol.* **2021**, *21*, 83–100. [CrossRef]
7. Russell, M.W.; Moldoveanu, Z.; Ogra, P.L.; Mestecky, J. Mucosal Immunity in COVID-19: A Neglected but Critical Aspect of SARS-CoV-2 Infection. *Front. Immunol.* **2020**, *11*, 611337. [CrossRef] [PubMed]
8. Reljic, R.; Sibley, L.; Huang, J.-M.; Pepponi, I.; Hoppe, A.; Hong, H.A.; Cutting, S.M. Mucosal Vaccination against Tuberculosis Using Inert Bioparticles. *Infect. Immun.* **2013**, *81*, 4071–4080. [CrossRef] [PubMed]
9. Jones, A.T.; Shen, X.; Walter, K.L.; LaBranche, C.C.; Wyatt, L.S.; Tomaras, G.D.; Montefiori, D.C.; Moss, B.; Barouch, D.H.; Clements, J.D.; et al. HIV-1 vaccination by needle-free oral injection induces strong mucosal immunity and protects against SHIV challenge. *Nat. Commun.* **2019**, *10*, 798. [CrossRef]
10. Loenenbach, A.D.; Poethko-Müller, C.; Pawlita, M.; Thamm, M.; Harder, T.; Waterboer, T.; Schröter, J.; Deleré, Y.; Wichmann, O.; Wiese-Posselt, M. Mucosal and cutaneous Human Papillomavirus seroprevalence among adults in the prevaccine era in Germany—Results from a nationwide population-based survey. *Int. J. Infect. Dis.* **2019**, *83*, 3–11. [CrossRef]
11. Matić, Z.; Šantak, M. Current view on novel vaccine technologies to combat human infectious diseases. *Appl. Microbiol. Biotechnol.* **2022**, *106*, 25–56. [CrossRef]
12. Cho, C.-S.; Hwang, S.-K.; Gu, M.-J.; Kim, C.-G.; Kim, S.-K.; Ju, D.-B.; Yun, C.-H.; Kim, H.-J. Mucosal Vaccine Delivery Using Mucoadhesive Polymer Particulate Systems. *Tissue Eng. Regen. Med.* **2021**, *18*, 693–712. [CrossRef]

13. Dmour, I.; Islam, N. Recent advances on chitosan as an adjuvant for vaccine delivery. *Int. J. Biol. Macromol.* **2022**, *200*, 498–519. [CrossRef]
14. Barbieri, S.; Buttini, F.; Rossi, A.; Bettini, R.; Colombo, P.; Ponchel, G.; Sonvico, F.; Colombo, G. Ex vivo permeation of tamoxifen and its 4-OH metabolite through rat intestine from lecithin/chitosan nanoparticles. *Int. J. Pharm.* **2015**, *491*, 99–104. [CrossRef] [PubMed]
15. Feng, C.; Wang, Z.; Jiang, C.; Kong, M.; Zhou, X.; Li, Y.; Cheng, X.; Chen, X. Chitosan/o-carboxymethyl chitosan nanoparticles for efficient and safe oral anticancer drug delivery: In vitro and in vivo evaluation. *Int. J. Pharm.* **2013**, *457*, 158–167. [CrossRef]
16. Liu, S.; Yang, S.; Ho, P.C. Intranasal administration of carbamazepine-loaded carboxymethyl chitosan nanoparticles for drug delivery to the brain. *Asian J. Pharm. Sci.* **2018**, *13*, 72–81. [CrossRef] [PubMed]
17. Cao, Y.; Tan, Y.F.; Wong, Y.S.; Liew, M.W.J.; Venkatraman, S. Recent Advances in Chitosan-Based Carriers for Gene Delivery. *Mar. Drugs* **2019**, *17*, 381. [CrossRef] [PubMed]
18. Boroumand, H.; Badie, F.; Mazaheri, S.; Seyedi, Z.S.; Nahand, J.S.; Nejati, M.; Baghi, H.B.; Abbasi-Kolli, M.; Badehnoosh, B.; Ghandali, M.; et al. Chitosan-Based Nanoparticles Against Viral Infections. *Front. Cell. Infect. Microbiol.* **2021**, *11*, 643953. [CrossRef]
19. Pearson, J.P.; Brownlee, I.A. Structure and Function of Mucosal Surfaces. In *Colonization of Mucosal Surfaces*; Nataro, J.P., Cohen, P.S., Mobley, H.L.T., Weiser, J.N., Eds.; ASM Press: Washington, DC, USA, 2014; pp. 1–16. ISBN 978-1-68367-201-2.
20. Kumar, R.; Srivastava, V.; Baindara, P.; Ahmad, A. Thermostable vaccines: An innovative concept in vaccine development. *Expert Rev. Vaccines* **2022**, *21*, 811–824. [CrossRef]
21. Cook, I.F. Evidence based route of administration of vaccines. *Hum. Vaccin.* **2008**, *4*, 67–73. [CrossRef]
22. Mohanan, D.; Slütter, B.; Henriksen-Lacey, M.; Jiskoot, W.; Bouwstra, J.A.; Perrie, Y.; Kündig, T.M.; Gander, B.; Johansen, P. Administration routes affect the quality of immune responses: A cross-sectional evaluation of particulate antigen-delivery systems. *J. Control. Release* **2010**, *147*, 342–349. [CrossRef] [PubMed]
23. Rosenbaum, P.; Tchitchek, N.; Joly, C.; Rodriguez Pozo, A.; Stimmer, L.; Langlois, S.; Hocini, H.; Gosse, L.; Pejoski, D.; Cosma, A.; et al. Vaccine Inoculation Route Modulates Early Immunity and Consequently Antigen-Specific Immune Response. *Front. Immunol.* **2021**, *12*, 645210. [CrossRef] [PubMed]
24. Yu, C.; Walter, M. *Needleless Injectors for the Administration of Vaccines: A Review of Clinical Effectiveness*; CADTH Rapid Response Reports; Canadian Agency for Drugs and Technologies in Health: Ottawa, ON, Canada, 2020.
25. Matchett, W.E.; Anguiano-Zarate, S.S.; Barry, M.A. Comparison of systemic and mucosal immunization with replicating Single cycle Adenoviruses. *Glob. Vaccines Immunol.* **2018**, *3*, 1–5. [CrossRef] [PubMed]
26. Criscuolo, E.; Caputo, V.; Diotti, R.A.; Sautto, G.A.; Kirchenbaum, G.A.; Clementi, N. Alternative Methods of Vaccine Delivery: An Overview of Edible and Intradermal Vaccines. *J. Immunol. Res.* **2019**, *2019*, 8303648. [CrossRef] [PubMed]
27. Holmgren, J.; Czerkinsky, C. Mucosal immunity and vaccines. *Nat. Med.* **2005**, *11*, S45–S53. [CrossRef]
28. Mitragotri, S. Immunization without needles. *Nat. Rev. Immunol.* **2005**, *5*, 905–916. [CrossRef]
29. Luczo, J.M.; Bousse, T.; Johnson, S.K.; Jones, C.A.; Pearce, N.; Neiswanger, C.A.; Wang, M.-X.; Miller, E.A.; Petrovsky, N.; Wentworth, D.E.; et al. Intranasal powder live attenuated influenza vaccine is thermostable, immunogenic, and protective against homologous challenge in ferrets. *Npj Vaccines* **2021**, *6*, 59. [CrossRef]
30. Sterlin, D.; Mathian, A.; Miyara, M.; Mohr, A.; Anna, F.; Claër, L.; Quentric, P.; Fadlallah, J.; Devilliers, H.; Ghillani, P.; et al. IgA dominates the early neutralizing antibody response to SARS-CoV-2. *Sci. Transl. Med.* **2021**, *13*, eabd2223. [CrossRef]
31. Wang, Z.; Lorenzi, J.C.C.; Muecksch, F.; Finkin, S.; Viant, C.; Gaebler, C.; Cipolla, M.; Hoffmann, H.-H.; Oliveira, T.Y.; Oren, D.A.; et al. Enhanced SARS-CoV-2 neutralization by dimeric IgA. *Sci. Transl. Med.* **2021**, *13*, eabf1555. [CrossRef]
32. Dalle Carbonare, L.; Valenti, M.T.; Bisoffi, Z.; Piubelli, C.; Pizzato, M.; Accordini, S.; Mariotto, S.; Ferrari, S.; Minoia, A.; Bertacco, J.; et al. Serology study after BTN162b2 vaccination in participants previously infected with SARS-CoV-2 in two different waves versus naïve. *Commun. Med.* **2021**, *1*, 38. [CrossRef]
33. Focosi, D.; Maggi, F.; Casadevall, A. Mucosal Vaccines, Sterilizing Immunity, and the Future of SARS-CoV-2 Virulence. *Viruses* **2022**, *14*, 187. [CrossRef] [PubMed]
34. Darwich, A.; Pozzi, C.; Fornasa, G.; Lizier, M.; Azzolini, E.; Spadoni, I.; Carli, F.; Voza, A.; Desai, A.; Ferrero, C.; et al. BNT162b2 vaccine induces antibody release in saliva: A possible role for mucosal viral protection? *EMBO Mol. Med.* **2022**, *14*, e15326. [CrossRef]
35. Lycke, N. Recent progress in mucosal vaccine development: Potential and limitations. *Nat. Rev. Immunol.* **2012**, *12*, 592–605. [CrossRef]
36. Luria-Pérez, R.; Sánchez-Vargas, L.A.; Muñoz-López, P.; Mellado-Sánchez, G. Mucosal Vaccination: A Promising Alternative Against Flaviviruses. *Front. Cell. Infect. Microbiol.* **2022**, *12*, 887729. [CrossRef] [PubMed]
37. Brandtzaeg, P. Secretory IgA: Designed for Anti-Microbial Defense. *Front. Immunol.* **2013**, *4*, 222. [CrossRef] [PubMed]
38. Strugnell, R.A.; Wijburg, O.L.C. The role of secretory antibodies in infection immunity. *Nat. Rev. Microbiol.* **2010**, *8*, 656–667. [CrossRef]
39. Rogier, E.; Frantz, A.; Bruno, M.; Kaetzel, C. Secretory IgA is Concentrated in the Outer Layer of Colonic Mucus along with Gut Bacteria. *Pathogens* **2014**, *3*, 390–403. [CrossRef]
40. Rakhra, K.; Abraham, W.; Wang, C.; Moynihan, K.D.; Li, N.; Donahue, N.; Baldeon, A.D.; Irvine, D.J. Exploiting albumin as a mucosal vaccine chaperone for robust generation of lung-resident memory T cells. *Sci. Immunol.* **2021**, *6*, eabd8003. [CrossRef]

41. Schenkel, J.M.; Masopust, D. Tissue-Resident Memory T Cells. *Immunity* **2014**, *41*, 886–897. [CrossRef]
42. Bartolomé-Casado, R.; Landsverk, O.J.B.; Chauhan, S.K.; Sætre, F.; Hagen, K.T.; Yaqub, S.; Øyen, O.; Horneland, R.; Aandahl, E.M.; Aabakken, L.; et al. CD4+ T cells persist for years in the human small intestine and display a TH1 cytokine profile. *Mucosal Immunol.* **2021**, *14*, 402–410. [CrossRef]
43. Yuan, R.; Yu, J.; Jiao, Z.; Li, J.; Wu, F.; Yan, R.; Huang, X.; Chen, C. The Roles of Tissue-Resident Memory T Cells in Lung Diseases. *Front. Immunol.* **2021**, *12*, 710375. [CrossRef]
44. Beverley, P.C.L.; Sridhar, S.; Lalvani, A.; Tchilian, E.Z. Harnessing local and systemic immunity for vaccines against tuberculosis. *Mucosal Immunol.* **2014**, *7*, 20–26. [CrossRef]
45. Shim, S.; Yoo, H.S. The Application of Mucoadhesive Chitosan Nanoparticles in Nasal Drug Delivery. *Mar. Drugs* **2020**, *18*, 605. [CrossRef]
46. Gong, X.; Gao, Y.; Shu, J.; Zhang, C.; Zhao, K. Chitosan-Based Nanomaterial as Immune Adjuvant and Delivery Carrier for Vaccines. *Vaccines* **2022**, *10*, 1906. [CrossRef] [PubMed]
47. Anthonsen, M.W.; Vårum, K.M.; Smidsrød, O. Solution properties of chitosans: Conformation and chain stiffness of chitosans with different degrees of N-acetylation. *Carbohydr. Polym.* **1993**, *22*, 193–201. [CrossRef]
48. Lieder, R.; Gaware, V.S.; Thormodsson, F.; Einarsson, J.M.; Ng, C.-H.; Gislason, J.; Masson, M.; Petersen, P.H.; Sigurjonsson, O.E. Endotoxins affect bioactivity of chitosan derivatives in cultures of bone marrow-derived human mesenchymal stem cells. *Acta Biomater.* **2013**, *9*, 4771–4778. [CrossRef] [PubMed]
49. Naskar, S.; Sharma, S.; Kuotsu, K. Chitosan-based nanoparticles: An overview of biomedical applications and its preparation. *J. Drug Deliv. Sci. Technol.* **2019**, *49*, 66–81. [CrossRef]
50. Vasiliev, Y.M. Chitosan-based vaccine adjuvants: Incomplete characterization complicates preclinical and clinical evaluation. *Expert Rev. Vaccines* **2015**, *14*, 37–53. [CrossRef]
51. Bakshi, P.S.; Selvakumar, D.; Kadirvelu, K.; Kumar, N.S. Chitosan as an environment friendly biomaterial—A review on recent modifications and applications. *Int. J. Biol. Macromol.* **2020**, *150*, 1072–1083. [CrossRef] [PubMed]
52. Yanat, M.; Schroën, K. Preparation methods and applications of chitosan nanoparticles; with an outlook toward reinforcement of biodegradable packaging. *React. Funct. Polym.* **2021**, *161*, 104849. [CrossRef]
53. Illum, L.; Jabbal-Gill, I.; Hinchcliffe, M.; Fisher, A.N.; Davis, S.S. Chitosan as a novel nasal delivery system for vaccines. *Adv. Drug Deliv. Rev.* **2001**, *51*, 81–96. [CrossRef]
54. Zhao, J.; Li, J.; Jiang, Z.; Tong, R.; Duan, X.; Bai, L.; Shi, J. Chitosan, N,N,N-trimethyl chitosan (TMC) and 2-hydroxypropyltrimethyl ammonium chloride chitosan (HTCC): The potential immune adjuvants and nano carriers. *Int. J. Biol. Macromol.* **2020**, *154*, 339–348. [CrossRef]
55. Norpi, A.S.M.; Nordin, M.L.; Ahmad, N.; Katas, H.; Fuaad, A.A.-H.A.; Sukri, A.; Marasini, N.; Azmi, F. New modular platform based on multi-adjuvanted amphiphilic chitosan nanoparticles for efficient lipopeptide vaccine delivery against group A streptococcus. *Asian J. Pharm. Sci.* **2022**, *17*, 435–446. [CrossRef]
56. Mehrabi, M.; Mohamadpour Dounighi, N.; Montazeri, H.; Vakili-Ghartavol, R.; Rashti, A. Chitosan based nanoparticles in mucosal vaccine delivery. *Arch. Razi Inst.* **2018**, *73*, 165–176. [CrossRef] [PubMed]
57. Dyer, A.M.; Hinchcliffe, M.; Watts, P.; Castile, J.; Jabbal-Gill, I.; Nankervis, R.; Smith, A.; Illum, L. Nasal delivery of insulin using novel chitosan based formulations: A comparative study in two animal models between simple chitosan formulations and chitosan nanoparticles. *Pharm. Res.* **2002**, *19*, 998–1008. [CrossRef]
58. Kammona, O.; Kiparissides, C. Recent advances in nanocarrier-based mucosal delivery of biomolecules. *J. Control. Release* **2012**, *161*, 781–794. [CrossRef]
59. Ways, T.M.M.; Filippov, S.K.; Maji, S.; Glassner, M.; Cegłowski, M.; Hoogenboom, R.; King, S.; Lau, W.M.; Khutoryanskiy, V.V. Mucus-penetrating nanoparticles based on chitosan grafted with various non-ionic polymers: Synthesis, structural characterisation and diffusion studies. *J. Colloid Interface Sci.* **2022**, *626*, 251–264. [CrossRef]
60. Torres, F.G.; Troncoso, O.P.; Pisani, A.; Gatto, F.; Bardi, G. Natural Polysaccharide Nanomaterials: An Overview of Their Immunological Properties. *Int. J. Mol. Sci.* **2019**, *20*, 5092. [CrossRef]
61. Nantachit, N.; Sunintaboon, P.; Ubol, S. Responses of primary human nasal epithelial cells to EDIII-DENV stimulation: The first step to intranasal dengue vaccination. *Virol. J.* **2016**, *13*, 142. [CrossRef] [PubMed]
62. Zhuo, S.-H.; Wu, J.-J.; Zhao, L.; Li, W.-H.; Zhao, Y.-F.; Li, Y.-M. A chitosan-mediated inhalable nanovaccine against SARS-CoV-2. *Nano Res.* **2022**, *15*, 4191–4200. [CrossRef] [PubMed]
63. Malik, A.; Gupta, M.; Gupta, V.; Gogoi, H.; Bhatnagar, R. Novel application of trimethyl chitosan as an adjuvant in vaccine delivery. *Int. J. Nanomed.* **2018**, *13*, 7959–7970. [CrossRef]
64. Liu, Q.; Zheng, X.; Zhang, C.; Shao, X.; Zhang, X.; Zhang, Q.; Jiang, X. Antigen-conjugated N-trimethylaminoethylmethacrylate Chitosan Nanoparticles Induce Strong Immune Responses After Nasal Administration. *Pharm. Res.* **2015**, *32*, 22–36. [CrossRef]
65. Slütter, B.; Bal, S.; Keijzer, C.; Mallants, R.; Hagenaars, N.; Que, I.; Kaijzel, E.; van Eden, W.; Augustijns, P.; Löwik, C.; et al. Nasal vaccination with N-trimethyl chitosan and PLGA based nanoparticles: Nanoparticle characteristics determine quality and strength of the antibody response in mice against the encapsulated antigen. *Vaccine* **2010**, *28*, 6282–6291. [CrossRef] [PubMed]
66. Hagenaars, N.; Mania, M.; de Jong, P.; Que, I.; Nieuwland, R.; Slütter, B.; Glansbeek, H.; Heldens, J.; van den Bosch, H.; Löwik, C. Role of trimethylated chitosan (TMC) in nasal residence time, local distribution and toxicity of an intranasal influenza vaccine. *J. Control. Release* **2010**, *144*, 17–24. [CrossRef] [PubMed]

67. Chadhar, V.; Parashar, A.K.; Patel, M.; Nema, R.K. Development and Characterization of Tetanus Toxoid Loaded Trimethylchitosan Chloride Nanoparticles for Nasal Immunization. *Curr. Res. Pharm. Sci.* **2012**, *2*, 197–202.
68. Tafaghodi, M.; Saluja, V.; Kersten, G.F.A.; Kraan, H.; Slütter, B.; Amorij, J.-P.; Jiskoot, W. Hepatitis B surface antigen nanoparticles coated with chitosan and trimethyl chitosan: Impact of formulation on physicochemical and immunological characteristics. *Vaccine* **2012**, *30*, 5341–5348. [CrossRef] [PubMed]
69. Abkar, M.; Fasihi-Ramandi, M.; Kooshki, H.; Sahebghadam Lotfi, A. Oral immunization of mice with Omp31-loaded N-trimethyl chitosan nanoparticles induces high protection against Brucella melitensis infection. *Int. J. Nanomed.* **2017**, *12*, 8769–8778. [CrossRef] [PubMed]
70. Zhao, K.; Sun, B.; Shi, C.; Sun, Y.; Jin, Z.; Hu, G. Intranasal immunization with O-2′-Hydroxypropyl trimethyl ammonium chloride chitosan nanoparticles loaded with Newcastle disease virus DNA vaccine enhances mucosal immune response in chickens. *J. Nanobiotechnol.* **2021**, *19*, 240. [CrossRef]
71. Peng, Y.; Yao, W.; Wang, B.; Zong, L. Mannosylated Chitosan Nanoparticles Based Macrophage-Targeting Gene Delivery System Enhanced Cellular Uptake and Improved Transfection Efficiency. *J. Nanosci. Nanotechnol.* **2015**, *15*, 2619–2627. [CrossRef]
72. Park, I.Y.; Kim, I.Y.; Yoo, M.K.; Choi, Y.J.; Cho, M.-H.; Cho, C.S. Mannosylated polyethylenimine coupled mesoporous silica nanoparticles for receptor-mediated gene delivery. *Int. J. Pharm.* **2008**, *359*, 280–287. [CrossRef]
73. Wu, M.; Zhao, H.; Li, M.; Yue, Y.; Xiong, S.; Xu, W. Intranasal Vaccination with Mannosylated Chitosan Formulated DNA Vaccine Enables Robust IgA and Cellular Response Induction in the Lungs of Mice and Improves Protection against Pulmonary Mycobacterial Challenge. *Front. Cell. Infect. Microbiol.* **2017**, *7*, 445. [CrossRef] [PubMed]
74. Bedford, J.G.; Caminschi, I.; Wakim, L.M. Intranasal Delivery of a Chitosan-Hydrogel Vaccine Generates Nasal Tissue Resident Memory CD8+ T Cells That Are Protective against Influenza Virus Infection. *Vaccines* **2020**, *8*, 572. [CrossRef] [PubMed]
75. Johansson, E.-L.; Wassén, L.; Holmgren, J.; Jertborn, M.; Rudin, A. Nasal and Vaginal Vaccinations Have Differential Effects on Antibody Responses in Vaginal and Cervical Secretions in Humans. *Infect. Immun.* **2001**, *69*, 7481–7486. [CrossRef]
76. Bergquist, C.; Johansson, E.L.; Lagergård, T.; Holmgren, J.; Rudin, A. Intranasal vaccination of humans with recombinant cholera toxin B subunit induces systemic and local antibody responses in the upper respiratory tract and the vagina. *Infect. Immun.* **1997**, *65*, 2676–2684. [CrossRef]
77. Mills, K.H.G.; Cosgrove, C.; McNeela, E.A.; Sexton, A.; Giemza, R.; Jabbal-Gill, I.; Church, A.; Lin, W.; Illum, L.; Podda, A.; et al. Protective Levels of Diphtheria-Neutralizing Antibody Induced in Healthy Volunteers by Unilateral Priming-Boosting Intranasal Immunization Associated with Restricted Ipsilateral Mucosal Secretory Immunoglobulin A. *Infect. Immun.* **2003**, *71*, 726–732. [CrossRef]
78. McNeela, E.A.; Jabbal-Gill, I.; Illum, L.; Pizza, M.; Rappuoli, R.; Podda, A.; Lewis, D.J.M.; Mills, K.H.G. Intranasal immunization with genetically detoxified diphtheria toxin induces T cell responses in humans: Enhancement of Th2 responses and toxin-neutralizing antibodies by formulation with chitosan. *Vaccine* **2004**, *22*, 909–914. [CrossRef] [PubMed]
79. Huo, Z.; Sinha, R.; McNeela, E.A.; Borrow, R.; Giemza, R.; Cosgrove, C.; Heath, P.T.; Mills, K.H.G.; Rappuoli, R.; Griffin, G.E.; et al. Induction of Protective Serum Meningococcal Bactericidal and Diphtheria-Neutralizing Antibodies and Mucosal Immunoglobulin A in Volunteers by Nasal Insufflations of the *Neisseria meningitidis* Serogroup C Polysaccharide-CRM197 Conjugate Vaccine Mixed with Chitosan. *Infect. Immun.* **2005**, *73*, 8256–8265. [CrossRef]
80. El-Kamary, S.S.; Pasetti, M.F.; Mendelman, P.M.; Frey, S.E.; Bernstein, D.I.; Treanor, J.J.; Ferreira, J.; Chen, W.H.; Sublett, R.; Richardson, C.; et al. Adjuvanted Intranasal Norwalk Virus-Like Particle Vaccine Elicits Antibodies and Antibody-Secreting Cells That Express Homing Receptors for Mucosal and Peripheral Lymphoid Tissues. *J. Infect. Dis.* **2010**, *202*, 1649–1658. [CrossRef]
81. Ruiz, G.A.M.; Corrales, H.F.Z. Chitosan, Chitosan Derivatives and their Biomedical Applications. In *Biological Activities and Application of Marine Polysaccharides*; Shalaby, E.A., Ed.; InTech: London, UK, 2017; ISBN 978-953-51-2859-5.
82. Rebouças, J.S.A.; Oliveira, F.P.S.; Araujo, A.C.D.S.; Gouveia, H.L.; Latorres, J.M.; Martins, V.G.; Prentice Hernández, C.; Tesser, M.B. Shellfish industrial waste reuse. *Crit. Rev. Biotechnol.* **2023**, *43*, 50–66. [CrossRef]
83. Bellich, B.; D'Agostino, I.; Semeraro, S.; Gamini, A.; Cesàro, A. "The Good, the Bad and the Ugly" of Chitosans. *Mar. Drugs* **2016**, *14*, 99. [CrossRef]
84. Paliwal, R.; Babu, R.J.; Palakurthi, S. Nanomedicine Scale-up Technologies: Feasibilities and Challenges. *AAPS PharmSciTech* **2014**, *15*, 1527–1534. [CrossRef] [PubMed]
85. Van Der Lubben, I.M.; Verhoef, J.C.; Borchard, G.; Junginger, H.E. Chitosan for mucosal vaccination. *Adv. Drug Deliv. Rev.* **2001**, *52*, 139–144. [CrossRef] [PubMed]

vaccines MDPI

Review

Nanoparticle-Based Adjuvants and Delivery Systems for Modern Vaccines

Brankica Filipić [1], Ivana Pantelić [2,*], Ines Nikolić [2,3], Dragomira Majhen [4], Zorica Stojić-Vukanić [1], Snežana Savić [2] and Danina Krajišnik [2]

[1] Department of Microbiology and Immunology, University of Belgrade-Faculty of Pharmacy, 11000 Belgrade, Serbia; brankica.filipic@pharmacy.bg.ac.rs (B.F.); zorica.stojic-vukanic@pharmacy.bg.ac.rs (Z.S.-V.)
[2] Department of Pharmaceutical Technology and Cosmetology, University of Belgrade-Faculty of Pharmacy, 11000 Belgrade, Serbia; ines.nikolic@pharmacy.bg.ac.rs (I.N.); snezana.savic@pharmacy.bg.ac.rs (S.S.); danina.krajisnik@pharmacy.bg.ac.rs (D.K.)
[3] Section of Pharmaceutical Sciences, University of Geneva, 1206 Geneva, Switzerland
[4] Division of Molecular Biology, Ruđer Bošković Institute, 10000 Zagreb, Croatia; dragomira.majhen@irb.hr
* Correspondence: ivana.pantelic@pharmacy.bg.ac.rs

Abstract: Ever since the development of the first vaccine, vaccination has had the great impact on global health, leading to the decrease in the burden of numerous infectious diseases. However, there is a constant need to improve existing vaccines and develop new vaccination strategies and vaccine platforms that induce a broader immune response compared to traditional vaccines. Modern vaccines tend to rely on certain nanotechnology platforms but are still expected to be readily available and easy for large-scale manufacturing and to induce a durable immune response. In this review, we present an overview of the most promising nanoadjuvants and nanoparticulate delivery systems and discuss their benefits from tehchnological and immunological standpoints as well as their objective drawbacks and possible side effects. The presented nano alums, silica and clay nanoparticles, nanoemulsions, adenoviral-vectored systems, adeno-associated viral vectors, vesicular stomatitis viral vectors, lentiviral vectors, virus-like particles (including bacteriophage-based ones) and virosomes indicate that vaccine developers can now choose different adjuvants and/or delivery systems as per the requirement, specific to combatting different infectious diseases.

Keywords: nanoadjuvants; nanoemulsions; viral-vectored system; virus-like particles

1. Introduction

The severe acute respiratory syndrome coronavirus 2 (SARS-CoV-2) pandemic served as a global reminder of how important the vaccinology field is. Modern vaccines are expected to combat a range of infectious and non-infectious diseases in a safe and efficacious manner while being readily available to developing countries as well [1]. However, conventional vaccine formulations generally favor humoral immunity, i.e., antibody-mediated responses, but fail to elicit strong cellular immune responses [2]. After decades of research in the nanotechnology field, nanostructured vaccine formulations (nanovaccines) were able to support different (alternative) routes of administration and offer facilitated uptake by the targeted cells while even modulating the immune response [3]. Moreover, lipid-based nanoparticles' successful translation from other drug targets (e.g., patisiran story) to the vaccinology field (namely, mRNA coronavirus disease 2019 (COVID-19) vaccines) implied the potential of other carrier platforms to be correspondingly successfully used in vaccines [4,5].

This review aims to offer a summarized description of nanotechnology advances that resulted in readily used nanoadjuvants and/or nanocarriers able to enhance a nanovaccine's immunization effect. According to the authors' opinions, certain inorganic nanoadjuvants, nanoemulsions, viral-vectored systems, virus-like particles and virosomes were focused on as nanoplatforms of special interest for the potentially successful translation to emerging targets. Novel insights into the design of the abovementioned nanoadjuvants and nanocarriers are given, along with distinct strengths and weaknesses, due to the fact that their inherent composition may often be subjected to certain modifications in order to better suit the intended purpose.

A critical search of the publications was performed through PubMed (Rockwille Pike Bethesda, USA), Scopus (London, UK), Web of Science (Philadelphia, PA, USA) and institutional libraries (namely, Serbian Library Consortium for Coordinated Acquisition—KoBSON, Belgrade, Serbia and National and University Library in Zagreb—NSK, Zagreb, Croatia). Both review and original scientific papers were considered, published between January 2000 and January 2023 (papers published before 2000 were considered, when necessary). Initial, broader search served to identify information and knowledge gaps in already published review papers dealing with nanosized adjuvants and nanostructured vaccine formulations. Although the presented paper is not a systematic review, a description of the mode of data collection and selection may facilitate other authors in critically assessing and further updating given information.

2. Nanoadjuvants

Adjuvants have been widely used within vaccine formulations, for decades even, to improve the immune response to target vaccine antigens. Recently, so-called nanoadjuvants have tended to prevail in modern vaccine formulation strategies [2,6,7]. For successful vaccine development, it is of high importance to select an appropriate adjuvant and, subsequently, a suitable adjuvant-antigen combination in order to obtain stable, immunogenic and safe vaccines [8]. A common classification of vaccine adjuvants considers that an adjuvant can act either as an immunostimulant (directly interacting with the immune system, increasing the response to antigens; e.g., bacterial toxins, cytokines, saponins, Toll-like receptor (TLR) ligands) or as a vehicle (which introduces the antigen to the immune system, providing a controlled delivery of antigens or immunostimulants; e.g., aluminium salts, nanoemulsions, liposomes, virosomes, etc.). A combination of an immunostimulant and a vehicle is marked as an adjuvant system [9].

Despite the widespread use of adjuvants, the mechanisms of action of the currently available adjuvants in licensed vaccines, at least in humans, are not yet well understood. Nevertheless, today, it is known that innate immune cells, especially dendritic cells (DCs), which are the major antigen-presenting cells (APCs), can sense pathogen- and damage-associated molecular patterns (PAMP and DAMP, respectively) by different receptors (such as TLR, NOD- and RIG-like receptors). Many of these molecules, which could be part of adjuvants or components released from injured or dying cells on the site of the vaccine injection, induce inflammation, activate DCs and induce adaptive immunity mediated by vaccine antigen-specific T and B lymphocytes [10–14]. Briefly, during the recognition of vaccine antigens presented on DCs by the CD4+ T helper (Th) cell, DC-derived cytokines and costimulatory molecules on their cell surface influence Th cell differentiation towards three major subsets called Th1, Th2 and Th17 [10]. These Th cell subsets secrete different sets of cytokines and function in the host defense against distinct types of infectious pathogens. Th1 cells secrete interferon (IFN)-γ, and this type of immunity is associated with an increased macrophage effector function (killing of phagocytosed microbes) and protection against intracellular pathogens (bacteria and viruses). In mice, IFN-γ induces the production of IgG antibody isotype (IgG2a) by effector B cells, with the ability to neutralize pathogens, enhance phagocytosis of opsonized microbes and activate complement. However, there is no evidence that in humans generation of IgG antibody isotypes (IgG1/3) with same functions involves IFN-γ. Cytokines produced by DCs and Th1 cells during CD8+ T cell activation also promote the proliferation of CD8+ T cells and their differentiation into cytotoxic T cells, which eliminate intracellular pathogens by killing infected host cells. Th2 cells generally induce an antibody-predominant response whose physiological function is a defense against extracellular microbes. IL-4, the signature cytokine of the Th2 subset, stimulates B cells to enhance the production of IgG antibodies that can neutralize pathogens (IgG4 in humans and IgG1 in mice) but do not bind to phagocyte Fc receptors or activate the complement. Th17 cells play a critical role in host defense against a variety of bacteria and fungi that can survive outside cells, primarily by producing the IL-17 cytokine, but data regarding Th17-inducing adjuvants are still scarce. Having all of this in mind, a major goal of the next-generation vaccines is the selective use of adjuvants and delivery systems to enhance optimal immune responses in humans [11–14].

In the era of modern vaccinology, research has been refocused towards novel adjuvants, and antigen administration is usually performed by applying nanoparticulated delivery systems, rendering more effective vaccines [15,16]. Therefore, in this section, novel inorganic nanoadjuvants and nanoemulsions as lipid-based adjuvants/adjuvant systems will be specially addressed. Their main immunological properties and clinical application are summarized in Table 1.

2.1. Inorganic Nanomaterials as Vaccine Adjuvants and/or Carriers

2.1.1. Conventional Alum Adjuvants

Aluminum-containing adjuvants (Alums) have historically served as immunostimulants in vaccines and continue to be the most widely used adjuvants generally regarded as safe [17]. Several aluminium compounds (aluminum hydroxide, aluminum phosphate and amorphous aluminum hydroxyphosphate sulfate) are used in the licensed vaccines (for intramuscular or subcutaneous applications) [18]; however, their distinctive physicochemical properties could have important implications for their immunomodulatory effects [19] (Figure 1).

The primary nanoparticles of aluminium adjuvants (aluminum hydroxide and aluminum phosphate) are fibers with an average size of 4.5 × 2.2 × 10 nm that form loosely connected porous aggregates. These aggregates present the functioning units in vaccines, and their size varies between 1 and about 20 μm [20] related to the adjuvant, particle size measurement method, and testing conditions. Along with the particle size, aluminium adjuvant's surface charges could differ notably, namely, the isoelectric point (IEP) of these adjuvants varies from 4.6 to 11.1 depending on the salt [21,22], resulting in different charges in the physiological environment, which can be of great significance for the interaction with the antigen [19,23,24].

The formulation of vaccines with aluminium salts requires the careful selection of excipients such as buffers, tonicity-adjusting agents, surfactants and other stabilizers [24–26]. The thermostability of vaccines containing aluminium adjuvants is of great importance, since exposure to both high and low temperatures (particularly freezing) can affect the stability of vaccines and potentially cause irreversible coagulation/agglomeration [22,27]. Although several technologies have been proposed to overcome sensitivity to freezing conditions [28–31], this remains one of the main drawbacks of these adjuvants. Aluminium-containing adjuvants can potentiate the immune response by: (i) forming a depot at the site of injection, which is associated with slow release of antigen, (ii) targeting antigens to APCs and enhancing their uptake, and (iii) direct stimulation of NOD-like re-ceptor protein 3 (NLRP3) and formation of the NLRP3 inflammasome, which generate the biologically active form of the inflammatory cytokines interleukin (IL)-1β and IL-18 [22,32–35]. Traditionally, alums enhance Th2 responses to protein antigens and promote antibody responses but have minimal effect on cell-mediated immunity, especially in mice. However, mixed Th1/Th2 immune responses can be primed in humans when aluminium-based adjuvants are injected intramuscularly [33,34].

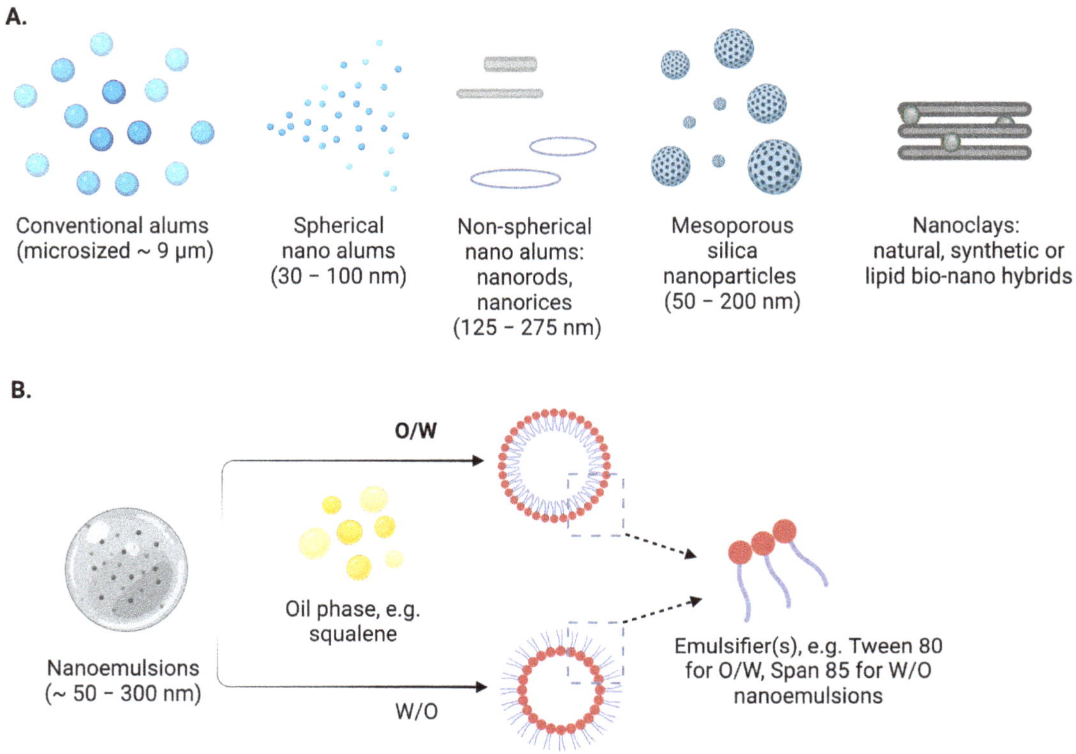

Figure 1. Schematic representation of the selected (**A**) inorganic nanomaterials (micro- to nanosized) and (**B**) nanoemulsions as adjuvants and/or carriers in contemporary vaccine formulations. Created with BioRender.com.

It has been apparent for some time now that satisfactory immunization outcomes are highly dependent on both objective side effects and the ones subjectively perceived for certain vaccines' components (e.g., effects emphasized by the patient's apprehensions). Moreover, it needs to be accepted that enhancing immunity inevitably leads to some undesired effects [36]. Alum adjuvants are generally deemed safe, and usually held responsible for local adverse reactions such as pain/tenderness at the injection site, accompanied by various degrees of inflammation, erythema, subcutaneous nodules and/or granuloma [37]. Additionally, in susceptible patients, alum adjuvants may trigger allergic reactions of various intensities, which appear to be alum concentration-dependent. However, intramuscularly administered alum vaccines may lead to more serious macrophagic myofasciitis-like local tissue impairments [38].

2.1.2. Nano Alum Adjuvants

Inorganic materials facilitate sustained and targeted release of antigens, enhance immunogenicity, and elicit a specific immune response. Over the past decade, inorganic materials have been intensively studied as vaccine adjuvants on the basisof nanoscale synthesis, and control of structural and functional properties. Moreover, it is possible to control their structural, optical, electrical or magnetic properties by engineering design, which allows the choise of the most suitable adjuvant formulations by means of in vitro and in vivo investigations [39–41].

It was confirmed by Sun and co-workers [41] that changes in alumina nanoparticles related to hydroxyl content, their shape and crystallinity had a significant influence in the activation of the NLRP3 inflammosome and improvement of antigen-specific immune responses. Particle size reduction of the conventional alum adjuvant to a nanometer scale (~112 nm) compared to mi-croscopic particles (~9 μm) elicited a stronger antigen-specific antibody response, which was likely related to their ability to effectively facilitate the uptake of adsorbed antigens by APCs [32]. A stronger adjuvant activity of insoluble aluminium oxyhydroxide nanoparticles (~30–100 nm) compared to microparticles (X50, 9.43 μm) was attributed to their stronger ability to stimulate uric acid production and consequently indirectly activate the inflammasome [42]. An evaluation of the specific immune response after the administration of aluminum hydroxide nanoparticles containingthe EsxV antigen of *Mycobacterium tuberculosis* (~200 nm) demonstrated strong activity in stimulating a Th1-type cellular immune response making them suitable adjuvant against *M. tuberculosis* infection [43].

Recently, researchers synthesized a series of non-spherical Alum nanoadjuvants (so-called nanorods or nanorices), showing satisfactory suspension stability upon model antigens' adsorption [6,44]. These nanoadjuvants are attributed with different aspect ratios but are generally characterized by hydrodynamic sizes ranging from 125 to 275 nm. Improved suspension stability indices were demonstrated both in water, saline and buffers, compared to commercially available Alums.

The effects of the surface coating of aluminium adjuvants were also investigated. Phospholipid bilayer-coated aluminium nanoparticles (PLANs; 50 nm in size) are formed via chemisorption and prepared by reverse ethanol injection–lyophilization. The anhydrous antigen-loaded PLANs showed satisfactory stability under the controlled-temperature-chain instead of the integrated cold-chain for distribution. Additionally, the coated adjuvants may stimulate powerful antigen-specific humoral and cellular immune responses with less local inflammation because they were more readily taken up by APCs [45]. Of course, the fact that higher positive zeta potentials of nanoadjuvants favor the resulting suspension stability was considered during Alum nanoparticles' engineering [46]. The cationic surface functionalization of aluminum oxyhydroxide nanorods (ALNRs) showed that NH2-functionalized ALNRs had higher levels of cellular uptake, lysosomal damage, oxidative stress, and NLRP3 inflammasome activation in cells than untreated and SO3H-functionalized ALNRs [47]. Similarly, Ren et al. [48] found that coating conventional Alums with polyethyleneimine could boost antigen cross-presentation by DCs to CD8+ cytotoxic T cells in a clinically relevant manner.

The findings available so far suggest that nano alums may trigger the abovementioned adverse reactions faster than the conventional alum adjuvants. On one hand, this may be regarded as another benefit of nano alums, considering the fact that certain side effects may be observed early by the healthcare provider upon vaccine administration. On the other hand, some researchers fear that nano alums may provoke more serious subcutaneous granuloma and other local inflammatory reactions [34]. Undoubtedly, the safety of nano-sized adjuvants will be continually scrutinized by both the researchers and regulatory authorities.

2.1.3. Silica Nanoparticles

The intrinsic structural characteristics of designed mesoporous silica nanoparticles (MSNs) (with an average size of 50 to 200 nm) such as the tailorable mesoporous structure, high specific surface area, large pore volume, low density, good biocompatibility, thermal and chemical stability and easy chemical functionalization, contributed to their various biomedical applications [45,49,50]. The in vivo behavior of MSNs is highly dependent on their preparation methods, particle sizes, geometries, surface chemistry, dosing regimens, and even routes of administration, as shown by systematic biosafety assessments [51–53]. In animal models, the use of MSN as an immune adjuvant has been shown to effectively enhance both cellular and humoral immunity [54,55]. As a representative, bovine serum albumin (BSA) encapsulated/adsorbed SBA-15 nanostructured silica (a mesopore diameter of ~10 nm) had the ability to increase the immunogenicity and stimulate mutually Th1 and

Th2 immune responses via intramuscular or oral administration in mice [56,57], while the intraperitoneal injection of ovalbumin and amorphous silica nanoparticles (diameter of 33 nm) exerts an adjuvant effect for Th1, Th2 and Th17 immune responses [58]. Additionally, SBA-15 was demonstrated to be an efficient carrier for the hepatitis B surface antigen (HBsAg) as well as diphtheria anatoxin for oral and/or subcutaneous vaccination [56–61]. Therefore, MSNs could be a valuable adjuvant for divalent and trivalent vaccine formulations. In the future, significant progress is expected in the rational design and tailored functionalization of MSN-based protein delivery systems, as well as in the evaluation of biocompatibility, stability, efficacy, and biological [62,63].

2.1.4. Clay Nanoparticles

Clay nanoparticles (or nanoclays), such as layered double hydroxides and hectorite nanoparticles, exhibited strong adjuvant activity leading to immune responses significantly potent than those obtained by adjuvants which are available on the market (such as Quil-A® and alum) [64]. Although natural clays are also investigated, the majority of research groups work with synthetic nanoclays [65]. Surface modifications through clay–organic interactions resulted in formation of hybrid organic–inorganic materials, where the added organic functionality leads to complementary qualities to those intrinsic to the silicate substrate [66]. Wicklein et al. [67] synthetized lipid-based bio-nano hybrids as carriers of viral particles in vaccines against influenza A, using sepiolite as a biocompatibile nanoclay modified with phosphatidylcholine. This clay-lipid formulated influenza A vaccine exhibited increased thermal stability in functional tests at elevated temperatures up to 48 °C and showed high immunogenicity in mice with strong induction of specific antibodies associated with a Th1 cell cytokine profile [67].

A variety of other representatives of inorganic nanomaterials such as gold and carbon nanoparticles, oxides and hydroxides (e.g., zinc oxide, iron oxide and iron hydroxide), quantum dots, nanodiamonds, etc. have been investigated as potential vaccine adjuvants [40,68]. A somewhat broader range of metal-based nanomaterials are being evaluated as potential adjuvants in nanovaccines for cancer immunotherapy [69].

3. Nanoemulsions

Nanoscale emulsions (nanoemulsions) represent heterogeneous systems consisting of two liquids of opposite polarity (therefore, imisscible), where one of them is dispersed in the other one, forming something very small—usually 50–300 nm in diameter. These dispersions can be water-in-oil (W/O) or oil-in-water (O/W) (Figure 1). Their physical stability is ensured through the appropriate selection of components and the preparation technique [70]. Nanoemulsions are characterized by numerous properties that are attractive for biomedical applications. These colloidal systems have been explored as carriers/vehicles for hydrophobic compounds, covering various routes of administration: parenteral, dermal, oral, ocular and nasal [71–74]. In the context of vaccine development, nanoemulsions have been used in both human and animal vaccines as lipid-based adjuvants [74]. In addition to their role in the presentation of the antigens to the immune system, emulsions promote slow antigen release and protection from swift elimination [75].

In approved vaccines, nanoemulsions were presented in the 1990s, when the pharmaceutical company Novartis, Basel, Switzerland registered the first emulsion-based adjuvant in Europe [76]. Indeed, from a historical point of view, the adjuvant effect of emulsions (precisely, paraffin oil in a W/O emulsion) was known more than eighty years ago, having been demonstrated by Freund. However, these first attempts to boost the immune response by applying Freund's complete adjuvant (W/O emulsion with thermally killed mycobacteria) and, afterwards, Freund's incomplete adjuvant (W/O emulsion without bacterial cells) were discouraged, as such systems proved to be too reactogenic for general applications [76,77]. Even so, insights into the adjuvant efficacy of (nano)emulsions were not dismissed. The subsequent experience and improvement of the raw materials used for nanoemulsion preparation segregated essential physical properties for an efficient and safe nanoemulsion-based adjuvant, such as: the droplet size and surface properties, O/W ratio and rheological behavior [26,78–80].

With a view to optimize the safety/immunogenicity ratio, it has been established that tolerability can be ameliorated by replacing W/O with O/W nanoemulsions. A reduction in the oil concentration is followed by a decrease in the droplet size and viscosity. Such shift enabled an easier injection and eliminated the tendency of the emulsion to stay at the injection site, causing reactogenity [22,81]. In addition, O/W emulsions offer the advantage of the simple mixing of aquous antigen solutions with preformed emulsion. Therefore, antigens and adjuvants may be stored separately and combined afterwards, just before the administration [82,83]. The investigation of other types of oil acceptable for adjuvant nanoemulsion formulations has also led to significant progress: mineral oil has been replaced by squalene-based formulations (Figure 2), as it is fully metabolizable and less toxic [74,84,85].

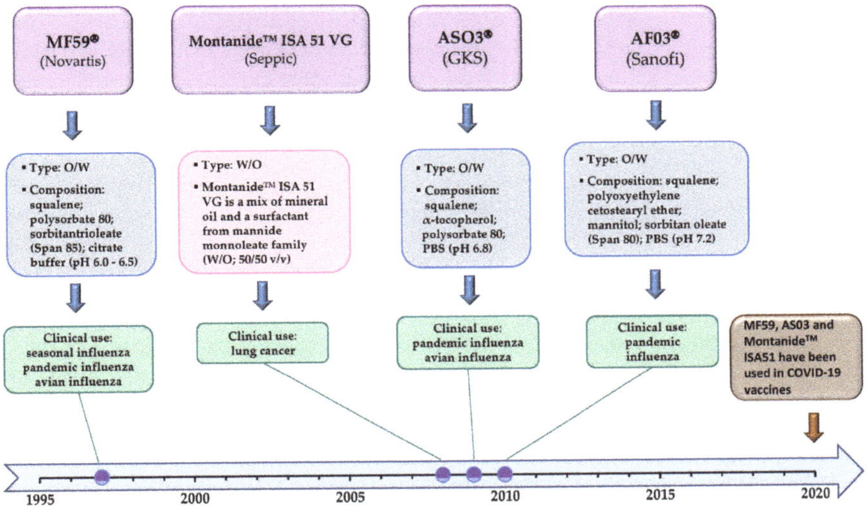

Figure 2. Timeline of nanoemulsion-based adjuvants' research and application.

One of the most widely used nanoemulsion-based adjuvants is MF59® (created by Novartis in 1997), containing 5% v/v of squalene as the oil phase and showing good performances in influenza vaccines [86]. Through extensive in vitro cell assays, Seubert et al. demonstrated that MF59® may increase immune cell migration to the injection site, promote monocyte differentiation towards DCs and consequently stimulate their maturation and antigen uptake while enhancing migration to the lymph nodes [87]. MF59 stimulates both Th1/Th2 cells, leading to humoral and cellular immune responses [18]. Another example of squalene-based nanoemulsion is the DETOX adjuvant system, developed by Ribi ImmunoChem Research, Hamilton, MO, USA (later acquired by Glaxo Smith Kline, London, UK (GSK)), containing a purified mycobacterium cell wall skeleton and monophospohoryl lipid A (as an immunostimulant; TLR4 ligand) dissolved in squalene. It is a specific adjuvant system used in a therapeutic melanoma vaccine called Melacine® [85,88,89]. Good illustrations of nanoemulsion-based adjuvants with squalene are ASO3® (containing alpha-tocopherol dissolved in squalene as an immunostimulant; developed by GSK) [90] and AFO3® (with sorbitan oleate and polyoxyethilenecetostearyl ether as stabilizers; created by Sanofi Pasteur, Lyon, France) [82]. AS03 has a similar mechanism of action to that of MF59 [18]. Additionallly, in the study of Morel et al. [90], it was depicted that the ability of ASO3® to induce an antigen-specific antibody response is likely linked to its capacity to activate innate immunity. The obtained results confirmed the good safety profile of ASO3®-containing vaccines. However, some findings imply that ASO3®

may be responsible for the reported narcolepsy cases after pediatric influenza vaccine administration [91]. On the other hand, the manufacturing peculiarity of AFO3® is that, unlike for other submicron emulsions, the phase inversion temperature procedure—a low-energy emulsification technique—was applied [82]. This technique is basically driven by the inherent physicochemical properties of the ingredients, not requiring sophisticated and expensive equipment [70]. This approach has effectively minimized both the development timelines and costs in industrial settings [84]. In response to the H1N1 influenza outbreak in 2009, European regulatory authorities granted approval to three pandemic influenza vaccines containing O/W emulsions (MF59®, AS03® and AF03®) as adjuvants [22]. Expectedly, the years that followed resulted in the new versions of W/O adjuvant nanoemulsions, among which the Seppic's (Courbevoie, France) Montanide® adjuvants (e.g., ISA 720, ISA 51), also squalene-based, can be underlined [92–94].

Unsurprisingly, the literature describes more examples of nanoemulsion-based adjuvants that are under investigation, such as CoVaccine HT® and Stable emulsion® (SE). CoVaccine HT® represents a combination of a sucrose fatty acid sulfate ester and squalane in a nanoemulsion for a single-dose pandemic influenza vaccine that is able to induce efficient Th1-biased cellular and humoral immune responses [95]. The study recently published by Lai et al. [96] demonstrated that CoVaccine HT® significantly contributes to robust immune responses against several SARS-CoV-2 variants.

Along with immunological behavior, physicochemical stability is one of the primary concerns for the nanoemulsion formulators. Both the physical instability (manifested either as droplet flocculation, Ostwald ripening, coalescence, creaming) and/or chemical degradation of nanoemulsion components jeopardize the adjuvant efficacy and safety. In addition, the factors related to potential freeze drying, which represents a common technique that promotes stability, facilitates transport and prolongs the product's shelf-life, ought to be addressed [97,98]. These issues should be overcome by the careful selection of aqueous phase components, oils, stabilizers and their concentration and mutual ratio, but the preparation technique must not be neglected [70,99]. Apparently, the majority of the approved nanoemulsions-based adjuvants have been formulated with squalene—a naturally derived and metabolizable oil—due to the good human safety track record of MF59® [82,100]. There has been some research covering the possibility of using other metabolizable oils, such as soybean, peanut or coconut oil. However, squalene over-performed their ability to provide nanoemulsions with an effective immune response and desirable physical properties [84,85,101]. Likewise, Tween 80®, as a well-established emulsifier, is applied as a stabilizer in most cases [84,102].

Last but not least, the COVID-19 vaccine development has also profited from the possibilities of lipid-based adjuvants. MF59, AS03, Freund's adjuvant and Montanide ISA51 have been used in COVID-19 vaccines [103,104]. GSK has been sharing its AS03 adjuvant with COVID-19 vaccine developers, and VidPrevtyn Beta, a recombinant COVID-19 vaccine developed by Sanofi, contains the ASO3 adjuvant and is approved by the European Commission as a booster dose vaccine for preventing COVID-19 in people aged 18 years and older [105,106].

In addition, the lipid nanoparticles, as another lipid-based nanosystem, especially underlined in the context of COVID-19 vaccines [107], have also been identified as a potent immunostimulatory component of mRNA and protein subunit vaccines (eliciting IL-6 production that is essential for antibody responses) [108,109].

Numerous studies point out a complex interplay among the droplet/particle size, surfactant concentration and immunostimulant components that have a significant effect on the biological/immunological performances of vaccine adjuvants [110–113]. Over the past two decades, noteworthy progress in the development of novel vaccine adjuvants has been made. Still, adjuvant activity is a result of multiple factors, and an enhanced immune response obtained with one antigen cannot be taken as a rule and extrapolated to another antigen [75]. A deeper understanding of the relationship among formulation science and relevant physical, chemical and physiological challenges related to specific pathological conditions, routes of administration and patient populations (children, adults, elderly) is required for rational and knowledge-based adjuvant design. Consequently, more potent adjuvants and improved vaccination compliance are envisaged. Taken all together, following their intensive human use worldwide, the benefits and safety profiles of nanoemulsion-adjuvanted vaccines are now very well established [78].

4. Virus-Based Nanocarriers as Vaccine Delivery Systems

The use of nanocarriers originated from natural sources that have specific properties such as the ability to escape host immune response and to enter the host cell in order to deliver specific genetic material, has become an important tool in the vaccine field [14]. Good example of naturaly originated nanocarriers that may have these characteristics are viruses that are an important class of naturally occurring bioinspired nanosystems for vaccine delivery [114]. Viruses can be rendered non-infectious and non-replicable without losing their ability to enter host cells, and since viral proteins exhibit ideal properties such as biocompatibility, biodegradability and uniform size distribution, all these characteristics underscore and encourage the use of viruses as vaccine delivery systems [115]. Traditional bio-based nanoparticle delivery systems such as liposomes present several challenges such as: less specific target delivery, poor stability in biological fluids, difficulties in mass production and in vivo toxicity that can be overcome with viral nanoparticles [14]. In addition, adjuvants currently used in humans have been observed to enhance humoral immunity; however, virus-based nanocarriers have advantage comparing to traditional adjuvants as they can induce both humoral and broad spectrum of cellular immune responses [116,117]. Different viral vectors with desired genetic information inserted into viral genome are used as vehicles: adenoviral, adeno-associated viral, vesicular stomatitis viral, lentiviral vector, etc. These vaccine platforms can be made using replicating or non-replicating viral vectors. Moreoften, the viral vectors are modified to be nonreplicating, as it is considered that nonreplicating viral vectors are safer than replicating viral vectors and have reduced reactogenicity. On the other hand, nonreplicating viral vectors may have lower immunogenicity profile requiring higher dose or repeated doses, and that's why in some cases the use of replicating viral vectors (e.g. vesicular stomatitis viral vector) is preferred [118,119].

Additionally, virus-derived features such as virus-like particles (VLPs) and virosomes are very attractive platforms for vaccine development as they are safe and stable, morphologically similar to the infectious virus with the ability to enter the host cell and induce both humoral and cellular immunity (Th1, Th2 and CD8+ T cell immune response) (Table 1) [120]. So, in this section, viral-vectored based delivery systems, as well as VLPs and virosomes, will be described in more detail as important virus-based bionanocarriers for vaccine delivery (Figure 3).

4.1. *Viral-Vectored-Based Delivery Systems*

Recombinant viral vectors have been shown to be profoundly effective vehicles for bringing foreign nucleic acids into target cells. Furthermore, infections naturally induce host immune reactions, which might be beneficial for vaccination purposes. Both of these features, joined by broad knowledge on techniques for manipulating the viral genome, make recombinant viral vectors appealing candidates for vaccine vector development. Recombinant viral vectors have been and are being examined as vaccines focusing on different viral, bacterial and protozoan microbes. They are especially utilized in disease areas where traditional vaccination approaches have been demonstrated to be ineffectual, troublesome or technically impossible. So far, viral vectors vaccines have been used as an approach in treating different infectious diseases including COVID-19, human immunodeficiency virus (HIV), Malaria, Ebola and others. Likewise, different viruses such as adenovirus, adeno-associated virus, vesicular stomatitis virus or lentivirus have been used for constructing recombinant viral vectors [118].

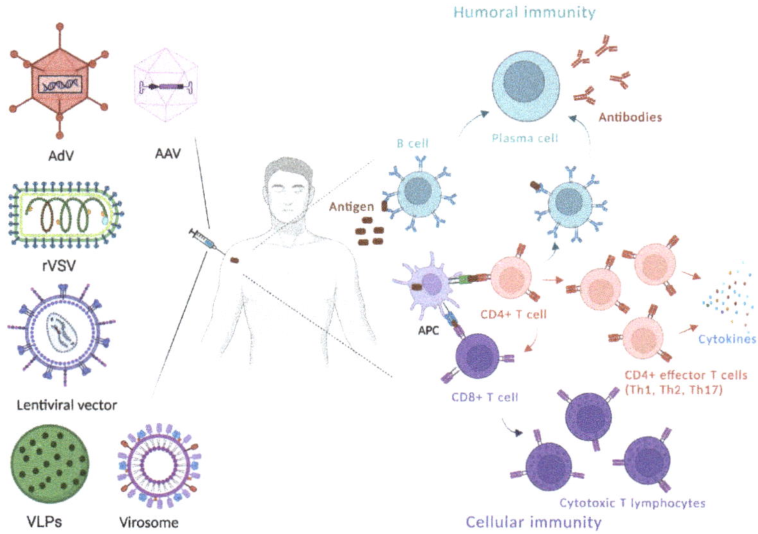

Figure 3. Different virus-based nanocarriers as vaccine delivery systems which function by activating cellular and humoral immunity to varying degrees. Apart from other presented virus-based nanocarriers, VLPs (including bacteriophage-based VLPs) and virosomes deliver antigens instead of a genetic material coding antigen structure. AdV—adeno-viral vector; AAV—adeno-associated viral vector; rVSV—recombinant vesicular stomatitis virus; VLPs—virus-like particles, APC—antigen-presenting cells. Created with BioRender.com.

4.1.1. Adenoviral Vector-Based Delivery System

Adenoviral vectors possess the ability to trigger strong immune responses on cellular, humoral, and mucosal levels, rendering them highly appealing for the development of vaccines.Compared to traditional vaccines, a diverse range of antigens can be incorporated into adenovirus vaccines to prompt an immune response that targets a wide array of pathogens while also acting as an adjuvant for an antigen-encoding vector.

By employing uncommon serotypes of both human and non-human origin, the issue of pre-existing immunity is effectively overcome,, while booster regimens with alternative adenoviral vectors have been shown to be safe and immunogenicVaccination strategies utilizing recombinant adenovirus-based vectors show great potential, especially in the field of infectious diseases where cellular immune responses are crucial for protection. Adenovirus-based vectors indeed stimulate robust and durable cellular responses, particularly in CD8+ T cells. [121].

So far, different adenovirus serotypes have been employed as vaccine vectors, the most prominent being human adenovirus type 5 (HAdV-C5), human adenovirus type 26 (HadV-D26) and chimpanzee adenovirus, ChAdOx1. Due to the fact that it has the capability to trigger immune responses involving both humoral and cell-mediated components , HAdV-C5 has been the most common choice for preclinical studies of adenovirus-based vector vaccines. In one of the first HAdV-C5 based vector vaccine studies, an efficacy assessment of a cell-mediated immunity to the HIV-1 vaccine (the Step Study) was performed, and, surprisingly, an increase in the number of HIV-1 infections in male recipients of the vaccine was noted [122]. Even though this could have been considered as a failure, the development of HAdV-C5-based vaccine vectors continued and peaked during the COVID-19 pandemic. A single vaccination with the replication defect HAdV-C5 encoding the spike protein of SARS-CoV-2 (Ad5-nCoV) was shown to completely protect mice from infection of the upper and lower respiratory tract with mouse-adapted SARS-CoV-2. This suggests that mucosal vaccination may provide the desired protective effect; thus, this route of administration requires further investigation in human clinical trials [123]. Later on, the Ad5-nCoV entered the clinical trials [124,125], and it was approved for use in 2021 in China, Hungary, Mexico and Pakistan.

Due to itslow occurrence in human populations and its ability to elicit a favorable immune response to inserted transgenes, HAdV-D26 has emerged as a promising framework for the development of vaccine vectors. Consequently, it has been the subject of numerous ongoing and completed clinical studies. HAdV-D26-based vaccine vectors have been assessed as potential interventions for diseases associated with HIV, respiratory syncytial virus (RSV), Ebola virus, Zika virus and SARS-CoV-2. Very recently, the European Medicines Agency (EMA) approved two HAdV-D26-based vector vaccines, namely, Ad26.ZEBOV against Ebola [126] and Ad26.COV2.S against COVID-19 [127]. The simian adenovirus-based vaccine vector ChAdOx1 against COVID-19 [128] has also been approved by regulatory agencies.

Rare occurrences of thrombocytopenia have been documented following the administration of adenoviral vector-based COVID-19 vaccines. A recently published study revealed that three adenoviruses utilized as vaccination vectors against SARS-CoV-2, namely HAdV-C5, HAdV-D26, and ChAdOx1, exhibit binding affinity towards platelet factor 4 (PF4), a protein implicated in the development of heparin-induced thrombocytopenia. The findings presented by these researchers suggest that stable complexes formed between PF4 and clinically relevant adenoviruses may play a significant role in elucidating the mechanisms underlying thrombosis with thrombocytopenia syndrome. [129]. These results underscore the necessity for further investigations into adenovirus-based vaccine vectors.

4.1.2. Adeno-Associated Viral Vector as a Delivery System for Vaccines

Adeno-associated virus (AAV) vectors have been extensively investigated as gene therapy vectors and have only recently gained attention in the field of vaccination. However, with the discovery that AAV can elicit a robust humoral and cellular immune response against the glycoprotein B of herpes simplex virus (HSV) type 1 [130], there has been a growing interest in exploring the application of AAV vectors for vaccination. Numerous studies have since delved into the utilization of AAV vectors for this purpose.Thus, it has been shown that a single injection of an AAV vector induced a potent and long-lasting antibody response without the need for an adjuvant [131,132]. Even though in some cases, compared to other vaccination strategies, AAV vectors have demonstrated the ability to generate a higher or longer-lasting antibody response. Additionally, they are regarded as having a low immunogenic profile in comparison to other viral vectors.. In addition to the low immunogenicity, AAV vectors suffer also from a very limited insertion space, namely, only 4.8 kb. However, AAV vectors are employed as vaccination vectors for delivering antigens. They facilitate the production of foreign proteins with a structure and arrangement that closely resemble those of natural pathogens [133], or as vehicles for delivering monoclonal antibodies and derivatives of immunoglobulins as a means of passive immunizations [134].

4.1.3. Vesicular Stomatitis Viral Vector for Vaccines

For more than a decade, Vesicular stomatitis virus (VSV) has been acknowledged as a robust vaccine vector for infectious diseases. Recent clinical trials examining the VSV-EBOV vaccine candidate against the Ebola virus (EBOV) have contributed a substantial amount of clinical data, indicating that VSV holds great potential as an ideal vaccine candidate for epidemic pathogens. [135]. The safety and immunogenicity of the VSV-based recombinant viral vector (rVSV) have been demonstrated in pre-clinical and clinical trials against Ebola, HIV and SARS-CoV-2. An extensive pre-clinical and clinical evaluation of the VSV-based Ebola vaccine demonstrated remarkable safety and efficacy, which ultimately led to the U.S. Food and Drug Administration (FDA) licence (also known by its brand name, Ervebo, Merck & Co., Rahway, NJ, USA) for use in humans against EBOV infection [136]. A recent study has demonstrated that a replicating VSV-SARS-CoV-2 vaccine triggers the production of potent neutralizing antibodies at high levels. This finding provides strong support for advancing the development of VSV-SARS-CoV-2 as an attenuated, replication-competent vaccine against SARS-CoV-2. [137].

4.1.4. Lentiviral Vector for Vaccines

Lentiviral vectors have become attractive as vaccine vectors in many preclinical animal models of [138] infection and oncology because of their high efficiency in transducing DCs and inducing long-term humoral and CD8+ immunity, as well as their potent protection. Recently, clear prophylactic effects of lentivirus-based vaccinations against SARS-CoV-2 have been described, indicating that intranasal vaccination is an effective approach against COVID-19. Namely, an intranasal strategy triggered a localized upper airway immune response that conferred disease protection in COVID-19 mice and hamster models [139]. Lentivirus vaccines have been tested in multiple preclinical studies in mice and rhesus monkeys, including against HIV, Zika and malaria, and have demonstrated high immunogenicity and sustained protective immune responses [140–142].

4.2. Virus-like Particles and Virosomes

4.2.1. Virus-like Particles (VLPs)

Virus-like particles (VLPs) are empty (without viral genetic material), non-replicating and non-infectious structures that have ability to self-assemble [143]. VLPs are nanomaterials, and the expression and self-assembly of the viral structural proteins can take place in various expression systems (including mammalian cell lines, plants, insects, yeasts and bacteria), after which the VLPs self-assemble into the structure similar to original virus, but without pathogenic potential [144–146]. First VLPs derived from Hepatitis B virus were introduced in 1968, after which different types of VLPs have been developed and with history of more than fifty years VLP-based vaccines are considered as safe [147,148].

The main building unit of VLPs is surface protein that is antigen of interest displayed in high levels after self-assembling process [149]. After phagocytosis, these antigens may be presented by DCs leading consequently to the induction of both humoral and cellular immune responses [117,150]. The most common VLPs size is within the range of 20~200 nm in diameter, which is optimal for particles to be recognized by DCs and other APCs including B cells, which process and present peptides of interest to T cells in lymph nodes and stimulate T cell-dependent antibody immune responses [144,151,152]. However, as VLPs may display multimeric epitopes, they can promote the cross-linking of antigen receptors on B cells, which can be strong enough for activating them and also induce the antibodies production but without the T cell help (T cell-independent antibody immune response) (Figure 3). Therefore, in most cases, adjuvants are not needed in VLP-based vaccines, but the use of an adjuvant within the VLP-based vaccine formulation may improve their immunogenicity [153].

Hundreds of VLPs are in different stages of preclinical and clinical studies for numerous infectious diseases, including HIV, influenza, hepatitis B, hepatitis E, malaria, Ebola, COVID-19, Zika virus, Dengue and foot and mouth disease, among others [154]. Several VLPs vaccines are licensed and are already in use, such as Engerix-BR (GSK) and Recombivax HBR (Merck & Co) against hepatitis B virus, Gardasil (Merck & Co) and Cervarix (GSK) against human papilloma viruses (HPV), Hecolin (Xiamen Innovax Biotech Co., Xiamen, China) against hepatitis E virus [155] and MosquirixTM (GSK) against malaria [156]. VLPs are classified as subunit vaccines, as they are made of non-infectious viral components.

Bacteriophage-Based VLPs

Bacteriophages or phages are viruses that infect and replicate within bacteria. VLPs from a variety of RNA bacteriophages, particularly VLPs derived from Leviviridae, are the most studied bacteriophage-based VLPs with the great potential for being used as vaccine platforms for the antigen delivery [157]. Bacteriophages belonging to the family Leviviridae possess positive-sense single-stranded RNA (ssRNA) packed inside an icosahedral protein shell made of 178 copies of coat protein (CP), with the diameter of about 30 nm [158,159]. The recombinant expression of ssRNA phage CPs in bacteria or yeasts most often result in the efficient assembly of VLPs [159]. Bacteriphage-based VLPs can display wide range of different antigens, and there are two main mechanisms for these antigens to be produced: a) genetic fusion with CPs or b) chemical conjugation to the surface of VLPs [160]. Several bacteriophage-based VLPs have been created using phage CPs of PP7 (infects *Pseudomonas aeruginosa*), Qβ (infects *Escherichia coli*), MS2 (infects *Escherichia coli*) and AP205 (infects *Acinetobacter* bacteria) bacteriophages [159,160]. Some examples of bacteriophage-based VLPs that are in different stages of clinical studies are: the PP7-based HPV vaccine [161], Qβ-based COVID-19 vaccine [162], MS2-based foot and mouth disease vaccine and HPV vaccines [157,163] and AP205-based flu and COVID-19 vaccines [164,165], but there is no approved bacteriophage-based vaccine from EMA or FDA yet.

The size (about 30 nm) of bacteriophage-based VLPs is important, as it allows for the direct dranage of VLPs to the lymph nodes and optimal B-cell activation. Moreover, the important advantages are that both their surfaces and their interior contents can be modified, which enables that VLPs can display different epitopes on their surface or interior of the VLPs can be packed with TLR ligands, such as RNA or DNA sequences which can activate TLR7/8 and TLR9, respectively [166].

4.2.2. Virosomes

Other delivery systems, also classified as subunit vaccines, similar to VLPs are virosomes. Virosomes, spherical, unilamellar vesicles (60–200 nm) comprising lipid nanomaterials, emerged as FDA-approved nanocarriers reconstituted of viral envelopes phospholipids with a removed nucleocapsid. Virosomes hold promising bioinspiration and biomimetic potency against viral infections [167]. The virosomes have a mono- or bilayer phospholipid envelope to which components/antigens originating from viruses or other pathogens can be attached or inserted and are preffered to VLPs because of the protein-based nature of VLPs [117]. The important advantage of virosomes is their ability to bind to the target cells through membrane fusion proteins retained from the native virus, which forms the basis for their enhanced immunogenicity. The virosomal surface contains antigen molecules that interact with receptors on B lymphocytes, which triggers a strong antibody response. Additionaly, virosomes efficiently interact with antigen-presenting cells, such as DCs, leading to the activation of the T lymphocites-mediated immune response [168] (Figure 3). Virosomes have shown promising results in vaccine development against various diseases including influenza and hepatitis A, for which the virosome-based formulations Inflexal® V and Epaxal® have been licenced, respectively. However, there is ongoing research to develop a potent vaccine against HIV-1, HPV and SARS-CoV-2 [167].

5. Conclusions

The application of nanotechnologies has led to great progress in the field of vaccine research and has become an important pillar of vaccinology. The great growth in this field enabled the development of new nano-sized vaccines and the design of nanoadjuvants and nanoparticulate delivery systems that are more reliable and/or effective and that can be applied in new medical treatments. The current findings imply that nanostructured vaccine formulations provide certain advantages in the elicited immune response and are suitable for a variety of administration routes and dose regimens. Some formulations benefit from the inclusion of a single nanoadjuvant, resulting in an improvement of the overall vaccination effect. Apart from most adjuvants currently used in humans, many new adjuvants (such as liposomes) or delivery systems (e.g., viral-vectored vaccines) elicit both humoral and specific T-cell responses. So far, certain inorganic nanoadjuvants (Alum-based, MSNs and metal-based nanomaterials), nanoemulsions (e.g., MF59®, ASO3®, AFO3®), viral-vectored systems, virus-like particles and virosomes have been of special interest for the potentially successful translation to emerging targets to deal with challenging diseases such as HIV infection, malaria, etc. All of these efforts enabled vaccine developers to choose different adjuvants and/or delivery systems as per the requirement, which was extremely important for the development of COVID-19 vaccines. It is expected that intense multidisciplinary research in the fields of nanomaterials, immunology, bacteriology, virology, oncology and pharmaceutical development will provide new perspectives in this field in the near future.

Table 1. Summary of various nanoadjuvants/delivery systems and their effects, safety aspects and clinical applications.

Type of Nanoadjuvant/Nanoparticulate Delivery System		Effect	Safety	Immune Target	Examples of Clinical Use	References
Inorganic nanomaterials	Nano-alum	NLRP3 inflammasome activation	Generally deemed safe Local adverse reactions (subcutaneous granuloma and other local inflammatory reactions)	Th1	/	[34,41–43]
	Silica nanoparticles	Delivery system with controlled antigen release	Safety is strongly dependent on preparation technique, physicochemical properties and administration route	Th1/Th2 Th17	/	[51–58]
	Clay nanoparticles	Depot effect	High immunogenicity; much higher compared to that of the FDA-approved adjuvants (e.g., Alum)	Th1	/	[64,67]
Lipid-based nanoadjuvants	Nanoemulsions	Sustained antigen release TLR4 ligand	O/W nanoemulsions (e.g., MF59®, AS03® and AF03®) are less reactogenic and less toxic compared to W/O (e.g., Montanide®) nanoemulsions Narcolepsy (AS03®)	Th1/Th2	Vaccines against seasonal influenza, pandemic influenza and avian influenza; therapeutic vaccines against lung cancer and melanoma	[18,76–80,82–84,91]
	Lipid nanoparticles	Antigen delivery system (especially for mRNA) Antigen protection from degradation	Highly dependent on lipid components Local reactions and the injection site	Th1/Th2 CD8+ T cells	Vaccines against SARS-CoV-2	[107–109,169]

Table 1. *Cont.*

Type of Nanoadjuvant/Nanoparticulate Delivery System		Effect	Safety	Immune Target	Examples of Clinical Use	References
Virus-based nanoparticles	Adenoviral vector	Antigen delivery system	Local reactions and the injection site Fatigue, fever and headache Rare cases of thrombotic thrombocytopaenia	Th1/ Th2 CD8+ T cells	Vaccines against SARS-CoV-2 and Ebola	[118,121,129]
	Adeno-associated viral vector	Delivery system for antigens as well as for monoclonal antibodies and derivatives of immunoglobulins for passive immunization	Generally safe, but with low immunogenicity Transient and asymptomatic hepatitis	Th1/ Th2 CD8+ T cells	/	[118,133,134,170]
	Vesicular stomatitis viral vector	Delivery system for antigens	Fatigue, fever and headache	Th1/ Th2 CD8+ T cells	Vaccine against Ebola virus	[135–137,171]
	Lentiviral vector	Delivery system for antigens	Safety is still under evaluation	Th1/ Th2 CD8+ T cells	/	[138,172]
	Virus-like particles (VLPs)	Delivery system for antigens	Improved safety compared to viral-vectored vaccines Generally recognized as safe	Th1/ Th2 CD8+ T cells	Vaccines against hepatitis B virus, vaccines against human papilloma viruses (HPV), vaccine against hepatitis E virus and vaccine against malaria	[120,155,156]
	Virosomes	Delivery system for antigens	Improved safety profile compared to VLPs	Th1/Th2 CD8+ T cells	Vaccines against influenza and hepatitis A viruses	[167,168]

Author Contributions: Conceptualization, B.F., I.N., D.M. and D.K.; Writing—Original Draft Preparation, B.F., I.P., I.N. and D.M.; Writing—Review & Editing, Z.S.-V. and S.S.; Visualization, B.F., I.P. and D.K.; Supervision, Z.S.-V. and S.S.; Funding Acquisition, B.F. and S.S. All authors have read and agreed to the published version of the manuscript.

Funding: This research was funded by the Ministry of Science, Technological Development and Innovation, Republic of Serbia through a Grant Agreement with University of Belgrade, Faculty of Pharmacy No: 451-03-47/2023-01/ 200161, the Croatian Science Foundation, grant number IP-2019-04-6048, and the COST Action CA16231 European Network of Vaccine Adjuvants (ENOVA).

Institutional Review Board Statement: Not applicable.

Informed Consent Statement: Not applicable.

Data Availability Statement: Not applicable.

Conflicts of Interest: The authors declare no conflict of interest. The funders had no role in the design of the study, in the collection, analyses or interpretation of the data, in the writing of the manuscript or in the decision to publish the results.

References

1. Lemoine, C.; Thakur, A.; Krajišnik, D.; Guyon, R.; Longet, S.; Razim, A.; Górska, S.; Pantelic, I.; Ilic, T.; Nikolic, I.; et al. Technological Approaches for Improving Vaccination Compliance and Coverage. *Vaccines* **2020**, *8*, 304. [CrossRef] [PubMed]
2. Candela, F.; Quarta, E.; Buttini, F.; Ancona, A.; Bettini, R.; Sonvico, F. Recent Patents on Nasal Vaccines Containing Nanoadjuvants. *Recent. Adv. Drug. Deliv.* **2022**, *16*, 103–121. [CrossRef]
3. Zhang, N.; Li, M.; Hou, Z.; Ma, L.; Younas, A.; Wang, Z.; Jiang, X.; Gao, J. From vaccines to nanovaccines: A promising strategy to revolutionize rheumatoid arthritis treatment. *J. Control. Release* **2022**, *350*, 107–121. [CrossRef] [PubMed]
4. Etter, E.; Mei, K.-C.; Nguyen, J. Delivering more for less: Nanosized, minimal-carrier and pharmacoactive drug delivery systems. *Adv. Drug. Deliv. Rev.* **2021**, *179*, 113994. [CrossRef]
5. Yuba, E.; Kado, Y.; Kasho, N.; Harada, A. Cationic lipid potentiated the adjuvanticity of polysaccharide derivative-modified liposome vaccines. *J. Control. Release*, 2022; *Online ahead of print*. [CrossRef]
6. Bi, S.; Li, M.; Liang, Z.; Li, G.; Yu, G.; Zhang, J.; Chen, C.; Yang, C.; Xue, C.; Zuo, Y.Y.; et al. Self-assembled aluminum oxyhydroxide nanorices with superior suspension stability for vaccine adjuvant. *J. Colloid. Interface Sci.* **2022**, *627*, 238–246. [CrossRef]
7. Carmona-Ribeiro, A.M. Supramolecular Nanostructures for Vaccines. *Biomimetics* **2021**, *7*, 6. [CrossRef]
8. Wack, A.; Rappuoli, R. Vaccinology at the beginning of the 21st century. *Curr. Opin. Immunol.* **2005**, *17*, 411–418. [CrossRef]
9. Reed, S.G.; Bertholet, S.; Coler, R.N.; Friede, M. New horizons in adjuvants for vaccine development. *Trends Immunol.* **2009**, *30*, 23–32. [CrossRef]
10. Ruterbusch, M.; Pruner, K.B.; Shehata, L.; Pepper, M. In Vivo CD4+ T Cell Differentiation and Function: Revisiting the Th1/Th2 Paradigm. *Annu. Rev. Immunol.* **2020**, *38*, 705–725. [CrossRef] [PubMed]
11. Rapaka, R.R.; Cross, A.S.; McArthur, M.A. Using Adjuvants to Drive T Cell Responses for Next-Generation Infectious Disease Vaccines. *Vaccines* **2021**, *9*, 820. [CrossRef]
12. Verma, S.K.; Mahajan, P.; Singh, N.K.; Gupta, A.; Aggarwal, R.; Rappuoli, R.; Johri, A.K. New-age vaccine adjuvants, their development, and future perspective. *Front. Immunol.* **2023**, *14*, 1043109. [CrossRef]
13. Pulendran, B.S.; Arunachalam, P.; O'Hagan, D.T. Emerging concepts in the science of vaccine adjuvants. *Nat. Rev. Drug. Discov.* **2021**, *20*, 454–475. [CrossRef] [PubMed]
14. Ojha, S.K.; Pattnaik, R.; Singh, P.K.; Dixit, S.; Mishra, S.; Pal, S.; Kumar, S. Virus as a Nanocarrier for Drug Delivery Redefining Medical Therapeutics—A Status Report. *Comb. Chem. High. Throughput Screen* **2022**, *25*, 1619–1629. [CrossRef] [PubMed]
15. Mamo, T.; Poland, G.A. Nanovaccinology: The next generation of vaccines meets 21st century materials science and engineering. *Vaccine* **2012**, *30*, 6609–6611. [CrossRef]
16. Yasamineh, S.; Kalajahi, H.G.; Yasamineh, P.; Yazdani, Y.; Gholizadeh, O.; Tabatabaie, R.; Afkhami, H.; Davodabadi, F.; Farkhad, A.K.; Pahlevan, D.; et al. An overview on nanoparticle-based strategies to fight viral infections with a focus on COVID-19. *J. Nanobiotechnol.* **2022**, *20*, 440. [CrossRef] [PubMed]
17. Lindblad, E.B.; Duroux, L. *Immunopotentiators in Modern Vaccines*; Schijns, V., O'Hagan, D., Eds.; Academic Press: Cambridge, MA, USA, 2017; Chapter 18; p. 347.
18. Shi, S.; Zhu, H.; Xia, X.; Liang, Z.; Ma, X.; Sun, B. Vaccine adjuvants: Understanding the structure and mechanism of adjuvanticity. *Vaccine* **2019**, *37*, 3167–3178. [CrossRef]
19. Hem, S.L.; Hogen Esch, H. *Vaccine Adjuvants and Delivery Systems*; Singh, M., Ed.; John Wiley & Sons, Inc.: Hoboken, NJ, USA, 2007; pp. 81–114.
20. Hem, S.L.; Hogenesch, H. Relationship between physical and chemical properties of aluminum-containing adjuvants and immunopotentiation. *Expert. Rev. Vaccines* **2007**, *6*, 685–698. [CrossRef]
21. Sheskey, P.J.; Cook, W.G.; Cable, C.T. *Handbook of Pharmaceutical Excipients*, 8th ed.; Pharmaceutical Press: London, UK, 2017; p. 67.
22. Garçon, N.; Friede, M. *Plotkin's Vaccines*; Plotkin, S.A., Orenstein, W.A., Offit, P.A., Edwards, K.M., Eds.; Elsevier: Philadelphia, PA, USA, 2018; Section 1; Chapter 6; pp. 61–74.
23. Caulfield, M.J.; Shi, L.; Wang, S.; Wang, B.; Tobery, T.W.; Mach, H.; Ahl, P.L.; Cannon, J.L.; Cook, J.C.; Heinrichs, J.H.; et al. Effect of alternative aluminum adjuvants on the absorption and immunogenicity of HPV16 L1 VLPs in mice. *Hum. Vaccines* **2007**, *3*, 139–145. [CrossRef]
24. HogenEsch, H.; O'Hagan, D.T.; Fox, C.B. Optimizing the utilization of aluminum adjuvants in vaccines: You might just get what you want. *NJP Vaccines* **2018**, *3*, 51. [CrossRef]
25. Morefield, G.L. A rational, systematic approach for the development of vaccine formulations. *AAPS J.* **2011**, *13*, 191–200. [CrossRef]
26. Krajišnik, D.; Ilifá, T.; Nikolifá, I.; Savifá, S. Established and advanced adjuvants in vaccines' formulation: Mineral adsorbents, nanoparticulate carriers and microneedle delivery systems. *Arh. Farm.* **2019**, *69*, 420–451. [CrossRef]
27. Galazka, A.M.; Milstien, J.B.; Zaffran, M. *Thermostability of Vaccines*; World Health Organization: Geneva, Switzerland, 1998; Available online: https://apps.who.int/iris/handle/10665/64980 (accessed on 20 April 2023).
28. Maa, Y.F.; Zhao, L.; Payne, L.G.; Chen, D. Stabilization of alum-adjuvanted vaccine dry powder formulations: Mechanism and application. *J. Pharm. Sci.* **2003**, *92*, 319–332. [CrossRef] [PubMed]
29. Clapp, T.; Siebert, P.; Chen, D.; Jones Braun, L. Vaccines with aluminum-containing adjuvants: Optimizing vaccine efficacy and thermal stability. *J. Pharm. Sci.* **2011**, *100*, 388–401. [CrossRef] [PubMed]

30. Tonnis, W.F.; Amorij, J.P.; Vreeman, M.A.; Frijlink, H.W.; Kersten, G.F.; Hinrichs, W.L. Improved storage stability and immunogenicity of hepatitis B vaccine after spray-freeze drying in presence of sugars. *Eur. J. Pharm. Sci.* **2014**, *55*, 36–45. [CrossRef]
31. Li, X.; Thakkar, S.G.; Ruwona, T.B.; Williams, R.O., 3rd; Cui, Z. A method of lyophilizing vaccines containing aluminum salts into a dry powder without causing particle aggregation or decreasing the immunogenicity following reconstitution. *J. Control. Release* **2015**, *204*, 38–50. [CrossRef]
32. Li, X.; Aldayel, A.M.; Cui, Z. Aluminum hydroxide nanoparticles show a stronger vaccine adjuvant activity than traditional aluminum hydroxide microparticles. *J. Control. Release* **2014**, *173*, 148–157. [CrossRef] [PubMed]
33. Hogenesch, H. Mechanism of immunopotentiation and safety of aluminum adjuvants. *Front. Immunol.* **2013**, *3*, 406. [CrossRef]
34. He, P.; Zou, Y.; Hu, Z. Advances in aluminum hydroxide-based adjuvant research and its mechanism. *Hum. Vaccines Immunother.* **2015**, *11*, 477–488. [CrossRef]
35. Quandt, D.; Rothe, K.; Baerwald, C.; Rossol, M. GPRC6A mediates Alum-induced Nlrp3 inflammasome activation but limits Th2 type antibody responses. *Sci. Rep.* **2015**, *5*, 16719. [CrossRef] [PubMed]
36. Mitchell, T.C.; Casella, C.R. No pain no gain? Adjuvant effects of alum and monophosphoryl lipid A in pertussis and HPV vac-cines. *Curr. Opin. Immunol.* **2017**, *47*, 17–25. [CrossRef]
37. Baylor, N.W.; Egan, W.; Richman, P. Aluminum salts in vaccines--US perspective. *Vaccine* **2002**, *20*, S18–S23. [CrossRef] [PubMed]
38. Nanishi, E.; Dowling, D.J.; Levy, O. Toward precision adjuvants: Optimizing science and safety. *Curr. Opin. Pediatr.* **2020**, *32*, 125–138. [CrossRef] [PubMed]
39. Hess, K.L.; Medintz, I.L.; Jewell, C.M. Designing inorganic nanomaterials for vaccines and immunotherapies. *Nano Today* **2019**, *27*, 73–98. [CrossRef]
40. Li, X.; Wang, X.; Ito, A. Tailoring inorganic nanoadjuvants towards next-generation vaccines. *Chem. Soc. Rev.* **2018**, *47*, 4954–4980. [CrossRef]
41. Sun, B.; Ji, Z.; Liao, Y.P.; Wang, M.; Wang, X.; Dong, J.; Chang, C.H.; Li, R.; Zhang, H.; Nel, A.E.; et al. Engineering an effective immune adjuvant by designed control of shape and crystallinity of aluminum oxyhydroxide nanoparticles. *ACS Nano* **2013**, *7*, 10834–10849. [CrossRef] [PubMed]
42. Thakkar, S.G.; Xu, H.; Li, X.; Cui, Z. Uric acid and the vaccine adjuvant activity of aluminum (oxy)hydroxide nanoparticles. *J. Drug Target.* **2018**, *26*, 474–480. [CrossRef]
43. Amini, Y.; Moradi, B.; Fasihi-Ramandi, M. Aluminum hydroxide nanoparticles show strong activity to stimulate Th-1 immune response against tuberculosis. *Artif. Cells Nanomed. Biotechnol.* **2017**, *45*, 1331–1335. [CrossRef] [PubMed]
44. Liang, Z.; Wang, X.; Yu, G.; Li, M.; Shi, S.; Bao, H.; Chen, C.; Fu, D.; Ma, W.; Xue, C.; et al. Mechanistic understanding of the aspect ratio-dependent adjuvanticity of engineered aluminum oxyhydroxide nanorods in prophylactic vaccines. *Nano Today* **2022**, *43*, 101445. [CrossRef]
45. Wang, T.; Zhen, Y.; Ma, X.; Wei, B.; Wang, N. Phospholipid bilayer-coated aluminum nanoparticles as an effective vaccine adjuvant-delivery system. *ACS Appl. Mater. Interfaces* **2015**, *7*, 6391–6396. [CrossRef]
46. Liang, Z.; Yang, Y.; Yu, G.; Zhu, H.; Xia, X.; Chen, C.; Fu, D.; Li, M.; Cheng, G.; Xue, C.; et al. Engineering aluminum hydroxyphosphate nanoparticles with well-controlled surface property to enhance humoral immune responses as vaccine adjuvants. *Biomaterials* **2021**, *275*, 120960. [CrossRef]
47. Sun, B.; Ji, Z.; Liao, Y.P.; Chang, C.H.; Wang, X.; Ku, J.; Xue, C.; Mirshafiee, V.; Xia, T. Enhanced Immune Adjuvant Activity of Aluminum Oxyhydroxide Nanorods through Cationic Surface Functionalization. *ACS Appl. Mater. Interfaces* **2017**, *9*, 21697–21705. [CrossRef] [PubMed]
48. Ren, H.; Mou, Y.; Lin, L.; Wang, L.; Hu, H. Efficient antigen cross-presentation through coating conventional aluminum adjuvant particles with PEI. *Am. J. Transl. Res.* **2021**, *13*, 4092–4102. [PubMed]
49. Ghasemi, M.; Bakhshi, B.; Khashei, R.; Soudi, S. Mesoporous silica nano-adjuvant triggers pro-inflammatory responses in Caco-2/peripheral blood mononuclear cell (PBMC) co-cultures. *Nanobiomedicine* **2022**, *9*, 18495435221088374. [CrossRef]
50. Jafari, S.; Derakhshankhah, H.; Alaei, L.; Fattahi, A.; Varnamkhasti, B.S.; Saboury, A.A. Mesoporous silica nanoparticles for therapeutic/diagnostic applications. *Biomed. Pharm.* **2019**, *109*, 1100–1111. [CrossRef]
51. Chen, Y.; Chen, H.; Shi, J. In vivo bio-safety evaluations and diagnostic/therapeutic applications of chemically designed mesoporous silica nanoparticles. *Adv. Mater.* **2013**, *25*, 3144–3176. [CrossRef] [PubMed]
52. Tang, F.; Li, L.; Chen, D. Mesoporous silica nanoparticles: Synthesis, biocompatibility and drug delivery. *Adv. Mater.* **2012**, *24*, 1504–1534. [CrossRef]
53. Zhang, L.; Yang, W.; Hu, C.; Wang, Q.; Wu, Y. Properties and applications of nanoparticle/microparticle conveyors with adjuvant characteristics suitable for oral vaccination. *Int. J. Nanomed.* **2018**, *13*, 2973–2987. [CrossRef]
54. Mahony, D.; Cavallaro, A.S.; Stahr, F.; Mahony, T.J.; Qiao, S.Z.; Mitter, N. Mesoporous silica nanoparticles act as a self-adjuvant for ovalbumin model antigen in mice. *Small* **2013**, *9*, 3138–3146. [CrossRef]
55. Mody, K.T.; Popat, A.; Mahony, D.; Cavallaro, A.S.; Yu, C.; Mitter, N. Mesoporous silica nanoparticles as antigen carriers and adjuvants for vaccine delivery. *Nanoscale* **2013**, *5*, 5167–5179. [CrossRef]
56. Carvalho, L.V.; Ruiz Rde, C.; Scaramuzzi, K.; Marengo, E.B.; Matos, J.R.; Tambourgi, D.V.; Fantini, M.C.; Sant'Anna, O.A. Immunological parameters related to the adjuvant effect of the ordered mesoporous silica SBA-15. *Vaccine* **2010**, *28*, 7829–7836. [CrossRef]

57. Wang, T.; Jiang, H.; Zhao, Q.; Wang, S.; Zou, M.; Cheng, G. Enhanced mucosal and systemic immune responses obtained by porous silica nanoparticles used as an oral vaccine adjuvant: Effect of silica architecture on immunological properties. *Int. J. Pharm.* **2012**, *436*, 351–358. [CrossRef] [PubMed]
58. Toda, T.; Yoshino, S. Enhancement of ovalbumin-specific Th1, Th2, and Th17 immune responses by amorphous silica nanoparticles. *Int. J. Immunopathol. Pharm.* **2016**, *29*, 408–420. [CrossRef] [PubMed]
59. Rasmussen, M.K.; Kardjilov, N.; Oliveira, C.L.P.; Watts, B.; Villanova, J.; Botosso, V.F.; Sant'Anna, O.A.; Fantini, M.C.A.; Bordallo, H.N. 3D visualisation of hepatitis B vaccine in the oral delivery vehicle SBA-15. *Sci. Rep.* **2019**, *9*, 6106. [CrossRef]
60. Rasmussen, M.K.; Bordallo, H.N.; Bordenalli, M.A.; Akamatsu, M.A.; Trezena, A.G.; Tino-De-Franco, M.; Sant'Anna, O.A.; da Silva Martins, T.; de Souza Lopes, H.; Fantini, M.C.; et al. Assessing the efficiency of SBA-15 as a nanocarrier for diphtheria anatoxin. *Microporous Mesoporous Mater.* **2021**, *312*, 110763. [CrossRef]
61. Scaramuzzi, K.; Tanaka, G.D.; Neto, F.M.; Garcia, P.R.; Gabrili, J.J.; Oliveira, D.C.; Tambourgi, D.V.; Mussalem, J.S.; Paixao-Cavalcante, D.; D'Azeredo Orlando, M.T.; et al. Nanostructured SBA-15 silica: An effective protective vehicle to oral hepatitis B vaccine immunization. *Nanomedicine* **2016**, *12*, 2241–2250. [CrossRef]
62. Xu, C.; Lei, C.; Yu, C. Mesoporous silica nanoparticles for protein protection and delivery. *Front. Chem.* **2019**, *7*, 290. [CrossRef]
63. Zhou, S.; Zhong, Q.; Wang, Y.; Hua, P.; Zhong, W.; Huang, C.-B.; Yu, Z.-Q.; Ding, C.-D.; Liu, H.; Fu, J. Chemically engineered mesoporous silica nanoparticles-based intelligent delivery systems for theranostic applications in multiple cancerous/non-cancerous diseases. *Coord. Chem. Rev.* **2022**, *452*, 214309. [CrossRef]
64. Chen, W.; Zuo, H.; Li, B.; Duan, C.; Rolfe, B.; Zhang, B.; Mahony, T.J.; Xu, Z.P. Clay nanoparticles elicit long-term immune responses by forming biodegradable depots for sustained antigen stimulation. *Small* **2018**, *14*, e1704465. [CrossRef]
65. Govea-Alonso, D.O.; García-Soto, M.J.; Betancourt-Mendiola, L.; Padilla-Ortega, E.; Rosales-Mendoza, S.; González-Ortega, O. Nanoclays: Promising Materials for Vaccinology. *Vaccines* **2022**, *10*, 1549. [CrossRef]
66. Pilar, A.; Darder, M.; Wicklein, B.; Rytwo, G.; Ruiz-Hitzky, E. *Hybrid Organic–Inorganic Interfaces: Towards Advanced Functional Materials*; Delville, M.H., Taubert, A., Eds.; Wiley-VCH Verlag GmbH & Co. KgaA: Weinheim, Germany, 2017; pp. 1–84.
67. Wicklein, B.; Martín del Burgo, M.Á.; Yuste, M.; Darder, M.; Llavata, E.C.; Aranda, P.; Ortín, J.; del Real, G.; Ruiz-Hitzky, E. Lipid-based bio-nanohybrids for functional stabilisation of influenza vaccines. *Eur. J. Inorg. Chem.* **2012**, *32*, 5186–5191. [CrossRef]
68. Chauhan, N.; Tiwari, S.; Iype, T.; Jain, U. An overview of adjuvants utilized in prophylactic vaccine formulation as immunomodulators. *Expert. Rev. Vaccines* **2017**, *16*, 491–502. [CrossRef] [PubMed]
69. Li, J.; Ren, H.; Zhang, Y. Metal-based nano-vaccines for cancer immunotherapy. *Coord. Chem. Rev.* **2022**, *455*, 214345. [CrossRef]
70. Tadros, T.; Izquierdo, P.; Esquena, J.; Solans, C. Formation and stability of nano-emulsions. *Adv. Colloid. Int. Sci.* **2004**, *108–109*, 303–318. [CrossRef]
71. Bonferoni, M.C.; Rossi, S.; Sandri, G.; Ferrari, F.; Gavini, E.; Rassu, G.; Giunchedi, P. Nanoemulsions for "Nose-to-Brain" Drug Delivery. *Pharmaceutic* **2019**, *11*, 84. [CrossRef]
72. Đorđević, S.M.; Santrač, A.; Cekić, N.D.; Marković, B.D.; Divović, B.; Ilić, T.M.; Savić, M.; Savić, S.D. Parenteral nanoemulsions of risperidone for enhanced brain delivery in acute psychosis: Physicochemical and in vivo performances. *Int. J. Pharm.* **2017**, *533*, 421–430. [CrossRef]
73. Nikolic, I.; Mitsou, E.; Pantelic, I.; Randjelovic, D.; Markovic, B.; Papadimitriou, V.; Xenakis, A.; Lunter, D.; Zugic, A.; Savic, S. Microstructure and biopharmaceutical performances of curcumin-loaded low-energy nanoemulsions containing eucalyptol and pinene: Terpenes' role overcome penetration enhancement effect? *Eur. J. Pharm. Sci.* **2020**, *142*, 105135. [CrossRef]
74. Tayeb, H.H.; Sainsbury, F. Nanoemulsions in drug delivery: Formulation to medical application. *Nanomedicine* **2018**, *13*, 2507–2525. [CrossRef]
75. The European Medicines Agency, Committee for Medicinal Products for Human Use. Guideline on Adjuvants in Vaccines for Human Use. 2005. Available online: https://www.ema.europa.eu/en/adjuvants-vaccines-human-use (accessed on 11 March 2023).
76. O'Hagan, D.T.; Ott, G.S.; Nest, G.V.; Rappuoli, R.; Giudice, G.D. The history of MF59® adjuvant: A phoenix that arose from the ashes. *Expert. Rev. Vaccines* **2013**, *12*, 13–30. [CrossRef] [PubMed]
77. Leroux-Roels, G. Unmet needs in modern vaccinology: Adjuvants to improve the immune response. *Vaccine* **2010**, *28*, C25–C36. [CrossRef] [PubMed]
78. Del Giudice, G.; Rappuoli, R.; Didierlaurent, A.M. Correlates of adjuvanticity: A review on adjuvants in licensed vaccines. *Semin. Immunol.* **2018**, *39*, 14–21. [CrossRef]
79. Oyewumi, M.O.; Kumar, A.; Cui, Z. Nano-microparticles as immune adjuvants: Correlating particle sizes and the resultant immune responses. *Expert. Rev. Vaccines* **2010**, *9*, 1095–1107. [CrossRef]
80. Wallis, J.; Shenton, D.P.; Carlisle, R.C. Novel approaches for the design, delivery and administration of vaccine technologies. *Clin. Exp. Immunol.* **2019**, *196*, 189–204. [CrossRef]
81. Aucouturier, J.; Dupuis, L.; Ganne, V. Adjuvants designed for veterinary and human vaccines. *Vaccine* **2001**, *19*, 2666–2672. [CrossRef]
82. Klucker, M.F.; Dalençon, F.; Probeck, P.; Haensler, J. AF03, an alternative squalene emulsion-based vaccine adjuvant prepared by a phase inversion temperature method. *J. Pharm. Sci.* **2012**, *101*, 4490–4500. [CrossRef]
83. Vogel, F.R.; Caillet, C.; Kusters, I.C.; Haensler, J. Emulsion-based adjuvants for influenza vaccines. *Expert. Rev. Vaccine* **2009**, *8*, 483–492. [CrossRef]

84. Allison, A.C. Squalene and squalane emulsions as adjuvants. *Methods* **1999**, *19*, 87–93. [CrossRef]
85. Fox, C. Squalene emulsions for parenteral vaccine and drug delivery. *Molecules* **2009**, *14*, 3286–3312. [CrossRef]
86. O'Hagan, D.T.; Podda, A. *Influenza Vaccines for the Future*; Rappuoli, R., Del Giudice, G., Eds.; Birkhauser Basel: Siena, Italy, 2008; Chapter 10; pp. 221–244. [CrossRef]
87. Seubert, A.; Monaci, E.; Pizza, M.; O'Hagan, D.T.; Wack, A. The adjuvants aluminum hydroxide and MF59 induce monocyte and granulocyte chemoattractants and enhance monocyte differentiation toward dendritic cells. *J. Immunol.* **2008**, *180*, 5402–5412. [CrossRef] [PubMed]
88. Ozao-Choy, J.; Lee, D.J.; Faries, M.B. Melanoma vaccines: Mixed past, promising future. *Surg. Clin.* **2014**, *94*, 1017–1030. [CrossRef]
89. Sosman, J.A.; Sondak, V.K. Melacine®: An allogeneic melanoma tumor cell lysate vaccine. *Expert. Rev. Vaccines* **2003**, *2*, 353–368. [CrossRef] [PubMed]
90. Morel, S.; Didierlaurent, A.; Bourguignon, P.; Delhaye, S.; Baras, B.; Jacob, V.; Planty, C.; Elouahabi, A.; Harvengt, P.; Carlsen, H.; et al. Adjuvant System AS03 containing α-tocopherol modulates innate immune response and leads to improved adaptive immunity. *Vaccine* **2011**, *29*, 2461–2473. [CrossRef] [PubMed]
91. Dye, T.J.; Simakajornboon, N. Narcolepsy in Children: Sleep disorders in children, A rapidly evolving field seeking consensus. *Pediatr. Pulmonol.* **2022**, *57*, 1952–1962. [CrossRef] [PubMed]
92. Aucouturier, J.; Dupuis, L.; Deville, S.; Ascarateil, S.; Ganne, V. Montanide ISA 720 and 51: A new generation of water in oil emulsions as adjuvants for human vaccines. *Expert. Rev. Vaccines* **2002**, *1*, 111–118. [CrossRef]
93. Mirzaee, M.; Haghighat, S.; Golkaran, B.; Asgarhalvaei, F.; Mirzaee, R.; Taghizadeh, M.; Savoji, M.A.; Esfandiari, B.; Mahdavi, M. Montanide ISA-720 and Naloxone in HBsAg Vaccine Formulation: Cytokine Profiling and Monitoring of Long-Lasting Humoral Immune Responses. *Biomed. Env. Sci.* **2022**, *35*, 792–803. [CrossRef]
94. Properties of MontanideTM Range of Ingredients. Available online: https://www.seppic.com/en/montanide_pharma (accessed on 15 March 2023).
95. Bodewes, R.; Kreijtz, J.H.C.M.; Van Amerongen, G.; Geelhoed-Mieras, M.M.; Verburgh, R.J.; Heldens, J.G.; Bendwell, J.; Van Der Brand, J.M.; Kuiken, T.; Van Baalen, C.A.; et al. A single immunization with CoVaccine HT-adjuvanted H5N1 influenza virus vaccine induces protective cellular and humoral immune responses in ferrets. *J. Virol.* **2010**, *84*, 7943–7952. [CrossRef] [PubMed]
96. Lai, C.Y.; To, A.; Ann, S.; Wong, T.; Lieberman, M.M.; Clements, D.E.; Senda, J.T.; Ball, A.H.; Pessaint, L.; Andersen, H.; et al. Recombinant protein subunit SARS-CoV-2 vaccines formulated with CoVaccine HT adjuvant induce broad, Th1 biased, humoral and cellular immune responses in mice [published online ahead of print, 2021 Nov 5]. *Vaccine X* **2021**, *9*, 100126. [CrossRef]
97. Anamur, C.; Winter, G.; Engert, J. Stability of collapse lyophilized influenza vaccine formulations. *Int. J. Pharm.* **2015**, *483*, 131–141. [CrossRef]
98. Do Vale Morais, A.R.; Do Nascimento Alencar, É.; Júnior, F.H.X.; De Oliveira, C.M.; Marcelino, H.R.; Barratt, G.; Fessi, H.; Do Egito, E.S.T.; Elaissari, A. Freeze-drying of emulsified systems: A review. *Int. J. Pharm.* **2016**, *503*, 102–114. [CrossRef]
99. Yuan, L.; Gao, X.D.; Xia, Y. Optimising the oil phases of aluminium hydrogel-stabilised emulsions for stable, safe and efficient vaccine adjuvant. *Front. Chem. Sci. Eng.* **2022**, *16*, 973–984. [CrossRef]
100. Schultze, V.D.; Agosto, V.; Wack, A.; Novicki, D.; Zorn, J.; Hennig, R. Safety of MF59 adjuvant. *Vaccine* **2008**, *26*, 3209–3222. [CrossRef]
101. Fox, C.B.; Baldwin, S.L.; Duthie, M.S.; Reed, S.G.; Vedvick, T.S. Immunomodulatory and physical effects of oil composition in vaccine adjuvant emulsions. *Vaccine* **2011**, *29*, 9563–9572. [CrossRef]
102. Assis, B.R.D.; da Silva, C.D.; Santiago, M.G.; Ferreira, L.A.M.; Goulart, G.A.C. Nanotechnology in adjuvants and vaccine development: What should we know? *Nanomedicine* **2021**, *16*, 2565–2568. [CrossRef]
103. O'Hagan, D.T.; van der Most, R.; Lodaya, R.N.; Coccia, M.; Lofano, G. "World in motion"—Emulsion adjuvants rising to meet the pandemic challenges. *NPJ Vaccines* **2021**, *6*, 158. [CrossRef]
104. Thanh Le, T.; Andreadakis, Z.; Kumar, A.; Gómez Román, R.; Tollefsen, S.; Saville, M.; Mayhew, S. The COVID-19 vaccine development landscape. *Nat. Rev. Drug. Discov.* **2020**, *19*, 305–306. [CrossRef] [PubMed]
105. European Medicines Agency, VidPrevtyn Beta Product Information File. 2022. Available online: https://www.ema.europa.eu/en/medicines/human/EPAR/vidprevtyn-beta#product-information-section (accessed on 4 April 2023).
106. Glaxo Smith Kline. Press. Release. 2022. Available online: https://www.gsk.com/en-gb/media/press-releases/sanofi-and-gsk-s-next-generation-covid-19-booster-vaccine-vidprevtyn-beta-approved-by-ec/ (accessed on 4 April 2023).
107. Schoenmaker, L.; Witzigmann, D.; Kulkarni, J.A.; Verbeke, R.; Kersten, G.; Jiskoot, W.; Crommelin, D.J.A. mRNA-lipid nanoparticle COVID-19 vaccines: Structure and stability. *Int. J. Pharm.* **2021**, *601*, 120586. [CrossRef] [PubMed]
108. Alameh, M.G.; Tombácz, I.; Bettini, E.; Lederer, K.; Sittplangkoon, C.; Wilmore, J.R.; Gaudette, B.T.; Soliman, O.Y.; Pine, M.; Hicks, P.; et al. Lipid nanoparticles enhance the efficacy of mRNA and protein subunit vaccines by inducing robust T follicular helper cell and humoral responses. *Immunity* **2021**, *54*, 2877–2892.e7. [CrossRef] [PubMed]
109. Kobiyama, K.; Ishii, K.J. Making innate sense of mRNA vaccine adjuvanticity. *Nat. Immunol.* **2022**, *23*, 474–476. [CrossRef] [PubMed]
110. Iyer, V.; Cayatte, C.; Guzman, B.; Schneider-Ohrum, K.; Matuszak, R.; Snell, A.; Rajani, G.M.; McCarthy, M.P.; Muralidhara, B. Impact of formulation and particle size on stability and immunogenicity of oil-in-water emulsion adjuvants. *Hum. Vaccin. Immunother.* **2015**, *11*, 1853–1864. [CrossRef]

111. Niikura, K.; Matsunaga, T.; Suzuki, T.; Kobayashi, S.; Yamaguchi, H.; Orba, Y.; Kawaguchi, A.; Hasegawa, H.; Kajino, K.; Ninomiya, T.; et al. Gold nanoparticles as a vaccine platform: Influence of size and shape on immunological responses in vitro and in vivo. *ACS Nano* **2013**, *7*, 3926–3938. [CrossRef]
112. Shah, R.R.; Taccone, M.; Monaci, E.; Brito, L.A.; Bonci, A.; O'Hagan, D.T.; Amiji, M.M.; Seubert, A. The droplet size of emulsion adjuvants has significant impact on their potency, due to differences in immune cell-recruitment and-activation. *Sci. Rep.* **2019**, *9*, 11520. [CrossRef] [PubMed]
113. Shah, R.R.; O'Hagan, D.T.; Amiji, M.M.; Brito, L.A. The impact of size on particulate vaccine adjuvants. *Nanomedicine* **2014**, *9*, 2671–2681. [CrossRef]
114. Somiya, M.; Kuroda, S. Development of a virus-mimicking nanocarrier for drug delivery systems: The bio-nanocapsule. *Adv. Drug. Deliv. Rev.* **2015**, *95*, 77–89. [CrossRef]
115. Chung, Y.H.; Cai, H.; Steinmetz, N.F. Viral nanoparticles for drug delivery, imaging, immunotherapy, and theranostic applications. *Adv. Drug. Deliv. Rev.* **2020**, *156*, 214–235. [CrossRef]
116. Fan, J.; Jin, S.; Gilmartin, L.; Toth, I.; Hussein, W.M.; Stephenson, R.J. Advances in Infectious Disease Vaccine Adjuvants. *Vaccines* **2022**, *10*, 1120. [CrossRef]
117. Petkar, K.C.; Patil, S.M.; Chavhan, S.S.; Kaneko, K.; Sawant, K.K.; Kunda, N.K.; Saleem, I.Y. An Overview of Nanocarrier-Based Adjuvants for Vaccine Delivery. *Pharmaceutics* **2021**, *27*, 455. [CrossRef]
118. Travieso, T.; Li, J.; Mahesh, S.; Mello, J.D.F.R.E.; Blasi, M. The use of viral vectors in vaccine development. *NPJ Vaccines* **2022**, *7*, 75. [CrossRef]
119. Luiz, M.T.; Dutra, J.A.P.; Tofani, L.B.; de Araújo, J.T.C.; Di Filippo, L.D.; Marchetti, J.M.; Chorilli, M. Targeted Liposomes: A Nonviral Gene Delivery System for Cancer Therapy. *Pharmaceutics* **2022**, *14*, 821. [CrossRef] [PubMed]
120. Plummer, E.M.; Manchester, M. Viral nanoparticles and virus-like particles: Platforms for contemporary vaccine design. *Wiley Interdiscip. Rev. Nanomed. Nanobiotechnol* **2011**, *3*, 174–196. [CrossRef]
121. Suleman, M.; Galea, S.; Gavard, F.; Merillon, N.; Klonjkowski, B.; Tartour, E.; Richardson, J. Antigen encoded by vaccine vectors derived from human adenovirus serotype 5 is preferentially presented to CD8+ T lymphocytes by the CD8α+ dendritic cell subset. *Vaccine* **2011**, *29*, 5892–5903. [CrossRef] [PubMed]
122. Buchbinder, S.P.; Mehrotra, D.V.; Duerr, A.; Fitzgerald, D.W.; Mogg, R.; Li, D.; Gilbert, P.B.; Lama, J.R.; Marmor, M.; Del Rio, C.; et al. Step Study Protocol Team. Efficacy assessment of a cell-mediated immunity HIV-1 vaccine (the Step Study): A double-blind, randomised, placebo-controlled, test-of-concept trial. *Lancet* **2008**, *372*, 1881–1893. [CrossRef] [PubMed]
123. Wu, S.; Zhong, G.; Zhang, J.; Shuai, L.; Zhang, Z.; Wen, Z.; Wang, B.; Zhao, Z.; Song, X.; Chen, Y.; et al. A single dose of an adenovirus-vectored vaccine provides protection against SARS-CoV-2 challenge. *Nat. Commun.* **2020**, *11*, 4081. [CrossRef] [PubMed]
124. Zhu, F.C.; Guan, X.H.; Li, Y.H.; Huang, J.Y.; Jiang, T.; Hou, L.H.; Li, J.X.; Yang, B.F.; Wang, L.; Wang, W.J.; et al. Immunogenicity and safety of a recombinant adenovirus type-5-vectored COVID-19 vaccine in healthy adults aged 18 years or older: A randomised, double-blind, placebo-controlled, phase 2 trial. *Lancet* **2020**, *396*, 479–488. [CrossRef]
125. Zhu, F.-C.; Li, Y.-H.; Guan, X.-H.; Hou, L.-H.; Wang, W.-J.; Li, J.-X.; Wu, S.-P.; Wang, B.-S.; Wang, Z.; Wang, L.; et al. Safety, tolerability, and immunogenicity of a recombinant adenovirus type-5 vectored COVID-19 vaccine: A dose-escalation, open-label, non-randomised, first-in-human trial. *Lancet* **2020**, *395*, 1845–1854. [CrossRef] [PubMed]
126. Milligan, I.D.; Gibani, M.M.; Sewell, R.; Clutterbuck, E.A.; Campbell, D.; Plested, E.; Nuthall, E.; Voysey, M.; Silva-Reyes, L.; McElrath, M.J.; et al. Safety and Immunogenicity of Novel Adenovirus Type 26- and Modified Vaccinia Ankara-Vectored Ebola Vaccines: A Randomized Clinical Trial. *JAMA* **2016**, *315*, 1610–1623. [CrossRef] [PubMed]
127. Sadoff, J.; Le Gars, M.; Shukarev, G.; Heerwegh, D.; Truyers, C.; de Groot, A.M.; Stoop, J.; Tete, S.; Van Damme, W.; Leroux-Roels, I.; et al. Interim Results of a Phase 1-2a Trial of Ad26.COV2.S Covid-19 Vaccine. *N. Engl. J. Med.* **2021**, *384*, 1824–1835. [CrossRef] [PubMed]
128. Voysey, M.; Clemens, S.A.C.; Madhi, S.A.; Weckx, L.Y.; Folegatti, P.M.; Aley, P.K.; Angus, B.; Baillie, V.L.; Barnabas, S.L.; Bhorat, Q.E.; et al. Oxford COVID Vaccine Trial Group. Safety and efficacy of the ChAdOx1 nCoV-19 vaccine (AZD1222) against SARS-CoV-2: An interim analysis of four randomised controlled trials in Brazil, South Africa, and the UK. *Lancet* **2021**, *397*, 99–111. [CrossRef]
129. Baker, A.T.; Boyd, R.J.; Sarkar, D.; Teijeira-Crespo, A.; Chan, C.K.; Bates, E.; Waraich, K.; Vant, J.; Wilson, E.; Truong, C.D. ChAdOx1 interacts with CAR and PF4 with implications for thrombosis with thrombocytopenia syndrome. *Sci. Adv.* **2021**, *7*, eabl8213. [CrossRef]
130. Manning, W.C.; Paliard, X.; Zhou, S.; Pat Bland, M.; Lee, A.Y.; Hong, K.; Walker, C.M.; Escobedo, J.A.; Dwarki, V. Genetic immunization with adeno-associated virus vectors expressing herpes simplex virus type 2 glycoproteins B and D. *J. Virol.* **1997**, *71*, 7960–7962. [CrossRef]
131. Kuck, D.; Lau, T.; Leuchs, B.; Kern, A.; Müller, M.; Gissmann, L.; Kleinschmidt, J.A. Intranasal vaccination with recombinant adeno-associated virus type 5 against human papillomavirus type 16 L1. *J. Virol.* **2006**, *80*, 2621–2630. [CrossRef]
132. Ploquin, A.; Szécsi, J.; Mathieu, C.; Guillaume, V.; Barateau, V.; Ong, K.C.; Wong, K.T.; Cosset, F.L.; Horvat, B.; Salvetti, A. Protection against henipavirus infection by use of recombinant adeno-associated virus-vector vaccines. *J. Infect. Dis.* **2013**, *207*, 469–478. [CrossRef]

133. Zhan, W.; Muhuri, M.; Tai, P.W.L.; Gao, G. Vectored Immunotherapeutics for Infectious Diseases: Can rAAVs Be The Game Changers for Fighting Transmissible Pathogens? *Front. Immunol.* **2021**, *2021*, 673699. [CrossRef] [PubMed]
134. Martinez-Navio, J.M.; Fuchs, S.P.; Pantry, S.N.; Lauer, W.A.; Duggan, N.N.; Keele, B.F.; Rakasz, E.G.; Gao, G.; Lifson, J.D.; Desrosiers, R.C. Adeno-Associated Virus Delivery of Anti-HIV Monoclonal Antibodies Can Drive Long-Term Virologic Suppression. *Immunity* **2019**, *50*, 567–575. [CrossRef] [PubMed]
135. Fathi, A.; Dahlke, C.; Addo, M.M. Recombinant vesicular stomatitis virus vector vaccines for WHO blueprint priority pathogens. *Hum. Vaccines Immunother.* **2019**, *15*, 2269–2285. [CrossRef] [PubMed]
136. Khurana, S.; Fuentes, S.; Coyle, E.M.; Ravichandran, S.; Davey, R.T., Jr.; Beigel, J.H. Human antibody repertoire after VSV-Ebola vaccination identifies novel targets and virus-neutralizing IgM antibodies. *Nat. Med.* **2016**, *22*, 1439–1447. [CrossRef]
137. Case, J.B.; Rothlauf, P.W.; Chen, R.E.; Kafai, N.M.; Fox, J.M.; Smith, B.K.; Shrihari, S.; McCune, B.T.; Harvey, I.B.; Keeler, S.P.; et al. Replication-Competent Vesicular Stomatitis Virus Vaccine Vector Protects against SARS-CoV-2-Mediated Pathogenesis in Mice. *Cell Host Microbe* **2020**, *28*, 465–474. [CrossRef]
138. Ku, M.W.; Charneau, P.; Majlessi, L. Use of lentiviral vectors in vaccination. *Expert. Rev. Vaccines* **2021**, *20*, 1571–1586. [CrossRef]
139. Ku, M.W.; Bourgine, M.; Authié, P.; Lopez, J.; Nemirov, K.; Moncoq, F.; Noirat, A.; Vesin, B.; Nevo, F.; Blanc, C.; et al. Intranasal vaccination with a lentiviral vector protects against SARS-CoV-2 in preclinical animal models. *Cell Host Microbe* **2021**, *29*, 236–249. [CrossRef]
140. Blasi, M.; Wescott, E.C.; Baker, E.J.; Mildenberg, B.; LaBranche, C.; Rountree, W.; Haynes, B.F.; Saunders, K.O.; Moody, M.A.; Negri, D.; et al. Therapeutic vaccination with IDLV-SIV-Gag results in durable viremia control in chronically SHIV-infected macaques. *NPJ Vaccines* **2020**, *5*, 36. [CrossRef]
141. Coutant, F.; Sanchez David, R.Y.; Félix, T.; Boulay, A.; Caleechurn, L.; Souque, P.; Thouvenot, C.; Bourgouin, C.; Beignon, A.S.; Charneau, P. A nonintegrative lentiviral vector-based vaccine provides long-term sterile protection against malaria. *PLoS ONE* **2012**, *7*, e48644. [CrossRef]
142. Ku, M.W.; Anna, F.; Souque, P.; Petres, S.; Prot, M.; Simon-Loriere, E.; Charneau, P.; Bourgine, M.A. Single Dose of NILV-Based Vaccine Provides Rapid and Durable Protection against Zika Virus. *Molecules* **2020**, *28*, 1772–1782. [CrossRef]
143. Yang, C.; Huang, K. Clinical applications of virus-like particles: Opportunities and challenges. *Curr. Protein Pept. Sci.* **2019**, *20*, 488–489. [CrossRef] [PubMed]
144. Nooraei, S.; Bahrulolum, H.; Hoseini, Z.S.; Katalani, C.; Hajizade, A.; Easton, A.J.; Ahmadian, G. Virus-like particles: Preparation, immunogenicity and their roles as nanovaccines and drug nanocarriers. *J. Nanobiotechnology* **2021**, *19*, 59. [CrossRef] [PubMed]
145. Mejía-Méndez, J.L.; Vazquez-Duhalt, R.; Hernández, L.R.; Sánchez-Arreola, E.; Bach, H. Virus-like Particles: Fundamentals and Biomedical Applications. *Int. J. Mol. Sci.* **2022**, *23*, 8579. [CrossRef] [PubMed]
146. Qian, C.; Liu, X.; Xu, Q.; Wang, Z.; Chen, J.; Li, T.; Zheng, Q.; Yu, H.; Gu, Y.; Li, S.; et al. Recent Progress on the Versatility of Virus-Like Particles. *Vaccines* **2020**, *8*, 139. [CrossRef] [PubMed]
147. Bayer, M.E.; Blumberg, B.S.; Werner, B. Particles associated with Australia antigen in the sera of patients with leukaemia, Down's Syndrome and hepatitis. *Nature* **1968**, *218*, 1057–1059. [CrossRef]
148. Thomas, D.L.; Astemborski, J.; Rai, R.M.; Anania, F.A.; Schaeffer, M.; Galai, N.; Nolt, K.; Nelson, K.E.; Strathdee, S.A.; Johnson, L. The natural history of hepatitis C virus infection: Host, viral, and environmental factors. *JAMA* **2000**, *284*, 450–456. [CrossRef]
149. Metz, S.W.; Thomas, A.; White, L. Dengue virus-like particles mimic the antigenic properties of the infectious dengue virus envelope. *Virol. J.* **2018**, *15*, 60. [CrossRef]
150. Perotti, M.; Perez, L. Virus-Like Particles and Nanoparticles for Vaccine Development against HCMV. *Viruses* **2019**, *12*, 35. [CrossRef]
151. Mohsen, M.O.; Gomes, A.C.; Vogel, M. Interaction of Viral Capsid-Derived Virus-Like Particles (VLPs) with the Innate Immune System. *Vaccines* **2018**, *6*, 37. [CrossRef]
152. Anzaghe, M.; Schülke, S.; Scheurer, S. Virus-like particles as carrier systems to enhance immunomodulation in allergen immunotherapy. *Curr. Allergy Asthma Rep.* **2018**, *18*, 71. [CrossRef]
153. Zepeda-Cervantes, J.; Ramírez-Jarquín, J.O.; Vaca, L. Interaction Between Virus-Like Particles (VLPs) and Pattern Recognition Receptors (PRRs) From Dendritic Cells (DCs): Toward Better Engineering of VLPs. *Front. Immunol.* **2020**, *11*, 1100. [CrossRef] [PubMed]
154. Tariq, H.; Batool, S.; Asif, S.; Ali, M.; Abbasi, B.H. Virus-Like Particles: Revolutionary Platforms for Developing Vaccines Against Emerging Infectious Diseases. *Front. Microbiol.* **2022**, *12*, 790121. [CrossRef]
155. Dai, S.; Wang, H.; Deng, F. Advances and challenges in enveloped virus-like particle (VLP)-based vaccines. *J. Immunol. Sci.* **2018**, *2*, 36–41.
156. Mohsen, M.O.; Zha, L.; Cabral-Miranda, G.; Bachmann, M.F. Major findings and recent advances in virus–like particle (VLP)-based vaccines. *Semin. Immunol.* **2017**, *34*, 123–132. [CrossRef] [PubMed]
157. Peabody, D.S.; Peabody, J.; Bradfute, S.B.; Chackerian, B. RNA Phage VLP-Based Vaccine Platforms. *Pharmaceuticals* **2021**, *4*, 764. [CrossRef]
158. Tars, K. ssRNA Phages: Life Cycle, Structure and Applications. *Biocommunication Phages* **2020**, *30*, 261–292. [CrossRef]
159. Liekniņa, I.; Černova, D.; Rūmnieks, J.; Tārs, K. Novel ssRNA phage VLP platform for displaying foreign epitopes by genetic fusion. *Vaccine* **2020**, *27*, 6019–6026. [CrossRef]
160. Palma, M. Aspects of Phage-Based Vaccines for Protein and Epitope Immunization. *Vaccines* **2023**, *11*, 436. [CrossRef]

161. Tumban, E.; Peabody, J.; Peabody, D.S.; Chackerian, B. A pan-HPV vaccine based on bacteriophage PP7 VLPs displaying broadly cross-neutralizing epitopes from the HPV minor capsid protein, L2. *PLoS ONE* **2011**, *6*, e23310. [CrossRef]
162. Guo, Y.; Guo, R.; Ma, Y.; Chang, W.; Ming, S.; Yang, G.; Guo, Y. Chimeric Virus-like Particles of Universal Antigen Epitopes of Coronavirus and Phage Qβ Coat Protein Trigger the Production of Neutralizing Antibodies. *Curr. Top. Med. Chem.* **2021**, *5*, 1235–1250. [CrossRef]
163. Kamel, M.; El-Sayed, A.; Castañeda Vazquez, H. Foot-and-mouth disease vaccines: Recent updates and future perspectives. *Arch. Virol.* **2019**, *164*, 1501–1513. [CrossRef] [PubMed]
164. Schmitz, N.; Beerli, R.R.; Bauer, M.; Jegerlehner, A.; Dietmeier, K.; Maudrich, M.; Pumpens, P.; Saudan, P.; Bachmann, M.F. Universal vaccine against influenza virus: Linking TLR signaling to anti-viral protection. *Eur. J. Immunol.* **2012**, *42*, 863–869. [CrossRef] [PubMed]
165. Liu, X.; Chang, X.; Rothen, D.; Derveni, M.; Krenger, P.; Roongta, S.; Wright, E.; Vogel, M.; Tars, K.; Mohsen, M.O.; et al. AP205 VLPs Based on Dimerized Capsid Proteins Accommodate RBM Domain of SARS-CoV-2 and Serve as an Attractive Vaccine Candidate. *Vaccines* **2021**, *19*, 403. [CrossRef] [PubMed]
166. Klimek, L.; Kündig, T.; Kramer, M.F.; Guethoff, S.; Jensen-Jarolim, E.; Schmidt-Weber, C.B.; Palomares, O.; Mohsen, M.O.; Jakob, T.; Bachmann, M. Virus-like particles (VLP) in prophylaxis and immunotherapy of allergic diseases. *Allergo J. Int.* **2018**, *27*, 245–255. [CrossRef] [PubMed]
167. Asadi, K.; Gholami, A. Virosome-based nanovaccines; a promising bioinspiration and biomimetic approach for preventing viral diseases: A review. *Int. J. Biol. Macromol.* **2021**, *182*, 648–658. [CrossRef] [PubMed]
168. Huckriede, A.; Bungener, L.; Stegmann, T.; Daemen, T.; Medema, J.; Palache, A.M.; Wilschut, J. The virosome concept for influenza vaccines. *Vaccine* **2005**, *23*, S26–S38. [CrossRef]
169. Kedmi, R.; Ben-Arie, N.; Peer, D. The systemic toxicity of positively charged lipid nanoparticles and the role of Toll-like receptor 4 in immune activation. *Biomaterials* **2010**, *31*, 6867–6875. [CrossRef]
170. Nidetz, N.F.; McGee, M.C.; Tse, L.V.; Li, C.; Cong, L.; Li, Y.; Huang, W. Adeno-associated viral vector-mediated immune responses: Understanding barriers to gene delivery. *Pharm.* **2020**, *207*, 107453. [CrossRef]
171. Pinski, A.N.; Messaoudi, I. To B or Not to B: Mechanisms of Protection Conferred by rVSV-EBOV-GP and the Roles of Innate and Adaptive Immunity. *Microorganisms* **2020**, *25*, 1473. [CrossRef]
172. Nemirov, K.; Bourgine, M.; Anna, F.; Wei, Y.; Charneau, P.; Majlessi, L. Lentiviral Vectors as a Vaccine Platform against Infectious Diseases. *Pharmaceutics* **2023**, *15*, 846. [CrossRef]

vaccines

MDPI

Article

A Thermal-Stable Protein Nanoparticle That Stimulates Long Lasting Humoral Immune Response

Ten-Tsao Wong [1], Gunn-Guang Liou [2,3] and Ming-Chung Kan [4,*]

[1] Department of Marine Biotechnology & Institute of Marine and Environmental Technology, University of Maryland Baltimore County, Baltiomre, MD 21202, USA
[2] Institute of Biological Chemistry, Academia Sinica, Taipei 11529, Taiwan
[3] Office of Research and Development, College of Medicine, National Taiwan University, Taipei 10051, Taiwan
[4] Vaxsia Biomedical Inc., Taipei 11503, Taiwan
* Correspondence: mingchung.kan@vaxsia.com

Abstract: A thermally stable vaccine platform is considered the missing piece of vaccine technology. In this article, we reported the creation of a novel protein nanoparticle and assessed its ability to withstand extended high temperature incubation while stimulating a long-lasting humoral immune response. This protein nanoparticle was assembled from a fusion protein composed of an amphipathic helical peptide derived from the M2 protein of the H5N1 influenza virus (AH3) and a superfolder green fluorescent protein (sfGFP). Its proposed structure was modeled according to transmission electronic microscope (TEM) images of protein nanoparticles. From this proposed protein model, we created a mutant with two gain-of-function mutations that work synergistically on particle stability. A protein nanoparticle assembled from this gain-of-function mutant is able to remove a hydrophobic patch from its surface. This gain-of-function mutant also contributes to the higher thermostability of protein nanoparticles and stimulates a long lasting humoral immune response after a single immunization. This assembled nanoparticle showed increasing particle stability at higher temperatures and salt concentrations. This novel protein nanoparticle may serve as a thermally-stable platform for vaccine development.

Keywords: nanoparticle; thermal stable; self-assembled protein nanoparticle

1. Introduction

The thermostability of a vaccine is considered an important characteristic that is essential to fulfilling the global vaccination initiative. Current vaccine logistics practice demands a continuous cold-chain environment from the manufacturer to remote clinics to maintain vaccine efficacy. It is estimated that there are 775 million people living outside of the electrical grid [1] and beyond the reach of vaccine cold chains and vaccination, which leads to millions of children dying of vaccine-preventable diseases each year [2]. To fully cover the population beyond the electrical grid, a vaccine has to be able to withstand temperatures up to 40 °C for two months, i.e., the duration that the vaccine is stored in a health post [3].

Subunit vaccines are a safe alternative to traditional inactivated or attenuated vaccines, but their efficacy is often hindered by the low antigenicity of recombinant proteins. Different approaches have been utilized to resolve this issue; among them, virus-like particles (VLP) and self-assembled protein nanoparticles (SAPN) are considered the best platforms for subunit vaccine development [4]. The size of VLPs ranges between 20 and 200 nm, which facilitates both efficient draining to lymph node and also uptake by antigen-presenting cells such as dendritic cells and macrophages [5]. The other benefit of a VLP-based vaccine is the induction of B cell receptor clustering when repetitive antigen is presented to B cell, a function that can activate antibody class-switching and somatic hypermutation in a T-cell-dependent mechanism [6]. Not only can the virus-like particles be directly used for vaccines, but heterologous antigens can also be applied to the particle surface through genetic fusion [7]. A universal flu vaccine candidate, ectopic M2 domain (M2e), has been genetically fused with Hepatitis B core antigen (HBc) and assembled into a nanoparticle that provides full protection to heterologous flu strains [8]. VLP purified from bacterially expressed capsid protein often encloses bacterial RNA that is required for particle stability [9]. Although the enclosed bacterial RNA could potentially act as an adjuvant during immunization [10], the labile nature of RNA also contributes to the low stability of VLP. A SAPN that is assembled strictly from protein components may provide better overall vaccine stability. The SAPN based on iron-transporting protein, ferritin, has been chosen as vaccine carrier for being constituted strictly by protein subunit [11,12]. It has good thermostability when exposed to high temperatures but long-term storage data for ferritin-based vaccines have not been reported yet. Recent advancements in the computer modeling of protein structures have enabled the de novo design and assembly of protein nanoparticle from protein subunits that are known to form either pentamer or trimer [13]. When fused with an RNA-binding motif, these protein nanoparticles are able to enclose their own RNA genomes and form artificial nucleocapsid that can be used for in vitro or in vivo selection for variants with higher stability [14].

The green fluorescent protein belongs to a fluorescent protein family which is structurally conserved and emits fluorescent light from a chromophore when excited by photons of shorter wavelength [15]. The shared features of fluorescent proteins include a sturdy barrel-shaped structure constituted by 11 β-sheets and an enclosed chromophore that emits fluorescent light when excited [16]. The function of the barrel shell is to provide a well-organized chemical environment to ensure the maturation of the chromophore and protect it from hostile elements [17]. Therefore, it is conceivable that the protein sequences among fluorescent protein family members in the barrel shell are highly variable and fluorescent proteins possess desirable biophysical properties that can be selected using directed evolution [15,18–21]. The applications of fluorescent proteins have been expanded into multiple areas beyond live imaging, which includes serving as biological sensors [22,23] or detectors for protein–protein interaction or protein folding [24,25].

Amphipathic α-helical peptide (AHP) forms hydrophilic and hydrophobic faces when folded and is often identified in proteins related to phospholipid membrane interaction. The N-terminal amphipathic helical peptide is required for membrane anchorage of the Hepatitis C virus NS3 protein and the protease function of the NS3/NS4a complex [26,27]. Several anti-microbial peptides also possess amphipathic properties and function by forming membrane pores or causing membrane disruption [28]. The amphipathic α-helical peptide of type-A influenza virus M2 protein is required for M2 protein anchorage and induces membrane curvature required for virus budding [29,30]. This report described the identification and design of a protein nanoparticle assembled from a fusion protein composing an amphipathic α-helical peptide (AH3) from the M2 protein of type-A influenza strain H5N1 and an sfGFP. We built a protein structure model based on TEM images using Ascalaph Designer. The model suggests AH3 serves as a polymerization module, and it induces fusion protein assembly into a corn-on-a-cob structure. The sturdy barrel structure of sfGFP was explored as an antigen-presenting module for peptide antigen presentation. From the protein model, we generated a gain-of-function mutant that provides higher stability to the protein nanoparticle. We showed biophysical and immunological evidence suggesting that the protein nanoparticle assembled from a fusion protein containing a gain-of-function mutant is able to withstand extended high-temperature exposure and stimulates a long-lasting antibody response to an inserted antigen in a single immunization without adjuvant.

2. Materials and Methods

2.1. Peptide Information, Expression and Purification of Recombinant Protein

The AH3 peptide sequence is MDRLFFKCIYRRLKYGLKRG. The sequence of peptide for anti-hM2e antibody titer ELISA is SLLTEVETPIRNEWGSRSNGSSDC. The sequence of the peptide in the insertion site of sfGFP is SLLTEVETPIRNEWGSRSNGSSDSSGGSLL-TEVETPIRNEWGSRSNGSSD. The protein expression vectors encoding target proteins were transformed into E coli competent cells using a heat-shock transformation. Colonies of transformed bacteria with the desired vector were scraped from the plate and inoculated in LB culture with an antibiotic. The bacterial culture was then grown exponentially to OD600 between 0.5 and 0.7 before cooling down on an ice bath and protein expression was induced by 1 mM IPTG at 20 °C for 14–16 h with 250 rpm shaking. After protein induction, bacteria were harvested by 5000 rpm centrifugation in a Sorvall SLC3000 rotor for 10 min. Bacterial pellet from 400 mL LB culture was re-suspended in 40 mL lysis buffer containing 10 mM Imidazole in 1XGF buffer (20 mM $Na(PO_4)$ pH7.4 and 300 mM NaCl) for sonication. For Ni–NTA resin purification, 10 mM Imidazol was added in 1XGF buffer for bacteria lysis (Lysis buffer), 20 mM Imidazol was added in 1XGF buffer for column wash (Wash buffer), and 500 mM Imidazol was added in 1XGF buffer for protein elution (Elution buffer). Bacteria were lysed using an ultrasonic sonicator (Misonix 3000) at 10 s on/20 s off cycles for 5 min at output level 5 in icy water. Insoluble cell debris was removed by centrifugation at 10,000 rpm for 10 min using a Sorval SS34 rotor at 4 °C. Soluble fraction containing the target protein was then used for purification by Ni–NTA resin as described in the user manual or was used for sedimentation ultracentrifugation. Purified proteins were stored in elution buffer in a cold room before further processing. Before immunization or thermostability testing, protein buffer was changed into 0.5× GF buffer using Sephadex-25 resin (GE, PD-10 and NAP-5).

2.2. Biophysical Analysis of Protein Oligomerization and Hydrophobicity

The oligomerization state of AH3–GFP was first analyzed by protein concentration tube Vivaspin 2 with an MWCO of 100, 300, or 1000 kDa from Sartorius. Protein samples were centrifuged in Vivaspin 2 at 1000× *g* for 20 min. Filtrate was analyzed by SDS-PAGE and then coomassie blue staining was performed for analysis. TEM images of the AH3–GFP protein complex was obtained after negative staining with uranyl acetate. Images were taken using Tecnai G2 Spirit Twin. For AH3–sfGFP–2xhM2e and its variants, the purified fusion protein was first crosslinked with Sulfo-SMCC (sulfosuccinimidyl 4-[N-maleimidomethyl]cyclohexane-1-carboxylate) at 4 µg/mL for 30 min before been processed using negative staining protocol. A sucrose density gradient was used to analyze AH3–sfGFP–2xhM2e-related proteins' oligomerization status and hydrophobicity. In a 13 mL polypropylene tube (Beckman cat#14287), the bottom was layered first with 1 mL 65% (*w*/*v*) sucrose solution and then was topped with 2 mL 45% (*w*/*v*) sucrose solution and then followed by 7 mL 15% (*w*/*v*) sucrose solution. Sudan III stock solution was prepared as 0.5% in isopropanol. Staining of the bacterial membrane was by adding Sudan III stock solution into the bacterial suspension before sonication at a 1:100 ratio. One milliliter of the soluble fraction was layered on top of 15% sucrose solution. To test protein nanoparticle membrane-binding activity, 1.5 mL baterial lysate from empty vector-transformed BL21(DE3) was mixed with 0.3 mg purified protein nanoparticle and incubated in RT for 30 min before been layered on the top of the sucrose solution and performing ultracentrifugation. The centrifuge tubes were photographed in front of a dark background under exposure to 450 nm LED light. Images were processed in ImageJ by first extracting green channels, and then the fluorescent intensity of ultracentrifugation results were quantitated from top-to-bottom using Plot Profile and compiled using a ROI manager. The protein nanoparticle's hydrodynamic diameter was determined by dynamic light scattering (DLS) using Stunner (Unchained Lab).

2.3. Animal Immunization and Antibody Titer Determination

The animal protocols were approved by IACUC of Fu Jen Catholic University and mice were purchased from BioLASCO and housed in the experimental animal center of Fu Jen Catholic University following SPF standards. Mice used in immunization procedures (BALB/c strain) were between 8 and 9 weeks of age. The mice used in this study were the

BALB/c strain, which is resistant to inflammation and autoimmune responses induced by adjuvant added in the vaccine. Additionally, BALB/c strain mice tend to develop Th2-type immune responses and produce higher humoral immune responses compared with C57BL/6 strain mice. Mice were immunized through an intramuscular injection of purified protein preparation in 1 mg/mL concentration. For blood collection, mice were bled from a facial vein after being pricked by the lancet. Sera collected were stored at −80 °C before being analyzed by ELISA assay. For each immunization, 20 μg of purified protein was injected intramuscularly in the thigh of the hind limb. Blood was bled 14 days post immunization or on planned dates for analysis using ELISA. For ELISA, the antigen was diluted with coating buffer for coating on a high-binding ELISA plate (Greiner Bio-One MICROLON high binding ELISA plate), and the antigen concentration of hM2e peptide was 2.5 μg/mL and sfGFP was 1 μg/mL. The ELISA plate was incubated at 4 °C overnight and then washed and blocked with 200 μL of blocking buffer containing 1% BSA in washing buffer. The washing buffer was composed of 10 mM $Na(PO_4)$, 150 mM NaCl at pH 7.4 with 0.05% Tween 20. Antiserum was diluted starting from 1:100 and followed by 4-fold serial dilutions. The secondary antibody, Goat anti-mouse IgG with HRP conjugation, was diluted at 1:5000 in blocking buffer. Chromogenic development was carried out by adding 100μL TMB and incubated for 10 min and stopped by adding 100 μL 2 M of H_2SO_4. Antibody titer was determined as the reciprocal value of the highest dilution that gives an OD450 reading of 0.1 above background.

2.4. Protein Structure Modeling and Intermolecular Force Calculation

Protein structure models of AH3 peptide monomer, dimer, and tetramer were generated using AscalaphDesigner version 1.8.79 (Agile Molecule, Stockholm, Sweden) and manual movement. The intermolecular interaction forces between either monomers or dimers were calculated using the intermolecular energy command of AscalaphDesigner. Hydrogen bonds between the peptide subunit and hydrophobic patch were determined and illustrated using Deep View/Swiss PDB Viewer version 4.1.1. The solvent-accessible surface area of the AH3 dimer and GFP were calculated using Jmol with a radius of 1.2 angstroms.

2.5. Thermostability Determination

The purified protein nanoparticle that was stored in the elution buffer was changed to 0.5× GF buffer (pH 8.0) using Sephadex-25 resin. Then the NaCl concentration of protein solutions was adjusted to the designed concentration by adding 5 M NaCl. Protein samples were incubated at various temperatures, and samples were taken at designated time points for sample preparation for SDS-PAGE and coomassie blue staining was performed at the end.

3. Results

3.1. Identification of a Protein Complex with High Antigenicity and Stability

As described in our patent application filed in 2015, we tested the immunogenicity of fusion proteins composed of an AHP and a GFP [31]. The results showed an increase of anti-GFP IgG titer in a range between 2 and 3 log under a prime and boost regime (Figure S1). One of the peptides, AH3, which is derived from the M2 protein of type-A influenza strain H5N1, provides extended stability to the GFP fusion protein when compared with another peptide, AH1 (Figure S2) as well as other peptides in our study (data not shown). Since a stable protein is essential for a vaccine carrier, we were interested in the mechanism of AH3–GFP stability and antigenicity. To study the potential mechanisms that contribute to the above-mentioned properties of AH3–GFP fusion protein, we first checked the composition of the AH3–GFP protein post expression and purification. One clue that led us to study the composition of AH3–GFP fusion protein is the difficulties encountered during protein purification. Unlike other fusion proteins studied, most of the AH3–GFP and AH5–GFP fusion proteins are expressed as insoluble inclusion body and the remaining soluble protein did not bind to Ni–NTA resin under the normal condition of 300 mM NaCl. The AH3–GFP and AH5–GFP fusion proteins only started to bind to Ni–NTA resin after we decreased the NaCl concentration from 300 mM to 50 mM. Additionally, the resistance of AH3–GFP fusion protein to hydrolysis suggested the linker between AH3 peptide and GFP is kept in a water tight complex. The likely hypothesis to explain these observations is that when GFP was fused with the AH3 peptide, the AH3 peptide induces the assembly of a hydrophobicity-driven protein complex formation that hinders N-terminal His tag from binding to Ni–NTA ligand.

3.2. Characterization of AH3–GFP Protein Complex

To test the hypothesis that AH3–GFP or AH5–GFP fusion protein forms a protein complex, we first used protein concentration tubes with a different molecular weight cut off (MWCO) to determine the protein complex sizes. As shown in Figure 1A, GFP protein with a molecular weight of 27 kDa was able to pass through membranes with MWCO of 100 kDa, 300 kDa, and 1000 kDa freely, but AH3–GFP fusion protein was prevented from passing through the membranes with an MWCO up to 1000 kDa. With a molecular weight of 30 kDa, the purified AH3–GFP fusion protein needs to form a complex with more than 35 monomers to be excluded from passing a membrane with a 1000 kDa MWCO. To further explore the geometric composition of the AH3–GFP protein complex, we examined the fusion protein under a transmissive electronic microscope (TEM). The TEM images showed the AH3–GFP fusion protein forms a corn-on-a-cob structure with a length up to 60 nm and a diameter around 10 nm (Figure 1B). The difference in length suggests that the particle may be assembled along the long axis. When scanning the AH3–GFP particles along their long axis, there is a repetitive pattern of the two-three–two-three array of white dots. The predicted structure, according to TEM images, is shown in Figure 1D. We also examined the AH5–GFP protein complex under TEM, but there was no clear evidence of the formation of a higher-order protein complex, suggesting the AH5–GFP protein complex is not as stable as AH3–GFP to withstand the conditions during negative staining. To find the correlation between protein complex formation and antigenicity, we immunized mice with purified AH3–GFP fusion proteins and the GFP proteins. Proteins were prepared from LPS synthesis defective *E. coli* strain, ClearColi BL21(DE3), to avoid the interference of LPS contamination, a known TLR4 ligand and an adjuvant. The mice were immunized with purified proteins by single intramuscular injection, and sera were collected at day 7, 14, 30, and 182 to evaluate anti-GFP IgG titer by ELISA. Deoxycholate was added to test if it affect in the concentration of 0.2% affects AH3–GFP antigenicity and the related experiment was terminated at 30 days post immunization when it showed no effect on the antigenicity of either GFP or AH3–GFP. These results suggested GFP alone is a poor antigen and it only gains high antigenicity after being fused with the AH3 peptide (Figure 1C).

3.3. Modeling the AH3–GFP Protein Complex Structure

To understand the potential molecular mechanism leading to the assembly of the AH3–GFP nanoparticle, we built an AH3 model based on a known helical structure and the hypothesis that hydrophobic interaction drives the complex formation in addition to two observations from TEM images: first, the particle was assembled along the long axis and second, the AH3–GFP protein particle had a repetitive three-two pattern along its long axis after examining the protein particle images. We first assembled two AH3 peptides as anti-parallel helices with a 24-degree angle to make close contacts involving side chains of Phe5, Phe6, Ile9, Leu13, and Leu17 using protein modeling software, AscalaphDesigner (Figure 2A). The hydrophobic core contributes the major intermolecular force binding these two helices and the AH3 dimer is surrounded by hydrophilic side chains from multiple lysine and arginine. Using another protein modeling software, Deepview, a hydrophobic patch was identified on one face of the assembled dimer as marked by a red surface (Figure 2B). In a water-accessible surface model of the AH3 dimer, two connected hydrophobic pockets can be seen located within the hydrophobic patch (marked by a dashed line) that serve as binding sites for two Arg12 side chains extruding from the opposite face of the second AH3 dimer. After turning counter clockwise for 36° while looking down the hydrophobic patch, the second AH3 dimer made close contact in tandem with the first dimer and formed a tetramer (Figure 2C). The intermolecular energy between two dimers from this model was calculated to have a ΔG of −101 kcal/mol (Figure 2C). After adding GFP protein structures onto the AH3 tetramer model, the AH3–GFP fusion protein tetramer formed a scissor-shaped assembling unit, and the stacking of every AH3–GFP tetramer on top of another tetramer increased the particle length by 2.8 nm and turned the axis by 72°. Since the GFP protein barrel diameter ranged between 2.7 and 3.5 nm, the out-extending GFP from the AH3–GFP tetramer can spatially fit into this model (Figure 2D). According to this model, the protein nanoparticle extends continuously along the long axis, with a hydrophobic patch presenting on the growing end of the assembling particle and serving as a point for polymerization.

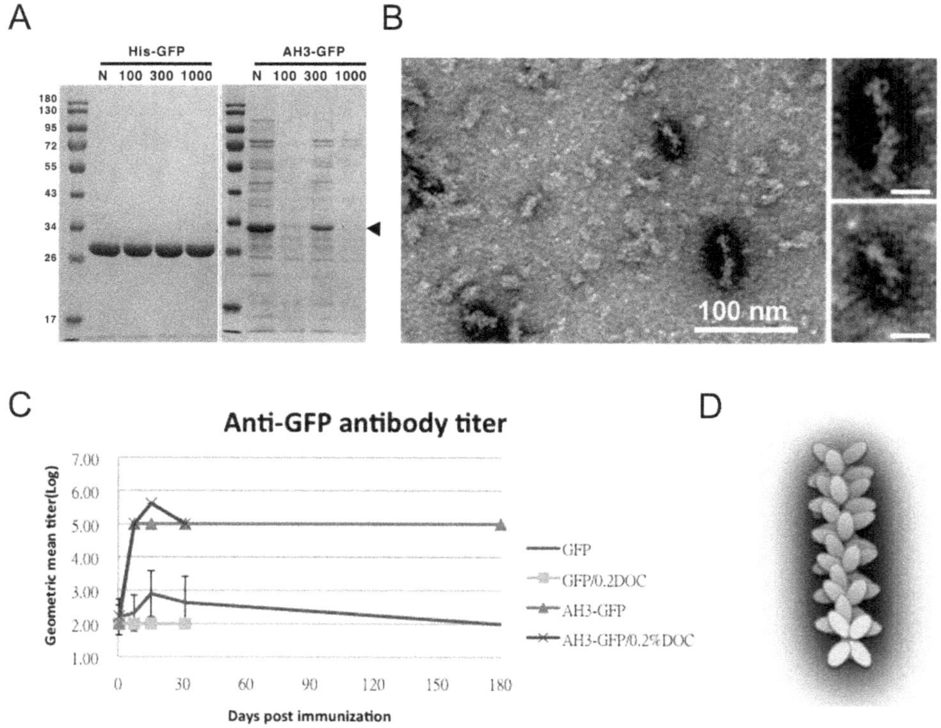

Figure 1. Characterization of GFP fusion protein oligomerization. (**A**) GFP and AH3–GFP fusion proteins purified and reconstituted in PBS were centrifuged through a size-exclusion membrane with a different molecular weight cut off (MWCO). The filtrate was then analyzed by 12% SDS-PAGE and coomassie blue staining. The position of AH3–GFP is marked by an arrow head. (**B**) The oligomerization of AH3–GFP protein was analyzed by negative staining using a transmissive electronic microscope (TEM). Scale bar represents 100 nm. Two of the representative particles are enlarged and shown in the right panels. (Scale bar = 20 nm) Images have been processed by Gaussian Blur filter using Photoshop. (**C**) The antigenicity of GFP and AH3–GFP fusion protein was evaluated by immunizing mice by single muscular immunization. The anti-GFP IgG titers were followed for 6 months by ELISA. (N = 5). Two samples (GFP/0.2% DOC and AH3–GFP/0.2% DOC) also included 0.2% deoxycholate when immunizing mice. (**D**) AH3–GFP protein nanoparticle model predicted from TEM images.

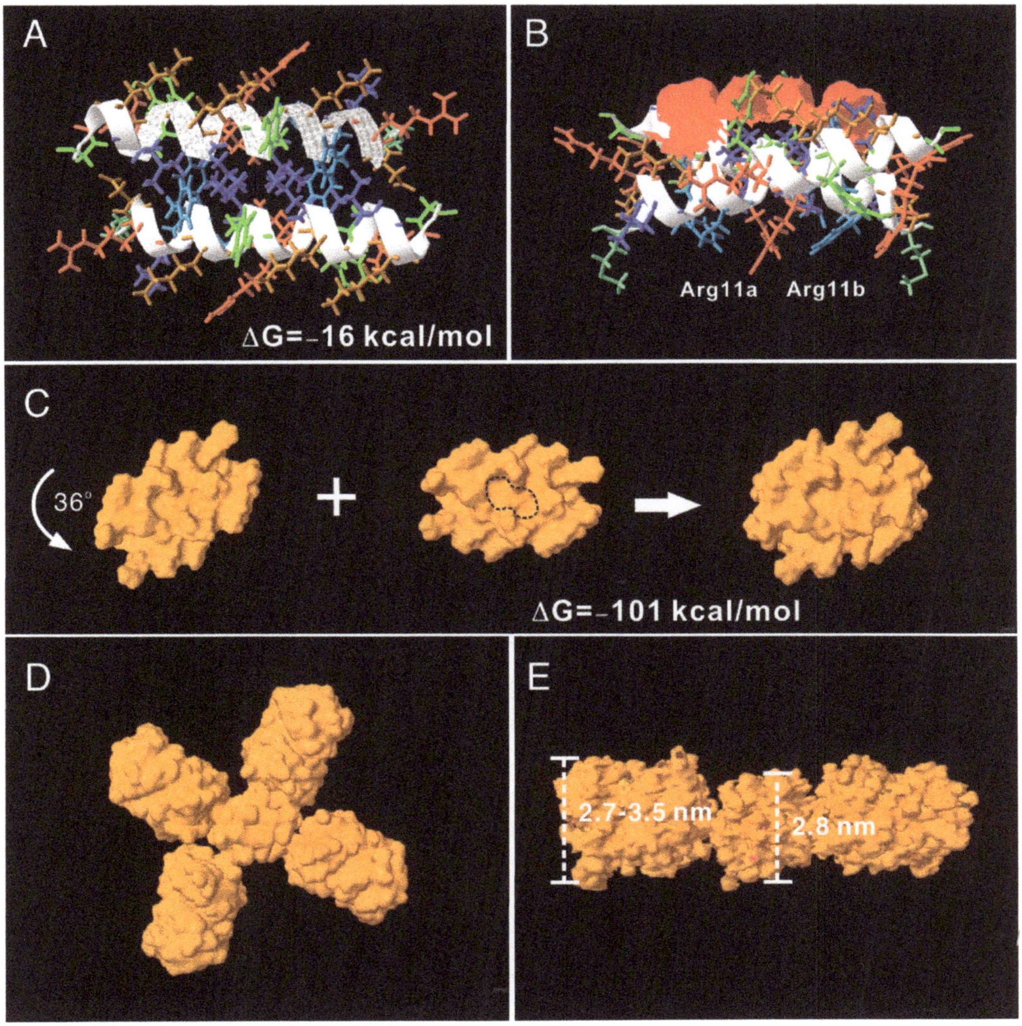

Figure 2. AH3–GFP protein nanoparticle structural modeling. (**A**) The assembling unit of AH3–GFP protein nanoparticles was modeled as a dimer interacting through hydrophobic interaction using AscalaphDesigner. Amino acid side chains are colored according to their hydrophobicity using SPDB v4.10. The most hydrophilic side chains are colored red, and the most hydrophobic side chains are colored blue. Side chains with hydrophobicity in between are colored transitionally. The two α-helix backbones are marked by a white ribbon. The intermolecular interaction force (ΔG) is shown in the bottom-right corner. (**B**) The bottom view of the AH3 dimer shows the location of the hydrophobic patch (marked by a red surface). The anti-parallel alpha helix fromed by two monomers forms a cross and has an angle of 24 degrees. (**C**) The water-accessible surface of the AH3 dimer is modeled using Jmol, and the stacking of the second dimer onto the first one is achieved after turning the second dimer counter clockwise by 36°. The two hydrophobic pockets are enclosed in a dashed line. The intermolecular force (ΔG) between two dimers is shown. (**D**) The AH3 tetramer with four fused GFP molecules is modeled to form a cross shape. (**E**) The front view of the AH3–GFP tetramer is shown with two GFP molecules removed for a clear view. The distance between two repeating atoms of the stacked tetramer is measured and shown as the thickness of the AH3 tetramer, and the thickness of the GFP monomer is measured as the distance across the protein barrel structure.

3.4. Designing a Vaccine Carrier That Enables Heterologous Antigen Insertion and High Stability

After proving that the AH3–GFP protein nanoparticle possesses high antigenicity, we decided to explore the potential of the AH3–GFP protein nanoparticle as a vaccine carrier. For the Hepatitis B core antigen, the amino acid 144 may serve as an insertion site for heterologous antigen fusion and expose the antigen to the immune system [32]. The GFP protein has a thermally stable structure that is constituted by 11 β-strands and 1 α-helix, and some of the loops connecting the β-strands have been explored as insertion sites for heterologous proteins [22,23,33]. Among those candidates, loop173 linking the strand 8 and the strand 9 was chosen because it has been shown to have a high capacity for foreign peptide insertion (Figure 3A) [31]. The original AH3–GFP recombinant protein was built in pET28a vector with an AH3 coding sequence inserted into the C-terminal of His-tag and thrombin cleavage site and followed immediately by GFP cloned from pEGFP–C2. This expression vector was low in soluble protein production and unable to express soluble recombinant protein when the peptide was inserted between D173 and G174. To resolve the expression and folding issues, we designed a new expression vector. First, we cloned AH3 peptide into the N-terminal following methionine in the pET27 vector, and then to its C-terminal we inserted a 6 a.a. linker (GTTSDV) followed by a synthetic sfGFP gene [34] with an antigen insertion site next to Ser175 of sfGFP. The length of the linker between the AH3 peptide and sfGFP is important for protein nanoparticle assembly as well since the fusion protein with a 2-amino-acid linker failed to assemble into nanoparticles. The antigen insertion site also contains an 8xHis tag to facilitate protein purification. To verify vaccine carrier function, we inserted two copies of a broad-spectrum flu vaccine candidate, the M2 ectopic peptide from PR8 strain (hM2e), separated by a 6 a.a. linker (Figure 3B). The newly constructed vector was proven to be efficient for expressing soluble AH3–sfGFP–2hM2e fusion proteins as a protein complex. However, the AH3–sfGFP–2xhM2e protein complex was not as stable as AH3–GFP because these nanoparticles fell apart during distilled water washing before TEM imaging. After few trials, we found the protein nanoparticles have to be crosslinked by sulfo-SMCC before TEM analysis. Following the established AH3–GFP protein complex model, we were seeking strategies to create a more stable AH3–sfGFP protein complex. First, using the protein modeling software AscalaphDesigner, we found the mutation of Ile9 to Leu increases the intermolecular force (ΔG) between two peptide helices from −16 kcal/mol to −39 kcal/mol (Figure 3C). Second, when we mutated Lys14 to Glu14, there were two hydrogen bonds formed between LYRRLE dimers, one between the side chain of Glu14c and Cys8a and the other one the side chain of Arg12a and the oxygen of the Tyr10c backbone (Figure 3D). The intermolecular force between the LYRRLE dimers increased from −101 kcal/mol to −161 kcal/mol. The K14E mutation not only enabled tighter binding between two adjacent dimers, but also resolved a key issue of AH3-mediated protein nanoparticle assembly, an exposed hydrophobic patch. From the protein modeling results, we found that the dimer assembled from the AH3 double mutant, LYRRLE, was able to bind to pre-formed particle in two orientations: either as a tandem dimer (ΔG = −161 kcal/mol) (Figure 3F) as an inverted dimer (ΔG = −59 kcal/mol). The formation of an inverted tetramer was able to enclose the hydrophobic patch inside the protein nanoparticle (Figure 3G).

Following the protein modeling result, we went further to verify whether the protein modeling results were correct using biophysical assays. We generated mutations in the AH3 peptide sequence in the context of AH3–sfGFP–2xhM2e (Figure 4A). The exposed hydrophobic patch on the end of the AH3–GFP protein nanoparticle, according to our hypothesis, will bind bacterial membrane and the co-sediment with the membrane during the sucrose step-gradient ultracentrifugation. This methodology was first verified by loading the bacterial-soluble fraction (containing bacterial membrane) prepared from homogenized bacterial culture in a centrifuge tube preloaded with 15% (w/v), 45% (w/v) and 65% (w/v) sucrose, as demonstrated in Figure 4B's left panel. The sedimentation of the bacterial membrane was marked by a lysochromic dye (Sudan III) that binds lipid. The control sample contained Sudan III with lysis buffer alone (Figure 4B, lane 1). After ultracentrifugation, the bacterial membrane was sedimented to the junction between 15% and 45% sucrose as marked by Sudan III staining (Figure 4B lane 2). Using the same protocol, AH3–GFP was found to co-sedimented with the bacterial membrane (Figure S3A) as well as the AH3–sfGFP–2xhM2e fusion protein but not a free GFP protein (Figure S3B) when the tubes were illuminated by 450 nm LED light. After protocol verification, the bacterial-soluble fractions from all six fusion protein expression cultures were prepared and analyzed using the same protocol. From the distribution of fluorescent protein between different percentages of sucrose, we made several observations: first, all of the expressed AH3–sfGFP–2xhM2e proteins bound to the bacterial membrane and were co-sedimented at the 15%/45% junction, as with AH3 variants LY and LYRLLK (Figure 4C, lane 1). Second, comparing tubes 4, 5, and 6 to tubes 1, 2, and 3, it is clear that when Lys14 was mutated to Glu, this mutation decreased the binding of expressed protein nanoparticles to bacterial membranes and about half of the protein complex remained on the top of the centrifuge tube. Third, the protein complex of AH3 variant LYRRLE and RRLE formed higher-order protein aggregates and was sedimented further to the 45%/65% junction (Figure 4C, lane 5 and 6); the identity of these high-order protein complexes is not known, but they were most likely aggregated protein nanoparticles. These results confirmed the predicted presence of a hydrophobic patch on the protein nanoparticles assembled from AH3–sfGFP–2xhM2e protein and LY, LYRLLK variants. Additionally, a portion of the protein nanoparticles assembled from AH3 variants containing a K14E mutation removed the hydrophobic patch from the surface when they were synthesized in cells. To verify if purified protein nanoparticles still have membrane-binding activity, the purified protein nanoparticle of AH3, LY, RRLE, and LYRRLE variants were first mixed with the bacterial-soluble fraction and incubated for 30 min before being evaluated by ultracentrifugation. A control tube with the same setup but no centrifugation was used to show the position of the protein nanoparticle before ultracentrifugation (Figure 4D). The data show that a fraction of the protein nanoparticles from AH3 and LY variants bound and were co-sedimented with the bacterial membrane during ultracentrifugation, as shown in the line plot that depicts the fluorescent protein distribution of all test samples (Figure 4E). None of the protein nanoparticles assembled by RRLE or LYRRLE variants co-sedimented with the bacterial membrane. These data are consistent with the protein modeling results suggesting that a protein nanoparticle assembled by the LYRRLE variant is able to eliminate hydrophobic patch through an inverted cap (Figure 3F).

To further evaluate the effect of AH3 mutants on protein nanoparticle formation, we examined the particle morphology using TEM and dynamic light scattering (DLS). The result is shown in the negative staining images (Figure 5A–D). Through visual observation and image refinement using ImageJ, we found some morphological evidence that support the protein modeling result. First, there were donut-shaped protein nanoparticles been observed in the AH3–sfGFP–2xhM2e protein preparation (Figure 5A). The donut- or disc-shape structure may be due to the collapse of the corn-on-a-cob structure because of low intermolecular interaction between AH3 monomers ($\Delta G = -16$ kcal/mol). This hypothesis is supported by the DLS data that measure the protein particle size. These data showed a 2.5-fold increase in particle size when Ile9 (AH3) was mutated to Leu9 (LY) and an increase of predicted intermolecular force between monomers from $\Delta G = -16$ kcal/mol to $\Delta G = -39$ kcal/mol (Table 1). The significance of Leu9 in stabilizing particle structure is also underscored by the change in morphology of protein nanoparticles assembled by RRLD and RRLE variants. Part of the fusion proteins containing RRLD or RRLE variants assembled into an eccentric ladder-like structure, suggesting an imbalance in the interaction forces within the protein nanoparticle.

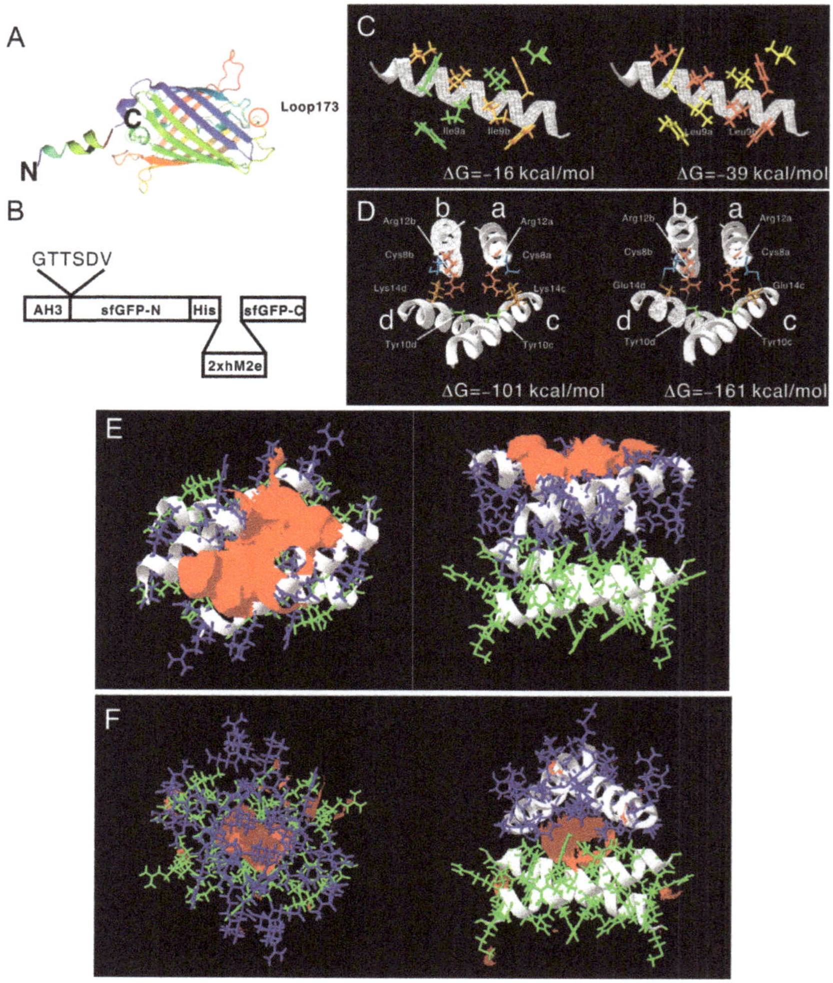

Figure 3. Construction of AH3–sfGFP–2xhM2e fusion protein and the protein model guided AH3 mutagenesis for variants with a higher nanoparticle stability. (**A**) The graphic presentation of the AH3–sfGFP fusion protein is shown and the antigen insertion site (loop173) is marked by a red circle. The AH3 peptide was fused to the N-terminal of sfGFP to mediate polymerization and a foreign antigen insertion site is genetically created in loop173 containing an 8xHis tag for efficient protein purification. The N-terminal and C-terminal of the fusion protein is marked by N and C respectively. (**B**) The graphic presentation of AH3–sfGFP–2xhM2e fusion protein is shown. (**C**) The comparison of AH3 and AH3 I9L (LY) mutant in dimer formation is shown. The calculated intermolecular force between two monomers are shown. (**D**) The hydrogen bonds that formed between AH3 variant LYRRLE dimers are shown in the right panel. AH3 tandem tetramer is shown in the left panel. Four monomers from each model are labeled from a to d. The Glu14 from monomers c and d form hydrogen bonds with Cys8 from monomers a and b respectively. Additionally, the side chain of Arg12 from monomers a and b forms a hydrogen bond with the backbone of monomers c and d, respectively. (**E**) The front view (left panel) and bottom view (right panel) of the LYRRLE tandem tetramer protein models. (**F**) The front view (left panel) and bottom view (right panel) of the LYRRLE inverted tetramer. The hydrophobic patch is marked as red surface, the first dimer is colored as green, and the second dimer is colored as blue. The calculated intermolecular interaction force is shown in the center of both graphs.

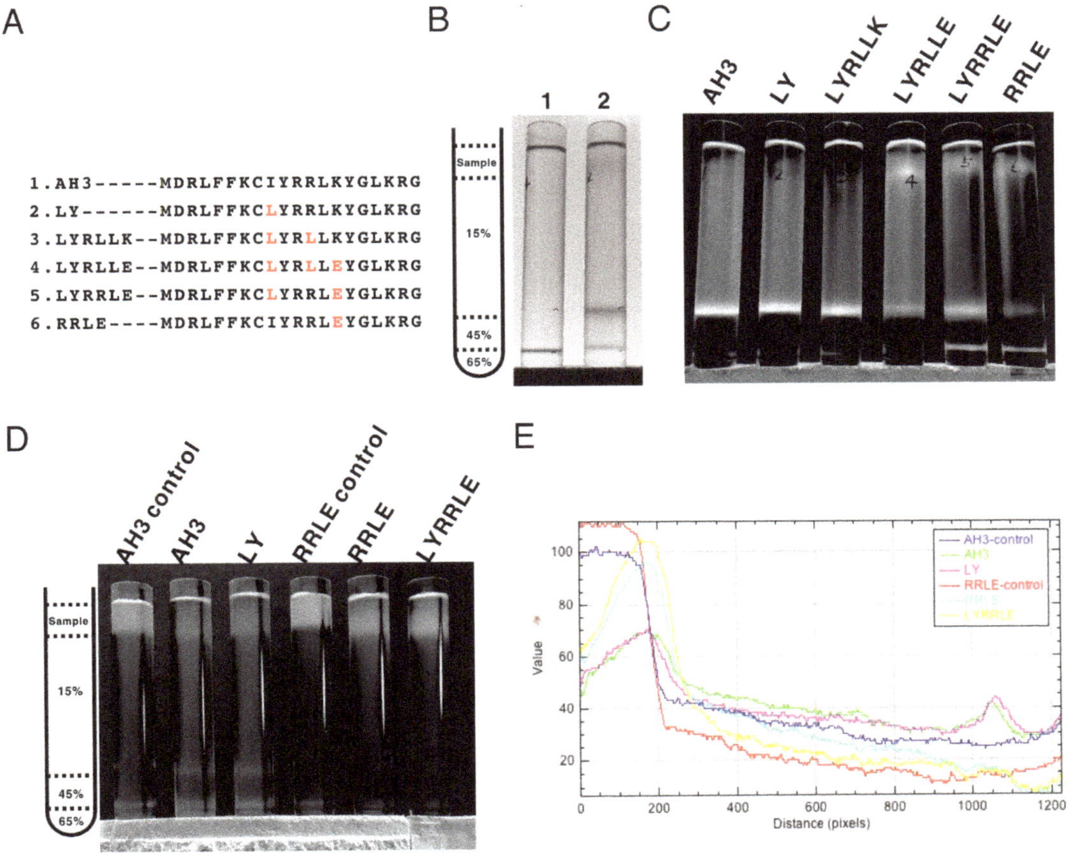

Figure 4. The co-sedimentation analysis of AH3–sfGFP–2xhM2e variants with bacterial membranes by analytic ultracentrifugation. (**A**) List of AH3 peptide variants introduced on the AH3–sfGFP–2xhM2e fusion protein for co-sedimentation analysis. (**B**) The right panel shows the graphic presentation of the centrifuge tube distribution of step sucrose solutions. The left panel shows the bacterial membrane distribution after ultracentrifugation, as marked by Sudan III staining. Lane 1 is a Sudan III solution only control without bacterial lysate, lane 2 is topped with 1 mL bacterial lysate mixed with Sudan III solution. The ratio between Sudan III staining solution and bacterial lysate is 1:100. (**C**) The distribution of sfGFP fusion proteins post ultracentrifugation as illuminated by 450 nm LED light was recorded as photos and analyzed by ImageJ. Only the green channel is shown. (**D**) The ability of purified protein nanoparticles to bind bacterial membranes was tested by first mixing with bacterial lysate and then analyzed by ultracentrifugation on a step sucrose gradient. The distribution of sucrose solutions of different percentages is shown in the left panel. The protein nanoparticles purified from fusion proteins containing various AH3 mutants AH3, LY, RRLE, and LYRRLE were tested. The control samples were not ultracentrifuged to reveal the original position of loaded samples. The photo taken is split into three channels by ImageJ; only the green channel is shown. (**E**) The combined line plot data showed the fluorescent intensity along the centrifuge tube from top to bottom. The image pixels had been averaged by the "smooth" command of imageJ before line plot analysis.

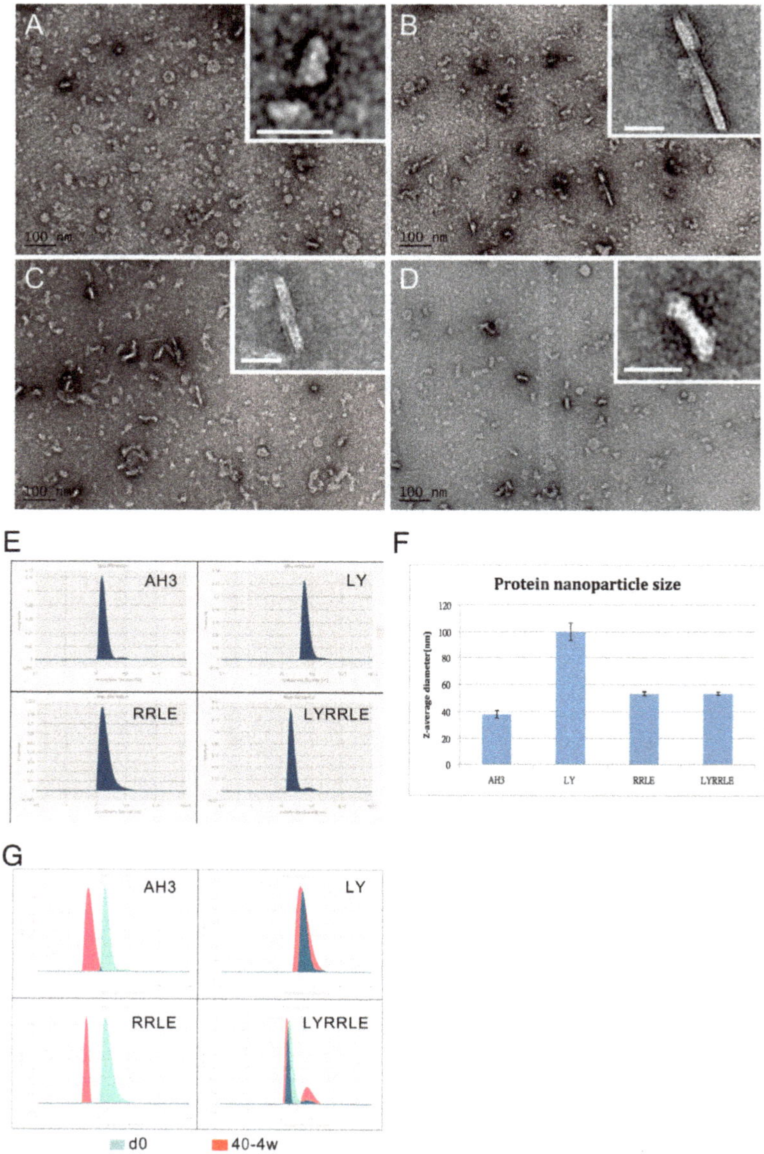

Figure 5. The TEM images of protein nanoparticles derived from AH3–sfGFP–2xhM2e and its variants. Purified protein nanoparticles were crosslinked with Sulfo-SMCC after buffer change from elution buffer using desalting resin (Sephadex 25). Protein samples were diluted to 0.1 mg/mL before being applied on the grid for TEM analysis. The representative images were presented in the sequence of (**A**) AH3–sfGFP–2xhM2e, (**B**) RRLD-sfGFP–2xhM2e, (**C**) RRLE–sfGFP–2xhM2e and (**D**) LYRRLE–sfGFP–2xhM2e. The enclosed window in the upper-right corner shows the magnified images of representative protein nanoparticles with size bar represent 20 nm. (**E**) The hydrodynamic diameter of protein nanoparticles derived from AH3–sfGFP–2xhM2e or its variant were analyzed by dynamic light scattering (DLS). (**F**) The Z average diameters of protein nanoparticles derived from AH3–sfGFP–2xhM2e or its variants were compared. (**G**) The merged hydrodynamic distribution graphs from protein nanoparticles prepared freshly or have been stored in 40 °C for 4 weeks.

Table 1. List of intermolecular forces and the access code of protein models deposited in ModelArchive.

	Assembling Unit	Intermolecular Interactions (kcal/mol)	Hydrophobic Patch Area (A^2)	Modelarchive Access Code
AH3	Monomer	na	na	ma-06g7k
	Monomer–Monomer	−16	285	ma-bhgiw
	Tandem dimer–dimer	−101	386	ma-xhrzb
	Inverted dimer–dimer	na	na	na
LY	Monomer	na	na	ma-0koys
	Monomer–Monomer	−39	389	ma-7x5gd
	Tandem dimer–dimer	na	na	na
	Inverted dimer–dimer	na	na	na
RRLE	Monomer	na	na	ma-xkn28
	Monomer–Monomer	−23	429	ma-ax78l
	Tandem dimer–dimer	−29	342	ma-vyybf
	Inverted dimer–dimer	na	na	na
LYRRLE	Monomer	na	na	ma-ivexo
	Monomer–Monomer	−36	393	ma-lptlj
	Tandem dimer–dimer	−161	433	ma-izbbn
	Inverted dimer–dimer	−59	111	ma-xomje

As shown in the sucrose step gradient ultracentrifugation, the presence of a hydrophobic patch enables nonspecific interaction of the AH3–GFP protein complex with the bacterial membrane, which may restrict the free movement of protein nanoparticles and keep them from reaching draining lymph nodes for stimulating immunity [6]. To compare the antigenicity of protein complexes derived from either AH3–sfGFP–2xhM2e or LYRRLE–sfGFP–2xhM2e, we immunized mice with a single injection of either recombinant protein. According to ImageJ analysis of SDS-PAGE gel, the purified protein nanoparticles had about 80% purity. Sera were collected on days 0, 16, 50, 92, 175, and 199 following immunization to evaluate anti-hM2e and anti-sfGFP–His IgG titer by ELISA (Figure 6A,B). The geometric mean titer of anti-hM2e IgG reached the highest point for the AH3–sfGFP–2xhM2e group and then declined afterward. However, in the LYRRLE–sfGFP–2xhM2e group, the GMT reached its highest point at day 50 and remained steady up to day 90. Surprisingly, the anti-sfGFP–His antibody did not decrease in the later stages but the IgG titer increased over time until the end of the experiment (Figure 6B). When the individual mouse serum results were observed separately, only one out of five mice from the AH3–sfGFP–2xhM2e group had a higher anti-hM2e IgG titer at day 199 than at day 16. However, four out of five mice from the LYRRLE–sfGFP–2xhM2e group showed a higher antibody titer on day 199 compared with day 16 (Figure 6C). These results suggest that the two point mutations of AH3 in I9L and K14E enabled the formation of a stable, high-antigenic protein complex that stimulates long-lasting immune responses in a single immunization.

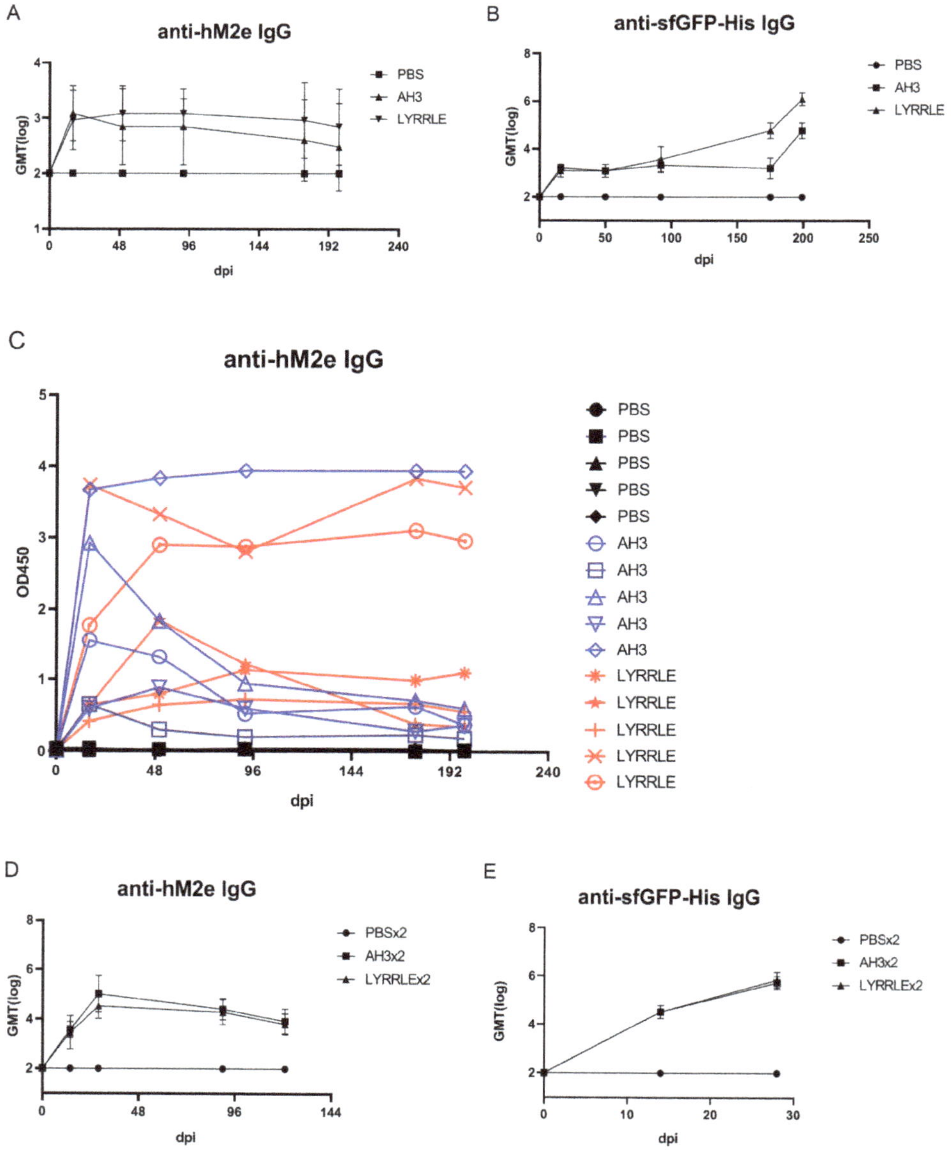

Figure 6. Immunization of mice with AH3–sfGFP–2xhM2e and LYRRLE–sfGFP–2xhM2e and detection of anti-hM2e IgG and anti-sfGFP IgG. The mice were immunized once at day 0 with 20 μg of purified protein nanoparticles assembled by the wild-type AH3 peptide, or the LYRRLE mutant, or PBS. Sera were collected at days 0, 16, 50, 91, 175, and 199 for analysis to detect (**A**) total anti-hM2e IgG and (**B**) total anti-sfGFP IgG by ELISA. (**C**) Anti-hM2e total IgG titers are presented as optical densities at OD450 using immune serum from each mouse diluted 1:100. (**D**) The anti-hM2e total IgG of mice immunized twice, 14 days apart, in a prime-boost regime was followed for 3 months and analyzed by ELISA (N = 5). (**E**) The anti-sfGFP–His total IgG of mice immunized twice with AH3–sfGFP–2xhM2e or LYRRLE–sfGFP–2xhM2e were evaluated 14 days after each immunization by ELISA (N = 5).

The pre-existing antibody against VLP may reduce the efficacy of booster doses due to carrier-induced epitope suppression [35]. Although GFP is a protein of low antigenicity, the fusion with AH3 strongly enhances its antigenicity, as shown in Figure 1C. To test if sfGFP backbone competes with inserted hM2e peptide for immune machinery, we immunized mice in a prime-boost protocol using the same protein preparations. The two consecutive injections were carried out 14 days apart and the sera that collected at day 14, 28, and 90 were subject to ELISA assay using either hM2e peptide or sfGFP–His protein as coating antigen. The results showed the IgG titer against hM2e elevated continuously after consecutive immunizations for both proteins as well as anti-sfGFP–His IgG titer. The result suggests that although carrier protein AH3–sfGFP also has high antigenicity, it does not interfere with the immune response against the heterologous protein, hM2e (Figure 6D,E).

One possible explanation of the long-lasting antibody response is the continuous stimulation of the immune system by the protein nanoparticles assembled by LYRRLE peptide but not AH3 peptide. To test this hypothesis, we incubated the purified protein nanoparticle in phosphate buffer containing 150 mM NaCl at 40 °C or 50 °C for 4 weeks. The integrity of protein nanoparticles was examined by SDS-PAGE and DLS. The SDS-PAGE data suggest protein nanoparticles assembled by LYRRLE and RRLE have better thermostability than those assembled by AH3 and LY peptides. The thermostability of LYRRLE-based nanoparticles is also supported by the DLS result (Figure 5G). Based on the data presented above, we conclude that the protein nanoparticle assembled by LYRRLE peptide is more suitable in serving as a vaccine carrier for two reasons. First, it is more stable in the host's body temperature, so it can endure high-temperature storage; second, the removal of the hydrophobic patch from the surface enables the nanoparticle to move freely through vessels to reach the target site.

While working with LYRRLE protein nanoparticles, it was observed that the protein nanoparticles often degraded after the buffer was changed from high-salt elution buffer (300 mM NaCl, 500 mM Imidazole) to isotonic buffer (150 mM NaCl). This suggests a possibility of salt concentration-dependent nanoparticle stabilization. To evaluate the effect of NaCl concentration on the stability of LYRRLE-assembled protein nanoparticles, the LYRRLE–sfGFP–2xhM2e protein nanoparticles were incubated in 4 °C, 22 °C, and 37 °C for up to 3 months, and the protein stability was evaluated by SDS-PAGE.

When the protein nanoparticles were stored at 4 °C, the protein particle stored in buffer containing 150 mM NaCl was stable during the first week, but hydrolyzed into a lower molecular weight before the fourth week (Figure 7B). Additionally, the protein nanoparticle in buffer containing 300 mM NaCl was stable for one month but it was later hydrolyzed before the end of the third month (Figure 7C). When the storage temperature was raised to 22 °C, the protein was hydrolyzed within a week in buffer containing 150 mM NaCl and within 4 weeks in buffer containing 150, 300, and 550 mM NaCl. The protein nanoparticle was only stable in buffer containing 800 mM NaCl (Figure 7D). When the temperature was shifted to 37 °C, the 300 mM NaCl was sufficient to keep the protein nanoparticle intact in the end of 3 months of incubation (Figure 7E,F). To be noted is that the protein samples that contained hydrolyzed protein nanoparticle were still fluorescent, an indication that sfGFP was still functionally intact. According to the size of hydrolysis product (20 kDa), the protein hydrolysis likely happened to the inserted loop that contains the 2xhM2e peptide. These results suggest that the stability of LYRRLE protein nanoparticle is salt dependent and it has good stability at high temperatures (37 °C). This salt-dependent nanoparticle stabilization is consistent with our assumption that the hydrophobic interaction is the major force that mediates monomer–monomer interaction.

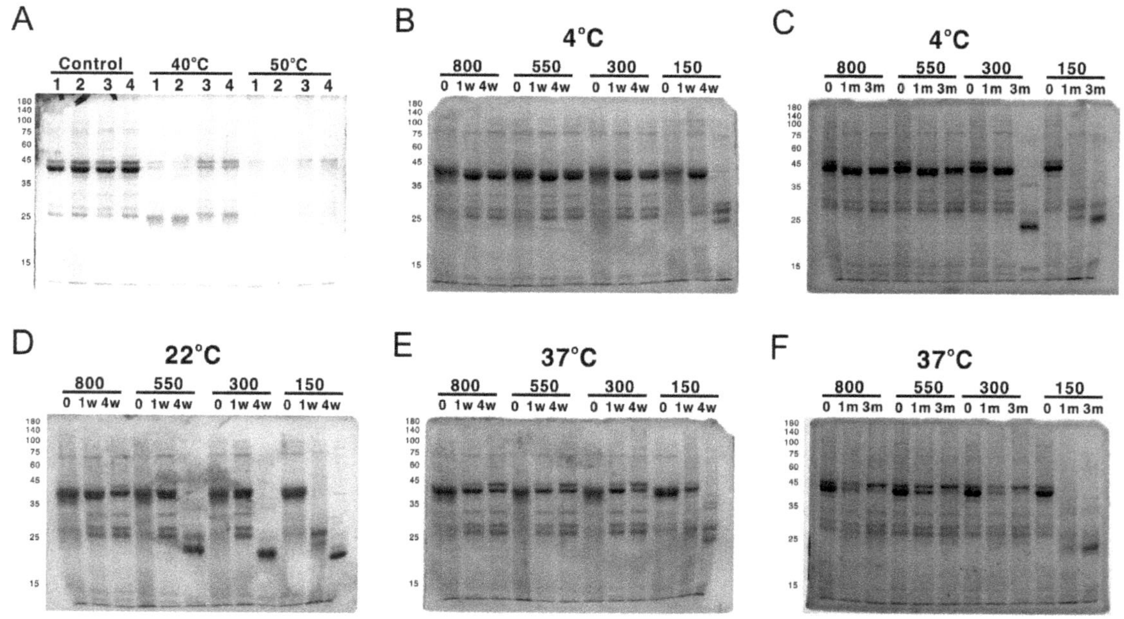

Figure 7. The analysis of protein nanoparticle stability using SDS-PAGE. The protein nanoparticle stability was analyzed by both SDS-PAGE and DLS. (**A**) Protein samples of 1) AH3–sfGFP–2xhM2e, 2) LY-sfGFP–2xhM2e, 3) RRLE–sfGFP–2xhM2e, or 4) LYRRLE–sfGFP–2xhM2e were incubated at either 40 °C or 50 °C for 4 weeks and analyzed in 12% SDS-PAGE for protein stability. LYRRLE–sfGFP–2xhM2e protein nanoparticles were first reconstituted in phosphate buffer containing 150 mM NaCl, and then the NaCl concentration was adjusted to either 300 mM, 550 mM, or 800 mM before being stored in 4 °C (**B**,**C**), 25 °C (**D**), or 37 °C (**E**,**F**) for the period of time indicated. Protein samples were removed and processed, and then stored before being analyzed by SDS-PAGE.

4. Discussion

Vaccines as a tool to prevent infectious diseases are the most cost-effective strategy, especially for attenuated viral vaccines such as Vaccinia, MMR, or the oral polio vaccine; they produce long-lasting and even lifetime-protective immune responses, but these attenuated viral vaccines took decades to develop. Apparently, this strategy will not be able to be used to develop a vaccine in time to ward off an emerging global pandemic such as that caused by COVID-19. Although new vaccine technologies such as DNA vaccines, mRNA vaccines, or adenovirus-based vaccines can be used to quickly develop a subunit vaccine after the genomic information of a pathogen becomes available, the immune responses generated are often reduced to a baseline level within a year [36–39]. This short-lived immune response may expose vaccinated people to the risk of a breakthrough infection and the need of a booster shot. In this study, we created a self-assembled protein nanoparticle composed of a gain-of-function mutant of the AH3 peptide that enables temperature- and salt-dependent protein nanoparticle thermostability. Protein nanoparticles assembled by this gain-of-function mutant are able to stimulate a long-lasting humoral immune response that is correlated with the thermostability of the gain-of-function mutant. The surprising continued increase of anti-sfGFP IgG titer at 3 months post-immunization is distinct from anti-hM2e IgG titer. This result suggests the continued presence of functional sfGFP protein without an inserted hM2e peptide in immunized mice for 3 to 6 months. These sfGFPs were likely presented to the immune system as loop-hydrolyzed nanoparticles since individual sfGFPs have low antigenicity. The results in Figure 5G also support this possibility; when the LYRRLE–FP fusion proteins were hydrolyzed into smaller products in SDS-PAGE, the nanoparticle size was maintained. The binding and removal of water molecules from proteins in the presence of high NaCl concentrations causes an increase in hydrophobic interaction. Although the physiological NaCl concentration is maintained at 150 mM, there are other electrolytes in the body fluid that can serve the same role without disturbing the electrical potential across the cell membrane. Therefore, the in vitro stability results shown in this study may not represent what happened in injection site during immunization. The mechanism of this long-lasting immune response may be mediated by the presence of thermal stable protein nanoparticle that remains intact in the injection site and stimulates the expression of long-lasting plasma cells through continuous exposure, the same mechanism that accounts for the life-long protection of attenuated viral vaccines [40].

The fluorescent protein (FP) family is a group of proteins with a conserved, sturdy barrel structure that encloses a fluorophore. Because the function of this beta-strand-constituted barrel is to provide a tightly controlled environment for fluorophore maturation and function, directed selection against fluorescent for mutants with high tolerance to structure destabilization can improve the thermostability of the FP barrel [15]. The variability and stability of FPs make them a powerful tool as vaccine carriers. First of all, the thermostability of FP partly contributes to the thermostability of LYRRLE–FP–based protein nanoparticles, a desired vaccine property for achieving the global vaccination initiative. Second, because of the diverse origins of FP [16], the LYRRLE–FP nanoparticle platform may be expanded into a line of vaccine carriers that share a common format but each have distinct serum types. This feature of the LYRRLE–FP platform may avoid the effects of pre-existing carrier antibodies when LYRRLE–FP carrier is used in several vaccines. Third, the extensive studies on the structure of FPs and their applications have provided useful tools to incorporate heterologous antigens onto LYRRLE–FP-based protein nanoparticles. With the features mentioned above, the LYRRLE–FP format is suitable for the commercialization of a nanoparticle-based subunit vaccine.

An AH3 peptide derived from the M2 protein of type-A influenza strain H5N1 was found to induce nanoparticle formation when fused with a GFP in this study. Previous studies about the M2 amphipathic helical peptide focused on its roles in virus budding and proton pump anchorage. It mediates virus budding by generating membrane curvature by embedding its hydrophobic face in the membrane bilayer [30]. Additionally, it anchors the M2 ectopic domain on the viral envelope, which serves as a proton pump that triggers envelope membrane fusion during infection [41]. Here we discovered a novel application of AH3, which serves as a nucleating center for protein nanoparticle assembly. This novel application is restricted to conditions when an AH3 peptide is fused with a fast-folding protein such as a fluorescent protein. Fusing amphipathic helical peptide to other slow-folding proteins causes fusion protein misfolding, which appears to be due to aggregation of slow-folding fusion proteins in the presence of AH3 (unpublished results). Therefore, the fast folding ability of sfGFP is the key to AH3–FP protein nanoparticle formation. Through protein modeling, we went further to identify two point mutations in the AH3 peptide, I9L and K14E, which act synergistically to stabilize protein nanoparticles. The I9L mutation increases monomer–monomer intermolecular interaction, and the K14E mutation enables hydrogen bond formation that holds the tandem dimers together and enables the removal of hydrophobic patches from surfaces. The synergistic effect of these two AH3 mutations contributes to a more stable protein nanoparticle and restricts it from binding cellular membranes, most likely resulting in extending the lifetime of antibody responses. Although there is a limit on the size of the peptide that can be inserted into the sfGFP loop and it may restrict the scope of vaccines that this protein nanoparticle can be applied to, an alternative strategy of tethering large proteins onto the LYRRLE–sfGFP protein nanoparticle has been developed that can expand the application of this protein nanoparticle to other important pathogenic antigens (unpublished result). Even at the peptide size limit, many short peptides have been studied and found to be good vaccine candidates in addition to M2e, for example, the neoantigen for personalized tumor vaccines or peptide vaccines against various infectious diseases [42,43].

The stabilization of the LYRRLE–FP nanoparticle by a high NaCl concentration is consistent with the assumption made during structure modeling. In clinical trials involving the study of neuromuscular pain, hypertonic saline up to 5.4% (6-fold concentration of normal saline, 0.9%) is routinely used. Intramuscular injection of hypertonic saline into muscle causes intense but short-lived muscular pain with no detectable damage to the injection site tissue [44]. The use of hypertonic saline (3.6% NaCl) in aluminum hydroxide formulated antigen was able to stimulate higher cellular and humoral immune responses than normal saline in mice [45]. Other electrolytes or ingredients such as sorbitol, sucrose, histidine, or recombinant human serum albumin may be added to decrease the use of NaCl and minimize interfering membrane potential post-immunization. These electrolytes can distract water molecules from solvating proteins and keep protein nanoparticles bound by hydrophobic interaction intact. The choice of antigen can also impact the stability of LYRRLE-FP-based protein nanoparticles. An LYRRLE–FP NP that incorporated a maltose-binding protein on the surface was able to maintain both protein integrity and particle size after 2 months of incubation at 40 °C (unpublished results). The fact that hypertonic saline is safe for human administration and has the potential benefit of further boosting immune responses is encouraging to apply LYRRLE–FP protein nanoparticles for the benefits of both stimulating long-lasting immune responses and having high thermostability during storage.

5. Conclusions

We designed a protein nanoparticle self-assembled from a fusion protein combining two functionally distinct domains: a polymerization domain and an antigen-presentation domain. In the absence of additional adjuvant, this protein nanoparticle is able to stimulate long lasting humoral immune response in single dose. This activity is correlated with the thermostability of protein nanoparticles at 37 °C.

Supplementary Materials: The following are available online at https://www.mdpi.com/article/10.3390/vaccines11020426/s1, Figure S1: Antigenicity of GFP fusion proteins with 7 different amphipathic helical peptides, Figure S2: Protein stability of AH1-GFP and AH3–GFP fusion proteins in ambient temperature. Figure S3: The hydrophobicity of the AH3-GFP protein nanoparticle was analyzed using density gradient ultracentrifugation. Figure S4: Densitometry quantification of protein bands from Figure 7. Table S1: The relative percentages of protein bands being quantifed from SDS-PAGE gels A-F of Figure S4.

Author Contributions: Conceptualization, M.-C.K. and T.-T.W.; methodology, M.-C.K. and G.-G.L.; software, M.-C.K.; validation, M.-C.K.; formal analysis, M.-C.K.; investigation, M.-C.K.; resources, M.-C.K. and T.-T.W.; data curation, M.-C.K.; writing—original draft preparation, M.-C.K.; writing—review and editing, M.-C.K., T.-T.W. and G.-G.L.; visualization, M.-C.K.; supervision, M.-C.K.; project administration, M.-C.K.; funding acquisition, M.-C.K. All authors have read and agreed to the published version of the manuscript.

Funding: This research was funded by SBIR from MOEA, Taiwan, grant number 1Z1060282, and University of Maryland Baltimore County STRT7TEN-1113 to T.-T.W.

Institutional Review Board Statement: The animal protocols had been approved by IACUC of Fu Jen Catholic University and mice were housed in the experimental animal center of Fu Jen Catholic University following SPF standards. (protocol code P10622 and P10809).

Informed Consent Statement: Not applicable.

Data Availability Statement: Protein model created in this study have been deposited in ModelArchive.org under access codes listed in Table 1.

Acknowledgments: We thank Image Core of Institute of Molecular Biology at Academia Sinica for kindly providing an FEI Tecnai G2 spirit EM to be used in this study.

Conflicts of Interest: M.-C.K. is the founder of Vaxsia biomedical Inc. that co-owns the patent right to the intellectual property from this research with the University of Maryland Baltimore County, employer of T.-T.W.

References

1. IEA. For the First Time in Decades, the Number of People without Access to Electricity is Set to Increase in 2022, IEA, Paris. Available online: https://www.iea.org/commentaries/for-the-first-time-in-decades-the-number-of-people-without-access-to-electricity-is-set-to-increase-in-2022 (accessed on 21 December 2022).
2. Kumar, R.; Srivastava, V.; Baindara, P.; Ahmad, A. Thermostable vaccines: An innovative concept in vaccine development. *Expert Rev. Vaccines* **2022**, *21*, 811–824. [CrossRef] [PubMed]
3. Karp, C.L.; Lans, D.; Esparza, J.; Edson, E.B.; Owen, K.E.; Wilson, C.B.; Heaton, P.M.; Levine, O.S.; Rao, R. Evaluating the value proposition for improving vaccine thermostability to increase vaccine impact in low and middle-income countries. *Vaccine* **2015**, *33*, 3471–3479. [CrossRef] [PubMed]
4. Lopez-Sagaseta, J.; Malito, E.; Rappuoli, R.; Bottomley, M.J. Self-assembling protein nanoparticles in the design of vaccines. *Comput. Struct. Biotechnol. J.* **2016**, *14*, 58–68. [CrossRef]
5. Ahsan, F.; Rivas, I.P.; Khan, M.A.; Torres Suárez, A.I. Targeting to macrophages: Role of physicochemical properties of particulate carriers-liposomes and microspheres-on the phagocytosis by macrophages. *J. Control. Release* **2002**, *79*, 29–40. [CrossRef]
6. Bachmann, M.F.; Jennings, G.T. Vaccine delivery: A matter of size, geometry, kinetics and molecular patterns. *Nat. Rev. Immunol.* **2010**, *10*, 787–796. [CrossRef]
7. Lu, Y.; Chan, W.; Ko, B.Y.; VanLang, C.C.; Swartz, J.R. Assessing sequence plasticity of a virus-like nanoparticle by evolution toward a versatile scaffold for vaccines and drug delivery. *Proc. Natl. Acad. Sci. USA* **2015**, *112*, 12360–12365. [CrossRef]
8. Ravin, N.V.; Blokhina, E.A.; Kuprianov, V.V.; Stepanova, L.A.; Shaldjan, A.A.; Kovaleva, A.A.; Tsybalova, L.M.; Skryabin, K.G. Development of a candidate influenza vaccine based on virus-like particles displaying influenza M2e peptide into the immunodominant loop region of hepatitis B core antigen: Insertion of multiple copies of M2e increases immunogenicity and protective efficiency. *Vaccine* **2015**, *33*, 3392–3397. [PubMed]
9. Manolova, V.; Flace, A.; Bauer, M.; Schwarz, K.; Saudan, P.; Bachmann, M.Ä. Nanoparticles target distinct dendritic cell populations according to their size. *Eur. J. Immunol.* **2008**, *38*, 1404–1413. [CrossRef]
10. Gomes, A.C.; Roesti, E.S.; El-Turabi, A.; Bachmann, M.F. Type of RNA Packed in VLPs Impacts IgG Class Switching-Implications for an Influenza Vaccine Design. *Vaccines* **2019**, *7*, 47. [CrossRef]
11. Johnston, S.C.; Ricks, K.M.; Lakhal-Naouar, I.; Jay, A.; Subra, C.; Raymond, J.L.; King, H.A.D.; Rossi, F.; Clements, T.L.; Fetterer, D.; et al. A SARS-CoV-2 Spike Ferritin Nanoparticle Vaccine Is Protective and Promotes a Strong Immunological Response in the Cynomolgus Macaque Coronavirus Disease 2019 (COVID-19) Model. *Vaccines* **2022**, *10*, 717. [CrossRef]
12. Kim, S.A.; Kim, S.; Kim, G.B.; Goo, J.; Kim, N.; Lee, Y.; Nam, G.H.; Lim, S.; Kim, T.; Chang, K.H.; et al. A Multivalent Vaccine Based on Ferritin Nanocage Elicits Potent Protective Immune Responses against SARS-CoV-2 Mutations. *Int. J. Mol. Sci.* **2022**, *23*, 6123. [CrossRef] [PubMed]
13. Bale, J.B.; Gonen, S.; Liu, Y.; Sheffler, W.; Ellis, D.; Thomas, C.; Cascio, D.; Yeates, T.O.; Gonen, T.; King, N.P.; et al. Accurate design of megadalton-scale two-component icosahedral protein complexes. *Science* **2016**, *353*, 389–394. [CrossRef] [PubMed]
14. Butterfield, G.L.; Lajoie, M.J.; Gustafson, H.H.; Sellers, D.L.; Nattermann, U.; Ellis, D.; Bale, J.B.; Ke, S.; Lenz, G.H.; Yehdego, A.; et al. Evolution of a designed protein assembly encapsulating its own RNA genome. *Nature* **2017**, *552*, 415–420. [CrossRef]
15. Tsien, R.Y. THE GREEN FLUORESCENT PROTEIN. *Annu. Rev. Biochem.* **1998**, *67*, 509–544. [CrossRef] [PubMed]
16. Dove, S.G.; Hoegh-Guldberg, O.; Ranganathan, S. Major colour patterns of reef-building corals are due to a family of GFP-like proteins. *Coral Reefs* **2001**, *19*, 197–204. [CrossRef]
17. Cody, C.W.; Prasher, D.C.; Westler, W.M.; Prendergast, F.G.; Ward, W.W. Chemical structure of the hexapeptide chromophore of the Aequorea green-fluorescent protein. *Biochemistry* **1993**, *32*, 1212–1218. [CrossRef]
18. Close, D.W.; Paul, C.D.; Langan, P.S.; Wilce, M.C.; Traore, D.A.; Halfmann, R.; Rocha, R.C.; Waldo, G.S.; Payne, R.J.; Rucker, J.B.; et al. Thermal green protein, an extremely stable, nonaggregating fluorescent protein created by structure-guided surface engineering. *Proteins* **2014**, *83*, 1225–1237. [CrossRef] [PubMed]
19. Kiss, C.; Temirov, J.; Chasteen, L.; Waldo, G.S.; Bradbury, A.R. Directed evolution of an extremely stable fluorescent protein. *Protein Eng. Des. Sel.* **2009**, *22*, 313–323. [CrossRef]
20. Cormack, B.P.; Valdivia, R.H.; Falkow, S. FACS-optimized mutants of the green fluorescent protein (GFP). *Gene* **1996**, *173*, 33–38. [CrossRef]
21. Crameri, A.; Whitehorn, E.A.; Tate, E.; Stemmer, W.P.C. Improved Green Fluorescent Protein by Molecular Evolution Using DNA Shuffling. *Nat. Biotechnol.* **1996**, *14*, 315–319. [CrossRef] [PubMed]
22. Pavoor, T.V.; Cho, Y.K.; Shusta, E.V. Development of GFP-based biosensors possessing the binding properties of antibodies. *Proc. Natl. Acad. Sci. USA* **2009**, *106*, 11895–11900. [CrossRef]
23. Wang, R.; Xiang, S.; Zhang, Y.; Chen, Q.; Zhong, Y.; Wang, S. Development of a functional antibody by using a green fluorescent protein frame as the template. *Appl. Environ. Microbiol.* **2014**, *80*, 4126–4137. [CrossRef]
24. Cabantous, S.p.; Terwilliger, T.C.; Waldo, G.S. Protein tagging and detection with engineered self-assembling fragments of green fluorescent protein. *Nat. Biotechnol.* **2005**, *23*, 102–107. [CrossRef] [PubMed]
25. Romei, M.G.; Boxer, S.G. Split Green Fluorescent Proteins: Scope, Limitations, and Outlook. *Annu. Rev. Biophys.* **2019**, *48*, 19–44. [CrossRef] [PubMed]
26. He, Y.; Weng, L.; Li, R.; Li, L.; Toyoda, T.; Zhong, J. The N-terminal helix alpha(0) of hepatitis C virus NS3 protein dictates the subcellular localization and stability of NS3/NS4A complex. *Virology* **2012**, *422*, 214–223. [CrossRef] [PubMed]

27. Horner, S.M.; Park, H.S.; Gale, M., Jr. Control of innate immune signaling and membrane targeting by the Hepatitis C virus NS3/4A protease are governed by the NS3 helix alpha0. *J. Virol.* **2012**, *86*, 3112–3120. [CrossRef]
28. Tossi, A.; Sandri, L.; Giangaspero, A. Amphipathic, alpha-helical antimicrobial peptides. *Pept. Sci.* **2000**, *55*, 4–30. [CrossRef]
29. Rossman, J.S.; Jing, X.; Leser, G.P.; Lamb, R.A. Influenza virus M2 protein mediates ESCRT-independent membrane scission. *Cell* **2010**, *142*, 902–913. [CrossRef]
30. Roberts, K.L.; Leser, G.P.; Ma, C.; Lamb, R.A. The amphipathic helix of influenza A virus M2 protein is required for filamentous bud formation and scission of filamentous and spherical particles. *J. Virol.* **2013**, *87*, 9973–9982. [CrossRef]
31. Kan, M.C. Novel Protein Structure Used for Efficient Antibody Production in Immunization. TWI683668B, 1 February 2018.
32. Borisova, G.P.; Berzins, I.; Pushko, P.M.; Pumpen, P.; Gren, E.J.; Tsibinogin, V.V.; Loseva, V.; Ose, V.; Ulrich, R.; Siakkou, H.; et al. Recombinant core particles of hepatitis B virus exposing foreign antigenic determinants on their surface. *FEBS Lett.* **1989**, *259*, 121–124. [CrossRef]
33. Kobayashi, T.; Morone, N.; Kashiyama, T.; Oyamada, H.; Kurebayashi, N.; Murayama, T. Engineering a novel multifunctional green fluorescent protein tag for a wide variety of protein research. *PLoS ONE* **2008**, *3*, e3822. [CrossRef] [PubMed]
34. Pédelacq, J.-D.; Cabantous, S.; Tran, T.; Terwilliger, T.C.; Waldo, G.S. Engineering and characterization of a superfolder green fluorescent protein. *Nat. Biotechnol.* **2006**, *24*, 79–88. [CrossRef] [PubMed]
35. Jegerlehner, A.; Wiesel, M.; Dietmeier, K.; Zabel, F.; Gatto, D.; Saudan, P.; Bachmann, M.F. Carrier induced epitopic suppression of antibody responses induced by virus-like particles is a dynamic phenomenon caused by carrier-specific antibodies. *Vaccine* **2010**, *28*, 5503–5512. [CrossRef]
36. Feldman, R.A.; Fuhr, R.; Smolenov, I.; Ribeiro, A.; Panther, L.; Watson, M.; Senn, J.J.; Smith, M.; Almarsson, R.; Pujar, H.S.; et al. mRNA vaccines against H10N8 and H7N9 influenza viruses of pandemic potential are immunogenic and well tolerated in healthy adults in phase 1 randomized clinical trials. *Vaccine* **2019**, *37*, 3326–3334. [CrossRef] [PubMed]
37. Li, J.X.; Hou, L.H.; Meng, F.Y.; Wu, S.P.; Hu, Y.M.; Liang, Q.; Chu, K.; Zhang, Z.; Xu, J.J.; Tang, R.; et al. Immunity duration of a recombinant adenovirus type-5 vector-based Ebola vaccine and a homologous prime-boost immunisation in healthy adults in China: Final report of a randomised, double-blind, placebo-controlled, phase 1 trial. *Lancet Glob. Health* **2016**, *5*, e324–e334. [CrossRef]
38. Modjarrad, K.; Roberts, C.C.; Mills, K.T.; Castellano, A.R.; Paolino, K.; Muthumani, K.; Reuschel, E.L.; Robb, M.L.; Racine, T.; Oh, M.-D.; et al. Safety and immunogenicity of an anti-Middle East respiratory syndrome coronavirus DNA vaccine: A phase 1, open-label, single-arm, dose-escalation trial. *Lancet Infect. Dis.* **2019**, *19*, 1013–1022. [CrossRef] [PubMed]
39. Zhu, F.C.; Wurie, A.H.; Hou, L.H.; Liang, Q.; Li, Y.H.; Russell, J.B.; Wu, S.P.; Li, J.X.; Hu, Y.M.; Guo, Q.; et al. Safety and immunogenicity of a recombinant adenovirus type-5 vector-based Ebola vaccine in healthy adults in Sierra Leone: A single-centre, randomised, double-blind, placebo-controlled, phase 2 trial. *Lancet* **2016**, *389*, 621–628. [CrossRef]
40. Amanna, I.J.; Carlson, N.E.; Slifka, M.K. Duration of humoral immunity to common viral and vaccine antigens. *N. Engl. J. Med.* **2007**, *357*, 1903–1915. [CrossRef]
41. Hu, B.; Siche, S.; Möller, L.; Veit, M. Amphipathic Helices of Cellular Proteins Can Replace the Helix in M2 of Influenza A Virus with Only Small Effects on Virus Replication. *J. Virol.* **2020**, *94*, e01605–e01619. [CrossRef] [PubMed]
42. Ma, M.; Liu, J.; Jin, S.; Wang, L. Development of tumour peptide vaccines: From universalization to personalization. *Scand. J. Immunol.* **2020**, *91*, e12875. [CrossRef]
43. Li, W.; Joshi, M.D.; Singhania, S.; Ramsey, K.H.; Murthy, A.K. Peptide Vaccine: Progress and Challenges. *Vaccines* **2014**, *2*, 515–536. [CrossRef]
44. Svendsen, O.; Edwards, C.N.; Lauritzen, B.; Rasmussen, A.D. Intramuscular injection of hypertonic saline: In vitro and in vivo muscle tissue toxicity and spinal neurone c-fos expression. *Basic Clin. Pharmacol. Toxicol.* **2005**, *97*, 52–57. [CrossRef]
45. Luo, M.; Shao, B.; Yu, J.Y.; Liu, T.; Liang, X.; Lu, L.; Ye, T.H.; He, Z.Y.; Xiao, H.Y.; Wei, X.W. Simultaneous enhancement of cellular and humoral immunity by the high salt formulation of Al(OH)3 adjuvant. *Cell Res.* **2017**, *27*, 586–589. [CrossRef]

vaccines

MDPI

Article

The Development of Surface-Modified Liposomes as an Intranasal Delivery System for Group A Streptococcus Vaccines

Jieru Yang [1], Jennifer C. Boer [2], Mattaka Khongkow [3], Sarunya Phunpee [3], Zeinab G. Khalil [4], Sahra Bashiri [1], Cyril Deceneux [2], Georgia Goodchild [2], Waleed M. Hussein [1], Robert J. Capon [4], Uracha Ruktanonchai [3], Magdalena Plebanski [2], Istvan Toth [1,4,5] and Mariusz Skwarczynski [1,*]

1 School of Chemistry and Molecular Biosciences, The University of Queensland, St. Lucia, QLD 4072, Australia
2 School of Health and Biomedical Sciences, RMIT University, Bundoora, Melbourne, VIC 3083, Australia
3 National Nanotechnology Center (NANOTEC), National Science and Technology Development Agency (NSTDA), 111 Thailand Science Park, Phahonyothin Road, Klong 1, Pathumthani 12120, Thailand
4 Institute for Molecular Bioscience, The University of Queensland, St. Lucia, QLD 4072, Australia
5 School of Pharmacy, The University of Queensland, Woolloongabba, QLD 4102, Australia
* Correspondence: m.skwarczynski@uq.edu.au; Tel.: +61-73-346-9894

Abstract: Intranasal vaccine administration can overcome the disadvantages of injectable vaccines and present greater efficiency for mass immunization. However, the development of intranasal vaccines is challenged by poor mucosal immunogenicity of antigens and the limited availability of mucosal adjuvants. Here, we examined a number of self-adjuvanting liposomal systems for intranasal delivery of lipopeptide vaccine against group A Streptococcus (GAS). Among them, two liposome formulations bearing lipidated cell-penetrating peptide KALA and a new lipidated chitosan derivative (oleoyl-quaternized chitosan, OTMC) stimulated high systemic antibody titers in outbred mice. The antibodies were fully functional and were able to kill GAS bacteria. Importantly, OTMC was far more effective at stimulating antibody production than the classical immune-stimulating trimethyl chitosan formulation. In a simple physical mixture, OTMC also enhanced the immune responses of the tested vaccine, without the need for a liposome delivery system. The adjuvanting capacity of OTMC was further confirmed by its ability to stimulate cytokine production by dendritic cells. Thus, we discovered a new immune stimulant with promising properties for mucosal vaccine development.

Keywords: group A Streptococcus; oleoyl-quaternized chitosan; adjuvant; intranasal vaccine; multilamellar liposome; cell-penetrating peptide

1. Introduction

Intranasal delivery is one of the most advantageous routes for vaccine administration [1–3]. Mucosal surfaces are well-vascularized, enabling rapid antigen absorption into the lymphatic system. The activity of proteolytic enzymes in the nasal cavity, which can potentially destroy antigens, is low in comparison to the oral delivery pathway. Intranasal immunization (a) can trigger systemic immune responses; (b) is needle-free and has high patient compliance; (c) can be self-administered and, therefore, the cost of vaccine administration can be greatly reduced (no need for specialized personnel); and (d) has greater capacity for mass immunizations. Furthermore, nasal-associated lymphoid tissue, the main tissue involved in intranasal immunity, is similar in humans and rodents, which greatly simplifies the translation of animal study into clinical trials. Two intranasally delivered vaccines (live-attenuated virus-based) have already been approved and licensed for influenza (FluMist/Fluenz™ and Nasovac™). However, before any vaccine can be approved for intranasal delivery, it must overcome a variety of challenges. These include (a) fast mucosal clearance and, therefore, limited interaction time of a vaccine with the mucosal surface; (b) poor mucosal permeability; (c) efficacy even with the limited dosage that can be administered via the nose; and (d) a lack of effective approved mucosal adjuvants (immune stimulators). These limitations can all lead to suboptimal vaccine efficacy. Consequently, the need exists for an effective mucosal vaccine self-adjuvanting delivery system. Furthermore, self-adjuvanting delivery systems provide additional advantages by circumventing the requirement of toxic adjuvants, such as complete Freund's adjuvant (CFA) [4–6], while still inducing potent immune responses.

Group A Streptococcus (*Streptococcus pyogenes*, GAS) is a gram-positive bacteria that causes a variety of diseases, from common pharyngitis (strep throat) to deadly rheumatic heart disease (RHD) [7]. RHD, alone, results in hundreds of thousands of deaths per year, worldwide [8]. There is no vaccine available to prevent this infection. Importantly, the traditional live-attenuated pathogen-based strategy cannot be used for GAS vaccine development, as the whole bacterium induces autoimmune reactions. Thus, a major virulent factor of GAS, M protein, was selected as the most promising vaccine antigen. Since similarities of M protein sequence with human proteins were detected, the development of the GAS vaccine has been purely focused on short peptide epitopes derived from GAS proteins. Significantly, all clinically tested vaccine candidates in the last two decades have been composed of peptide epitopes derived from M-protein [9]. Among them, J8 peptide (QAEDKVKQSREAKKQVEKALKQLEDKVQ) is the only conserved epitope recognized by a variety of GAS strains that reached clinical trials [10]. Thus, we selected J8 as an antigen for mucosal vaccine development and investigated a variety of intranasal delivery platforms for one of our lead GAS vaccine candidates, LCP−1 (Figure 1a) [11,12].

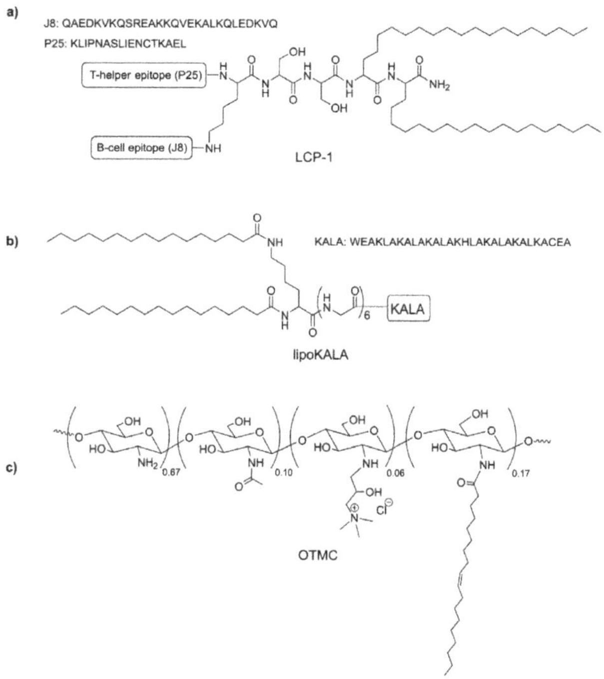

Figure 1. The chemical structure of (**a**) LCP−1, (**b**) lipoKALA, and (**c**) OTMC.

Liposome-based systems are one the most popular intranasal delivery platforms [2,13,14], especially as the liposomes' immunological properties can be modified by polyelectrolyte-based coatings [15,16]. These coatings can stabilize liposomes, and more effectively protect encapsulated antigens and improve their mucoadhesive properties. For example, we have demonstrated that the immunogenicity of liposomes bearing lipopeptide-based vaccines was enhanced by coating liposomes with alginate and trimethyl chitosan

(TMC) [17]. Moreover, when such vaccine was coated with dextran and TMC and used for intranasal immunization of mice, higher antibody titers were observed compared to antigen delivered with a commercial mucosal adjuvant, cholera toxin subunit B [18]. Interestingly, once negatively charged dextran was converted to its cationic derivative, diethylaminoethyl(DEAE)-dextran, it provided adjuvanting activity on its own [19,20]. We also recently showed that cell-penetrating peptides (CPPs), such as lipidated KALA (WEAKLAKALAKALAKHLAKALAKALKACEA) and polyethylenimine (PEI), can improve vaccine efficacy upon intranasal immunization (Figure 1b) [21,22]. While TMC is a widely investigated mucoadhesive polymer, Ruktanonchai and co-workers recently demonstrated that oleoyl-quaternized chitosan (OTMC, Figure 1c) also has mucoadhesive properties [23]. Therefore, we hypothesized that OTMC may also improve the immunogenicity of mucosal vaccines.

As a leading vaccine candidate against GAS, lipopeptide LCP−1 was selected for this study [11,24]. LCP−1 carries universal P25 T-helper epitope, J8, GAS M protein-derived B-cell epitope, and two lipidic moieties (Figure 1). Here, we designed several delivery systems for the LCP−1 vaccine to determine which has the greatest capacity to induce strong and effective antibody responses following intranasal immunization. These included five liposome-based vaccine candidates: LCP−1 encapsulated into liposomes carrying lipoKALA peptide (**L1**); [21] OTMC (**L2**); lipoPEI (**L3**); [22] and LCP−1 coated with alginate/DEAE-dextran (**L4**); or alginate/TMC (**L5**) (Figure 2). Moreover, three corresponding physical mixtures were also investigated: LCP−1/OTMC; LCP−1/DEAE-dextran; and LCP−1/TMC, which contained equivalent amounts of LCP−1 and polymers to the liposomal formulations. The vaccine candidates were evaluated in outbred mice following intranasal administration. A common key feature that defines many vaccine adjuvants is an ability to stimulate immune-stimulatory cytokines from the main immune-activating cell of the immune system, the dendritic cell (DC) [25]. Indeed, DC are uniquely capable of priming the immune system to recognize new pathogens, given their ability to endocytose, process and present antigens to lymphocytes, and release secondary signals, such as pro-inflammatory cytokines, which further result in their production and activation. Therefore, the ability of vaccines to stimulate cytokine production by DC has also been evaluated.

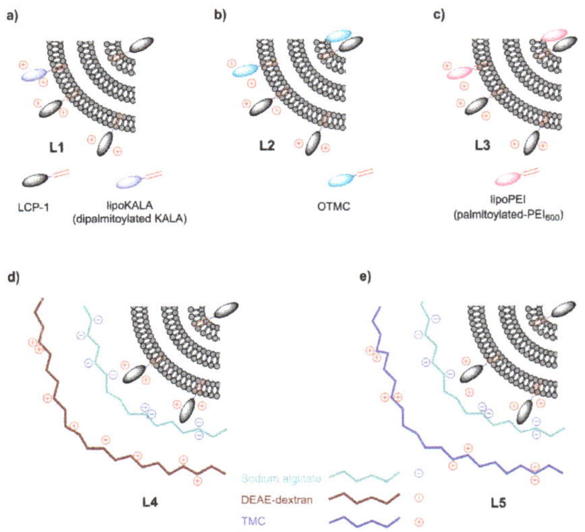

Figure 2. The multilamellar liposome−based LCP−1 vaccine delivery system bearing immunity enhancers, (**a**) lipoKALA (**L1**), (**b**) OTMC (**L2**), (**c**) lipoPEI (**L3**), (**d**) DEAE-dextran (**L4**), and (**e**) TMC (**L5**).

2. Materials and Methods

All chemicals used in this study were analytical-grade or equivalent, unless stated otherwise. Protected Fmoc/Boc-amino acids were obtained from Novabiochem (Läufelfingen, Switzerland). 1-[Bis(dimethylamino)methylene]-1*H*-1,2,3-triazolo[4,5-b]pyridinium 3-oxid hexafluorophosphate (HATU) was purchased from Mimotopes (Melbourne, Australia). *p*-MBHA·HCl resin was purchased from Peptides International (Louisville, KY, USA). Rink amide 4-methylbenzhydrylamine (MBHA) resin was purchased from Novabiochem (Hohenbrunn, Germany). Methanol, dichloromethane (DCM), *N,N*′-dimethylformamide (DMF), chloroform, *N,N*-diisopropylethylamine (DIPEA), HPLC-grade acetonitrile, trifluoroacetic acid (TFA), and piperidine were purchased from Merck (Hohenbrunn, Germany). Triisopropylsilane (TIPS), copper wire, phenylmethylsulfonyl fluoride (PMSF), and phosphate-buffered saline (PBS) tablets were purchased from Gibco (Paisly, UK). *O*-phenylenediamine dihydrochloride (OPD) was purchased from SIGMAFAST^{TM}. Goat anti-mouse IgG conjugated to horseradish peroxidase and its substrate were purchased from Bio-Rad (Hercules, CA, USA). Dipalmitoylphosphatidylcholine (DPPC), cholesterol (CH), an Avanti mini extruder, PC membranes, and filter supports were purchased from Avanti (Alabaster, AL, USA).

Analytical reverse-phase high-performance liquid chromatography (RP-HPLC) was performed on a Shimadzu LCMS-2020 instrument (Kyoto, Japan) with a Vydac analytical C-4 (214TP; 10 μm, 250 × 4.6 mm) or C-18 column (218TP; 10 μm, 250 × 4.6 mm) at a flow rate of 1 mL/min. Detection was conducted at 214 nm. Preparative RP-HPLC was performed on a Shimadzu instrument using either a Vydac or Altima preparative C-18 column (218TP; 10 mm, 250 × 22 mm) and C-4 column (214TP; 10 mm, 250 × 22 mm), or a semi-preparative column in linear gradient mode using a flow rate of 10–20 mL/min. Detection was conducted at 214 nm. Compounds were synthesized as previously reported: LCP−1 [26], lipoKALA [21], OTMC [23], lipoPEI [22], and TMC (degree of quaternization 65%) [27].

2.1. Preparation of LCP−1-Loaded Multilamellar Liposomes

All liposomal formulations were prepared to contain 1 mg/mL of LCP−1.

Liposomes **L1** were formulated with DPPC, CH, LCP−1, and lipoKALA at a molar ratio of 2:1:0.05:0.01. DPPC (4 mg in 1 mL chloroform), CH (1.05 mg in 1 mL chloroform), LCP−1 (1 mg in 0.5 mL methanol), and lipoKALA (0.135 mg in 0.5 mL methanol) were mixed together in a 5 mL round bottom flask. The solvents were very slowly removed under reduced pressure to produce a dry lipid film. The flask was stored under a high vacuum (in a freeze dryer) overnight to remove residual solvent. The formed film was rehydrated with Milli-Q water (1 mL) to produce multilamellar liposomes.

Liposomes **L2** and **L3** were formulated in an identical manner to **L1**, but with DPPC, CH, LCP−1, and OTMC at a molar ratio of 2:1:0.05:0.01 (**L2**); and DPPC, CH, LCP−1, and lipoPEI at a molar ratio of 2:1:0.05:0.01 (**L3**).

Liposomes **L4** and **L5** were formulated with DPPC, CH, and LCP−1 at a molar ratio of 2:1:0.05 in a similar manner to **L1**, except an additional coating was performed. The liposome coating procedure was performed after thin-film hydration and was based on electrostatic interactions. The optimal amount of polymer in the coating (sodium alginate, DEAE-dextran, and TMC) was determined based on the least amount required for full-surface coating, as detected by changes in zeta potential (Supplementary Figure S1). The rehydrated liposome solution (1 mL) was split into 10 individual vials (100 μL liposome solution in each, containing 100 μg LCP−1), while three polymer stock solutions (sodium alginate, TMC, and DEAE-dextran) were prepared to a 3 mg/mL concentration in Milli-Q water. Stock solutions were stirred for 2 h prior to the coating process. Various amounts of alginate stock solution (0, 5, 10, 20, 30, 40, 60, 65, 70, 75, 80, 100, or 200 μg of alginate) were added drop-by-drop to 100 μL liposome solutions and incubated for 1 h with gentle stirring at room temperature. The zeta potential of each solution was monitored by DLS, a stable negative zeta potential indicated that the coating was complete. The saturation quantity of sodium alginate to coat 100 μL of liposome solution (containing 100 μg LCP−1) was identified as 75 μg. Rehydrated liposome solution (1 mL) with 750 μg of coating was prepared and split into 10 individual vials. DEAE-dextran stock solution (0, 80, 110, 115, 125, 130, 150, 200, or 240 μg) was added drop-by-drop to individual alginate-coated liposome solutions (containing 100 μg LCP−1); the solutions were then incubated for 1 h with gentle stirring at room temperature. The optimal amount of DEAE-dextran coating was identified as 125 μg/100 μL liposomes (containing 100 μg LCP−1); this was used to produce **L4**. The amount of TMC for coating (150 μg/100 μL) of **L5** was determined in the same way.

2.2. Characterization of Vaccine Candidates by Dynamic Light Scattering

Average particle size, zeta potential, and the polydispersity index (PDI) were determined by dynamic light scattering (DLS) at a back-scattering angle of 173° at 25 °C in folded capillary cuvettes, using a Zetasizer Nano ZP instrument (Malvern, UK) with Malvern Zetasizer Analyser 6.2 software. The formulations were tested at 0.1 mg/mL concentration.

2.3. Immunization Study

Outbred female Swiss (CD-1) mice (9–10 weeks old) obtained from the Animal Resource Centre (Perth, Western Australia) were used for the immunization study. Mice were immunized intranasally with 30 μL (15 μL/nare) of PBS (negative control group), **L1**, **L2**, **L3**, **L4**, or **L5** (five mice per group). Each liposomal formulation, **L1–L5**, contained 30 μg of LCP−1. Mice were also immunized with physical mixtures of LCP−1/OTMC (containing 30 μg of LCP−1, and 45 μg of OTMC), LCP−1/DEAE-dextran (30 μg of LCP−1, and 37.5 μg of DEAE-dextran), LCP−1/TMC (30 μg of LCP−1, and 45 μg of TMC), and LCP−1 alone (30 μg). Three boosts were performed on days 14, 28, and 42. Blood was collected via tail bleed on days −1, 13, 27, and 41 and by cardiac puncture on day 52. The clear supernatant serum was collected after centrifugation for 10 min at 956× *g* (3600 rpm). Serum samples were stored at −80 °C.

2.4. Determination of IgG Titres

ELISA was used to measure J8-specific IgG antibody titers, as previously described [21]. Briefly, plates were coated with J8 (50 μg/plate) in carbonate coating buffer and then blocked with a 5% skim milk/PBS-Tween 20 buffer. Serial dilutions of two-fold serum samples were implemented in a 0.5% skim milk/PBS-Tween 20 buffer, with an initial serum dilution of 1:200. Horseradish-conjugated goat anti-mouse IgG (H + L) was used as a secondary antibody and O-phenylenediamine as substrate. An antibody titer was described as the lowest concentration producing an absorbance (at 450 nm) that was higher than 3 standard deviations (STD) above the average of serum from PBS-immunized mice (negative control).

2.5. Opsonization Assays

Opsonization assays were performed as described previously [28], using clinical isolates of D3840 (nasopharynx swab) and GC2203 (wound swab) donated by the Princess Alexandra Hospital (Brisbane, Australia). Briefly, bacterial isolates were streaked onto Todd–Hewitt broth supplemented with 5% yeast extract agar plates, then incubated for 24 h at 37°C. Single colonies were transferred to Todd–Hewitt broth (5 mL) supplemented with 5% yeast extract and incubated for 24 h at 37 °C to produce approximately 4.6×10^6 colony-forming units (CFU)/mL. The culture was serially diluted (×100) in PBS. Aliquots (10 μL) were mixed with horse blood (80 μL) and heat inactivated sera (10 μL, inactivated using a water bath at 50 °C for 15 min). Bacteria were incubated with the sera in a 96-well plate for 3 h at 37 °C. Bacterial survival was examined by plating a 10 μL aliquot of the culture material onto Todd–Hewitt agar plates supplemented with 5% horse blood and 5% yeast extract. Plates were incubated for 24 h at 37 °C and colonies were counted as CFU. Antibody opsonic activity (%) was calculated as (1 − [CFU in the presence of immunized mouse serum]/[mean CFU in the presence of untreated wells]) × 100%. Assays were performed from two independent cultures, in duplicate.

2.6. Ex vivo Cytokine Profiling in Dendritic Cells

Bone marrow-derived dendritic cells (BMDCs) were obtained from C57BL/6 mice (6–8 week-old) in triplicate, for each condition. Briefly, mice were culled by CO_2 asphyxiation. The femur and tibia of both legs were extracted and soaked in ethanol (70%) for 1 min, then washed and soaked in sterile Roswell Park Memorial Institute (RPMI) complete media (supplemented with 10% fetal bovine serum (FBS), 20 mM HEPES, 2 mM L-glutamine, 0.1 mM 2-mercaptoethanol, 100 µg of streptomycin and 100 units/mL penicillin; complete media (CM)). To extract bone marrow cells, bones were flushed and cells were then dissociated with a pipette, filtered through a cell strainer (100 µm, Millipore, Billerica, MA, USA) into a centrifuge tube (10 mL) and centrifuged at 372× *g* at room temperature for 5 min. The supernatant was removed and the cells were re-suspended in ammonium-chloride-potassium (ACK) lysis buffer (1 mL) for 1 min for erythrocyte lysis. The lysis buffer reaction was stopped with CM (9 mL). The cells were centrifuged again (372× *g*) at room temperature for 5 min, the supernatant was removed, and bone marrow cells were re-suspended in CM (10 mL). The bone marrow cells were then adjusted to 5×10^5 cells/mL in CM, plated in 6-well plates (Corning) and granulocyte and monocyte colony-stimulating factor (GM-CSF) (PeproTech, Rocky Hill, NJ, USA) was added to the cell suspension to reach a concentration of 10 ng/mL. Bone marrow cells were adjusted to a concentration of 5×10^5 cells/mL in CM and incubated for 3 days in 5% CO_2 at 37 °C. On day 3, the cells were incubated with lipopolysaccharide (LPS), alone (negative control), OTMC, LCP−1, or OTMC mixed with LCP−1. Three mice were used for biological replicates, and each condition was plated in duplicate. As a control, to ensure that cytokine upregulation was indeed directly stimulated by the compounds tested, the cells were plated in identical conditions and incubated with liposomes for 24 h at 4 °C. Following 24 h incubation with the compounds, the cell supernatant was collected to analyze mouse cytokine expression by ELISA for TNFα (cat. 558534, BD), IL1β (cat. 432601 Biolgened), IL6 (cat. 555240, BD), IL4 (cat. 555232, BD), IL12 (cat. 555256, BD), and IL23 (cat. 433704, Biolegend, CA, USA). ELISA was performed following the manufacturer's instructions. Absorbance was read on a plate reader (Multiscan GO, Thermo Fisher, Waltham, MA, USA) at 450 nm, within 30 min of adding the stop solution. Background absorbance was subtracted from all data points. Standard curves were generated using either linear regression or four-parameter logistic regression.

2.7. Ethics Statement

This study was performed according to the regulations set by the National Health and Medical Research Council (NHMRC) of Australia (Australian Code of Practice for the Care and Use of Animals for Scientific Purposes, 8th edition, 2013). All animal procedures and protocols were approved by The University of Queensland Animal Ethics Committee (AEC), AEC Approval Number SCMB/AIBN/069/17. In addition, C57BL/6 mice 6–8 weeks old were obtained from the Animal Resource Centre (ARC) in Western Australia. All ex vivo experiments had ethics approval by the Research Animal Facility (RAF) of RMIT university under AEC-approved project number 1917.

2.8. Statistical Analysis

GraphPad Prism® 7 software (GraphPad Software, Inc., San Diego, CA, USA) was used for all statistical analysis. One-way ANOVA followed by Tukey's multiple comparison test was applied for statistical analysis, or multiple comparison Kruskal–Wallis tests with $p < 0.05$ were considered statistically significant.

3. Results

Liposomes **L1–L5** were prepared without extrusion as multilamellar vesicles. Liposomes were composed of neutral lipids (composed of DPPC and CH) and the positively charged (due to high cationic amino acid content) lipopeptide vaccine, LCP−1. They had an overall positive surface charge (50 ± 2 mV). These liposomes carried additional anchored moieties: lipoKALA (**L1**), OTMC (**L2**), lipoPEI (**L3**), or were coated with alginate/DEAE-dextran (**L4**), or alginate/TMC (**L5**). All were positively charged according to DLS analysis. OTMC-bearing liposomes (**L2**) had a higher positive charge than TMC-bearing liposomes (**L5**), which can be explained by the presence of negatively charged alginate in **L5. L1–L5** formed a variety of sizes, ranging from 60–5000 nm, and were highly polydisperse (PDI), which is typical for multilamellar liposomes (Supplementary Figure S2). Physical mixtures of LCP−1/OTMC, LCP−1/DEAE-dextran, and LCP−1/TMC were also prepared.

The new liposome- and physical mixture-based formulations were examined in outbred Swiss mice for the ability to induce antibody production upon intranasal immunization, and compared to our previous lead formulation, **L1** [21], bearing lipoKALA as an immune enhancer. Again, **L1** elicited a high J8-specific IgG titer, which was significantly higher than those induced by LCP−1 alone (Figure 3a). Interestingly, a comparable J8-specific IgG titer was produced by mice vaccinated with **L2,** which incorporated OTMC. Moreover, when mice were immunized with OTMC and physically mixed with LCP−1, the produced IgG titers were not significantly different from **L1**. In contrast, significantly lower antibody titers were produced by mice vaccinated with the two coated liposomes (**L4** and **L5**). **L3**, which contained lipoPEI, produced a higher IgG titer compared to the negative control (PBS) group, but it was not significantly different from that of mice immunized with LCP−1 alone. Similarly, the two physical mixtures, LCP−1/DEAE-dextran and LCP−1/TMC, did not stimulate higher IgG expression than LCP−1. The serum isolated from mice immunized with the three most effective formulations (**L1**, **L2**, and LCP−1/OTMC). The weakly performing **L4** and LCP−1 (control) were analyzed for their ability to kill GAS bacteria. **L1**, and especially **L2**, had the highest opsonization potentials against both GAS clinical isolates (Figure 3b,c).

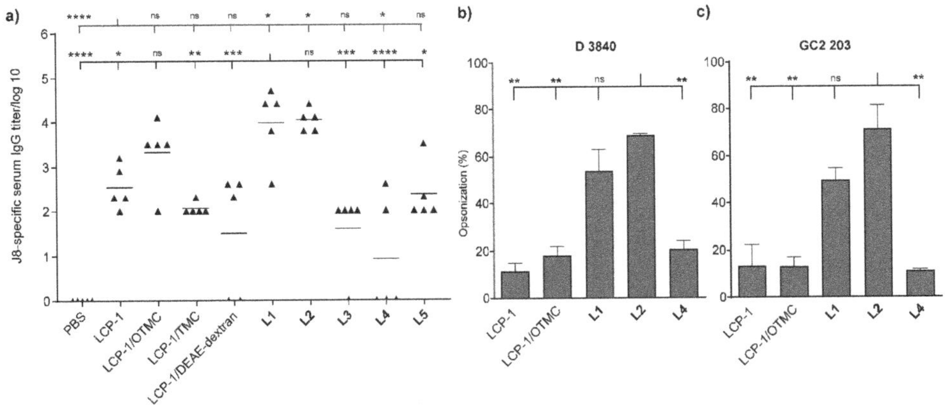

Figure 3. (**a**) J8-specific antibody responses (log10) following intranasal administration of LCP−1-based liposomes and controls in Swiss mice (n = 5), as determined by ELISA. Serum was collected on day 52. The average opsonization percentage of group A Streptococcus strains (**b**) D3840 and (**c**) GC2 203). Statistical analysis was performed using one-way ANOVA followed by Tukey's post-hoc test. All groups were compared with **L1** or LCP−1. Not significant (ns), $p > 0.05$; *; $p < 0.05$; **; $p < 0.01$; ***; $p < 0.001$, and ****; $p < 0.0001$.

Cytokine profiling was performed using C57BL/6 mouse-derived BMDCs upon treatment with OTMC and controls, as this chitosan derivative had not been previously tested for its adjuvanting capabilities (Figure 4). Secretion of pro-inflammatory cytokine TNFα by BMDCs was significantly higher for LCP−1/OTMC and OTMC than for the negative control (Figure 4a). OTMC also induced significantly higher levels of expression of pro-inflammatory cytokine IL6 and IL1β, while its mixture with LCP−1 did not (Figure 4b,c). While the production of these cytokines was higher in cells treated with OTMC than LCP−1/OTMC, the difference was not statistically significant. Despite its reported self-adjuvanting properties [29], LCP−1, alone, did not induce significant cytokine release with the concentration of tested lipopeptide. Production of IL4 and IL12 was slightly upregulated by LCP−1/OTMC and OTMC; however, this did not differ significantly from the negative control (Figure 4d,e). None of the compounds stimulated the production of IL23 (Figure 4f). The upregulation of cytokine production was not observed under cold conditions (Supplementary Figure S3). Interestingly, TNFα and IL1β production triggered by OTMC was higher than with the gold-standard stimulant, LPS, in the experimental conditions used here.

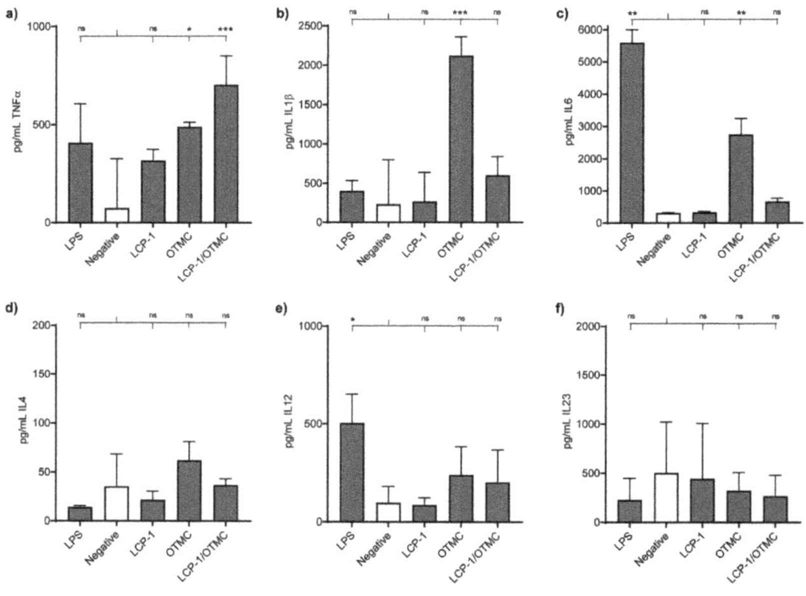

Figure 4. Cytokine production by BMDCs following stimulation with LPS, alone (negative control), OTMC, LCP−1, and OTMC mixed with LCP−1. Cytokine production was expressed as pg/mL ± SD. Production of (**a**) TNFα; (**b**) IL-1β; (**c**) IL6; (**d**) IL4; (**e**) IL12; and (**f**) IL23 was detected following BMDC stimulation by LPS (1 µg/mL). Statistical analysis was performed using multiple comparison Kruskal–Wallis test (ns, $p > 0.05$; * $p < 0.05$; ** $p < 0.01$; and *** $p < 0.001$).

4. Discussion

Liposomal antigen delivery has been recognized as a promising strategy for vaccine development. Liposome-based vaccine formulations have been approved for human use against hepatitis A (Epaxal) and influenza (Inflexal V) [17,30]. Liposomes can carry antigens encapsulated inside their aqueous core or attached to their membrane via lipid-based anchoring. Furthermore, a variety of lipidated moieties can be anchored to liposomes to modify their properties. One such moiety, CPPs, has been used to enhance vaccine efficacy [31]. We recently screened a variety of lipidated CPPs anchored to liposomes for their ability to enhance humoral immune responses against GAS. Among the tested formulations, multilamellar liposomes were the most effective. Of the CPPs tested, lipoKALA (Figure 1) stimulated the production of the highest level of opsonic IgG titers. Thus, the optimized formulation, **L1** carrying lipoKALA, was chosen as a control for the further development of intranasal vaccine delivery systems. For this study, we selected a variety of known and previously examined immune-stimulating moieties (TMC, lipoPEI, DEAE-dextran) [22,26,32], as well as untested OTMC. Chitosan derivatives (e.g., TMC) are known immune stimulators [33], while lipidated sugars often demonstrate adjuvanting abilities, especially once incorporated into liposomes [34–38]. Therefore, we hypothesized that a lipidated analog of TMC (OTMC) may also be an effective immune stimulator.

Multilamellar liposomes were chosen as a vaccine carrier based on our previous studies demonstrating their (i.e., **L1**) ability to induce stronger immune responses compared to unilamellar liposomes (120 nm) [21]. Indeed, we also demonstrated that smaller unilamellar liposomes induced higher antibody titers than larger (70 nm > 140 nm > 400 nm) in mice following intranasal administration; however, multilamellar liposomes (150–1000 nm) were as effective as the smallest unilamellar liposomes [39]. Thus, four new liposomal formulations (**L2**–**L5**) were produced and examined for their ability to stimulate antibody production upon intranasal administration in outbred mice. As expected, liposome **L1** induced high-level antigen-specific IgG titers. Surprisingly, coated liposome **L5** failed to improve the immunogenicity of LCP−1, despite the TMC/alginate system having been reported as effective previously [26,28,40,41]. DEAE-dextran and lipidated PEI were even less effective than TMC.

It is important to note that the polymers mentioned above were examined previously only in unilamellar liposomal formulation, not multilamellar, as reported here. The new chitosan derivative, OTMC, showed a remarkable ability to induce IgG production once formulated into liposomes (**L2**) and was effective even as a simple mixture with LCP−1. However, once antibodies were tested for their GAS opsonization efficacy, only the liposomal formulation, **L2**, triggered the production of clearly opsonic antibodies at the tested concentrations. The high efficacy of **L2** was related to the special properties of OTMC, rather than the high positive charge of the liposomes (e.g., **L2** = +63 mV vs. **L5** = +36 mV), as OTMC was also very immunogenic in the physical mixture (LCP−1/OTMC vs. LCP−1/TMC, Figure 3a). In addition, OTMC/LCP−1 was more immunogenic than TMC/LCP−1, suggesting that more than the liposome-anchoring ability of OTMC was responsible for its activity. Thus, to further analyze the ability of OTMC to act as an adjuvant, we analyzed its ability to stimulate cytokine production. Pro-inflammatory cytokines play an important role in the control of adaptive immune responses. Several different cytokines were tested (TNFα, IL1β, IL6, IL4, IL12, and IL23). Cytokines, such as IL6 and TNFα, have the ability to enhance immune responses against viral infections/vaccines [42–44]. Both IL6 and IL12 play important roles in the induction of immune responses against influenza [42]. IL6 is also involved in the terminal differentiation of B-cells [42]. while IL12 and IL23 regulate the differentiation of CD4+ T follicular helper cells, providing support for B-cell generation of high-affinity antibodies and differentiation of B-cells into memory B-cells [45,46]. Three pro-inflammatory cytokines, TNFα, IL1β, and IL6, were secreted in significantly higher amounts upon dendritic cell stimulation with OTMC. OTMC stimulated higher cytokine production alone, compared to LCP−1, suggesting a lack of synergistic effect between these two lipidated compounds. IL6 and TNFα's ability to stimulate plasma cell longevity was also reported [47], and OTMC clearly upregulated the production of these cytokines. Moreover, IL1β produced at higher levels by OTMC than even the positive control LPS, and has a critical non-dispensable role in stimulating and priming naïve T cell activation [48], a useful new feature for this potential new adjuvant. In addition, cytokines IL12 and IL23 have been associated with autoimmunity and extensive inflammation [49]. Harmful effects associated with the overexpression of IL-4 have also been reported [50]. OTMC did not greatly enhance the level of IL4, IL12, or IL23 produced by BMDCs, suggesting no undesired immunity would be stimulated. Finally, overexpression of IL-1β can be associated with pain, inflammation, and autoimmune reactions [51]. Although OTMC significantly overexpressed this cytokine, we did not observe any adverse effects in mice immunized with OTMC. This could mean that, when administered together with LCP−1, it does not overstimulate IL-1β production.

In summary, both liposomal formulations, **L1** (bearing lipidated CPP KALA) and **L2** (bearing lipidated TMC), stimulated the production of fully functional opsonic antibodies. A newly discovered adjuvant (OTMC) showed the ability to trigger cytokine release by dendritic cells, as well as IgG production, even when administered as a physical mixture with the antigen. However, it was more effective in liposomal formulation.

5. Conclusions

We examined a variety of liposomal intranasal vaccine delivery systems. While we confirmed that the lipoKALA liposome delivery strategy is efficient in inducing antibody production, we also discovered the remarkable adjuvanting capacity of OTMC. OTMC was not only effective in enhancing antibody production upon anchoring to liposomes, but also on its own when mixed with the LCP−1 vaccine. OTMC triggered cytokine release by dendritic cells when administered with or without LCP−1, further confirming its adjuvanting activity. TMC/DEAE dextran-coated and lipoPEI-incorporated liposomes were much less immunogenic than expected. In summary, liposomes bearing lipoKALA and OTMC are promising self-adjuvanting platforms for intranasal peptide-based vaccine delivery.

Supplementary Materials: The following are available online at https://www.mdpi.com/article/10.3390/vaccines11020305/s1, Figure S1: Coating of polyelectrolyte-coated liposomes. Effect of concentration of sodium alginate, TMC, and DEAE-dextran on zeta potentials of LCP−1 loaded liposome formulation. Zeta potential changes during the coating of liposome with (a) sodium alginate; (b) DEAE-dextran; and (c) with TMC. The red box marks the values selected for the final liposome coating. Figure S2: DLS spectra of particles of (a) **L1**, (b) **L2**, (c) L3, (d) **L4**, and (e) **L5**. Size distributions by intensity, PDI, and zeta potential. Figure S3: Production of (a) TNFα; (b) IL-1β; (c) IL6; (d) IL4; (e) IL12; and IL23 cytokines following stimulation of BMDC in cold-controlled conditions (4 °C). Stimulation was performed with LPS, none (negative control), OTMC, LCP−1, and OTMC mixed with LCP−1. No significant difference between the groups was detected.

Author Contributions: M.S. and I.T. designed the project. J.Y. performed the experiments and wrote the first draft of the manuscript. J.C.B., C.D., G.G. and M.P. performed and analyzed the cytokine profiling in dendritic cells. M.K., S.P. and U.R. synthesized OTMC. S.B. assisted in the TMC synthesis. Z.G.K., R.J.C. and W.M.H. performed the opsonization study and data analysis. M.S., W.M.H. and I.T. reviewed and edited the manuscript. All authors have read and agreed to the published version of the manuscript.

Funding: This research was funded by the National Health and Medical Research Council, Australia (NHMRC, APP1132975) and The University of Queensland, Australia.

Institutional Review Board Statement: The study was conducted in compliance with the guidelines from the Australian National Health and Medical Research Council (NHMRC). All immunization protocols were approved by The University of Queensland Ethics Committee (Animal Ethics Unit, Office of Research Ethics, The University of Queensland; approval number SCMB/AIBN/069/17).

Informed Consent Statement: Not applicable.

Data Availability Statement: The data presented in this study are available in the article and Supplementary Materials.

Acknowledgments: M.P. is a recipient of an NHMRC SRFB Fellowship.

Conflicts of Interest: The authors declare no conflict of interest.

References

1. Skwarczynski, M.; Toth, I. Non-invasive mucosal vaccine delivery: Advantages, challenges and the future. *Expert Opin. Drug Deliv.* **2020**, *17*, 435–437. [CrossRef] [PubMed]
2. Marasini, N.; Skwarczynski, M.; Toth, I. Intranasal delivery of nanoparticle-based vaccines. *Ther. Deliv.* **2017**, *8*, 151–167. [CrossRef] [PubMed]
3. Yusuf, H.; Kett, V. Current prospects and future challenges for nasal vaccine delivery. *Hum. Vaccin. Immunother.* **2017**, *13*, 34–45. [CrossRef] [PubMed]

4. Oscherwitz, J.; Hankenson, F.C.; Yu, F.; Cease, K.B. Low-dose intraperitoneal Freund's adjuvant: Toxicity and immunogenicity in mice using an immunogen targeting amyloid-beta peptide. *Vaccine* **2006**, *24*, 3018–3025. [CrossRef]
5. Powers, J.G.; Nash, P.B.; Rhyan, J.C.; Yoder, C.A.; Miller, L.A. Comparison of immune and adverse effects induced by AdjuVac and Freund's complete adjuvant in New Zealand white rabbits (Oryctolagus cuniculus). *Lab. Anim.* **2007**, *36*, 51–58. [CrossRef]
6. Petrovsky, N. Comparative Safety of Vaccine Adjuvants: A Summary of Current Evidence and Future Needs. *Drug Saf.* **2015**, *38*, 1059–1074. [CrossRef]
7. Sanyahumbi, A.S.; Colquhoun, S.; Wyber, R.; Carapetis, J.R. Global disease burden of group A Streptococcus. In *Streptococcus pyogenes: Basic Biology to Clinical Manifestations*; University of Oklahoma Health Sciences Center: Oklahoma City, OK, USA, 2016.
8. Watkins, D.A.; Johnson, C.O.; Colquhoun, S.M.; Karthikeyan, G.; Beaton, A.; Bukhman, G.; Forouzanfar, M.H.; Longenecker, C.T.; Mayosi, B.M.; Mensah, G.A.; et al. Global, Regional, and National Burden of Rheumatic Heart Disease, 1990–2015. *N. Engl. J. Med.* **2017**, *377*, 713–722. [CrossRef]
9. Azuar, A.; Jin, W.; Mukaida, S.; Hussein, W.M.; Toth, I.; Skwarczynski, M. Recent advances in the development of peptide vaccines and their delivery systems against group a streptococcus. *Vaccines* **2019**, *7*, 58. [CrossRef]
10. Sekuloski, S.; Batzloff, M.R.; Griffin, P.; Parsonage, W.; Elliott, S.; Hartas, J.; O'Rourke, P.; Marquart, L.; Pandey, M.; Rubin, F.A. Evaluation of safety and immunogenicity of a group A streptococcus vaccine candidate (MJ8VAX) in a randomized clinical trial. *PLoS ONE* **2018**, *13*, e0198658. [CrossRef]
11. Alharbi, N.; Skwarczynski, M.; Toth, I. The influence of component structural arrangement on peptide vaccine immunogenicity. *Biotechnol. Adv.* **2022**, *60*, 108029. [CrossRef]
12. Marasini, N.; Ghaffar, K.A.; Giddam, A.K.; Batzloff, M.R.; Good, M.F.; Skwarczynski, M.; Toth, I. Highly Immunogenic Trimethyl Chitosan-based Delivery System for Intranasal Lipopeptide Vaccines against Group A Streptococcus. *Curr. Drug Deliv.* **2017**, *14*, 701–708. [CrossRef] [PubMed]
13. Yang, J.; Azuar, A.; Toth, I.; Skwarczynski, M. Liposomes for the Delivery of Lipopeptide Vaccines. In *Vaccine Design: Methods and Protocols, Volume 3. Resources for Vaccine Development*; Thomas, S., Ed.; Springer: New York, NY, USA, 2022; pp. 295–307.
14. Marasini, N.; Ghaffar, K.A.; Skwarczynski, M.; Toth, I. Liposomes as a Vaccine Delivery System. In *Micro- and Nanotechnology in Vaccine Development*; Skwarczynski, M., Toth, I., Eds.; William Andrew Inc.: Norwich, UK, 2017; pp. 221–239.
15. Zhao, L.; Skwarczynski, M.; Toth, I. Polyelectrolyte-Based Platforms for the Delivery of Peptides and Proteins. *ACS Biomater. Sci. Eng.* **2019**, *5*, 4937–4950. [CrossRef] [PubMed]
16. Volodkin, D.V.; Schaaf, P.; Mohwald, H.; Voegel, J.-C.; Ball, V. Effective embedding of liposomes into polyelectrolyte multilayered films: The relative importance of lipid-polyelectrolyte and interpolyelectrolyte interactions. *Soft Matter* **2009**, *5*, 1394–1405. [CrossRef]
17. Ghaffar, K.A.; Marasini, N.; Giddam, A.K.; Batzloff, M.R.; Good, M.F.; Skwarczynski, M.; Toth, I. Liposome-based intranasal delivery of lipopeptide vaccine candidates against group A streptococcus. *Acta Biomater.* **2016**, *41*, 161–168. [CrossRef]
18. Marasini, N.; Giddam, A.K.; Khalil, Z.G.; Hussein, W.M.; Capon, R.J.; Batzloff, M.R.; Good, M.F.; Toth, I.; Skwarczynski, M. Double adjuvanting strategy for peptide-based vaccines: Trimethyl chitosan nanoparticles for lipopeptide delivery. *Nanomedicine* **2016**, *11*, 3223–3235. [CrossRef] [PubMed]
19. Houston, W.E.; Crabbs, C.L.; Kremer, R.J.; Springer, J.W. Adjuvant effects of diethylaminoethyl-dextran. *Infect. Immun.* **1976**, *13*, 1559–1562. [CrossRef]
20. Piedrafita, D.; Preston, S.; Kemp, J.; de Veer, M.; Sherrard, J.; Kraska, T.; Elhay, M.; Meeusen, E. The Effect of Different Adjuvants on Immune Parameters and Protection following Vaccination of Sheep with a Larval-Specific Antigen of the Gastrointestinal Nematode, Haemonchus contortus. *PLoS ONE* **2013**, *8*, e78357. [CrossRef] [PubMed]
21. Yang, J.; Firdaus, F.; Azuar, A.; Khalil, Z.G.; Marasini, N.; Capon, R.J.; Hussein, W.M.; Toth, I.; Skwarczynski, M. Cell-Penetrating Peptides-Based Liposomal Delivery System Enhanced Immunogenicity of Peptide-Based Vaccine against Group A Streptococcus. *Vaccines* **2021**, *9*, 499. [CrossRef] [PubMed]
22. Dai, C.C.; Yang, J.; Hussein, W.M.; Zhao, L.; Wang, X.; Khalil, Z.G.; Capon, R.J.; Toth, I.; Stephenson, R.J. Polyethylenimine: An Intranasal Adjuvant for Liposomal Peptide-Based Subunit Vaccine against Group A Streptococcus. *ACS Infect. Dis.* **2020**, *6*, 2502–2512. [CrossRef] [PubMed]
23. Yostawonkul, J.; Surassmo, S.; Iempridee, T.; Pimtong, W.; Suktham, K.; Sajomsang, W.; Gonil, P.; Ruktanonchai, U.R. Surface modification of nanostructure lipid carrier (NLC) by oleoyl-quaternized-chitosan as a mucoadhesive nanocarrier. *Colloids Surf. B Biointerfaces* **2017**, *149*, 301–311. [CrossRef]
24. Bartlett, S.; Skwarczynski, M.; Toth, I. Lipids as Activators of Innate Immunity in Peptide Vaccine Delivery. *Curr. Med. Chem.* **2020**, *27*, 2887–2901. [CrossRef] [PubMed]
25. Wilson, K.L.; Flanagan, K.L.; Prakash, M.D.; Plebanski, M. Malaria vaccines in the eradication era: Current status and future perspectives. *Expert Rev. Vaccines* **2019**, *18*, 133–151. [CrossRef] [PubMed]
26. Zhao, L.; Jin, W.; Cruz, J.G.; Marasini, N.; Khalil, Z.G.; Capon, R.J.; Hussein, W.M.; Skwarczynski, M.; Toth, I. Development of polyelectrolyte complexes for the delivery of peptide-based subunit vaccines against group A streptococcus. *Nanomaterials* **2020**, *10*, 823. [CrossRef] [PubMed]

27. Zhao, L.; Bashiri, S.; Toth, I.; Skwarczynski, M. Preparation of Trimethyl Chitosan-Based Polyelectrolyte Complexes for Peptide Subunit Vaccine Delivery. In *Bacterial Vaccines: Methods and Protocols*; Bidmos, F., Bossé, J., Langford, P., Eds.; Springer: New York, NY, USA, 2022; pp. 141–149.
28. Marasini, N.; Giddam, A.K.; Ghaffar, K.A.; Batzloff, M.R.; Good, M.F.; Skwarczynski, M.; Toth, I. Multilayer engineered nanoliposomes as a novel tool for oral delivery of lipopeptide-based vaccines against group A Streptococcus. *Nanomedicine* **2016**, *11*, 1223–1236. [CrossRef] [PubMed]
29. Zaman, M.; Abdel-Aal, A.B.; Phillipps, K.S.; Fujita, Y.; Good, M.F.; Toth, I. Structure-activity relationship of lipopeptide Group A streptococcus (GAS) vaccine candidates on toll-like receptor 2. *Vaccine* **2010**, *28*, 2243–2248. [CrossRef]
30. Tenchov, R.; Bird, R.; Curtze, A.E.; Zhou, Q. Lipid Nanoparticles—From Liposomes to mRNA Vaccine Delivery, a Landscape of Research Diversity and Advancement. *ACS Nano* **2021**, *15*, 16982–17015. [CrossRef]
31. Yang, J.R.; Luo, Y.C.; Shibu, M.A.; Toth, I.; Skwarczynski, M. Cell-Penetrating Peptides: Efficient Vectors for Vaccine Delivery. *Curr. Drug Deliv.* **2019**, *16*, 430–443. [CrossRef]
32. Zhao, L.; Yang, J.; Nahar, U.J.; Khalil, Z.G.; Capon, R.J.; Hussein, W.M.; Skwarczynski, M.; Toth, I. A dual-adjuvanting strategy for peptide-based subunit vaccines against group A Streptococcus: Lipidation and polyelectrolyte complexes. *Bioorg. Med. Chem.* **2020**, *28*, 115823. [CrossRef]
33. Bashiri, S.; Koirala, P.; Toth, I.; Skwarczynski, M. Carbohydrate Immune Adjuvants in Subunit Vaccines. *Pharmaceutics* **2020**, *12*, 965. [CrossRef]
34. Nahar, U.J.; Toth, I.; Skwarczynski, M. Mannose in vaccine delivery. *J. Control. Release* **2022**, *351*, 284–300. [CrossRef]
35. Reintjens, N.R.M.; Tondini, E.; de Jong, A.R.; Meeuwenoord, N.J.; Chiodo, F.; Peterse, E.; Overkleeft, H.S.; Filippov, D.V.; van der Marel, G.A.; Ossendorp, F.; et al. Self-Adjuvanting Cancer Vaccines from Conjugation-Ready Lipid A Analogues and Synthetic Long Peptides. *J. Med. Chem.* **2020**, *63*, 11691–11706. [CrossRef] [PubMed]
36. Norpi, A.S.M.; Nordin, M.L.; Ahmad, N.; Katas, H.; Fuaad, A.A.A.; Sukri, A.; Marasini, N.; Azmi, F. New modular platform based on multi-adjuvanted amphiphilic chitosan nanoparticles for efficient lipopeptide vaccine delivery against group A streptococcus. *Asian J. Pharm. Sci.* **2022**, *17*, 435–446. [CrossRef]
37. Stanisic, D.I.; Ho, M.-F.; Nevagi, R.; Cooper, E.; Walton, M.; Islam, M.T.; Hussein, W.M.; Skwarczynski, M.; Toth, I.; Good, M.F. Development and Evaluation of a Cryopreserved Whole-Parasite Vaccine in a Rodent Model of Blood-Stage Malaria. *mBio* **2021**, *12*, e0265721. [CrossRef]
38. Al-Nazal, H.A.; Cooper, E.; Ho, M.F.; Eskandari, S.; Majam, V.; Giddam, A.K.; Hussein, W.M.; Islam, M.T.; Skwarczynski, M.; Toth, I.; et al. Pre-clinical evaluation of a whole-parasite vaccine to control human babesiosis. *Cell Host Microbe* **2021**, *29*, 894–903.e5. [CrossRef]
39. Ghaffar, K.A.; Marasini, N.; Giddam, A.K.; Batzloff, M.R.; Good, M.F.; Skwarczynski, M.; Toth, I. The Role of Size in Development of Mucosal Liposome-Lipopeptide Vaccine Candidates against Group A Streptococcus. *Med. Chem.* **2017**, *13*, 22–27. [CrossRef] [PubMed]
40. Li, X.; Kong, X.; Shi, S.; Zheng, X.; Guo, G.; Wei, Y.; Qian, Z. Preparation of alginate coated chitosan microparticles for vaccine delivery. *BMC Biotechnol.* **2008**, *8*, 89. [CrossRef] [PubMed]
41. Slütter, B.; Plapied, L.; Fievez, V.; Sande, M.A.; des Rieux, A.; Schneider, Y.-J.; Van Riet, E.; Jiskoot, W.; Préat, V. Mechanistic study of the adjuvant effect of biodegradable nanoparticles in mucosal vaccination. *J. Control. Release* **2009**, *138*, 113–121. [CrossRef]
42. Fernández-Ruiz, M.; Humar, A.; Baluch, A.; Keshwani, S.; Husain, S.; Kumar, D. Baseline serum interleukin-6 to interleukin-2 ratio is associated with the response to seasonal trivalent influenza vaccine in solid organ transplant recipients. *Vaccine* **2015**, *33*, 7176–7182. [CrossRef] [PubMed]
43. Su, B.; Wang, J.; Wang, X.; Jin, H.; Zhao, G.; Ding, Z.; Kang, Y.; Wang, B. The effects of IL-6 and TNF-alpha as molecular adjuvants on immune responses to FMDV and maturation of dendritic cells by DNA vaccination. *Vaccine* **2008**, *26*, 5111–5122. [CrossRef]
44. Ovsyannikova, I.G.; Reid, K.C.; Jacobson, R.M.; Oberg, A.L.; Klee, G.G.; Poland, G.A. Cytokine production patterns and antibody response to measles vaccine. *Vaccine* **2003**, *21*, 3946–3953. [CrossRef] [PubMed]
45. Schmitt, N.; Morita, R.; Bourdery, L.; Bentebibel, S.E.; Zurawski, S.M.; Banchereau, J.; Ueno, H. Human dendritic cells induce the differentiation of interleukin-21-producing T follicular helper-like cells through interleukin-12. *Immunity* **2009**, *31*, 158–169. [CrossRef]
46. Schmitt, N.; Bustamante, J.; Bourdery, L.; Bentebibel, S.E.; Boisson-Dupuis, S.; Hamlin, F.; Tran, M.V.; Blankenship, D.; Pascual, V.; Savino, D.A.; et al. IL-12 receptor β1 deficiency alters in vivo T follicular helper cell response in humans. *Blood* **2013**, *121*, 3375–3385. [CrossRef] [PubMed]
47. Cassese, G.; Arce, S.; Hauser, A.E.; Lehnert, K.; Moewes, B.; Mostarac, M.; Muehlinghaus, G.; Szyska, M.; Radbruch, A.; Manz, R.A. Plasma cell survival is mediated by synergistic effects of cytokines and adhesion-dependent signals. *J. Immunol.* **2003**, *171*, 1684–1690. [CrossRef] [PubMed]
48. Plebanski, M.; Elson, C.J.; Billington, W.D. Dependency on interleukin-1 of primary human in vitro T cell responses to soluble antigens. *Eur. J. Immunol.* **1992**, *22*, 2353–2358. [CrossRef]
49. Schurich, A.; Raine, C.; Morris, V.; Ciurtin, C. The role of IL-12/23 in T cell–related chronic inflammation: Implications of immunodeficiency and therapeutic blockade. *Rheumatology* **2018**, *57*, 246–254. [CrossRef] [PubMed]

50. Junttila, I.S. Tuning the Cytokine Responses: An Update on Interleukin (IL)-4 and IL-13 Receptor Complexes. *Front. Immunol.* **2018**, *9*, 888. [CrossRef]
51. Ren, K.; Torres, R. Role of interleukin-1beta during pain and inflammation. *Brain Res. Rev.* **2009**, *60*, 57–64. [CrossRef] [PubMed]

vaccines

MDPI

Review

Vaccination Strategies Based on Bacterial Self-Assembling Proteins as Antigen Delivery Nanoscaffolds

Félix Lamontagne [1,2,3,4,5,†], Vinay Khatri [1,2,3,4,5,†], Philippe St-Louis [1,3,5], Steve Bourgault [2,3,4,5,*] and Denis Archambault [1,3,5,*]

1 Department of Biological Sciences, Université du Québec à Montréal, C.P.8888, Succursale Centre-Ville, Montréal, QC H3C 3P8, Canada
2 Department of Chemistry, Université du Québec à Montréal, C.P.8888, Succursale Centre-Ville, Montréal, QC H3C 3P8, Canada
3 The Swine and Poultry Infectious Diseases Research Centre (CRIPA), Saint-Hyacinthe, QC J2S 2M2, Canada
4 Quebec Network for Research on Protein Function, Engineering and Applications (PROTEO), Quebec, QC G1V 0A6, Canada
5 The Center of Excellence in Research on Orphan Diseases—Fondation Courtois (CERMO-FC), Montréal, QC H3C 3P8, Canada
* Correspondence: bourgault.steve@uqam.ca (S.B.); archambault.denis@uqam.ca (D.A.)
† These authors contributed equally to this work.

Abstract: Vaccination has saved billions of human lives and has considerably reduced the economic burden associated with pandemic and endemic infectious diseases. Notwithstanding major advancements in recent decades, multitude diseases remain with no available effective vaccine. While subunit-based vaccines have shown great potential to address the safety concerns of live-attenuated vaccines, their limited immunogenicity remains a major drawback that still needs to be addressed for their use fighting infectious illnesses, autoimmune disorders, and/or cancer. Among the adjuvants and delivery systems for antigens, bacterial proteinaceous supramolecular structures have recently received considerable attention. The use of bacterial proteins with self-assembling properties to deliver antigens offers several advantages, including biocompatibility, stability, molecular specificity, symmetrical organization, and multivalency. Bacterial protein nanoassemblies closely simulate most invading pathogens, acting as an alarm signal for the immune system to mount an effective adaptive immune response. Their nanoscale architecture can be precisely controlled at the atomic level to produce a variety of nanostructures, allowing for infinite possibilities of organized antigen display. For the bottom-up design of the proteinaceous antigen delivery scaffolds, it is essential to understand how the structural and physicochemical properties of the nanoassemblies modulate the strength and polarization of the immune responses. The present review first describes the relationships between structure and the generated immune responses, before discussing potential and current clinical applications.

Keywords: vaccines; antigen delivery systems; nanostructures; bacterial self-assembling proteins; immunomodulation; self-assembly

1. Introduction

Vaccination has saved billions of lives and has significantly reduced the economic burden of many infectious diseases [1]. In addition to protecting the immunized individual from severe symptoms, vaccination confers protection to the community by limiting the spread of the targeted pathogen [1]. Despite the numerous advances in vaccine technologies over the last two centuries (Figure 1), a multitude of infectious diseases remain without any available vaccine. The human immunodeficiency virus (HIV), the respiratory syncytial virus (RSV), and many others have no clinically approved vaccine [2,3]. Moreover, several currently available vaccine formulations need to be optimized, including the Bacille

Calmette-Guérin (BCG) vaccine and the seasonal flu vaccine. The BCG is still used to prevent infection with *Mycobacterium tuberculosis*, which causes tuberculosis. This formulation is effective in children for the first dose; however, the induced immunity diminishes over time, and it is not effective in booster doses [4]. Furthermore, there is a constant need to adjust the formulation of the influenza vaccine each year to accommodate antigenic drift and/or the major circulating strains, which can be challenging to predict [5]. This was exemplified by the 2014–2015 formulation, which showed an effectiveness of only 19% against infection [6].

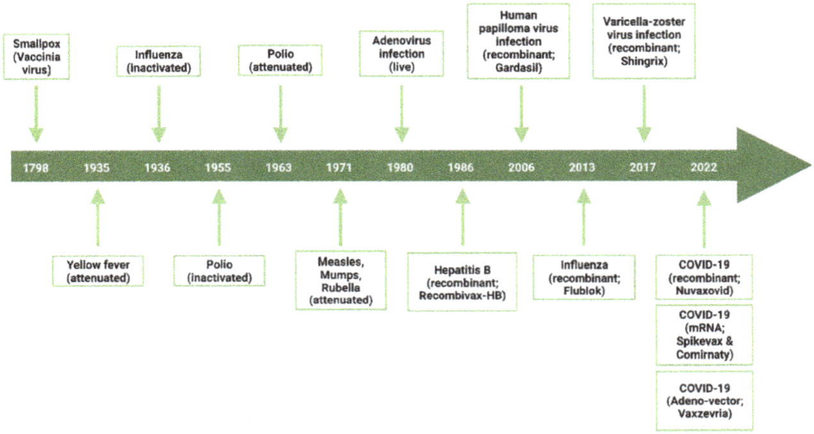

Figure 1. Timeline exposing major breakthroughs related to vaccine development for human uses.

Historically, vaccines were composed of whole pathogens that were either attenuated or inactivated. Attenuated vaccines are usually produced by a series of cell passages under suboptimal conditions to select variants with lower virulence in humans, whereas inactivated vaccines are produced by physical and/or chemical treatments [7]. Although live-attenuated vaccines induce a robust immune response, the pathogen can still replicate and mutate within the host, which risks its reverting to the original virulent form [8]. In contrast, inactivated vaccines are safer but tend to induce a moderate immune response, particularly in children and the elderly [9]. To address these limitations, vaccine technologies have shifted towards nucleic acid and protein-based vaccines (Figure 2) that aim to induce a targeted and safe immune response [10–13]. For instance, large-scale vaccination against SARS-CoV-2 has shown the efficacy and safety of nucleic acid-based vaccines. Cominarty (Pfizer-BioNTech) and Spikevax (Moderna, Cambridge, MA, USA), as well as Covishield/Vaxzevria (Oxford-AstraZeneca), have been administered to more than 69% of the global population (received at least one dose) [14]. This type of vaccine formulation delivers the antigenic coding sequence(s) to the host organism that expresses the protein(s) against which an immune response is to be induced [15]. To date, several nucleic acid-based vaccine technologies have been developed, such as mRNA, self-replicating RNA and plasmid DNA vaccines as well as various viral vectors [15]. Although they proved to be effective, these vaccine formulations usually require cold environments for long-term storage (from −20 to −80 °C), limiting its worldwide distribution [16]. In contrast, the protein-based subunit vaccines that have been used in clinics for many decades are stable formulations, although they tend to be weakly immunogenic on their own, requiring the use of adjuvants and/or immunomodulating delivery systems. The present review focuses on self-assembling bacterial proteins as nanoscaffold for antigen delivery in subunit vaccines.

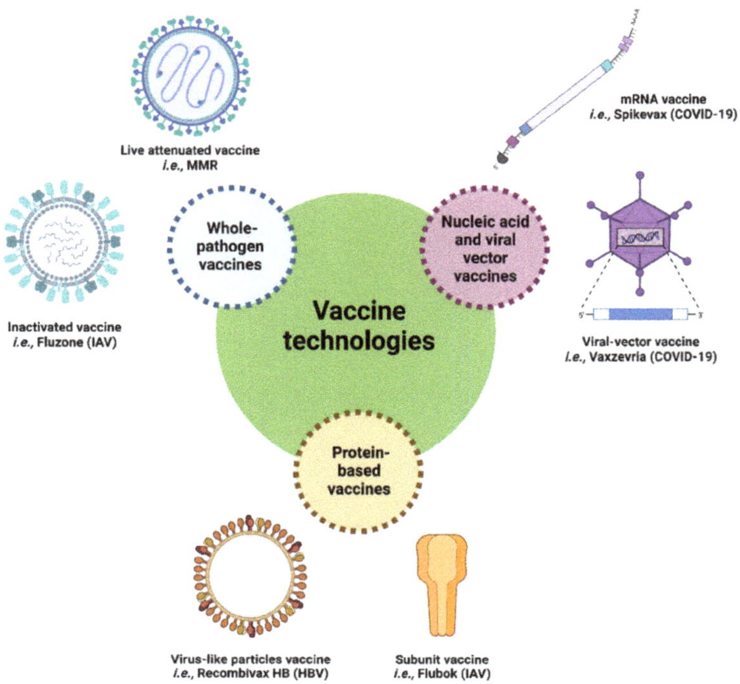

Figure 2. Schematic representation of vaccine technologies and examples of commercialized vaccines for human use. HBV: hepatitis B virus; IAV: influenza A virus; MMR: measles, mumps, and rubella.

2. Protein-Based Subunit Vaccines

Subunit protein vaccines are composed of one or more protein antigens to which a specific immune response is desired. The first subunit-based vaccine, Recombivax-HB (Merck, Rahway, NJ, USA), was approved in 1986 for use in humans [17]. It is composed of the hepatitis B virus (HBV) surface antigen, HBsAg, produced in the yeast Saccharomyces cerevisiae, which is mixed with aluminum salts (Alum), a commonly used adjuvant [17]. This vaccine formulation has demonstrated a protective efficacy of nearly 95% [18]. Since then, several protein-based subunit vaccines against viruses have been approved worldwide. Two other HBV vaccines, Heplisav-B (Dynavax, Emeryville, CA, USA) and Engerix-B (GlaxoSmithKline (GSK), Brentford, UK) have subsequently been licensed, with a similar efficacy to Recombivax-HB [18,19]. Other examples of protein subunit vaccine formulations approved for human use include Gardasil (Merck) and Cervarix (GSK) human papillomavirus (HPV) vaccine, Shingrix (GSK) varicella-zoster virus (VZV) vaccine, and Flublok (Protein Sciences Corporation) influenza A virus (IAV) vaccine [20]. Protein subunit vaccines are safe since they do not contain the original pathogen and they do not require highly specialized infrastructures for production or specialized equipment for long-term storage [21]. In order to produce the antigen, the coding gene is cloned into a vector, which is then transferred into a host cell for expression [22]. To date, five main expression systems have been utilized to produce recombinant proteins for vaccine purposes: bacterial, yeast, insect and mammalian cells, and more recently plants [22,23], with each of them having their advantages and limitations, as shown in Table 1.

Table 1. Expression systems to produce protein subunit vaccines.

Expression System	Advantages	Limitations	Vaccines	Antigens
Bacteria	Simple, well established, low cost, large-scale production	No PTM, inclusions body	Bexsero (against *Neisseria meningitidis*)	fHbp, NadA, NHBA & PorA [24]
Yeast	Simple, low cost, large scale production	Low PTM, hyperglycosylation	All HBV vaccines Gardasil (against HPV) Corbevax (against SARS-CoV-2)	HBs-Ag L1 RBD [24,25]
Insect cells	Human-like PTM, transient Expression	High costs, longer than bacteria and yeast and lower yield	Cervarix (against HPV) Flublok (against IAV) Nuvaxovid (against SARS-CoV-2)	L1 HA S protein [11,26]
Mammalian cells	Human identical PTM, stable expression	High cost, time-consuming to generate stable lines and lower yields	Several candidates against SARS-CoV-2	S, S1 & RBD [21]
Plants	Large scale production, easily modified genome, transient expression	New technology, high time required for implementation	Covifenz (against SARS-CoV-2)	S [27]

PTM: Post-translational modification; fHbp: factor H binding protein; NadA: Neisseria adhesin A; NHBA: Neisserial heparin binding antigen; HBV: Hepatitis B virus; HPV: Human papillomavirus; HBs-Ag: Hepatitis B surface antigen; L1: major capsid protein; IAV: Influenza A virus; S: spike protein; S1: S1 domain of spike protein; RBD: Receptor-binding domain; HA: Haemagglutinin; PorA: Porin A.

3. Cellular and Molecular Mechanisms of Immune Responses to Subunit-Based Vaccines

The immune response to subunit vaccines is multifaceted. After the administration of the antigen(s) into the host organism, antigen-presenting cells (APCs), including dendritic cells, macrophages and monocytes, internalize the proteins and display them at their surface with the aim of activating the adaptive immune system (T and B lymphocytes) to mount an antigen-specific adaptive immune response. The activation of naive T and B cells occurs in secondary lymphoid organs (SLOs) such as the spleen or lymph nodes. To mount an effective immune response following vaccination, antigens must accumulate at these sites via cell-mediated or lymphatic transport [28].

Following antigen uptake, APCs are activated via a multitude of immune receptors (i.e., toll-like receptors (TLR) or cytokine receptors) [29]. They progressively gain the ability to present a high density of antigen-derived peptides and also upregulate co-stimulatory molecules that are necessary for the efficient activation of the adaptive immune response [30]. APCs migrate from peripheral tissues to SLOs and enter the T-cell zone (Figure 3), seeking to engage with a T cell receptor (TCR) that can recognize and bind to antigen-derived peptide loaded on type I or type II major histocompatibility complex (MHC) molecules presented at its surface.

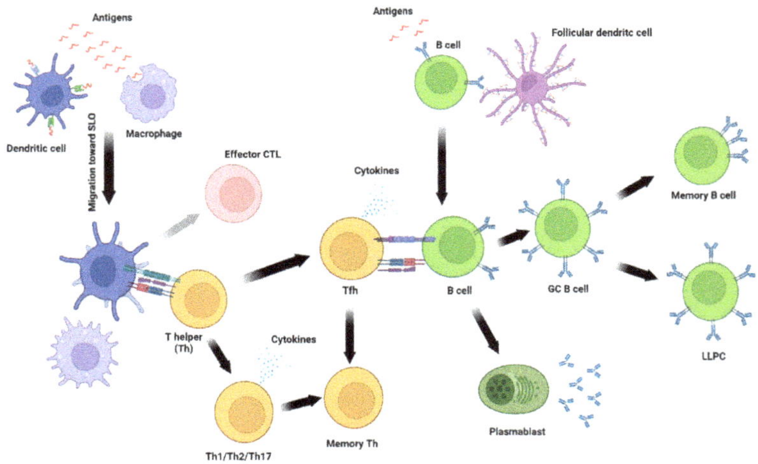

Figure 3. Immune response to subunit vaccines. In brief, antigen presenting cells, including dendritic cells and macrophages, internalize antigens at the site of injection and migrate toward secondary lymphoid organs (SLO) where they present antigen-derived peptides loaded on major histocompatibility complex (MHC) molecules to T helper (Th) and cytotoxic T cells (CTL). Activated Th cellsproliferate and differentiate into Th1, Th2, or Th17 and secrete cytokines and modulate the activity of other immune cells. An activated T helper can also differentiate into T follicular helper (Tfh) which provides direct help for B cell activation and germinal center reaction. Following the contraction of the immune response, some activated T helpers remain as memory T helpers. On the other hand, CTL can be reactivated (grey arrow) following antigen presentation on MHC-I molecules. Simultaneously, antigen-specific B cells are activated by soluble or membrane-bound antigens immobilized on follicular dendritic cells. Activated B cells proliferate and then differentiate into plasmablasts that secrete low-avidity antibodies or become germinal center B cells (GC B cells) where their BCR undergoes somatic hypermutation to increase antibodies' avidity. GC B cells then differentiate into long-lived plasma cells (LLPC) that constitutively secrete antibodies and reside in the bone marrow or memory B cells that patrol secondary lymphoid organs waiting for subsequent exposure to the same antigen.

Once a T cell is activated, it proliferates and produces a multitude of clones to help defend the organism against potential invaders. CD4 T cells or helper T cells (Th cells) bind to MHC-II-loaded peptides and secrete cytokines and express co-stimulatory molecules that help to activate other immune cells while polarizing the immune response [31]. Th cells orchestrate and direct the immune response toward a specific type of pathogen. Accordingly, there are multiple Th subsets: Th1 cells are associated with intracellular pathogens (i.e., viruses or certain bacteria), Th2 cells are mainly associated with helminth or allergen response, while Th17 cells induce protection against extracellular bacteria or fungi. T follicular helper cells (Tfh) are a specialized subset of CD4 T cells that localize to the B cell follicles in SLOs and promote B cell activation, germinal center (GC) reaction, antibody affinity maturation and isotype switching. The nature and the quality of the activating signals during antigen presentation greatly affect the polarization of Th cells and the resulting immune response [32–34]. Another type of T cells, CD8 T cells or cytotoxic T cells (CTL), recognize MHC-I-loaded peptides and are involved in killing infected cells. Although subunit vaccines are not efficient at promoting primary CTL response, they can effectively stimulate secondary CD8 T cells responses to common pathogens [35].

Simultaneously, B cells recognize, via their B cell receptor (BCR), membrane-bound antigens on follicular dendritic cells (FDCs) and subcapsular macrophages or soluble antigens that are drained to the follicles of the SLOs. After binding to antigens, the BCR and its bound antigen are internalized and digested before the antigen-derived peptide is docked on an MHC-II molecule at the surface of the B cell. This permits the interaction with an antigen-specific CD4 T cell that provides the co-stimulatory signal for full activation of the B lymphocyte. B cells can also be directly activated by antigens in a T-cell-independent manner following the strong signaling of co-receptors (i.e., TLR) or cross-linking of BCR by multivalent antigens [36].

Activated B cells start to proliferate and can take one of three differentiation paths. They can become short-lived plasma cells (SLPCs), or plasmablasts, a subset of B cells that secrete relatively low-affinity antigen-specific antibodies of switched or unswitched isotypes that defend the organism against immediate danger. Antigen-experienced B cells can also become GC B cells, which could enter the GCs, where they proliferate and undergo somatic hypermutation (SMH) to give rise to high-affinity antigen-specific BCR. Ultimately, GC B cells differentiate into long-lived plasma cells (LLPCs), and memory B cells (MBCs), or re-enter the GC for another round of proliferation and SMH. It is important to note that MBCs and LLPCs can arise through GC-independent mechanisms but with relatively low-affinity BCR [37,38]. After the initial proliferation burst following antigen recognition and GC reaction, LLPCs mostly migrate to the bone marrow, where they will secrete antibodies of high affinity following extensive SHM. On the other hand, MBCs take up residency in SLOs and other tissues in which antigen encounter is promoted. In these strategic locations, MBCs are in a quiescent state, ready to react to eventual re-exposure to the antigens [37].

4. Strategies to Enhance the Immune Response to Subunit Vaccines

Although protein-based subunit vaccines have shown efficacy in clinics, they remain poorly immunogenic when composed solely of soluble antigens. In fact, low-molecular-weight polypeptidic antigens are readily eliminated from the organism and generate little to no immune response. To circumvent these issues, different strategies have been developed, including (i) the use of adjuvants in the vaccine formulation, (ii) the addition of TLR agonists, and (iii) the conjugation of the antigen(s) to nanoparticles.

4.1. Adjuvants and Recruitment of Immune Cells at the Injection Site

Adjuvants are substances that can enhance, or modulate, the immune response directed toward an antigen. Traditionally, adjuvants were developed empirically, without a clear understanding of the molecular mechanisms involved in their immunostimulating properties [39]. Insoluble aluminum salts, or alum, remains the most widely used adjuvant in human vaccine formulations. Although its mechanisms of action are not fully understood, it is known that alum mainly works through the adsorption of antigens on particles of aluminum and the induction of an inflammatory milieu following the release of danger signals by affected cells. The pro-inflammatory environment caused by alum enhances the recruitment of immune cells, mostly neutrophils, at the site of injection and the antigen adsorption to particles increases uptake by APCs. This leads to an increase in antigen-directed antibodies. Alum also induces a CD4 T cell response that is Th2-biased in mice, while this bias is less clear in humans [40–42].

Oil-in-water emulsions are another type of adjuvant that is approved in humans and has been extensively used in vaccination. MF59 and AS03 are two proprietary adjuvants of Novartis and GlaxoSmithKline, respectively. Following the administration of antigen with an oil-in-water adjuvant, immune cells such as macrophages, dendritic cells and granulocytes are recruited at the site of injection. This leads to an increase in antigen uptake by APCs and greatly enhances the antibody and cellular immune response toward the antigen. Oil-in-water emulsions usually lead to a broader immune response and a more balanced Th1-Th2 cellular immune response compared to alum [39,41,43].

4.2. Stimulation of Immune Cells via the Activation of TLRs

Another strategy developed to increase the immune response emerged following the discovery of pattern recognition receptors (PRRs), which are germline-encoded immune receptors that bind to the pathogen-associated molecular pattern (PAMP) and danger-associated molecular patterns (DAMP). The PRRs are expressed on innate immune cells, B cells and some epithelial and fibroblastic cells and promote their activation following the binding of ligands to their cognate receptor [44,45].

The PRR-targeting adjuvants licensed for human use are 3-O-desacyl-4′-monophosphoryl lipid A (MPLA) and cytosine phosphoguanosine (CpG) 1018 [43]. MPLA is a purified form of lipopolysaccharides (LPS) from the bacteria *Salmonella minnesota* and activates the TLR4, a type of PRR. It was first used in combination with alum under the proprietary name AS04 (GSK) in HPV and HBV subunit vaccines. The combination of MPLA and alum induces higher antibody titers compared to alum alone in both vaccines while increasing the breadth of protection against multiple strains of HPV [41].

Bacterial proteins with the ability to activate membrane-expressed TLR2, TLR4, or TLR5 also show great potential as immune-enhancing components in vaccines. Flagellin was the first known agonist for TLR5. Its adjuvant properties were first evaluated thirty years ago, and it has been extensively used in the context of vaccine research in pre-clinical and clinical settings [46]. Flagellin can promote the maturation of APCs, inducing the secretion of pro-inflammatory cytokines, and increasing the level of antibodies directed toward an antigen when injected in a co-mixture, as a fusion protein, or incorporated into nanoparticles. Since TLR5 is highly expressed in the airway epithelial and immune cells, flagellin shows potential as a mucosal adjuvant for intranasal or oral vaccines [46]. A recent study has identified the protein P97c from Mycoplasma hyopneumonia as a novel TLR5 agonist capable of inducing concentration-dependent cytokine production in HEK-blue mTLR5 (mouse TLR5) cells. Furthermore, its efficacy as an adjuvant was demonstrated by conjugating the ectodomain matrix 2 protein (M2e) of influenza A, leading to higher M2e-specific antibody titer following mice immunization [47]. Another membrane PRR, TLR2 is more promiscuous than TLR5 and recognizes lipoproteins and a wide variety of hydrophobic proteins [48–50]. It forms a heterodimer with TLR1 or TLR6. A multitude of bacterial proteins have TLR2 agonist activity, mostly cell surface-expressed proteins like porins or outer membrane proteins (OMPs). Bacterial proteins with TLR2 binding properties have been shown to promote leukocyte recruitment, induce the production of pro-inflammatory cytokines, the maturation of APCs and the production of antigen-specific antibodies and cellular response. While most bacterial proteins with TLR2-activation capacities induce a Th1-skewed immune response, some have reported a balanced Th1-Th17 immune response, mainly in the lungs [51,52]. Interestingly, some TLR2-targeting proteins can activate TLR4 [51]. While LPS and its derivatives are the most characterized TLR4 agonists, it is becoming increasingly clear that a myriad of bacterial proteins have the same ability. These proteins have been shown to promote the maturation and migration of DC to lymph nodes, the production of pro-inflammatory cytokines and a robust B and T cell activation. Similar to TLR2 signaling by bacterial proteins, TLR4 agonists induce a Th1-Th17 immune response [51]. Moreover, bacterial components that activate the immune system in a TLR-independent manner have also been studied for their potential use as adjuvants, for example, the heat-labile enterotoxin (LT) and the cholera toxin (CT). Early studies on their underlying mechanism point to the accumulation of cAMP as adjuvant effects and the activation of the Nod-like receptor Pyrin 3 (NLRP3) [53,54]. The LT protein induces strong Th17 responses and increases sIgA titers when used as a mucosal adjuvant. However, adverse side effects were reported in clinical studies and linked the adjuvant to Bell's Palsy symptoms, including facial paralysis, in early 2000 [55]. This prompted research on genetically modified LT with reduced toxicity in humans for adjuvant uses (reviewed in [56]).

4.3. Conjugation to Nanoscale Antigen Delivery Systems

Since the immune system has evolved to recognize pathogens under their particulate forms, mimicking this structure in subunit vaccines could potentially increase the immune response generated following administration. Accordingly, the conjugation of antigens with particles increases their immunogenicity and physical and metabolic stability, while limiting potential toxicity. Antigens under particulate forms enhance uptake by APCs, lymph node trafficking and persistence, which leads to a stronger cellular and humoral immune responses being directed against the antigen [28,42,57–65].

The biodistribution of antigens following administration is critical for vaccine effectiveness. Particle size is a parameter that greatly affects the pharmacokinetics of vaccines. Nanoparticles under 5 nm readily diffuse in the blood where they circulate systemically, while the ones between 10 nm and 100 nm drain in the lymphoid system toward LN [61,66–71]. Particles over 100–200 nm are mainly trapped in the extracellular matrix at the site of injection and must be brought to LN via APCs [42,72,73]. Shape, rigidity, and surface chemistry are also important factors influencing the biodistribution of particles [42,57,64]. Limited circulation is a clear advantage of nanoparticles because the systemic dissemination of small molecules limits the efficacy of vaccination while potentially increasing the side effects [28,74]. Once in the LN, particles have been shown to persist for a longer period compared to smaller molecules. While nanoparticles smaller than 15 nm are rapidly found in the follicles of the LN following immunization, they are also rapidly eliminated. On the contrary, particles of 50–100 nm tend to take more time to reach the follicles but can persist for a few weeks when immobilized on FDCs [28,75]. It is interesting to note that this phenomenon seems to be dependent on a complement, which opsonizes the particles and promotes their retention by the FDCs via their complementary receptors [75,76]. The key advantages of conjugating antigenic materials on nanoparticles for subunit vaccines are summarized in Figure 4.

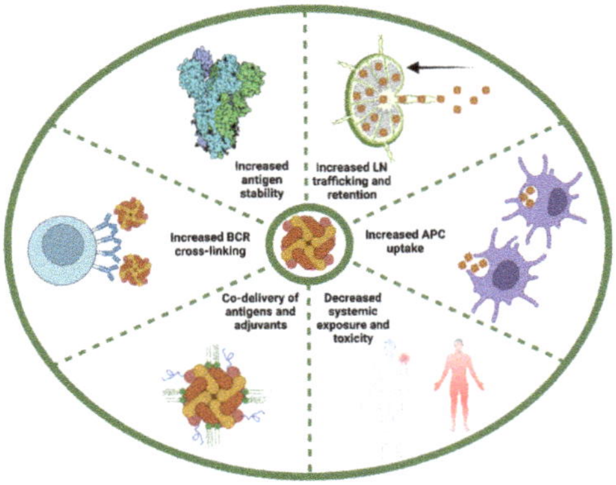

Figure 4. Key advantages of conjugating antigenic determinants on nanoparticle for protein subunit vaccines. Nanoparticle-associated antigens diffuse more efficiently to the draining lymph node (LN) from the injection site and are retained for a longer period of time compared to soluble antigens. The repetitive nature of antigens arrayed on nanoparticles also enhances internalization by APCs via multiple mechanisms. Nanoparticles have also shown to offer limited systemic toxicity. The use of nanoparticles can favor the co-delivery of antigen and adjuvant to the same immune cell, which enhances adjuvant effects and limits off-target effects. Multivalent antigens efficiently induce B cell receptor (BCR) cross-linking, which greatly enhances uptake and presentation by high- and low-affinity B cells. The conjugation of antigen on nanoparticles stabilizes them and allows for the display of antigens under "locked" conformation.

Nanoparticle-associated antigens can also be brought to the LN from the site of injection by APCs. Particles that are over 20 nm are more efficiently internalized by DC and macrophages compared to soluble antigens [66,77–79]. The repetitive nature of nanoparticles also enhances internalization by APCs via multiple mechanisms that are not fully understood. Amongst others, the binding of natural antibodies to repetitive patterns triggers the recruitment of the complement which, in turn, interacts with Fc receptors (FcR) and promotes internalization of the opsonized material by APCs [57].

To enhance the presentation of antigen-derived peptides on MHC molecules, APCs must be activated. The co-delivery of adjuvant and antigens by particles promotes internalization, antigen processing, maturation of the APC and presentation of the antigen-derived peptide at the surface of the cell. Danger signals also promote the cross-presentation of exogenous peptides on MHC-I molecules by DCs [80,81]. Since the inflammation generated by activated immune cells is usually the source of vaccine side effects, targeting adjuvants to antigen-experienced cells offers the possibility to decrease systemic toxicity while promoting vaccine effectiveness [28]. The co-delivery of antigen and adjuvant also promotes B cell engagement and enhances GC formation and the humoral immune response generated toward the antigen [82].

Moreover, nanoparticles can present a high density of antigens at their surface. This allows for a repetitive display, similar to virus particles [83]. Multivalent antigens efficiently induce BCR cross-linking, which greatly enhances uptake and presentation by high- and low-affinity B cells, compared to the soluble protein, which is mostly uptaken by B cells with high-affinity BCR. The engagement of low-affinity B cells could be useful in the generation of broadly neutralizing antibodies (Nabs) [58]. It was also shown that multivalent antigens can enhance T cell activation by promoting presentation by B cells while increasing antigen-specific antibodies by up to 10-fold [84–86].

In summary, the display of antigens on nanoparticles offers many advantages over soluble antigens (Figure 4). Notably, nanoparticles are preferentially uptaken by APCs, and efficiently drain to lymph nodes, where they are retained for a longer period and offer limited systemic toxicity. Furthermore, the use of nanoparticles favors the co-delivery of antigen and adjuvant to the same cell, which enhances adjuvant effects and limits off-target effects. The conjugation of antigen on nanoparticles stabilizes them and allows for the display of antigens under "locked" conformation (i.e., pre-fusion stabilized GP). Moreover, antigen multivalency on nanoparticles can facilitate BCR cross-linking and antibody production [87]. It has also been demonstrated that the multivalent display of an antigen on nanoparticles not only significantly improves their immunogenicity, but also induces potent immune responses at relatively low immunogen dose when compared to animals immunized with the soluble antigen alone [88,89].

Recent developments have made it possible to design nanoparticles with distinctive physicochemical characteristics. One may construct nanoparticles with specific biological features by tuning and controlling factors including size, shape, solubility, surface chemistry, and hydrophilicity. These characteristics suggest that nanoparticles are promising immune cell stimulators and antigen carriers for immunization. These nanoscale materials can be conceived de novo or derived from living organisms. A wide range of particles, including inorganic and polymeric nanoparticles, virus-like particles (VLPs), liposomes, and self-assembling protein-based nanoparticles, have been evaluated as antigen carriers [71]. Since many bacterial proteins can self-assemble into nanoparticles of defined structure, they are interesting candidates for the generation of nanoparticle-based subunit vaccines with enhanced immunogenicity. They self-assemble into highly symmetric, stable nanoparticles with diameters of 10–150 nm [71,83], which is an ideal size range ideal for interacting with different immune cells [83]. Since they can be used as nanoplatforms for the organized display of specific immunogen, these nanoparticles are of particular interest in the context of vaccine design because they can mimic the repetitive surface architecture of most naturally occurring pathogens.

5. Self-Assembling Bacterial Proteins as Nanoscaffolds for Antigen Delivery in Subunit Vaccines

In recent decades, bacterial protein-based nanoparticles have been evaluated in numerous subunit vaccine formulations [57,90–93]. The unique particulate nature and repetitive subunit organization of these assemblies make them ideal candidates for antigen display, which, in turn, would provide a robust antigen-specific immune response. These self-assembling bacterial protein-based nanoparticles for the delivery of antigenic determinants are summarized in detail in Table 2 and Figure 5.

Table 2. *Cont.*

Self-Assembling Protein	Organism	Antigen	Expression System	Assembly Structure	Size	Method for Conjugating the Antigen	Animal Model/ Administration Route	Studied Immunity	Comments	Reference
Ferritin	*H. pylori*	V1V2, gp120 and gp140 trimers (HIV)	N-acety lglucosaminy ltransferase I-negative (GnTI/) HEK293S, HEK293F and ExpiCHO cells			Genetic	in vitro antibody binding and B cell activation	N/A	Ferritin nanoparticles displaying trimeric V1V2, gp120 and gp140. Demonstrated high-yield gp140 nanoparticle production and robust stimulation of B cells carrying cognate VRC01 receptors by gp120 and gp140 nanoparticles	[102]
Ferritin	*H. pylori*	Envelope trimers (BG505 SOSIP.664) (HIV)	293F cells		30–40 nm in diameter	Genetic	Balb/c mice and New Zealand White Rabbits IM	Humoral	HIV-1 envelope GP trimers (BG505 SOSIP.664) -bearing nanoparticles were significantly more immunogenic than trimers in both mice and rabbits Adjuvant: MPLA liposomes	[103]
Ferritin	*H. pylori*	Envelope trimers (ConM) (HIV)	293F cells		30–40 nm in diameter	Genetic	New Zealand White Rabbits and Rhesus macaques IM	Humoral	The ConM trimers elicited strong NAb responses against the autologous virus in rabbits and macaques that are significantly enhanced when it is presented on Ferritin nanoparticles Adjuvant: Iscomatrix (Isco) or MF59	[104]
Ferritin	*H. pylori*	Envelope trimers (ConM SOSIP.v7) (HIV)	293F cells			Genetic	New Zealand White Rabbits IM	Humoral	Stronger NAbs responses were elicited when the ConM SOSIP trimers were presented on Ferritin nanoparticles Adjuvant: Squalene emulsion and MPLA liposomes	[105]
Ferritin	*H. pylori*	VP6 (Rotavirus A)	*E. coli* BL21 (DE3) cells and transgenically expressed in the milk of mice		~ 20 nm	Genetic	Balb/c mice PO	Humoral	Recombinant VP6–ferritin nanoparticle vaccine efficiently prevented the death and malnutrition induced by the rotavirus infection in pups Adjuvant: Cholera toxin subunit B (CTB)	[106]

Table 2. *Cont.*

Self-Assembling Protein	Organism	Antigen	Expression System	Assembly Structure	Size	Method for Conjugating the Antigen	Animal Model/ Administration Route	Studied Immunity	Comments	Reference
Ferritin	*H. pylori*	GP5 (PRRSV)	Sf9 cells			Genetic	Pigs IM	Humoral and cellular	Immunization with PRRSV modified GP5 protein coupled to ferritin elicited improved protective immunity against PRRSV compared to inactivated vaccine	[107]
Ferritin	*H. pylori*	VP1 & G-H loop (FMDV)	Sf9 cells			Genetic	C57BL/6 mice IM	Humoral and cellular	Ferritin nanoparticles carrying recombinant proteins exhibited good immunogenicity with 66.7% survival rate but less than inactivated vaccine Adjuvant: Montanide ISA201VG	[108]
Ferritin	*H. pylori*	E2 (CFSV)	Sf9 cells			Genetic	Rabbit IM	Humoral and cellular	E2-expressing ferritin nanoparticles induced stronger immune responses than E2 alone Adjuvant: Montanide gel 02	[109]
Ferritin	*H. pylori*	Spike (SARS-CoV-2)	Expi 293 cells		15–19 nm	Spy tag and Spy catcher	Balb/C mice IM	Humoral	Recombinant expression of ferritin with a N-linked glycan increased yield in mammalian expression systems and Increased S-directed Nabs Adjuvant: SAS	[89]
Ferritin	*H. pylori*	RBD & Spike (SARS-CoV-2)	ExpiCHO cells		47.9 nm	Spy tag and Spy catcher	Balb/C mice IP	Cellular and Humoral	The 24-meric RBD-ferritin and spike-ferritin elicited a more potent Nab response than the RBD or Spike alone Adjuvant: MF59 or Alum	[110]
Ferritin	*Escherichia coli (E. coli)*	RBD (MERS-CoV)	*E. coli* strain SHuffle® T7		20–40 nm	Genetic	Balb/c mice IM	Humoral	ChapeRNA-mediated folding of RBD-ferritin controlled the overall kinetic network of the antigen folding pathway in favor of enhanced assemblage of NPs into highly regular and immunologically relevant conformations Adjuvant: MF59 or Alum	[111]

Table 2. *Cont.*

Self-Assembling Protein	Organism	Antigen	Expression System	Assembly Structure	Size	Method for Conjugating the Antigen	Animal Model/ Administration Route	Studied Immunity	Comments	Reference
Ferritin	*Pyrococcus furiosus* (*P. furiosus*)	MD39 env trimer (HIV)	FreeStyle™ 293-F Cells		~40 nm diameter	Genetic	Balb/c mice SC	Humoral	Nanoparticles with heavily glycosylated antigens were accumulated and were retained on FDCs in a mannose-binding lectin- and complement-dependent manner	[112]
Ferritin	*P. furiosus*	preS1 domain of HBV	BL21 (DE3) competent *E. coli*		-	SpyTag and SpyCatcher	Balb/c mice SC	Humoral	preS1-Ferritin nanoparticle targets SIGNR1+ APC, which are involved in Tfh and B cell activation. The vaccine induced a high-level and persistent anti-preS1 response that resulted in efficient viral clearance and partial serological conversion in a chronic HBV mouse model offering a promising translatable vaccination strategy for the functional cure of chronic hepatitis B	[113]
Ferritin	*P. furiosus*	RBD (SARS-CoV-2)	BL21 (DE3) competent *E. coli* and 293F cells			SpyTag and SpyCatcher	C57BL/6 mice SC	Humoral	Vaccine generated an effective antibody response and long-term MBCs in mice that was sustained for at least 7 months after inoculation	[114]
Ferritin	*P. furiosus*	HPV minor capsid protein L2	Sf9 and High Five insect cells	an octahedral structure composed by 24 protomers		Genetic	Balb/c mice and Guinea Pig IM	Humoral	The ferritin-Trx-L2 trimer induced a broadly Nab response covering 14 oncogenic and two non-oncogenic HPV types, which lasted for at least one year Adjuvant: MF59 or Alum	[115]
Ferritin	*Thermotoga maritima* (*T. maritima*)	GPΔMUC trimer (EBOV)	ExpiCHO	24-subunit protein icosahedron	34.6 nm	Genetic	Balb/C mice IPNew Zealand white rabbit IM	Humoral and Cellular	GP trimers and nanoparticles elicited cross-ebolavirus NAbs, as well as non-NAbs that enhanced pseudovirus infection Adjuvant: MF59 or Alum	[116]

Table 2. *Cont.*

Self-Assembling Protein	Organism	Antigen	Expression System	Assembly Structure	Size	Method for Conjugating the Antigen	Animal Model/ Administration Route	Studied Immunity	Comments	Reference
Lumazine synthase	*Aquifex aelocus (A. aelocus)*	gp120 (HIV)	N-acetylglucosaminyltransferase I-negative (GnTI/) HEK293S, HEK293F and ExpiCHO cells	Self-assembles into a 60-mer		Genetic	in vitro BCR expressing cell stimulation	Humoral	Demonstrated high yield gp140 nanoparticle production and robust stimulation of B cells carrying cognate VRC01 receptors by gp120 and gp140 nanoparticles	[102]
Lumazine syn-thase	*A. aelocus*	gp120 (HIV)	FreeStyle™ 293-F Cells		~32 nm diameter	Genetic	Balb/c mice SC	Humoral	The findings highlighted how the innate immune system recognizes HIV nanoparticles and the importance of antigen glycosylation in the design of next-generation nano-based vaccines	[112]
Lumazine syn-thase	*A. aelocus*	gp120 (HIV)	FreeStyle™ 293-F Cells			Genetic	Balb/c mice IP or SC	Humoral	The results suggested that rational epitope design can prime rare B cell precursors for affinity maturation to desired targets Adjuvant: Ribi, Alum or Isco	[117]
Lumazine syn-thase	*A. aelocus*	gp120 (HIV)	FreeStyle™ 293-F Cells			Genetic	Balb/c mice IP or SC	Humoral	When multimerized on nanoparticles, the immunogen (eOD-GT6) activated germline and mature VRC01-class B cells Adjuvant: Ribi, Alum or Isco	[118]
Lumazine syn-thase	*A. aelocus*	Spike (SARS-CoV-2)	Expi 293 cells		15–19 nm	Spy tag and Spy catcher	Balb/C mice IM	Humoral	SARS-CoV-2-spike nanoparticles elicited substantially higher Nab responses than spike alone Adjuvant: SAS	[89]
Lumazine syn-thase	*A. aelocus*	GP350 (EBV)	High-Five cells		~20 nm	Genetic	Balb/C mice SC non-human primate IM	Humoral	Nanoparticle vaccine elicited potent Nab antibody responses against EBV infection Adjuvant: MF59 or Alum	[119]

Table 2. *Cont.*

Self-Assembling Protein	Organism	Antigen	Expression System	Assembly Structure	Size	Method for Conjugating the Antigen	Animal Model/ Administration Route	Studied Immunity	Comments	Reference
Lumazine syn-thase	*A. aelocus*	Gc env (SBV)	*E. coli* BL21 (DE3) and Drosophila S2 Cells		15 nm	Spy tag and Spy catcher	C57BL/6 mice and cattle (German domestic cow breed) SC	Humoral	Even a single-shot vaccination protected about 80% of mice from an otherwise lethal dose of SBV and induced a virtually sterile immunity in cattle Adjuvant: Emulsigen (mice) or Ploygen (Cattle)	[120]
Lumazine syn-thase	*A. aelocus*	Gn (RVFV)	*E. coli* BL21 (DE3) and High FIve cells	Assembles via 12 pentamers into an icosahedral particle	15 nm	Spy tag and Spy catcher	Balb/C mice and Texel-German lamb IM	Humoral	Lumazine synthase-based nanoparticles, prevented mortality in a lethal mouse model and protected lambs Adjuvant: Stimune (mice) and TS6 (lamb)	[121]
Encapsulin	*T. maritima*	M2e (IAV)	*E. coli* BL21 (DE3)	60-mer (T = 1) icosohedral capsid-like particles	24 nm	Genetic	Balb/C mice SC	Humoral	Nanoparticle immunization elicited antibody responses against both the surface epitope and the loaded cargo protein Adjuvant: Freund's adjuvant	[93]
Encap-sulin	*T. maritima*	GP350 (EBV)	Expi293F		~20–30 nm	Genetic	Balb/C Mice and Rhesus macaques IM	Humoral	The structurally designed nanoparticle vaccine increased neutralization from 10- to 100-fold compared to soluble gp350 by targeting a functionally conserved site of vulnerability, improving vaccine-induced protection in a EBV mouse experimental challenge Adjuvant: SAS	[99]
small Heat shock protein (sHsp) 16.5	*Methanocaldoccus jannaschii (archea)*	Model antigen, ovalbumin (OVA)	*E. coli* BL21 DE3	24 repeating subunits self-assemble to produce cage-like nanoparticles	30–41 nm	Chemically	Balb/C and C57BL/6 mice SC	Humoral	sHsp nanoparticles elicited quick and intense antibody responses, and these accelerated responses could similarly be targeted toward antigens chemically conjugated to the sHsp Adjuvant: Alum	[122]

Table 2. *Cont.*

Self-Assembling Protein	Organism	Antigen	Expression System	Assembly Structure	Size	Method for Conjugating the Antigen	Animal Model/ Administration Route	Studied Immunity	Comments	Reference
small Heat shock protein (sHsp) 16.5	*Methanocaldoccus jannaschii*	-	*E. coli*			-	C57BL/6 (CD45.2), BALB/c, and μMT (B10.129S2(B6)-Igh-6^{tmlCgn}) mice IN	-	Bronchus-associated lymphoid tissue elicited by a protein cage nanoparticle enhanced protection in mice against diverse respiratory viruses	[123]
P22	Bacteriophage	Model antigen, ovalbumin (OVA)	*E. coli* BL21 DE3	Non-infectious empty viral capsid	30–41 nm	Priming agent	Balb/C and C57BL/6 mice SC	Humoral	Pretreatment of mice with P22 further accelerated the onset of the antibody response to OVA–sHsp, demonstrating the utility of conjugating antigens to VLPs for pre-, or possibly post-exposure prophylaxis of lung, all without the need for adjuvant Adjuvant: Alum	[122]
P22	Bacteriophage	Conserved nucleoprotein (NP) from influenza (H1N1 and H3N2)	*E. coli* BL21 DE3		29–54 nm	Priming agent	BALB/c mice IN	Humoral and Cellular	P22 encapsulating NP (truncated and full-length constructs) elicited a strong protective immune response in mice against challenge with both H1N1 and H3N2 (IAV), without the addition of adjuvants	[124]
P22	Bacteriophage	HA (PR8 IAV)	ClearColi BL21 (λDE3)		26 nm	Spy tag and Spy catcher	C57BL/6 (CD45.2) mice IN	Humoral	P22 VLPs can be rapidly modified in a modular fashion, resulting in an effective vaccine construct that can generate protective immunity without the need for additional adjuvants	[125]
BP26	*Brucella abortus*	M2e (IAV)	*E. coli* BL21 (DE3)	Nanobarrels (forms a barrel-like structure with a hollow center through self-assembly of 16 monomeric proteins)	11–22 nm	Genetic	Balb/C mice SC	Humoral and Cellular	BP26-M2e nanobarrels effectively protected mice from (IAV) infection-associated death, even without the use of a conventional adjuvant Adjuvant: Alum	[126]

Table 2. *Cont.*

Self-Assembling Protein	Organism	Antigen	Expression System	Assembly Structure	Size	Method for Conjugating the Antigen	Animal Model/ Administration Route	Studied Immunity	Comments	Reference
Flagellin	*Bacillus subtilis*	M2e (IAV)	*E. coli* Rosetta DE3	Ring-like nanostructures	10–15 nm	Genetic	Balb/C Mice IN	Humoral and Cellular	Flagellin ring-like nanostructures were efficiently internalized by APCs, and avidly activated the TLR5 in vitro as well as the innate and adaptive immune responses	[127]
Flagellin	Salmonella serovar enterica typhimurium (*S. typhimurium*)	Viral envelope protein from Dengue virus (DENV2)	SF9 & SF21 insect cells	Filaments	35 nm	Genetic	C57BL/6J, B10.D2, 6.5-TCR (Tg(Tcra/Tcrb)1Vbo) mice IN or IP	Humoral and Cellular	Reengineered hybrid FliC enhanced T-cell-dependent and possibly induced T-independent antibody responses from B-1 B cells	[128]
E2p	*Bacillus stearothermophilus (B. stearothermophilus)*	gp120/gp140 (HIV)	N-acetylglucosaminyl transferase I-negative (GnTI/) HEK293S, HEK293F and ExpiCHO cells	60-mer assembles into a pentagonal dodecahedral scaffold		Genetic	in vitro BCR expressing cell stimulation	Humoral	Demonstrated high-yield gp140 nanoparticle production and robust stimulation of B cells carrying cognate VRC01 receptors by gp120 and gp140 nanoparticles	[102]
E2p	*B. stearothermophilus*	GP (EBOV)	HEK293F and ExpiCHO cells		45.9 nm	Genetic	Balb/C mice IP New Zealand white rabbit IM	Humoral and Cellular	GP trimers and nanoparticles elicited cross-ebolavirus NAbs, as well as non-NAbs that enhanced pseudovirus infection Adjuvant: MF59 or Alum	[116]
E2p	*B. stearothermophilus*	RBD & Spike (SARS-CoV-2)	ExpiCHO cells		55.9 nm	Spy tag and Spy catcher	Balb/C mice IP	Humoral and Cellular	E2 elicited up to 10-fold higher NAb titers. Adjuvant: MF59 or Alum	[110]
E2p	*B. stearothermophilus*	Gn (RVFV)	*E. coli* BL21 (DE3) and High FIve cells		27 nm	Spy tag and Spy catcher	Balb/C mice and Texel-German lamb IM	Humoral	Geobacillus stearothermophilus E2p or a modified KDPG Aldolase provided complete protection in lambs from RVFV challenge Adjuvant: Stimune (mice) and TS6 (lamb)	[121]
I3-01	*T. maritima*	GP (EBOV)	ExpiCHO	60-subunit protein icosahedron	49.2 nm	Genetic	Balb/C mice IP and New Zealand white rabbit IM	Humoral and Cellular	GP trimers and nanoparticles elicited cross-ebolavirus NAbs, as well as non-NAbs that enhanced pseudovirus infection Adjuvant: MF59 or Alum	[116]

Table 2. *Cont.*

Self-Assembling Protein	Organism	Antigen	Expression System	Assembly Structure	Size	Method for Conjugating the Antigen	Animal Model/ Administration Route	Studied Immunity	Comments	Reference
I3-01	*T. maritima*	Gn (RVFV)	*E. coli* BL21 (DE3) and High FIve cells		25 nm	Spy tag and Spy catcher	Balb/C mice and Texel-German lamb IM	Humoral	I3-01 modified KDPG Aldolase provided complete protection in lambs from RVFV challenge Adjuvant: Stimune (mice) and TS6 (lamb)	[121]
I3-01	*T. maritima*	GP350 (EBV)	High-Five cells		~25 nm	Genetic	Balb/C mice SC	Humoral	The self-assembled nanoparticle vaccine elicited potent Nabs responses against EBV infection Adjuvant: MF59 or Alum	[119]
I3-01	*T. maritima*	RBD and Spike (SARS-CoV-2)	ExpiCHO cells		59.3 nm	Spy tag and Spy catcher	Balb/C mice IP	Humoral and Cellular	I3-01v9 60-mers elicited up to 10-fold higher NAb titers. I3-01v9 SApNP also induced critically needed T cell immunity Adjuvant: MF59 or Alum	[110]
I3-01	*T. maritima*	GP (EBOV)	HEK293F and ExpiCHO cells		49.2 nm	Genetic	Balb/C mice and New Zealand white rabbit IM	Humoral and Cellular	GP trimers and nanoparticles elicited cross-ebolavirus NAbs, as well as non-NAbs that enhanced pseudovirus infection Adjuvant: MF59 or Alum	[116]
I53-50	-	Fusion (RSV)	HEK293F and Expi293 cells	Icosahedral assembly	55 nm	Genetic	Balb/C mice and Indian Rhesus macaques IM	Humoral and Cellular	Computationally designed self-assembling nanoparticle that displayed 20 copies of a trimeric viral protein induced potent Nab responses Adjuvant: MF59 (mice) and Squalene emulsion (macaques)	[129]
I53-50	-	RBD of Spike (SARS-Cov-2)	Expi293F		28 nm	Genetic	Balb/C mice and Pigtail Macaque IM	Humoral	The nanoparticle vaccine exhibits 60 copies of the RBD of the spike protein and induced strong humoral response in mice and non-human primates Adjuvant: MF59	[88]

Table 2. *Cont.*

Self-Assembling Protein	Organism	Antigen	Expression System	Assembly Structure	Size	Method for Conjugating the Antigen	Animal Model/ Administration Route	Studied Immunity	Comments	Reference
I53-50	-	RBD of Spike (SARS-Cov-2)	Expi293F		28 nm	Genetic	Rhesus Macaque IM	Humoral and Cellular	Follow-up study to Walls et al., 2020 [88] where the platform was combined to different adjuvant. All combination showed potent humoral response, but combination with AS03 was the only one that elicited mixed TH1/TH2 cellular response Adjuvant: AS03 or AS37	[130]
I53-50	-	Spike (SARS-Cov-2)	HEK293F		30 nm	Genetic	Balb/C mice SC, New Zealand White rabbit IM and Cynomolgus macaque IN	Humoral and Cellular	The I53-50 nanoparticle construct exposed several Spike proteins and induced strong Nabs in all animal models. Furthermore, the vaccine generated IFN-γ secreting T cells in non-human primate Adjuvant: Poly-IC (mice), Squalene emulsion (rabbit) or MPLA liposomes (macaque)	[131]

Administration routes: IM: intramuscular, IN: intranasal, IP: intraperitoneal, PO: oral, SC: subcutaneous.

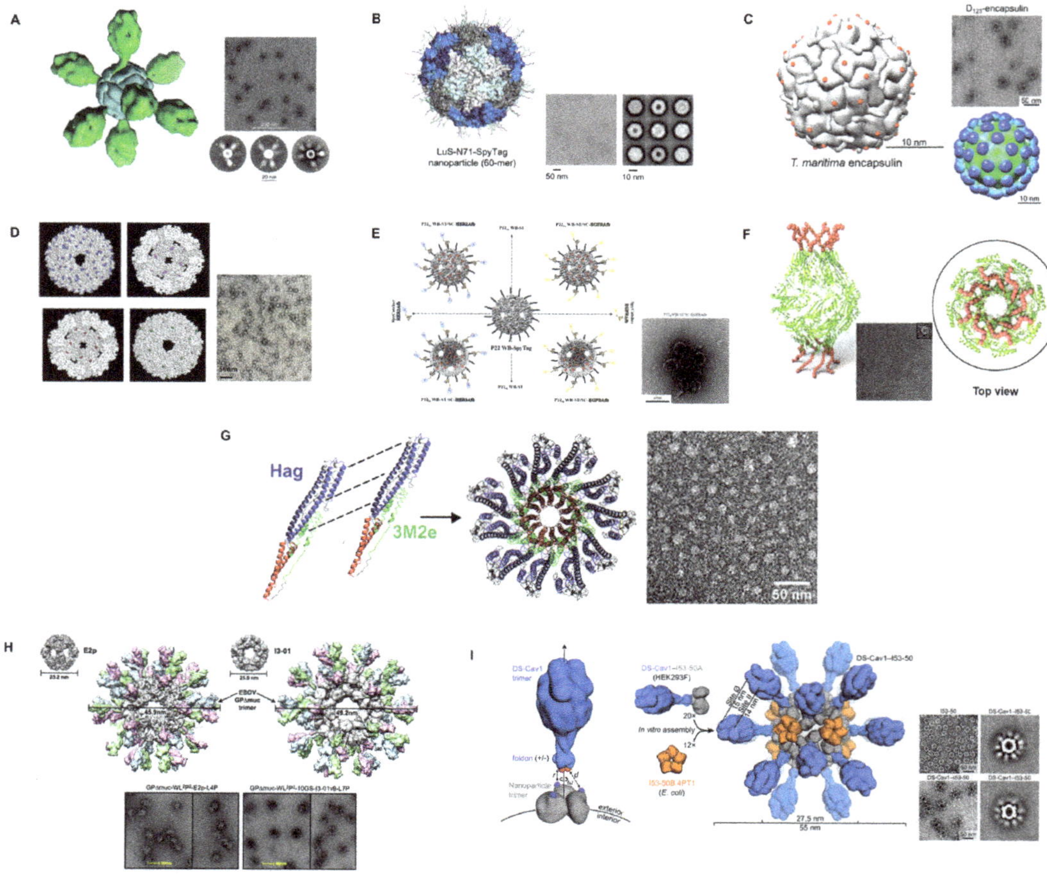

Figure 5. Selected illustrations of self-assembling bacterial proteins as nanoscaffolds for antigen delivery. (**A**) ferritin nanoparticle with the Fusion (RSV) antigen [100]. (**B**) lumazine synthase-N71-SpyTag nanoparticles [89]. (**C**) encapsulin with the insertion sites for exogenous antigens indicated as red spheres [99]. (**D**) Space-filling models of the 24 subunit sHsp cages [132]. (**E**) Schematic representation of various configurations of target-tunable and cargo-loadable P22-based delivery nanoplatforms [133]. (**F**) *Brucella* outer membrane protein BP26-derived nanoarchitecture displaying the influenza extracellular domain of matrix protein-2 (M2e) [126]. (**G**) Hag-3M2e$_{Ct}$ ring-like nanostructures [127]. (**H**) 60-meric E2p and I3-01 with GP (EBOV) trimer [116]. (**I**) Structural model of DS-Cav 1-I53-50 and schematic representation of the self-assembly process. Each nanoparticle comprises 20 trimeric and 12 pentameric building blocks for a total of 60 copies of each subunit [129]. (**A–I**) Reprinted with permission from [89,99,100,116,126,127,129,132,133].

5.1. Ferritin

Almost all living species, including bacteria, fungi, plants, and mammals, expressed the protein ferritin [134,135]. Its main physiological function is to store iron in an insoluble, non-toxic form while making it intracellularly accessible by converting it to a soluble form [136], playing a crucial role in iron homeostasis. It also protects against free-iron-related toxicity, such as the production of reactive oxygen species, which can damage cellular machinery and cause cell death [137]. Structurally, ferritin nanoparticles have a hollow core with inner and outer diameters of 8 and 12 nm that may internalize up to 4500 iron atoms in the form of ferric oxyhydroxide [138], as well as varying quantities of phosphate [139]. Each ferritin particle consists of 24 identical or homologous subunits that self-assemble in an octahedral (432) symmetry [135]. Ferritin has lately emerged as a promising antigen-displaying platform [140]. In addition to its ability to self-assemble, the ferritin protein complex exhibits remarkable thermal and pH stability, biocompatibility, biodegradability and is significantly cost-effective for large-scale production, hollow cavity with reversible assembly/disassembly, and amenability to surface conjugation via chemical or genetic approaches [141]. Ferritin has been employed in nanobiotechnology for drug delivery, biomimetic synthesis, bioimaging, and cell targeting [135,141–145].

Helicobacter pylori, *E. coli* and *Pyrococcus furiosus* 24-mer non-heam self-assembling ferritin have been most widely used for the development of bacterial protein-based nanoparticle subunit vaccines (Figure 5A), notably against SARS-CoV-2 (using the spike (S protein), RBD and heptad repeat (HR)) [89,94–96,110,114], Middle East respiratory syndrome-coronavirus (MERS-CoV) [111], IAV [97], IAV subtype H5N1 [98], Ebola virus (EboV) [116], RSV [100], hepatitis C virus (HCV) [101], HIV [102–105,112], rotavirus A [106], porcine reproductive and respiratory syndrome virus (PRRSV) [107], foot-and-mouth disease virus (FMDV) [108], Epstein-Barr virus (EBV) [99], classical swine fever virus (CFSV) [109], HBV [113] and HPV [115]. These self-assembling ferritins have been produced in a variety of hosts including bacteria (E. coli BL21 (DE3)), insects (Sf9, High-Five and Drosophila S2 cells) and mammalian cells (Expi293F, ExpiCHO, FreeStyle™ 293-F, FreeStyle™ HEK293F, HEK293S and FreeStyle™ CHO-S). The antigens were fused to the ferritin using either a genetic fusion, chemical modification, or the SpyTag/SpyCather system. Upon self-assembling, ferritin nanoparticle-based vaccine candidates yielded nanoparticles with 20–50 nm in diameter. The advantage of self-assembling in 24-mer is that 24 copies of antigen/epitope can be genetically or covalently conjugated. These self-assembling vaccine candidates have been used in various model animals, such as mice, rabbits, pigs, non-human primates and many others, to study humoral and cellular immunity.

Using ferritin nanoparticles, Ma et al. (2020) [95] exposed that ferritin-based nanoparticle vaccines induce potent Nabs and cellular immune responses against SARS-CoV-2. Furthermore, the vaccination of transgenic hACE2 mice with RBD and/or RBD-HR nanoparticles demonstrated a significantly reduced viral load and elevated protection following SARS-CoV-2 infection. Moreover, nanoparticles also assisted in inducing Nabs and cellular immune responses against other coronaviruses. Additionally, the authors have exhibited that the use of these nanoparticles also induced Nabs, as well as T cell responses in Rhesus Macaques, which persisted for more than three months. Of note, they also assessed that the nanoparticles did not induce ADE. In another study, Wang et al. (2021) [114] showed that a ferritin nanoparticle-based SARS-CoV-2 RBD vaccine generated an effective antibody response in mice that sustained for at least 7 months after inoculation. Moreover, they also reported that, upon antigen exposure, a large proportion of MBCs were preserved and considerably recalled and induced persistent antibody response. They have also exhibited that ferritin nanoparticle vaccine with preS1 domain of HBV elicited a therapeutic antibody response against chronic hepatitis B in a mouse model [113]. In nonhuman primates, Joyce et al. (2021) [96] designed and tested the SARS-CoV-2 spike ferritin nanoparticle (SpFN) vaccine, which produced a Th1-biased CD4 T cell response and Nabs against wild-type SARS-CoV-2, variants, as well as SARS-CoV-1. Following a high-dose SARS-CoV-2 respiratory challenge, these powerful humoral and cell-mediated immune responses resulted in the fast clearance of replicating virus in the upper and lower airways, as well as the lung parenchyma of nonhuman primates. In another example, an SpFN nanoparticle vaccine candidate was paired with two distinct adjuvants, Alhydrogel® or Army Liposome Formulation containing QS-21 (ALFQ) [94]. They exhibited that SpFN-ALFQ efficiently activates innate immune cells and improves SARS-CoV-2-specific long-lasting adaptive immune T cell responses. Innate immune cell activation was linked to robust antigen-specific polyfunctional cytokine responses and cytolytic activity. The role of the adjuvant ALFQ, together with ferritin nanoparticles, was responsible for directing the interaction of innate and adaptive immune responses.

For reliable and repeatable manufacturing of nanoparticle-based vaccines, monomeric antigen folding and subsequent assembly into highly ordered structures are critical. Despite substantial breakthroughs in silico design and structure-based assembly, most engineered nanoparticle-based vaccines are resistant to soluble expression and fail to assemble as intended, posing severe manufacturing problems in nanoparticle-based vaccine development. Therefore, the RNA-interaction domain (RID) was used by Kim et al. (2018) [111] as a reliable protein-folding vehicle to assemble NPs in bacterial hosts. They genetically fused the MERS-CoV RBD with Helicobacter pylori ferritin and RID to produce nanoparticles in a soluble form in Escherichia coli. The results exhibited that MERS-CoV RBD binding to the cellular receptor hDPP4 was efficiently inhibited by mice sera following vaccination. In conclusion, their findings demonstrated that RID regulates the antigen folding pathway's overall kinetic network in favor of improved nanoparticles assembling into highly regular and immunologically relevant conformations. Moreover, the concentration of the Fe^{2+}, salt, and fusion linker all played a role in in vitro assembly and stability of these nanoparticles.

Kanekiyo et al. (2013) [97] genetically linked the haemagglutinin (HA) of IAV (H1N1) to ferritin nanoparticles. Immunization with this influenza nanoparticle vaccine resulted in haemagglutination inhibition antibody titers that were more than tenfold greater than those obtained with the approved inactivated vaccine. It also evoked Nabs to the stem and the receptor-binding site on the head of HA, two highly conserved sites that are universal vaccine targets. As a result, these self-assembling nanoparticles induced a broader and more powerful protection than standard influenza vaccinations. As mentioned above, the subdominant stem region of HA is highly conserved and identified by antibodies that can bind different HA subtypes. Therefore, Yassine et al. (2015) [98] developed a ferritin nanoparticle-based HA H1 stem-only GP vaccine candidate using a structural-based design approach. Despite the lack of detectable H5N1 neutralizing activity in vitro, the vaccination of mice and ferrets with ferritin nanoparticle-based stem-only HA resulted in widely cross-reactive antibodies that totally protected mice and partially protected ferrets against deadly heterosubtypic H5N1 IAV challenge. Moreover, the passive transfer of antibodies from immunized mice to naïve mice protected them from the H5N1 challenge, demonstrating that the vaccine-elicited HA stem-specific antibodies can protect against a variety of group 1 influenza viruses.

To enhance antigen presentation and target antibody responses to important epitopes of the F protein of RSV, Swanson et al. (2020) [100] used a structure-based rational design and fused a stabilized pre-F protein to ferritin nanoparticles (pre-F-NP) (Figure 5A). They also concealed non-neutralizing epitopes with glycans. The multimeric pre-F-NP induced long-lasting specific Nabs in nonhuman primates and mice over 150 days.

Chen et al. (2020) [108] investigated ferritin nanoparticles that were recombinantly linked to VP1 and G-H loop subunits of FMDV, an acute, febrile, and highly contagious infectious disease common in cloven-hoofed animals. Their findings indicated that ferritin nanoparticles containing recombinant proteins were immunogenic. Recombinant FMDV subunit vaccinations boosted FMDV-specific IgG (IgG1 and IgG2a) antibody titers, as well as IL-4 and IFN-γproduction. The ferritin nanoparticles also provided partial protection in mice.

Li et al. (2019) [106] proposed an experimental rotavirus vaccine transgenically expressed in the milk of mice. It was tested for immunological protection in a mouse model in order to create a rotavirus vaccine that would be suited to both mammary-gland-based manufacturing and milk-based administration. Their findings implied that the recombinant VP6–ferritin nanoparticle vaccine can effectively prevent the mortality and malnutrition caused by rotavirus infection in pups, making it a promising candidate vaccination for rotavirus.

Trimers of the HIV-1 envelope (Env) are typically poorly immunogenic. Therefore, Sliepen et al. (2021) [105] compared the efficacy of several adjuvants (squalene emulsion, ISCOMATRIX, GLA-LSQ, and MPLA liposomes) to promote Nab responses in rabbits using the clinically relevant ConM SOSIP.v7 trimer fused to ferritin nanoparticles. When the ferritin nanoparticles were delivered with ISCOMATRIX, stronger Nab responses were evoked. In conclusion, they exhibited that the ferritin nanoparticle's immune response could be enhanced with the combination of adjuvant, but the nature of the antigens and nanoparticles must be taken into consideration. In other studies, they reported that HIV-1 Env GP trimers bearing ferritin nanoparticles were significantly more immunogenic than trimers in both mice and New Zealand white rabbits [103] and elicited a strong Nabs response against the autologous virus in New Zealand white rabbits and Rhesus macaques [104].

Zhao et al. (2021) [109] reported a complete vaccination method for displaying the E2 GP of the CSFV on the surface of self-assembling ferritin nanocages. In vivo data showed that rabbits inoculated with ferritin nanoparticles triggered both humoral and cellular immunity as indicated by Nab titers and expression of IL-4 and IFN-γ.

In an in vitro study, He et al. (2015) [101] developed and characterized epitope vaccine antigens targeting the antigenic locations of HCV envelope GP E1 (residues 314–324) and E2 (residues 412–423). They then used a "multivalent scaffolding" strategy to improve antibody binding avidity by displaying 24 copies of an epitope scaffold on a ferritin self-assembling nanoparticle. Their research shows the value of a multi-scale scaffolding technique in epitope vaccine development and identifies prospective HCV immunogens for in vivo testing. The nanoparticles had delayed off-kinetics for both antigenic sites, indicating a substantial avidity impact owing to multivalent antibody binding.

Due to its capacity to elicit responses against a greater variety of distinct HPV strains, the HPV minor capsid protein L2 has been studied as a possible antigen candidate substitute for major capsid protein L1. In this regard, the ferritin-L2 antigen nanoparticles caused a broadly Nab response in guinea pigs and mice that covered 14 oncogenic and two non-oncogenic HPV strains. The immune response lasted for at least one year and provided protection in a cervico-vaginal mouse model of HPV infection [115].

In contrast to inactivated PRRSV, Ma et al. (2021) [107] showed that a modified envelope glycoprotein 5 (GP5)-ferritin nanoparticle vaccine produced greater serum antibody titers in pigs. Additionally, the nanoparticle vaccine boosted a Th1-dominant cellular immune response and enhanced specific T lymphocyte immune responses. In comparison to unvaccinated pigs, GP5-ferritin-vaccinated pigs had significantly lower mean rectal temperatures, respiratory scores, viremia, and scores for both macroscopic and microscopic lung lesions after the challenge. These findings demonstrated the potential of ferritin subunit vaccines to induce protective immune responses and serve as vaccine candidates.

5.2. Lumazine Synthase

Lumazine synthase (LS, EC 2.5.1.78) is an enzyme that catalyzes the penultimate step in the biosynthesis of riboflavin, widely known as vitamin B2 [146]. LS from *Bacillus subtilis*, hyperthermophilic bacterium *Aquifex aeolicus* and a variety of other bacteria and archaea produce icosahedral capsids with triangulation (Figure 5B). The icosahedral compound has a triangulation number (T) = 1 [147–150]. The capsids have an outside diameter of roughly 15–16 nm and are made up of 12 pentameric units, totaling 60 identical subunits linked by symmetry axes of twofold, threefold, and fivefold of ~960,000 Daltons. In an alkaline medium (pH > 8), *Bacillus subtilis'* LS transforms from T = 1 state to a T = 3 state, which is composed of 180 identical subunits with a diameter of roughly 29 nm [149]. This transformation is caused by the loss of a phosphate ion per monomer, which stabilizes the T = 1 state [146,149]. LS's symmetric nanoparticle carriers have been shown to display a structurally ordered array of immunogens, which are summarized here.

In this regard, using the SpyTag/SpyCatcher system, Zhang et al. (2020) [89] created a modular 60-subunit *Aquifex aeolicus* LS-based self-assembling nanoparticle (in parallel with the 24-subunit Helicobacter pylori ferritin) platform that enables the plug-and-play display of trimeric viral GP on nanoparticle surfaces (Figure 5B). This technique was tested using three viral trimeric GP that were pre-fusion (preF)-stabilized: RSV fusion (RSV F) GP, human parainfluenza virus type 3 fusion (PIV3 F) GP, and SARS-CoV-2 S GP. The higher antigenicity of the apical epitopes of trimeric viral GP attached to LS nanoparticles resulted in improved immunogenicity, especially at lower doses. The vaccination of mice with 0.08 μg of SARS-CoV-2 spike-LS nanoparticle induced identical neutralizing reactions as 2.0 μg of the spike, which was 25-fold greater in terms of weight-per-weight. Aebischer et al. (2021) [120] also established a durable and adaptable self-assembling multimeric protein scaffold particle (MPSP) vaccination platform using LS from Aquifex aeolicus. They used the SpyTag/SpyCatcher-mediated plug-and-display on the pre-assembled particles of LS-MPSPs to display two model antigens of Schmallenberg virus (SBV) and studied their efficacy in mouse and cattle models. In both models, they showed that the nanoparticles improved immunogenicity and vaccine efficacy. For example, a single dose of this vaccine protected roughly 80% of mice and gave cattle an almost sterile immunity against an otherwise deadly dosage of SBV.

Furthermore, Tokatlian et al. (2019) [112] evaluated the fate of two different extensively glycosylated HIV antigens, gp120 and gp140. To generate protein nanoparticles, the gp120 antigen was fused to a bacterial protein LS, and, on the other hand, archaeal ferritin was used to fuse the gp140 antigen. Unlike monomeric antigens, multivalent glycosylated antigens displayed on nanoparticles stimulate mannose-binding lectin-mediated innate immune recognition in vivo, resulting in fast complement-dependent transport to FDCs and their subsequent accumulation in GCs. This focused trafficking was linked to improved antibody responses, suggesting that immunogen glycosylation may be a significant design requirement for future nanoparticulate vaccines. These findings are particularly intriguing in the context of HIV vaccine development, where the thick envelope "glycan shield" is frequently seen as a barrier to eliciting effective antibody responses. Designing immunogens that can trigger broadly Nabs that bind to the HIV-1 viral Env GP is one of the challenging aspects of HIV-1 vaccine research. LS nanoparticles, when multimerized with a rationally designed epitope (HIV-1 gp-120), showed that formulated nanoparticles activate germline and mature VRC01-class B cells [118] and could prime a broadly Nabs response to HIV-1 [117].

5.3. Encapsulin

Encapsulin is a nanocarrier particle that contains virus capsid-like nanocompartments found in a variety of bacteria and archaea [93,151–154]. It has been demonstrated that these nanocompartments may store iron and protect bacteria from oxidative stress [155]. Structurally, encapsulin nanocompartments differ from organism to organism. For example, *Rhodococcus erytropolis* [156], *Mycobacterium tuberculosis* [157], and *Thermotoga maritima* [153] encapsulin are made up of 60 identical subunits that form T = 1 icosahedral capsid-like particles with a diameter of 20–24 nm. On the other hand, *Myxococcus xanthus* [155] and *Pyrococcus furiosus* [151] encapsulin nanocompartments comprise of 180 protein subunits in a T = 3 icosahedral particles with a diameter of from 30 to 32 nm. Encapsulin nanocompartments package functional enzymes such as ferritin-like proteins and Dyp-type peroxidases in bacteria [93,153,155,157]. A specific C-terminus sequence of these encapsulated proteins causes them to bind to the internal surface of the encapsulin [153]. Taking advantage of this, many non-native cargo proteins (i.e., antigens) were packaged using the unique cargo-loading ability of the encapsulin nanocompartments (Figure 5C) [158,159].

Lagoutte et al. (2018) [93] demonstrated the ability of *Thermotoga maritima's* encapsulin nanoparticles for simultaneous surface display of M2e epitope of IAV and packaging of a green fluorescent protein (GFP) reporter into the internal cavity. In this study, the researchers successfully demonstrated that the engineered encapsulin nanoparticles facilitated both surface display and packaging properties. This surface engineering of encapsulin nanoparticles also enhanced the cargo-loading capacity of the heterologous reporter protein. Furthermore, the immunogenicity study in mice revealed strong antibody responses against both the surface epitope and the loaded cargo protein without compromising the booster immune response to the targeted epitope. Their study represents the enormous potential of encapsulin nanoparticles for rational vaccine design and antigen delivery.

Kanekiyo et al. (2015) [99] have developed an EBV vaccine candidates based on encapsulin and ferritin self-assembling nanoparticles (Figure 5C). These platforms elicited potent and long-lasting virus-Nabs in mice and nonhuman primates that target the receptor-binding site on the viral envelope protein gp350. More specifically, they blocked gp350's CR2-binding site and exhibited enhanced vaccine-induced protection in a mouse model. Of note, the ferritin-based nanoparticles provided an 80% survival rate versus 20% for the encapsulin-based nanoparticle in the mice model.

5.4. sHSP and P22

The small heat shock protein 16.5 (sHsp) from *Methanocaldococcus jannaschii* (a hyperthermophilic archaeon) [122,160,161] is composed of 24 repeating subunits. Due to the high symmetry and quaternary structure, these subunits self-assemble to form an empty cage-like shape, similar to that of a viral capsid (Figure 5D) [123,132,162]. Flenniken et al. (2003) and Abedin et al. (2009) [132,163] previously exhibited that sHsp may be genetically modified to contain cysteine residues, therefore enabling attachment sites for bioconjugation [132,163,164], which has been used for the display of a foreign protein [165,166]. On the other hand, P22 is a bacteriophage capsid that infects *Salmonella typhimurium* in the presence of intact tail fibers [167,168]. The P22 employed for vaccine developments lacks both genetic material and tail fibers, leaving just the non-infectious empty viral capsid (Figure 5E). P22 forms an icosahedral procapsid (58 nm in diameter) from 415 copies of the 46.6 kDa coat protein and roughly 300 copies of the 33.6 kDa scaffolding protein [122]. Richert et al. (2012) [122] used sHsp and P22 phage-derived VLPs as immunomodulatory platforms in both nonspecific pre-priming scenarios and as the delivery of particular antigens to the lung. Specifically, they exhibited that ovalbumin (OVA) could be chemically fused to the outside of an sHsp cage. When naive mice were given OVA–sHsp intranasally, the immune response to OVA was expedited and strengthened and OVA-specific IgG1 responses were observed 5 days after a single dose, demonstrating its potential for vaccine delivery platform. Furthermore, the research group showed a strong mucosal sIgA titer alongside GCs for B and Tfh cell accumulation. As a result, they demonstrated that sHsp and P22 VLPs may be employed as both immunomodulatory agents and antigen carriers, allowing for local immunization of the lower respiratory tract against pathogens with a single dose. In another study, Wiley et al. (2009) [123] generated protein cage nanoparticles (PCNs) derived from the sHsp 16.5, which comprised 24 identical subunits that spontaneously self-assemble into hollow, spherical protein cages that were 12 nm in diameter. They showed that mice pre-treated with PCN, independent of any viral antigens, were protected from both sub-lethal and fatal dosages of two distinct IAV, a mouse-adapted SARS-coronavirus, or mouse pneumovirus. Treatment with PCN markedly improved viral clearance, expedited the induction of viral-specific antibody production, reduced morbidity and lung damage and greatly increased survival.

To study the response of P22 VLP to IAV variability, Sharma et al. (2020) [125] used a mouse model to assess the immunogenic potential and protective effects of P22 VLPs conjugated with multiple copies of the globular head domain of the HA protein from the PR8 strain of IAV. The HA globular head was coupled to preassembled P22 VLPs via a covalent attachment technique (SpyTag/SpyCatcher). Mice immunized with this P22-HA head combination were completely protected from morbidity and mortality after infection with a homologous IAV strain. The authors also exhibited that P22 VLPs may be quickly modified in a modular way to design potent vaccine(s) that can produce protective immunity without the use of additional adjuvants. Similarly, Patterson et al. (2013) [124] developed a vaccine strategy where the antigenically conserved nucleoprotein from influenza was fused on the interior of P22-derived VLP. They reported that the P22-derived vaccine protected mice against many strains of IAV (H1N1 and H3N2) in a nucleoprotein-specific CD8+ T cell-dependent manner. Their findings also highlighted the P22 system's ability to integrate large antigenic proteins, which is frequently a limiting element in other VLP systems.

5.5. BP26

Through the self-assembly of 16 monomeric proteins, BP26, an OMP of the zoonotic pathogen *Brucella abortus*, creates a barrel-like shape with a hollow core (Protein Data Bank ID: 4HVZ) (Figure 5F) [126,169]. As BP26 is a common immunodominant antigen in *Brucella* bacteria, its barrel-like nanostructure exhibits potent adjuvant efficacy [170]. As a result, it has been hypothesized that engineered BP26 monomers containing an antigen could self-assemble into antigen-displaying BP26-based nanobarrels. These nanobarrels have been shown to improve recognition by BCRs, resulting in increased antigen-specific antibody production [59,62,171]. Kang et al. (2021) [126] created a universal IAV vaccine platform that is cross-protective using the nanobarrel self-assembly of BP26 to build a protein nanoarchitecture displaying the extracellular domain from M2e of IAV (BP26-M2e) (Figure 5F). Mice immunized with BP26-M2e nanobarrel vaccines produced high levels of anti-M2e antibodies with antibody-dependant cell cytotoxicity capacity. Furthermore, their platform induced T-cell responses and effectively protected mice from IAV infection-associated death, even without the use of a conventional adjuvant.

5.6. Flagellin

The flagellum of bacteria is made up of the protein flagellin. Due to its capacity to trigger elements of the innate immune system, flagellin has shown considerable promise as a vaccine adjuvant [172–175]. This protein has attracted interest for its use as an adjuvant in vaccine formulations as a powerful TLR5 agonist. TLR5 activation increases the synthesis of IL-6 through the MyD88-dependent pathway [172]. Flagellin can also activate the NAIP5 and NLRC4 NOD-like receptors (NLRs), which, in turn, causes the NLRC4 inflammasome to assemble and activates caspase-1, resulting in proinflammatory signals [176–178]. Although flagellin is capable of self-assembling into linear filaments (nanotubes), the majority of research to date has only employed the protein's monomeric form as an adjuvant [128]. However, monomeric flagellin-based vaccines have been proven to cause visible and long-lasting side effects, such as discomfort at the injection site, exhaustion, and muscular pains due to the sustained activation of TLR5, which has compromised their use in clinical trials [127,179]. In contrast, the use of nanotubes as an adjuvant cannot be effectively internalized by APCs or passively drained to the lymph nodes due to their aspect ratio [57]. Additionally, studies have demonstrated that when compared to their monomeric counterpart, flagellin nanotubes cause a significantly lower activation of TLR5 and reduced stimulation of the innate immunity [180,181]. To avoid using µm-long nanotubes as innately immunostimulatory antigen delivery platforms, it may be possible to manipulate flagellin's ability to self-assemble into lower-aspect-ratio supramolecular nanostructures [127].

Flagellins have from two to four domains, with D0 and D1 playing a role in the TLR5 interaction [182,183]. On the other hand, the D2 and D3 domains, display considerable sequence variation between species and are involved in flagellated pathogen immune evasion [184]. Evidence shows that the D0 and D1 domains are involved in the self-assembly aspect of this protein [185]. Côté-Cyr et al. (2022) [179] fused an IAV M2e epitope to self-assembling flagellins Hag from *Bacillus subtilis* (having only D0 and D1 domains) and FljB from *S. Typhimurium* (containing D0, D1, D2 and D3 domains) [182] to investigate their immunostimulatory and pro-inflammatory characteristics. Both flagellins activated TLR5, but FljB was 2.5 times more potent than Hag. However, mice inoculated with FljB or Hag elicited a strong M2e-specific antibody response, with Hag showing reduced production of pro-inflammatory markers and weight loss. Therefore, the study showed that flagellin Hag was a powerful immunoadjuvant with minimal side effects. In another study, Côté-Cyr et al. (2022) [127] also described a method for controlling the self-assembly of the *Bacillus subtilis* flagellin protein Hag into reduced aspect ratio nanoparticles by preventing the non-covalent contacts that cause the protein to elongate into nanotubes. They exhibited that adding three repetitions of the M2e antigenic sequence from the IAV to the C-terminus of flagellin prevented filament elongation and led to low-aspect-ratio ring-like nanostructures during salting-out-induced self-assembly (Figure 5G). APCs successfully absorbed flagellin-ring-like nanostructures, which, in turn, triggered TLR5 response in vitro as well as innate and adaptive immune responses. In summary, the authors reported that these nanostructures have the potential to act as antigen carriers because they are intrinsically immunostimulatory. The intranasal vaccination of mice with these nanostructures led to the potentiation of the antigen-specific antibody response and protection against a deadly IAV infection.

5.7. Other Self-Assembling Proteins with Potential Usage in Subunit Vaccines

A new development in bacterial protein-based self-assembling nanoparticles has been the emergence of computationally designed protein nanoparticles as a robust and versatile platform for multivalent antigen presentation [129,186–191]. In preclinical studies, vaccine candidates based on designed protein nanoparticles have significantly improved the potency or breadth of antibody responses against numerous antigens, including prefusion RSV F [129], Env from HIV-1 [192], HA from IAV [193], and *P. falciparum* cysteine-rich protective antigen (CyRPA) [194], relative to either soluble antigen or commercial vaccine comparators. Among these platforms, I3-01 (derived from *Thermotoga maritima*) [187], E2p (derived from *Geobacillus stearothermophilus*) and I53-50 [186] are the most extensively used, computationally designed, self-assembling subunit vaccines.

I3-01 is derived from *Thermotoga maritima's* 2-Keto-3-deoxy-6-phosphogluconate (KDPG) aldolase, which is pyruvate aldolase central to the Entner–Doudoroff glycolytic pathway [119,187,195,196]. In this metabolic pathway, glucose and galactose are converted to the corresponding 3-deoxy-6-phosphohexulosonate and then cleaved enzymatically to pyruvate and d-glyceraldehyde-3-phosphate [195,196]. The KDPG aldolase protein forms a capsule-like protein that resembles a virus capsid. Therefore, the KDPG structure has been studied and computationally modified to generate I3-01 (24-kDa), which assembles into a 25 nm dodecahedron (Figure 5H) [119]. This ability to design proteins that self-assemble into precisely specified, highly ordered icosahedral structures has opened the door to a new generation of protein 'containers' that could exhibit properties custom-made for various vaccine and drug-delivery applications.

Furthermore, *Geobacillus stearothermophilus* E2p (26-kDa), which self-assembles into a pentagonal dodecahedral scaffold of 27 nm, is an icosahedral scaffold formed by the acyltransferase component (E2 protein) of the multienzyme pyruvate dehydrogenase complex (PDH) [197]. E2p has been reported for its ability to display peptides in a highly immunogenic form [198,199]. The core C-terminal catalytic domain of E2p self-assembles into trimers, which, in turn, aggregate to generate a 60-chain core with an icosahedral symmetry [200–202]. Moreover, this 60-meric icosahedral structure can be regenerated with high efficiency from denaturing conditions in vitro, without the need for chaperonins [203,204]. The robustness of this peptide-based VLP rendered it an attractive macromolecular scaffold for the presentation of exogenous molecules on its surface [198–200] and for molecular encapsulation in its cavity [205,206]. Efficient refolding of E2p to the 60-mer is also possible, with foreign peptides replacing the natural peripheral domains, as N-terminal fusions to the core domain. Thus, a suitably engineered E2p core (E2DISP) can display 60 copies of heterologous peptides on the surface of a high-molecular-mass scaffold [198–200], which could be exploited for vaccine design (Figure 5H).

I53-50 is another computationally designed, two-component protein complex comprising 20 trimeric "A" components and 12 pentameric "B" components for a total of 120 subunits with icosahedral symmetry that is 28 nm broad (Figure 5I) [131,186]. I53 stands for icosahedral assembly constructed from pentamers and trimers. One of the major advantages of using I53-50 is that these nanoparticles could be easily constructed in vitro by simply combining I53-50A and I53-50B that have been produced and purified separately, a characteristic that has aided in its usage as a platform for multivalent antigen presentation [131]. Altogether, such robustly designed nanostructures of I3-01, I53-50 and E2p were studied to have considerable utility for targeted vaccine design, which are discussed and summarized here.

He et al. (2021) [116] demonstrated a rationally designed GPΔmuc trimer to 60-mer E2p and I3-01 (as well as 24-mer ferritin) protein nanoparticles (Figure 5H). These nanoparticles were re-engineered in such a way that a dimeric locking domain (LD) is fused to the C terminus of a nanoparticle subunit and a Th cell epitope. These GP trimers and nanoparticles were observed to induce cross-EBV Nabs in mice and rabbits. In another study, He et al. (2016) [102] exhibited that trimeric HIV-1 antigens gp120 and gp140 trimer based on 60-meric E2p with 20 spikes (in parallel with ferritin) resemble VLPs. They found that gp120 and gp140 nanoparticles provide a strong activation of B cells with cognate VRC01 receptors in vitro. Kang et al. (2021) [119] created, described and assessed the immunogenicity of gp350-based (gp350D_{123}) self-assembled nanoparticles of EBV. They used I3-01 derived from KDPG aldolase and *Aquifex aeolicus's* LS as self-assembling nanoparticles to display 60 copies of gp350 in a repeating pattern. Nanoparticles carrying the gp350D_{123} generated a greater titer (65 to 133-fold higher) of binding and Nabs in mice and nonhuman primates' serum than the soluble gp350 monomer. Moreover, they also induced a Th2-biased response in mice. Additionally, gp350D_{123}-based nanoparticle vaccinations produced long-lasting antibody responses due to their excellent retention in lymph nodes and thermal stability.

Wichgers Schreur et al. (2021) [121] investigated whether linking antigens to E2p and I3-01 (in parallel to LS), could increase the immunogenicity of a potential subunit vaccine against the zoonotic Rift Valley fever virus (RVFV). To test immunogenicity and effectiveness in vivo, the head domain of RVFV GP Gn, a known target for Nabs, was coupled to these nanoparticles. The Gn head domain, when bound to the LS, decreased mortality in a fatal mouse model and protected lambs from viremia and clinical symptoms following vaccination and experimental challenge. Furthermore, in lambs, the LS nanoparticles, in combination with E2p or I3-01, offered complete protection.

I53-50 with 20 copies of prefusion RSV F GP (a trimeric viral protein (DS-Cav1)) has been investigated as the main target of Nabs (Figure 5I) [129]. These nanoparticles created stable, highly ordered, monodisperse immunogens that show DS-Cav1 at a controlled density and generated 10-fold greater NAbs responses in mice and nonhuman primates than trimeric DS-Cav1. Moreover, Walls et al., 2020 [88], Brouwer et al., 2021 [131] and Arunachalam et al., 2021 [130] exhibited the structure-based design of self-assembling protein nanoparticle immunogens that cause mice to develop strong and protective antibody responses against SARS-CoV-2. In mice, rabbits, and cynomolgus macaques, the nanoparticle vaccinations generated Nab titers that were significantly greater than the prefusion-stabilized spike despite a lower dosage, because they feature 60 copies of monomeric SARS-CoV-2 spike RBDs in a highly immunogenic array (currently in phase 1/2 clinical trials: NCT04742738 and NCT04750343) [88]. They also suggested that these assembled nanoparticles' excellent yield and stability imply that nanoparticle vaccine production could be scalable to suit mass-production. Interestingly, a recent study evaluated the impact of scaffold-specific antibodies on the antigen-specific immune response using I53-50 [207]. Four different viral antigens, HA (IAV), F (RSV), RBD (SARS-CoV-2) and Env (HIV), were used to monitor the antigen specific response. The scaffold-specific immune response did not affect antigen-specific immune response for HA, F and RBD but reduced Env-specific immune response, since Env is immunologically subdominant to the scaffold. For dominant antigens, prior scaffold immunity does not seem to affect antigen-specific immune responses. Nonetheless, the impact of scaffold immunity will still have to be considered if multiple doses of the same vaccine are administered to the same individuals.

Apart from the self-assembling bacterial proteins described above, supramolecular architectures consisting of different proteins are also attracting attention as candidates for the development of vaccine nanoplatforms. In this regard, a self-assembling foldon, the natural trimerization domain of bacteriophage T4 fibritin and GCN4-based isoleucine zipper from *Saccharomyces cerevisiae,* has been used to facilitate the trimerization of viral structural proteins, such as IAV, HIV-1, RSV, MERS-CoV and SARS-CoV-2 antigens, to mimic their native viral structure [208–211]. The inclusion of these trimerization domain promotes a stable trimeric structure and has been shown to increase the immunogenicity of antigens, supporting the importance of mimicking native viral trimeric structures [210,212–214]. Furthermore, a recent study showed that genetically fusing MERS-CoV RBD with a foldon domain elicited potent RBD-specific neutralizing antibodies and protected hDPP4 transgenic mice from viral infection. Moreover, it has been exhibited that these commonly used protein trimerization domains can be highly immunogenic, but they can be further immunosilenced by the addition of *N*-glycans [215].

6. Conclusions

The need to develop safe, scalable, effective, and affordable vaccination strategies is emphasized by the constant appearance and evolution of new pathogens and the lack of vaccines for many infectious diseases. Despite their limited immunogenicity and prerequisite for adjuvants, subunit vaccines offer a safe alternative to vaccine formulations based on live-attenuated and inactivated whole pathogens. Numerous adjuvants have largely non-specific and poorly understood mechanisms of action, and these substances are often associated with undesirable side effects. Therefore, bacterial protein supramolecular assemblies provide an interesting alternative to traditional vaccine adjuvants by combining the characteristics of an immunostimulant with those of an antigen delivery system. Compared to their soluble equivalents, antigenic determinants repeatedly presented on bacterial protein-based nanoparticles show strong immune-activating properties. Important aspects of the immune response, such as uptake by APCs, biodistribution, antigen retention, and activation of TLRs, can be modulated by engineering the self-recognition process at the molecular level. This allows for control of the size, morphology, surface chemistry, and symmetry of the nanoassemblies. Vaccination with bacterial protein-based nanoparticles was shown to promote the production of high levels of antigen-specific antibodies and, often, of a robust cellular immune response, while being safe. Future studies are still required to precisely elucidate how nanoparticles' shape, supramolecular architecture and surface chemistry affect the interactions with the immune system to fully take advantage of bacterial self-assembling proteins for the delivery of subunit antigens.

Author Contributions: F.L., V.K. and P.S.-L. made substantial contributions to the concept, design, drafting and writing of this review. D.A. and S.B. contributed to critical revision and approval of the submitted version. All authors have read and agreed to the published version of the manuscript.

Funding: This work was supported by funding from the Natural Sciences and Engineering Research Council of Canada (NSERC) grants RGPN-2016-06532 (D.A.) and RGPIN-2018-06209 (S.B.). F.L. acknowledges fellowships from the FRQNT strategic cluster PROTEO.

Institutional Review Board Statement: Not applicable.

Informed Consent Statement: Not applicable.

Data Availability Statement: No new data reported.

Acknowledgments: The authors would like to thank Mélanie Côté-Cyr for fruitful discussions in the preparation of this manuscript.

Conflicts of Interest: The authors declare no conflict of interest.

References

1. Orenstein, W.A.; Ahmed, R. Simply put: Vaccination saves lives. *Proc. Natl. Acad. Sci. USA* **2017**, *114*, 4031–4033. [CrossRef] [PubMed]
2. Mascola, J.R.; Fauci, A.S. Novel vaccine technologies for the 21st century. *Nat. Rev. Immunol.* **2020**, *20*, 87–88. [CrossRef] [PubMed]
3. Svoboda, E. Research round-up: Vaccines. *Nature* **2019**, *575*, S46–S47. [CrossRef] [PubMed]
4. Delany, I.; Rappuoli, R.; De Gregorio, E. Vaccines for the 21st century. *EMBO Mol. Med.* **2014**, *6*, 708–720. [CrossRef]

5. Tenforde, M.W.; Kondor, R.J.G.; Chung, J.R.; Zimmerman, R.K.; Nowalk, M.P.; Jackson, M.L.; Jackson, L.A.; Monto, A.S.; Martin, E.T.; Belongia, E.A.; et al. Effect of Antigenic Drift on Influenza Vaccine Effectiveness in the United States-2019–2020. *Clin. Infect. Dis.* **2021**, *73*, e4244–e4250. [CrossRef]
6. Zimmerman, R.K.; Nowalk, M.P.; Chung, J.; Jackson, M.L.; Jackson, L.A.; Petrie, J.G.; Monto, A.S.; McLean, H.Q.; Belongia, E.A.; Gaglani, M. 2014–2015 influenza vaccine effectiveness in the United States by vaccine type. *Clin. Infect. Dis.* **2016**, *2016*, ciw635. [CrossRef]
7. Ghattas, M.; Dwivedi, G.; Lavertu, M.; Alameh, M.G. Vaccine Technologies and Platforms for Infectious Diseases: Current Progress, Challenges, and Opportunities. *Vaccines* **2021**, *9*, 1490. [CrossRef]
8. Hanley, K.A. The double-edged sword: How evolution can make or break a live-attenuated virus vaccine. *Evol. Educ. Outreach* **2011**, *4*, 635–643. [CrossRef]
9. Beyer, W.E.; Palache, A.M.; de Jong, J.C.; Osterhaus, A.D. Cold-adapted live influenza vaccine versus inactivated vaccine: Systemic vaccine reactions, local and systemic antibody response, and vaccine efficacy. A meta-analysis. *Vaccine* **2002**, *20*, 1340–1353. [CrossRef]
10. El Sahly, H.M.; Baden, L.R.; Essink, B.; Doblecki-Lewis, S.; Martin, J.M.; Anderson, E.J.; Campbell, T.B.; Clark, J.; Jackson, L.A.; Fichtenbaum, C.J.; et al. Efficacy of the mRNA-1273 SARS-CoV-2 Vaccine at Completion of Blinded Phase. *N. Engl. J. Med.* **2021**, *385*, 1774–1785. [CrossRef]
11. Heath, P.T.; Galiza, E.P.; Baxter, D.N.; Boffito, M.; Browne, D.; Burns, F.; Chadwick, D.R.; Clark, R.; Cosgrove, C.; Galloway, J.; et al. Safety and Efficacy of NVX-CoV2373 Covid-19 Vaccine. *N. Engl. J. Med.* **2021**, *385*, 1172–1183. [CrossRef] [PubMed]
12. Markowitz, L.E.; Drolet, M.; Perez, N.; Jit, M.; Brisson, M. Human papillomavirus vaccine effectiveness by number of doses: Systematic review of data from national immunization programs. *Vaccine* **2018**, *36*, 4806–4815. [CrossRef] [PubMed]
13. Voysey, M.; Clemens, S.A.C.; Madhi, S.A.; Weckx, L.Y.; Folegatti, P.M.; Aley, P.K.; Angus, B.; Baillie, V.L.; Barnabas, S.L.; Bhorat, Q.E.; et al. Safety and efficacy of the ChAdOx1 nCoV-19 vaccine (AZD1222) against SARS-CoV-2: An interim analysis of four randomised controlled trials in Brazil, South Africa, and the UK. *Lancet* **2021**, *397*, 99–111. [CrossRef]
14. COVID-19 Tracker. COVID-19 Vaccination Tracker. Available online: https://covid19tracker.ca/vaccinationtracker.html (accessed on 25 March 2022).
15. Gebre, M.S.; Brito, L.A.; Tostanoski, L.H.; Edwards, D.K.; Carfi, A.; Barouch, D.H. Novel approaches for vaccine development. *Cell* **2021**, *184*, 1589–1603. [CrossRef]
16. Uddin, M.N.; Roni, M.A. Challenges of storage and stability of mRNA-based COVID-19 vaccines. *Vaccines* **2021**, *9*, 1033. [CrossRef]
17. Venters, C.; Graham, W.; Cassidy, W. Recombivax-HB: Perspectives past, present and future. *Expert Rev. Vaccines* **2004**, *3*, 119–129. [CrossRef]
18. Coates, T.; Wilson, R.; Patrick, G.; Andre, F.; Watson, V. Hepatitis B vaccines: Assessment of the seroprotective efficacy of two recombinant DNA vaccines. *Clin. Ther.* **2001**, *23*, 392–403. [CrossRef]
19. Jackson, S.; Lentino, J.; Kopp, J.; Murray, L.; Ellison, W.; Rhee, M.; Shockey, G.; Akella, L.; Erby, K.; Heyward, W.L.; et al. Immunogenicity of a two-dose investigational hepatitis B vaccine, HBsAg-1018, using a toll-like receptor 9 agonist adjuvant compared with a licensed hepatitis B vaccine in adults. *Vaccine* **2018**, *36*, 668–674. [CrossRef]
20. Food and Drug Administration. Vaccines Licensed for Use in the United States. Available online: https://www.fda.gov/vaccines-blood-biologics/vaccines/vaccines-licensed-use-united-states (accessed on 25 March 2022).
21. Pollet, J.; Chen, W.H.; Strych, U. Recombinant protein vaccines, a proven approach against coronavirus pandemics. *Adv. Drug Deliv. Rev.* **2021**, *170*, 71–82. [CrossRef]
22. Hansson, M.; Nygren, P.A.; Stahl, S. Design and production of recombinant subunit vaccines. *Biotechnol. Appl. Biochem.* **2000**, *32*, 95–107. [CrossRef]
23. Chung, Y.H.; Church, D.; Koellhoffer, E.C.; Osota, E.; Shukla, S.; Rybicki, E.P.; Pokorski, J.K.; Steinmetz, N.F. Integrating plant molecular farming and materials research for next-generation vaccines. *Nat. Rev. Mater.* **2021**, *7*, 372–388. [CrossRef] [PubMed]
24. Bill, R.M. Recombinant protein subunit vaccine synthesis in microbes: A role for yeast? *J. Pharm. Pharm.* **2015**, *67*, 319–328. [CrossRef] [PubMed]
25. Pino, M.; Abid, T.; Pereira Ribeiro, S.; Edara, V.V.; Floyd, K.; Smith, J.C.; Latif, M.B.; Pacheco-Sanchez, G.; Dutta, D.; Wang, S.; et al. A yeast expressed RBD-based SARS-CoV-2 vaccine formulated with 3M-052-alum adjuvant promotes protective efficacy in non-human primates. *Sci. Immunol.* **2021**, *6*, eabh3634. [CrossRef] [PubMed]
26. Felberbaum, R.S. The baculovirus expression vector system: A commercial manufacturing platform for viral vaccines and gene therapy vectors. *Biotechnol. J.* **2015**, *10*, 702–714. [CrossRef]
27. Ward, B.J.; Gobeil, P.; Seguin, A.; Atkins, J.; Boulay, I.; Charbonneau, P.Y.; Couture, M.; D'Aoust, M.A.; Dhaliwall, J.; Finkle, C.; et al. Phase 1 randomized trial of a plant-derived virus-like particle vaccine for COVID-19. *Nat. Med.* **2021**, *27*, 1071–1078. [CrossRef]
28. Irvine, D.J.; Aung, A.; Silva, M. Controlling timing and location in vaccines. *Adv. Drug Deliv. Rev.* **2020**, *158*, 91–115. [CrossRef]
29. Guermonprez, P.; Valladeau, J.; Zitvogel, L.; Thery, C.; Amigorena, S. Antigen presentation and T cell stimulation by dendritic cells. *Annu. Rev. Immunol.* **2002**, *20*, 621–667. [CrossRef]
30. Banchereau, J.; Steinman, R.M. Dendritic cells and the control of immunity. *Nature* **1998**, *392*, 245–252. [CrossRef]
31. Abbas, A.K.; Lichtman, A.H.; Pillai, S. *Cellular and Molecular Immunology*, 9th ed.; Elsevier: Philadelphie, PA, USA, 2018.

32. Hilligan, K.L.; Ronchese, F. Antigen presentation by dendritic cells and their instruction of CD4+ T helper cell responses. *Cell Mol. Immunol.* **2020**, *17*, 587–599. [CrossRef]
33. Leon, B.; Lund, F.E. Compartmentalization of dendritic cell and T-cell interactions in the lymph node: Anatomy of T-cell fate decisions. *Immunol. Rev.* **2019**, *289*, 84–100. [CrossRef]
34. Bhattacharyya, N.D.; Feng, C.G. Regulation of T helper cell fate by TCR signal strength. *Front. Immunol.* **2020**, *11*, 624. [CrossRef] [PubMed]
35. Beijnen, E.M.S.; van Haren, S.D. Vaccine-Induced CD8(+) T Cell Responses in Children: A Review of Age-Specific Molecular Determinants Contributing to Antigen Cross-Presentation. *Front. Immunol.* **2020**, *11*, 607977. [CrossRef] [PubMed]
36. Cyster, J.G.; Allen, C.D.C. B Cell Responses: Cell Interaction Dynamics and Decisions. *Cell* **2019**, *177*, 524–540. [CrossRef] [PubMed]
37. Akkaya, M.; Kwak, K.; Pierce, S.K. B cell memory: Building two walls of protection against pathogens. *Nat. Rev. Immunol.* **2020**, *20*, 229–238. [CrossRef] [PubMed]
38. Elsner, R.A.; Shlomchik, M.J. Germinal Center and Extrafollicular B Cell Responses in Vaccination, Immunity, and Autoimmunity. *Immunity* **2020**, *53*, 1136–1150. [CrossRef]
39. Coffman, R.L.; Sher, A.; Seder, R.A. Vaccine adjuvants: Putting innate immunity to work. *Immunity* **2010**, *33*, 492–503. [CrossRef]
40. Kool, M.; Soullie, T.; van Nimwegen, M.; Willart, M.A.; Muskens, F.; Jung, S.; Hoogsteden, H.C.; Hammad, H.; Lambrecht, B.N. Alum adjuvant boosts adaptive immunity by inducing uric acid and activating inflammatory dendritic cells. *J. Exp. Med.* **2008**, *205*, 869–882. [CrossRef]
41. Pulendran, B.; Arunachalam, S.; O'Hagan, D.T. Emerging concepts in the science of vaccine adjuvants. *Nat. Rev. Drug. Discov.* **2021**, *20*, 454–475. [CrossRef]
42. Roth, G.A.; Picece, V.; Ou, B.S.; Luo, W.; Pulendran, B.; Appel, E.A. Designing spatial and temporal control of vaccine responses. *Nat. Rev. Mater.* **2021**, *7*, 174–195. [CrossRef]
43. Ong, G.H.; Lian, B.S.X.; Kawasaki, T.; Kawai, T. Exploration of Pattern Recognition Receptor Agonists as Candidate Adjuvants. *Front. Cell Infect. Microbiol.* **2021**, *11*, 745016. [CrossRef]
44. Iwasaki, A.; Medzhitov, R. Toll-like receptor control of the adaptive immune responses. *Nat. Immunol.* **2004**, *5*, 987–995. [CrossRef] [PubMed]
45. Gong, T.; Liu, L.; Jiang, W.; Zhou, R. DAMP-sensing receptors in sterile inflammation and inflammatory diseases. *Nat. Rev. Immunol.* **2020**, *20*, 95–112. [CrossRef] [PubMed]
46. Cui, B.; Liu, X.; Fang, Y.; Zhou, P.; Zhang, Y.; Wang, Y. Flagellin as a vaccine adjuvant. *Expert Rev. Vaccines* **2018**, *17*, 335–349. [CrossRef]
47. Gauthier, L.; Babych, M.; Segura, M.; Bourgault, S.; Archambault, D. Identification of a novel TLR5 agonist derived from the P97 protein of Mycoplasma hyopneumoniae. *Immunobiology* **2020**, *225*, 151962. [CrossRef] [PubMed]
48. Basto, A.P.; Leitao, A. Targeting TLR2 for vaccine development. *J. Immunol. Res.* **2014**, *2014*, 619410. [CrossRef]
49. Zottig, X.; Al-Halifa, S.; Cote-Cyr, M.; Calzas, C.; Le Goffic, R.; Chevalier, C.; Archambault, D.; Bourgault, S. Self-assembled peptide nanorod vaccine confers protection against influenza A virus. *Biomaterials* **2021**, *269*, 120672. [CrossRef] [PubMed]
50. Al-Halifa, S.; Zottig, X.; Babych, M.; Cote-Cyr, M.; Bourgault, S.; Archambault, D. Harnessing the Activation of Toll-Like Receptor 2/6 by Self-Assembled Cross-beta Fibrils to Design Adjuvanted Nanovaccines. *Nanomaterilas* **2020**, *10*, 1981. [CrossRef]
51. Kumar, S.; Sunagar, R.; Gosselin, E. Bacterial Protein Toll-Like-Receptor Agonists: A Novel Perspective on Vaccine Adjuvants. *Front. Immunol.* **2019**, *10*, 1144. [CrossRef]
52. Chatterjee, S.; Dwivedi, V.P.; Singh, Y.; Siddiqui, I.; Sharma, P.; Van Kaer, L.; Chattopadhyay, D.; Das, G. Early secreted antigen ESAT-6 of Mycobacterium tuberculosis promotes protective T helper 17 cell responses in a toll-like receptor-2-dependent manner. *PLoS Pathog.* **2011**, *7*, e1002378. [CrossRef]
53. Cheng, E.; Cárdenas-Freytag, L.; Clements, J.D. The role of cAMP in mucosal adjuvanticity of Escherichia coli heat-labile enterotoxin (LT). *Vaccine* **1999**, *18*, 38–49. [CrossRef]
54. Larena, M.; Holmgren, J.; Lebens, M.; Terrinoni, M.; Lundgren, A. Cholera toxin, and the related nontoxic adjuvants mmCT and dmLT, promote human Th17 responses via cyclic AMP-protein kinase A and inflammasome-dependent IL-1 signaling. *J. Immunol.* **2015**, *194*, 3829–3839. [CrossRef] [PubMed]
55. Mutsch, M.; Zhou, W.; Rhodes, P.; Bopp, M.; Chen, R.T.; Linder, T.; Spyr, C.; Steffen, R. Use of the inactivated intranasal influenza vaccine and the risk of Bell's palsy in Switzerland. *N. Engl. J. Med.* **2004**, *350*, 896–903. [CrossRef] [PubMed]
56. Clements, J.D.; Norton, E.B. The mucosal vaccine adjuvant LT (R192G/L211A) or dmLT. *MSphere* **2018**, *3*, e00215–e00218. [CrossRef] [PubMed]
57. Bachmann, M.F.; Jennings, G.T. Vaccine delivery: A matter of size, geometry, kinetics and molecular patterns. *Nat. Rev. Immunol.* **2010**, *10*, 787–796. [CrossRef] [PubMed]
58. Fries, C.N.; Curvino, E.J.; Chen, J.L.; Permar, S.R.; Fouda, G.G.; Collier, J.H. Advances in nanomaterial vaccine strategies to address infectious diseases impacting global health. *Nat. Nanotechnol.* **2021**, *16*, 1–14. [CrossRef]
59. Irvine, D.J.; Read, B.J. Shaping humoral immunity to vaccines through antigen-displaying nanoparticles. *Curr. Opin. Immunol.* **2020**, *65*, 1–6. [CrossRef]
60. Irvine, D.J.; Swartz, M.A.; Szeto, G.L. Engineering synthetic vaccines using cues from natural immunity. *Nat. Mater.* **2013**, *12*, 978–990. [CrossRef]

61. Moyer, T.J.; Zmolek, A.C.; Irvine, D.J. Beyond antigens and adjuvants: Formulating future vaccines. *J. Clin. Investig.* **2016**, *126*, 799–808. [CrossRef]
62. Singh, A. Eliciting B cell immunity against infectious diseases using nanovaccines. *Nat. Nanotechnol.* **2021**, *16*, 16–24. [CrossRef]
63. Ding, Y.; Li, Z.; Jaklenec, A.; Hu, Q. Vaccine delivery systems toward lymph nodes. *Adv. Drug Deliv. Rev.* **2021**, *179*, 113914. [CrossRef]
64. Ke, X.; Howard, G.P.; Tang, H.; Cheng, B.; Saung, M.T.; Santos, J.L.; Mao, H.-Q. Physical and chemical profiles of nanoparticles for lymphatic targeting. *Adv. Drug Deliv. Rev.* **2019**, *151*, 72–93. [CrossRef] [PubMed]
65. Babych, M.; Bertheau-Mailhot, G.; Zottig, X.; Dion, J.; Gauthier, L.; Archambault, D.; Bourgault, S. Engineering and evaluation of amyloid assemblies as a nanovaccine against the Chikungunya virus. *Nanoscale* **2018**, *10*, 19547–19556. [CrossRef]
66. Manolova, V.; Flace, A.; Bauer, M.; Schwarz, K.; Saudan, P.; Bachmann, M.F. Nanoparticles target distinct dendritic cell populations according to their size. *Eur. J. Immunol.* **2008**, *38*, 1404–1413. [CrossRef] [PubMed]
67. Miller, N.E.; Michel, C.C.; Nanjee, M.N.; Olszewski, W.L.; Miller, I.P.; Hazell, M.; Olivecrona, G.; Sutton, P.; Humphreys, S.M.; Frayn, K.N. Secretion of adipokines by human adipose tissue in vivo: Partitioning between capillary and lymphatic transport. *Am. J. Physiol. Endocrinol. Metab.* **2011**, *301*, E659–E667. [CrossRef] [PubMed]
68. Reddy, S.T.; Rehor, A.; Schmoekel, H.G.; Hubbell, J.A.; Swartz, M.A. In vivo targeting of dendritic cells in lymph nodes with poly(propylene sulfide) nanoparticles. *J. Control Release* **2006**, *112*, 26–34. [CrossRef] [PubMed]
69. Roth, G.A.; Saouaf, O.M.; Smith, A.A.A.; Gale, E.C.; Hernandez, M.A.; Idoyaga, J.; Appel, E.A. Prolonged Codelivery of Hemagglutinin and a TLR7/8 Agonist in a Supramolecular Polymer-Nanoparticle Hydrogel Enhances Potency and Breadth of Influenza Vaccination. *ACS Biomater. Sci. Eng.* **2021**, *7*, 1889–1899. [CrossRef]
70. Supersaxo, A.; Hein, W.R.; Steffen, H. Effect of molecular weight on the lymphatic absorption of water-soluble compounds following subcutaneous administration. *Pharm. Res.* **1990**, *7*, 167–169. [CrossRef]
71. Al-Halifa, S.; Gauthier, L.; Arpin, D.; Bourgault, S.; Archambault, D. Nanoparticle-Based Vaccines Against Respiratory Viruses. *Front. Immunol.* **2019**, *10*, 22. [CrossRef]
72. Moon, J.J.; Suh, H.; Li, A.V.; Ockenhouse, C.F.; Yadava, A.; Irvine, D.J. Enhancing humoral responses to a malaria antigen with nanoparticle vaccines that expand Tfh cells and promote germinal center induction. *Proc. Natl. Acad. Sci. USA* **2012**, *109*, 1080–1085. [CrossRef]
73. Yus, C.; Irusta, S.; Sebastian, V.; Arruebo, M. Controlling Particle Size and Release Kinetics in the Sustained Delivery of Oral Antibiotics Using pH-Independent Mucoadhesive Polymers. *Mol. Pharm.* **2020**, *17*, 3314–3327. [CrossRef]
74. Smirnov, D.; Schmidt, J.J.; Capecchi, J.T.; Wightman, P.D. Vaccine adjuvant activity of 3M-052: An imidazoquinoline designed for local activity without systemic cytokine induction. *Vaccine* **2011**, *29*, 5434–5442. [CrossRef] [PubMed]
75. Zhang, Y.N.; Lazarovits, J.; Poon, W.; Ouyang, B.; Nguyen, L.N.M.; Kingston, B.R.; Chan, W.C.W. Nanoparticle Size Influences Antigen Retention and Presentation in Lymph Node Follicles for Humoral Immunity. *Nano Lett.* **2019**, *19*, 7226–7235. [CrossRef] [PubMed]
76. Mohsen, M.O.; Gomes, A.C.; Vogel, M.; Bachmann, M.F. Interaction of Viral Capsid-Derived Virus-Like Particles (VLPs) with the Innate Immune System. *Vaccines* **2018**, *6*, 37. [CrossRef] [PubMed]
77. Foged, C.; Brodin, B.; Frokjaer, S.; Sundblad, A. Particle size and surface charge affect particle uptake by human dendritic cells in an in vitro model. *Int. J. Pharm.* **2005**, *298*, 315–322. [CrossRef]
78. Petithory, T.; Pieuchot, L.; Josien, L.; Ponche, A.; Anselme, K.; Vonna, L. Size-Dependent Internalization Efficiency of Macrophages from Adsorbed Nanoparticle-Based Monolayers. *Nanomaterials* **2021**, *11*, 1963. [CrossRef]
79. Snapper, C.M. Distinct Immunologic Properties of Soluble Versus Particulate Antigens. *Front. Immunol.* **2018**, *9*, 598. [CrossRef]
80. Desch, A.N.; Gibbings, S.L.; Clambey, E.T.; Janssen, W.J.; Slansky, J.E.; Kedl, R.M.; Henson, P.M.; Jakubzick, C. Dendritic cell subsets require cis-activation for cytotoxic CD8 T-cell induction. *Nat. Commun.* **2014**, *5*, 4674. [CrossRef]
81. Sporri, R.; Reis e Sousa, C. Inflammatory mediators are insufficient for full dendritic cell activation and promote expansion of CD4+ T cell populations lacking helper function. *Nat. Immunol.* **2005**, *6*, 163–170. [CrossRef]
82. Hou, B.; Saudan, P.; Ott, G.; Wheeler, M.L.; Ji, M.; Kuzmich, L.; Lee, L.M.; Coffman, R.L.; Bachmann, M.F.; DeFranco, A.L. Selective utilization of Toll-like receptor and MyD88 signaling in B cells for enhancement of the antiviral germinal center response. *Immunity* **2011**, *34*, 375–384. [CrossRef]
83. Zottig, X.; Cote-Cyr, M.; Arpin, D.; Archambault, D.; Bourgault, S. Protein Supramolecular Structures: From Self-Assembly to Nanovaccine Design. *Nanomaterilas* **2020**, *10*, 1008. [CrossRef]
84. Bennett, N.R.; Zwick, D.B.; Courtney, A.H.; Kiessling, L.L. Multivalent Antigens for Promoting B and T Cell Activation. *ACS Chem. Biol.* **2015**, *10*, 1817–1824. [CrossRef] [PubMed]
85. Hong, S.; Zhang, Z.; Liu, H.; Tian, M.; Zhu, X.; Zhang, Z.; Wang, W.; Zhou, X.; Zhang, F.; Ge, Q.; et al. B Cells Are the Dominant Antigen-Presenting Cells that Activate Naive CD4(+) T Cells upon Immunization with a Virus-Derived Nanoparticle Antigen. *Immunity* **2018**, *49*, 695–708.e694. [CrossRef]
86. Pompano, R.R.; Chen, J.; Verbus, E.A.; Han, H.; Fridman, A.; McNeely, T.; Collier, J.H.; Chong, A.S. Titrating T-cell epitopes within self-assembled vaccines optimizes CD4+ helper T cell and antibody outputs. *Adv. Health Mater.* **2014**, *3*, 1898–1908. [CrossRef] [PubMed]
87. Nguyen, B.; Tolia, N.H. Protein-based antigen presentation platforms for nanoparticle vaccines. *Npj Vaccines* **2021**, *6*, 70. [CrossRef] [PubMed]

88. Walls, A.C.; Fiala, B.; Schäfer, A.; Wrenn, S.; Pham, M.N.; Murphy, M.; Longping, V.T.; Shehata, L.; O'Connor, M.A.; Chen, C. Elicitation of potent neutralizing antibody responses by designed protein nanoparticle vaccines for SARS-CoV-2. *Cell* **2020**, *183*, 1367–1382.e17. [CrossRef]
89. Zhang, B.; Chao, C.W.; Tsybovsky, Y.; Abiona, O.M.; Hutchinson, G.B.; Moliva, J.I.; Olia, A.S.; Pegu, A.; Phung, E.; Stewart-Jones, G.B.E.; et al. A platform incorporating trimeric antigens into self-assembling nanoparticles reveals SARS-CoV-2-spike nanoparticles to elicit substantially higher neutralizing responses than spike alone. *Sci. Rep.* **2020**, *10*, 18149. [CrossRef]
90. Rohovie, M.J.; Nagasawa, M.; Swartz, J.R. Virus-like particles: Next-generation nanoparticles for targeted therapeutic delivery. *Bioeng. Transl. Med.* **2017**, *2*, 43–57. [CrossRef]
91. Yan, D.; Wei, Y.-Q.; Guo, H.-C.; Sun, S.-Q. The application of virus-like particles as vaccines and biological vehicles. *Appl. Microbiol. Biotechnol.* **2015**, *99*, 10415–10432. [CrossRef]
92. Zabel, F.; Kündig, T.M.; Bachmann, M.F. Virus-induced humoral immunity: On how B cell responses are initiated. *Curr. Opin. Virol.* **2013**, *3*, 357–362. [CrossRef]
93. Lagoutte, P.; Mignon, C.; Stadthagen, G.; Potisopon, S.; Donnat, S.; Mast, J.; Lugari, A.; Werle, B. Simultaneous surface display and cargo loading of encapsulin nanocompartments and their use for rational vaccine design. *Vaccine* **2018**, *36*, 3622–3628. [CrossRef]
94. Carmen, J.M.; Shrivastava, S.; Lu, Z.; Anderson, A.; Morrison, E.B.; Sankhala, R.S.; Chen, W.H.; Chang, W.C.; Bolton, J.S.; Matyas, G.R.; et al. SARS-CoV-2 ferritin nanoparticle vaccine induces robust innate immune activity driving polyfunctional spike-specific T cell responses. *NPJ Vaccines* **2021**, *6*, 151. [CrossRef] [PubMed]
95. Ma, X.; Zou, F.; Yu, F.; Li, R.; Yuan, Y.; Zhang, Y.; Zhang, X.; Deng, J.; Chen, T.; Song, Z. Nanoparticle vaccines based on the receptor binding domain (RBD) and heptad repeat (HR) of SARS-CoV-2 elicit robust protective immune responses. *Immunity* **2020**, *53*, 1315–1330.e9. [CrossRef]
96. Joyce, M.G.; King, H.A.; Elakhal-Naouar, I.; Ahmed, A.; Peachman, K.K.; Macedo Cincotta, C.; Subra, C.; Chen, R.E.; Thomas, P.V.; Chen, W.-H. A SARS-CoV-2 ferritin nanoparticle vaccine elicits protective immune responses in nonhuman primates. *Sci. Transl. Med.* **2021**, *14*, eabi5735. [CrossRef] [PubMed]
97. Kanekiyo, M.; Wei, C.-J.; Yassine, H.M.; McTamney, P.M.; Boyington, J.C.; Whittle, J.R.; Rao, S.S.; Kong, W.-P.; Wang, L.; Nabel, G.J. Self-assembling influenza nanoparticle vaccines elicit broadly neutralizing H1N1 antibodies. *Nature* **2013**, *499*, 102–106. [CrossRef] [PubMed]
98. Yassine, H.M.; Boyington, J.C.; McTamney, P.M.; Wei, C.-J.; Kanekiyo, M.; Kong, W.-P.; Gallagher, J.R.; Wang, L.; Zhang, Y.; Joyce, M.G. Hemagglutinin-stem nanoparticles generate heterosubtypic influenza protection. *Nat. Med.* **2015**, *21*, 1065–1070. [CrossRef] [PubMed]
99. Kanekiyo, M.; Bu, W.; Joyce, M.G.; Meng, G.; Whittle, J.R.; Baxa, U.; Yamamoto, T.; Narpala, S.; Todd, J.-P.; Rao, S.S. Rational design of an Epstein-Barr virus vaccine targeting the receptor-binding site. *Cell* **2015**, *162*, 1090–1100. [CrossRef] [PubMed]
100. Swanson, K.A.; Rainho-Tomko, J.N.; Williams, Z.P.; Lanza, L.; Peredelchuk, M.; Kishko, M.; Pavot, V.; Alamares-Sapuay, J.; Adhikarla, H.; Gupta, S. A respiratory syncytial virus (RSV) F protein nanoparticle vaccine focuses antibody responses to a conserved neutralization domain. *Sci. Immunol.* **2020**, *5*, eaba6466. [CrossRef] [PubMed]
101. He, L.; Cheng, Y.; Kong, L.; Azadnia, P.; Giang, E.; Kim, J.; Wood, M.R.; Wilson, I.A.; Law, M.; Zhu, J. Approaching rational epitope vaccine design for hepatitis C virus with meta-server and multivalent scaffolding. *Sci. Rep.* **2015**, *5*, 12501. [CrossRef]
102. He, L.; De Val, N.; Morris, C.D.; Vora, N.; Thinnes, T.C.; Kong, L.; Azadnia, P.; Sok, D.; Zhou, B.; Burton, D.R. Presenting native-like trimeric HIV-1 antigens with self-assembling nanoparticles. *Nat. Commun.* **2016**, *7*, 12041. [CrossRef]
103. Sliepen, K.; Ozorowski, G.; Burger, J.A.; van Montfort, T.; Stunnenberg, M.; LaBranche, C.; Montefiori, D.C.; Moore, J.P.; Ward, A.B.; Sanders, R.W. Presenting native-like HIV-1 envelope trimers on ferritin nanoparticles improves their immunogenicity. *Retrovirology* **2015**, *12*, 82. [CrossRef]
104. Sliepen, K.; Han, B.W.; Bontjer, I.; Mooij, P.; Garces, F.; Behrens, A.-J.; Rantalainen, K.; Kumar, S.; Sarkar, A.; Brouwer, P.J. Structure and immunogenicity of a stabilized HIV-1 envelope trimer based on a group-M consensus sequence. *Nat. Commun.* **2019**, *10*, 2355. [CrossRef] [PubMed]
105. Sliepen, K.; Schermer, E.; Bontjer, I.; Burger, J.A.; Lévai, R.F.; Mundsperger, P.; Brouwer, P.J.; Tolazzi, M.; Farsang, A.; Katinger, D. Interplay of diverse adjuvants and nanoparticle presentation of native-like HIV-1 envelope trimers. *Npj Vaccines* **2021**, *6*, 103. [CrossRef] [PubMed]
106. Li, Z.; Cui, K.; Wang, H.; Liu, F.; Huang, K.; Duan, Z.; Wang, F.; Shi, D.; Liu, Q. A milk-based self-assemble rotavirus VP6-ferritin nanoparticle vaccine elicited protection against the viral infection. *J. Nanobiotechnol.* **2019**, *17*, 13. [CrossRef] [PubMed]
107. Ma, H.; Li, X.; Li, J.; Zhao, Z.; Zhang, H.; Hao, G.; Chen, H.; Qian, P. Immunization with a recombinant fusion of porcine reproductive and respiratory syndrome virus modified GP5 and ferritin elicits enhanced protective immunity in pigs. *Virology* **2021**, *552*, 112–120. [CrossRef] [PubMed]
108. Chen, Y.; Hu, Y.; Chen, H.; Li, X.; Qian, P. A ferritin nanoparticle vaccine for foot-and-mouth disease virus elicited partial protection in mice. *Vaccine* **2020**, *38*, 5647–5652. [CrossRef] [PubMed]
109. Zhao, Z.; Chen, X.; Chen, Y.; Li, H.; Fang, K.; Chen, H.; Li, X.; Qian, P. A Self-Assembling Ferritin Nanoplatform for Designing Classical Swine Fever Vaccine: Elicitation of Potent Neutralizing Antibody. *Vaccines* **2021**, *9*, 45. [CrossRef] [PubMed]
110. He, L.; Lin, X.; Wang, Y.; Abraham, C.; Sou, C.; Ngo, T.; Zhang, Y.; Wilson, I.A.; Zhu, J. Single-component, self-assembling, protein nanoparticles presenting the receptor binding domain and stabilized spike as SARS-CoV-2 vaccine candidates. *Sci. Adv.* **2021**, *7*, eabf1591. [CrossRef]

111. Kim, Y.-S.; Son, A.; Kim, J.; Kwon, S.B.; Kim, M.H.; Kim, P.; Kim, J.; Byun, Y.H.; Sung, J.; Lee, J. Chaperna-mediated assembly of ferritin-based Middle East respiratory syndrome-coronavirus nanoparticles. *Front. Immunol.* **2018**, *9*, 1093. [CrossRef]
112. Tokatlian, T.; Read, B.J.; Jones, C.A.; Kulp, D.W.; Menis, S.; Chang, J.Y.; Steichen, J.M.; Kumari, S.; Allen, J.D.; Dane, E.L. Innate immune recognition of glycans targets HIV nanoparticle immunogens to germinal centers. *Science* **2019**, *363*, 649–654. [CrossRef]
113. Wang, W.; Zhou, X.; Bian, Y.; Wang, S.; Chai, Q.; Guo, Z.; Wang, Z.; Zhu, P.; Peng, H.; Yan, X. Dual-targeting nanoparticle vaccine elicits a therapeutic antibody response against chronic hepatitis B. *Nat. Nanotechnol.* **2020**, *15*, 406–416. [CrossRef]
114. Wang, W.; Huang, B.; Zhu, Y.; Tan, W.; Zhu, M. Ferritin nanoparticle-based SARS-CoV-2 RBD vaccine induces a persistent antibody response and long-term memory in mice. *Cell. Mol. Immunol.* **2021**, *18*, 749–751. [CrossRef] [PubMed]
115. Yang, F.; Mariz, F.C.; Zhao, X.; Spagnoli, G.; Ottonello, S.; Müller, M. Broad neutralization responses against oncogenic human papillomaviruses induced by a minor capsid l2 polytope genetically incorporated into bacterial ferritin nanoparticles. *Front. Immunol.* **2020**, *11*, 606569. [CrossRef] [PubMed]
116. He, L.; Chaudhary, A.; Lin, X.; Sou, C.; Alkutkar, T.; Kumar, S.; Ngo, T.; Kosviner, E.; Ozorowski, G.; Stanfield, R.L. Single-component multilayered self-assembling nanoparticles presenting rationally designed glycoprotein trimers as Ebola virus vaccines. *Nat. Commun.* **2021**, *12*, 2633. [CrossRef] [PubMed]
117. Jardine, J.G.; Ota, T.; Sok, D.; Pauthner, M.; Kulp, D.W.; Kalyuzhniy, O.; Skog, P.D.; Thinnes, T.C.; Bhullar, D.; Briney, B. Priming a broadly neutralizing antibody response to HIV-1 using a germline-targeting immunogen. *Science* **2015**, *349*, 156–161. [CrossRef] [PubMed]
118. Jardine, J.; Julien, J.-P.; Menis, S.; Ota, T.; Kalyuzhniy, O.; McGuire, A.; Sok, D.; Huang, P.-S.; MacPherson, S.; Jones, M. Rational HIV immunogen design to target specific germline B cell receptors. *Science* **2013**, *340*, 711–716. [CrossRef]
119. Kang, Y.-F.; Zhang, X.; Yu, X.-H.; Zheng, Q.; Liu, Z.; Li, J.-P.; Sun, C.; Kong, X.-W.; Zhu, Q.-Y.; Chen, H.-W. Immunization with a self-assembled nanoparticle vaccine elicits potent neutralizing antibody responses against EBV infection. *Nano Lett.* **2021**, *21*, 2476–2486. [CrossRef]
120. Aebischer, A.; Wernike, K.; König, P.; Franzke, K.; Wichgers Schreur, P.J.; Kortekaas, J.; Vitikainen, M.; Wiebe, M.; Saloheimo, M.; Tchelet, R. Development of a modular vaccine platform for multimeric antigen display using an orthobunyavirus model. *Vaccines* **2021**, *9*, 651. [CrossRef]
121. Wichgers Schreur, P.J.; Tacken, M.; Gutjahr, B.; Keller, M.; van Keulen, L.; Kant, J.; van de Water, S.; Lin, Y.; Eiden, M.; Rissmann, M. Vaccine efficacy of self-assembled multimeric protein scaffold particles displaying the glycoprotein Gn Head Domain of Rift Valley Fever Virus. *Vaccines* **2021**, *9*, 301. [CrossRef]
122. Richert, L.E.; Servid, A.E.; Harmsen, A.L.; Rynda-Apple, A.; Han, S.; Wiley, J.A.; Douglas, T.; Harmsen, A.G. A virus-like particle vaccine platform elicits heightened and hastened local lung mucosal antibody production after a single dose. *Vaccine* **2012**, *30*, 3653–3665. [CrossRef]
123. Wiley, J.A.; Richert, L.E.; Swain, S.D.; Harmsen, A.; Barnard, D.L.; Randall, T.D.; Jutila, M.; Douglas, T.; Broomell, C.; Young, M. Inducible bronchus-associated lymphoid tissue elicited by a protein cage nanoparticle enhances protection in mice against diverse respiratory viruses. *PLoS ONE* **2009**, *4*, e7142. [CrossRef]
124. Patterson, D.P.; Rynda-Apple, A.; Harmsen, A.L.; Harmsen, A.G.; Douglas, T. Biomimetic antigenic nanoparticles elicit controlled protective immune response to influenza. *ACS Nano* **2013**, *7*, 3036–3044. [CrossRef] [PubMed]
125. Sharma, J.; Shepardson, K.; Johns, L.L.; Wellham, J.; Avera, J.; Schwarz, B.; Rynda-Apple, A.; Douglas, T. A self-adjuvanted, modular, antigenic VLP for rapid response to influenza virus variability. *ACS Appl. Mater. Interfaces* **2020**, *12*, 18211–18224. [CrossRef] [PubMed]
126. Kang, S.; Kim, Y.; Shin, Y.; Song, J.J.; Jon, S. Antigen-Presenting, Self-Assembled Protein Nanobarrels as an Adjuvant-Free Vaccine Platform against Influenza Virus. *ACS Nano* **2021**, *15*, 10722–10732. [CrossRef] [PubMed]
127. Cote-Cyr, M.; Zottig, X.; Gauthier, L.; Archambault, D.; Bourgault, S. Self-Assembly of Flagellin into Immunostimulatory Ring-like Nanostructures as an Antigen Delivery System. *ACS Biomater. Sci. Eng.* **2022**, *8*, 694–707. [CrossRef] [PubMed]
128. Bennett, K.M.; Gorham, R.D.; Gusti, V.; Trinh, L.; Morikis, D.; Lo, D.D. Hybrid flagellin as a T cell independent vaccine scaffold. *BMC Biotechnol.* **2015**, *15*, 71. [CrossRef] [PubMed]
129. Marcandalli, J.; Fiala, B.; Ols, S.; Perotti, M.; de van der Schueren, W.; Snijder, J.; Hodge, E.; Benhaim, M.; Ravichandran, R.; Carter, L. Induction of potent neutralizing antibody responses by a designed protein nanoparticle vaccine for respiratory syncytial virus. *Cell* **2019**, *176*, 1420–1431.e17. [CrossRef]
130. Arunachalam, P.S.; Walls, A.C.; Golden, N.; Atyeo, C.; Fischinger, S.; Li, C.; Aye, P.; Navarro, M.J.; Lai, L.; Edara, V.V. Adjuvanting a subunit COVID-19 vaccine to induce protective immunity. *Nature* **2021**, *594*, 253–258. [CrossRef]
131. Brouwer, P.J.; Brinkkemper, M.; Maisonnasse, P.; Dereuddre-Bosquet, N.; Grobben, M.; Claireaux, M.; de Gast, M.; Marlin, R.; Chesnais, V.; Diry, S. Two-component spike nanoparticle vaccine protects macaques from SARS-CoV-2 infection. *Cell* **2021**, *184*, 1188–1200.e19. [CrossRef]
132. Flenniken, M.L.; Willits, D.A.; Brumfield, S.; Young, M.J.; Douglas, T. The small heat shock protein cage from Methanococcus jannaschii is a versatile nanoscale platform for genetic and chemical modification. *Nano Lett.* **2003**, *3*, 1573–1576. [CrossRef]
133. Kim, H.; Choi, H.; Bae, Y.; Kang, S. Development of target-tunable P22 VLP-based delivery nanoplatforms using bacterial superglue. *Biotechnol. Bioeng.* **2019**, *116*, 2843–2851. [CrossRef]
134. Munro, H.N.; Linder, M.C. Ferritin: Structure, biosynthesis, and role in iron metabolism. *Physiol. Rev.* **1978**, *58*, 317–396. [CrossRef] [PubMed]

135. Rodrigues, M.Q.; Alves, P.M.; Roldão, A. Functionalizing ferritin nanoparticles for vaccine development. *Pharmaceutics* **2021**, *13*, 1621. [CrossRef] [PubMed]
136. Bhaskar, S.; Lim, S. Engineering protein nanocages as carriers for biomedical applications. *NPG Asia Mater.* **2017**, *9*, e371. [CrossRef] [PubMed]
137. Pantopoulos, K.; Porwal, S.K.; Tartakoff, A.; Devireddy, L. Mechanisms of mammalian iron homeostasis. *Biochemistry* **2012**, *51*, 5705–5724. [CrossRef] [PubMed]
138. Chiancone, E.; Ceci, P.; Ilari, A.; Ribacchi, F.; Stefanini, S. Iron and proteins for iron storage and detoxification. *Biometals* **2004**, *17*, 197–202. [CrossRef]
139. Theil, E.C. Ferritin: Structure, gene regulation, and cellular function in animals, plants, and microorganisms. *Annu. Rev. Biochem.* **1987**, *56*, 289–315. [CrossRef]
140. Li, C.Q.; Soistman, E.; Carter, D.C. Ferritin nanoparticle technology ... A new platform for antigen presentation and vaccine development. *Ind. Biotechnol.* **2006**, *2*, 143–147. [CrossRef]
141. Khoshnejad, M.; Parhiz, H.; Shuvaev, V.V.; Dmochowski, I.J.; Muzykantov, V.R. Ferritin-based drug delivery systems: Hybrid nanocarriers for vascular immunotargeting. *J. Control Release* **2018**, *282*, 13–24. [CrossRef]
142. Chiou, B.; Connor, J.R. Emerging and dynamic biomedical uses of ferritin. *Pharmaceuticals* **2018**, *11*, 124. [CrossRef]
143. He, D.; Marles-Wright, J. Ferritin family proteins and their use in bionanotechnology. *New Biotechnol.* **2015**, *32*, 651–657. [CrossRef]
144. Uchida, M.; Kang, S.; Reichhardt, C.; Harlen, K.; Douglas, T. The ferritin superfamily: Supramolecular templates for materials synthesis. *Biochim. Biophys. Acta (BBA)-Gen. Subj.* **2010**, *1800*, 834–845.
145. Wang, Z.; Xu, L.; Yu, H.; Lv, P.; Lei, Z.; Zeng, Y.; Liu, G.; Cheng, T. Ferritin nanocage-based antigen delivery nanoplatforms: Epitope engineering for peptide vaccine design. *Biomater. Sci.* **2019**, *7*, 1794–1800. [CrossRef] [PubMed]
146. Ladenstein, R.; Morgunova, E. Second career of a biosynthetic enzyme: Lumazine synthase as a virus-like nanoparticle in vaccine development. *Biotechnol. Rep.* **2020**, *27*, e00494. [CrossRef]
147. Caspar, D.L.; Klug, A. Physical principles in the construction of regular viruses. *Cold Spring Harb. Symp. Quant. Biol.* **1962**, *27*, 1–24. [CrossRef]
148. Ladenstein, R.; Schneider, M.; Huber, R.; Bartunik, H.-D.; Wilson, K.; Schott, K.; Bacher, A. Heavy riboflavin synthase from Bacillus subtilis: Crystal structure analysis of the icosahedral β60 capsid at 3· 3 Å resolution. *J. Mol. Biol.* **1988**, *203*, 1045–1070. [CrossRef]
149. Zhang, X.; Konarev, P.V.; Petoukhov, M.V.; Svergun, D.I.; Xing, L.; Cheng, R.H.; Haase, I.; Fischer, M.; Bacher, A.; Ladenstein, R. Multiple assembly states of lumazine synthase: A model relating catalytic function and molecular assembly. *J. Mol. Biol.* **2006**, *362*, 753–770. [CrossRef] [PubMed]
150. Zhang, X.; Meining, W.; Fischer, M.; Bacher, A.; Ladenstein, R. X-ray structure analysis and crystallographic refinement of lumazine synthase from the hyperthermophile Aquifex aeolicus at 1.6 Å resolution: Determinants of thermostability revealed from structural comparisons. *J. Mol. Biol.* **2001**, *306*, 1099–1114. [CrossRef] [PubMed]
151. Akita, F.; Chong, K.T.; Tanaka, H.; Yamashita, E.; Miyazaki, N.; Nakaishi, Y.; Suzuki, M.; Namba, K.; Ono, Y.; Tsukihara, T. The crystal structure of a virus-like particle from the hyperthermophilic archaeon Pyrococcus furiosus provides insight into the evolution of viruses. *J. Mol. Biol.* **2007**, *368*, 1469–1483. [CrossRef]
152. Rosenkrands, I.; Rasmussen, P.B.; Carnio, M.; Jacobsen, S.; Theisen, M.; Andersen, P. Identification and characterization of a 29-kilodalton protein from Mycobacterium tuberculosis culture filtrate recognized by mouse memory effector cells. *Infect. Immun.* **1998**, *66*, 2728–2735. [CrossRef]
153. Sutter, M.; Boehringer, D.; Gutmann, S.; Günther, S.; Prangishvili, D.; Loessner, M.J.; Stetter, K.O.; Weber-Ban, E.; Ban, N. Structural basis of enzyme encapsulation into a bacterial nanocompartment. *Nat. Struct. Mol. Biol.* **2008**, *15*, 939–947. [CrossRef]
154. Valdes-Stauber, N.; Scherer, S. Nucleotide sequence and taxonomical distribution of the bacteriocin gene lin cloned from Brevibacterium linens M18. *Appl. Environ. Microbiol.* **1996**, *62*, 1283–1286. [CrossRef]
155. McHugh, C.A.; Fontana, J.; Nemecek, D.; Cheng, N.; Aksyuk, A.A.; Heymann, J.B.; Winkler, D.C.; Lam, A.S.; Wall, J.S.; Steven, A.C. A virus capsid-like nanocompartment that stores iron and protects bacteria from oxidative stress. *EMBO J.* **2014**, *33*, 1896–1911. [CrossRef] [PubMed]
156. Rahmanpour, R.; Bugg, T.D. Assembly in vitro of Rhodococcus jostii RHA 1 encapsulin and peroxidase DypB to form a nanocompartment. *FEBS J.* **2013**, *280*, 2097–2104. [CrossRef] [PubMed]
157. Contreras, H.; Joens, M.S.; McMath, L.M.; Le, V.P.; Tullius, M.V.; Kimmey, J.M.; Bionghi, N.; Horwitz, M.A.; Fitzpatrick, J.A.; Goulding, C.W. Characterization of a Mycobacterium tuberculosis nanocompartment and its potential cargo proteins. *J. Biol. Chem.* **2014**, *289*, 18279–18289. [CrossRef]
158. Rurup, W.F.; Snijder, J.; Koay, M.S.; Heck, A.J.; Cornelissen, J.J. Self-sorting of foreign proteins in a bacterial nanocompartment. *J. Am. Chem. Soc.* **2014**, *136*, 3828–3832.
159. Tamura, A.; Fukutani, Y.; Takami, T.; Fujii, M.; Nakaguchi, Y.; Murakami, Y.; Noguchi, K.; Yohda, M.; Odaka, M. Packaging guest proteins into the encapsulin nanocompartment from Rhodococcus erythropolis N771. *Biotechnol. Bioeng.* **2015**, *112*, 13–20. [CrossRef]
160. Kim, R.; Kim, K.K.; Yokota, H.; Kim, S.-H. Small heat shock protein of Methanococcus jannaschii, a hyperthermophile. *Proc. Natl. Acad. Sci. USA* **1998**, *95*, 9129–9133. [CrossRef]
161. Kim, K.K.; Kim, R.; Kim, S.-H. Crystal structure of a small heat-shock protein. *Nature* **1998**, *394*, 595–599. [CrossRef] [PubMed]

162. Kaiser, C.R.; Flenniken, M.L.; Gillitzer, E.; Harmsen, A.L.; Harmsen, A.G.; Jutila, M.A.; Douglas, T.; Young, M.J. Biodistribution studies of protein cage nanoparticles demonstrate broad tissue distribution and rapid clearance in vivo. *Int. J. Nanomed.* **2007**, *2*, 715.
163. Abedin, M.J.; Liepold, L.; Suci, P.; Young, M.; Douglas, T. Synthesis of a cross-linked branched polymer network in the interior of a protein cage. *J. Am. Chem. Soc.* **2009**, *131*, 4346–4354. [CrossRef]
164. Liepold, L.O.; Abedin, M.J.; Buckhouse, E.D.; Frank, J.A.; Young, M.J.; Douglas, T. Supramolecular protein cage composite MR contrast agents with extremely efficient relaxivity properties. *Nano Lett.* **2009**, *9*, 4520–4526. [CrossRef] [PubMed]
165. Chatterji, A.; Ochoa, W.; Shamieh, L.; Salakian, S.P.; Wong, S.M.; Clinton, G.; Ghosh, P.; Lin, T.; Johnson, J.E. Chemical conjugation of heterologous proteins on the surface of cowpea mosaic virus. *Bioconjug. Chem.* **2004**, *15*, 807–813. [CrossRef] [PubMed]
166. Gupta, S.S.; Kuzelka, J.; Singh, P.; Lewis, W.G.; Manchester, M.; Finn, M. Accelerated bioorthogonal conjugation: A practical method for the ligation of diverse functional molecules to a polyvalent virus scaffold. *Bioconjug. Chem.* **2005**, *16*, 1572–1579. [CrossRef] [PubMed]
167. Kang, S.; Uchida, M.; O'Neil, A.; Li, R.; Prevelige, P.E.; Douglas, T. Implementation of p22 viral capsids as nanoplatforms. *Biomacromolecules* **2010**, *11*, 2804–2809. [CrossRef]
168. Prevelige, P.E., Jr.; Thomas, D.; King, J. Scaffolding protein regulates the polymerization of P22 coat subunits into icosahedral shells in vitro. *J. Mol. Biol.* **1988**, *202*, 743–757. [CrossRef]
169. Kim, D.; Park, J.; Kim, S.J.; Soh, Y.-M.; Kim, H.M.; Oh, B.-H.; Song, J.-J. Brucella immunogenic BP26 forms a channel-like structure. *J. Mol. Biol.* **2013**, *425*, 1119–1126. [CrossRef]
170. Yang, X.; Walters, N.; Robison, A.; Trunkle, T.; Pascual, D.W. Nasal immunization with recombinant Brucella melitensis bp26 and trigger factor with cholera toxin reduces B. melitensis colonization. *Vaccine* **2007**, *25*, 2261–2268. [CrossRef]
171. Chattopadhyay, S.; Chen, J.-Y.; Chen, H.-W.; Hu, C.-M.J. Nanoparticle vaccines adopting virus-like features for enhanced immune potentiation. *Nanotheranostics* **2017**, *1*, 244. [CrossRef]
172. Hayashi, F.; Smith, K.D.; Ozinsky, A.; Hawn, T.R.; Yi, E.C.; Goodlett, D.R.; Eng, J.K.; Akira, S.; Underhill, D.M.; Aderem, A. The innate immune response to bacterial flagellin is mediated by Toll-like receptor 5. *Nature* **2001**, *410*, 1099–1103. [CrossRef]
173. Huleatt, J.W.; Nakaar, V.; Desai, P.; Huang, Y.; Hewitt, D.; Jacobs, A.; Tang, J.; McDonald, W.; Song, L.; Evans, R.K. Potent immunogenicity and efficacy of a universal influenza vaccine candidate comprising a recombinant fusion protein linking influenza M2e to the TLR5 ligand flagellin. *Vaccine* **2008**, *26*, 201–214. [CrossRef]
174. Lockner, J.W.; Eubanks, L.M.; Choi, J.L.; Lively, J.M.; Schlosburg, J.E.; Collins, K.C.; Globisch, D.; Rosenfeld-Gunn, R.J.; Wilson, I.A.; Janda, K.D. Flagellin as carrier and adjuvant in cocaine vaccine development. *Mol. Pharm.* **2015**, *12*, 653–662. [CrossRef] [PubMed]
175. Stepanova, L.A.; Mardanova, E.S.; Shuklina, M.A.; Blokhina, E.A.; Kotlyarov, R.Y.; Potapchuk, M.V.; Kovaleva, A.A.; Vidyaeva, I.G.; Korotkov, A.V.; Eletskaya, E.I. Flagellin-fused protein targeting M2e and HA2 induces potent humoral and T-cell responses and protects mice against various influenza viruses a subtypes. *J. Biomed. Sci.* **2018**, *25*, 33. [CrossRef] [PubMed]
176. Lightfield, K.L.; Persson, J.; Brubaker, S.W.; Witte, C.E.; Von Moltke, J.; Dunipace, E.A.; Henry, T.; Sun, Y.-H.; Cado, D.; Dietrich, W.F. Critical function for Naip5 in inflammasome activation by a conserved carboxy-terminal domain of flagellin. *Nat. Immunol.* **2008**, *9*, 1171–1178. [CrossRef] [PubMed]
177. Matusiak, M.; Van Opdenbosch, N.; Vande Walle, L.; Sirard, J.-C.; Kanneganti, T.-D.; Lamkanfi, M. Flagellin-induced NLRC4 phosphorylation primes the inflammasome for activation by NAIP5. *Proc. Natl. Acad. Sci. USA* **2015**, *112*, 1541–1546. [CrossRef]
178. Zhao, Y.; Yang, J.; Shi, J.; Gong, Y.-N.; Lu, Q.; Xu, H.; Liu, L.; Shao, F. The NLRC4 inflammasome receptors for bacterial flagellin and type III secretion apparatus. *Nature* **2011**, *477*, 596–600. [CrossRef]
179. Cote-Cyr, M.; Gauthier, L.; Zottig, X.; Bourgault, S.; Archambault, D. Recombinant Bacillus subtilis flagellin Hag is a potent immunostimulant with reduced proinflammatory properties compared to Salmonella enterica serovar Typhimurium FljB. *Vaccine* **2022**, *40*, 11–17. [CrossRef]
180. Skountzou, I.; del Pilar Martin, M.; Wang, B.; Ye, L.; Koutsonanos, D.; Weldon, W.; Jacob, J.; Compans, R.W. Salmonella flagellins are potent adjuvants for intranasally administered whole inactivated influenza vaccine. *Vaccine* **2010**, *28*, 4103–4112. [CrossRef]
181. Smith, K.D.; Andersen-Nissen, E.; Hayashi, F.; Strobe, K.; Bergman, M.A.; Barrett, S.L.R.; Cookson, B.T.; Aderem, A. Toll-like receptor 5 recognizes a conserved site on flagellin required for protofilament formation and bacterial motility. *Nat. Immunol.* **2003**, *4*, 1247–1253. [CrossRef]
182. Song, W.S.; Jeon, Y.J.; Namgung, B.; Hong, M.; Yoon, S.-i. A conserved TLR5 binding and activation hot spot on flagellin. *Sci. Rep.* **2017**, *7*, 40878. [CrossRef]
183. Zhang, L.; Pan, Z.; Kang, X.; Yang, Y.; Kang, H.; Zhang, N.; Rosati, J.M.; Jiao, X. Amino acids 89–96 of Salmonella typhimurium flagellin represent the major domain responsible for TLR5-independent adjuvanticity in the humoral immune response. *Cell. Mol. Immunol.* **2015**, *12*, 625–632. [CrossRef]
184. Ikeda, J.S.; Schmitt, C.K.; Darnell, S.C.; Watson, P.R.; Bispham, J.; Wallis, T.S.; Weinstein, D.L.; Metcalf, E.S.; Adams, P.; O'Connor, C.D. Flagellar phase variation of Salmonella enterica serovar Typhimurium contributes to virulence in the murine typhoid infection model but does not influence Salmonella-induced enteropathogenesis. *Infect. Immun.* **2001**, *69*, 3021–3030. [CrossRef] [PubMed]

185. Biedma, M.E.; Cayet, D.; Tabareau, J.; Rossi, A.H.; Ivičak-Kocjan, K.; Moreno, G.; Errea, A.; Soulard, D.; Parisi, G.; Jerala, R. Recombinant flagellins with deletions in domains D1, D2, and D3: Characterization as novel immunoadjuvants. *Vaccine* **2019**, *37*, 652–663. [CrossRef] [PubMed]
186. Bale, J.B.; Gonen, S.; Liu, Y.; Sheffler, W.; Ellis, D.; Thomas, C.; Cascio, D.; Yeates, T.O.; Gonen, T.; King, N.P. Accurate design of megadalton-scale two-component icosahedral protein complexes. *Science* **2016**, *353*, 389–394. [CrossRef]
187. Hsia, Y.; Bale, J.B.; Gonen, S.; Shi, D.; Sheffler, W.; Fong, K.K.; Nattermann, U.; Xu, C.; Huang, P.-S.; Ravichandran, R. Design of a hyperstable 60-subunit protein icosahedron. *Nature* **2016**, *535*, 136–139. [CrossRef]
188. King, N.P.; Bale, J.B.; Sheffler, W.; McNamara, D.E.; Gonen, S.; Gonen, T.; Yeates, T.O.; Baker, D. Accurate design of co-assembling multi-component protein nanomaterials. *Nature* **2014**, *510*, 103–108. [CrossRef] [PubMed]
189. King, N.P.; Sheffler, W.; Sawaya, M.R.; Vollmar, B.S.; Sumida, J.P.; André, I.; Gonen, T.; Yeates, T.O.; Baker, D. Computational design of self-assembling protein nanomaterials with atomic level accuracy. *Science* **2012**, *336*, 1171–1174. [CrossRef] [PubMed]
190. Stupka, I.; Heddle, J.G. Artificial protein cages–inspiration, construction, and observation. *Curr. Opin. Struct. Biol.* **2020**, *64*, 66–73. [CrossRef]
191. Ueda, G.; Antanasijevic, A.; Fallas, J.A.; Sheffler, W.; Copps, J.; Ellis, D.; Hutchinson, G.B.; Moyer, A.; Yasmeen, A.; Tsybovsky, Y. Tailored design of protein nanoparticle scaffolds for multivalent presentation of viral glycoprotein antigens. *Elife* **2020**, *9*, e57659. [CrossRef]
192. Brouwer, P.J.; Antanasijevic, A.; Berndsen, Z.; Yasmeen, A.; Fiala, B.; Bijl, T.P.; Bontjer, I.; Bale, J.B.; Sheffler, W.; Allen, J.D. Enhancing and shaping the immunogenicity of native-like HIV-1 envelope trimers with a two-component protein nanoparticle. *Nat. Commun.* **2019**, *10*, 4272. [CrossRef]
193. Boyoglu-Barnum, S.; Hutchinson, G.B.; Boyington, J.C.; Moin, S.M.; Gillespie, R.A.; Tsybovsky, Y.; Stephens, T.; Vaile, J.R.; Lederhofer, J.; Corbett, K.S. Glycan repositioning of influenza hemagglutinin stem facilitates the elicitation of protective cross-group antibody responses. *Nat. Commun.* **2020**, *11*, 791. [CrossRef]
194. Bruun, T.U.; Andersson, A.-M.C.; Draper, S.J.; Howarth, M. Engineering a rugged nanoscaffold to enhance plug-and-display vaccination. *ACS Nano* **2018**, *12*, 8855–8866. [CrossRef] [PubMed]
195. Fullerton, S.W.; Griffiths, J.S.; Merkel, A.B.; Cheriyan, M.; Wymer, N.J.; Hutchins, M.J.; Fierke, C.A.; Toone, E.J.; Naismith, J.H. Mechanism of the Class I KDPG aldolase. *Bioorg. Med. Chem.* **2006**, *14*, 3002–3010. [CrossRef] [PubMed]
196. Griffiths, J.S.; Wymer, N.J.; Njolito, E.; Niranjanakumari, S.; Fierke, C.A.; Toone, E.J. Cloning, isolation and characterization of the Thermotoga maritima KDPG aldolase. *Bioorg. Med. Chem.* **2002**, *10*, 545–550. [CrossRef]
197. Caivano, A.; Doria-Rose, N.A.; Buelow, B.; Sartorius, R.; Trovato, M.; D'Apice, L.; Domingo, G.J.; Sutton, W.F.; Haigwood, N.L.; De Berardinis, P. HIV-1 Gag p17 presented as virus-like particles on the E2 scaffold from Geobacillus stearothermophilus induces sustained humoral and cellular immune responses in the absence of IFNγ production by CD4+ T cells. *Virology* **2010**, *407*, 296–305. [CrossRef] [PubMed]
198. Berardinis, P.D.; Sartorius, R.; Caivano, A.; Mascolo, D.; Domingo, G.J.; Pozzo, G.D.; Gaubin, M.; Perham, R.N.; Piatier-Tonneau, D.; Guardiola, J. Use of fusion proteins and procaryotic display systems for delivery of HIV-1 antigens: Development of novel vaccines for HIV-1 infection. *Curr. HIV Res.* **2003**, *1*, 441–446. [CrossRef] [PubMed]
199. Domingo, G.J.; Caivano, A.; Sartorius, R.; Barba, P.; Bäckström, M.; Piatier-Tonneau, D.; Guardiola, J.; De Berardinis, P.; Perham, R.N. Induction of specific T-helper and cytolytic responses to epitopes displayed on a virus-like protein scaffold derived from the pyruvate dehydrogenase multienzyme complex. *Vaccine* **2003**, *21*, 1502–1509. [CrossRef]
200. Domingo, G.J.; Perham, R.N. Multiple display of peptides and proteins on a macromolecular scaffold derived from a multienzyme complex. *J. Mol. Biol.* **2001**, *305*, 259–267. [CrossRef]
201. Henderson, C.E.; Perham, R.N.; Finch, J.T. Structure and symmetry of B. stearothermophilus pyruvate dehydrogenase multienzyme complex and implications for eucaryote evolution. *Cell* **1979**, *17*, 85–93. [CrossRef]
202. Perham, R.N. Swinging arms and swinging domains in multifunctional enzymes: Catalytic machines for multistep reactions. *Annu. Rev. Biochem.* **2000**, *69*, 961. [CrossRef]
203. Allen, M.D.; Perham, R.N. The catalytic domain of dihydrolipoyl acetyltransferase from the pyruvate dehydrogenase multienzyme complex of Bacillus stearothermophilus: Expression, purification and reversible denaturation. *FEBS Lett.* **1997**, *413*, 339–343. [CrossRef]
204. Lessard, I.A.; Domingo, G.J.; Borges, A.; Perham, R.N. Expression of genes encoding the E2 and E3 components of the Bacillus stearothermophilus pyruvate dehydrogenase complex and the stoichiometry of subunit interaction in assembly in vitro. *Eur. J. Biochem.* **1998**, *258*, 491–501. [CrossRef] [PubMed]
205. Dalmau, M.; Lim, S.; Chen, H.C.; Ruiz, C.; Wang, S.W. Thermostability and molecular encapsulation within an engineered caged protein scaffold. *Biotechnol. Bioeng.* **2008**, *101*, 654–664. [CrossRef] [PubMed]
206. Dalmau, M.; Lim, S.; Wang, S.-W. Design of a pH-dependent molecular switch in a caged protein platform. *Nano Lett.* **2009**, *9*, 160–166. [CrossRef]
207. Kraft, J.C.; Pham, M.N.; Shehata, L.; Brinkkemper, M.; Boyoglu-Barnum, S.; Sprouse, K.R.; Walls, A.C.; Cheng, S.; Murphy, M.; Pettie, D.; et al. Antigen- and scaffold-specific antibody responses to protein nanoparticle immunogens. *Cell Rep. Med.* **2022**, *3*, 100780. [CrossRef]

208. Guo, L.; Bi, W.; Wang, X.; Xu, W.; Yan, R.; Zhang, Y.; Zhao, K.; Li, Y.; Zhang, M.; Cai, X. Engineered trimeric ACE2 binds viral spike protein and locks it in "Three-up" conformation to potently inhibit SARS-CoV-2 infection. *Cell Res.* **2021**, *31*, 98–100. [CrossRef] [PubMed]
209. Lu, Y.; Welsh, J.P.; Swartz, J.R. Production and stabilization of the trimeric influenza hemagglutinin stem domain for potentially broadly protective influenza vaccines. *Proc. Natl. Acad. Sci. USA* **2014**, *111*, 125–130. [CrossRef]
210. Kim, E.; Erdos, G.; Huang, S.; Kenniston, T.W.; Balmert, S.C.; Carey, C.D.; Raj, V.S.; Epperly, M.W.; Klimstra, W.B.; Haagmans, B.L. Microneedle array delivered recombinant coronavirus vaccines: Immunogenicity and rapid translational development. *EBioMedicine* **2020**, *55*, 102743. [CrossRef]
211. Li, S.; Wang, B.; Jiang, S.; Pan, Y.; Shi, Y.; Kong, W.; Shan, Y. Surface-Functionalized Silica-Coated Calcium Phosphate Nanoparticles Efficiently Deliver DNA-Based HIV-1 Trimeric Envelope Vaccines against HIV-1. *ACS Appl. Mater. Interfaces* **2021**, *13*, 53630–53645. [CrossRef]
212. Grundner, C.; Li, Y.; Louder, M.; Mascola, J.; Yang, X.; Sodroski, J.; Wyatt, R. Analysis of the neutralizing antibody response elicited in rabbits by repeated inoculation with trimeric HIV-1 envelope glycoproteins. *Virology* **2005**, *331*, 33–46. [CrossRef]
213. Krammer, F.; Margine, I.; Tan, G.S.; Pica, N.; Krause, J.C.; Palese, P. A carboxy-terminal trimerization domain stabilizes conformational epitopes on the stalk domain of soluble recombinant hemagglutinin substrates. *PLoS ONE* **2012**, *7*, e43603. [CrossRef]
214. Li, J.; Ulitzky, L.; Silberstein, E.; Taylor, D.R.; Viscidi, R. Immunogenicity and protection efficacy of monomeric and trimeric recombinant SARS coronavirus spike protein subunit vaccine candidates. *Viral Immunol.* **2013**, *26*, 126–132. [CrossRef] [PubMed]
215. Sliepen, K.; van Montfort, T.; Melchers, M.; Isik, G.; Sanders, R.W. Immunosilencing a highly immunogenic protein trimerization domain. *J. Biol. Chem.* **2015**, *290*, 7436–7442. [CrossRef] [PubMed]

vaccines

MDPI

Article

Immune Enhancement of Nanoparticle-Encapsulated Ginseng Stem-Leaf Saponins on Porcine Epidemic Diarrhea Virus Vaccine in Mice

Fei Su [1], Lihua Xu [1], Yin Xue [2], Wei Xu [3], Junxing Li [1], Bin Yu [1], Shiyi Ye [1] and Xiufang Yuan [1,*]

1 Institute of Animal Husbandry and Veterinary Science, Zhejiang Academy of Agricultural Sciences, Hangzhou 310002, China
2 Zhejiang Center of Animal Disease Control, Hangzhou 310020, China
3 Zhejiang Provincial Key Laboratory of Preventive Veterinary Medicine, MOA Key Laboratory of Animal Virology, Center for Veterinary Sciences, College of Animal Sciences, Zhejiang University, Hangzhou 310058, China
* Correspondence: yuanxf@zaas.ac.cn or xiufangyuan@126.com; Tel.: +86-571-8640-4121

Abstract: Porcine epidemic diarrhea virus (PEDV) causes severe enteric disease in pigs, particularly neonatal piglets. Current vaccines do not provide complete protection against PEDV. Ginseng stem-leaf saponins (GSLS), a promising oral adjuvant candidate, can improve intestinal immune responses in poultry and mice. However, its low stability limits further use. Poly lactic-co-glycolic acid (PLGA), a biocompatible and biodegradable nanoparticle, has been widely used in biomedicine for stable and targeted drug delivery. In this study, we developed GSLS-PLGA nanoparticles (GSLS-NPs) and evaluated the mucosal adjuvant efficacy in vitro and in vivo. GSLS-NPs significantly enhanced antigen internalization and pro-inflammatory cytokine secretion by DC2.4 cells. Mice orally administered GSLS-NPs before intramuscular inoculation generated $CD11b^{+}CD8\alpha^{-}$ and $CD11b^{-}CD103^{+}$ dendritic cells in the spleen and draining mesenteric lymph nodes, respectively, which are the types mainly responsible for antigen presentation. Additionally, enhanced neutralizing and non-neutralizing antibody responses and expanded activities of specific effector and memory $CD4^{+}$ and $CD8^{+}$ T cells were also observed in mice immunized with PEDV vaccines plus GSLS-NPs compared to mice receiving the vaccines alone. Furthermore, GSLS-NPs showed a good safety profile and presented great advantages over GSLS aqueous solution. Collectively, our results highlight the potential of GSLS-NPs as a mucosal adjuvant and provide an attractive vaccination strategy for combatting PEDV. Further study is required to evaluate the efficacy of this mucosal adjuvant in swine.

Keywords: ginseng stem-leaf saponins; nanoparticles; mucosal adjuvant; porcine epidemic diarrhea virus

1. Introduction

Porcine epidemic diarrhea virus (PEDV), a member of the genus *Alphacoronavirus* in the family *Coronaviridae*, causes acute enteric symptoms characterized by severe diarrhea, vomiting, and dehydration in pigs, causing high mortality in neonatal piglets that reaches nearly 100% [1]. In 2013–2014, a PEDV outbreak led to approximately 8 million pigs dying in the United States and cost between USD 900 million and 1.8 billion in annual losses [2]. PEDV is now considered one of the most devastating porcine viruses that has emerged or re-emerged and presents a significant threat to the worldwide swine industry [3]. The main PEDV transmission route is the fecal-oral route, although airborne transmission via the fecal-nasal route is also frequent within or between farms. As PEDV mainly infects and replicates in the villus epithelium of the small intestine, especially in the jejunum and ileum, locally produced virus-specific immunoglobulin A (IgA) antibodies are the most important factor in the primary adaptive defense at mucosal surfaces [4]. However, repeated outbreaks of PEDV on large farms and the emergence of highly pathogenic strains indicate that the effectiveness of current parenteral vaccination is not sufficient. Therefore, more effective vaccination strategies for PEDV prevention are urgently needed. One alternative solution is to develop potent adjuvants to improve mucosal immunity.

Ginseng saponins, consisting of steroid or triterpene aglycone linked to one or more sugars, are believed to be the major bioactive constituent of *Panax ginseng* C. A. Meyer. The total ginseng saponins extracted from the stems and leaves (GSLS) are composed of the ginsenosides Re, Rg1, Rb2, Rd and so on, which exhibit pharmacological activities similar to those from the roots and are available at a lower cost [5]. GSLS are widely used as an oral adjuvant for improving vaccination in poultry and mice. Zhai et al. reported that oral administration of GSLS could significantly enhance the efficacy of an infectious bursa vaccine in chickens, increasing the populations of intestinal intraepithelial lymphocytes (iIELs) and lamina propria IgA-secreting cells from the duodenum, jejunum and ileum mucosa when compared with the vaccine alone, and provide more effective protection against the virulent strain of chicken infectious bursal virus [5]. In addition, oral administration of GSLS could recover antibody and cellular responses to an inactive bivalent vaccine for Newcastle disease and avian influenza in chickens immunosuppressed by cyclophosphamide [6]. Similar results were also observed in Li's study [7], in which oral administration of GSLS before immunization of mice with a foot-and-mouth disease vaccine improved specific serum immunoglobulin G (IgG) antibody levels and proliferative splenocyte responses, and significantly increased the numbers of iIELs in the duodenal mucosa and IgA-secreting cells in the lamina propria. These activities were associated with the chemical structures of saponins, in which the carbohydrate moieties might interact with the receptors on the surface of antigen-presenting cells (APCs), promoting antigen phagocytosis and cytokine secretion, while acyl chain domains assist in the delivery of exogenous antigens into APCs and facilitated adaptive immunity [8]. However, the major limitation of using GSLS to improve immunity is their low oral bioavailability and rapid clearance in the gastrointestinal tract, especially for the components with poor water solubility [9]. Therefore, GSLS need to be biochemically stabilized to avoid degradation and enhance absorption in the gastrointestinal tract to achieve maximal beneficial effects.

In recent years, nanoparticle (NP) adjuvant delivery systems have offered attractive features that dictate and direct an immune response, including protection of associated drugs from degradation, increased drug bioavailability, and facilitation of antigen uptake and processing by APCs [10–12]. Poly lactic-co-glycolic acid (PLGA), a biodegradable and nontoxic copolymer approved by the United States Food and Drug Administration (FDA) for clinical use [13], has been extensively used to prepare NPs for drug and vaccine delivery. Here, we utilized PLGA to form a nanocarrier for GSLS by the double emulsion/solvent evaporation method. The immunological effects of GSLS nanoparticles (GSLS-NPs) on the phagocytosis and activation of DC2.4 cells were preliminarily investigated in vitro. Then, GSLS-NPs were orally administered to mice before PEDV vaccination, and the adjuvant effects on systemic and intestinal PEDV-specific IgG and IgA antibody responses, neutralizing antibody titers, dendritic cell (DC) activation, and effector and memory T-cell induction in the spleen and mesenteric lymph nodes (MLNs) were further investigated and compared with aqueous solution of GSLS. Moreover, the biosafety of GSLS-NPs was also evaluated in vitro and in vivo. Our results demonstrate that GSLS-NPs may be a potent mucosal adjuvant for the efficient induction of robust and long-term sustained adaptive immune responses against PEDV.

2. Material and Methods

2.1. Reagents

Standardized GSLS were purchased from Hongjiu Ginseng Industry Co. Ltd. (Jilin, China) and mainly contained Re (29.34%), Rg1 (12.94%), Rd (12.23%), Rb2 (12.22%) and Rc (7.36%), according to HPLC analysis. PLGA (DG-50DLG025, copolymer composition of 50:50, 15–24 kDa, ester terminated) was purchased from Jinan Daigang Biomaterial Co.,

Ltd. (Jinan, China). All solutions were passed through a gel endotoxin-removing column (Pierce, Shanghai, China), and the endotoxin level of solutions was below 0.5 EU/mL.

2.2. *Cells and Virus Stock*

Vero cells were obtained from the National Animal Gene Research Center of China Agricultural University. Dulbecco's modified Eagle's medium (DMEM), RPMI-1640 medium, trypsin (0.25%), fetal bovine serum (FBS), penicillin and streptomycin were purchased from Gibco (Shanghai, China). Cells were maintained in DMEM supplemented with 10% (v/v) heat-inactivated FBS and antibiotics (100 U/mL penicillin and 8 μg/mL streptomycin) in a humidified 5% CO_2 incubator at 37 °C.

The murine DC line DC2.4 was obtained from the cell bank at the Chinese Academy of Science (Shanghai, China). DC2.4 cells were cultured in complete RPMI-1640 medium supplemented with 10% (v/v) FBS and antibiotics (100 U/mL penicillin and 8 μg/mL streptomycin) in a 5% CO_2 incubator at 37 °C.

The PEDV strain ZJ-ZX2018-C10 (GenBank Accession No. MK250953) was isolated in our laboratory. The virus ($1 \times 10^{6.8}$ $TCID_{50}$/mL) was inactivated with 0.05% β-propiolactone (Solarbio, Beijing, China) and used as an inactivated PEDV vaccine. An attenuated PEDV vaccine was obtained by propagating the virus on Vero cells for more than 130 passages (1×10^5 $TCID_{50}$/mL).

2.3. *Preparation and Characterization of Nanoparticles*

GSLS-NPs were prepared by the double emulsion/solvent evaporation method [14]. Briefly, 10 mg of GSLS was dissolved in 1 mL of phosphate-buffered saline (PBS) and emulsified with 10 mL of 20 mg/mL PLGA dichloromethane solution under sonication in an ice bath at 190 W for 2 min. Subsequently, the primary emulsion was diluted with 110 mL of 2% (w/v) PVA aqueous solution and sonicated on ice at 190 W for 5 min, followed by evaporation of the organic solvents under magnetic stirring overnight. GSLS-NPs were collected from the supernatant colloidal suspension by centrifugation at 3000 rpm for 30 min and lyophilized in a freeze dryer (SCIENTZ-18N, SCIENTZ, Ningbo, China) for further use. Empty PLGA nanoparticles were prepared using a similar method. NP size and the polydispersity index (PDI) were determined with a laser particle size analyzer (Litesizer^{TM} 500, Anton Paar, Graz, Austria). Morphological characterization of NPs was performed using transmission electron microscopy (JEM-1200EX, JEOL, Tokyo, Japan).

A total of 100 mg of freeze-dried NPs was resuspended in 1 mL of acetonitrile to disrupt encapsulation and then diluted in 3 mL of methanol to extract the drugs. The obtained solution was subsequently filtered and subjected to high-performance liquid chromatography (HPLC, Ultra Violet detector, Shimadzu, Kyoto, Japan) with a C18 reversed-phase column (Dikma, Foothill Ranch, CA, USA) at a wavelength of 203 nm. The mobile phase was a mixture of acetonitrile and water (81:19, v/v) pumped at a rate of 1 mL/min. The HPLC system was controlled using LCSolution software. The encapsulation efficiency (EE) was calculated according to Equation (1):

$$\text{EE (\%)} = \frac{\text{weight of the drug in nanoparticles}}{\text{weight of the total drug}} \times 100\% \quad (1)$$

2.4. *Cytotoxicity Assay*

The cytotoxicity of GSLS-NPs to DC2.4 cells was tested using a CCK-8 assay. Briefly, DC2.4 cells were plated in a 96-well plate at a density of 1×10^5 cells/mL and incubated overnight to allow cell adhesion. The cells were then treated with GSLS-NPs, GSLS aqueous solution (G-Solution) or empty NPs at different concentrations ranging from 50 to 400 μg/mL. Treatment with LPS (Sigma, Shanghai, China) was used as a positive control. PBS was used to replace the drugs in the negative control (NC) group. After 48 h of incubation, 10 μL of CCK-8 (Dojindo Laboratories, Shanghai, China) was added to each well and incubated for 4 h at 37 °C. The absorbance at 450 nm was evaluated with a microplate reader (Thermo Multiskan MK3, Thermo Fisher Scientific, Shanghai, China). Cell viability was calculated using Equation (2):

$$\text{Cell viability } (\%) = \frac{\text{OD value of treated cultures}}{\text{OD value of non} - \text{treated cultures}} \times 100\% \quad (2)$$

2.5. Cellular Uptake

DC2.4 cells were seeded in a 6-well plate at a density of 1×10^5 cells/mL and then incubated in complete RPMI-1640 medium overnight. After pre-treatment with GSLS-NPs, G-Solution or empty NPs at a final concentration of 100 μg/mL for 4 h, the cells were incubated with the inactivated PEDV antigen for 24 h. Untreated cells served as the NC. After removing the supernatant, the cells were washed three times with PBS and fixed with Immunol Staining Fix Solution (P0098, Beyotime, Shanghai, China). A monoclonal antibody (1:100) against the PEDV nucleocapsid protein was then added and incubated for 1.5 h at 37 °C. DyLight 488-conjugated goat anti-mouse IgG (E032210-01, EARTHOX, San Francisco, CA, USA) (1:200) was used as the secondary antibody. LysoTracker Red (C1046, Beyotime, Shanghai, China) and DAPI (C1005, Beyotime, Shanghai, China) were used to stain the lysosomes and nuclei, respectively. The cells were observed under an inverse confocal laser scanning microscope (FV1000, Olympus, Tokyo, Japan). The images were acquired using Olympus confocal software (FV10-ASW 3.1). In addition, a monoclonal antibody against the PEDV nucleocapsid protein was labeled with the fluorescent dye APC by using an APC conjugation kit (ab201807, Abcam, Shanghai, China). The relative fluorescence intensity of intracellular PEDV was quantified on a FACSCanto™ (BD Biosciences, San Diego, CA, USA).

2.6. Animal Studies

Female BALB/c mice aged 6–8 weeks were purchased from the Shanghai Laboratory Animal Center (Shanghai, China). Mice were housed under specific pathogen-free conditions at a stable temperature (24 ± 1 °C) and randomly divided into 10 groups, with 6 mice in each group. All experiments related to animal care and use were approved by the Animal Care Committee of the Zhejiang Academy of Agricultural Sciences in accordance with the recommendations of the National Institutes of Health Guide for the Care and Use of Laboratory Animals (Ethics protocol No. 2020001).

In one experiment, mice were gavaged daily for 7 days according to their group: GSLS-NPs 1 (2.5 mg/kg), GSLS-NPs 2 (5 mg/kg), GSLS-NPs 3 (10 mg/kg), G-Solution 1 (2.5 mg/kg), G-Solution 2 (5 mg/kg), G-Solution 3 (10 mg/kg), empty NPs (10 mg/kg), LPS (50 μg/kg) or saline. Four hours after the final gavage, the mice received 1×10^4 $TCID_{50}$ of attenuated PEDV vaccine via i.m. injection (day 0), and a booster injection of the inactivated vaccine ($1 \times 10^{5.8}$ $TCID_{50}$) was administered on day 21. The mice treated with saline alone served as the NC. On day 35, blood and feces were collected for antibody determination. The blood was centrifuged at 1000× *g* for 10 min to separate the serum. Three to four fresh fecal pellets were collected into a micro-tube and weighed, and a 10× volume (per gram of feces) of extraction buffer (5% FBS and 0.02% sodium azide in PBS) was added to each tube to normalize the variation in the fecal weight among different mice. The fecal samples were vortexed vigorously and then centrifuged at 10,000× *g* for 10 min. The supernatant was collected for further use.

In the other experiment, mice were gavaged daily with GSLS-NPs, G-Solution or empty NPs at a dose of 5 mg/kg for 7 days. Four hours after the final gavage, the mice were immunized with 1×10^4 $TCID_{50}$ of attenuated PEDV vaccine via the i.m. route (day 0); then, they received a booster immunization with $1 \times 10^{5.8}$ $TCID_{50}$ of inactivated vaccine on day 21. Blood was harvested on days 2 and 23 for biochemical analysis and collected on day 35 for antibody determination. Feces and small intestines were collected on day 35 for immunological testing. Three-centimeter sections of the duodenum, jejunum and ileum were extracted for immunohistochemical and hematoxylin-eosin (H.E.) staining. In addition, intestinal samples were cut into pieces and vortexed with 400 μL of extraction buffer. The supernatant was harvested after centrifugation for antibody detection. Spleens and MLNs were harvested on day 23 for analysis of DC activation and collected on day 35 for evaluation of cytokine responses and T-cell activities. The spleens and MLNs were homogenized and passed through a steel mesh to obtain a cell suspension. After centrifugation at 380× *g* for 10 min, the cell pellets were washed in Hank's balanced salt solution and re-suspended in complete RPMI-1640 medium for further use.

2.7. Enzyme-Linked Immunosorbent Assay (ELISA) for Antibody Determination

Blood collected on day 35 was centrifuged at 1000× *g* for 10 min to separate the serum. PEDV-specific IgA, IgG, IgG1 and IgG2a antibodies were measured by ELISA. Briefly, polyvinyl 96-well microplates pre-coated with the PEDV antigen were obtained from PEDV antibody detection kits (HomSun Bio-Technology, Hangzhou, China). After adding either 100 μL of diluted serum samples (1:100) or fecal fluids (1:2), the plates were incubated at 37 °C for 1 h. After washing, 100 μL of horseradish peroxidase (HRP)-conjugated goat anti-mouse IgA (1:2000), IgG (1:5000), IgG1 (1:2000) or IgG2a (1:2000) was added, and the plate was incubated at 37 °C for 1 h. After another round of washing, 100 μL of 3,3′,5,5′-tetramethyl benzidine substrate solution was added to each well, and the reaction was stopped using a stop solution after 15 min. The optical density (OD) of the wells was read at 450 nm by a microplate reader (Thermo Multiskan MK3, Thermo Fisher Scientific, Shanghai, China).

2.8. Immunohistochemistry and Haematoxylin-Eosin Staining

IgA^+ cells were identified with an immunohistochemical staining method described previously [15]. Briefly, fixed intestinal samples were embedded in paraffin, sectioned at a thickness of 6 μm, and mounted onto glass slides. Endogenous peroxidase activity was blocked by incubation with 3% H_2O_2 in methanol for 20 min. After rinsing three times with PBS, the slides were successively incubated with a goat anti-mouse IgA antibody (1:400) and HRP-conjugated rabbit anti-goat IgG antibody (1:250). 3, 3′-Diaminobenzidine (DAB) (BOSTER, Wuhan, China) detection was performed according to the manufacturer's instructions. Images were captured with a DS-U3 camera interface (Nikon, Tokyo, Japan) and evaluated using the image analysis software Image-Pro Plus 6.0. Three fields per slide were randomly selected at 400× magnification, and the integrated optical density (IOD) was analyzed.

After hematoxylin–eosin staining, the number of iIELs in five different fields of intestinal villi on each slide was counted to statistically analyze the data.

2.9. Neutralization Assay

Serum and fecal fluids were inactivated at 56 °C for 30 min and serially diluted 1:2. The serial dilutions of the samples were mixed with equal volumes of 100 $TCID_{50}$ of PEDV virus and incubated at 37 °C for 1 h. The mixtures were then transferred to Vero cell monolayers on 96-well plates. The plates were incubated for 7 days at 37 °C in a 5% CO_2 atmosphere, and the cytopathic effect (CPE) was recorded. Neutralizing antibody titer that prevented CPE in 50% of the wells was calculated by the Reed and Muench method [16].

2.10. Cytokine Release

DC2.4 cells were plated at a density of 1×10^5 cells/mL in a 24-well plate and treated with GSLS-NPs, G-Solution or empty NPs at a final concentration of 100 μg/mL for 72 h. LPS- and PBS-treated cells served as the positive control and NC, respectively. The supernatant was collected to determine the concentrations of IL-1β, TNF-α and IL-6. Mononuclear cells from the spleen or MLNs of mice were re-stimulated with the inactivated PEDV antigen (10 μg/mL) for 72 h at 37 °C in a 5% CO_2 atmosphere. The supernatant was harvested to detect the levels of IFN-γ, IL-10 and IL-6. Cytokine levels were measured with commercial ELISA kits (Neobioscience, Shanghai, China) according to the manufacturer's instructions.

2.11. Flow Cytometry

To detect the activation of DCs, mononuclear cells from the spleen or MLNs were stained with PE/Cy7-conjugated anti-CD11c (clone N418), PE-conjugated anti-MHC II (clone M5/114.15.2), APC-conjugated anti-PEDV, FITC-conjugated anti-CD11b (clone M1/70), PerCP-Cy5.5-conjugated anti-CD8α (clone 53-6.7) and PerCP-Cy5.5-conjugated anti-CD103 (clone 2E7). To evaluate memory T-cell responses, mononuclear cells from the spleen or MLNs were stained with FITC-conjugated anti-CD3e (clone 145-2C11), APC-conjugated anti-CD4 (clone GK1.5), PerCP-Cy5.5-conjugated anti-CD8a (clone 53-6.7), PE-Cy7-conjugated anti-CD44 (clone IM7) and PE-conjugated anti-CD62L (clone MEL-14). The PE/Cy7-conjugated anti-CD11c and PerCP-Cy5.5-conjugated anti-CD103 antibodies were purchased from BioLegend (San Diego, CA, USA). The APC-conjugated anti-PEDV antibody was prepared in our laboratory. The PE-conjugated anti-CD62L antibody was purchased from Southern Biotech (Birmingham, AL, USA). The other antibodies were obtained from MultiSciences (Hangzhou, China). All data were collected on a FACSCanto™ (BD Biosciences, San Diego, CA, USA) and analyzed using FlowJo (Version 10.0).

2.12. Blood Biochemistry Analysis

Serum was assayed for the levels of alanine aminotransferase (ALT), aspartate aminotransferase (AST), lactate dehydrogenase (LDH), alkaline phosphatase (ALP), total bilirubin (TBIL) and blood urea nitrogen (BUN) with appropriate assay kits (Nanjing Jiancheng Bioengineering Institute, Nanjing, China) following the manufacturer's instructions.

2.13. Statistical Analysis

Data are expressed as the mean ± standard deviation (SD). All statistical analyses were performed with GraphPad Prism 7.0 software (San Diego, CA, USA); an unpaired Student's t test was used for two groups and one-way ANOVA with Tukey's multiple comparisons test was used for three or more groups. A p value < 0.05 was considered statistically significant. All experiments were conducted a minimum of three times.

3. Results

3.1. Synthesis and Characterization of Nanoparticles

As shown in Figure 1A, GSLS-NPs and empty NPs exhibited a smooth surface and spherical morphology. The center of empty NPs was bright, while that of GSLS-NPs showed a dark area, which suggested that GSLS was successfully loaded into the PLGA shell. The average diameter of GSLS-NPs was 230.7 ± 5.3 nm, with a PDI of 0.1348 ± 0.0276 (Figure 1B,C). There were no significant differences in particle size or PDI between GSLS-NPs and empty NPs. As determined by HPLC analysis, the encapsulation efficiencies of GSLS-NPs were 44.32 ± 6.93% for Re, 51.20 ± 6.25% for Rg1, 69.34 ± 7.94% for Rd, 36.63 ± 4.36% for Rb2, and 58.56 ± 4.36% for Rc (Figure 1D).

3.2. Viability and Activation of DCs Induced by GSLS-NPs In Vitro

DCs are highly specialized in processing and presenting antigens to activate lymphocytes. An in vitro cytotoxicity experiment was conducted with the murine DC line DC2.4 using a CCK-8 assay. Cells were cultured with GSLS-NPs at concentrations of 50–400 μg/mL, and no significant cell death was observed over 48 h (Figure 2A). To investigate the potential of GSLS-NPs to enhance the phagocytic capacity of DCs in vitro, the uptake of a PEDV antigen by DC2.4 cells was first evaluated by laser confocal microscopy detection. As shown in Figure 2B, GSLS-NPs effectively induced antigen internalization by DC2.4 cells, and the fluorescence intensity of the intracellular antigen achieved with GSLS-NPs was obviously stronger than that achieved by treatment with G-Solution, empty PLGA NPs or PBS. G-Solution had a slight enhancing effect on antigen internalization when compared with empty PLGA NPs or PBS. A negligible effect was found for empty PLGA NPs. These results were further confirmed by flow cytometric analysis. As shown in Figure 2C, when compared with empty NPs or PBS, GSLS-NPs ($p < 0.001$) and G-Solution

($p < 0.05$) significantly promoted the antigen uptake ratio of DC2.4 cells. Surprisingly, from the results of the flow cytometric analysis, no significant differences in the antigen uptake ratio were observed between the GSLS-NP and G-Solution treatments.

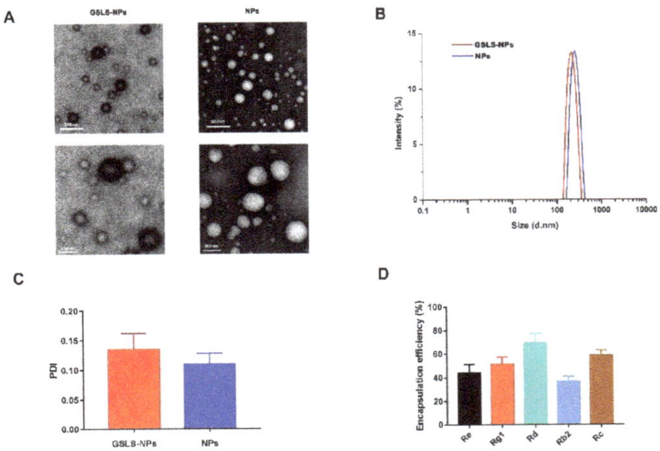

Figure 1. Characterization of nanoparticles. (**A**) Transmission electron microscopy images of GSLS-NPs and empty NPs. Low magnification, scale bar = 500 nm. High magnification, scale bar = 200 nm. (**B**) Particle Size. (**C**) PDI value. (**D**) Encapsulation efficiency.

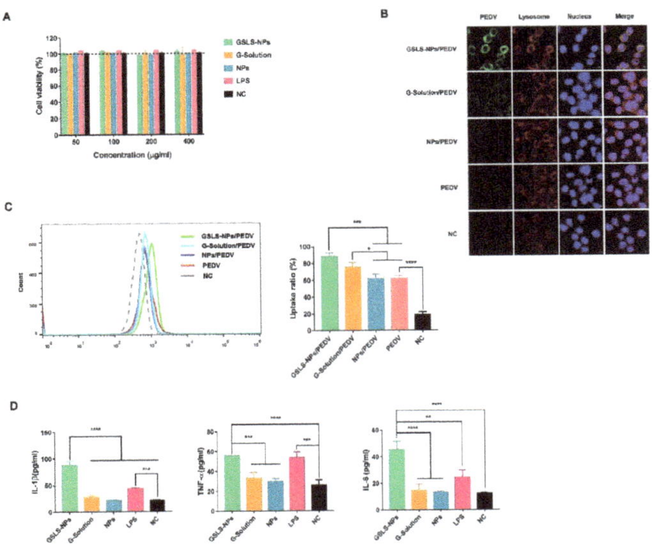

Figure 2. Effect of GSLS-NPs on the activation of DC2.4 cells. (**A**) Viability of DC2.4 cells measured with a CCK-8 assay after incubation with GSLS-NPs, G-Solution or empty NPs for 48 h. LPS was used as a positive control, and untreated cells served as a negative control (NC). (**B**,**C**) Representative fluorescence microscopy images and flow cytometric analysis of antigen uptake by DC2.4 cells after pre-treatment with GSLS-NPs, G-Solution, empty NPs or PBS for 4 h and incubation with the PEDV antigen for the following 24 h. (**D**) Production of cytokines by DC2.4 cells after 72 h of incubation. The experiments were performed in triplicate. Data represent the averages of triplicate samples from three identical experiments and are presented as the mean ± SD. Statistical analysis was performed using an unpaired Student's t test. * $p < 0.05$, ** $p < 0.01$, *** $p < 0.001$, **** $p < 0.0001$.

Next, we analyzed the pro-inflammatory cytokines expressed by DC2.4 cells in response to stimulation with GSLS-NPs, G-Solution or empty NPs. As shown in Figure 2D, DC2.4 cells treated with GSLS-NPs secreted significantly higher levels of IL-1β, TNF-α and IL-6 than untreated cells ($p < 0.0001$) or cells treated with G-Solution or empty NPs ($p < 0.0001$ for IL-1β and IL-6, $p < 0.001$ for TNF-α). Interestingly, the production of IL-1β and IL-6 in the GSLS-NP group was even higher than that in the lipopolysaccharide (LPS)-treated group ($p < 0.0001$ for IL-1β, $p < 0.01$ for IL-6). However, little effect was found for treatment with G-Solution or empty NPs.

3.3. Oral Administration of GSLS-NPs Enhances PEDV-Specific Antibody Responses in the Serum and Gastrointestinal Tract

To evaluate the adjuvant effect of GSLS-NPs on PEDV vaccination, PEDV-specific antibody responses in mice were analyzed 14 days after a booster immunization. When compared with the vaccine-only control, the addition of the adjuvant GSLS-NPs at a dose of 5 or 10 mg/kg significantly enhanced the IgG responses in the serum ($p < 0.05$), but only the dose of 5 mg/kg could strongly increase the titer of specific IgA antibodies in the feces ($p < 0.05$) (Figure 3A). G-Solution or empty NPs showed little enhancing effect on the vaccine-induced serum IgG and fecal IgA antibody responses. As a positive control, LPS was effective only in elevating serum IgG responses ($p < 0.05$), not in increasing fecal IgA antibody levels.

As shown in Figure 3B, the prominent IgG isotypes elevated in the serum by vaccination were IgG1 and IgG2a, which were significantly higher in the adjuvant group treated with GSLS-NPs than in the vaccine-only control group ($p < 0.001$ for IgG1 and IgG2a) and vaccine-plus-empty-NPs group ($p < 0.001$ for IgG1, $p < 0.01$ for IgG2a). Mice immunized with the vaccine plus G-Solution generated similar levels of serum IgG1 but lower levels of IgG2a ($p < 0.05$) than mice immunized with the vaccine supplemented with GSLS-NPs. The serum IgG1 isotype titer induced by addition of G-Solution was apparently higher than that in the vaccine-only control ($p < 0.01$) and the addition of empty NPs ($p < 0.05$). Furthermore, the IgA-secreting cells in each region of the small intestine were analyzed using immunohistochemistry. As shown in Figure 3C,D, the addition of GSLS-NPs significantly increased the numbers of IgA-secreting cells in the duodenum, jejunum and ileum when compared with the vaccine-only control, vaccine plus G-Solution or vaccine plus empty NPs ($p < 0.0001$). The number of IgA-secreting cells in the whole small intestine was similar among mice receiving the vaccine alone or vaccine plus G-Solution or empty NPs. Notably, vaccination alone increased only IgA-secreting cells in the duodenum when compared to no immunization ($p < 0.05$). As expected, the GSLS-NPs adjuvant also elevated PEDV-specific IgA antibody levels in segments of the duodenum ($p < 0.0001$), jejunum ($p < 0.01$) and ileum ($p < 0.05$) compared with the vaccine alone (Figure 3E), which was consistent with the increased populations of IgA-secreting cells. However, compared to the addition group of G-Solution or empty NPs, GSLS-NPs elicited stronger IgA antibody responses in only the duodenum ($p < 0.0001$) and jejunum ($p < 0.01$), not in the ileum, which was different from the IgA-secreting cell population results.

The neutralizing antibody titers in serum and feces were further analyzed. As shown in Figure 3F, mice receiving the vaccines alone generated a higher serum neutralizing antibody titer than unimmunized mice ($p < 0.0001$), but no difference was observed in the feces. GSLS-NPs significantly promoted neutralizing antibody responses in both the serum and the feces when compared with the vaccine-only control ($p < 0.0001$), vaccine plus G-Solution ($p < 0.001$) or empty NPs ($p < 0.0001$). However, the G-Solution and empty NPs had little effect on the neutralizing antibody response in either the serum or the feces compared to the vaccine alone.

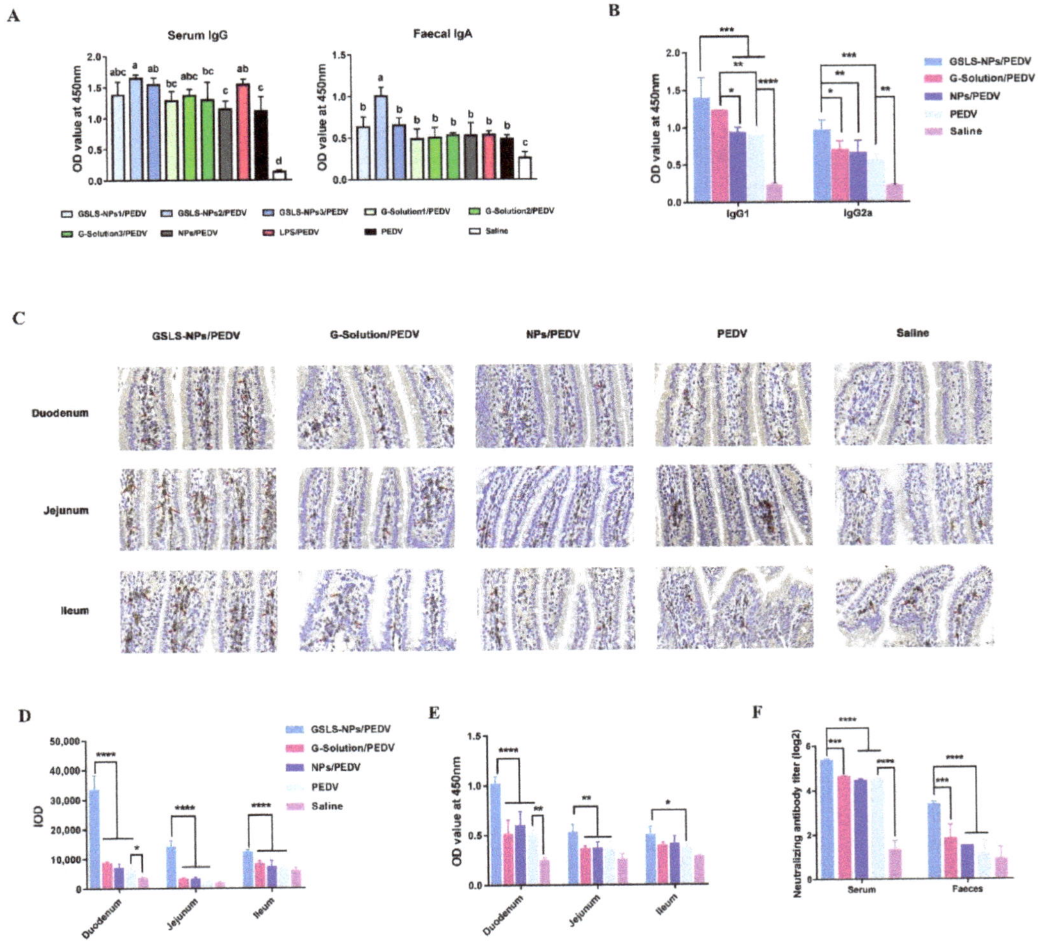

Figure 3. GSLS-NPs enhanced PEDV-specific antibody responses in the blood and intestinal mucosa. (**A**) PEDV-specific serum IgG and fecal IgA antibodies. Mice (n = 6) were gavaged daily with saline, empty NPs (10 mg/kg), G-Solution (2.5–10 mg/kg) or GSLS-NPs (2.5–10 mg/kg) for 7 days. LPS (50 μg/kg) was used as a positive control. After the final gavage, the mice were intramuscularly immunized with an attenuated PEDV vaccine (1×10^4 $TCID_{50}$) on day 0, followed by a booster immunization with an inactivated vaccine ($1 \times 10^{5.8}$ $TCID_{50}$) on day 21. Non-immunized mice served as the negative control (NC). Serum and feces were collected on day 35 for antibody analysis by ELISA. Data are expressed as the mean ± SD. Statistical analysis was performed using one-way ANOVA with Tukey's multiple comparisons test. Different letters represent significant differences ($p < 0.05$). (**B**) Serum IgG isotypes. Mice (n = 6) were gavaged daily with saline, empty NPs (5 mg/kg), G-Solution (5 mg/kg) or GSLS-NPs (5 mg/kg) for 7 days and immunized as described above. The mice were euthanized on day 35. (**C**) IgA-secreting cells in the duodenum, jejunum and ileum. (**D**) Integrated optical density (IOD) of IgA-secreting cells. (**E**) PEDV-specific IgA antibody responses in the duodenum, jejunum and ileum. (**F**) Neutralizing antibody titer. Experiments were performed in triplicate. Data are presented as the mean ± SD. Stars denote significant differences: * $p < 0.05$, ** $p < 0.01$, *** $p < 0.001$, **** $p < 0.0001$. Unpaired Student's t test.

3.4. Oral Administration of GSLS-NPs Increases iIEL Proliferation

iIELs are small round cells with densely stained nuclei located within the epithelial cell layer in the small intestine. As shown in Figure 4, intramuscular (i.m.) inoculation of PEDV vaccines had little enhancing effect on the number of iIELs. Mice receiving the GSLS-NPs adjuvant generated more iIELs in the intestinal mucosa than mice immunized with the vaccine alone ($p < 0.01$), vaccine plus G-Solution ($p < 0.05$) or vaccine plus empty NPs ($p < 0.01$). However, no significant increases in iIELs were observed in the G-Solution or empty-NPs addition group when compared with the vaccine-only control group.

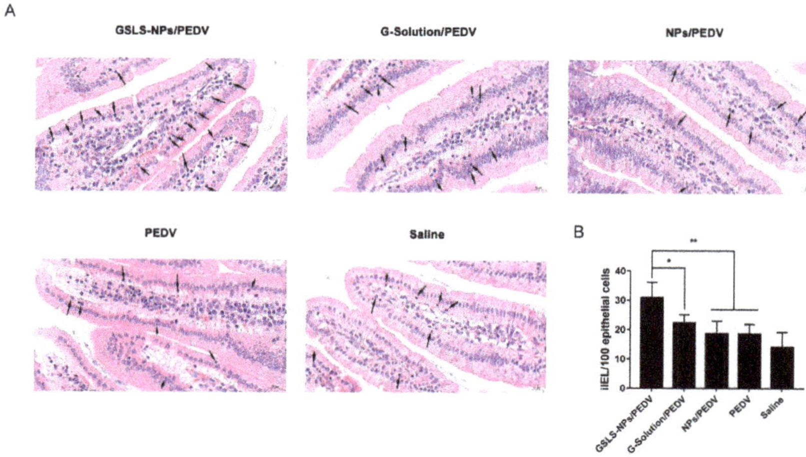

Figure 4. GSLS-NPs increased intestinal intraepithelial lymphocytes (iIELs) in the mucosa of the small intestine. Mice (n = 6) were gavaged daily with saline, empty NPs, G-Solution or GSLS-NPs for 7 days. After the final gavage, the mice were immunized as described above and euthanized on day 35. Non-immunized mice served as the NC. (**A**) Representative images of iIELs in the epithelial cell layer of the small intestine identified by hematoxylin-eosin staining (indicated as arrows). Scale bar = 20 μm. (**B**) iIEL count per 100 epithelial cells in five different fields of intestinal villi on each slide. Data are presented as the mean ± SD. * $p < 0.05$, ** $p < 0.01$. Unpaired Student's t test.

3.5. GSLS-NPs Regulate Distinct Subsets of Splenic and Intestinal DCs

DCs are critical initiators and modulators of adaptive immune responses, and as such, targeting DCs directly is an approach to enhance responses to vaccines. In this study, approximately 40% of mature CD11c$^+$ DCs from the spleen (Figure 5A,B) and MLNs (Figure 6A,B) highly expressed major histocompatibility complex (MHC) class II and acquired the PEDV antigen after immunization with the vaccine plus GSLS-NPs, which was significantly higher than the rates in the vaccine-only control ($p < 0.01$ in the spleen and MLNs) and empty-NPs-addition ($p < 0.05$ in the spleen, $p < 0.01$ in the MLNs) groups. Moreover, the population of PEDV-positive (PEDV$^+$) DCs in the MLNs of mice receiving G-Solution was significantly lower than that in mice treated with GSLS-NPs ($p < 0.05$), which was similar to those in the vaccine-only control and vaccine-plus-empty-NPs groups. Notably, i.m. inoculation of PEDV vaccines enhanced the antigen uptake by DCs only in the spleen ($p < 0.05$).

Multiple subsets of DCs with distinct functions and roles in the priming of T-cell responses exist [14]. Thus, we analysed various subsets of MHCII$^+$ CD11c$^+$ DCs to determine which specific subsets were involved in the uptake of the PEDV antigen. As shown in Figure 5C,D, the CD11b$^+$ CD8α$^-$ subset was the most frequent PEDV$^+$ DC population in the spleen of immunized mice, and mice receiving the vaccine plus GSLS-NPs showed an apparently higher proportion of CD11b$^+$ CD8α$^-$ DCs than mice immunized with the vaccine alone ($p < 0.01$), vaccine plus G-Solution ($p < 0.05$) or vaccine plus empty NPs

($p < 0.01$). Furthermore, the PEDV antigen was mainly associated with CD11b$^-$ CD103$^+$ DCs in the MLNs of mice receiving GSLS-NPs adjuvant (Figure 6C,D), and this population was significantly larger in these mice than in those treated with the vaccine alone ($p < 0.01$), vaccine supplemented with G-Solution ($p < 0.05$) or empty NPs ($p < 0.01$), suggesting that GSLS-NPs may have a more profound impact on migratory CD11b$^-$ CD103$^+$ DCs than on other cells. However, negligible differences in the intestinal DC subsets were found among the vaccine-only, vaccine-plus-G-Solution, vaccine-plus-empty-NPs and non-immunized groups.

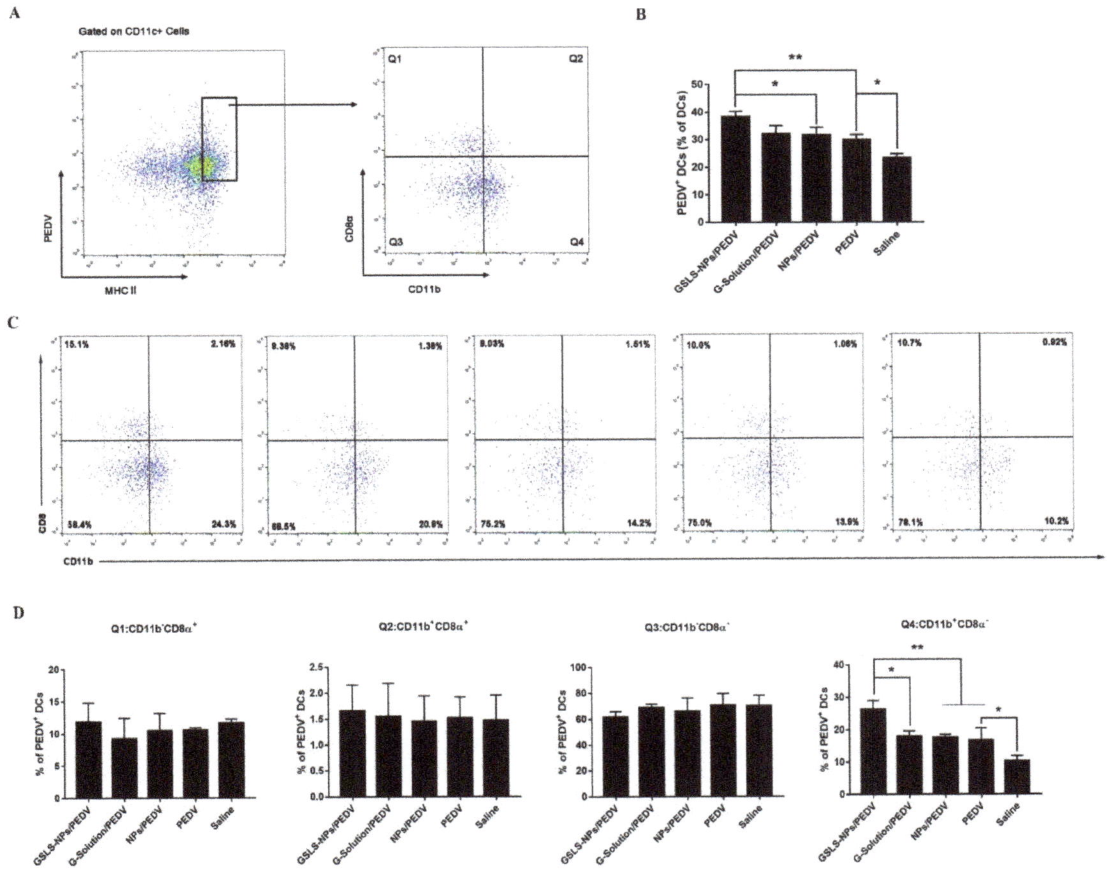

Figure 5. Responses of dendritic cell subsets in the spleen. Mice (n = 6) were gavaged and immunized as described above, and splenocytes were collected after euthanasia on day 23. (**A**,**B**) Frequency of PEDV-positive DCs. (**C**) Representative dot plot of the DC subset distribution. (**D**) Frequencies of PEDV-positive DC subsets. Data are expressed as the mean ± SD. * $p < 0.05$, ** $p < 0.01$. Unpaired Student's t test.

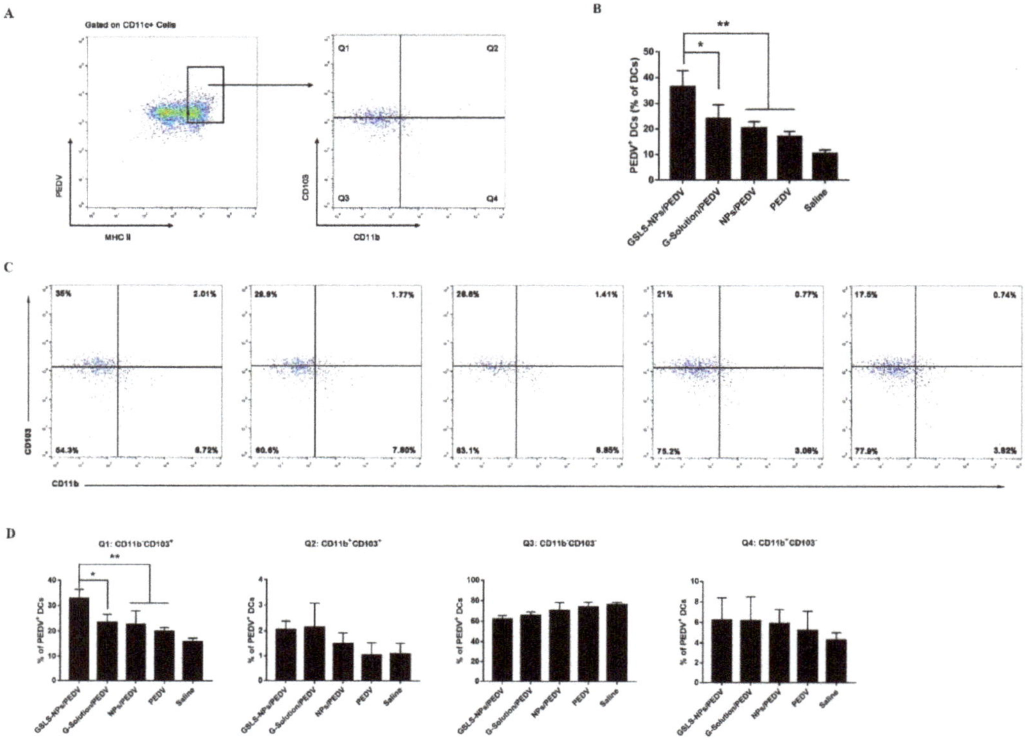

Figure 6. Responses of dendritic cell subsets in the MLNs. Mice (n = 6) were gavaged and immunized as described above, and mononuclear cells were isolated from the MLNs after euthanasia on day 23. (**A**,**B**) Proportion of PEDV-positive DCs. (**C**) Representative flow cytometry plot of the DC subset distribution. (**D**) Frequencies of PEDV-positive DC subsets. Data are presented as the mean ± SD. * $p < 0.05$, ** $p < 0.01$. Unpaired Student's t test.

3.6. GSLS-NPs Expand Memory and Effector T-Cell Responses

T lymphocyte ($CD3^+$) subpopulations in the spleen and MLNs were analyzed 14 days after a secondary immunization. As shown in Figure 7A,B, the GSLS-NPs adjuvant was more effective than G-Solution and empty NPs in triggering T-cell differentiation when delivered orally. The proportion of the $CD4^+$ subpopulation was significantly higher in spleens and MLNs from mice immunized with the vaccine supplemented with GSLS-NPs than in those from mice immunized with the vaccine alone ($p < 0.001$ in the spleen, $p < 0.0001$ in the MLNs), vaccine plus G-Solution ($p < 0.05$ in the spleen and MLNs) or vaccine plus empty NPs ($p < 0.01$ in the spleen, $p < 0.0001$ in the MLNs). Increased $CD8^+$ T cells were observed only in the MLNs, not the spleen, in the GSLS-NPs adjuvant group compared to other immunized groups ($p < 0.05$). Notably, parenteral inoculation increased only $CD4^+$ ($p < 0.001$) cells, not $CD8^+$ T cells, in the spleen but had little impact on the T-cell subpopulation in the MLNs.

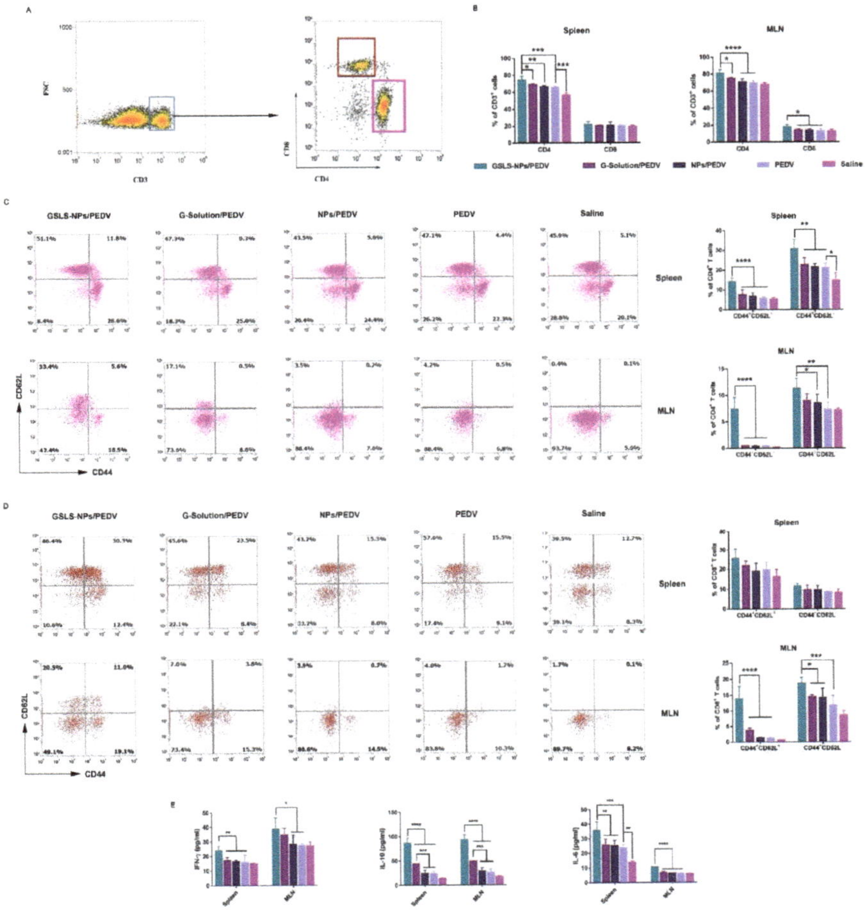

Figure 7. GSLS-NPs expanded PEDV-specific T-cell responses in the spleen and MLNs. Mice (n = 6) were gavaged and immunized as described above, and mononuclear cells were isolated from the spleen and MLNs after euthanasia on day 35. (**A**,**B**) Frequencies of $CD4^+$ and $CD8^+$ T cells. (**C**,**D**) Representative dot plots and percentages of memory ($CD44^{hi}$ $CD62L^{hi}$) and effector ($CD44^{hi}$ $CD62L^{lo}$) $CD4^+/CD8^+$ T cells. (**E**) Cytokine production in response to re-stimulation with the PEDV antigen. Data are presented as the mean ± SD. Statistical analysis was performed using an unpaired Student's *t* test. * $p < 0.05$, ** $p < 0.01$, *** $p < 0.001$, **** $p < 0.0001$.

Effective activation of memory T cells contributes to long-lasting immunity, which helps to strengthen adaptive immune responses [17]. As shown in Figure 7C, GSLS-NPs increased specific memory ($CD44^{hi}$ $CD62L^{hi}$) and effector ($CD44^{hi}$ $CD62L^{lo}$) $CD4^+$ T cells in both the spleen ($p < 0.0001$ for memory, $p < 0.01$ for effector) and the MLNs ($p < 0.0001$ for memory, $p < 0.01$ for effector) compared to the vaccine-only control. Moreover, the proportions of memory and effector $CD4^+$ T cells in the spleen ($p < 0.0001$ for memory, $p < 0.01$ for effector) and the MLNs ($p < 0.0001$ for memory, $p < 0.05$ for effector) were higher in the GSLS-NPs adjuvant group than in the empty NPs addition group. When compared with treatment of the G-Solution, the addition of GSLS-NPs showed elevated numbers of memory $CD4^+$ T cells in both the spleen and MLNs ($p < 0.0001$), while a larger effector $CD4^+$ T-cell subpopulation was observed only in the spleen ($p < 0.01$), not in the MLNs. Notably, parenteral inoculation increased only effector $CD4^+$ T cells in the spleen, not in the MLNs ($p < 0.05$), and had little enhancing effect on memory $CD4^+$ T cells in either the spleen or the MLNs. Furthermore, the results for specific memory (CD44^{hi} CD62L^{hi}) and effector (CD44^{hi} CD62L^{lo}) CD8$^+$ T cells are presented in Figure 7D. In the spleen, rare significant differences were found among all the immunized groups in either memory or effector CD8$^+$ T cells. In the MLNs, mice immunized with the vaccine plus GSLS-NPs exhibited significantly higher percentages of memory and effector CD8$^+$ T cells than mice receiving the vaccine alone ($p < 0.0001$ for memory, $p < 0.001$ for effector) or vaccine plus G-Solution or empty NPs ($p < 0.0001$ for memory, $p < 0.05$ for effector).

We further evaluated mononuclear cells isolated from the spleen or MLNs of immunized mice for cytokine production upon PEDV antigen re-stimulation in vitro. As shown in Figure 7E, the vaccine alone increased the level of IL-6 only in the spleen compared to no immunization ($p < 0.01$). Oral administration of GSLS-NPs significantly increased the levels of Th1 (IFN-γ), Th2 (IL-10) and Th17 (IL-6) cytokines in both the spleen ($p < 0.01$ for IFN-γ, $p < 0.0001$ for IL-10 and $p < 0.001$ for IL-6) and the MLNs ($p < 0.05$ for IFN-γ, $p < 0.0001$ for IL-10 and IL-6) compared to vaccine-only immunization. When compared with the administration of empty NPs, treatment with GSLS-NPs remarkably elevated the production of IFN-γ, IL-10 and IL-6 in the spleen ($p < 0.01$ for IFN-γ, $p < 0.0001$ for IL-10 and $p < 0.01$ for IL-6) and MLNs ($p < 0.05$ for IFN-γ, $p < 0.0001$ for IL-10 and IL-6). Moreover, compared with treatment of the G-Solution, the addition of GSLS-NPs exhibited higher levels of IL-10 and IL-6 in both the spleen ($p < 0.0001$ for IL-10 and $p < 0.01$ for IL-6) and the MLNs ($p < 0.0001$ for IL-10 and IL-6), while a higher IFN-γ level was observed only in the spleen ($p < 0.01$), not in the MLNs. Notably, oral administration of the G-Solution increased the production of Th2-type cytokines (IL-10) in both the spleen and the MLNs compared to treatment with empty NPs or the vaccine alone ($p < 0.001$).

3.7. *Biosafety of GSLS-NPs as an Oral Adjuvant*

To evaluate the safety of oral delivery of GSLS-NPs, blood biochemistry analysis was conducted 2 days after the primary and secondary immunizations. As shown in Figure 8, no significant treatment-related differences in the serum levels of ALT, AST, LDH, ALP, TBIL or BUN were recorded among all the groups, indicating no overall adverse effects on hepatic or renal function.

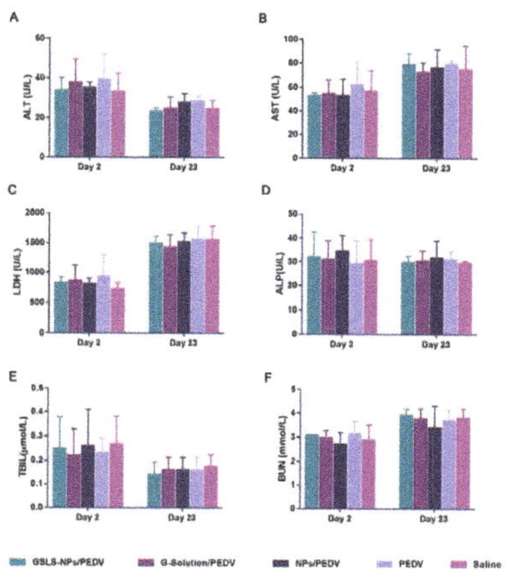

Figure 8. Serum levels of alanine aminotransferase (ALT), aspartate aminotransferase (AST), lactate dehydrogenase (LDH), alkaline phosphatase (ALP), total bilirubin (TBIL) and blood urea nitrogen (BUN). Mice (n = 6) were gavaged and immunized as described above, and serum was collected on days 2 and 23. Data are presented as the mean ± SD. Statistical differences were determined by an unpaired Student's *t* test.

4. Discussion

To date, parenteral vaccination to prevent PEDV infection has achieved little success, as the current vaccines fail to induce sufficient mucosal immunity, which plays a crucial role in immune defenses against PEDV infection. One attractive way to circumvent this issue would be the use of an effective mucosal adjuvant to sensitize the immune status of the intestinal mucosa before vaccination. In this study, we prepared a nano-based delivery system for GSLS and explored its mucosal adjuvant potential in the context of PEDV vaccine.

The GSLS-NPs exhibited a spherical shape with an average diameter of approximately 230 nm and a low PDI value, suggesting the narrow size distribution of the NPs. According to the HPLC analysis results, the encapsulation efficiency was high for the hydrophilic saponins of GSLS, including Re, Rg1, Rd and Rc, but low for Rb2, which possesses lipophilic properties, indicating that the water-in-oil-in-water (w/o/w) method is more suitable to encapsulate water-soluble agents, as previously reported [18]. Based on the results of the in vitro viability assay and in vivo blood biochemistry analysis, no apparent cytotoxicity or hepatic or renal dysfunction was observed after treatment with GSLS-NPs, indicating that GSLS-NPs have a good biosafety profile.

DCs, as one of the most important professional APC populations, have a specific function in antigen presentation, which is distinct from that of macrophages and B cells. DCs can present extracellular and intracellular antigens by direct presentation and cross-presentation, which promote the differentiation and activation of $CD4^+$ and $CD8^+$ T cells, respectively [19]. In this work, we initially found that GSLS-NPs significantly enhanced antigen internalization and cytokine responses by DC2.4 cells, suggesting that GSLS-NPs could promote the activation of DCs to enhance immune activities, which was further proven in animals. Distinct DC subsets may be programmed for different functions depending on the nature of the stimulus [20]. Oral administration of GSLS-NPs tended to promote the differentiation of splenic DCs into the $CD11b^+$ $CD8\alpha^-$ phenotype, which is the population specifically responsible for antigen presentation in the spleen. $CD11b^+$ $CD8\alpha^-$ DCs can further present antigens exclusively to $CD4^+$ T cells and prime antigen-specific immune responses [19], which can be seen from the biased increase in $CD4^+$ T cells and balanced improvements in Th1, Th2 and Th17 cytokine responses in spleen from the mice immunized with the vaccine plus GSLS-NPs compared to those from mice in other immunized groups. Notably, splenic $CD4^+$ T cells from mice immunized with the vaccine supplemented with GSLS-NPs further differentiated into specific memory and effector $CD4^+$ T-cell subsets, both of which were more abundant in this group than in the other immunized groups. Memory T cells form the cellular basis for accelerated protection upon re-exposure to the same pathogen, which is a hallmark of adaptive immunity [21,22]. In contrast, splenic $CD11b^+$ $CD8\alpha^-$ DCs activated by parenteral inoculation alone preferentially primed effector $CD4^+$ T cells but not memory $CD4^+$ T cells and mainly promoted the Th17 immune response. It was surprising to note that none of the treatment groups showed an effect on $CD8\alpha^+$ DCs, resulting in little activation of CD8+ T cells in the spleen. In contrast, in the draining MLNs, GSLS-NPs guided the recruitment of intestinal $CD11b^-$ $CD103^+$ DCs, which can cross-present antigens to $CD8^+$ T cells and promote the differentiation of $CD4^+$ T cells into Th1 and Foxp3+ regulatory T (Treg) cells [23,24]. The activated $CD4^+$ and $CD8^+$ T cells in the MLNs of mice immunized with the vaccine plus GSLS-NPs further differentiated into memory and effector T-cell subsets and expressed higher levels of Th1, Th2 and Th17 cytokines than those in other immunized groups, indicating that robust cytotoxic T-cell activities and that balanced Th1/Th2/Th17 responses were induced in the draining MLNs by the additional administration of GSLS-NPs. It has been reported that $CD11b^-$ $CD103^+$ DCs can activate the gut-homing receptors CCR9 and integrin $\alpha4\beta7$ on the surface of T cells to promote the homing of effector T cells to the effector site in the intestinal mucosa and protect against pathogens [25]. Correlatively, iIELs, which form one of the first lines of immune defense in gut tissue, were also significantly increased in the intestinal epithelium of mice immunized with the vaccine plus GSLS-NPs compared with that of other immunized mice, suggesting that intestinal defense against PEDV invasion was established. However, parenteral inoculation alone had little promotive effect on cellular immunity in the intestine. Except for the induction of Th2-biased activities, the administration of the mixture containing major saponins of GSLS elicited cellular immune responses in the spleen and MLNs similar to those in the vaccine-only group.

B cells are an essential part of the adaptive immune system, acting as drivers of humoral immune responses through T-cell-dependent and T-cell-independent pathways, and humoral immunity, especially intestinal IgA responses, plays an important role in immune protection against PEDV infection [26]. Our antibody results demonstrated that the ability of GSLS-NPs to improve antibody responses was not dose-dependent, and the optimal GSLS-NPs dose for adjuvant activities appeared to be 5 mg/kg, which elicited a stronger serum IgG antibody response, specifically responses involving the IgG1 and IgG2a isotypes, 14 days after a booster immunization than vaccination alone. The predominant IgG subclass changes progressively during an immune response, which is dependent on the pattern of cytokines secreted by $CD4^+$ T cells [15]. In mice, IL-10 favors IgG1 isotype (Th2 type) production by activated B cells, while IFN-γ promotes IgG2a production (Th1 type). The results indicated that GSLS-NPs stimulated unbiased Th1 and Th2 immune responses in mice, consistent with the cytokine responses. Furthermore, the numbers of IgA-secreting plasma cells in the mucosa of the duodenum, jejunum and ileum were increased by the additional administration of GSLS-NPs, a result corresponding to the finding of more abundant IgA antibodies in all segments of the small intestine. Surprisingly, the large number of IgA-secreting cells induced by GSLS-NPs in the ileum did not produce sufficient IgA antibodies, the level of which was higher than only that in the vaccine-only group but similar to those in other immunized mice. As expected, the additional administration of GSLS-NPs compared to the other treatments enhanced the levels of neutralizing antibodies in the serum and intestine, which are critical for protection against PEDV infection, especially those in the intestine. Interestingly, parenteral inoculation alone could induce a PEDV-specific serum IgG response with balanced IgG1 and IgG2a subtypes and an intestinal IgA response limited to the duodenum but had no effect on IgA antibodies in the jejunum and ileum, which are the main invasion and replication sites of PEDV. Moreover, parenteral inoculation induced neutralizing antibodies only in the serum, not in the local mucosa. The aqueous solution of GSLS preferentially enhanced the PEDV-specific IgG1 isotype response in the serum, which correlated with the increased induction of IL-10 in the spleen. However, it was surprising that the aqueous solution had a rare enhancing effect on the serum total IgG antibody response and local IgA antibody response compared to the vaccine alone, which led to a limited effect on serum and intestinal neutralizing antibodies.

5. Conclusions

In this study, we successfully developed a safe and efficient well-characterized nanoparticle delivery system for GSLS that offers a promising strategy for improving the efficacy of parenteral vaccines against PEDV. In addition to enhancing systemic cellular and humoral immune responses, oral administration of GSLS-NPs at the dose of 5 mg/kg before parenteral inoculation potently enhanced the phagocytic function and activation of DCs, expanded the activities of effector and memory T cells, and increased PEDV-specific IgA and neutralizing antibodies in the local intestinal mucosa, showing great advantages over the aqueous solution of GSLS. Therefore, GSLS-NPs could be an attractive mucosal adjuvant for parenteral vaccines commonly used in the field to prevent PEDV infection.

Author Contributions: Conceptualization, F.S. and X.Y.; methodology, F.S.; software, F.S.; validation, X.Y.; formal analysis, L.X., Y.X. and W.X.; investigation, F.S., L.X., Y.X., J.L. and S.Y.; resources, F.S. and B.Y.; data curation, Y.X. and B.Y.; writing—original draft preparation, F.S.; writing—review and editing, F.S. and X.Y.; visualization, F.S.; supervision, X.Y.; project administration, F.S. and X.Y.; funding acquisition, F.S. All authors have read and agreed to the published version of the manuscript.

Funding: This work was supported by the Zhejiang Province Public Welfare Technology Application Research Project (Agriculture and Rural Project) (Grant No. LGN21C180001).

Institutional Review Board Statement: All experiments related to animal care and use were approved by the Animal Care Committee of the Zhejiang Academy of Agricultural Sciences in accordance with the recommendations of the National Institutes of Health Guide for the Care and Use of Laboratory Animals (Ethics protocol No. 2020001).

Informed Consent Statement: Not applicable.

Data Availability Statement: All data generated or analyzed during the current study are available from the corresponding author on reasonable request.

Conflicts of Interest: The authors declare no conflict of interest.

References

1. Jung, K.; Saif, L.J.; Wang, Q. Porcine epidemic diarrhea virus (PEDV): An update on etiology, transmission, pathogenesis, and prevention and control. *Virus Res.* **2020**, *286*, 198045. [CrossRef] [PubMed]
2. Langel, S.N.; Paim, F.C.; Lager, K.M.; Vlasova, A.N.; Saif, L.J. Lactogenic immunity and vaccines for porcine epidemic diarrhea virus (PEDV): Historical and current concepts. *Virus Res.* **2016**, *226*, 93–107. [CrossRef]
3. Wang, Q.; Vlasova, A.N.; Kenney, S.P.; Saif, L.J. Emerging and re-emerging coronaviruses in pigs. *Curr. Opin. Virol.* **2019**, *34*, 39–49. [CrossRef]
4. Suda, Y.; Miyazaki, A.; Miyazawa, K.; Shibahara, T.; Ohashi, S. Systemic and intestinal porcine epidemic diarrhea virus-specific antibody response and distribution of antibody-secreting cells in experimentally infected conventional pigs. *Vet. Res.* **2021**, *52*, 2. [CrossRef]
5. Zhai, L.; Wang, Y.; Yu, J.; Hu, S. Enhanced immune responses of chickens to oral vaccination against infectious bursal disease by ginseng stem-leaf saponins. *Poult. Sci.* **2014**, *93*, 2473–2481. [CrossRef]
6. Yu, J.; Shi, F.S.; Hu, S. Improved immune responses to a bivalent vaccine of Newcastle disease and avian influenza in chickens by ginseng stem-leaf saponins. *Vet. Immunol. Immunopathol.* **2015**, *167*, 147–155. [CrossRef] [PubMed]
7. Li, R.; Ma, Y.; Zhai, L.; Lu, Y.; Chi, X.; Wu, J.; Hu, S. Enhanced immune response to foot-and-mouth disease vaccine by oral administration of ginseng stem-leaf saponins. *Immunopharmacol. Immunotoxicol.* **2016**, *38*, 257–263. [CrossRef] [PubMed]
8. Marciani, D.J. Vaccine adjuvants: Role and mechanisms of action in vaccine immunogenicity. *Drug Discov. Today* **2003**, *8*, 934–943. [CrossRef]
9. Zhang, W.; Wang, X.; Zhang, M.; Xu, M.; Tang, W.; Zhang, Y.; Lu, L.; Gao, J.; Gao, S. Intranasal Delivery of Microspheres Loaded with 20 (R)-ginsenoside Rg3 Enhances Anti-Fatigue Effect in Mice. *Curr. Drug Deliv.* **2017**, *14*, 867–874. [CrossRef]
10. Canadas, C.; Alvarado, H.; Calpena, A.C.; Silva, A.M.; Souto, E.B.; Garcia, M.L.; Abrego, G. In vitro, ex vivo and in vivo characterization of PLGA nanoparticles loading pranoprofen for ocular administration. *Int. J. Pharm.* **2016**, *511*, 719–727. [CrossRef]
11. Du, G.; Hathout, R.M.; Nasr, M.; Nejadnik, M.R.; Tu, J.; Koning, R.I.; Koster, A.J.; Slutter, B.; Kros, A.; Jiskoot, W.; et al. Intradermal vaccination with hollow microneedles: A comparative study of various protein antigen and adjuvant encapsulated nanoparticles. *J. Control. Release* **2017**, *266*, 109–118. [CrossRef] [PubMed]
12. Gu, P.; Wusiman, A.; Wang, S.; Zhang, Y.; Liu, Z.; Hu, Y.; Liu, J.; Wang, D. Polyethylenimine-coated PLGA nanoparticles-encapsulated Angelica sinensis polysaccharide as an adjuvant to enhance immune responses. *Carbohydr. Polym.* **2019**, *223*, 115128. [CrossRef] [PubMed]
13. Su, Y.; Zhang, B.; Sun, R.; Liu, W.; Zhu, Q.; Zhang, X.; Wang, R.; Chen, C. PLGA-based biodegradable microspheres in drug delivery: Recent advances in research and application. *Drug Deliv.* **2021**, *28*, 1397–1418. [CrossRef] [PubMed]
14. Mueller, S.N. Spreading the load: Antigen transfer between migratory and lymph node-resident dendritic cells promotes T-cell priming. *Eur. J. Immunol.* **2017**, *47*, 1798–1801. [CrossRef] [PubMed]
15. Su, F.; Wu, Y.; Li, J.; Huang, Y.; Yu, B.; Xu, L.; Xue, Y.; Xiao, C.; Yuan, X. Escherichia coli Heat-Labile Enterotoxin B Subunit Combined with Ginsenoside Rg1 as an Intranasal Adjuvant Triggers Type I Interferon Signaling Pathway and Enhances Adaptive Immune Responses to an Inactivated PRRSV Vaccine in ICR Mice. *Vaccines* **2021**, *9*, 266. [CrossRef] [PubMed]
16. Reed, L.J.; Muench, H. A simple method of estimating fifty per cent endpoints. *Am. J. Epidemiol.* **1938**, *27*, 493–497. [CrossRef]
17. Sahu, R.; Dixit, S.; Verma, R.; Duncan, S.A.; Coats, M.T.; Giambartolomei, G.H.; Singh, S.R.; Dennis, V.A. A nanovaccine formulation of Chlamydia recombinant MOMP encapsulated in PLGA 85:15 nanoparticles augments CD4(+) effector (CD44(high) CD62L(low)) and memory (CD44(high) CD62L(high)) T-cells in immunized mice. *Nanomedicine* **2020**, *29*, 102257. [CrossRef] [PubMed]
18. Jain, R.A. The manufacturing techniques of various drug loaded biodegradable poly(lactide-co-glycolide) (PLGA) devices. *Biomaterials* **2000**, *21*, 2475–2490. [CrossRef]
19. Zhang, W.; Xu, L.; Zhang, X.; Xu, J.; Jin, J.O. CD8alpha(-) conventional dendritic cells control Vbeta T-cell immunity in response to Staphylococcus aureus infection in mice. *Immunology* **2020**, *159*, 404–412. [CrossRef] [PubMed]
20. Macri, C.; Pang, E.S.; Patton, T.; O'Keeffe, M. Dendritic cell subsets. *Semin. Cell Dev. Biol.* **2018**, *84*, 11–21. [CrossRef] [PubMed]

21. Song, N.; Sengupta, S.; Khoruzhenko, S.; Welsh, R.A.; Kim, A.; Kumar, M.R.; Sonder, S.U.; Sidhom, J.W.; Zhang, H.; Jie, C.; et al. Multiple genetic programs contribute to CD4 T cell memory differentiation and longevity by maintaining T cell quiescence. *Cell Immunol.* **2020**, *357*, 104210. [CrossRef] [PubMed]
22. Wang, T.; Shen, Y.; Luyten, S.; Yang, Y.; Jiang, X. Tissue-resident memory CD8(+) T cells in cancer immunology and immunotherapy. *Pharmacol. Res.* **2020**, *159*, 104876. [CrossRef] [PubMed]
23. Esterhazy, D.; Loschko, J.; London, M.; Jove, V.; Oliveira, T.Y.; Mucida, D. Classical dendritic cells are required for dietary antigen-mediated induction of peripheral T(reg) cells and tolerance. *Nat. Immunol.* **2016**, *17*, 545–555. [CrossRef] [PubMed]
24. Mayer, J.U.; Brown, S.L.; MacDonald, A.S.; Milling, S.W. Defined Intestinal Regions Are Drained by Specific Lymph Nodes That Mount Distinct Th1 and Th2 Responses Against Schistosoma mansoni Eggs. *Front. Immunol.* **2020**, *11*, 592325. [CrossRef] [PubMed]
25. Tezuka, H.; Ohteki, T. Regulation of IgA Production by Intestinal Dendritic Cells and Related Cells. *Front. Immunol.* **2019**, *10*, 1891. [CrossRef]
26. Guo, H.; Yao, Z.; Chen, L.; Li, L.; Li, Y.; Wang, Y.; Li, T.; Wang, H.; Sun, L.; Hao, D.; et al. Humoral immune responses in piglets experimentally infected with a field strain of porcine epidemic diarrhea virus. *Vet. Microbiol.* **2020**, *246*, 108742. [CrossRef] [PubMed]

vaccines

Review

Innovation in mRNA Vaccines and RNAi via Protein Nanocages

Sohrab Ahmadivand

Faculty of Veterinary Medicine, Ludwig-Maximilians-Universität München, 80539 Munich, Germany; sohrab.ahmadivand@lmu.de

Abstract: Self-assembling protein nanocages (SAPNs) are distinct natural structures formed by the self-assembly of identical subunits, providing a highly efficient platform and a novel strategy for vaccine development and RNAi therapy. Their internal cavity allows for precise cargo encapsulation, while the externally modifiable surface supports multivalent antigen presentation, thereby enhancing stability, targeted delivery, and immune activation. In addition to serving as stable subunit vaccines with multivalent antigen display, SAPNs can be incorporated into mRNA vaccines (SAPN-RNA vaccines) by pre-fusing with the antigen. This strategy stabilizes secreted antigenic proteins with prolonged presentation to the immune system, and improves vaccine efficacy while reducing off-target effects and minimizing required doses. Additionally, SAPNs can overcome cellular uptake barriers, enhance DNA vaccine efficacy, and enable the co-delivery of antigens and adjuvants. Functionalization with adjuvants or targeting ligands further improves their immunostimulatory properties and specificity. The SAPN-RNAi strategy optimizes siRNA delivery by promoting lysosomal escape, enhancing targeted uptake, and protecting siRNA from degradation through SAPN encapsulation. This review examines the structural and functional properties of protein nanocages and their applications in vaccine design and RNAi delivery, emphasizing their synergistic effects, and exploring current progress, challenges, and future directions. In conclusion, SAPNs represent a versatile multifunctional platform with broad applicability across subunit, mRNA and DNA vaccines, adjuvant co-delivery, and RNAi therapeutics, with significant potential against viral infections.

Keywords: protein nanocages; siRNA delivery; mRNA vaccines; vaccine platform; ferritin; virus

1. Introduction

RNA biopharmaceuticals are an emerging class of therapies showing significant promise for the treatment and prevention of a wide range of diseases. mRNA vaccines, in particular, offer great potential for both human and animal diseases [1,2]. However, to maximize their efficacy, it is crucial to modulate immunogenicity, developing safe and efficient delivery systems, and stabilizing RNA against degradation. Additionally, cold-chain requirements and biocompatibility concerns hinder large-scale clinical deployment [3, 4].

Similarly, RNA interference (RNAi) offers a targeted and highly specific therapeutic strategy to manage viral infections and other diseases by silencing target gene expression. This flexibility provides a distinct advantage over traditional antiviral treatments. Nevertheless, RNAi therapies face significant barriers in clinical translation, primarily related to toxicity, off-target effects, and inefficient delivery to target cells [5,6].

The multivalent display of structurally defined antigenic epitopes on the surface of self-assembling nanostructures such as virus-like particles (VLPs) and self-assembling

 https://doi.org/10.3390/vaccines13060653

protein nanocages (SAPNs), constitutes a highly effective strategy for enhancing antigen stability, immunogenicity, and the induction of robust, targeted immune responses. VLPs, derived from viral structural proteins, mimic the size, shape, and surface organization of native virions but are rendered non-infectious due to the absence of viral genomic material. In contrast, SAPNs are naturally occurring, highly ordered, biocompatible protein nanostructures formed through the spontaneous self-assembly of identical subunits. The protein nanocages exhibit exceptional physicochemical properties, including high thermal and pH stability, intrinsic biodegradability, and reversible assembly and disassembly under physiological conditions [7,8]. These characteristics underpin the broad applicability of SAPNs across various biomedical fields, including targeted drug delivery, vaccine development, molecular diagnostics, and therapeutic interventions, particularly when integrated with advanced molecular imaging techniques [9].

Their unique architecture features an internal cavity capable of precise cargo encapsulation, coupled with a modifiable surface that enables multivalent display of ligands or antigenic epitopes [10,11]. This dual functionality facilitates targeted delivery, controlled release, and potent immune activation, positioning SAPNs as versatile platforms for vaccine development and RNAi therapeutics [11–13]. Unlike VLPs, which mimic viral structures but require lipid membranes leading to structural variability, SAPNs serve as robust scaffolds that can replace viral envelopes, overcoming challenges associated with VLP assembly and characterization [14,15].

SAPNs were initially explored for subunit vaccines to improve antigen stability and immunogenicity, while supporting DIVA (Differentiating Infected from Vaccinated Animals) strategies and rapid vaccine design [16,17]. Fusing antigens to protein nanocages enhances their solubility in prokaryotic expression systems, likely by promoting proper folding, preventing aggregation, and shielding hydrophobic regions [15]. This improved structural presentation not only facilitates expression but also contributes to enhanced vaccine safety, as the biocompatible nature of SAPNs helps mitigate the antigen-associated toxicity often observed with unstable or poorly expressed proteins [15,18].

In mRNA vaccines, SAPNs fusion (referred to here as SAPN-RNA vaccines) allow for multivalent antigen presentation, which stabilizes antigenic proteins, improves intracellular trafficking, and potentially prolongs antigen expression. These effects lead to more potent and sustained immune responses, including high-affinity neutralizing antibodies at lower doses [19,20]. Similarly, these protein nanocages address challenges related to poor antigen expression and inefficient cellular uptake in DNA vaccines, thereby enhancing their overall immunogenicity [21].

Clinical trials of SAPN-based subunit vaccines targeting influenza, HIV, EBV, and SARS-CoV-2 have demonstrated strong immunogenicity and favorable safety profiles [16,22]. SAPN-RNA vaccines, such as the mRNA (NCT05001373 and NCT05414786) and saRNA HIV candidates utilizing lumazine synthase nanocages, further underscore the promise of this approach in next-generation vaccine development [23,24].

Beyond structural antigen display, SAPNs enable co-delivery of antigens and adjuvants [25–27]. This feature enhances immune responses while minimizing off-target effects and systemic toxicity. The multivalent surface display of adjuvant proteins, such as Flagellin, a TLR5 agonist, potentiates immune activation, while encapsulating small/molecule adjuvants within the internal cavity ensures sustained release, thereby improving vaccine efficacy [10,25–27].

SAPNs also enhance RNAi delivery by overcoming biological barriers such as instability, nuclease degradation, and endosomal entrapment [28]. Their reversible disassembly (e.g., ferritin) and internal targeting domains (e.g., encapsulin TPs) allow for efficient cargo loading [8,10,11]. Recent studies have shown that SAPNs can protect siRNAs from degradation, promote lysosomal escape, and enhance targeted delivery, resulting in improved gene silencing efficiency [13,29–31].

These multifunctional capabilities position SAPNs as a novel and promising platform for RNAi therapy and vaccine development, particularly for mRNA vaccines. This review comprehensively analyzes the structural features, functional mechanisms, and biomedical applications of SAPNs in mRNA, DNA, and subunit vaccines, RNAi delivery, and adjuvant functionalization. It also addresses key design principles, technical challenges, and innovative strategies to facilitate the clinical translation of SAPNs, with a focus on viral infections.

2. Self-Assembling Protein Nanocages (SAPNs)

2.1. Characteristics

Self-assembling protein nanocages (SAPNs), naturally occurring in various organisms, have emerged as promising platforms for vaccine development and therapeutic protein delivery, including RNAi. These nanoscale structures (10–100 nm) closely mimic the size and shape of viruses, enabling the multivalent display of antigens in highly ordered architectures with up to 240 identical subunits [12,20]. They exhibit exceptional thermal and pH stability, monodispersity, uniform size, biodegradability, biocompatibility, and reversible spontaneous assembly/disassembly [16,22]. Therapeutic molecules can be encapsulated within SAPNs, while engineered surfaces enable antigen display and targeted delivery. Modifying intersubunit interactions allows for modulation of self-assembly and controlled molecular release [32].

The current studied protein nanocages include ferritin, lumazine synthase (LuS), Encapsulin, E2 protein (E2p), and small heat-shock proteins (sHSP), each possessing distinct structural and physiochemical characteristics. They are generally non-immunogenic, and those derived from bacterial sources are preferred to mitigate the risk of autoimmunity [15,33].

Ferritin is a conserved intracellular iron-storage protein and a well-characterized SAPN used in vaccine and drug delivery [17,34]. In mammals, it consists of heavy (21 kD) and light (19 kD) chains, while teleost fish have an additional medium chain (20 kD). Bacterial and plant ferritins contain a single subunit resembling the vertebrate heavy chain [35,36]. Despite up to 80% sequence variation, ferritin maintains a conserved structure, assembling into an octahedral (432) symmetry nanocage of 24 subunits with internal and external diameters of 8 nm and 12 nm [34]. Toxicological studies show ferritin nanoparticles are biocompatible, with no cytotoxicity observed in zebrafish liver (ZFL) cells up to 100 μg/mL [15] and no adverse effects in vivo at imaging doses [35]. Ferritin also exhibits exceptional thermal and pH stability, tolerating temperatures up to 100 °C and pH ranges from 3 to 10, with reversible disassembly below pH 3.4 [8,36]. This structure provides dual functionality: encapsulating internal cargo and displaying multivalent antigens on the external surface via the N-terminus, with *Helicobacter pylori* being the most studied origin of some in clinical trial [17].

Lumazine synthase (LuS) is a self-assembling protein nanocage involved in riboflavin biosynthesis, forming capsids composed of 60 subunits arranged into 12 pentameric units [37,38]. With both N- and C-termini exposed on the outer surface and a greater number of subunits compared to ferritin, LuS facilitates efficient antigen attachment, potentially enhancing antigen presentation and immunogenicity. Among the LuS variants, those from *Aquifex aeolicus* (AaLuS) and *Bacillus subtilis* (BsLuS) have been most studied. AaLuS, one of the most thermostable proteins known, has a melting temperature of 120 °C, making it ideal for high-temperature applications [37]. While BsLuS forms smaller capsids, AaLuS can assemble larger shells composed of 180 or more subunits, offering greater versatility in encapsulation [38]. Despite these promising features, LuS remains underexplored, with research primarily focused on pathogens like HIV, SBV, SARS-CoV-2, and RVFV [12,20].

Encapsulins are another self-assembling, icosahedral protein nanocage found in bacteria and archaea, typically composed of 60 subunits, with some variants, like those in *Quasibacillus thermotolerans*, forming larger cages of up to 240 subunits. These protein shells are highly stable under varying pH and temperature conditions, ranging in size from 20 to 42 nm, and can incorporate diverse cargos, including proteins, therapeutic RNA, and adjuvants [39,40]. Encapsulins are classified into four families based on cargo type and operon structure, with computational analysis, of which Family 1 is the most extensively characterized [41]. Encapsulin shell protomers share structural similarity with the HK97 bacteriophage capsomere gp5 and comprise three conserved domains: the axial (A-domain), peripheral (P-domain), and extended loop (E-loop), with the E-loop regulating subunit arrangement and triangulation. Cargo proteins are attached via targeting peptides (TPs) at the C-terminus or, in some cases, the N-terminus, as seen in Firmicute bacteria [42]. Encapsulins can undergo reversible disassembly and reassembly, with the smallest T = 1 (60-mer) T. maritima encapsulin nanocage demonstrating the most stable and effective reassembly after disassembly. It can withstand up to 3 M GuHCl, pH 12, and 90 °C, with reversible unfolding induced by basic pH or GuHCl, making it suitable for encapsulation applications. However, repeated cycles of disassembly and reassembly may result in significant loss of structural integrity [8]. Their simplicity enables efficient production in *E. coli*, and by fusing antigen-encoding genes to the C-terminal region, they have shown potential as a vaccine platform against infectious pathogens such as influenza [43], SARS-CoV-2 [44], and rotavirus [45].

Dihydrolipoyl Acetyltransferase (E2 protein; E2p) from *Geobacillus stearothermophilus* is another protein nanocage that forms a 27 nm icosahedral structure with 60 subunits and 532-fold symmetry [46]. However, some variant forms of octahedral E2p consist of 24 identical polypeptide chains arranged with 432 symmetries [46].

As a key component of the pyruvate dehydrogenase complex, E2p transfers the acetyl group to CoA. Gram-negative bacteria like *E. coli* form 24-subunit structures, while yeast and mammals form the 60-subunit pentagonal dodecahedron, with catalytic domains creating cavities for binding lipoyl groups and CoA [47]. E2p nanocages offer high stability, ease of expression, and molecular encapsulation. Dalmau et al. [48] engineered E2p for expression in E. coli, demonstrating exceptional thermostability (unfolding onset: 81.1 $\pm$ 0.9 °C, midpoint: 91.4 $\pm$ 1.4 °C) while preserving its dodecahedral structure. Peng et al. [49] also developed a pH-responsive E2p nanocage incorporating a switchable GALA peptide, which disassembles at pH 7.0 and reassembles at pH 4.0 in a reversible manner. These modifications enable non-native functionality and cargo encapsulation. The N-terminus of truncated E2p, retaining only the catalytic domain, remains accessible for antigen fusion, facilitating the presentation of target antigens. E2p has been explored as a platform for biomedical applications, successfully displaying multivalent antigens from infectious viruses such HIV-1 gp140, Ebola glycoprotein, modified SARS-CoV-2 RBD, and has even shown promise in malaria transmission-blocking vaccines [50]. Its unique structure and larger size enhance E2p versatility as both an antigen presentation and therapeutic cargo delivery platform.

Small heat-shock proteins (sHSPs) are found in a wide variety of organisms and are unique for being ATP-independent. sHSPs consist of three domains: the N-terminal domain (NTD), the α-crystallin domain (ACD), and the C-terminal domain (CTD), which regulate the assembly of monomers (13–43 kDa) into large oligomers (up to 40 subunits) [51]. They form nanoparticles ranging from 12 to 20 nm in diameter, typically composed of 12 or 24 subunits [52]. They can be easily modified for antigen display by exposing either the N-terminus (e.g., M. tuberculosis HSP16.3) or the C-terminus (e.g., *M. jannaschii* HSP16.5) [53].

These nanocages also exhibit intrinsic protective effects against respiratory viruses, making them promising candidates for vaccine platforms [54]. The sHsp nanocages have eight pores (3 nm and 1.7 nm) for molecular trafficking and exhibit high stability [55]. The archaeal sHsp from *M. jannaschii* forms a 12 nm nanocage, tolerating up to 70 °C and pH 5–11 [56]. sHSP nanocages have diverse biomedical applications. However, their intrinsic immunogenicity requires careful evaluation to avoid autoimmunity, particularly with homologous applications [57].

Given these characteristics, SAPNs represent an efficient strategy to enhance the immunity and stability of subunit and nucleic acid-based vaccines, as well as for therapeutic delivery (Tables 1 and 2). Further studies are essential to explore and characterize additional protein nanocages with optimized designs, to assess long-term efficacy and facilitate regulatory approval, ensuring their widespread adoption in disease control strategies.

Table 1. Characteristics of main Self-Assembling Protein Nanocages (SAPNs) used in Vaccine Development and RNAi Delivery.

SAPNs	Biological Function	Origin	Cavity Size (nm)	Outer Size (nm)	Subunits	Subunit MW	Stability	Ref.
Ferritin	Iron storage and transport	*Helicobacter pylori*	8	12	24	20 kDa	pH: 3–12; 80–100 °C	[17]
Lumazine Synthase (LuS)	Riboflavin biosynthesis	*Aquifex aeolicus*	9	15.4	60	17.8 kDa	120 °C	[38]
		Bacillus subtilis	16	30	180	16.3 kDa	Up to pH: 10 and 80 °C	[37]
Encapsulin	Oxidative stress response through encapsulation of other related proteins	*Thermotoga maritima*	20	24	60	31 kDa	pH: 12, <3 M GuHCl; 90 °C	[8,58]
		Pyrococcus furiosus	22	32	180	39 kDa	<4 M GuHCl; 80 °C	[8,58]
		Quasibacillus thermotolerance	36	42	240	32.2 kDa	<1 M GuHCl; 40 °C	[8,58]
E2p	Enzyme complex for metabolic pathways	*Geobacillus stearothermophilus*	12 pores of 5 nm	27	60	41 kDa	80 °C	[48]
sHSP	Response to cellular stress	*Methanococcus jannaschii*	6	12	24	16.5 kDa	70 °C and pH 5–11	[56]

Table 2. Innovative Strategies Employing Self-Assembling Protein Nanocages (SAPNs) in Vaccine Design and RNAi Therapy.

Approach	Developing Method	Key Advantages Compared to Conventional Method
SAPN-Subunit Vaccines	Fuse the antigen to the N- or C-termini of SAPN, express and purify the fusion protein to devlop the subunit vaccine.	Multivalent antigen display, enhanced solubility, stability and biocompatibility, cellular uptake, heightened immunogenicity, targeted delivery, and customized vaccine design.
SAPN-RNA Vaccines	Fuse the antigen to the N- or C-termini of SAPN, produce the antigen via RNA vaccine platforms (mRNA, saRNA, taRNA).	Promotes multivalent antigen display, enhances stability and immunogenicity of secreted proteins with prolonged expression, reduces off-target effects, and minimizes required doses.
SAPN-DNA Vaccines	Fuse the antigen to the N- or C-termini of SAPN, clone the fusion construct into a eukaryotic DNA plasmid.	Multivalent antigen display that stabilizes the antigenic protein and enhances expression, cellular uptake, antigen presentation, and immune responses.
SAPN-Adjuvant	Fuse a protein adjuvant to the N- or C-termini of SAPN, co-deliver with the antigen or encapsulate within SAPN.	Enables co-delivery of adjuvant with antigen to same cell (e.g., APC), targeted and enhanced biocompatibility, stability and immune responses with multivalent display of protein adjuvants, protection and controlled release of encapsulated small-molecule adjuvants.
SAPN-RNAi Delivery	Encapsulate siRNA or miRNA within SAPN using the disassembly/reassembly of SAPN.	Enables targeted gene silencing, controlled release, enhanced delivery efficiency, reduced immunogenicity, and scalable production.

2.2. Production

Protein nanocages provide a versatile platform for antigen presentation, with most architectures orienting the C- and N-termini toward the interior and exterior, respectively. Antigen display on the nanocage surface has been successfully implemented through genetic fusion, chemical conjugation, and protein ligation systems [17]. While genetic fusion to these termini enables precise antigen incorporation, optimizing structural design is crucial to maintaining nanoparticle stability and assembly integrity, particularly for larger or more complex antigens. Chemical conjugation, though effective, is a complex procedure that requires specific conditions and reagents, potentially resulting in heterogeneous decoration that hinders efficient immune presentation [59].

The SpyTag-SpyCatcher system, a novel technology for vaccine development, offers a solution by enabling highly efficient coupling of antigens to the platform while preserving their native conformation. This system utilizes the Streptococcus pyogenes fibronectin-binding protein FbaB, split into SpyCatcher (113 aa) and SpyTag (13 aa), to form an irreversible peptide bond between the NP and antigen, regardless of antigen size [60].

The expression and purification of the protein nanocages are critical steps to ensure high-quality nanoparticles with consistent structural and immunological properties. SAPNs can be produced using bacterial, mammalian, or insect cells, depending on the complexity of the protein and required post-translational modifications. Mammalian systems, such as HEK293F and ExpiCHO, ensure proper folding and glycosylation, which are crucial for antigenicity in ferritin-based nanoparticles displaying viral antigens like HIV gp120 or SARS-CoV-2 spike proteins [12]. Baculovirus expression systems provide high yields with appropriate modifications, enabling the successful development of SAPN-based vaccines, including ferritin, Lus, and E2p [61,62]. Bacterial expression systems offer a cost-effective, efficient platform for SAPN production, with high yields and rapid synthesis. Genetic fusion with nanocage platforms like ferritin enhances solubility and stability, while optimization strategies such as molecular chaperones and RNA-Interaction Domains (RIDs) improve protein folding and assembly efficiency, making bacteria an ideal choice for scalable SAPN vaccine development [63].

Purification typically involves affinity chromatography, size-exclusion chromatography, or ultracentrifugation to isolate fully assembled nanoparticles. For instance, ferritin NPs displaying SARS-CoV-2 spike proteins are purified via nickel-affinity chromatography followed by size-exclusion to remove unassembled or aggregated particles, ensuring a homogenous population with optimal antigen presentation capabilities [12]. In bacterial systems, recovery from inclusion bodies may require denaturation and refolding, followed by dialysis to remove salts and stabilize nanoparticles for downstream applications. Optimizing factors like protein concentration, pH, ionic strength, and temperature during expression and purification is essential to achieve high-yield assembly of SAPNs, as demonstrated in ferritin nanoparticles, which require divalent cations like Fe^{2+} for proper self-assembly [14].

Recent advances focus on optimizing the design and production of SAPN-based vaccines to enhance their scalability and efficacy. Tyrosinase-mediated bioconjugation enables site-specific attachment of antigens such as the SARS-CoV-2 spike receptor binding domain to ferritin nanoparticles while preserving antigen integrity and allowing controlled multivalent display [64]. Concurrently, the strategic insertion of polyhistidine tags into structurally permissive loops of *H. pylori* ferritin facilitates efficient immobilized metal affinity chromatography (IMAC)-based purification, with purification yield dependent on tag location and antigen size—the His-4 site proving optimal for spike protein constructs [65]. These innovations represent critical steps toward the streamlined development and manufacturing of SAPN-based vaccines.

Beyond subunit vaccines (SAPN-Subunit), SAPNs can also be incorporated into nucleic acid vaccines, such as RNA and DNA vaccines, by genetically fusing the antigen, resulting in the secretion of stable antigenic proteins from host cells in a multivalent display manner that enhances immune responses. SAPNs further enable the co-delivery of protein antigens, adjuvants, RNA, and other cargo, either within the internal cavity or on the external surface.

Early strategies for internal cavity loading involved engineering the nanocage's C-terminus for cargo encapsulation. Encapsulins exemplify drug delivery by enabling in vivo co-packaging of proteins through selective mechanisms such as targeting domains (TDs), transporter peptides (TPs), or RNA-binding domains [11]. Other SAPN loading methods include biomineralization, passive diffusion, and disassembly/reassembly [66]. Biomineralization is effective for synthesizing metal oxide nanoparticles but has limited applicability for other drugs. Passive diffusion allows small molecules to enter ferritin without compromising structural integrity, whereas disassembly/reassembly enables loading of larger compounds by dissociating nanocages at pH 2 and reassembling at pH 7. Alternative methods, such as urea-induced (8M urea) or hydrostatic pressure (450 MPa) disassembly, provide controlled loading and release [66].

However, reassembly rates vary among nanocages, with ferritin showing the highest efficiency, while others, such as encapsulin (T = 3), may exhibit lower reassembly or lack the ability to reassemble completely [8]. Moreover, the recovery of reassembled proteins can be suboptimal [7].

The external surface of SAPNs can be modified for targeted delivery, allowing precise control over the location and timing of payload release or immune activation. Surface modifications such as the KALA peptide and L17E peptide have been employed to functionalize SAPNs for tumor therapy, thereby enhancing targeted delivery and cellular uptake [13,67]. To significantly enhance vaccine efficacy, one effective strategy is the incorporation of dendritic cell-targeting ligands such as CD40 and lymphotactin (XCL1). These ligands can be conjugated to SAPNs using the SpyTag-SpyCatcher system or chemical methods to facilitate targeted delivery. Together, these strategies highlight the versatility of SAPNs as multifunctional platforms for vaccine and therapeutic delivery (Figure 1).

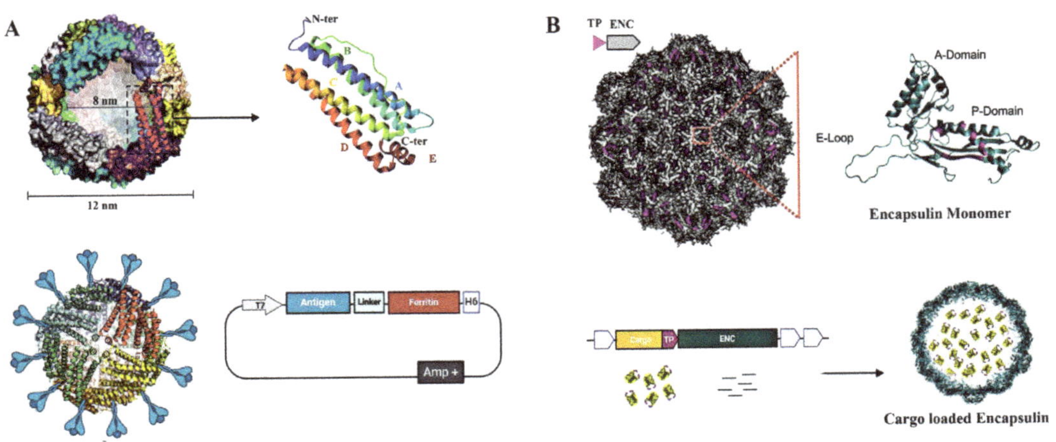

Figure 1. Self-assembling protein nanocages (SAPNs) for multivalent antigen display and cargo encapsulation. (A) Schematic representation of ferritin nanocages and a monomeric subunit. Ferritin assembles into a 24-subunit spherical cage with inner and outer diameters of approximately 8 nm and 12 nm, respectively. Multivalent antigen display can be achieved via genetic fusion of antigen-encoding sequences to the C-terminal region of ferritin subunits. (B) Schematic illustration of a core encapsulin operon and targeting peptide (TP)-mediated cargo encapsulation. Encapsulin monomers

self-assemble into icosahedral protein cages ranging from 20 to 42 nm in diameter, composed of 60, 180, or 240 identical subunits. A domain-level view of the encapsulin monomer is shown. Antigen display on the encapsulin surface can also be achieved through C-terminal genetic fusion. Created with BioRender.com.

3. Self-Assembling Protein Nanocages (SAPNs) as Vaccine Platform

3.1. SAPN-Subunit Vaccines

Subunit vaccines offer a safer alternative to traditional whole-virus vaccines by utilizing purified viral antigens, reducing the risks associated with live virus replication. Despite their safety, subunit vaccines can have variable efficacy, necessitating strong adjuvants and repeated administration. Additionally, due to GI instability, injection is required, making mass administration costly, labor-intensive, and stressful, particularly in veterinary medicine, where the vaccination of large animal populations presents significant challenges [68,69].

Through the SAPN-Subunit vaccine approach, self-assembling protein nanocages can be engineered for subunit vaccines by modifying the external surface for multivalent antigen display via genetic fusion, chemical conjugation, or protein ligation, with the potential for targeted delivery [16,17,70]. These vaccines are DIVA-compatible and can be produced in a variety of expression systems, including both prokaryotic and eukaryotic systems, for large-scale production. They outperform subunit vaccines by efficiently displaying complex antigenic proteins from enveloped viruses and rescuing insoluble antigens [15]. The approach also enhances the immunogenicity and stability of subunit vaccines through their repetitive nanostructure, facilitating both oral delivery and mass administration. The unique nanoscale size (<50 nm) and multivalent structure of SAPN-subunit vaccines mimic viral particles, promoting enhanced immune activation (Figure 2A). SAPNs improve antigen uptake by antigen-presenting cells (APCs) and the direct activation of naïve B cells, and enable efficient cross-linking of B-cell receptors (BCRs), inducing innate immune responses via pattern recognition receptors (PRRs) [71,72].

They engage MHC-I and MHC-II pathways, triggering both cellular and humoral responses with Tfh cell activation, which drive germinal center responses essential for generating durable, high-affinity antibody responses [19].

SAPN-Subunit vaccines have emerged as a promising approach with significant potential for human and veterinary applications. Ferritin nanocages, first used in H1N1 influenza vaccines in 2013, have since been developed for influenza, HIV, EBV, and SARS-CoV-2, with several candidates currently in human clinical trials, particularly following the SARS-CoV-2 pandemic in 2021 [17]. Notably, three ferritin-based influenza vaccines (NCT03186781, NCT04579250 and NCT03814720) have reached phase I trials, demonstrating favorable safety profiles and the ability to induce broadly neutralizing antibody responses [73–75].

For veterinary use, ferritin vaccines have been explored for economically significant swine respiratory viruses, including swine influenza virus, classical swine fever virus (CSFV) [76], porcine deltacoronavirus (PDCoV) [62], and porcine reproductive and respiratory syndrome virus (PRRSV) [77]. These vaccines offer strain cross-protection and induce stronger humoral and cellular immune responses compared to commercially available inactivated or subunit vaccines [17]. A ferritin-based nanovaccine displaying the H1HA stem region induced broad cross-reactive antibodies, providing full protection against lethal H5N1 in mice and partial protection in ferrets [78]. In comparison to conventional swine influenza vaccines, ferritin-based vaccines triggered stronger immune responses, including higher hemagglutinin inhibition titers and robust IgG production. This vaccine offered complete protection against H1N1 and reduced lung damage in mice and piglets, even against a heterologous strain (81% homology), with cross-protection against H1N1 and H3N2 in mice [79]. A self-assembled nanoparticle African swine fever (ASF) vaccine was designed to target dendritic cells by coupling a 24-mer ferritin scaffold with dominant ASFV epitopes (p72, CD2v, pB602L, p30) and fusing it to the chemokine XCL1 receptor via the SpyTag/SpyCatcher system. Compared to monomeric protein forms, the vaccine elicited significantly stronger T and B cell responses in mice, with ASFV-specific antibody responses sustained for over 231 days [80].

Recently, in the first study for non-mammalian viruses, the ferritin platform was evaluated against infectious hematopoietic necrosis virus (IHNV), an enveloped WOAH-notifiable rhabdovirus affecting aquaculture, demonstrating potential for controlling fish viruses and its high stability for oral mass delivery [15].

Lumazine synthase (LuS) and E2 protein nanocages have been investigated as platforms for presenting complex viral antigens in vaccines, enhancing neutralizing responses. For instance, an AaLuS-based Schmallenberg virus (SBV) vaccine provided 80% protection in mice and induced sterile immunity in cattle with a single dose [81]. When conjugated with the SARS-CoV-2 spike protein, the LuS vaccine elicited stronger neutralizing responses compared to the spike protein alone [82]. Additionally, the LuS vaccine activated VRC01-class B cells, fostering a broadly neutralizing response against HIV-1 [83]. The IAVI G001 trial (NCT03547245) is an important early-stage study in the HIV germline-targeting vaccine strategy, utilizing LuS nanocages. SARS-CoV-2 vaccines using ferritin and E2p nanocages displaying HR2-deleted glycine-capped spike (S2GΔHR2) proteins generated high neutralizing antibody titers and robust T cell responses, including Th1 and memory CD4+ cells. E2p-based vaccines, in particular, demonstrated improved immunogenicity at lower doses [84]. Furthermore, a LuS-subunit vaccine using the Gn head domain provided protection against viremia and clinical signs in mice and lambs infected with Rift Valley fever virus (RVFV), with complete protection achieved in lambs when conjugated to E2p nanocages [85].

Encapsulin, another SAPN, has shown promise for vaccines against rotavirus, influenza, and SARS-CoV-2. Encapsulin from Myxococcus xanthus, engineered with the SpyTag-SpyCatcher system to display 180 copies of the SARS-CoV-2 spike RBD, demonstrates strong antigenicity and thermotolerance. When adjuvanted with SWE, it elicits potent neutralizing antibody titers against multiple variants, including Omicron [44]. Encapsulin-based influenza vaccines (180 copies) outperformed ferritin formulations (24 copies), emphasizing the impact of nanocage architecture on immunogenicity [43]. For rotavirus, a 60-mer encapsulin vaccine induced strong VP8*-specific IgG responses and neutralizing antibodies, surpassing subunit vaccines in mice and balancing Th1/Th2 immune responses [45].

The study by Choi et al. [86] demonstrated that *T. maritima* encapsulin nanocages fused with the OT-1 peptide enabled effective antigen delivery to dendritic cells, resulting in the activation of antigen-specific cytotoxic $CD8^+$ T cells. In a melanoma mouse model, vaccination with OT-1-Encaps led to significant tumor suppression.

Small heat shock proteins (sHSPs) have also shown potential for durable antibody production. In mice, intranasal OVA-sHSP vaccines accelerated IgG1 and mucosal IgA responses, improving lung immunity and influenza challenge outcomes [54].

Overall, SAPN-subunit vaccines have demonstrated significant potential in both human and veterinary medicine, offering enhanced antigen presentation, strong immune responses, cross-protection, and stability for oral delivery. Beyond subunit vaccines, SAPNs can also be integrated into nucleic acid vaccine platforms (mRNA and DNA) and utilized for RNAi delivery, with multiple advantages (Figure 2 and Table 3).

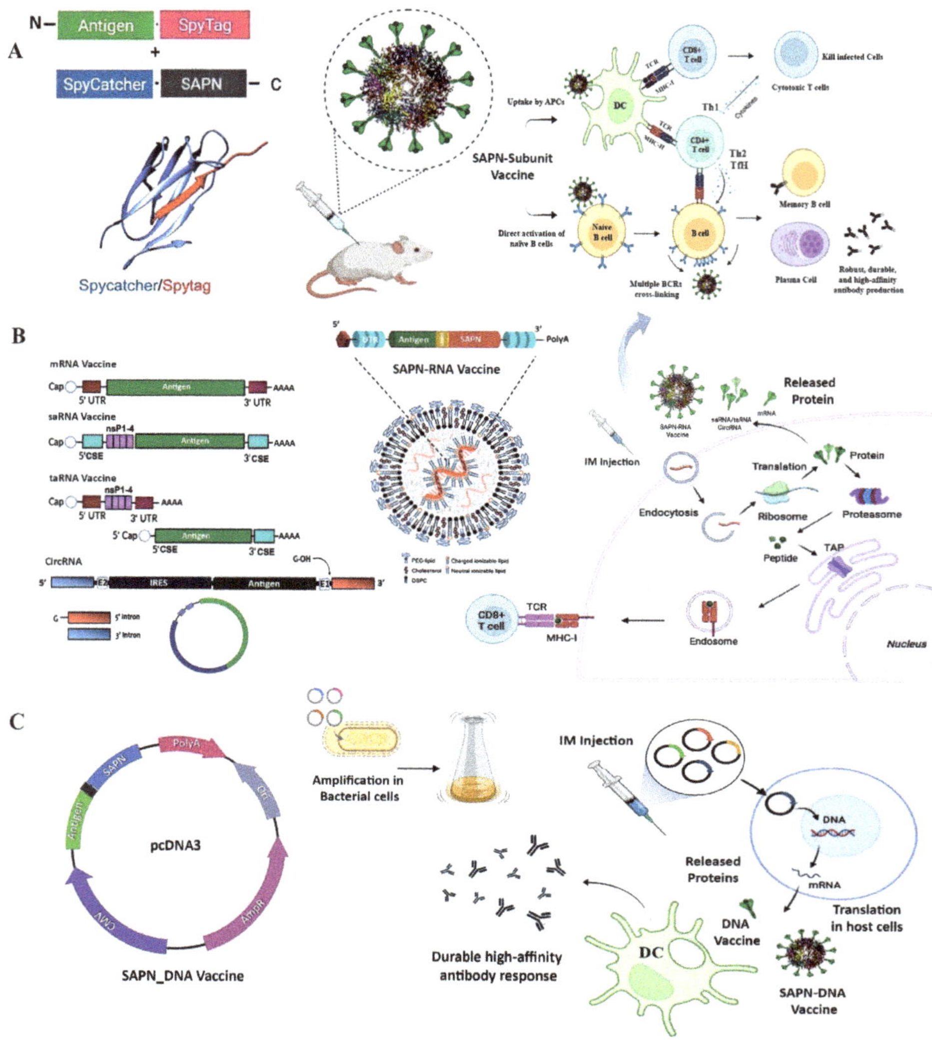

Figure 2. Self-assembling protein nanocages (SAPNs) as versatile platforms for subunit, RNA, and DNA vaccines. (**A**) SAPN-Subunit Vaccines: antigens are covalently linked to protein nanocages using the SpyTag/SpyCatcher system, enabling a highly ordered, multivalent display with enhanced stability. This promotes robust immune responses by directly activating naïve B cells through efficient BCR cross-linking and facilitating antigen uptake by antigen-presenting cells (APCs) such as such as dendritic cells (DCs), leading to strong cellular immunity and durable, high-affinity antibody production. (**B**) SAPN-RNA Vaccines: RNA platforms, including conventional mRNA, self-amplifying RNA (saRNA), trans-amplifying RNA (taRNA), and circular RNA (circRNA), encode SAPN-antigen fusion proteins. Delivered via lipid nanoparticles (LNPs), these RNAs are translated in situ, generating stabilized antigens that enhance immune presentation, cellular uptake, and overall vaccine efficacy while reducing off-target effects and required doses. (**C**) SAPN-DNA Vaccines: plasmid DNA encoding SAPN–antigen fusions is delivered intramuscularly (IM), leading to in situ antigen production within host cells. This forms multivalent antigen displays with enhanced stability, cellular uptake, and immunogenicity. The SAPN platform is based on lumazine synthase, which self-assembles into 60-subunit icosahedral nanocages (~16 nm in diameter; PDB ID: 8F25). Created with BioRender.com.

Table 3. Self-Assembling Protein Nanocages (SAPNs) for Nucleic Acid Vaccine (mRNA and DNA) Development and RNAi Delivery.

Platform	SAPN Type	Mode of Incorporation	Diseases/Pathogen	Species/Cells	Key Finding	Ref.
mRNA vaccine	Ferritin	Genetic Surface Fusion	MERS-CoV	Mice; NHPs and Alpacas	This vaccine is temperature- and pH-stable, eliciting robust antibody titers in BALB/c mice and NHPs (non-human primates) with cross-clade neutralization (A, B, C). Complete protection in alpacas.	[19]
mRNA vaccine	Ferritin	Genetic Surface Fusion	HIV-1 Env	Mice	Induced neutralizing antibody responses in mice capable of neutralizing heterologous HIV-1 isolates.	[87]
mRNA vaccine	Ferritin	Genetic Surface Fusion	SARS-CoV-2 (RBD)	Mice and Hamsters	Elevated specific IgG levels and significantly higher neutralizing antibody titers correlated with improved protection in hamsters.	[88]
mRNA vaccine	LuS	Genetic Surface Fusion	Rotavirus (P2-VP8*)	Mice and Guinea Pigs	Induced the highest specific IgG titers compared to conventional mRNA and subunit vaccine.	[89]
mRNA vaccine	LuS	Genetic Surface Fusion	SARS-CoV-2 (RBD)	Mice	Strong neutralizing antibody responses and protection against the Delta variant.	[90]
mRNA vaccine	LuS	Genetic Surface Fusion	HIV (gp120)	Human	The vaccines were safe and induce early maturation of HIV broadly neutralizing antibody (bnAb) precursors in humans. Phase 1 clinical trial (NCT05001373 and NCT05414786).	[23]
saRNA vaccine	LuS	Genetic Surface Fusion	HIV (gp120)	Mice	Elicited enhanced Tfh cell activation, increased B cell responses, and stronger antibody responses in mice compared to protein vaccines	[24]
saRNA vaccine	Ferritin	Genetic Surface Fusion	Yersinia pestis	Mice	Induced specific IgG responses and neutralizing antibodies, conferring protection in mice against *Yersinia pestis* challenge.	[91]
DNA vaccine	Ferritin	Genetic Surface Fusion	SARS-CoV-2 (RBD)	Rat	Generates more potent neutralizing responses compared to the subunit vaccine.	[21]
siRNA	Ferritin	Reversible dissassembly	Cancer	hMSC and PBMCs cells	Efficient encapsulation enabling gene silencing in tumor cells and PBMCs at low concentrations without inducing immune activation.	[29]
siRNA	Ferritin	Reversible dissassembly	Acute myeloid leukemia	HL-60 and K-562	The siPD-L1/HFn nanocarrier effectively silenced PD-L1 expression in HL-60 and K-562 cells.	[30]
siRNA	Ferritin	Reversible dissassembly	Cancer	A549 and HeLa cells	L17E-modified ferritin boosts stability, uptake, and BCL-2 silencing in A549 and HeLa cells.	[13]
siRNA	Ferritin (tHFn(+))	Reversible dissassembly	Glioblastoma	Mice	Efficiently crosses the blood–brain barrier, targets glioblastoma, and demonstrates therapeutic efficacy with siTERT and siEGFR in mice.	[31]

Note: In vaccines, SAPNs are genetically fused to antigens for multivalent display; in RNAi therapies, they encapsulate siRNA via reversible disassembly.

3.2. *SAPN-RNA Vaccines: Synergizing mRNA Vaccines with Protein Nanocages*

3.2.1. mRNA Vaccines and Mechanisms

Messenger RNA (mRNA) vaccines are an innovative platform for vaccine development, offering rapid production, scalability, and the ability to target a wide range of pathogens and diseases as demonstrated during the SARS-CoV-2 pandemic. LNP delivery systems effectively protect mRNA, facilitate cellular uptake, and enhance immune responses through Tfh cell activation, supporting high-affinity antibody production [20,92]. They are categorized into four main types: conventional mRNA vaccines, self-amplifying RNA (saRNA) vaccines, trans-amplifying RNA (taRNA) vaccines, and circular RNA (circRNA) vaccines (Figure 2B).

Conventional mRNA vaccines are based on synthetic mRNA transcribed in vitro, incorporating stabilizing features such as 5′ capping, untranslated regions (UTRs), and poly(A) tails to optimize translation and prevent degradation. Advances such as anti-reverse cap analogs (ARCAs) and modified nucleotides like N1-methylpseudouridine have improved mRNA stability and reduced unintended immune activation, as demonstrated in COVID-19 vaccines [93].

In contrast, saRNA vaccines incorporate replication machinery derived from alphaviruses, including nonstructural proteins (nsP1–4), enabling intracellular RNA amplification [94]. This self-amplifying feature reduces the required mRNA doses and lowers production costs while prolonging the expression of the encoded antigen, lasting up to 86 days [95], while mRNA vaccines show reduced expression within two weeks [4]. saRNA vaccines also elicit type I interferon (IFN-I) responses, enhancing adaptive immunity without integrating into the host genome, ensuring a safer profile [96]. However, the longer RNA sequence of saRNA presents challenges in stability and manufacturing. The taRNA vaccines address this by separating the replicase and antigen-encoding sequences onto distinct mRNA molecules, improving translation efficiency while reducing the challenges posed by long RNA constructs [97].

Meanwhile, circRNA vaccines, with their covalently closed structure, offer enhanced stability and prolonged antigen expression, as they are resistant to exonucleases. Despite lacking a 5′ cap and poly(A) tail, translation in circRNA vaccines is facilitated by internal ribosome entry sites (IRESs), leading to sustained protein production with reduced immune activation, positioning circRNA as a promising platform for future vaccine development [98].

Recent studies have revealed the potential of mRNA vaccines as a promising strategy for immunization against diverse human and zoonotic pathogens beyond SARS-CoV-2, such as influenza, dengue virus, HSV-2, rabies, and Zika virus, with distinct advantages compared to other platforms [2].

mRNA vaccines elicit strong humoral and cellular immunity by directing antigen translation in the cytosol. Antigen presentation via MHC-I activates cytotoxic CD8+ T cells, while MHC-II processing stimulates CD4+ T cells and enhances B cell antibody production, ensuring a robust and durable immune response [4].

mRNA vaccines are also gaining attention in veterinary medicine as innovative solutions for managing viral infections in farm animals [1]. They elicit strong humoral and cellular immunity, often surpassing conventional vaccines in swine, poultry, and companion animals. Studies on foot-and-mouth disease virus (FMDV), bovine viral diarrhea virus (BVDV), ASFV, porcine reproductive and respiratory syndrome virus (PRRSV), and avian influenza virus (AIV), demonstrate high neutralizing antibody titers and durable protection [99].

mRNA vaccines are also emerging as a promising approach to control viral diseases in aquaculture, a major and ongoing industry challenge [100–102], showing encouraging results despite the unique and simpler immune systems of aquatic species [103,104].

By targeting zoonotic pathogens, mRNA vaccines provide a proactive approach to reducing the risk of animal-to-human transmission, supporting One Health strategies [1].

While mRNA vaccines have shown success against viruses and cancers, they face challenges in producing bacterial proteins, particularly due to issues with folding, transport, and post-translational modifications. Their inability to produce complex antigens, such as polysaccharides, limits their competitiveness with traditional bacterial vaccines [105].

Overall, mRNA vaccines show great promise for human and animal diseases, but optimizing immunogenicity, delivery, and stability is essential to improve efficacy and address challenges like cold-chain storage and biocompatibility.

3.2.2. Self-Assembling Protein Nanocages for mRNA Vaccines Development

mRNA vaccines elicit strong humoral and cellular immunity by mimicking a viral infection. Once inside host cells, the mRNA is translated into protein, which can either function intracellularly and/or be secreted. Cytoplasmic proteins are degraded by proteasomes into peptides that are presented on MHC class I molecules, activating CD8+ T cells. Secreted proteins are taken up by APCs, processed, and presented via MHC class II [4] (Figure 2B). Moreover, cross-presentation allows exogenous antigens to be processed and displayed on MHC class I, further enhancing CD8+ T cell responses [106]. The mRNA-LNP formulation also activates innate immunity by engaging PRRs like TLR7/8, triggering cytokine and Tfh cell activation essential for high-affinity neutralizing antibodies [20,92].

Self-assembling protein nanocages can be engineered by fusing antigens to the N- or C-terminus, a strategy compatible with mRNA, saRNA, taRNA, and circRNA vaccines (SAPN-RNA vaccines). This approach synergistically enhances vaccine efficacy by promoting higher and longer expression of antigenic proteins, stabilizing secreted proteins, improving their uptake by antigen-presenting cells, and increasing immunogenicity. SAPNs mimic viral structures without genetic material, enhancing B cell activation through highly repetitive antigen display. Their nanoscale size enables efficient APC uptake, prolonged antigen retention in lymph nodes, and Tfh cell activation, driving germinal center responses which enhances the generation of long-lived plasma cells and memory B cells [20]. As a result of these synergistic mechanisms of protein nanocages and the mRNA platform, SAPN-RNA vaccines can induce higher neutralizing antibody titers with cross-reactivity against heterogeneous strains, making them particularly beneficial for viruses undergoing rapid antigenic changes, such as influenza.

Incorporating SAPNs into RNA vaccine design not only enhances immunogenicity and dose efficiency, but also offers higher biocompatibility, due to their natural protein composition and favorable kinetic properties [15,35].

SAPNs such as ferritin have shown considerable promise in enhancing the immunogenicity of RNA vaccines, boosting both humoral and cellular immune responses. In the case of HIV-1, multivalent ferritin nanovaccines, such as those displaying the glycosylated RBD, have been shown to induce robust neutralizing antibody responses in mice, which effectively neutralized diverse HIV-1 strains [87].

The approach has demonstrated notable advantages, such as enhancing antigen expression and eliciting strong neutralizing responses [19]. Incorporating ferritin into an MERS-CoV RNA vaccine enabled broad cross-clade neutralization, including a pangolin-derived strain, and sustained immune responses for over five months post booster, underscoring their potential for durable, broad-spectrum protection [19].

Furthermore, SAPNs have been successfully applied in the development of mRNA vaccines for SARS-CoV-2. Ferritin-RBD fusion vaccines, in particular, elicited significantly higher neutralizing antibody levels compared to RBD alone, demonstrating a stronger humoral immune response in mouse models, with improved protection in hamsters [88].

LuS self-assembling nanocages enhance mRNA vaccine immunogenicity by presenting antigens in a virus-like 60-mer structure [38]. Phase 1 trials IAVI-G002 (NCT05001373) and IAVI-G003 (NCT05414786) demonstrated that LuS-mRNA vaccines were safe, and effectively induced broadly neutralizing antibody precursors with improved maturation and neutralization, validating this strategy in high-HIV-burden regions [23]. An mRNA vaccine encoding the Rotavirus P2-VP8 antigen displayed on LuS nanocages elicited the highest IgG and neutralizing antibody titers in mice and guinea pigs compared to conventional mRNA and subunit vaccines [89]. For SARS-CoV-2, the LuS-displayed multivalent RBD of the Delta variant induced potent neutralizing responses and conferred protection in mice [90]. The approach also enables bivalent vaccine designs, producing antibody responses comparable to monovalent formulations, highlighting its versatility [90].

Both non-replicating and self-amplifying mRNA vaccines rely on host-cell machinery to produce antigens and induce adaptive immunity. Non-replicating mRNA encodes only the target protein, whereas self-amplifying mRNA includes additional replication machinery, reducing the required dose for effective immunity [3]. Incorporating SAPNs can further enhance this by multivalent antigen display, improving cellular uptake, and enhancing expression and antigen presentation by APCs.

Besides conventional mRNA vaccines, this approach has been applied to self-amplifying RNA (saRNA) vaccines. Incorporation of LuS in saRNA induced robust anti-HIV Env antibody production and significantly enhanced antigen-specific B cell activation in mice compared to protein immunization, eliciting germinal center B and T follicular helper-cell responses with a balanced IgG1/IgG2 profile [24].

For bacteria, saRNA with ferritin-incorporation constructs encoding the F1 and V antigens of Yersinia pestis provided significant protection in mice against subsequent challenges. Mice immunized with 1 μg or 5 μg of saRNA in LNP showed survival (5/7 mice) against both low and high bacterial doses. This saRNA vaccine induced specific IgG responses and neutralizing antibodies, with the potential for reduced boosting and rapid production, making it a promising candidate for plague vaccine development and outbreak response [91]. Therefore, the SAPN-RNA approach can be applied to both non-replicating vaccines and self-replicating RNA vaccines. However, despite the potential of the approach, there is no data on taRNA and circRNA vaccines, and further exploration of these could synergize their efficacy.

Overall, the SAPN-RNA vaccine approach, i.e., integrating self-assembling protein nanocages into mRNA vaccines, enables multivalent antigen display, prolongs protein expression, reduces off-target effects, lowers dosing requirements, and enhances antigen stability, immunogenicity, and biocompatibility. Careful design is essential to prevent excessive innate immune activation and potential autoimmunity.

3.3. SAPN-DNA Vaccines: Integration with DNA Plasmids for Multivalent Antigen Display

DNA immunization is a promising vaccination strategy due to its stability, ease of design, and efficient plasmid production. However, unlike mRNA vaccines that act directly in the cytoplasm, DNA vaccines require nuclear entry, raising biosafety concerns under GMO regulations (Figure 2C). Moreover, DNA vaccines often exhibit lower immunogenicity compared to traditional platforms and typically require IM administration to achieve optimal efficacy, which has limited their broader use [107,108]. While immune modulators, cytokines, and costimulatory molecules have been investigated to enhance immune responses, these efforts have yielded only modest improvements [109].

The key challenge in optimizing DNA vaccines lies in enhancing transfection efficiency, as effective delivery and expression are crucial for inducing robust immune responses [110]. Technologies such as gene guns and electroporation have demonstrated potential for improving transfection rates and enhancing immune responses. However, these methods are not yet fully optimized for large-scale, routine applications [111].

SAPNs enhance DNA vaccine efficacy by fusing antigens to nanocages in DNA constructs, leveraging their nanostructure and multivalent surface display to improve cellular uptake, boost protein expression, stabilization and immunogenicity, and with potential for targeted delivery functionalization. For example, the SARS-CoV-2 glycosylated receptor-binding domain (gRBD) fused to *H. pylori* ferritin 24-mer elicited stronger neutralizing antibodies than the mi3 60-mer, highlighting the potential to enhance antibody responses and reduce vaccine doses and production costs [21].

Although nanoencapsulation techniques, such as chitosan nanoparticles [69,112] and live vectors [113], have proven effective for DNA vaccine delivery, early attempts to use protein nanocages for encapsulation of DNA likely encountered challenges [114,115]. The negative charge and large size of DNA plasmids seems to have hindered their efficient encapsulation and delivery, with electrostatic repulsion between negatively charged ferritin and the cell membrane limiting effective transfection.

Overall, SAPNs represent a promising approach to enhancing DNA vaccine efficacy. Their unique structural properties, such as the ability to multivalent display of antigens and the fact that they may encapsulate DNA for targeted delivery, allow for superior immunogenicity. As research progresses and challenges related to large-scale production and regulatory approval are addressed, SAPN-DNA vaccines could become a key component of modern vaccination strategies, offering substantial improvements in infectious disease prevention and cancer immunotherapy.

3.4. SAPN-Adjuvant: Multivalent Display of Adjuvant and Co-Delivery with Antigen

Beyond structural antigen presentation, SAPNs provide a versatile platform for the co-delivery of antigens and adjuvants, significantly enhancing APC activation and promoting robust immune responses [25–27]. The multivalent surface display of adjuvant proteins potentiates immune stimulation, while encapsulation of small-molecule adjuvants within the internal cavity of the nanocage allows for sustained release, prolonging vaccine efficacy and minimizing off-target effects [10,25–27]. These combined strategies enable precise immune modulation tailored to diverse pathogens.

The SAPN-Adjuvant approach utilizes the SAPNs to enhance immune responses by presenting multiple copies of protein-based adjuvants, such as interleukin cytokines, TLR agonists (e.g., flagellin), or other immune modulators, through various design strategies, including genetic fusion, chemical conjugation, and SpyTag/SpyCatcher [17]. However, reversible disassembly and genetic engineering of the platform's internal cavity also enable encapsulation and co-delivery of these or small-molecule adjuvants [10].

Approaches include (1) multivalent surface display, (2) internal encapsulation, (3) a dual strategy where one adjuvant is surface-displayed and another encapsulated, and (4) co-delivery of an antigenic protein with an adjuvant, either both on the surface or with the antigen on the surface and adjuvant encapsulated. These configurations enable fine-tuning immune responses for Th1- or Th2-biased immunity or a balanced immune profile, depending on the pathogen. Additionally, the encapsulated or surface-displayed adjuvant using SAPNs can be easily mixed with an SAPN-based antigenic protein, further enhancing immune responses through combined administration. Protein-based adjuvants (PBAs) are potent immunomodulators, particularly for viral vaccines [116]. Flagellin, a TLR5 agonist and example of a PBA, stimulates APCs, promoting cytokine secretion and enhanced antigen uptake [117].

Conjugation of flagellin with influenza hemagglutinin on the exterior of ferritin nanocages significantly enhances both humoral and antigen-specific T cell responses, including Th1 cytokine production, and confers protection against lethal viral challenge in mice [26]. Similarly, genetic fusion of A-type flagellin (FliC) to ferritin nanocages generated the ReFliC-FN vaccine, which elicited a robust Th1-biased immune response and improved protection against *Pseudomonas aeruginosa*, while maintaining biocompatibility and safety [27]. Furthermore, immunization with Infectious Bronchitis Virus (IBV)-Flagellin SAPNs in chickens induced strong antibody responses, increased peripheral blood mononuclear-cell proliferation, and reduced viral shedding, demonstrating the approach's efficacy across species [25].

Controlled release of encapsulated adjuvants enhances vaccine safety by reducing toxicity and promoting localized immune stimulation. SAPNs such as ferritin nanocages are effective delivery vehicles, due to their reversible disassembly; however, their small internal cavity (~8 nm) limits cargo capacity [17]. In contrast, larger encapsulin nanocages (20–42 nm) offer improved co-delivery of proteins, including antigens and adjuvants, with selective cargo loading mediated by targeting domains, thereby optimizing immune responses [118,119].

Effective incorporation of adjuvants into nanocage platforms requires careful consideration of their biological targets and activation mechanisms. For example, since flagellin is recognized by TLR5 on the extracellular membrane, internal encapsulation may limit direct receptor engagement. Truncated flagellin variants, such as the D0/D1 domains, have also been proposed to preserve TLR5 activation, while allowing efficient fusion and surface display on nanocage platforms due to their smaller size [120].

Overall, SAPN-Adjuvants provide a versatile approach for vaccine development, enabling precise immune modulation through tailored adjuvant presentation and delivery. By co-delivery with antigens through multivalent surface display or internal encapsulation, they hold potential to enhance immunogenicity, safety, and efficacy of vaccines.

4. SAPN-RNAi: Self-Assembling Protein Nanocages for RNAi Delivery

4.1. RNA Interference (RNAi) and Mechanisms

RNAi is a cellular mechanism that regulates gene expression by binding to complementary RNA, targeting mRNA, transposon intermediates, RNA viruses, and other types for degradation or translational repression [121]. The main RNAi pathways are mediated by small interfering RNAs (siRNAs) and microRNAs (miRNAs), typically 19–25 nucleotides in length with negative charge and molecular weight of ~14 kDa [122].

siRNAs are typically exogenous, double-stranded RNA molecules that are processed into smaller fragments by the enzyme Dicer. In contrast, miRNAs are endogenous, single-stranded RNAs that are derived from primary transcripts. Both siRNAs and miRNAs play significant roles in post-transcriptional regulation, though they function differently. siRNAs are highly specific and typically target a single mRNA for degradation, making them suitable for precise silencing, especially in therapeutic applications and viral infections. miRNAs, on the other hand, generally regulate the translation of multiple mRNAs by binding with partial complementarity, leading to broader modulation of gene expression [121–123].

The RNAi mechanism is conserved across many organisms and involves two main steps: initiation and effector. In siRNA-mediated RNAi, exogenous double-stranded RNA (dsRNA), which can originate from repetitive sequences or long hairpin structures, is processed by Dicer into 21–24 nucleotide siRNA duplexes. These duplexes, possessing 3′-OH, 5′ phosphate, and 3′ dinucleotide overhangs, then associate with Argonaute proteins to form the precursor RNA-induced silencing complex (pre-RISC). One strand of the siRNA duplex, the passenger strand, is cleaved by Argonaute, while the remaining guide strand directs the mature RISC complex to complementary mRNAs for cleavage and degradation [28,124].

In miRNA-mediated RNAi, the process initiates with the transcription of primary miRNAs (pri-miRNAs). These pri-miRNAs are processed by the Microprocessor complex, consisting of the enzyme Drosha and its cofactor DGCR8, into precursor miRNAs (pre-miRNAs). The pre-miRNAs are then exported from the nucleus to the cytoplasm by Exportin-5 in a Ran-GTP-dependent manner. In the cytoplasm, the RNase III enzyme Dicer further processes the pre-miRNAs into miRNA duplexes. One strand of the duplex, known as the guide strand, is incorporated into the RISC. The guide strand directs RISC to target mRNAs with partial complementarity, leading to translational repression or mRNA degradation, thereby enabling broader modulating gene activity [125].

Another type of RNAi molecule, short hairpin RNA (shRNA), is a synthetic RNA structure that mimics endogenous RNA precursors and is commonly used in stable gene knockdown experiments [126]. They form a hairpin structure, target specific genes, and are processed into siRNA by the RNAi machinery when introduced into cells via vectors, with first-generation shRNAs having off-target effects, while second-generation shRNAs offer greater adaptability and stability but remain more complex [127]. Additionally, piwi-interacting RNAs (piRNAs), 21–35 nt in length, regulate gene expression through PIWI proteins and are involved in transposon silencing and maintaining genome stability in germline cells by mediating both transcriptional and post-transcriptional silencing, with therapeutic potential [128].

A key aspect of RNAi-based therapies is effective delivery. Various strategies have been developed to enhance delivery and minimize degradation [127]. LNPs can protect RNA from degradation and facilitate cellular uptake [129]. Viral vectors also offer efficient delivery but may raise concerns regarding immunogenicity and potential genome integration [130]. Synthetic polymers enhance stability and targeting with less toxicity, while physical methods like electroporation allow direct delivery but are invasive [127]. Despite these advancements, RNAi delivery remains a major challenge. The SAPN-RNAi delivery approach for antiviral therapy is shown in Figure 3.

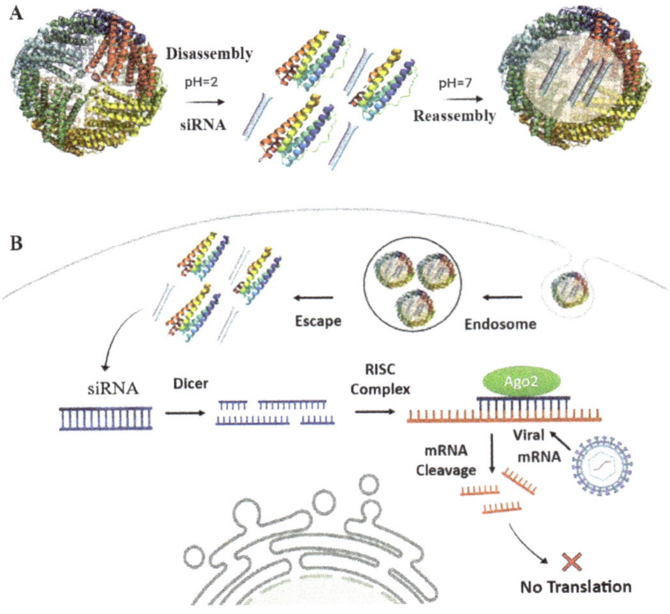

Figure 3. Schematic of SAPN-RNAi delivery for antiviral therapy. (**A**) Self-assembling protein nanocages (SAPNs) for RNAi encapsulation: siRNA is encapsulated within ferritin via reversible disassembly at pH 2 and reassembly at pH 7. (**B**) Antiviral mechanism of SAPN-RNAi: SAPN protects siRNA from degradation and facilitates cellular uptake via endocytosis. After endosomal escape, Dicer processes long dsRNA into siRNAs that guide Ago2-RISC to degrade viral mRNA and inhibit protein synthesis. Targeting-ligands on the SAPN surface ensure cell-specific delivery. Created with BioRender.com.

4.2. Self-Assembling Protein Nanocages for RNAi Delivery Against Viral Infection

RNA interference provides a targeted therapeutic strategy for managing viral infections, offering greater adaptability than traditional antiviral approaches. siRNAs are particularly effective due to their high specificity, binding to and degrading viral RNA, thereby disrupting replication and limiting disease progression [6].

Once introduced into cells, siRNAs incorporate into RISC, where the guide strand directs cleavage of complementary viral mRNAs, preventing protein synthesis and viral propagation [5]. Additionally, synthetic siRNAs bypass the Dicer-mediated processing step and directly integrate into RISC to initiate gene silencing [131,132].

The efficacy of siRNAs in targeting chronic and lethal viruses, including HBV, HIV, influenza, and SARS-CoV, has been demonstrated, with promising results in inhibiting viral replication [124]. Currently, five siRNA-based drugs including patisiran, givosiran, inclisiran, lumasiran, and vurtisiran, have been approved, with others in late-stage clinical trials [133]. However, efficient delivery of RNAi molecules into target cells remains a major challenge, as effective delivery systems are needed to ensure sufficient RNA concentrations while minimizing off-target effects and immune responses [127].

Nanoparticles have proven to be effective siRNA delivery systems with minimal toxicity and off-target effects [134]. Among them, protein nanocages such as ferritin, offer a highly efficient and versatile platform for RNAi therapy. Their internal cavity enables siRNA encapsulation, while surface engineering allows targeted delivery. With excellent biosafety, rapid dispersion, and controlled release, SAPNs enhance therapeutic stability and efficacy [66,70]. Their well-defined nanostructures, structural stability, and functionalization capacity protect RNA from degradation and facilitate cellular uptake [13,29–31]. SAPNs are well-suited for delivering small RNA molecules such as siRNA and miRNA, rather than larger mRNA or DNA vaccines, as their size (5–8 nm) aligns with the protein nanocage cavity size and loading capacity [28,115].

The ability of SAPNs to undergo controlled disassembly and reassembly enhances their utility for RNAi delivery. For example, ferritin, a naturally occurring SAPN, can disassemble at acidic pH 2, allowing RNA loading, and reassemble at physiological pH (~7.4), encapsulating the RNA payload and protecting it from degradation [66]. This method enables flexible RNAi delivery while maintaining structural integrity. Alternative techniques, such as urea gradients and hydrostatic pressure, further facilitate RNA loading [9], providing versatile approaches for developing SAPN- RNAi delivery systems.

Encapsulation of siRNA within human apoferritin has demonstrated superior protection against enzymatic degradation and efficient cellular uptake. This system achieves high transfection efficiency in diverse cell types, including peripheral blood mononuclear cells (PBMCs), without inducing immune activation [29].

Such characteristics make SAPN-based RNAi delivery systems promising for antiviral therapies aimed at suppressing viral gene expression and modulating immune responses. Insights from ferritin-based antitumor delivery systems can inform their application in antiviral contexts. For instance, ferritin modified with the cell-penetrating peptide L17E demonstrated enhanced cellular uptake and stability relative to wild-type ferritin. When loaded with BCL-2 siRNA, this platform (Fn-L17E) effectively silenced the anti-apoptotic gene BCL-2 in A549 and HeLa cells [13]. Similarly, ferritin fused with the KALA peptide effectively targeted dendritic cells and enhanced antigen presentation. Co-delivery of SOCS1 siRNA further promoted DC–T cell interactions, resulting in durable antitumor immunity [67].

Additionally, siPD-L1/HFn ferritin nanocarriers have exhibited significant PD-L1 silencing in HL-60 and K-562 cancer cell lines, further supporting their promise in RNAi-based therapeutic strategies [30]. In targeted RNAi therapy for glioblastoma, the positively charged ferritin variant, tHFn(+), enhances lysosomal escape, facilitating the release of siRNA into the cytoplasm. Both in vitro and in vivo studies confirm that tHFn(+) efficiently crosses the blood–brain barrier (BBB) and specifically targets glioblastoma, with siTERT and siEGFR showing therapeutic efficacy in mouse models [31].

Encapsulins, with their larger size and superior cargo loading capacity, outperform other protein nanocages in delivering larger RNA molecules or multiple siRNA sequences for advanced antiviral therapies [118]. Unlike traditional systems requiring in vitro cargo loading, encapsulins enable in vivo co-packaging of RNA and proteins via selective loading mechanisms and targeting domains, preventing conditions that could destabilize RNA [11]. However, integrating a non-protein cargo like siRNA or mRNA requires additional engineering, such as RNA-binding domains or charge-based modifications like a cationic RKRK domain, to overcome electrostatic repulsion with the cell membrane and ensure effective transfection [11,115,135].

A key advantage of co-delivering protein and RNA within a single nanocage is the ability to target multiple intracellular pathways, thereby enhancing immune responses through antigen presentation and silencing of immune-suppressive pathways. Nanocages also protect RNA from degradation, with pH-sensitive disassembly facilitating efficient release. In vivo assembly further simplifies production and minimizes non-specific loading [11,42].

However, challenges such as optimizing loading capacity, preventing RNA degradation during pH modulation, and preserving the native structure of the protein nanocages remain critical. To address these issues, various ferritin mutants have been engineered to enhance pH-responsive disassembly/reassembly for drug delivery. Approaches include C-terminal modifications (e.g., GALA fusion, DE helix deletion), loop truncations (AB, DE loops), and histidine substitutions, to enable structural reversibility and cargo encapsulation [136]. Future advancements may improve RNA selectivity via sequence-specific RNA-binding domains. Notably, functionalizing SAPNs with viral-targeting ligands and refining assembly mechanisms could enable these systems for effective antiviral siRNA delivery.

Overall, the SAPN-RNAi approach shows promise for treating viral infections by targeting conserved viral genes or modulating host factors critical for viral replication. Advancements in delivery strategies, including the evaluation of various protein nanocages and loading techniques, could lead to next-generation antiviral therapies with improved specificity, stability, and efficacy.

5. Challenges and Future Direction

Self-assembling protein nanocages (SAPNs) occur naturally in various organisms, providing a versatile platform for vaccine development and therapeutic delivery, including RNAi. Their modular architecture enables precise immune modulation through multivalent antigen display, while the inner cavity supports cargo loading for enhanced stability, controlled release, with targeted delivery. The unique nanostructure improves cellular uptake and antigen presentation, and facilitates the co-delivery of antigens and adjuvants, optimizing immune responses. Incorporating SAPNs into nucleic acid vaccines platforms (mRNA and DNA) offers a novel strategy to synergize their immune responses at lower doses while enhancing stability, versatility and cross-protection against evolving pathogens. Encapsulation and protection of therapeutic RNAi with targeted delivery is another potential of this novel strategy. Approaches for vaccine design and RNAi delivery, along with offered corresponding terms, are summarized in Table 2.

Despite their potential, several challenges must be addressed to fully harness the benefits of SAPNs in vaccine development and RNAi therapeutics. Antigen subunits can be genetically fused to the C- or N-terminus of proteins; however, increasing antigen size and structural complexity may compromise nanoparticle assembly and integrity. To optimize antigen presentation, regardless of size, SAPNs designed for subunit vaccines may require modular engineering approaches, such as SpyTag-SpyCatcher systems. Larger protein nanocages or multi-epitopic/short transduced antigens, combined with suitable linkers, may be preferable to full-length antigens for nucleic acid-based vaccines (RNA and DNA vaccines). Excessive innate immune activation or autoimmunity must be considered, as they could impact antigen translation and RNAi delivery. Thus, biocompatibility studies and dose optimization using non-homologous protein nanocages, especially those derived from bacterial sources, are recommended. Importantly, protein nanocages should be screened for autoimmune epitopes and engineered to minimize risks.

Alongside multivalent antigen display, future research should focus on the encapsulation of mRNA vaccines using protein nanocages. Challenges related to charge properties, release kinetics, and structural integrity must be addressed. The negative charge of SAPNs presents a limitation, hindering RNA loading and delivery. This can be mitigated by engineering positively charged SAPN variants or incorporating nanoparticles like chitosan to enhance electrostatic interactions. Optimizing RNA encapsulation while maintaining controlled release is also critical. Advancements, such as pH- or enzyme-responsive disassembly mechanisms, could enhance RNA release at target sites. Furthermore, incorporating cell-specific ligands or fusion proteins could improve targeting and therapeutic efficacy. Hybrid delivery strategies, combining SAPNs with lipid or polymer nanocarriers, may also be applied to expand clinical applicability. Beyond RNAi delivery, SAPNs also hold promise for emerging nucleic acid therapeutics like antisense oligonucleotides (ASOs), which offer adaptable gene-silencing mechanisms

Selecting the appropriate SAPN scaffold for each application is crucial, as factors such as stability, flexibility, expression system compatibility, loading capacity, and subunit composition impact efficacy. Well-characterized nanocages, like ferritin, have proven useful for both antigen display and RNA encapsulation, while other protein nanocages, such as encapsulins, may offer advantages in size, subunit composition, and dual delivery potential. Nonetheless, further exploration of protein nanocages from diverse sources, along with extensive characterization, is necessary to validate this approach across a variety of pathogens and diseases in both humans and animals.

Comprehensive preclinical and clinical studies are thus essential to validate the safety and efficacy of SAPN platforms. Addressing regulatory challenges will be crucial to ensuring these innovative technologies meet stringent clinical approval standards. By optimizing SAPN design and production and overcoming these challenges, these platforms hold significant promise for advancing vaccines and RNAi therapeutics for viral infections and other diseases.

Funding: This research received no external funding.

Conflicts of Interest: The author declares no conflict of interest.

References

1. Le, T.; Sun, C.; Chang, J.; Zhang, G.; Yin, X. mRNA vaccine development for emerging animal and zoonotic diseases. *Viruses* **2022**, *14*, 401. [CrossRef]
2. Zhang, G.; Tang, T.; Chen, Y.; Huang, X.; Liang, T. mRNA vaccines in disease prevention and treatment. *Signal Transduct. Target. Ther.* **2023**, *8*, 365. [CrossRef] [PubMed]
3. Wadhwa, A.; Aljabbari, A.; Lokras, A.; Foged, C.; Thakur, A. Opportunities and Challenges in the Delivery of mRNA-Based Vaccines. *Pharmaceutics* **2020**, *12*, 102. [CrossRef] [PubMed]
4. Pardi, N.; Krammer, F. mRNA vaccines for infectious diseases—Advances, challenges and opportunities. *Nat. Rev. Drug Discov.* **2024**, *23*, 838–861. [CrossRef] [PubMed]
5. Tan, F.; Yin, J. RNAi, a new therapeutic strategy against viral infection. *Cell Res.* **2004**, *14*, 460–466. [CrossRef]

6. Kelleher, A.D.; Cortez-Jugo, C.; Cavalieri, F.; Qu, Y.; Glanville, A.R.; Caruso, F.; Symonds, G.; Ahlenstiel, C.L. RNAi therapeutics: An antiviral strategy for human infections. *Curr. Opin. Pharmacol.* **2020**, *54*, 121–129. [CrossRef]
7. Kim, M.; Rho, Y.; Jin, K.S.; Ahn, B.; Jung, S.; Kim, H.; Ree, M. pH-dependent structures of ferritin and apoferritin in solution: Disassembly and reassembly. *Biomacromolecules* **2011**, *12*, 1629–1640. [CrossRef]
8. Boyton, I.; Goodchild, S.C.; Diaz, D.; Elbourne, A.; Collins-Praino, L.E.; Care, A. Characterizing the Dynamic Disassembly/Reassembly Mechanisms of Encapsulin Protein Nanocages. *ACS Omega* **2022**, *7*, 823–836. [CrossRef]
9. Wang, Z.; Gao, H.; Zhang, Y.; Liu, G.; Niu, G.; Chen, X. Functional Ferritin Nanoparticles for Biomedical Applications. *Front. Chem. Sci. Eng.* **2017**, *11*, 633–646. [CrossRef]
10. Mohanty, A.; Parida, A.; Raut, R.K.; Behera, R.K. Ferritin: A Promising Nanoreactor and Nanocarrier for Bionanotechnology. *ACS Bio Med. Chem. Au* **2022**, *2*, 258–281. [CrossRef]
11. Kwon, S.; Giessen, T.W. Engineered Protein Nanocages for Concurrent RNA and Protein Packaging In Vivo. *ACS Synth Biol.* **2022**, *11*, 3504–3515. [CrossRef] [PubMed]
12. Lamontagne, F.; Khatri, V.; St-Louis, P.; Bourgault, S.; Archambault, D. Vaccination strategies based on bacterial self-assembling proteins as antigen delivery nanoscaffolds. *Vaccines* **2022**, *10*, 1920. [CrossRef] [PubMed]
13. Qin, L.-M.; Feng, M.-F.; Sun, B.; Yu, X.; Wu, Y. Targeted Delivery of BCL-2 siRNA Using Modified Ferritin Nanocarriers for Antitumor Therapy. *ACS Appl. Nano Mater.* **2024**, *7*, 22487–22496. [CrossRef]
14. Kim, Y.S.; Son, A.; Kim, J.; Kwon, S.B.; Kim, M.H.; Kim, P.; Kim, J.; Byun, Y.H.; Sung, J.; Lee, J.; et al. Chaperna-Mediated Assembly of Ferritin-Based Middle East Respiratory Syndrome-Coronavirus Nanoparticles. *Front. Immunol.* **2018**, *9*, 1093. [CrossRef]
15. Ahmadivand, S.; Krpetic, Z.; Martínez, M.M.; García-Ordoñez, M.; Roher, N.; Palić, D. Self-Assembling Ferritin Nanoplatform for the Development of Infectious Hematopoietic Necrosis Virus Vaccine. *Front. Immunol.* **2024**, *15*, 1346512. [CrossRef]
16. Hauser, H.; Lopez-Sagaseta, J.; Malito, E. *Self-Assembling Protein Nanoparticles in the Design of Vaccines:* 2022 Update. *Vaccines* **2022**, *10*, 1447. [CrossRef]
17. Ahmadivand, S.; Fux, R.; Palić, D. Ferritin Vaccine Platform for Animal and Zoonotic Viruses. *Vaccines* **2024**, *12*, 1112. [CrossRef]
18. Jiang, R.; Yuan, S.; Zhou, Y.; Wei, Y.; Li, F.; Wang, M.; Chen, B.; Yu, H. Strategies to overcome the challenges of low or no expression of heterologous proteins in *Escherichia coli* . *Biotechnol. Adv.* **2024**, *75*, 108417. [CrossRef]
19. Powell, A.E.; Caruso, H.; Park, S.; Chen, J.-L.; O'Rear, J.; Ferrer, B.J.; Walker, A.; Bruening, A.; Hartwig, A.; Ahyong, V.; et al. A stabilized MERS-CoV spike ferritin nanoparticle vaccine elicits robust and protective neutralizing antibody responses. *bioRxiv* **2024**. 2024.07.01.601243. [CrossRef]
20. Ahmadivand, S.; Fux, R.; Palić, D. Role of T Follicular Helper Cells in Viral Infections and Vaccine Design. *Cells* **2025**, *14*, 508. [CrossRef]
21. Guo, Y.; He, W.; Mou, H.; Zhang, L.; Chang, J.; Peng, S.; Ojha, A.; Tavora, R.; Parcells, M.S.; Luo, G.; et al. An engineered receptor-binding domain improves the immunogenicity of multivalent SARS-CoV-2 vaccines. *mBio* **2021**, *12*, e00930-21. [CrossRef] [PubMed]
22. Pandey, K.K.; Sahoo, B.R.; Pattnaik, A.K. Protein Nanoparticles as Vaccine Platforms for Human and Zoonotic Viruses. *Viruses* **2024**, *16*, 936. [CrossRef] [PubMed]
23. Willis, J.R.; Prabhakaran, M.; Muthui, M.; Naidoo, A.; Sincomb, T.; Wu, W.; Cottrell, C.A.; Landais, E.; deCamp, A.C.; Keshavarzi, N.R. Vaccination with mRNA-encoded nanoparticles drives early maturation of HIV bnAb precursors in humans. *Science* **2025**, *0*, eadr8382. [CrossRef] [PubMed]
24. Melo, M.; Porter, E.; Zhang, Y.; Silva, M.; Li, N.; Dobosh, B.; Liguori, A.; Skog, P.; Landais, E.; Menis, S.; et al. Immunogenicity of RNA replicons encoding HIV Env immunogens designed for self-assembly into nanoparticles. *Mol. Ther.* **2019**, *27*, 2080–2090. [CrossRef]
25. Li, J.; Helal, Z.H.; Karch, C.P.; Mishra, N.; Girshick, T.; Garmendia, A.; Burkhard, P.; Khan, M.I. A Self-Adjuvanted Nanoparticle-Based Vaccine Against Infectious Bronchitis Virus. *PLoS ONE* **2018**, *13*, e0203771. [CrossRef]
26. Lee, E.B.; Jeon, H.M.; Kim, C.U.; Park, S.M.; Cho, G.; Kim, H.J.; Kim, Y.; Kim, D.J.; Kim, Y.S.; Lee, H.; et al. Attachment of flagellin enhances the immunostimulatory activity of a hemagglutinin-ferritin nano-cage. *Nanomedicine* **2019**, *17*, 223–235. [CrossRef]
27. Wei, Y.; Cheng, X.; Liao, Y.; Zeng, S.; Li, Y.; Zhang, Y.; Gao, C.; Zhang, Y.; Wan, J.; Gu, J.; et al. Recombinant Pseudomonas aeruginosa flagellin delivered using ferritin nanoparticles provides enhanced cross-protection against lung infection in mice. *Mol. Immunol.* **2023**, *163*, 235–242. [CrossRef]
28. Hu, B.; Zhong, L.; Weng, Y.; Peng, L.; Huang, Y.; Zhao, Y.; Liang, X.J. Therapeutic siRNA: State of the Art. *Signal Transduct. Target. Ther.* **2020**, *5*, 101. [CrossRef]
29. Li, L.; Muñoz-Culla, M.; Carmona, U.; Lopez, M.P.; Yang, F.; Trigueros, C.; Otaegui, D.; Zhang, L.; Knez, M. Ferritin-mediated siRNA delivery and gene silencing in human tumor and primary cells. *Biomaterials.* **2016**, *98*, 143–151. [CrossRef]
30. Rajabinejad, M.; Valadan, R.; Tehrani, M.; Najafi, A.; Negarandeh, R.; Saeedi, M.; Asgarian-Omran, H. Effective delivery of anti-PD-L1 siRNA with human heavy chain ferritin (HFn) in acute myeloid leukemia cell lines. *Med Oncol.* **2024**, *41*, 149. [CrossRef]

31. Jin, Y.; Zhang, B.; Li, J.; Guo, Z.; Zhang, C.; Chen, X.; Ma, L.; Wang, Z.; Yang, H.; Li, Y.; et al. Bioengineered protein nanocarrier facilitating siRNA escape from lysosomes for targeted RNAi therapy in glioblastoma. *Sci Adv.* **2025**, *11*, eadr9266. [CrossRef]
32. Bhaskar, S.; Lim, S. Engineering protein nanocages as carriers for biomedical applications. *NPG Asia Mater.* **2017**, *9*, e371. [CrossRef] [PubMed]
33. Kanekiyo, M.; Wei, C.-J.; Yassine, H.M.; McTamney, P.M.; Boyington, J.C.; Whittle, J.R.R.; Rao, S.S.; Kong, W.-P.; Wang, L.; Nabel, G.J. Self-assembling influenza nanoparticle vaccines elicit broadly neutralizing H1N1 antibodies. *Nature* **2013**, *499*, 102–106. [CrossRef] [PubMed]
34. Rodrigues, M.Q.; Alves, P.M.; Roldão, A. Functionalizing Ferritin Nanoparticles for Vaccine Development. *Pharmaceutics* **2021**, *13*, 1621. [CrossRef] [PubMed]
35. Charlton, J.R.; Pearl, V.M.; Denotti, A.R.; Lee, J.B.; Swaminathan, S.; Scindia, Y.M.; Charlton, N.P.; Baldelomar, E.J.; Beeman, S.C.; Bennett, K.M. Biocompatibility of ferritin-based nanoparticles as targeted MRI contrast agents. Nanomedicine: Nanotechnology. *Biol. Med.* **2016**, *12*, 1735–1745.
36. Zhang, Y.; Orner, B.P. Self-Assembly in the Ferritin Nano-Cage Protein Superfamily. *Int. J. Mol. Sci.* **2011**, *12*, 5406–5421. [CrossRef]
37. Ladenstein, R.; Morgunova, E. Second career of a biosynthetic enzyme: Lumazine synthase as a virus-like nanoparticle in vaccine development. *Biotechnol. Rep.* **2020**, *27*, e00494. [CrossRef]
38. Ladenstein, R.; Fischer, M.; Bacher, A. The lumazine synthase/riboflavin synthase complex: Shapes and functions of a highly variable enzyme system. *FEBS J.* **2013**, *280*, 2537–2563. [CrossRef]
39. Putri, R.M.; Allende-Ballestero, C.; Luque, D.; Klem, R.; Rousou, K.-A.; Liu, A.; Traulsen, C.H.-H.; Rurup, W.F.; Koay, M.S.T.; Castón, J.R.; et al. Structural Characterization of Native and Modified Encapsulins as Nanoplatforms for in Vitro Catalysis and Cellular Uptake. *ACS Nano* **2017**, *11*, 12796–12804. [CrossRef]
40. Lagoutte, P.; Mignon, C.; Stadthagen, G.; Potisopon, S.; Donnat, S.; Mast, J.; Lugari, A.; Werle, B. Simultaneous Surface Display and Cargo Loading of Encapsulin Nanocompartments and Their Use for Rational Vaccine Design. *Vaccine* **2018**, *36*, 3622–3628. [CrossRef]
41. Wiryaman, T.; Toor, N. Recent Advances in the Structural Biology of Encapsulin Bacterial Nanocompartments. *J. Struct. Biol. X* **2022**, *6*, 100062. [CrossRef] [PubMed]
42. Chmelyuk, N.S.; Oda, V.V.; Gabashvili, A.N.; Abakumov, M.A. Encapsulins: Structure, Properties, and Biotechnological Applications. *Biochemistry* **2023**, *88*, 35–49. [CrossRef] [PubMed]
43. Kar, U.; Khaleeq, S.; Garg, P.; Bhat, M.; Reddy, P.; Vignesh, V.S.; Upadhyaya, A.; Das, M.; Chakshusmathi, G.; Pandey, S.; et al. Comparative Immunogenicity of Bacterially Expressed Soluble Trimers and Nanoparticle Displayed Influenza Hemagglutinin Stem Immunogens. *Front. Immunol.* **2022**, *13*, 890622. [CrossRef] [PubMed]
44. Khaleeq, S.; Sengupta, N.; Kumar, S.; Patel, U.R.; Rajmani, R.S.; Reddy, P.; Pandey, S.; Singh, R.; Dutta, S.; Ringe, R.P.; et al. Neutralizing Efficacy of Encapsulin Nanoparticles against SARS-CoV2 Variants of Concern. *Viruses* **2023**, *15*, 346. [CrossRef]
45. Jung, H.G.; Jeong, S.; Kang, M.J.; Hong, I.; Park, Y.S.; Ko, E.; Kim, J.O.; Choi, D.Y. Molecular Design of Encapsulin Protein Nanoparticles to Display Rotavirus Antigens for Enhancing Immunogenicity. *Vaccines* **2024**, *12*, 1020. [CrossRef]
46. Allen, M.D.; Perham, R.N. The Catalytic Domain of Dihydrolipoyl Acetyltransferase from the Pyruvate Dehydrogenase Multienzyme Complex of Bacillus stearothermophilus: Expression, Purification and Reversible Denaturation. *FEBS Lett.* **1997**, *413*, 339–343. [CrossRef]
47. Izard, T.; Aevarsson, A.; Allen, M.D.; Westphal, A.H.; Perham, R.N.; de Kok, A.; Hol, W.G. Principles of Quasi-Equivalence and Euclidean Geometry Govern the Assembly of Cubic and Dodecahedral Cores of Pyruvate Dehydrogenase Complexes. *Proc. Natl. Acad. Sci. USA.* **1999**, *96*, 1240–1245. [CrossRef]
48. Dalmau, M.; Lim, S.; Chen, H.C.; Ruiz, C.; Wang, S.W. Thermostability and molecular encapsulation within an engineered caged protein scaffold. *Biotechnol. Bioeng.* **2008**, *101*, 654–664. [CrossRef]
49. Peng, T.; Lee, H.; Lim, S. Design of a reversible inversed pH-responsive caged protein. *Biomater. Sci.* **2015**, *3*, 627–635. [CrossRef]
50. Salinas, N.D.; Ma, R.; Dickey, T.H.; McAleese, H.; Ouahes, T.; Long, C.A.; Miura, K.; Lambert, L.E.; Tolia, N.H. A Potent and Durable Malaria Transmission-Blocking Vaccine Designed from a Single-Component 60-Copy Pfs230D1 Nanoparticle. *NPJ Vaccines* **2023**, *8*, 124. [CrossRef]
51. Van Montfort, R.; Slingsby, C.; Vierling, E. Structure and Function of the Small Heat Shock Protein/Alpha-Crystallin Family of Molecular Chaperones. *Adv. Protein Chem.* **2001**, *59*, 105–156. [PubMed]
52. Biswas, S.; Garg, P.; Dutta, S.; Suguna, K. Multiple nanocages of a cyanophage small heat shock protein with icosahedral and octahedral symmetries. *Sci. Rep.* **2021**, *11*, 21023. [CrossRef]
53. Kim, K.K.; Kim, R.; Kim, S.H. Crystal Structure of a Small Heat-Shock Protein. *Nature* **1998**, *394*, 595–599. [CrossRef] [PubMed]
54. Richert, L.E.; Servid, A.E.; Harmsen, A.L.; Rynda-Apple, A.; Han, S.; Wiley, J.A.; Douglas, T.; Harmsen, A.G. A virus-like particle vaccine platform elicits heightened and hastened local lung mucosal antibody production after a single dose. *Vaccine* **2012**, *30*, 3653–3665. [CrossRef] [PubMed]

55. Schreiber, A.; Schiller, S.M. Nanobiotechnology of protein-based compartments: Steps toward nanofactories. *Bioinspired Biomim. Nanobiomater.* **2013**, *2*, 154–172. [CrossRef]
56. Kim, D.R.; Lee, I.; Ha, S.C.; Kim, K.K. Activation Mechanism of HSP16.5 from Methanococcus jannaschii. *Biochem. Biophys. Res. Commun.* **2003**, *307*, 991–998. [CrossRef]
57. Carra, S.; Alberti, S.; Benesch, J.L.P.; Boelens, W.; Buchner, J.; Carver, J.A.; Cecconi, C.; Ecroyd, H.; Gusev, N.; Hightower, L.E.; et al. Small Heat Shock Proteins: Multifaceted Proteins with Important Implications for Life. *Cell Stress. Chaperones* **2019**, *24*, 295–308. [CrossRef]
58. Gabashvili, A.N.; Chmelyuk, N.S.; Efremova, M.V.; Malinovskaya, J.A.; Semkina, A.S.; Abakumov, M.A. Encapsulins—Bacterial Protein Nanocompartments: Structure, Properties, and Application. *Biomolecules* **2020**, *10*, 966. [CrossRef]
59. Brune, K.D.; Howarth, M. New Routes and Opportunities for Modular Construction of Particulate Vaccines: Stick, Click, and Glue. *Front. Immunol.* **2018**, *9*, 1432. [CrossRef]
60. Hatlem, D.; Trunk, T.; Linke, D.; Leo, J.C. Catching a SPY: Using the SpyCatcher-SpyTag and Related Systems for Labeling and Localizing Bacterial Proteins. *Int. J. Mol. Sci.* **2019**, *20*, 2129. [CrossRef]
61. Chen, Y.; Hu, Y.; Chen, H.; Li, X.; Qian, P. A ferritin nanoparticle vaccine for foot-and-mouth disease virus elicited partial protection in mice. *Vaccine* **2020**, *38*, 5647–5652. [CrossRef]
62. Wang, Y.; Song, J.; Deng, X.; Wang, J.; Zhang, M.; Liu, Y.; Tang, P.; Liu, H.; Zhou, Y.; Tong, G.; et al. Nanoparticle vaccines based on the receptor binding domain of porcine deltacoronavirus elicit robust protective immune responses in mice. *Front. Immunol.* **2024**, *15*, 1328266. [CrossRef] [PubMed]
63. Tang, S.; Zhao, C.; Zhu, X. Engineering Escherichia Coli-Derived Nanoparticles for Vaccine Development. *Vaccines* **2024**, *12*, 1287. [CrossRef] [PubMed]
64. Rodrigues, M.Q.; Patão, S.; Thomaz, M.; Nunes, T.; Alves, P.M.; Roldão, A. Tyrosinase-Mediated Conjugation for Antigen Display on Ferritin Nanoparticles. *Bioconjug. Chem.* **2024**, *35*, 1608–1617. [CrossRef] [PubMed]
65. Weidenbacher, P.; Musunuri, S.; Powell, A.E.; Tang, S.; Do, J.; Sanyal, M.; Kim, P.S. Simplified Purification of Glycoprotein-Modified Ferritin Nanoparticles for Vaccine Development. *Biochemistry* **2023**, *62*, 292–299. [CrossRef]
66. Tu, Z.; Timashev, P.; Chen, J.; Liang, X.J. Ferritin-based drug delivery system for tumor therapy. *BMEMat* **2023**, *1*, e12022. [CrossRef]
67. Wu, J.; Liang, J.; Li, S.; Lu, J.; Li, Y.; Zhang, B.; Gao, M.; Zhou, J.; Zhang, Y.; Chen, J. Cancer Vaccine Designed from Homologous Ferritin-Based Fusion Protein with Enhanced DC-T Cell Crosstalk for Durable Adaptive Immunity against Tumors. *Bioact. Mater.* **2025**, *30*, 1024–1042. [CrossRef]
68. Brun, A. (Ed.) Vaccines and Vaccination for Veterinary Viral Diseases: A General Overview. In *Vaccine Technologies for Veterinary Viral Diseases*; Methods in Molecular Biology; Humana Press: New York, NY, USA, 2016; Volume 1349, pp. 1–24.
69. Ahmadivand, S.; Soltani, M.; Behdani, M.; Evensen, Ø.; Alirahimi, E.; Hassanzadeh, R.; Soltani, E. Oral DNA vaccines based on CS-TPP nanoparticles and alginate microparticles confer high protection against infectious pancreatic necrosis virus (IPNV) infection in trout. *Dev. Comp. Immunol.* **2017**, *74*, 178–189. [CrossRef]
70. Cao, S.; Ma, D.; Ji, S.; Zhou, M.; Zhu, S. Self-Assembled Ferritin Nanoparticles for Delivery of Antigens and Development of Vaccines: From Structure and Property to Applications. *Molecules.* **2024**, *29*, 4221. [CrossRef]
71. Tesarova, B.; Musilek, K.; Rex, S.; Heger, Z. Taking advantage of cellular uptake of ferritin nanocages for targeted drug delivery. *J. Control Release.* **2020**, *325*, 176–190. [CrossRef]
72. Zepeda-Cervantes, J.; Ramírez-Jarquín, J.O.; Vaca, L. Interaction Between Virus-Like Particles (VLPs) and Pattern Recognition Receptors (PRRs) from Dendritic Cells (DCs): Toward Better Engineering of VLPs. *Front. Immunol.* **2020**, *11*, 1100. [CrossRef] [PubMed]
73. Houser, K.V.; Chen, G.L.; Carter, C.; Crank, M.C.; Nguyen, T.A.; Burgos Florez, M.C.; Berkowitz, N.M.; Mendoza, F.; Hendel, C.S.; Gordon, I.J.; et al. Safety and immunogenicity of a ferritin nanoparticle H2 influenza vaccine in healthy adults: A phase 1 trial. *Nat. Med.* **2022**, *28*, 383–391. [CrossRef] [PubMed]
74. Widge, A.T.; Hofstetter, A.R.; Houser, K.V.; Awan, S.F.; Chen, G.L.; Burgos Florez, M.C.; Berkowitz, N.M.; Mendoza, F.; Hendel, C.S.; Holman, L.A.; et al. An influenza hemagglutinin stem nanoparticle vaccine induces cross-group 1 neutralizing antibodies in healthy adults. *Sci. Transl. Med.* **2023**, *15*, eade4790. [CrossRef] [PubMed]
75. Casazza, J.P.; Hofstetter, A.R.; Costner, P.J.M.; Holman, L.A.; Hendel, C.S.; Widge, A.T.; Wu, R.L.; Whalen, W.R.; Cunningham, J.; Arthur, A.; et al. Phase 1 dose-escalation trial evaluating a group 2 influenza hemagglutinin stabilized stem nanoparticle vaccine. *npj Vaccines* **2024**, *9*, 171. [CrossRef]
76. Zhong, D.; Lu, Z.; Xia, Y.; Wu, H.; Zhang, X.; Li, M.; Song, X.; Wang, Y.; Moon, A.; Qiu, H.-J.; et al. Ferritin Nanoparticle Delivery of the E2 Protein of Classical Swine Fever Virus Completely Protects Pigs from Lethal Challenge. *Vaccines* **2024**, *12*, 629. [CrossRef]
77. Chang, X.; Ma, J.; Zhou, Y.; Xiao, S.; Xiao, X.; Fang, L. Development of a Ferritin Protein Nanoparticle Vaccine with PRRSV GP5 Protein. *Viruses.* **2024**, *16*, 991. [CrossRef]

78. Yassine, H.M.; Boyington, J.C.; McTamney, P.M.; Wei, C.J.; Kanekiyo, M.; Kong, W.P.; Gallagher, J.R.; Wang, L.; Zhang, Y.; Joyce, M.G.; et al. Hemagglutinin-stem nanoparticles generate heterosubtypic influenza protection. *Nat. Med.* **2015**, *21*, 1065–1070. [CrossRef]
79. Tang, P.; Cui, E.-H.; Chang, W.-C.; Yu, C.; Wang, H.; Du, E.-Q.; Wang, J.-Y. Nanoparticle-Based Bivalent Swine Influenza Virus Vaccine Induces Enhanced Immunity and Effective Protection against Drifted H1N1 and H3N2 Viruses in Mice. *Viruses* **2022**, *14*, 2443. [CrossRef]
80. Song, J.; Wang, M.; Zhou, L.; Tian, P.; Sun, Z.; Sun, J.; Wang, X.; Zhuang, G.; Jiang, D.; Wu, Y.; et al. A Candidate Nanoparticle Vaccine Comprised of Multiple Epitopes of the African Swine Fever Virus Elicits a Robust Immune Response. *J. Nanobiotechnol.* **2023**, *21*, 424. [CrossRef]
81. Aebischer, A.; Wernike, K.; König, P.; Franzke, K.; Wichgers Schreur, P.J.; Kortekaas, J.; Vitikainen, M.; Wiebe, M.; Saloheimo, M.; Tchelet, R.; et al. Development of a Modular Vaccine Platform for Multimeric Antigen Display Using an Orthobunyavirus Model. *Vaccines* **2021**, *9*, 651. [CrossRef]
82. Zhang, B.; Chao, C.W.; Tsybovsky, Y.; Abiona, O.M.; Hutchinson, G.B.; Moliva, J.I.; Olia, A.S.; Pegu, A.; Phung, E.; Stewart-Jones, G.B.E.; et al. A platform incorporating trimeric antigens into self-assembling nanoparticles reveals SARS-CoV-2-spike nanoparticles to elicit substantially higher neutralizing responses than spike alone. *Sci. Rep.* **2020**, *10*, 18149. [CrossRef]
83. He, L.; de Val, N.; Morris, C.D.; Vora, N.; Thinnes, T.C.; Kong, L.; Azadnia, P.; Sok, D.; Zhou, B.; Burton, D.R.; et al. Presenting native-like trimeric HIV-1 antigens with self-assembling nanoparticles. *Nat. Commun.* **2016**, *7*, 12041. [CrossRef]
84. He, L.; Lin, X.; Wang, Y.; Abraham, C.; Sou, C.; Ngo, T.; Zhang, Y.; Wilson, I.A.; Zhu, J. Single-component, self-assembling, protein nanoparticles presenting the receptor binding domain and stabilized spike as SARS-CoV-2 vaccine candidates. *Sci. Adv.* **2021**, *7*, eabf1591. [CrossRef]
85. Wichgers Schreur, P.J.; Tacken, M.; Gutjahr, B.; Keller, M.; van Keulen, L.; Kant, J.; van de Water, S.; Lin, Y.; Eiden, M.; Rissmann, M.; et al. Vaccine Efficacy of Self-Assembled Multimeric Protein Scaffold Particles Displaying the Glycoprotein Gn Head Domain of Rift Valley Fever Virus. *Vaccines* **2021**, *9*, 301. [CrossRef]
86. Choi, B.; Moon, H.; Hong, S.J.; Shin, C.; Do, Y.; Ryu, S.; Choi, K.; Kim, Y.S. Effective Delivery of Antigen-Encapsulin Nanoparticle Fusions to Dendritic Cells Leads to Antigen-Specific Cytotoxic T Cell Activation and Tumor Rejection. *ACS Nano* **2016**, *10*, 7339–7350. [CrossRef]
87. Mu, Z.; Wiehe, K.; Saunders, K.O.; Henderson, R.; Cain, D.W.; Parks, R.; Martik, D.; Mansouri, K.; Edwards, R.J.; Newman, A.; et al. mRNA-encoded HIV-1 Env trimer ferritin nanoparticles induce monoclonal antibodies that neutralize heterologous HIV-1 isolates in mice. *Cell Rep.* **2022**, *38*, 110514. [CrossRef]
88. Yu, T.; Zhang, C.; Xing, J.; Zhang, T.; Xu, Z.; Di, Y.; Yang, S.; Jiang, R.; Tang, J.; Zhuang, X.; et al. Ferritin-binding and ubiquitination-modified mRNA vaccines induce potent immune responses and protective efficacy against SARS-CoV-2. *Int. Immunopharmacol.* **2024**, *129*, 111630. [CrossRef]
89. Roier, S.; Mangala Prasad, V.; McNeal, M.M.; Lee, K.K.; Petsch, B.; Rauch, S. mRNA-based VP8* nanoparticle vaccines against rotavirus are highly immunogenic in rodents. *npj Vaccines* **2023**, *8*, 190. [CrossRef]
90. Brandys, P.; Montagutelli, X.; Merenkova, I.; Barut, G.T.; Thiel, V.; Schork, N.J.; Trüeb, B.; Conquet, L.; Deng, A.; Antanasijevic, A.; et al. A mRNA vaccine encoding for a RBD 60-mer nanoparticle elicits neutralizing antibodies and protective immunity against the SARS-CoV-2 Delta variant in transgenic K18-hACE2 mice. *Front. Immunol.* **2022**, *13*, 912898. [CrossRef]
91. Shattock, R.J.; Andrianaivoarimanana, V.; McKay, P.F.; Randriantseheno, L.N.; Murugaiah, V.; Samnuan, K.; Rogers, P.; Tregoning, J.S.; Rajerison, M.; Moore, K.M.; et al. A self-amplifying RNA vaccine provides protection in a murine model of bubonic plague. *Front. Microbiol.* **2023**, *14*, 1247041. [CrossRef]
92. Alameh, M.G.; Tombácz, I.; Bettini, E.; Lederer, K.; Sittplangkoon, C.; Wilmore, J.R.; Gaudette, B.T.; Soliman, O.Y.; Pine, M.; Hicks, P.; et al. Lipid nanoparticles enhance the efficacy of mRNA and protein subunit vaccines by inducing robust T follicular helper cell and humoral responses. *Immunity* **2022**, *54*, 2877–2892.e7. [CrossRef]
93. Sahin, U.; Muik, A.; Derhovanessian, E.; Vogler, I.; Kranz, L.M.; Vormehr, M.; Baum, A.; Pascal, K.; Quandt, J.; Maurus, D.; et al. COVID-19 vaccine BNT162b1 elicits human antibody and TH1 T cell responses. *Nature* **2020**, *586*, 594–599. [CrossRef]
94. Schmidt, C.; Schnierle, B.S. Self-amplifying RNA vaccine candidates: Alternative platforms for mRNA vaccine development. *Pathogens* **2023**, *12*, 138. [CrossRef]
95. Komori, M.; Nogimori, T.; Morey, A.L.; Sekida, T.; Ishimoto, K.; Hassett, M.R.; Masuta, Y.; Ode, H.; Tamura, T.; Suzuki, R.; et al. saRNA vaccine expressing membrane-anchored RBD elicits broad and durable immunity against SARS-CoV-2 variants of concern. *Nat. Commun.* **2023**, *14*, 2810. [CrossRef]
96. Blakney, A.K.; Ip, S.; Geall, A.J. An update on self-amplifying mRNA vaccine development. *Vaccines* **2021**, *9*, 97. [CrossRef]
97. Yıldız, A.; Răileanu, C.; Beissert, T. Trans-amplifying RNA: A journey from alphavirus research to future vaccines. *Viruses* **2024**, *16*, 503. [CrossRef]
98. Zhou, W.Y.; Cai, Z.R.; Liu, J.; Wang, D.S.; Ju, H.Q.; Xu, R.H. Circular RNA: Metabolism, functions and interactions with proteins. *Mol. Cancer* **2020**, *19*, 172. [CrossRef]

99. Fazel, F.; Doost, J.S.; Raj, S.; Boodhoo, N.; Karimi, K.; Sharif, S. The mRNA vaccine platform for veterinary species. *Vet. Immunol. Immunopathol.* **2024**, *274*, 110803. [CrossRef]
100. Ahmadivand, S.; Soltani, K.; Shokrpoor, S.; Rahmati-Holasoo, H.; El-Matbouli, M.; Taheri-Mirghaed, A. Cyprinid herpesvirus 3 (CyHV-3) transmission and outbreaks in Iran: Detection and characterization in farmed common carp. *Microb. Pathog.* **2020**, *149*, 104321. [CrossRef]
101. Soltani, M.; Zamani, M.; Taheri-Mirghaed, A.; Ahmadivand, S.; Mohamdian, S.; Abdi, K.; Soltani, E. Incidence and Genetic Analysis of White Spot Syndrome Virus (WSSV) in Farmed Shrimps (P. indicus and L. Vannamei) in Iran. *Bull. Eur. Assoc. Fish. Pathol.* **2018**, *38*, 24–34.
102. Ahmadivand, S.; Palić, D.; Weidmann, M. Molecular Epidemiology of Novirhabdoviruses Emerging in Iranian Trout Farms. *Viruses* **2021**, *13*, 448. [CrossRef]
103. Dahl, L.O.S.; Hak, S.; Braaen, S.; Molska, A.; Rodà, F.; Parot, J.; Wessel, Ø.; Fosse, J.H.; Bjørgen, H.; Borgos, S.E.; et al. Implementation of mRNA-Lipid Nanoparticle Technology in Atlantic Salmon (Salmo salar). *Vaccines* **2024**, *12*, 788. [CrossRef]
104. See, S.A.; Bhassu, S.; Tang, S.S.; Yusoff, K. Newly Developed mRNA Vaccines Induce Immune Responses in Litopenaeus vannamei Shrimps During Primary Vaccination. *Dev. Comp. Immunol.* **2024**, *162*, 105264. [CrossRef]
105. Ozdilek, A.; Paschall, A.V.; Dookwah, M.; Tiemeyer, M.; Avci, F.Y. Host protein glycosylation in nucleic acid vaccines as a potential hurdle in vaccine design for nonviral pathogens. *Proc. Natl. Acad. Sci. USA* **2020**, *117*, 1280–1282. [CrossRef]
106. Rijkers, G.T.; Weterings, N.; Obregon-Henao, A.; Lepolder, M.; Dutt, T.S.; van Overveld, F.J.; Henao-Tamayo, M. Antigen Presentation of mRNA-Based and Virus-Vectored SARS-CoV-2 Vaccines. *Vaccines* **2021**, *9*, 848. [CrossRef]
107. Li, L.; Petrovsky, N. Molecular Mechanisms for Enhanced DNA Vaccine Immunogenicity. *Expert. Rev. Vaccines* **2016**, *15*, 313–329. [CrossRef]
108. Ahmadivand, S.; Soltani, M.; Behdani, M.; Evensen, Ø.; Alirahimi, E.; Soltani, E.; Hassanzadeh, R.; Ashrafi-Helan, J. VP2 (PTA motif) encoding DNA vaccine confers protection against lethal challenge with infectious pancreatic necrosis virus (IPNV) in trout. *Mol. Immunol.* **2018**, *94*, 61–67. [CrossRef]
109. Lu, B.; Lim, J.M.; Yu, B.; Song, S.; Neeli, P.; Sobhani, N.; K, P.; Bonam, S.R.; Kurapati, R.; Zheng, J.; et al. The Next-Generation DNA Vaccine Platforms and Delivery Systems: Advances, Challenges and Prospects. *Front. Immunol.* **2024**, *15*, 1332939. [CrossRef]
110. Kozak, M.; Hu, J. DNA Vaccines: Their Formulations, Engineering and Delivery. *Vaccines* **2024**, *12*, 71. [CrossRef]
111. Ledesma-Feliciano, C.; Chapman, R.; Hooper, J.W.; Elma, K.; Zehrung, D.; Brennan, M.B.; Spiegel, E.K. Improved DNA Vaccine Delivery with Needle-Free Injection Systems. *Vaccines* **2023**, *11*, 280. [CrossRef]
112. Soltani, M.; Ahmadivand, S.; Behdani, M.; Hassanzadeh, R.; Rahmati-Holasoo, H.; Taheri-Mirghaed, A. Transcription of adaptive-immune genes upon challenge with infectious pancreatic necrosis virus (IPNV) in DNA vaccinated rainbow trout. *Int. J. Aquat. Biol.* **2016**, *4*, 353–359.
113. Naderi-Samani, M.; Soltani, M.; Dadar, M.; Taheri-Mirghaed, A.; Zargar, A.; Ahmadivand, S.; Hassanzadeh, R.; Moazami Goudarzi, L. *Oral immunization of trout fry with recombinant Lactococcus lactis* NZ3900 expressing G gene of viral hemorrhagic septicaemia virus (VHSV). *Fish Shellfish. Immunol.* **2020**, *105*, 62–70. [CrossRef]
114. Almeida, A.V.; Carvalho, A.J.; Calmeiro, T.; Jones, N.C.; Hoffmann, S.V.; Fortunato, E.; Pereira, A.S.; Tavares, P. Condensation and protection of DNA by the Myxococcus xanthus encapsulin: A novel function. *Int. J. Mol. Sci.* **2022**, *23*, 7829. [CrossRef]
115. Paiva, N.F.; Rios, W.M.; Antonio, G.P.; Umino, G.M.; Ferreira, M.P.; Brandão, I.T.; Del Ciampo, J.O.; Bentley, M.V.L.B.; Casagrande, R.; Verri, W.A.; et al. Ferritin-Based Systems as a Platform for DNA Construct Delivery: Preparation, Characterization, and In Vitro Evaluation. *BioNanoSci.* **2025**, *15*, 128. [CrossRef]
116. Díaz-Dinamarca, D.A.; Salazar, M.L.; Castillo, B.N.; Manubens, A.; Vasquez, A.E.; Salazar, F.; Becker, M.I. Protein-Based Adjuvants for Vaccines as Immunomodulators of the Innate and Adaptive Immune Response: Current Knowledge, Challenges, and Future Opportunities. *Pharmaceutics* **2022**, *14*, 1671. [CrossRef]
117. Li, X.; Cao, Y.; Mou, M.; Li, J.; Huang, S.; Zhang, E.; Yan, H.; Yang, J.; Zhong, M. Enhanced TLR5-dependent migration and activation of antigen-loaded airway dendritic cells by flagellin. *J. Leukoc. Biol.* **2023**, *113*, 567–576. [CrossRef]
118. Jones, J.A.; Benisch, R.; Giessen, T.W. Encapsulin cargo loading: Progress and potential. *J. Mater. Chem. B* **2023**, *11*, 4377–4388. [CrossRef]
119. Altenburg, W.J.; Rollins, N.; Silver, P.A.; Giessen, T.W. Exploring targeting peptide-shell interactions in encapsulin nanocompartments. *Sci. Rep.* **2021**, *11*, 4951. [CrossRef]
120. Afzal, H.; Murtaza, A.; Cheng, L.T. Structural Engineering of Flagellin as Vaccine Adjuvant: Quest for the Minimal Domain of Flagellin for TLR5 Activation. *Mol. Biol. Rep.* **2025**, *52*, 104. [CrossRef]
121. Kim, D.; Rossi, J. RNAi Mechanisms and Applications. *Biotechniques* **2008**, *44*, 613–616. [CrossRef]
122. Lam, J.K.; Chow, M.Y.; Zhang, Y.; Leung, S.W. siRNA Versus miRNA as Therapeutics for Gene Silencing. *Mol. Ther. Nucleic Acids* **2015**, *4*, e252. [CrossRef] [PubMed]
123. Agrawal, N.; Dasaradhi, P.V.; Mohmmed, A.; Malhotra, P.; Bhatnagar, R.K.; Mukherjee, S.K. RNA Interference: Biology, Mechanism, and Applications. *Microbiol. Mol. Biol. Rev.* **2003**, *67*, 657–685. [CrossRef] [PubMed]

124. Chokwassanasakulkit, T.; Oti, V.B.; Idris, A.; McMillan, N.A.J. SiRNAs as antiviral drugs—Current status, therapeutic potential and challenges. *Antivir. Res.* **2024**, *232*, 106024. [CrossRef] [PubMed]
125. Fabian, M.R.; Sonenberg, N.; Filipowicz, W. Regulation of mRNA translation and stability by microRNAs. *Annu. Rev. Biochem.* **2010**, *79*, 351–379. [CrossRef]
126. McAnuff, M.A.; Rettig, G.R.; Rice, K.G. Potency of siRNA versus shRNA mediated knockdown in vivo. *J. Pharm. Sci.* **2007**, *96*, 2922–2930. [CrossRef]
127. Isenmann, M.; Stoddart, M.J.; Schmelzeisen, R.; Gross, C.; Della Bella, E.; Rothweiler, R.M. Basic Principles of RNA Interference: Nucleic Acid Types and In Vitro Intracellular Delivery Methods. *Micromachines* **2023**, *14*, 1321. [CrossRef]
128. Ozata, D.M.; Gainetdinov, I.; Zoch, A.; O'Carroll, D.; Zamore, P.D. PIWI-interacting RNAs: Small RNAs with big functions. *Nat.Rev. Genet.* **2019**, *20*, 89–108. [CrossRef]
129. Akinc, A.; Querbes, W.; De, S.; Qin, J.; Frank-Kamenetsky, M.; Jayaprakash, K.N.; Jayaraman, M.; Rajeev, K.G.; Cantley, W.L.; Dorkin, J.R.; et al. Targeted delivery of RNAi therapeutics with endogenous and exogenous ligand-based mechanisms. *Mol. Ther.* **2010**, *18*, 1357–1364. [CrossRef]
130. Lee, C.S.; Bishop, E.S.; Zhang, R.; Yu, X.; Farina, E.M.; Yan, S.; Zhao, C.; Zheng, Z.; Shu, Y.; Wu, X.; et al. Adenovirus-Mediated Gene Delivery: Potential Applications for Gene and Cell-Based Therapies in the New Era of Personalized Medicine. *Genes. Dis.* **2017**, *4*, 43–63. [CrossRef]
131. Paddison, P.J.; Hannon, G.J. RNA interference: The new somatic cell genetics? *Cancer Cell* **2002**, *2*, 17–23. [CrossRef]
132. Qureshi, A.; Tantray, V.; Kirmani, A.G.; Ahangar, A.G. A review on current status of antiviral siRNA. *Rev Med Virol.* **2018**, *28*, e1976. [CrossRef] [PubMed]
133. Motamedi, H.; Ari, M.M.; Alvandi, A.; Abiri, R. Principle, application and challenges of development siRNA-based therapeutics against bacterial and viral infections: A comprehensive review. *Front Microbiol.* **2024**, *15*, 1393646. [CrossRef] [PubMed]
134. Moazzam, M.; Zhang, M.; Hussain, A.; Yu, X.; Huang, J.; Huang, Y. The landscape of nanoparticle-based siRNA delivery and therapeutic development. *Mol. Ther.* **2024**, *32*, 284–312. [CrossRef] [PubMed]
135. Charousova, M.; Kudlickova Peskova, M.; Takacsova, P.; Kapolkova, K.; Haddad, Y.; Bilek, J.; Sivak, L.; Bartejs, T.; Heger, Z.; Pekarik, V. Engineered human H-chain ferritin with reversed charge of the internal cavity exhibits RNA-mediated spongelike effect for loading RNA/DNA-binding molecules. *Biomater. Sci.* **2024**, *12*, 1249–1262. [CrossRef]
136. Zhu, Y.; Zhu, Y.; Cao, T.; Liu, X.; Liu, X.; Yan, Y.; Shi, Y.; Wang, J.-C. Ferritin-based nanomedicine for disease treatment. *Med. Rev.* **2023**, *3*, 49–74. [CrossRef]

MDPI AG
Grosspeteranlage 5
4052 Basel
Switzerland
Tel.: +41 61 683 77 34

Vaccines Editorial Office
E-mail: vaccines@mdpi.com
www.mdpi.com/journal/vaccines

www.ingramcontent.com/pod-product-compliance
Lightning Source LLC
Chambersburg PA
CBHW061546220926
51850CB00005B/267
9783725871223